The Economy of Nature

Sixth Edition

ABOUT THE AUTHOR

Robert E. Ricklefs is Curators' Professor of Biology at the University of Missouri–St. Louis, where he joined the faculty in 1995 after 27 years in the Biology Department at the University of Pennsylvania. Professor Ricklefs is a native of California and holds an undergraduate degree from Stanford University and a Ph.D. from the University of Pennsylvania. His interests include the energetics of reproduction in birds, evolutionary differentiation of life histories, biogeography, and the historical development of biological communities, including the generation and maintenance of large-scale patterns of biodiversity. His research has taken him to a wide variety of habitats from the lowland tropics to seabird islands in Antarctica. Professor Ricklefs is a Fellow of the American Association for the Advancement of Science and the American Academy of Arts and Sciences. He is also the coauthor of *Ecology*, now in its fourth edition, and *Aging: A Natural History*, both published by W. H. Freeman and Company.

The Economy of Nature

SIXTH EDITION

Robert E. Ricklefs
University of Missouri–St. Louis

W. H. Freeman and Company
New York

Acquisitions Editor: Jerry Correa
Developmental Editor: Susan Moran
Associate Director of Marketing: Debbie Clare
Supplements and Media Editor: Daniel Gonzalez
Senior Project Editor: Georgia Lee Hadler
Copy Editor: Norma Roche
Cover Designer: Paula Jo Smith
Text Designer: Victoria Tomaselli
Photo Editor: Cecilia Varas
Photo Researcher: Julie Tesser
Illustration Coordinator: Susan Timmons
Illustrations: Dragonfly Media Group
Production Manager: Julia DeRosa
Composition: Matrix Publishing Services
Printing and Binding: RR Donnelley

The chapter-opening images were provided by the following photographers:
Ch. 1: Luiz C. Marigo/Peter Arnold; Ch. 2: Francois Gohier/Photo Researchers; Ch. 3: Hanne and Jens Eriksen/Nature Picture Library; Ch. 4: Bill Brooks/Alamy; Ch. 5: Alan and Linda Detrick/ Photo Researchers; Ch. 6: Adrienne Gibson/Animals Animals – Earth Scenes; Ch. 7: Michel Roggo-Bios/Peter Arnold; Ch.8: Simon D. Pollard/ Photo Researchers; Ch. 9: John R. McGregor/Peter Arnold; Ch. 10: E. R. Degginger; Ch. 11: Images & Stories/Alamy; Ch. 12: Ernst Haas/Getty Images; Ch. 13: Stockbyte/Alamy; Ch. 14: Peggy Greb/Agricultural Research Services/U.S. Department of Agriculture; Ch. 15; Tom Brakefield/DRK Photo; Ch. 16: Boris Karpinski/Alamy; Ch. 17: Andy Rouse/DRK Photo; Ch. 18: Tom Bean/DRK Photo; Ch. 19: NASA; Ch. 20: Jim Edds/Photo Researchers; Ch. 21; John Cancalosi/DRK Photo; Ch. 22; R. E. Ricklefs; Ch. 23: Tom and Pat Leeson/Photo Researchers; Ch. 24: Will and Deni McIntyre/Photo Researchers; Ch. 25: James P. Blair/Corbis; Ch. 26: William Campbell/Peter Arnold; Ch. 27: Jim Wilson/*The New York Times*/Redux

Library of Congress Control Number: 2008932083

ISBN-13: 978-0-7167-8697-9
ISBN-10: 0-7167-8697-4
©2008, 2001, 1997, 1993 by W. H. Freeman and Company

Printed in the United States of America

Fifth printing

W. H. Freeman and Company
41 Madison Avenue
New York, NY 10010
Houndmills, Basingstoke RG21 6XS, England
www.whfreeman.com

BRIEF CONTENTS

CONTENTS

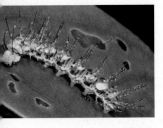

PREFACE

Enduring Vision

Since the first edition of *The Economy of Nature* appeared in 1976, it has maintained a consistent vision of the teaching of ecology based on three fundamental tenets:

• **First, a solid grounding in natural history.** The more we know about habitats and their resident organisms, the better we can understand how ecological and evolutionary processes have shaped the natural world.

• **Second, an appreciation of the organism as the fundamental unit of ecology.** The structures and dynamics of populations, communities, and ecosystems express the activities of, and interactions among, the organisms they comprise.

• **Third, the central position of evolutionary thinking in the study of ecology.** The qualities of all ecological systems express the evolutionary adaptations of their component species.

Readers familiar with the fifth edition of this book will find the same emphasis on field ecology in this edition. Most chapters contain one or more **Ecologists in the Field** essays highlighting the research of ecologists working in a variety of systems and on a variety of problems through field observation, experimentation, and laboratory research. These essays emphasize to students the importance of ecology as a living science.

Students will also have the opportunity to analyze datasets themselves in the **Data Analysis Modules** provided at the ends of several chapters and on the Companion Web Site at www.whfreeman.com/ricklefs6e. These modules introduce students to the importance of data analysis for interpreting patterns in the natural world as well as results of experimental manipulations, while providing some grounding in basic statistical procedures.

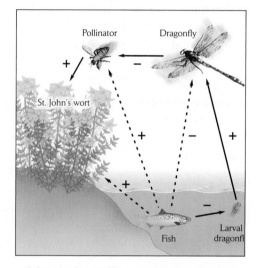

Fish have indirect effects on the populations of several species in and around ponds. The solid arrows represent direct effects, and the dashed arrows indirect effects; the nature of the effect is indicated by a + or −. Fish have indirect effects, through a trophic cascade, on several terrestrial species: dragonfly adults (−), pollinators (+), and plants (+). After T. M. Knight et al., *Nature* 437:880–883 (2005).

New to This Edition

The revision of this textbook has been guided by three overarching goals:

• **To apply the insights of ecology to understanding the impact of human activities on the environment.** As we continue to alter our surroundings, our effects on populations and ecosystems will depend on the particular responses of individual plants, animals, and microorganisms to changes in their environment.

• **To further emphasize the principles of evolution as a foundation of ecology, with repercussions that extend even to managing global change.** For example, the rate of speciation influences large-scale patterns of species richness over the surface of the earth, and understanding the dynamics of this process provides guidelines for preserving biodiversity.

• **To show how modern approaches to studying ecology are illuminating ecological structures and functions.** For example, the increasing availability of a

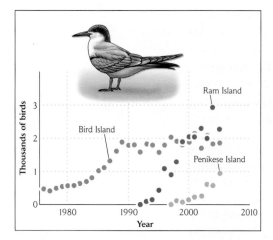

Populations of common terns are limited by nesting space. Populations of common terns (*Sterna hirundo*) on several islands in Buzzards Bay, Massachusetts, have grown rapidly, then leveled off as suitable nesting sites were occupied. Data courtesy of Ian C. T. Nisbet.

wide variety of markers of genetic variation now allows ecologists to take into account the history of movements of individuals and changes in the size of populations over time in analyzing population structure.

What's New . . .

Consolidated Coverage of Evolution. The newly reworked Chapter 6 presents Darwinian evolutionary principles, including natural selection, adaptation as a process, and relevant topics from population genetics. The chapter provides a more focused discussion of evolution by bringing together topics previously covered separately across several chapters. In a complementary way, at the end of the section on populations, Chapter 13 summarizes recent advances in the use of genetic markers to study population processes, including the estimation of effective population size, fitness effects of inbreeding in small populations, and historical changes in population size. These genetic tools have made significant contributions to the conservation and management of wild populations.

Increased Emphasis on Global Change. Five two-page spreads, all but the first written by Rick Relyea of the University of Pittsburgh, explore "global change" as an important ecological principle:

- Carbon Dioxide and Global Warming (Chapter 3, p. 52)

- Global Warming and Flowering Time (Chapter 7, p. 146)

- Changing Ocean Temperatures and Shifting Fish Distributions (Chapter 10, p. 206)

- Invasive Plant Species and the Role of Herbivores (Chapter 17, p. 364)

- Rising Carbon Dioxide Concentrations and the Productivity of Grasslands (Chapter 23, p. 492)

By considering the extent of human impacts on ecosystems in these spreads and elsewhere in the chapters, students will gain an understanding of the relationship between humans and their environment. In addition, they will learn about potential approaches to preventing future ecological crises, such as climatic warming, decreased crop production, emerging diseases, and species extinctions.

New Chapter on Landscape Ecology. To address an increasing interest in landscape ecology, Chapter 25, written by Rick Relyea of the University of Pittsburgh, presents a modern synthesis of large-scale ecology, including human influences on landscapes and the ways in which landscape structure affects individuals, populations, and communities. The chapter focuses on the way in which the scale of spatial heterogeneity in the environment matches the scale of organismal behavior, including foraging activity and dispersal between suitable habitat patches, a key to understanding ecological complexity.

New Organization. Coverage of ecosystems ecology has been moved to follow the material on community ecology, bringing the table of contents into closer alignment with the order in which ecology is taught in most courses. Thus, *The Economy of Nature* now follows a hierarchical organization scheme by sequentially addressing increasing levels of ecological complexity from organisms to populations, communities, and ecosystems.

Clear Connections Between Adaptations and the Physical Environment. To help students make a more meaningful connection between the physical environment and an organism's adaptations to it, Chapters 2 and 3 have been reconceived as a chapter on water (Chapter 2) and a chapter on energy (Chapter 3). Water and energy, including heat, are two of the most important drivers of ecological function and are becoming increasingly important to the study of ecology as emissions of carbon dioxide and other greenhouse gases cause our climate to heat up at a rate never before experienced over the history of the earth.

New Aquatic Examples. Significant advances in aquatic research are introduced throughout the book as Ecologists in the Field essays and elsewhere in the chapters, providing more balanced treatment of terrestrial and aquatic examples. *The Economy of Nature* has always provided students with a broad view of the diversity of organisms and natural systems, and this tradition is further expanded in the sixth edition. Rick Relyea, an aquatic ecologist at the University of Pittsburgh, has provided several of these new examples:

- Effects of predation on the evolution of guppy life histories (Chapter 7, p. 141)

- Effects of fishing on sex switching (Chapter 8, p. 168)

- The chytrid fungus and the global decline of amphibians (Chapter 15, p. 317)

- Apparent competition between corals and algae mediated by microbes (Chapter 16, p. 342)

- Mimicking the effects of ice scouring on the rocky coast of Maine (Chapter 18, p. 384)

- A trophic cascade from fish to flowers (Chapter 18, p. 388)

Up-to-Date Coverage. The new edition incorporates modern developments in ecology, both technical and conceptual, including applications of stable isotopes and phylogenetics, recent developments in macroecology, neutral theory, invasion biology, and global processes connected with human activities. Among the new topics included in this edition are the following:

- Ecological niche modeling (Chapter 10, p. 204)

- Macroecologic correlation between abundance and geographic distribution; inverse correlation between population size and body size (Chapter 10, p. 216)

- Use of genetic markers to study population processes (Chapter 13, p. 269)

- Pathogen-host dynamics (Chapter 15, p. 315)

- Apparent competition (Chapter 16, p. 342)

- Alternative stable community states (Chapter 18, p. 383)

- New measures of relative abundance and new beta diversity indices (Chapter 20, p. 413)

- Expanded discussion of Hubbell's neutral model (Chapter 20, p. 432)

- The influence of phylogenetic relationships on community assembly (Chapter 21, p. 450)

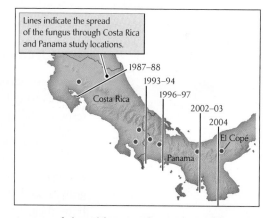

A wave of chytrid fungus infection spread from the northwest to the southeast through Costa Rica and Panama from 1987 to 2004. The red dots indicate locations sampled for infected amphibians. From Lips et al., *Proc. Natl. Acad. Sci. USA* 103:3165–3170 (2006).

- Using phylogenetic trees to test hypotheses explaining high species diversity in the tropics (Chapter 21, p. 457)

- The fossil record of diversity: studies of pollen morphotypes and fossil mammal assemblages (Chapter 21, p. 458)

- Stoichiometry and nutrient balances (Chapter 22, p. 476)

- Using isotopes to trace the fate of nitrates in rainwater (Chapter 23, p. 496)

- Large-scale mapping of habitats using satellites and GIS (Chapter 25, p. 532)

- Emerging diseases and their effects on rates of extinction (Chapter 26, p. 561)

- Using population viability analysis (PVA) to predict the probability that a population will avoid extinction (Chapter 26, p. 566)

New End-of-Chapter Review Questions. Each chapter now includes 8 to 10 questions that will help students review the most important material in the chapter.

Resources for the Student

The Companion Web Site (www.whfreeman.com/ricklefs6e) provides a place for students to enhance, test, and expand their knowledge of the material. The following resources are available to students online:

- **Living Graph Simulations** are interactive tutorials that allow students to practice manipulating variables on a graph and to master important quantitative concepts such as exponential growth, Lotka–Volterra predator–prey interactions, logistic growth, and the Hardy–Weinberg principle. Living Graph Simulations help students hone their data analysis and data interpretation skills. See page xviii for a complete list of Living Graph Simulations.

- **Data Analysis Modules** provide inquiry-based exercises to help students learn important quantitative topics in a step-by-step manner in the context of a real experiment. Five Data Analysis Modules are included in the text, while seven additional modules have been posted on the Companion Web Site. **Datasets** for all twelve modules are available to instructors online in Excel spreadsheets for student assignments. See page xviii for a complete list of Data Analysis Modules.

- **More on the Web** icons are found throughout the book to indicate supplementary topics now discussed on the Companion Web Site that will enhance those presented in the book. These topics include phenotypic plasticity and contrasting mechanisms of growth in animals and plants; sequential hermaphroditism; the origin of female choice; and environmental sex determination. See page xviii for a complete list of More on the Web topics.

- **Online Quizzes** that allow students to review for exams are available for each chapter. Students get instant feedback on their progress and can take the quizzes more than once for practice.

- **The eBook** allows instructors and students access to the entire textbook anywhere, anytime. It is also available as a download for use offline. The eBook is fully searchable and can be annotated with note-taking and highlighting features. The eBook is offered at about half the price of the printed textbook. For more information visit www.coursesmart.com (1-4292-3335-4).

Resources for the Instructor

The following resources for the instructor are available in both the Instructor view of the **Companion Web Site** (1-4292-3547-0) and on an **Instructor Resource CD-ROM** (1-4292-3549-7):

• **Fully optimized JPEGs** of every illustration, photograph, and table in the textbook are offered in labeled and unlabeled versions. The type size, configuration, and color saturation of every image have been individually treated for maximum clarity and visibility.

• **PowerPoint Image Set** includes fully optimized JPEGs of all the illustrations, photos, and tables in the textbook.

• **Test Bank** provides instructors with questions (and accompanying answers) for each chapter. Developed by Thomas Wentworth of North Carolina State University, questions test knowledge, comprehension, application, and analysis. Question formats include multiple choice, fill-in-the-blank, and essay. Among the essay questions are at least five interrelated application and analysis questions based on case histories drawn from actual experiments or hypothetical situations.

• **Living Graph Simulations** can be used in class to review important quantitative topics in an interactive fashion that is less intimidating to students.

• **Datasets** for each Data Analysis Module in the textbook and on the Companion Web Site are available for student assignment.

The following resources are also available for instructors:

• **Overhead Transparencies.** Available on demand, the transparency set contains over 200 figures from the textbook, formatted for maximum visibility in large lecture halls (1-4292-3676-0).

• **WebCT/Blackboard.** Cartridge downloads are available for instructors using either WebCT or Blackboard. Cartridges include the entire suite of student and instructor resources from the Companion Web Site (1-4292-3548-9).

Companion Web Resources

A number of Living Graph Simulations, Data Analysis Modules, and More on the Web topics are available to students and instructors on the **Companion Web Site** (www.whfreeman.com/ricklefs6e), in addition to practice tests and study aids. Page numbers indicate the location in the text of the icon referring to the module, simulation, or topic.

 Living Graph Simulations

 Data Analysis Modules

Data Analysis Modules available in the text

Acknowledgments

I particularly want to acknowledge the contributions of the two people with whom I
worked most closely on this edition: Jerry Correa, Acquisitions Editor, and Susan Moran,
Senior Development Editor. Jerry provided the overall direction for the new edition,
always giving support and encouragement, while Susan worked closely with me to
improve the organization, writing, and illustrations.

I appreciate the proficiency and professionalism of Georgia Lee Hadler, Senior Project Editor; Norma Sims Roche, Copy Editor; Julia DeRosa, Senior Production Manager; Victoria Tomaselli, Senior Designer; Cecilia Varas, Photo Editor; Julie Tesser, Photo Researcher; Susan Timmons, Illustration Coordinator; and Daniel Gonzalez, Supplements and Media Editor. Debbie Clare ably directed the marketing of the book.

I am particularly grateful to Matt Whiles, Southern Illinois University Carbondale, who drew on his own experience as a teacher to create most of the Data Analysis Modules. Jeff Ciprioni and especially Elaine Palucki shepherded the modules through production with enthusiasm and intelligence, making the process painless and enjoyable. I am also grateful to Rick Relyea, University of Pittsburgh, for enriching this textbook through his contributions of the Landscape of Ecology chapter and many Global Change and Ecologists in the Field essays.

Of particular importance to me are the many colleagues who read the manuscript and provided useful advice and guidance:

Jonathan M. Adams, Rutgers University
Loreen Allphin, Brigham Young University
Anthony H. Bledsoe, University of Pittsburgh
Chad E. Brassil, University of Nebraska
Robert S. Capers, Oklahoma State University
Walter P. Carson, University of Pittsburgh
Lisa M. Castle, Glenville State College
Samantha Chapman, Villanova University
Patricia Clark, Indiana University–Purdue University, Indianapolis
Kenneth Ede, Oklahoma State University–Tulsa
Llody Fitzpatrick, University of North Texas
Jason Fridley, Syracuse University
Jack Grubaugh, University of Memphis
Stephen J. Hecnar, Lakehead University
Tara Jo Holmberg, Northwestern Connecticut Community College
Claus Holzapfel, Rutgers University
Thomas R. Horton, SUNY College of Environmental Science and Forestry
R. Stephen Howard, Middle Tennessee State University
Anthony Ippolito, DePaul University
Thomas W. Jurik, Iowa State University
Jamie Kneitel, California State University, Sacramento
John L. Koprowski, University of Arizona
Dr. Mary E. Lehman, Longwood University
Patrick Mathews, Friends University
Dean G. McCurdy, Albion College
Rob McGregor, Institute of Urban Ecology, Douglas College
Bill McMillan, Malaspina University–College
Randall J. Mitchell, University of Akron
L. Maynard Moe, California State University, Bakersfield
Patrick L. Osborne, University of Missouri–St. Louis
Diane Post, University of Texas–Permian Basin
Mark Pyron, Ball State University
Rick Relyea, University of Pittsburgh
John P. Roche, Boston College
Steven J. Rothenberger, University of Nebraska–Kearney
Ted Schuur, University of Florida

Erik P. Scully, Towson University
William R. Teska, Pacific Lutheran University
Diana F. Tomback, University of Colorado–Denver
William Tonn, University of Alberta
Joseph von Fischer, Colorado State University
Diane Wagner, University of Alaska
William E. Walton, University of California–Riverside
Xianzhong Wang, Indiana University–Purdue University, Indianapolis
Thomas Wentworth, North Carolina State University
Bradley M. Wetherbee, University of Rhode Island
Susan K. Willson, St. Lawrence University
Mosheh Wolf, University of Illinois at Chicago
John A. Yunger, Governors State University

The following people provided valuable expertise and assisted with the development of the Data Analysis Modules written by Matt Whiles: Walter K. Dodds, Kansas State University; James E. Garvey, Southern Illinois University Carbondale; Alexander D. Huryn, University of Alabama; Clayton K. Nielson, Southern Illinois University Carbondale; John D. Reeve, Southern Illinois University Carbondale; and Eric M. Schauber, Southern Illinois University Carbondale.

Many thanks also to the readers who reviewed the Data Analysis Modules: Patricia Clark, Indiana University–Purdue University, Indianapolis; Robert Colwell, University of Connecticut; Theodore Fleming, University of Miami; Michael Ganger, Massachusetts College of Liberal Arts; Zachary Jones, Colorado College; Aaron King, University of Michigan; Timothy McCay, Colgate University; George Robinson, University at Albany-SUNY; John P. Roche, Boston College; Joseph von Fischer, Colorado State University; I. Michael Weis, University of Windsor; Thomas Wentworth, North Carolina State University; Peter White, University of North Carolina at Chapel Hill.

CHAPTER 1

Introduction

In his book *Uncommon Ground,* William Cronon challenged two common perceptions of nature and of humankind's relationship to nature. The first is the idea that nature tends toward a self-restoring equilibrium when left alone, a notion referred to as "the balance of nature." The second is the idea that in the absence of human interference, nature exists in a pristine state. Ecological studies present scientific evidence both for and against the idea of balance in nature, and they show us how humans have influenced ecological systems. However, Cronon goes beyond these issues to address the cultural foundations of the way we view our own relationship with nature. He advances the idea that the conservation movement and, to some extent, the scientific field of ecology regard pristine nature as an absolute against which there can be no appeal. The unspoiled Amazon rain forest, for example, is likened by many to the Garden of Eden before Adam and Eve, which embodies complete good and also the temptations of complete evil. Cronon suggested that, in the minds of some people, the extinction of species brings out a deep fear of losing paradise or of having to face the reality of our imperfect world.

Ecological studies paint a different picture. They show great variation in nature over time and demonstrate that the pervasive influence of human activities extends to the most remote regions of the earth. These findings challenge the notion of a pristine, balanced environment. Paradise never did exist, at least not in human experience. Where we humans fit in a less than perfect world is a judgment each of you must make, guided by your own sense of values and moral beliefs. Regardless of your own stand, it will be more useful to you and to humankind in general if your judgment is informed by a scientific understanding of how natural systems work and how humans function as a part of the natural world. The purpose of *The Economy of Nature* is to help you achieve that understanding.

- Ecological systems can be as small as individual organisms or as large as the biosphere
- Ecologists study nature from several perspectives
- Plants, animals, and microorganisms play different roles in ecological systems
- The habitat defines an organism's place in nature; the niche defines its functional role
- Ecological systems and processes have characteristic scales in time and space

- Ecological systems are governed by basic physical and biological principles
- Ecologists study the natural world by observation and experimentation
- Humans are a prominent part of the biosphere
- Human impacts on the natural world have increasingly become a focus of ecology

The English word *ecology* is taken from the Greek *oikos*, meaning "house," and thus refers to our immediate surroundings, or **environment.** In 1870, the German zoologist Ernst Haeckel gave the word a broader meaning:

> By ecology, we mean the body of knowledge concerning the economy of nature—the investigation of the total relations of the animal both to its organic and to its inorganic environment; including above all, its friendly and inimical relation with those animals and plants with which it comes directly or indirectly into contact—in a word, ecology is the study of all the complex interrelationships referred to by Darwin as the conditions of the struggle for existence.

Thus, **ecology** is the science by which we study how organisms interact in and with the natural world.

The word *ecology* came into general use only in the late 1800s, when European and American scientists began to call themselves ecologists. The first societies and journals devoted to ecology appeared in the early decades of the twentieth century. Since that time, ecology has grown and diversified, and professional ecologists now number in the tens of thousands. The science of ecology has produced an immense body of knowledge about the world around us. At the same time, the rapid growth of the human population and our increasing technology and materialism have accelerated change in our environment, frequently with dramatic consequences. Now more than ever, we need to understand how ecological systems function if we are to develop the best policies for managing the watersheds, agricultural lands, wetlands, and other areas—what are generally called environmental support systems—on which humanity depends for food, water, protection against natural catastrophes, and public health. Ecologists provide that

understanding through studies of population regulation by predators, the influence of soil fertility on plant growth, the evolutionary responses of microorganisms to environmental contaminants, the spread of organisms, including pathogens, over the surface of the earth, and a multitude of similar issues. Managing biotic resources to sustain a reasonable quality of human life depends on the wise application of ecological principles to solve or prevent environmental problems and to inform our economic, political, and social thought and practice.

This chapter will start you on the road to ecological thinking. We shall first view ecological knowledge and insight from several different vantage points—for example, as levels of complexity, varieties of organisms, types of habitat, and scales in time and space. We shall see how organisms, assemblages of organisms, and organisms together with their environment are united to form larger **ecological systems** by the regular interaction and interdependence of their parts. Although ecological systems vary in scale from a single microbe to the entire biosphere blanketing the surface of the earth, they all obey similar principles. Some of the most important of these principles concern their physical and chemical attributes, the regulation of their structure and function, and evolutionary change. Applying these principles to environmental issues can help us to meet the challenge of maintaining a supportive environment for natural systems—and for ourselves—in the face of increasing ecological stresses.

As we begin this journey of inquiry and exploration, we should be mindful of two points. First, ecology as a science is distinct from environmental science, applied ecology, conservation biology, and related fields. Those fields use an ecological understanding (obtained through scientific investigation) to solve problems concerning the environment and its inhabitants. Of course, science and applications of science are intimately connected, and information flows between them in both directions. Indeed, much of the science of ecology has developed through research on practical problems in pest management, species conservation, habitat restoration, and the like. Throughout this book, we shall see the connections between science and application, between the generation of knowledge and its use.

The second point concerns the nature of science itself. Science is a process, not the knowledge it generates. As we shall see later in this chapter, scientific investigation makes use of a variety of tools to develop an understanding of the workings of nature. This understanding is never complete or absolute, but constantly changes as scientists discover new ways of thinking. Much of our knowledge about the natural world is well established because it has withstood many tests and is consistent with a large body of observations and with the results of experimental manipulations. Our understanding of many issues, however, is incomplete and imperfect. For example, ecologists have yet to agree broadly on the factors that determine many patterns and processes, such as global patterns of species richness, how and where the biosphere sequesters carbon dioxide, the role of certain mineral nutrients in marine production, and the role of predators in controlling prey populations and shifting the character of natural communities. These are areas of active research in which ecologists are exploring alternative explanations for natural phenomena.

Ecological systems can be as small as individual organisms or as large as the biosphere

An ecological system may be an organism, a population, an assemblage of populations living together (often called a community), an ecosystem, or the entire biosphere. Each smaller ecological system is a subset of the next larger one, so that the different types of ecological systems form a hierarchy. This arrangement is shown diagrammatically in Figure 1.1, which represents the idea that a population is made up of many individual organisms, a community comprises many interacting populations, an ecosystem represents the linkage of many communities through their use of energy and nutrient resources, and the biosphere encompasses all the ecosystems on earth.

The **organism** is the most fundamental unit of ecology, the elemental ecological system. No smaller unit in biology, such as the organ, cell, or macromolecule, has a separate life in the environment. Every organism is bounded by a membrane or other covering across which it exchanges energy and materials with its environment. This boundary separates the "internal" processes and structures of the ecological system—in this case, an organism—from the "external" resources and conditions of the environment.

In the course of their lives, organisms transform energy and process materials. To accomplish this, organisms must acquire energy and nutrients from their surroundings and rid themselves of unwanted waste products. In doing so, they modify the conditions of the environment and the resources available for other organisms, and they contribute to energy fluxes and the cycling of chemical elements in the natural world. Assemblages of organisms together with their physical and chemical environments make up an **ecosystem.** Ecosystems are large and complex ecological systems, sometimes including many thousands of different kinds of organisms living in a great variety of individual surroundings. A warbler flitting among the leaves overhead searching for caterpillars and a bacterium decomposing the organic soil underfoot are both part of the same forest ecosystem. We may speak of a forest ecosystem, a prairie ecosystem, and an estuarine ecosystem as distinct units because relatively little energy and few substances are exchanged between these units compared with the innumerable transformations going on within each of them. We can think of an ecosystem, like an organism, as having "internal" processes and exchanges with the "external" surroundings. Thus, we can treat both organism and ecosystem as ecological systems.

Ultimately, all ecosystems are linked together in a single **biosphere** that includes all the environments and organisms on earth. The far-flung parts of the biosphere are linked together by exchanges of energy and nutrients carried by currents of wind and water and by the movements of organisms. A river flowing from its headwaters to an estuary connects the terrestrial and aquatic ecosystems of the watershed to those of the marine realm (Figure 1.2). The migrations of gray whales link the ecosystems of the Bering Sea and the Gulf of California because feeding conditions in the Bering Sea influence the numbers of migrating whales and the number of young they produce on their calving grounds in the Gulf of California. The whale population, in turn, influences both marine ecosystems by consuming vast numbers of marine invertebrates and churning up marine sediments in search of prey. Energy and materials also move between different types of ecosystems within the biosphere, for example, when grizzly bears capture salmon migrating from the ocean to their spawning areas in rivers and lakes. The biosphere is the ultimate ecological system. External to the biosphere,

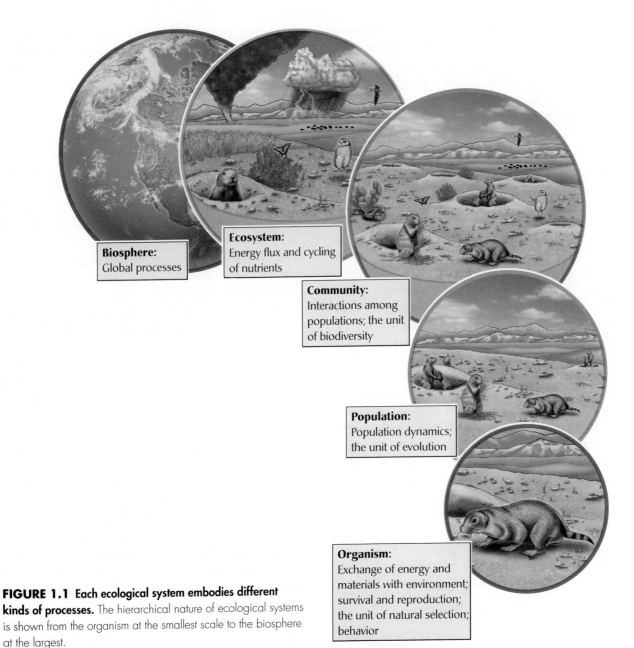

Biosphere: Global processes

Ecosystem: Energy flux and cycling of nutrients

Community: Interactions among populations; the unit of biodiversity

Population: Population dynamics; the unit of evolution

Organism: Exchange of energy and materials with environment; survival and reproduction; the unit of natural selection; behavior

FIGURE 1.1 Each ecological system embodies different kinds of processes. The hierarchical nature of ecological systems is shown from the organism at the smallest scale to the biosphere at the largest.

you will find only sunlight streaming toward the earth and the black coldness of space. Except for the energy arriving from the sun and the heat lost to the depths of space, all the transformations of the biosphere are internal. We have all the materials that we will ever have; our wastes have nowhere to go and must be recycled within the biosphere.

The concepts of ecosystems and the biosphere emphasize the transformation of energy and the synthesis and degradation of materials—ecological systems as physical machines and chemical laboratories. Another perspective emphasizes the uniquely biological properties of ecological systems that are embodied in populations.

A **population** consists of many organisms of the same kind living together. Populations differ from organisms in that they are potentially immortal, since their numbers are maintained over time by the births of new individuals that replace those that die. Populations also have properties that are not exhibited by individual organisms. These distinctive properties include geographic ranges, densities (number of individuals per unit of area), and variations in size or composition (for example, evolutionary responses to environmental change and periodic cycles of numbers).

Many populations of different kinds living in the same place make up an ecological **community.** The populations

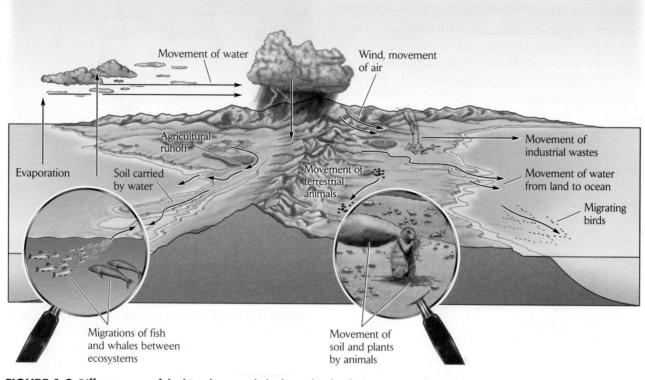

Movement of water

Wind, movement of air

Movement of industrial wastes

Agricultural runoff

Movement of water from land to ocean

Evaporation

Soil carried by water

Movement of terrestrial animals

Migrating birds

Migrations of fish and whales between ecosystems

Movement of soil and plants by animals

FIGURE 1.2 Different parts of the biosphere are linked together by the movement of air, water, and organisms.

within a community interact in various ways. For example, many species are predators that eat other kinds of organisms; almost all species are themselves prey. Some, such as bees and the plants whose flowers they pollinate, and many microorganisms living together with plants and animals, enter into cooperative interactions from which both parties benefit. All these interactions influence the numbers of individuals in populations. Unlike organisms, but like ecosystems, communities have no rigidly defined boundaries; no perceptible skin separates a community from what surrounds it. The interconnectedness of ecological systems means that interactions among populations spread across the globe as individuals and materials move between habitats and regions.

Ecologists study nature from several perspectives

Each level in the hierarchy of ecological systems is distinguished by unique structures and processes. Therefore, each level has given rise to a different approach to the study of ecology. Of course, all the approaches intersect. Within these areas of overlap, ecologists may bring several perspectives to the study of particular ecological problems.

The **organism approach** to ecology emphasizes the way in which an individual's form, physiology, and behavior help it to survive in its environment. This approach also seeks to understand why each type of organism is limited to some environments and not others, and why related organisms living in different environments have different characteristic appearances. For example, as we shall see later in this book, the dominant plants of warm, moist environments are trees, while regions with cool, wet winters and hot, dry summers typically support shrubs with small, tough leaves.

Ecologists who use the organism approach are often interested in studying the adaptations of organisms. Adaptations are modifications of structure and function that better suit an organism for life in its environment: enhanced kidney function to conserve water in deserts; cryptic coloration to avoid detection by predators; flowers shaped and scented to attract certain kinds of pollinators. Adaptations are the result of evolutionary change through the process of natural selection. Because evolution occurs through the replacement of one genetically distinct type of organism by another within a population, the study of adaptations represents a point of overlap between the organism and population approaches to ecology.

The **population approach** is concerned with variation in the numbers of individuals, the sex ratio, the relative sizes of age classes, and the genetic makeup of a population through time. Together these constitute the study of population dynamics. Changes in numbers reflect births and deaths within a population. These events may be influenced by physical conditions of the environment, such as temperature and the availability of water. In the process of evolution, genetic mutations may alter birth and death rates, new genetically distinct types of individuals may become common within a population, and the overall genetic makeup of the population may change. Organisms of other species, which might be food items, pathogens, or predators, also influence the births and deaths of individuals within a population. In some cases, interactions with other species can produce dramatic oscillations of population size or less predictable population changes. Interactions between different kinds of organisms are the common ground of the population and community approaches.

The **community approach** to ecology is concerned with understanding the diversity and relative abundances of different kinds of organisms living together in the same place. The community approach focuses on interactions between populations, which both promote and limit the coexistence of species. These interactions include feeding relationships, which are responsible for the movement of energy and materials through the ecosystem, providing a link between the community and ecosystem approaches. Community studies have considerably expanded their scale in recent years to consider the distribution of species over the surface of the earth and the history of change in community composition—or more generally, global patterns of biodiversity.

The **ecosystem approach** to ecology describes organisms and their activities in terms of common "currencies," principally amounts of energy and various chemical elements essential to life, such as oxygen, carbon, nitrogen, phosphorus, and sulfur. The study of ecosystems deals with movements of energy and materials and how these movements are influenced by climate and other physical factors. Ecosystem function reflects the activities of organisms as well as physical and chemical transformations of energy and materials in the soil, atmosphere, and water.

Plants, algae, and some bacteria transform the energy of sunlight into the stored chemical energy of carbohydrates through photosynthesis. In eating those photosynthetic organisms, animals transform some of the energy available in those carbohydrates into animal biomass. Thus, the activities of organisms as different as bacteria and birds can be compared by describing the energy transformations of populations in such units as watts per

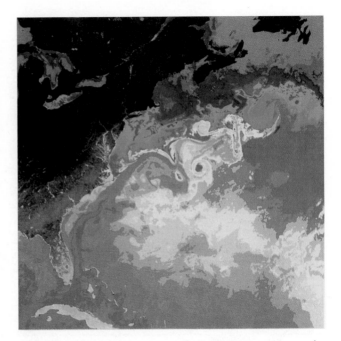

FIGURE 1.3 Ocean currents and winds carry moisture and heat over the earth. This satellite image of the North Atlantic Ocean during the first week of June, 1984, shows the Gulf Stream moving along the coast of Florida and breaking up into large eddies as it begins to cross the Atlantic toward northern Europe. Warm water is indicated by red and cold water by green or blue, and then by red at the top of the picture. Courtesy of Otis Brown, Robert Evans, and Mark Carle, University of Miami Rosenstiel School of Marine and Atmospheric Science.

square meter of habitat. In spite of their commonalities, however, community and ecosystem approaches to ecology provide different ways of looking at the natural world. We may speak of a forest ecosystem, or we may speak of the community of animals and plants that live in the forest, using different jargon and referring to different facets of the same ecological system.

The **biosphere approach** to ecology is concerned with the largest scale in the hierarchy of ecological systems. This approach tackles the movements of air and water, and the energy and chemical elements they contain, over the surface of the earth (see, for example, Figure 1.3). Ocean currents and winds carry the heat and moisture that define the climates at each location on earth, which in turn govern the distributions of organisms, the dynamics of populations, the composition of communities, and the productivity of ecosystems. Another important goal of the biosphere approach is to understand the ecological consequences of natural variations in climate, such as El Niño events, and human-abetted changes, including the formation of the ozone hole over the Antarctic, the conversion of grazing lands to desert over much of Africa, and the increase in atmospheric carbon dioxide, which is having a global impact on climate.

Plants, animals, and microorganisms play different roles in ecological systems

The largest and most conspicuous forms of life, plants and animals, perform a large share of the energy transformations within the biosphere, but no more so than the countless microorganisms in soils, water, and sediments. The characteristics that distinguish plants, animals, fungi, protists, and bacteria have important implications for the way we study and come to understand nature because different kinds of organisms have different functions in natural systems (Figure 1.4).

Early ecosystems were dominated by bacteria of various forms. Bacteria not only gave rise to all other forms of life, but modified the biosphere, making it possible for more complex forms of life to exist. Photosynthetic bacteria present over 3 billion years ago in the earliest ecosystems on earth produced oxygen as a by-product of carbon dioxide assimilation. The resulting increase in oxygen concentration in the atmosphere and oceans eventually permitted the evolution of complex, mobile life forms with high metabolic requirements, which have dominated the earth for the last 500 million years. As these new forms of life evolved, however, their simpler ancestors persisted because their unique biochemical capabilities allowed them to use resources and tolerate ecological conditions that their more complex descendants could not. Indeed, the characteristics of modern ecosystems depend on the activities of many varied forms of life, with each major group filling a unique and necessary role in the biosphere.

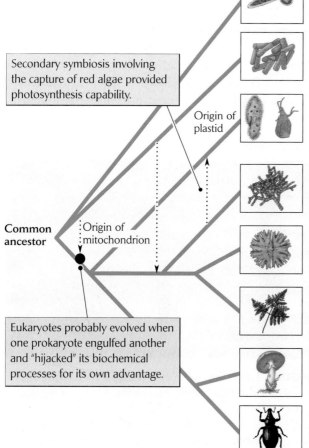

Archaebacteria
Simple prokaryotic organisms lacking an organized nucleus and other cell organelles. Adapted to living in extreme conditions of high salt concentration, high temperature, and pH (both acid and alkaline).

Eubacteria
Like archaebacteria, simple prokaryotic organisms having a wide variety of biochemical reactions of ecological importance in cycling elements through the ecosystem. Many forms are symbiotic or parasitic.

Various protists
An extremely diverse group of mostly single-celled eukaryotes–organisms having nuclear membranes and other cell organelles–ranging from slime molds and protozoans to the photosynthetic dinoflagellates, brown algae, and diatoms.

Red algae
Perhaps 6000 species of photosynthetic protists distinguished by several accessory photosynthetic pigments. Predominately coastal in distribution, the coralline algae are important reef builders.

Green algae
One of the lines of phototsynthetic protists, which are responsible for most of the biological production in aquatic systems, and thought to have been the ancestor of green plants.

Green plants
Complex, primarily terrestrial photosynthetic (photoautotrophic) organisms responsible for fixing most of the organic carbon in the biosphere.

Fungi
Primarily terrestrial heterotrophic organisms of great importance in recycling plant detritus in ecosystems. Many forms are pathogenic and others form important symbioses (lichens, mycorrhizae).

Animals
Aquatic and terrestrial heterotrophic organisms that feed on other forms of life or their remains. Complexity and mobility have led to remarkable diversification of animal life.

Secondary symbiosis involving the capture of red algae provided photosynthesis capability.

Origin of plastid

Common ancestor

Origin of mitochondrion

Eukaryotes probably evolved when one prokaryote engulfed another and "hijacked" its biochemical processes for its own advantage.

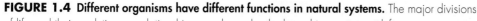

FIGURE 1.4 Different organisms have different functions in natural systems. The major divisions of life and their evolutionary relationships are shown by the branching pattern at left.

FIGURE 1.5 Epiphytic air plants grow high above the ground on the limbs of trees in tropical rain forests. Photo by R. E. Ricklefs.

Plants use the energy of sunlight to produce organic matter

All ecological systems depend on transformations of energy. For most systems, the ultimate source of that energy is sunlight. Plants and other photosynthetic organisms use the energy of sunlight to synthesize organic molecules from carbon dioxide and water. On land, most plants have structures with large exposed surfaces—their leaves—to capture the energy of sunlight. Their leaves are thin because surface area for light capture is more important than bulk. Rigid stems support their aboveground parts. To obtain carbon, terrestrial plants take up gaseous carbon dioxide from the atmosphere. At the same time, they lose prodigious amounts of water by evaporation from their leaf tissues to the atmosphere. Thus, plants need a steady supply to replace water lost during photosynthesis. Not surprisingly, most plants are firmly rooted in the ground, in constant touch with water in the soil. Those that are not, such as orchids and other tropical "air plants" (epiphytes), can be actively photosynthetic only in humid environments bathed in clouds (Figure 1.5).

Animals feed on other organisms or their remains

The organic carbon produced by photosynthesis provides food, either directly or indirectly, for the rest of the ecological community. Some animals consume plants; some consume animals that have eaten plants; others, such as fly larvae, consume the dead remains of plants or animals.

Animals and plants differ in many important ways besides their sources of energy (Figure 1.6). Animals, like plants, need large surfaces for exchanging substances with their environment. However, because they don't need to capture light as an energy source, their exchange surfaces can be enclosed within the body. A modest pair of human lungs has a surface area of about 100 square meters, which is half the size of a tennis court. By internalizing their exchange surfaces in lungs, gills, and guts, animals can achieve bulk and streamlined body shapes, and they can develop the skeletal and muscular systems that make mobility possible. In addition, the internalized exchange surfaces of terrestrial animals lose less water by evaporation than do the exposed leaves of plants, so that animals need not be continuously supplied with water.

Fungi are highly effective decomposers

The fungi assume unique roles in the ecosystem because of their distinctive growth form. Fungi, like plants and animals, are multicellular (except for the unicellular yeasts and their relatives). Most fungal organisms consist of threadlike structures called hyphae that are only a single cell in diameter. These hyphae may form a loose network, which can invade plant or animal tissues or dead leaves and wood on the soil surface, or they can grow together into the reproductive structures that we recognize as mushrooms (Figure 1.7). Because fungal hyphae can penetrate deeply, they readily decompose dead plant material, eventually

FIGURE 1.6 Plants derive their energy from sunlight, and animals derive their energy from plants. A mammal browsing on vegetation on a savanna in East Africa emphasizes the fundamental differences between plants, which assimilate the energy of sunlight and use it to convert atmospheric carbon dioxide into organic carbon compounds, and animals, which derive their energy ultimately from the production of plants. Photo by R. E. Ricklefs.

FIGURE 1.7 Fungi are effective decomposers. The mushrooms produced by this sulphur tuft fungus (*Hypholoma fasciculare*) in Belgium are fruiting bodies produced by the much larger, unseen masses of threadlike hyphae that penetrate decaying wood and leaf litter. Photo by Philippe Clement/Nature Picture Library.

making its nutrients available to other organisms. Fungi digest their foods externally, secreting acids and enzymes into their immediate surroundings, cutting through dead wood and dissolving recalcitrant nutrients from soil minerals. Fungi are the primary agents of rot—unpleasant to our senses and sensibilities, perhaps, but very important to ecosystem functioning.

Protists are the single-celled ancestors of more complex life forms

The protists are a highly diverse group of mostly single-celled eukaryotic organisms that includes the algae, slime molds, and protozoans. The bewildering variety of protists fills almost every ecological role. For instance, algae, including the single-celled diatoms, are the primary photosynthetic organisms in most aquatic systems. Algae can also form large plantlike structures—some seaweeds can be up to 100 meters in length (see, for example, Figure 1.16)—but their cells are not organized into the specialized tissues and organs that one sees in plants.

The other members of this group are not photosynthetic. Foraminifera and radiolarians are protozoans that feed on tiny particles of organic matter or absorb small dissolved organic molecules, and that secrete shells of calcite or silicate. Some of the ciliate protozoans are effective predators—on other microorganisms, of course. Many protists are commensal or parasitic, living in the guts or tissues of their host organisms. Some of these, such as

the *Plasmodium* organism responsible for human malaria, cause debilitating diseases.

Bacteria have a wide variety of biochemical mechanisms for energy transformations

Bacteria are the biochemical specialists of the ecosystem. Each bacterium consists of a simple, single cell without a nucleus to contain its DNA and lacking any other intracellular membranes and organelles. Nonetheless, the enormous range of metabolic capabilities of the diverse bacteria, as well as their minute size, enables them to accomplish many unique biochemical transformations and occupy parts of the ecosystem that larger organisms cannot. Some bacteria can assimilate molecular nitrogen (N_2, the common form found in the atmosphere), which they use to synthesize proteins and nucleic acids. Others can use inorganic compounds such as hydrogen sulfide (H_2S) as sources of energy. Plants, animals, fungi, and most protists cannot accomplish these feats. Furthermore, many bacteria live under anaerobic conditions (lacking free oxygen) in mucky soils and sediments, where their metabolic activities regenerate nutrients, making them available for plants. We will have much more to say about the special place of bacteria in the functioning of the ecosystem later in this book.

Many types of organisms cooperate in nature

Because each type of organism is specialized for a particular way of life, it is not surprising that many different types of organisms live together in close association. A close physical relationship between two different types of organisms is referred to as a **symbiosis.** When each partner in a symbiosis provides something that the other lacks, their relationship is called a **mutualism.** Some familiar examples include lichens, which comprise a fungus and an alga in one organism (Figure 1.8); bacteria that ferment plant material in the guts of cows; protozoans that digest wood in the guts of termites; fungi associated with the roots of plants that help them to extract mineral nutrients from the soil in return for carbohydrate energy from the plant; photosynthetic algae in the flesh of corals and giant clams; and nitrogen-fixing bacteria in the root nodules of legumes. The specialized organelles so characteristic of the eukaryotic cell—chloroplasts for photosynthesis, mitochondria for various oxidative energy transformations—originated as symbiotic bacteria living within the cytoplasm of host cells.

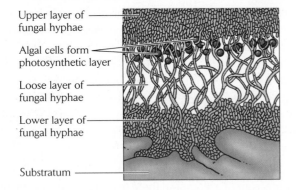

Upper layer of fungal hyphae

Algal cells form photosynthetic layer

Loose layer of fungal hyphae

Lower layer of fungal hyphae

Substratum

FIGURE 1.8 A lichen is a symbiotic association of a fungus and a green alga. Photo by R. E. Ricklefs.

Parasites live in all types of organisms

The line between mutualism and parasitism—that is, living off another organism without giving back in equal amount—is often crossed. Parasites are live-in predators. Because their futures depend on the survival of their hosts, they rarely kill the host organism outright, but rather consume small amounts of host tissue or nutrients. When parasites cause disease symptoms, they are called **pathogens.** From the standpoint of a parasite, organisms such as humans are walking grocery stores filled with an abundance of well-prepared food. Even tiny bacteria are beset by a plethora of even smaller viruses. Parasites are ecologically unique: living in relative ease within a host, they are able to shed many functions needed in the outside world, yet they must often adopt complicated life cycles to find new hosts. Parasites may be adapted to living in the external environment during one life stage or even to using other animals to transport themselves from one host to another. Malaria parasites, for example, infect mosquitoes during one life stage as a means of getting from one human to another. Or, perhaps, they use humans to get from one mosquito to the next.

(a)

(b)

(c)

(d)

FIGURE 1.9 Terrestrial habitats are distinguished by their dominant vegetation. (a) In the tropical rain forest, warm temperatures and abundant rainfall support the highest productivity and biodiversity on earth. (b) In tropical seasonal forest habitats, trees shed their leaves during the pronounced dry season to avoid water stress. (c) Tropical grasslands, which develop where rainfall is sparse, nonetheless support vast herds of grazing herbivores during the productive rainy season. (d) Freezing temperatures on the Antarctic ice cap preclude all life except for occasional bacteria in crevices of sun-warmed exposed rock. Photos by R. E. Ricklefs.

The habitat defines an organism's place in nature; the niche defines its functional role

Ecologists have found it useful to distinguish between where an organism lives and what it does. The **habitat** of an organism is the place, or physical setting, in which it lives. Habitats are distinguished by conspicuous structural features, often including the predominant form of plant life or, sometimes, animal life (Figure 1.9). Thus, we speak of forest habitats, desert habitats, and coral reef habitats. During ecology's early years, much effort was devoted to classifying habitats. For example, ecologists distinguish terrestrial and aquatic habitats; among aquatic habitats, freshwater and marine; among marine habitats, ocean and estuary; among ocean habitats, benthic (on or within the ocean bottom) and pelagic (in the open sea); and so on. However, as such classifications became more complex, they ultimately broke down, because habitat types overlap broadly and absolute distinctions between them rarely exist. The idea of habitat is nonetheless useful because it emphasizes the variety of conditions to which organisms are exposed. Inhabitants of abyssal ocean depths and tropical rain forest canopies experience altogether different conditions of light, pressure, temperature, oxygen concentration, moisture, viscosity, and salt concentrations, not to mention food resources and enemies.

An organism's **niche** represents the range of conditions it can tolerate and the ways of life it pursues—that is, its role in the ecological system. An important principle of ecology is that each species has a distinct niche (Figure 1.10). No two species are exactly the same, because each has distinctive attributes of form and function that determine the conditions it can tolerate, how it feeds, and how it escapes its enemies.

The variety of habitats holds the key to much of the diversity of living organisms. No one organism can live under all the conditions on earth; each must specialize with respect to both the range of habitats within which it can live and the niche that it can occupy within a habitat.

Ecological systems and processes have characteristic scales in time and space

Most properties of the environment, such as air temperature or the number of individuals in a population per unit of area, vary from one place to the next and from one moment to the next. As a result, each measurement exhibits highs and lows, and successive highs or lows are separated by short or long intervals in time or distances in space. Variation in each measurement exhibits a characteristic **scale,** which is the dimension in time or space over which the variation is perceived. It is important to select the appropriate scale of measurement to match the scale of variation of an ecological pattern in either time or space. For example, over time, air temperature can plunge dramatically in a matter of hours as a cold front passes through a region, whereas ocean water may require weeks or months to cool the same amount. Ecological studies focus on patterns and processes occurring on time scales of hours, weeks, months, and years and spatial scales of millimeters, meters, and kilometers. Biospheric processes and evolutionary change, however, occupy vastly larger scales. Only large networks of collaborating investigators can work on these scales, and they must use special technologies to probe deep time and vast areas and to process the mountains of data now available.

(a)

(b)

(c)

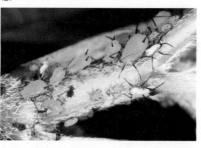

FIGURE 1.10 Each species has a distinct niche. (a) This Peruvian rhinoceros katydid (*Copiphora rhinoceros*) is specialized for chewing leaves. (b) These aphids are specialized for sucking juices from vessels in the stems and leaves of milkweed plants. (c) Ichneumonid wasps, such as this species of *Thalessa* from Ohio, lay their eggs in the larvae of beetles burrowing deep within wood. Photo (a) by Nature's Images/Photo Researchers; photo (b) by Scott Camazine/Photo Researchers; photo (c) by Gary Maszaros/Visuals Unlimited.

Temporal variation

We perceive temporal variation as our environment changes over time, for example, with the alternation of day and night and with the seasonal progression of temperature and precipitation. Superimposed on these more or less predictable cycles are irregular and unpredictable variations, such as droughts and fires, as well as long-term trends, such as the current warming of the earth's climate. The term *climate* refers to average atmospheric conditions over long periods, whereas *weather* refers to atmospheric phenomena that vary over periods of days or hours. Winter climates are generally cold and wet, but the weather at any particular time cannot be predicted much in advance; it varies perceptibly over intervals of a few hours or days with the passage of cold fronts and other atmospheric phenomena. Some irregularities in conditions, such as a string of especially wet or dry years, occur over long periods. Other events of great local ecological consequence, such as fires and tornadoes, strike a particular place only at very long intervals.

How organisms and populations respond to change in their environment depends on how often it occurs. In general, the more extreme the condition, the less frequent it is. However, both the severity and the frequency of events are relative measures, depending on the organism that experiences them. Forest fires may touch an individual tree many times, but skip dozens of generations of an insect population.

Patterns of temporal variation may be intrinsic to an ecological system or imposed by variation in external factors. In pine woodlands, for example, the probability of a destructive fire increases with time since the last such event. As litter and other fuels accumulate and burn, they produce a characteristic fire cycle. Similarly, a communicable disease may spread through a population at regular intervals, reappearing whenever there are enough young individuals lacking immunity gained from previous exposure. Such processes within ecological systems help to regulate their temporal dynamics.

Spatial variation

The environment also differs from place to place. Variations in climate, topography, and soil type cause large-scale heterogeneity (across meters to hundreds of kilometers; see the variation in water temperature in the western Atlantic Ocean illustrated in Figure 1.3). At smaller scales, heterogeneity is generated by the structures of plants, the activities of animals, and the content of soils. As with temporal variation, a particular scale of spatial variation may be important to one organism and not to another. The difference between the top and the underside of a leaf is important to an aphid, but not to a moose, which happily eats the whole leaf, aphid and all.

As an individual moves through an environment that varies in space, it encounters environmental variation as a sequence in time. In other words, a moving individual perceives spatial variation as temporal variation. The faster an individual moves, and the smaller the scale of spatial variation, the more quickly it encounters new environmental conditions, and the shorter the temporal scale of the variation. This principle applies to plants as well as to animals. Roots growing through soil may encounter new conditions if the scale of spatial variation in soil characteristics is small enough. Wind and animals disperse seeds, which may land in a variety of habitats depending on how far they travel relative to the scale of spatial variation in the habitat.

Correlation of spatial and temporal dimensions

With regard to ecologically important phenomena, duration in time usually increases with the extent of the space affected. For example, tornadoes last only a few minutes and affect small areas, whereas hurricanes inflict devastation across hundreds of kilometers over days or weeks. In the oceans, at one extreme, small eddy currents may last only a few days; at the other extreme, ocean gyres (circulating currents encompassing entire ocean basins) are stable over millennia.

Compared with marine and, especially, atmospheric phenomena, variations in landforms have very long temporal scales at any spatial scale. The reason is simply that topography and geology are transformed at a snail's pace by such processes as mountain building, volcanic eruptions, erosion, and even continental drift. In contrast, spatial heterogeneity in the open ocean results from physical processes in water, which are obviously more changeable than those in rock and soil. Because air is even more fluid than water, atmospheric processes have very short time scales at a given spatial scale.

The spatial and temporal scales of *patterns* that we measure in nature often match the scales of the *processes* that produce them. For example, the large-scale processes of species formation and extinction create a global pattern of increasing species richness in most groups of organisms from high latitudes to the equator. The formation of new species generally requires evolutionary periods of time and continental scales of space (new species do not form readily on individual small islands, for example), and the extinction of species under natural circumstances might result from millennial or even slower changes in climate and environment. At the other extreme, the distribution of individuals within a population depends on behavioral responses of individuals to variations in the environment and the presence of other individuals over periods of hours, minutes, and seconds.

Ecological systems are governed by basic physical and biological principles

With all of these ongoing patterns and processes, ecological systems are busy places, yet these dynamic and complex systems are governed by a small number of basic principles. A brief consideration of four of these principles will help to illustrate the underlying unity of ecology.

Ecological systems obey the laws of physics

Life builds on the physical properties and chemical reactions of matter. The diffusion of oxygen across body surfaces, the rates of chemical reactions, the resistance of vessels to the flow of fluids, and the transmission of nerve impulses all obey the laws of thermodynamics. Biological systems are powerless to alter these fundamental properties of matter and energy, but, within the broad limits imposed by these constraints, life can pursue many options, and it has done so with astounding invention.

Ecological systems exist in dynamic states

Whether we focus on the organism, the population, the community, the ecosystem, or the biosphere, each of these ecological systems continuously exchanges matter and energy with its surroundings. When gains and losses are balanced, ecological systems remain unchanged. This balance is the essence of a dynamic **steady state.** A warm-blooded animal continuously loses heat to a cold environment. This loss is balanced, however, by heat gained from the metabolism of foods, so body temperature remains constant. The proteins of our bodies are continuously broken down and replaced by newly synthesized proteins, yet our appearance does not change.

This principle of the steady state applies to all levels of ecological organization. For the individual organism, assimilated food and energy must balance energy expenditure and metabolic breakdown of tissues. For the population, gains and losses are births and deaths. The diversity

of a community decreases when species become extinct, and it increases when new species invade the community's habitat. Ecosystems and the biosphere itself could not exist without the energy received from the sun, yet this gain is balanced by heat energy radiated by the earth back out into space. How the steady states of ecological systems are maintained and regulated is one of the most important questions posed by ecologists, one to which we will return frequently throughout this book.

Of course, ecological systems also change. Organisms grow; populations cycle in abundance; abandoned fields revert to forest. Yet all ecological systems have mechanisms that tend to maintain their integrity.

Living systems must expend energy to maintain themselves

Because life is so special—consider that the molecules of life are rare or nonexistent in the physical world—living organisms exist out of equilibrium with the physical environment. What the organism loses to its surroundings, however, is not returned by the environment for free. If it were, life would be the equivalent of a perpetual motion machine. The organism must procure energy or materials to replace its losses. To do this, it must expend energy. Thus, it must replace the energy lost as heat and motion by metabolizing food or stored reserves, which it must expend energy to capture and assimilate. The price of maintaining a living system in a dynamic state is energy.

Ecological systems evolve over time

Over the history of life on earth, the attributes of organisms have changed and diversified dramatically through the process of **evolution.** Although the physical and chemical properties of matter and energy are immutable, what living systems do with matter and energy is as variable as all the forms of organisms that have existed in the past, are alive today, or might exist in the future. The structures and functions of those organisms are products of evolutionary change in populations in response to their particular environments. For example, prey animals are often colored in such a way that they blend in with their surroundings and escape the notice of predators (Figure 1.11). Many plants in hot, dry climates have small leaves with waxy surfaces to reduce water loss by evaporation. Such attributes of structure and function that suit an organism to the conditions of its environment are called **adaptations.**

The close correspondence of organisms to their environments is no accident. It derives from a process unique

(a)

(b)

FIGURE 1.11 Adaptations help organisms to survive in their environments. (a) The cryptic coloration of a Costa Rican mantid protects it from predators. (b) The waxy, succulent leaves of the South American wax agave (*Echeveria agavoides*; Crassulaceae) reduce water loss in its arid environment. Photo (a) by Michael Fogden/DRK; photo (b) by Peter Anderson, DK Limited/ CORBIS.

to biological systems: **natural selection.** Only individuals that are well suited to their environments survive and produce offspring. The favorable traits inherited by their progeny are preserved. Other individuals survive less well or produce fewer offspring, and their less suitable traits are not passed on. Charles Darwin recognized that this process allowed populations to respond, over many generations, to changes in their environment. A wonderful thing about natural selection and evolution is that as each species changes, new possibilities for further change are opened up for itself and for other species with which it interacts. In this way, the complexity of ecological communities and ecosystems builds on, and is fostered by, existing complexity. An important goal of ecology as a science is to understand how these complex ecological systems came into being and how they function in their environmental settings.

Ecologists study the natural world by observation and experimentation

Like other scientists, ecologists apply many methods to learn about nature. Most of these methods reflect three facets of scientific investigation, often referred to as the scientific method: (1) observation and description, (2) development of hypotheses or explanations, and (3) testing of these hypotheses, often with experiments.

Most research programs begin with a set of **observations** about nature that invite explanation or speculation. Usually these facts describe a consistent pattern. For example, measurements of plant production in various parts of the world show a strong positive relationship between plant growth and precipitation, which we would expect because we know that plants require water. In the wettest areas of the tropics, however, plant production decreases. This unexpected finding clearly cannot be accounted for by an explanation based on replenishment of the water lost to evaporation. We'll discuss some possibilities later in this book. Even a simple observation can stimulate speculation. For example, ecologists (as well as farmers and gardeners) have long known that insects feed on plants, typically removing about 10% of the leaf biomass of trees. Many years ago, ecologists wondered why insects and other herbivores didn't eat more than they do. We'll return to this question in a moment.

Hypotheses are ideas about how a system works—that is, they are explanations. If correct, a hypothesis may help us to understand the cause of an observed pattern. Suppose we observe that male frogs sing on warm nights after periods of rain. If a reasonable amount of observation produces few exceptions to this pattern, it may be regarded as a **generalization** that enables us to predict the behavior of frogs from the weather. Having established the existence of such a pattern, we may wish to understand it better. For example, we may wish to explain how a frog responds to temperature and rainfall; we may also wish to explain why a frog responds the way it does. The "how" part of this particular phenomenon involves details of sensory perception, the interplay between environmental stimuli and the frog's hormonal status, and the frog's nervous system and muscles—in other words, it involves physiological processes and the **proximate factors** in the environment that stimulate the frog's behavior. The "why" question addresses the costs and benefits of the behavior to the individual and deals with **ultimate factors** in the environment that guide evolution—factors such as predators and female frogs, both of which are attracted to singing males, but for different reasons. If we suspect that male frogs sing to attract females, perhaps males sing on warm nights after rains because that is when females look for mates. If frogs chorused at other times, they might attract few mates (low benefit) but still expose themselves to predation or other risks (high cost). We have now generated a number of hypotheses about frog behavior: (1) singing by males attracts females and leads to mating; (2) females actively search for males only on warm nights after rains (perhaps because those nights produce the best conditions for laying eggs); (3) singing imposes a cost, which compels males to save their singing for times when it will do them the most good.

Using experimental manipulations to test hypotheses

If we are to convince ourselves that a hypothesis is valid, we must put it to the test. Only rarely can a particular idea be proved beyond a doubt, but our confidence increases as we explore the implications of a hypothesis and find it to be consistent with the facts. Let's return to the question of insect herbivores and plants. We are surprised to observe that herbivores consume so little plant biomass. Two ideas come to mind. The first is that plants defend themselves against herbivores, not by running and hiding, but by synthesizing various chemicals to reduce their palatability. These defenses not only deter individual herbivores from feeding, but by doing so, they also check herbivore population growth. The second idea is that predators reduce herbivore populations and keep them from overeating their food plants. Both of these ideas are hypotheses about how an ecological system works.

Robert Marquis and Chris Whelan of the University of Missouri at St. Louis became interested in whether the predation hypothesis (which is also referred to as "top-down control" of herbivore populations) applied to insects feeding on oak trees in Missouri. They had observed that birds consume many insects on oak foliage, and they specu-

lated that the avian predators controlled the insect herbivore populations. They predicted that if their hypothesis was correct, then insect populations should increase and consume more leaf biomass if the birds were removed. A **prediction** is a statement that follows logically from a hypothesis. If Marquis and Whelan could confirm their prediction, their hypothesis would be strengthened; if not, their hypothesis would be weakened, or perhaps it might be rejected altogether.

Because many hypotheses are plausible, it is necessary to conduct investigations to determine which explanations best account for the facts. The strongest tests of hypotheses are often **experiments,** in which one or a small number of variables are manipulated independently of others to reveal their particular effects. To test their hypothesis, Marquis and Whelan constructed bird-proof cages around oak trees (Figure 1.12) that excluded birds from the foliage, but allowed insects to pass through freely. The number of insects and the amount of leaf damage inside the cages were then monitored through the summer growing season. Of course, insect populations could be influenced by variables other than predation, such as the weather. So Marquis and Whelan also monitored neighboring trees without exclusion cages to account for spatial and temporal variation in insect populations. Such a treatment, which reproduces all aspects of an experiment except the variable of interest (bird exclusion), is called a **control.** Similarly, because the cages might have effects on the foliage other than bird exclusion (shading, for example), the investigators also enclosed some trees within incomplete cages that allowed birds access to the foliage. This third

FIGURE 1.12 Experiments are the strongest tests of hypotheses. A cage has been placed around a white oak sapling to exclude bird predators that would otherwise consume caterpillars feeding on its leaves. Courtesy of C. Whelan, from R. J. Marquis and C. Whelan, *Ecology* 75:2007–2014 (1994).

set of trees provided a control for experimental effects. Finally, to make sure that their results were repeatable, the investigators applied the experimental treatments to several trees, with a similar number of trees designated as controls.

Marquis and Whelan found that the numbers of insects recorded on the trees from which birds were excluded were 70% higher than on the control trees, and that the percentage of leaf area missing at the end of the growing season jumped from 22% on the control trees to 35% on the experimental trees. These findings led them to conclude that avian predators do reduce the abundance of insect herbivores as well as the damage caused by herbivores to trees. Thus, the experiments confirmed the investigators' predictions and strengthened their hypothesis. It did not, however, address the alternative hypothesis that defensive chemicals produced by plants also reduce insect herbivory. This hypothesis will require separate testing, as we shall see later in this book. The finding that birds reduce insect herbivores suggests another question: Bird populations are declining in response to forest fragmentation in the eastern United States and elsewhere. Will insect damage to the remaining forest increase as a result?

Alternative approaches to hypothesis testing

Although the methods of acquiring scientific knowledge appear to be straightforward, many pitfalls exist. For example, a correlation between variables does not imply a causal relationship; the mechanism of causation must be determined independently by suitable investigation. In addition, many hypotheses cannot be tested by experimental methods because the scales of the relevant processes are too large, or the important variables cannot be isolated because suitable controls cannot be devised. These limitations become particularly restrictive with patterns that have evolved over long periods and with systems such as entire populations or ecosystems that are too large for practical manipulation.

Different hypotheses might explain a particular observation equally well, so investigators must make predictions that distinguish among the alternatives. Of course, more than one mechanism might produce a pattern, in which case more than one hypothesis would be supported. For example, the observed decrease in species richness at higher latitudes has many potential explanations. As one travels north from the equator, average temperature and precipitation decrease, sunlight and biological production decrease, and seasonality and other environmental variations increase. Each of these factors could interact with

ecological systems in ways that could affect the numbers of species that can coexist in a locality, and dozens of hypotheses based on these factors and invoking various ecological and evolutionary mechanisms have been proposed. Isolating the effect of each factor has proved difficult because each tends to vary in parallel with the others.

Faced with these difficulties, ecologists have resorted to several alternative approaches to hypothesis testing. One of these is the **natural experiment,** which relies on natural variation in the environment to create reasonably controlled experimental treatments. For example, the hypothesis that the number of species on an island is influenced by the rate at which new colonists arrive there from continental source areas has been "tested" by comparing species diversity on islands with similar sizes and habitats but at different distances from continental coasts. As predicted, diversity decreases with distance from a continent.

Another approach is the **microcosm** experiment, which attempts to replicate the essential features of an ecological system in a simplified laboratory or field setting (Figure 1.13). It assumes that an aquarium with five animal species will behave like the more complex natural system in a pond, or even like ecological systems more generally. If so, experimental manipulations of the microcosm may yield results that can be generalized to the larger system. For example, the hypothesis that diversity decreases as temporal variation in the environment increases might be approached in a microcosm experiment by varying

FIGURE 1.13 Microcosm experiments are designed to replicate the essential features of an ecological system. Communities of freshwater invertebrates are housed in cattle tanks at the Kellogg Biological Station of Michigan State University. Numerous tanks are used for replicates of different experimental treatments. Photo by R. E. Ricklefs.

temperature, light, acidity, or nutrient resource conditions and observing whether some species disappear from the system. It may be a stretch to generalize from an aquarium to a "real" ecological system, but if such variation consistently resulted in a loss of species in a variety of microcosms, the hypothesis would be strengthened.

Ecologists have also used **mathematical models** to explore the behavior of complex systems. The investigator represents such a system as a set of equations corresponding to the postulated relationships of each of the system's components to other components and to outside influences. In this sense, a mathematical model is a hypothesis; it provides an explanation of the observed structure and functioning of the system. We can test models by comparing the predictions they yield with observations. Most models make predictions about attributes of a system that have not been measured or about the response of the system to perturbation. Whether these predictions are consistent with observations determines whether the hypothesis on which they are based is supported or rejected. For example, epidemiologists have developed models to describe the spread of communicable diseases. These models include such factors as the proportions of a population that are susceptible, exposed, infectious, and recovered (and thus resistant because of acquired immunity), as well as rates of transmission and the virulence of the disease organism. Such models are able to make predictions about the frequency and severity of disease outbreaks, and these predictions can be compared with observations to test the models.

At a larger scale, ecologists have created global carbon balance models to investigate how the burning of fossil fuels affects the carbon dioxide content of the atmosphere. Understanding this relationship is critically important to managing human impacts on our environment. Global carbon balance models include, among other factors, equations for the uptake of carbon dioxide by plants and the dissolution of carbon dioxide in the oceans. The output of early versions of these models failed to match observations; specifically, the models overestimated the annual increase in atmospheric carbon dioxide concentrations. The real world evidently contains carbon dioxide "sinks" that remove the gas from the atmosphere but were not accurately represented in the model. This discrepancy has caused ecosystem modelers to look more closely at processes such as the regeneration of forests and the movement of carbon dioxide across the air–water interface. Such processes have been updated in the models to create more refined descriptions of the functioning of the biosphere and more accurate predictions of the future of atmospheric change.

Humans are a prominent part of the biosphere

Why do ecologists do all this? The wonders of the natural world summon our natural curiosity about life and our surroundings. For many of us, our curiosity about nature and the challenges of its study are reason enough. In addition, however, our need to understand nature is becoming more and more urgent as the growing human population stresses the capacity of natural systems to maintain their structure and functioning. Environments that human activities either dominate or have produced—including our urban and suburban living places, our agricultural breadbaskets, and our recreational areas, tree farms, and fisheries—are also ecological systems. The welfare of humanity depends on maintaining the functioning of these systems, whether they are natural or artificial. Virtually all of the earth's surface is, or soon will be, strongly influenced by people, if not fully under their control. Already, humans usurp nearly half of the biological production of the biosphere. We cannot take this responsibility lightly.

The human population is approaching the 7 billion mark, and it consumes energy and resources, and produces wastes, far in excess of needs dictated by biological metabolism. These activities have caused two related problems of global dimensions. The first is their impact on natural systems, including the disruption of ecological processes and the extermination of species. The second is the steady deterioration of humankind's own environment as we push the limits of what ecological systems can sustain. Understanding ecological principles is a necessary step in dealing with these problems. Two examples drive this point home.

ECOLOGISTS IN THE FIELD   **Introduction of the Nile perch into Lake Victoria.** During the 1950s and early 1960s, the Nile perch (*Lates niloticus*) and Nile tilapia (*Oreochromis niloticus*) were introduced into Lake Victoria, a large, shallow lake straddling the equator in East Africa. The introduction was intended to provide additional food for the people living in the area and additional income from exporting the surplus catch (Figure 1.14). During the 1980s, the population of Nile perch increased dramatically, and the burgeoning fishery attracted many people to the region surrounding Lake Victoria. Indeed, by 2003, the fishery was producing exports to the European Union valued at almost 170 million euros annually. However, because basic ecological principles were ignored, the introduction ended up destroying most of the lake's

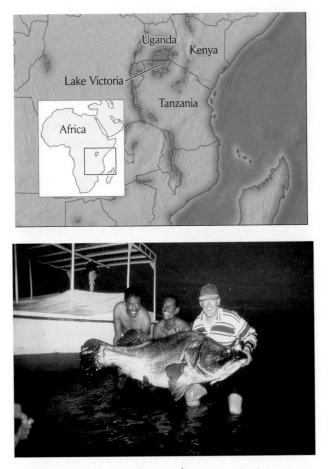

FIGURE 1.14 The introduction of a new species into an ecosystem can have drastic effects. Nile perch were introduced into Lake Victoria in the 1950s to improve the local fishery, but they have driven many native fishes to extinction and completely changed the ecosystem of the lake. Photo courtesy of Tim Baily/The African Angler and Joe Bucher Tackle Company.

traditional fishery and eventually led to the demise of the new fishery as well.

Until the introduction of the Nile perch, Lake Victoria supported a sustainable catch of a variety of local fishes, mostly species belonging to the family Cichlidae. One of these native species was a tilapia that feeds primarily on dead organic matter, plants, and small aquatic invertebrates. Nile perch are very large, and they eat vast quantities of other fish: the smaller cichlids, in this instance. Moreover, because energy is lost with each step in the food chain, predatory fish populations cannot be harvested at as high a rate as their prey species, even though they may be easier to catch. Because the Nile perch was alien to Lake Victoria, the local cichlids had few adaptations to help them escape this predator. Inevitably, the perch annihilated the cichlid populations, driving many unique species to extinction, destroying the native fishery, and severely reducing its own food supply. Consequently, the perch's voracious habits among defenseless prey brought

about its own demise as an exploitable fish species and completely changed the Lake Victoria ecosystem.

At the present time, the Nile perch fishery in Lake Victoria has collapsed. Introduction of the Nile perch had secondary consequences for the terrestrial ecosystems surrounding the lake as well. The flesh of the perch is oily and must be preserved by smoking rather than sun drying, so local forests were cut for firewood. To be sure, the native fishery was already precariously close to being overexploited as a result of the growing local human population and the use of advanced, nontraditional fishing technologies. However, rather than by introducing an efficient predator on the local fishes, these problems might have been more appropriately addressed by regulating fishing methods, enforcing limits on the total annual catch of local fishes, and developing alternative non-fish food sources. ▌

ECOLOGISTS IN THE FIELD   **The California sea otter.** Half a world away from Lake Victoria, efforts to save the sea otter (*Enhydra lutris*) along the coast of California illustrate the intricate intermingling of ecology and other human concerns (Figure 1.15). The sea otter was once widely distributed around the northern Pacific Rim from Japan to Baja California. In the 1700s and 1800s, intense hunting for otter pelts reduced the population nearly to extinction. Predictably, the fur industry collapsed as it overexploited its economic base. Following the rediscovery of a small population in the 1930s, the California sea otter population was placed under strict protection. It had increased to several thousand individuals by the 1990s, but it is now decreasing again.

Initially, the sea otter's recovery irked some California fishers, who claim that the otters—which do not need commercial fishing licenses—drastically reduce stocks of valuable abalone, sea urchins, and spiny lobsters. Matters deteriorated at one point to the marine equivalent of a range war between the fishing industry and conservationists, with the otter caught in the line of fire, often fatally. Ironically, the otters benefited one commercial marine enterprise: the harvesting of kelps. Kelps, which are large seaweeds often used in making fertilizer, grow in shallow water in stands called kelp forests, which provide refuge and feeding grounds for larval fish (Figure 1.16). Kelps are also grazed by sea urchins, which, when abundant, can denude an area. The sea otter is a principal predator of sea urchins. When the expanding otter population spread into new areas, the otters kept urchin populations in check, and kelp forests could regrow.

In recent years, the gill nets used to exploit a newly developed fishery along the California coast inadvertently killed otters in substantial numbers until new legislation moved the fishery farther offshore. Recent otter deaths have been attributed to the protozoan parasites *Toxoplasma gondii* and *Sarcocystis neurona*. These parasites

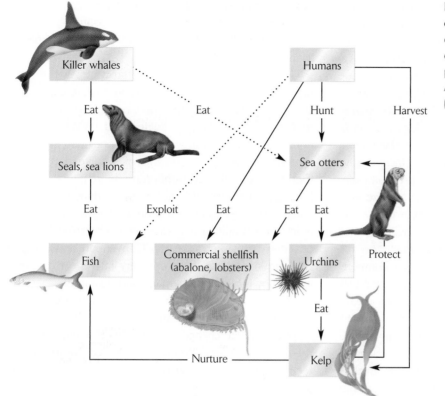

FIGURE 1.15 Human activities have complex effects on ecosystems. Several components of the otter–urchin–kelp ecosystem are altered when humans reduce populations of sea otters by hunting them. After J. A. Estes et al., *Science* 282:473–476 (1998).

normally infect cats, but might have entered the marine ecosystem by way of kitty litter flushed through the sewer system.

Elsewhere, where other factors are at work, the otter population is also dwindling. In a report published in 1998 in the journal *Science,* J. A. Estes and his colleagues at the University of California at Santa Cruz showed that popula-

tions of otters in the vicinity of the Aleutian Islands, Alaska, had declined precipitately during the 1990s. The reason? Killer whales, or orcas (*Orcinus orca*), which previously had not preyed on otters, have been coming close to shore and taking large numbers of them. A predictable result of the decline in otter populations has been a dramatic increase in urchins and decimation of kelps in affected areas. Why did killer whales adopt this new behavior? Estes pointed out that populations of the principal prey of killer whales—seals and sea lions—collapsed during the same period, perhaps inducing the whales to seek alternative food sources. Why did the seals and sea lions decline? One can only speculate at this point. However, intense human fisheries have reduced fish stocks exploited by the seals to levels low enough to seriously threaten seal populations.

FIGURE 1.16 The integrity of kelp forest habitat depends on the presence of sea otters. The kelp forest provides feeding grounds and refuge for many fishes and invertebrates. The otters eat sea urchins that would otherwise destroy young kelp. Photo by Jeff Rotman/Photo Researchers.

Human impacts on the natural world have increasingly become a focus of ecology

Although the plight of endangered species arouses us emotionally, ecologists increasingly realize that the only effective means of preserving natural resources is through the conservation of entire ecological systems and the management of large-scale ecological processes. Individual species, including the ones that humans rely on for food and other products, themselves depend on the maintenance of environmental support systems. We have already seen that predators such as the Nile perch and the sea otter may assume key roles in the functioning of ecological systems. Human activities that changed the abundances of these predators have altered entire ecosystems. Local impacts of human activities on ecological systems can often be managed once we understand the underlying mechanisms responsible for change. Increasingly, however, our activities have multiple, widespread impacts that are more difficult for scientists to characterize and for legislative and regulatory bodies to control. For this reason, a sound scientific understanding of environmental problems is a necessary prerequisite to action.

The daily newspapers are filled with environmental problems: disappearing tropical forests, depleted fish stocks, emerging diseases, global warming. Wars create staggering environmental catastrophe as well as human tragedy. But there are success stories as well. Many developed countries have made great strides in cleaning up their rivers, lakes, and air. Fish are once again migrating up the major rivers of North America and Europe to spawn. Acid rain has decreased, thanks to changes in the combustion of fossil fuels. The release of chlorofluorocarbons, which damage the ozone layer that shields the earth's surface from ultraviolet radiation, has decreased dramatically. The inevitability of global warming caused by increasing atmospheric carbon dioxide concentrations has set off an international research effort and provoked global concern. Conservation efforts, including the breeding of endangered species in captivity, have saved some animals and plants from certain extinction. They have also

heightened public awareness of environmental issues, and they have sometimes sparked public controversy. Without public concern and understanding, however, political action is impossible.

Particularly encouraging is the growing level of international cooperation exemplified in such organizations as the International Union for Conservation of Nature (IUCN) and the World Wildlife Fund (WWF). In addition, the nations of the world have concluded important agreements for the protection of wildlife and nature. One of these is the Convention on International Trade in Endangered Species (CITES), which forbids transport of endangered species or their products (hides, feathers, and ivory, for example) across international borders, depriving poachers of markets. The Rio Convention on Biological Diversity recognizes the proprietary interest of countries in their own biological heritage and guarantees fees and royalties for the exploitation of local plants and animals for uses such as pharmaceutical products. The Kyoto Protocol on climate change, an agreement designed to limit the emission of greenhouse gases, may prove to be ineffective, but it is an initial commitment to do something about global change in the biosphere.

These successes would not have been possible without general consensus founded on evidence produced by scientific study of the natural world. Understanding ecology will not by itself solve our environmental problems in all their political, economic, and social dimensions. However, as we contemplate the need for global management of natural systems, our effectiveness in this enterprise will hinge on our understanding of their structure and functioning—an understanding that depends on knowing the principles of ecology.

SUMMARY

1. Ecology is the scientific study of the natural environment and of the relationships of organisms to one another and to their surroundings.

2. An ecological system may be an organism, a population, a community, an ecosystem, or the entire biosphere. These ecological systems represent levels of organization of ecological structure and functioning and form a hierarchy of progressively more complex entities.

3. Ecologists use several different approaches to study nature, focusing on the interactions of organisms with their environment; the resulting transformations of energy and chemical elements in ecosystems and the biosphere;

the dynamics of populations, including evolutionary change; and the interactions of populations within ecological communities.

4. Different kinds of organisms play different roles in the functioning of ecosystems. Plants, algae, and some bacteria transform the energy of sunlight into stored chemical energy. Animals and protozoans consume these biological forms of energy. Fungi play an important role in decomposing biological materials and replenishing nutrients in the ecosystem. Bacteria are biochemical specialists, able to accomplish such transformations as the assimilation of nitrogen and the use of hydrogen sulfide as an energy source.

5. Different kinds of organisms may form mutually beneficial partnerships, as in the case of the algae and fungi that constitute lichens. Many organisms live parasitically on or within other organisms, feeding on their hosts' nutrients or tissues and often causing disease.

6. An individual's habitat is the place where it lives. The habitat concept emphasizes the structure and conditions of the environment. An individual's niche includes the range of conditions that it can tolerate and the ways of life that it can pursue—that is, its functional role in the natural system.

7. Ecological processes and structures have characteristic temporal and spatial scales. In general, the scales of patterns and processes in time and in space are correlated; large systems tend to change more slowly than small systems.

8. Ecological systems are governed by a small number of basic ecological principles. Ecological systems function within the physical and chemical constraints governing energy transformations. Furthermore, all ecological systems exchange materials and energy with their surroundings. When inputs and outputs are balanced, the system is said to be in a dynamic steady state.

9. All living systems must expend energy to maintain their integrity. Organisms must expend energy to replace the energy and materials they lose through natural processes.

10. All ecological systems are subject to evolutionary change, which results from the differential survival and reproduction within populations of individuals that exhibit different genetically determined traits. As a result of natural selection, organisms exhibit adaptations of structure and function that suit them to the conditions of their environment.

11. Ecologists employ a variety of techniques to study natural systems. The most important of these are observation, development of hypotheses to explain observations, and testing of those hypotheses. Experiments are an important tool for testing hypotheses. When natural systems do not lend themselves readily to experimentation, ecologists may work with microcosms or mathematical models.

12. Humans play a dominant role in the functioning of the biosphere, and human activities have created an environmental crisis of global proportions. Solving our acute environmental problems will require an understanding of the principles of ecology and their application within the framework of political, economic, and social action.

REVIEW QUESTIONS

1. Why do ecologists consider both organisms and ecosystems to be ecological systems?

2. What are the unique processes and structures that are examined when taking the organism, population, community, and ecosystem approaches to studying ecology?

3. How do the sources of energy acquired by plants, animals, and fungi differ?

4. Compare and contrast an organism's habitat and an organism's niche.

5. What is the relationship between the frequency of change in environmental conditions and the spatial extent of the change?

6. Describe how ecological systems are governed by general physical and biological principles.

7. In the Northern Hemisphere, many species of birds fly south during the autumn months. Propose a proximate and an ultimate cause for this behavior.

8. When experimental manipulations are conducted to test a hypothesis, what is the purpose of including a control?

9. In what ways do experimental manipulations differ from natural experiments and microcosm experiments?

10. How can our knowledge of ecological systems help humans to manage these systems?

SUGGESTED READINGS

Barel, C. D. N. et al. 1985. Destruction of fisheries in Africa's lakes. *Nature* 315:19–20.

Bartholomew, G. A. 1986. The role of natural history in contemporary biology. *BioScience* 36:324–329.

Berner, R. A. et al. 2003. Phanerozoic atmospheric oxygen. *Annual Review of Earth and Planetary Sciences* 31:105–134.

Booth, W. 1988. Reintroducing a political animal. *Science* 241: 156–158.

Cohn, J. P. 1998. Understanding sea otters. *BioScience* 48(3): 151–155.

Cronon, W. (ed.). 1996. *Uncommon Ground: Rethinking the Human Place in Nature.* W. W. Norton, New York.

Estes, J. A. et al. 1998. Killer whale predation on sea otters linking oceanic and nearshore systems. *Science* 282:473–476.

Franklin, J. F., C. S. Bledsoe, and J. T. Callahan. 1990. Contributions of the Long-Term Ecological Research Program. *BioScience* 40:509–523.

Goldschmidt, T., F. Witte, and J. Wanink. 1993. Cascading effects of the introduced Nile perch on the detritivorous/phytoplanktivorous species in the sublittoral areas of Lake Victoria. *Conservation Biology* 7:686–700.

Harley, J. L. 1972. Fungi in ecosystems. *Journal of Animal Ecology* 41:1–16.

Kitchell, J. F. et al. 1997. The Nile perch in Lake Victoria: Interactions between predation and fisheries. *Ecological Applications* 7(2): 653–664.

Leibold, M. A. 1995. The niche concept revisited—mechanistic models and community context. *Ecology* 76:1371–1382.

Marquis, R. J., and C. Whelan. 1994. Insectivorous birds increase growth of white oak through consumption of leaf-chewing insects. *Ecology* 75:2007–2014.

McIntosh, R. P. 1985. *The Background of Ecology: Concept and Theory.* Cambridge University Press, Cambridge.

Nichols, F. H. et al. 1986. The modification of an estuary. *Science* 231:567–573.

Price, P. W. 1996. Empirical research and factually based theory: What are their roles in entomology? *American Entomologist* 42(2): 209–214.

Reisewitz, S. E., J. A. Estes, and C. A. Simenstad. 2006. Indirect food web interactions: Sea otters and kelp forest fishes in the Aleutian archipelago. *Oecologia* 146(4):623–631.

Schneider, D. C. 2001. The rise of the concept of scale in ecology. *BioScience* 51(7):545–553.

Sinclair, A. R. E., and J. M. Frywell. 1985. The Sahel of Africa: Ecology of a disaster. *Canadian Journal of Zoology* 63:987–994.

Urban, D. L., R. V. O'Neill, and H. H. Shugart, Jr. 1987. Landscape ecology. *BioScience* 37:119–127.

Wilson, E. O. 1992. *The Diversity of Life.* W. W. Norton, New York.

Worster, D. 1994. *Nature's Economy.* Cambridge University Press, Cambridge.

CHAPTER 2

Adaptations to the Physical Environment: Water and Nutrients

Sperm whales routinely dive to depths of 500 meters, staying below the surface for more than an hour searching for fish, squid, and other food items that are found at these depths. Like all mammals, sperm whales breathe air to obtain oxygen. While diving, however, a sperm whale has to rely on oxygen stored in its body. It might surprise you to learn that very little of this oxygen resides in the lungs. Most of it is bound to hemoglobin in the blood or to a similar oxygen storage molecule, myoglobin, in the muscles. While under water, deep-diving mammals slow down their metabolism considerably by reducing blood flow to nonvital organs, such as the skin, viscera, lungs, kidneys, and muscles (which have their own oxygen supply bound to myoglobin); blood flow is maintained primarily to the brain and heart. Consequently, the temperature of all but a few key organs drops, the heart rate slows, and demand for oxygen drops to a minimum.

If having enough oxygen is one challenge faced by deep-diving mammals, another is keeping warm. Most deep-diving mammals living in cold waters, including the Weddell seal, a native of Antarctic waters (Figure 2.1), are well insulated by a thick layer of fat under the skin, which slows the conduction of heat generated by its internal organs to the surrounding water. Unlike mammals, diving birds, such as the Adélie penguin, are insulated against the cold by air trapped in their plumage. Although air makes great insulation, it creates a problem for diving animals because it increases buoyancy. Imagine trying to dive under water wearing a life jacket! Indeed, a penguin expends most of its energy during a dive overcoming its positive buoyancy. The air trapped in its plumage also enlarges a penguin's waistline and therefore increases the drag of the water on its body. As the penguin dives deeper, however, the pressure of the water compresses the air in the plumage. As its volume decreases, buoyancy and drag also decrease, until

(a)

(b)

FIGURE 2.1 The Weddell seal and the Adélie penguin are excellent divers. (a) The Weddell seal (*Leptonychotes weddellii*), though clumsy on land, can dive to over 500 meters depth and remain submerged for up to an hour and 20 minutes. (b) The Adélie penguin (*Pygoscelis adeliae*) must overcome the buoyancy of its plumage to descend, but that buoyancy also allows it to ascend rapidly. Photo (a) by R. E. Ricklefs; photo (b) by Roland Seitre/Peter Arnold.

at 60 meters depth (where the pressure is 6 times that at the surface), the penguin has the same density as seawater and is neutrally buoyant. At the end of a dive, as the penguin swims upward, the air trapped in the plumage expands, and its positive buoyancy now propels the penguin toward the surface, fast enough that it can jump onto an ice floe a meter or more above the surface.

These aerobic and hydrodynamic limits confronting diving mammals and birds illustrate some of the ways in which organisms are restricted by their physical environments. In this chapter, we shall explore ways in which the properties of water both support and constrain aquatic and terrestrial plants and animals. These properties also direct the evolution of adaptations. The tendency of organisms to adapt to their environments helps us to understand why they are found under the conditions to which they are best suited. After all, whales are pretty helpless out of water!

CHAPTER CONCEPTS

- Water has many properties favorable to life
- Many inorganic nutrients are dissolved in water
- Plants obtain water and nutrients from the soil by the osmotic potential of their root cells
- Forces generated by transpiration help to move water from roots to leaves

- Salt balance and water balance go hand in hand
- Animals must excrete excess nitrogen without losing too much water

We often speak of the living and the nonliving as opposites. But although we can easily distinguish the two, life does not exist in isolation from its abiotic environment. Life depends on the physical world. Water is the basic medium of life, and energy powers life processes. Indeed, living organisms themselves are physical as well as biological systems.

Organisms, in turn, affect the physical world: soils, the atmosphere, lakes and oceans, and many sedimentary rocks owe their properties in part to the activities of organisms. Many conditions favorable for the development and maintenance of life depend on the activities of living organisms. The oxygen in the atmosphere, for example, owes its existence to photosynthetic microorganisms early in the history of life and, later, to plants. Today, for the first time in the earth's history, a single species is capable of significantly changing the physical world within the lifetime of a single individual. This sobering realization that we humans have an overriding global impact emphasizes the close relationship between the physical and biological dominions.

Although they are distinct from purely abiotic systems, organisms nonetheless function within limits set by physical laws. The physical world provides the context for life, but also constrains its expression. Life exists out of equilibrium with the physical world. Biological systems must use energy to counteract the physical forces of gravity, heat flow, diffusion, and chemical reaction.

The most important distinction between biological and abiotic systems is that living organisms have a purposeful existence. Their structures, physiological processes, and behaviors, shaped by evolutionary responses to natural selection, are directed toward procuring energy and resources that are ultimately used to produce offspring. Certainly life is constrained by physics and chemistry, just as architecture is constrained by the properties of building materials. However, as in biological systems, the purpose of the design of a building is unrelated to, and transcends, the qualities of bricks and mortar.

In the final analysis, life is a unique part of the physical world, but it exists in a state of constant tension with its physical surroundings. Organisms ultimately receive their energy from sunlight and their nutrients from soil and water, and they must tolerate extremes of temperature, moisture, salinity, and the other physical factors of their surroundings. In this chapter and the next, we shall explore some of the attributes of the physical environment that are most consequential for life. Because life processes take place in an aqueous environment, and because water makes up the largest proportion of all organisms, water seems a logical place to start.

Water has many properties favorable to life

Water is abundant over most of the earth's surface, and within the temperature range usually encountered there, it is liquid. Because water also has an immense capacity to dissolve inorganic compounds, it is an excellent medium for the chemical processes of living systems. It is hard to imagine life having any other basis than water. No other common substance is liquid under most conditions at the earth's surface. Organisms can move only because the aqueous substrate of life is fluid. Gases are also fluid, but the higher density of water is needed to achieve the concentrations of molecules necessary for rapid chemical reactions. Try to imagine life based on a rigid solid or a thin gas.

The thermal properties of water

Water stays liquid over a broad range of temperatures because it resists changes in temperature. The temperature of water remains steady even when heat is removed

FIGURE 2.2 Water expands and becomes less dense as it freezes. Because the density of ice is 0.92 g per cm^3 (just slightly less than that of liquid water, which is 1 g per cm^3), this Antarctic iceberg floats, but more than 90% of its bulk lies below the sea surface. Photo by R. E. Ricklefs.

or added rapidly, as can happen at the air–water interface or at an organism's surface. Water also resists changes between its solid (ice), liquid, and gaseous (water vapor) states. Over 500 times as much energy must be added to evaporate a quantity of water as to raise its temperature by 1°C! Freezing requires the removal of 80 times as much heat as is needed to lower the temperature of the same quantity of water by 1°C. This property helps to keep large bodies of water from freezing solid during winter. In addition, water conducts heat rapidly, so thermal energy tends to spread evenly throughout a body of water, further slowing local changes in temperature.

Another curious, but fortunate, thermal property of water is that, whereas most substances become denser at colder temperatures, water becomes less dense as it cools below 4°C. Moreover, upon freezing, water expands farther and becomes even less dense. Consequently, ice floats (Figure 2.2). This property of water prevents the bottoms of lakes and oceans from freezing and enables aquatic plants and animals to find refuge there in winter.

The density and viscosity of water

Because water is dense (800 times denser than air), it provides support for organisms, which, after all, are themselves mostly water. But water is also viscous, meaning that it resists flow or the movement of a body through it. These physical properties of water create a favorable environment for life, but at the same time place limits on its form and function.

Organisms often deal with such limits by taking advantage of the physical properties of natural substances or by exploiting physical principles. For example, animals and plants contain bone, proteins, and other materials that are

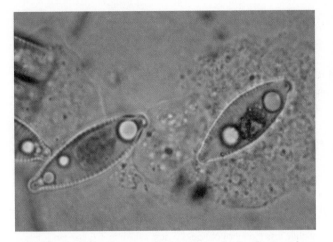

FIGURE 2.3 **The droplets of oil in these algal cells provide buoyancy.** Photo by Larry Jon Friesen/Saturdaze.

FIGURE 2.5 **Appendages prevent small aquatic animals from sinking.** These long, filamentous and feathery projections from the body of a tropical marine planktonic crustacean retard sinking. Photo by Image Quest 3-D.

denser than salt water and much denser than fresh water. These materials would cause aquatic organisms to sink, were it not for a variety of adaptations that reduce their density or retard their rate of sinking. Many fish species have a gas-filled swim bladder whose size can be adjusted to make the density of the fish's body equal to that of the surrounding water. Some large kelps have gas-filled bulbs that float their leaves to the sunlit surface waters (see Figure 1.16). At the other end of the size spectrum, many of the microscopic unicellular algae that float in great numbers in the surface waters of lakes and oceans use droplets of oil as flotation devices. With 90%–93% of the density of pure water, these droplets compensate for the natural tendency of the algal cells to sink (Figure 2.3).

The high viscosity of water hampers movement, so it is no surprise that fast-moving aquatic animals have evolved

streamlined shapes, which reduce the drag encountered in moving through a dense and viscous medium (Figure 2.4). As animals become smaller, the momentum of their movement decreases relative to the viscosity of water. A tiny water flea, viewed from a human perspective, seems to be swimming in molasses. But what impedes swimming also prevents sinking. Many tiny marine animals have evolved long, filamentous appendages that take advantage of the viscosity of water to retard sinking (Figure 2.5), just as a parachute slows the fall of a body through air.

Many inorganic nutrients are dissolved in water

Organisms require a variety of chemical elements to build necessary biological structures and maintain life processes (Table 2.1). The elements they require in the greatest quantities are hydrogen, carbon, and oxygen, which are the elements in carbohydrates. Organisms also require varying quantities of nitrogen, phosphorus, sulfur, potassium, calcium, magnesium, and iron. Certain organisms need other elements in abundance as well. For example, diatoms construct their glassy shells from silicates (Figure 2.6); tunicates, which are sessile marine invertebrates, accumulate vanadium in high concentrations, possibly as a defense against predators; nitrogen-fixing bacteria require molybdenum as a part of the enzyme they use to assimilate nitrogen from the atmosphere. Animals ultimately acquire these nutrients from plants, and plants acquire most of them from water.

FIGURE 2.4 **Streamlined shapes reduce the drag of water.** The sleek bodies of barracudas (*Sphyraena*) allow them to swim rapidly through water with a relatively low expenditure of energy. Photo by Larry Jon Friesen/Saturdaze.

TABLE 2.1	Major nutrients required by organisms, and some of their primary functions
Element	Function
Nitrogen (N)	Structural component of proteins and nucleic acids
Phosphorus (P)	Structural component of nucleic acids, phospholipids, and bone
Sulfur (S)	Structural component of many proteins
Potassium (K)	Major solute in animal cells
Calcium (Ca)	Structural component of bone and of material between woody plant cells; regulator of cell permeability
Magnesium (Mg)	Structural component of chlorophyll; involved in the function of many enzymes
Iron (Fe)	Structural component of hemoglobin and many enzymes
Sodium (Na)	Major solute in extracellular fluids of animals

(a)   (b)

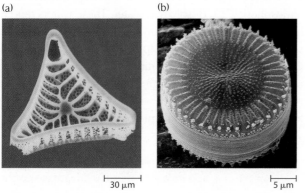

$\vdash$ 30 μm $\dashv$ $\vdash$ 5 μm $\dashv$

FIGURE 2.6 Diatoms use silicates to build their shells. The glassy outer shells of these photosynthetic protists take diverse forms, as shown by these scanning election micrographs of (a) *Entogonia* and (b) *Cyclotella*. Photo (a) by F. Rossi; photo (b) by Ann Smith/Photo Researchers.

The solvent capacity of water

Water has an impressive capacity to dissolve various substances, making them accessible to living systems and providing a medium within which they can react to form new compounds. Water is a powerful solvent because water molecules are strongly attracted to many solids. Some solid compounds consist of electrically charged atoms or groups of atoms called **ions.** For example, common table salt, sodium chloride (NaCl), contains positively charged sodium ions (Na^+) and negatively charged chloride ions (Cl^-). In its solid state, these ions are arranged in close proximity in a crystal lattice. In water, however, the crystal lattice dissolves. The charged sodium and chloride ions are powerfully attracted by water molecules, which them-

selves have both positive and negative charges. In fact, these forces of attraction are stronger than the forces that hold salt crystals together. Therefore, the crystals readily separate into their component ions when surrounded by water molecules—another way of saying that the salt dissolves.

The powerful solvent properties of water are responsible for the presence of minerals in streams, rivers, lakes, and oceans. Water vapor in the atmosphere condenses to form clouds and, eventually, precipitation (rain and snow). When it condenses, the water is nearly pure, except for dissolved atmospheric gases (principally nitrogen, oxygen, and carbon dioxide). Rainwater acquires some minerals from dust particles and droplets of ocean spray in the atmosphere as it falls. As it flows over and under the ground toward the ocean, it picks up additional minerals from rocks and soils. So-called "hard" water has high concentrations of dissolved calcium. The water in most lakes and rivers contains 0.01%–0.02% dissolved minerals, far less than the average concentration of the ocean (3.4% by weight), in which salts and other minerals have accumulated over several billion years.

The ocean functions like a large still, concentrating ions as mineral-laden water arrives via streams and rivers and as pure water evaporates from its surface. Here the concentrations of some elements, particularly calcium, reach limits set by the maximum solubility of the compounds they form. In the oceans, calcium ions (Ca^{2+}) readily combine with dissolved carbon dioxide to form calcium carbonate, which is soluble only to the extent of 0.014 grams per liter (g per L) of water, or 0.0014% by weight. Its concentration in the oceans reached this level eons ago, so the excess calcium carbonate formed from calcium ions washing into the oceans each year

FIGURE 2.7 Limestone is made of calcium carbonate. The limestone sediments that form many mountains represent calcium carbonate precipitated out of solution in shallow seas. Photo by Larry Jon Friesen/Saturdaze.

precipitates to form limestone sediments (Figure 2.7). This reaction serves as an important sink for carbon dioxide. At the other extreme, the solubility of sodium compounds, such as sodium chloride (360 g per L) and sodium bicarbonate (69 g per L), far exceeds the concentration of sodium in seawater (approximately 10 g per L at present). Most of the sodium chloride washing into ocean basins remains dissolved, so the concentration of this compound in seawater has continued to increase over geologic time.

Hydrogen ions in ecological systems

Among dissolved substances in water, hydrogen ions (H^+) deserve special mention because they are extremely reactive. At high concentrations, they affect the activities of most enzymes and have other, generally negative consequences for life processes. They also play a crucial role in dissolving minerals from rocks and soils.

The concentration of hydrogen ions in a solution is referred to as its **acidity.** The acidity is commonly measured as **pH,** which is the negative of the common logarithm of hydrogen ion concentration, measured in moles per liter (Figure 2.8). In pure water at any given time, a small fraction of the water molecules (H_2O) are dissociated into their hydrogen (H^+) and hydroxide (OH^-) ions. The pH of pure water, which is defined as neutral pH, is 7, which means that the concentration of hydrogen ions is 10^{-7} (0.0000001) moles per liter, or one ten-millionth of a gram per kilogram of water. In contrast, strong acids, such as sulfuric acid (H_2SO_4) and hydrochloric acid (HCl), dissociate almost completely when dissolved in water. At high concentrations, these acids can produce pH values approaching 0, which is 1 mole (the equivalent of 1 gram) of H^+ per liter. The acid in your stomach has a pH of 1 (0.1 mole per liter). Most natural waters contain

weak acids, such as carbonic acid (H_2CO_3), formed when atmospheric CO_2 dissolves in water. These waters tend to have pH values close to neutral. Some natural waters are somewhat basic, or alkaline (pH > 7), having an excess of OH^- over H^+. The normal range of pH in natural waters is between 6 and 9, although small ponds and streams in regions with acid rainfall, or which are polluted by sulfuric acid draining out of coal mining wastes, can reach pH values as low as 4.

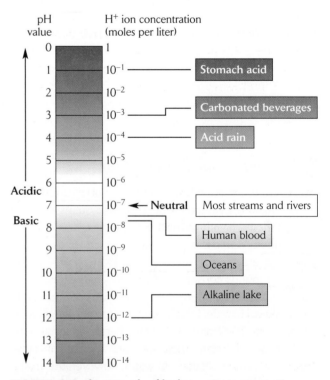

FIGURE 2.8 The pH scale of hydrogen ion concentration extends from 0 (highly acidic) to 14 (highly alkaline).

Hydrogen ions, because of their high reactivity, dissolve minerals from rocks and soils, enhancing the natural solvent properties of water. For example, in the presence of hydrogen ions, the calcium carbonate in limestone dissolves readily, according to the chemical equation

$$H^+ + CaCO_3 \rightarrow Ca^{2+} + HCO_3^-$$

Calcium ions are important to life processes, and their presence at high concentrations is vital to organisms, such as snails, that form shells made of calcium carbonate. Indeed, mollusks are less abundant and diverse in streams and lakes that are nutrient poor. Thus, hydrogen ions are essential for making certain nutrients available for life processes. However, this same reactivity of hydrogen ions helps to dissolve highly toxic heavy metals, such as arsenic, cadmium, and mercury. When made soluble in natural waters, these metals are detrimental to life processes.

Plants obtain water and nutrients from the soil by the osmotic potential of their root cells

Plants acquire the inorganic nutrients they need—other than oxygen, carbon, and some nitrogen—as ions dissolved in water in the soil around their roots. Nitrogen exists in soil as ammonium (NH_4^+) and nitrate ions (NO_3^-), phosphorus as phosphate ions (PO_4^{3-}), calcium and potassium as their elemental ions Ca^{2+} and K^+, respectively. The availability of these and other inorganic nutrients varies with their chemical form in the soil and with temperature, acidity, and the presence of other ions. The scarcity (relative to need) of inorganic nutrients often limits plant growth. Phosphorus, in particular, often limits plant production in terrestrial environments; even when phosphorus is abundant, most of the compounds it forms in soil do not dissolve easily. We shall have much more to say about nutrient uptake by plants in later chapters.

Soil structure and water-holding capacity

Most terrestrial plants obtain the water they need from the soil. The amount of water in soil and its availability to plants varies with the physical structure of the soil. Because of their electrical charges, water molecules cling to one another by hydrogen bonding (the basis for surface tension) and to the surfaces of soil particles (a tendency known as capillary attraction). This clinginess is the reason why soil is able to retain water. The more surface area a soil has per unit of volume, the more water that soil can hold.

Soils consist of particles of clay, silt, and sand, as well as particles of organic material, in varying proportions. Clay particles are the smallest, at less than 0.002 mm in diameter; silt particles vary from 0.002 to 0.05 mm in diameter; and sand particles are the largest, at more than 0.05 mm. Because the total surface area of particles in a given volume of soil increases as particle size decreases, soils with abundant clay and silt hold more water than coarse sands, through which water drains quickly (Figure 2.9). However, because clay particles are small and hold water tightly, less water is available to plants in a clay soil than in a soil with a mixture of particles of different sizes, commonly called a loam.

Plant roots easily take up water that clings loosely to soil particles. But close to the surfaces of soil particles, water adheres tightly by more powerful forces of attraction. The strength of the attractive forces holding water in the soil is called the **water potential** of the soil. Most of this water potential is generated by the attraction of water to the surfaces of soil particles—the soil matrix—so it is often referred to as the **matric potential.** As we shall see below, plants must also overcome the pull of the earth's gravity and the diffusion of water from the roots into the soil due to the presence of dissolved substances in soil water.

By convention, pure water is defined as having a water potential of 0. Water always moves from a higher to a lower water potential. Soil has a negative water potential because soil attracts water from a pure solution with a

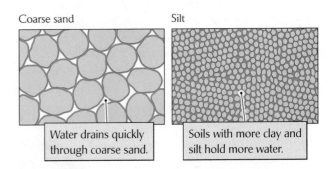

Coarse sand   Silt

Water drains quickly through coarse sand.   Soils with more clay and silt hold more water.

FIGURE 2.9 Soils with smaller particles hold more water. Soils with large particles have large spaces between them that are not completely filled with water at field capacity. Soils with very small particles hold more water, but they hold it so tightly that its availability to plants is reduced.

water potential of 0. Consequently, plants must develop a water potential lower than that of the soil to overcome the matric potential and extract water. Soil scientists quantify water potential in units of pressure, called megapascals (MPa). For reference, the standard atmospheric pressure at the surface of the earth is approximately one-tenth of a megapascal (0.1 MPa).

Matric potential is greatest right at the surfaces of soil particles and decreases with distance from them. Water held by a matric potential of less than about −0.01 MPa drains out of the soil under the pull of gravity and joins the groundwater in the crevices of the bedrock below. Water drains through the interstices between large soil particles as long as it is more than about 0.005 mm (five thousandths of a millimeter!) from their surfaces. The amount of water held against gravity by a matric potential of less than −0.01 MPa is called the **field capacity** of the soil. The field capacity represents the maximum amount of water available to a plant in well-drained soil. How much water is this? Imagine a particle of silt with a diameter of 0.01 mm enlarged to the size of this page (×25,000). The film of water held at field capacity would be as thick as half the width of the page. The volume of water in soil at field capacity varies from about 10% of the total soil volume for sandy soils to 50% for soils dominated by fine clay particles.

As soils dry out, the remaining water is held ever more tightly because a greater proportion of that water lies close to the surfaces of soil particles. Soils with water potentials as low as −10 MPa are very dry. Most crop plants can extract water from soils with water potentials down to about −1.5 MPa. At lower water potentials, these plants wilt, even though some water still remains in the soil. Agronomists and ecologists refer to a water potential of −1.5 MPa as the **wilting coefficient** or **wilting point** of the soil. This is only a general rule of thumb, however, because many drought-adapted species can extract water from even drier soils.

Osmotic potential and water uptake by plants

Water in the environment, and in organisms, contains dissolved substances, called **solutes,** which influence the diffusion of water molecules. Plants take advantage of the tendency of water to move from regions of low solute concentration to regions of high solute concentration (Figure 2.10). When the fluid in a cell has a high concentration of ions and other solutes (and thus a low water potential), water tends to move from the surrounding environment into the cell. This process is called **osmosis.** The force with which an aqueous solution attracts water by osmosis is known as its **osmotic potential.** Like the matric potential of soil, osmotic potential is expressed in units of pressure. It is the osmotic potential in the roots of trees that causes water to enter the roots from the soil

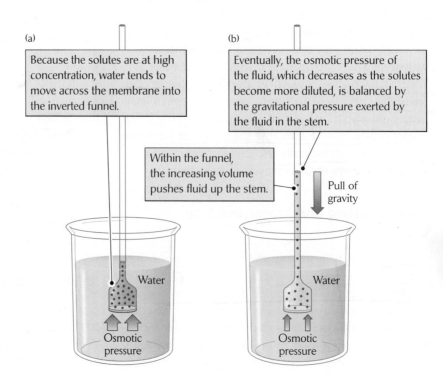

FIGURE 2.10 Solutes enclosed within a membrane that is permeable to water create an osmotic potential. Plant roots use this principle to draw water from the soil and to develop sufficient osmotic pressure to push water up into the stem.

(a) Because the solutes are at high concentration, water tends to move across the membrane into the inverted funnel.

Within the funnel, the increasing volume pushes fluid up the stem.

Water

Osmotic pressure

(b) Eventually, the osmotic pressure of the fluid, which decreases as the solutes become more diluted, is balanced by the gravitational pressure exerted by the fluid in the stem.

Pull of gravity

Water

Osmotic pressure

against the attraction of soil particles and the downward pull of gravity.

A complicating factor is that ions and other solutes diffuse through water from regions of high solute concentration to regions of low solute concentration. So, as water is coming into the cell, solutes tend to move out. Eventually, solute concentrations within the cell and in the surrounding water would come into equilibrium. At this point, the osmotic potentials of the cell and its surroundings would be equal, and there would be no net movement of water across the cell membrane. Cells prevent this equalization of osmotic potential in two ways. First, a cell membrane can be **semipermeable,** meaning that some small molecules and ions can diffuse across it, but larger ones cannot. Many carbohydrates and most proteins are too large to pass through the pores of a cell membrane, so they remain inside the cell and help to maintain its low water potential. Second, cell membranes can transport ions and small molecules actively against a concentration gradient to maintain their concentrations within the cell. This **active transport** requires expenditure of energy.

The osmotic potential generated by an aqueous solution depends on its solute concentration. More specifically, it depends on the number of solute molecules or ions per volume of solution. Thus, a given mass of a small solute molecule generates greater osmotic potential than the same mass of a larger molecule. You will remember from your introductory chemistry course that the concentration of molecules in solution is expressed in terms of gram molecular weights, or moles, per liter. For example, the sugar glucose ($C_6H_{12}O_6$) has a molecular weight of 180, and so a 1 molar solution contains 180 grams of glucose per liter of water. The amino acid alanine ($C_3H_7NO_2$) has a molecular weight of 89, and so the same mass of that substance per liter of water would contain twice as many molecules and have twice the osmotic potential of glucose.

Plants growing in deserts and salty environments can lower the water potential of their roots to as much as -6 MPa, thereby overcoming soil water potentials down to -6 MPa, by increasing the concentrations of amino acids, carbohydrates, or organic acids in their root cells. They pay a high metabolic price, however, to maintain high concentrations of dissolved substances.

Forces generated by transpiration help to move water from roots to leaves

Osmotic potential draws water from the soil into the cells of plant roots. But how does that water get from the roots to the leaves? Plants conduct water to their leaves through xylem elements, which are the empty remains of xylem cells in the cores of roots and stems, connected end-to-end to form the equivalent of water pipes. Osmotic potential in the roots, which draws water from the soil into the plant, creates a **root pressure** that forces water into the xylem elements. However, this pressure is counteracted by gravity and the osmotic potential of living root cells, and under the best circumstances it can raise water to a height of no more than about 20 meters, far short of the leaves of the tallest trees.

Leaves themselves generate water potential when water evaporates from leaf cell surfaces into the air spaces within the leaves, a process known as **transpiration.** The column of water in a xylem element is continuous from the roots to the leaves, since it is held together by hydrogen bonds between the water molecules. Thus, low water potentials in leaves can literally draw water upward through the xylem elements against the osmotic potential of the living root cells and the pull of gravity. Dry air at 20°C has a water potential of -133 MPa. The water potential in the air spaces within leaves is never this low because of retained water vapor, but it is low enough under most conditions to pull water through the roots, xylem, and leaves. Thus, transpiration creates a continuous gradient of water potential as high as -2 to -5 MPa from leaf surfaces in contact with the atmosphere to the surfaces of root hairs in contact with soil water. This explanation of the mechanism of water movement from roots to leaves is known as the **cohesion–tension theory** (Figure 2.11).

Although transpiration generates a powerful force, when the soil reaches the wilting point, water lost from the leaves of a plant can no longer be replaced by new water moving up from the roots. To prevent further water loss from the leaves, plants have various mechanisms for controlling transpiration. Most of the cells on the exterior of a leaf are coated with a waxy cuticle that retards water loss. Gas exchange between the atmosphere and the interior of the leaf occurs through small openings at the leaf surfaces, called **stomates** (Figure 2.12). (Many botanists prefer the term **stomata,** singular **stoma,** from the Latin for "mouth.") The stomates are the points of entry for CO_2 and the exits for water escaping to the atmosphere by transpiration. Plants can reduce water loss by closing their stomates. As leaf water potential decreases, the so-called guard cells bordering a stomate collapse slightly, which causes them to press together and shut off the opening. Closing of the stomates prevents further water from escaping, but it also prevents the carbon dioxide required for photosynthesis from entering the leaf. Such compromises are simply a fact of life.

FIGURE 2.11 The cohesion–tension theory explains the movement of water from the roots to the leaves of a plant. The water potential that draws water upward is generated by transpiration.

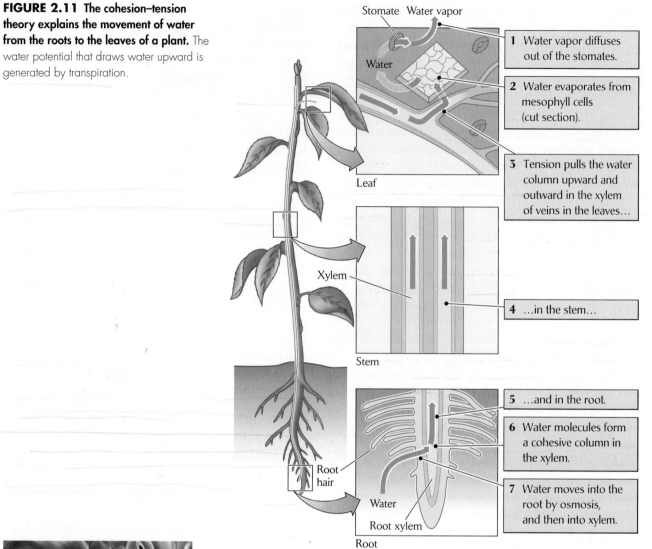

Leaf

Stomate Water vapor

Water

1 Water vapor diffuses out of the stomates.

2 Water evaporates from mesophyll cells (cut section).

3 Tension pulls the water column upward and outward in the xylem of veins in the leaves…

Xylem

Stem

4 …in the stem…

Root hair

Water

Root xylem

Root

5 …and in the root.

6 Water molecules form a cohesive column in the xylem.

7 Water moves into the root by osmosis, and then into xylem.

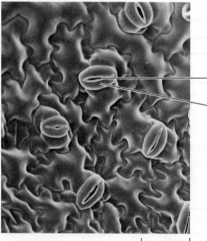

Stomate

Guard cell

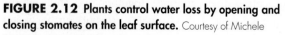

50 μm

FIGURE 2.12 Plants control water loss by opening and closing stomates on the leaf surface. Courtesy of Michele McCauley, from P. H. Raven, R. F. Evert, and S. E. Eichorn, *Biology of Plants*, 6th ed., W. H. Freeman and Company and Worth Publishers, New York (1999), p. 630.

Salt balance and water balance go hand in hand

To maintain the proper amounts of water and dissolved substances in their bodies, organisms must balance losses with intake. Often, organisms take in water with a solute concentration that differs from that of their bodies, so they must either acquire additional solutes to make up the deficit or rid themselves of excess solutes. When water evaporates from the surfaces of terrestrial organisms into the atmosphere, solutes are left behind, and their concentration in the body tends to increase. Under such circumstances, organisms must excrete excess salts to maintain the proper concentrations in their bodies. Salt concentrations that are too high can change the way proteins interact with other molecules and disrupt cell function.

Because solutes determine the osmotic potential of body fluids, the mechanisms that organisms use to maintain a proper salt balance are referred to as **osmoregulation.**

Management of salt balance by plants

Terrestrial plants transpire hundreds of grams of water for every gram of dry matter they accumulate in tissue growth, and they inevitably take up dissolved salts along with the water that passes into their roots. Where salt concentrations in soil water are high, plants pump excess salts back into the soil by active transport across their root surfaces, which therefore function as the plant's "kidneys." Mangroves are plants that grow on coastal mudflats that are inundated daily by high tides (Figure 2.13). Not only does this habitat impose a high salt load, but the high osmotic potential of the saltwater environment also makes it difficult for the roots to take up water. To counter these problems, many mangroves maintain high concentrations of organic solutes—various amino acids and small sugar molecules—in their roots and leaves to increase their osmotic potential. In addition, salt glands in the leaves secrete salt by active transport to the exterior leaf surface. Many mangrove species also exclude salts from their roots by active transport. Because many of these adaptations parallel those of plants from environments with scarce water, the mangrove habitat can be thought of as an osmotic desert, even though plant roots are frequently immersed in water.

Water balance and salt balance in terrestrial animals

Water is as important to animals as it is to plants. Terrestrial animals, with their internalized gas exchange surfaces, are less vulnerable to respiratory water loss than plants are, and because they are not immersed continuously in water like aquatic organisms, they have little trouble retaining ions. They acquire the mineral ions they need in the water they drink and the food they eat, and they use water to eliminate excess salts in their urine. Where fresh water abounds, animals can drink large quantities of water to flush out salts that would otherwise accumulate in the body. Where water is scarce, however, animals must produce concentrated urine to conserve water.

As one would expect, desert animals have champion kidneys. For example, whereas human kidneys can concentrate most solutes in their urine to about 4 times the levels in their blood plasma, the kangaroo rat's kidneys produce urine with solute concentrations as high as 14 times the levels in its blood. However, because sodium

(a)

(b)

FIGURE 2.13 Mangroves have adaptations for coping with a high salt load. (a) The roots of mangroves are immersed in salt water at high tide. (b) Specialized glands in the leaves of the button mangrove (*Conocarpus erecta*) excrete salt, which precipitates on their outer surfaces. Photos by R. E. Ricklefs.

and chloride ions participate in the mechanism by which the animal kidney retains water, the kidney does not excrete these ions efficiently. Hence, many animals lacking access to fresh water have specialized salt-secreting organs that work on a different principle than the kidney, more like the salt glands of mangrove plants. The "salt glands" of birds and reptiles, which are particularly well developed in marine species, are actually modified tear glands located in the orbit of the eye, capable of secreting a concentrated salt solution.

These adaptations help animals to balance their water budget on land, but even aquatic animals face challenges in the management of water.

Water exclusion in freshwater animals

The water balance of aquatic animals is closely tied to the concentrations of salts and other solutes in their body tissues and in the environment. The body fluids of vertebrate animals, which have an osmotic potential of about -0.3 to -0.5 MPa, occupy an intermediate position between fresh water (with an osmotic potential close to zero) and seawater (-1.2 MPa). Thus, the tissues of freshwater fish have higher salt concentrations than the surrounding water. Such organisms, which are referred to as **hyperosmotic,** tend to gain water from their surroundings and lose solutes.

Freshwater fish continuously gain water by osmosis across the surfaces of the mouth and gills, which are the most permeable of their tissues that are exposed to the freshwater environment, as well as in their food (Figure 2.14). To counter this influx, the fish eliminate excess water in their urine. If the fish did not also selectively retain solutes, however, they would soon become lifeless bags of water. The kidneys of freshwater fish retain salts by actively removing ions from the urine and infusing them back into the bloodstream. In addition, the gills can selectively absorb ions from the surrounding water and release them into the bloodstream.

Water retention in marine animals

Marine fish are surrounded by water with a salt concentration higher than that of their bodies; in other words, they are **hypo-osmotic.** As a result, they tend to lose water to the surrounding seawater and must drink seawater to replace it (see Figure 2.14). The salts that come in with that water and with their food, as well as the salts that diffuse in across their body surfaces, must be excreted by the gills and kidneys.

Some sharks and rays have found a unique solution to the problem of water balance. Sharks retain urea—a common nitrogenous by-product of protein metabolism in vertebrates—in the bloodstream, instead of excret-

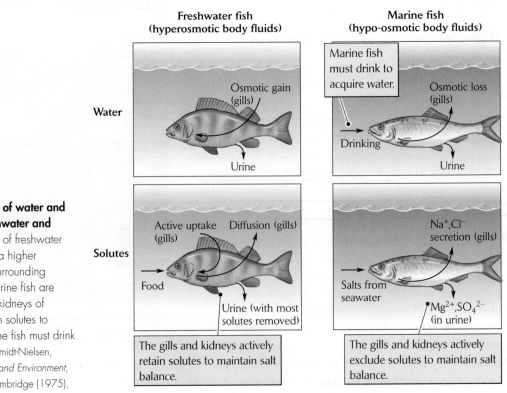

FIGURE 2.14 Exchanges of water and solutes differ between freshwater and marine fish. The body fluids of freshwater fish are hyperosmotic (have a higher salt concentration than the surrounding water), whereas those of marine fish are hypo-osmotic. The gills and kidneys of fish actively exclude or retain solutes to maintain salt balance. Marine fish must drink to acquire water. After K. Schmidt-Nielsen, *Animal Physiology: Adaptation and Environment,* Cambridge University Press, Cambridge (1975).

Freshwater fish (hyperosmotic body fluids)

Marine fish (hypo-osmotic body fluids)

Water

Osmotic gain (gills)

Urine

Marine fish must drink to acquire water.

Osmotic loss (gills)

Drinking

Urine

Solutes

Active uptake (gills)   Diffusion (gills)

Food

Urine (with most solutes removed)

The gills and kidneys actively retain solutes to maintain salt balance.

Na^+, Cl^- secretion (gills)

Salts from seawater

Mg^{2+}, SO_4^{2-} (in urine)

The gills and kidneys actively exclude solutes to maintain salt balance.

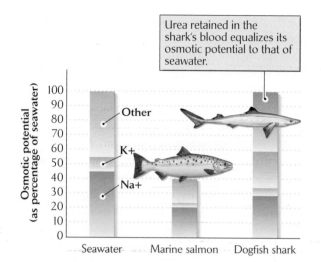

Urea retained in the shark's blood equalizes its osmotic potential to that of seawater.

FIGURE 2.15 Sharks match their total solute concentration to that of seawater. Sodium, potassium, urea, and other solutes (mostly chloride ions) contribute differently to the osmotic potential of seawater and of the body fluids of marine fish and sharks. From data in K. Schmidt-Nielsen, *Animal Physiology: Adaptation and Environment,* 5th ed., Cambridge University Press, London and New York (1997), Table 8.6.

ing it from the body in their urine as other animals do. Concentrations of urea up to 2.5% (compared with less than 0.03% in other vertebrates) raise the osmotic potential of their blood to the level of seawater without any increase in the concentrations of sodium and chloride ions (Figure 2.15). Consequently, the movement of water across a shark's body surfaces is balanced, with neither gain nor loss. This adaptation frees sharks and rays from having to drink extra salt-laden water to replace water lost by osmosis. The observation that freshwater species of rays do not accumulate urea in their blood empha-

sizes the importance of urea for osmoregulation in marine members of this group. The downside of retaining urea is that urea impairs protein function. Sharks and many other marine organisms that use urea to maintain their water balance also accumulate high concentrations of a compound called trimethylamine oxide to protect proteins from its negative effects.

ECOLOGISTS IN THE FIELD **Flip-flopping osmoregulation in a small marine invertebrate.** The small copepod *Tigriopus* is exposed to widely fluctuating salt concentrations over short periods and must adjust its physiology rapidly to compensate for these changes. *Tigriopus* lives in splash pools high in the intertidal zone along rocky coasts (Figure 2.16), which receive seawater infrequently from the splash of high waves. As the water evaporates, the salt concentration in these pools rises to high levels. However, a heavy rainfall can rapidly lower the salt concentration, causing a rapid reversal of environmental conditions.

Ron Burton, at the Scripps Institute of Oceanography, has shown that *Tigriopus,* like sharks and rays, manages its water balance by changing the osmotic potential of its body fluids. When the salt concentration in a pool is high, individuals synthesize large quantities of certain amino acids such as alanine and proline. These small molecules increase the osmotic potential of the body fluids to match that of the environment without the deleterious physiological consequences of high levels of salts or urea.

This response to excess salts in the environment is costly, however. In a laboratory experiment, individual *Tigriopus* were switched from 50% seawater to 100% seawater to mimic what might happen when waves at high tide filled a pool previously flushed with rainwater. In response to this change, the respiration rate of the copepods initially

(a)

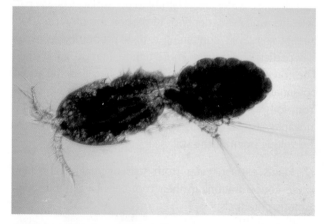

(b)

FIGURE 2.16 The tiny copepod *Tigriopus,* shown here with an attached egg mass, lives in splash pools high in the rocky intertidal zone in California. Courtesy of Ron Burton (a); R. E. Ricklefs (b).

declined, owing to salt stress, and then increased as they synthesized alanine and proline to restore their water balance. In a second experiment, the copepods were switched from 100% seawater to 50% seawater. In this case, the copepods' respiration rate immediately increased as they rapidly degraded and metabolized excess free amino acids to reduce their osmotic potential to that of their new environment.

Certain environments pose special osmotic challenges. The salt concentrations in some water-filled landlocked basins greatly exceed those of seawater and even splash pools, particularly in arid regions where evaporation outpaces precipitation. The Great Salt Lake in Utah contains 5%–27% salt—that is, up to 8 times more than normal seawater—depending on the water level. The osmotic potential of its water—well in excess of −10 MPa—would shrivel most organisms. However, a few aquatic creatures, such as brine shrimp (*Artemia*), thrive in the Great Salt Lake, providing an important food resource for birds and other creatures. Brine shrimp can survive in the Great Salt Lake because they excrete salt at a prodigious rate, and at a high energetic cost. They obtain the energy they need by feeding on the abundant photosynthetic bacteria that live in their hypersaline environment. ▌

Animals must excrete excess nitrogen without losing too much water

Most carnivores, whether they eat crustaceans, fish, insects, or mammals, consume excess nitrogen. This nitrogen is part of the proteins and nucleic acids in their diet, and it must be eliminated from the body when these compounds are metabolized. Most aquatic animals produce a simple metabolic by-product of nitrogen metabolism: ammonia (NH_3). Although ammonia is mildly poisonous to tissues, aquatic animals eliminate it rapidly in copious dilute urine, or directly across the body surface, before it reaches a dangerous concentration within the body.

Terrestrial animals cannot use large quantities of water to excrete excess nitrogen. Instead, they produce metabolic by-products that are less toxic than ammonia and which can therefore accumulate to higher levels in the blood and urine without danger. In mammals, this metabolic by-product is urea [$CO(NH_2)_2$], the same substance that sharks produce and retain to achieve osmotic balance in marine environments. Because urea dissolves in water, excreting it still requires some urinary water loss—how much depends on the concentrating power of the kidneys. Birds and reptiles have carried adaptation to terrestrial life one step further: they excrete nitrogen in the form of uric acid ($C_5H_4N_4O_3$), which crystallizes out of solution and can then be excreted as a highly concentrated paste in the urine.

Although excreting urea and uric acid conserves water, it is costly in terms of the energy lost in the organic carbon used to form these compounds. For each atom of nitrogen excreted, 0.5 and 1.25 atoms of organic carbon are lost in urea and uric acid, respectively. None are lost in the excretion of ammonia.

SUMMARY

1. Water is the basic medium of life. It is abundant over most of the earth's surface, it is liquid within the range of temperatures usually encountered there, and it is a powerful solvent. These properties of water make it an ideal medium for living systems.

2. Water conducts heat rapidly and resists changes in temperature and state. Temperatures are therefore relatively evenly distributed throughout bodies of water.

3. Water is denser, and provides more buoyancy, than air, but it is also more viscous and therefore impedes movement.

4. All natural waters contain dissolved substances picked up in the atmosphere or from soils and rocks through which water flows.

5. The concentration of hydrogen (H^+) ions in a solution is referred to as acidity and is expressed as pH. Most natural waters have pH values between 6 (slightly acidic) and 9 (slightly alkaline).

6. Because water clings tightly to the surfaces of soil particles, its availability depends in part on the physical structure of soil. Soils containing a high proportion of small clay particles hold water more tightly than do sandy soils. The force by which soil holds water is called the water potential of the soil. Most plants cannot remove water from soils with a water potential more negative than −1.5 megapascals (MPa). This water potential is referred to as the wilting point of the soil.

7. Plants extract water from soils by maintaining high solute concentrations in their root cells to generate high osmotic potentials.

8. According to the cohesion–tension theory, water is drawn from the roots to the leaves of a plant by a gradient in water potential generated by transpiration—the evaporation of water from leaf cell surfaces. When under water limitation, plants can reduce transpirational water losses by closing their stomates.

9. Terrestrial animals reduce their use of water for eliminating excess salts by concentrating salts in their urine or by excreting them through salt glands.

10. To maintain salt and water balance, freshwater animals, which are hyperosmotic, retain salts while excreting the water that continuously diffuses into their bodies.

11. Marine animals, which are hypo-osmotic, actively exclude salts. Some marine animals increase the concentrations of solutes, such as urea and amino acids, in their body fluids to match the osmotic potential of seawater and thus reduce the movement of water out of their bodies.

12. Nitrogenous by-products of protein metabolism are excreted as ammonia by most aquatic organisms, as urea by mammals, and as uric acid by birds and reptiles.

REVIEW QUESTIONS

1. For aquatic organisms, how can the viscosity of water both hinder and facilitate movement?

2. Describe how water changes in mineral content as one moves from rainwater to lake water and, eventually, to ocean water.

3. Why might bodies of water with low pH pose a danger to organisms that live in them?

4. Explain the relationship between soil particle size and the field capacity of soil.

5. Explain why the availability of water to plants is highest in soils with particle sizes intermediate between sand and clay.

6. How can we be sure that root pressure is not sufficient to explain the movement of water in trees?

7. For saltwater and freshwater fish, describe what would happen if they lacked their adaptations to control the movement of water and salts across their external surfaces.

8. Describe the costs and benefits associated with the different nitrogen products excreted by fish, mammals, and birds.

SUGGESTED READINGS

Canny, M. J. 1998. Transporting water in plants. *American Scientist* 86:152–159.

Chapin, F. S., III. 1991. Integrated responses of plants to stress. *BioScience* 41:29–36.

Feldman, L. J. 1988. The habits of roots. *BioScience* 38:612–618.

Hochachka, P. W., and G. N. Somero. 1984. *Biochemical Adaptation.* Princeton University Press, Princeton, N.J.

Koch, G., et al. 2004. The limits to tree height. *Nature* 428:851–854.

Kooyman, G. L., and P. J. Ponganis. 1997. The challenges of diving to depth. *American Scientist* 85:530–539.

Phleger, C. F. 1998. Buoyancy in marine fishes: Direct and indirect role of lipids. *American Zoologist* 38:321–330.

Ryan, M., and B. Yoder. 1997. Hydraulic limits to tree height and tree growth. *BioScience* 47:235–242.

Schenk, H. J., and R. B. Jackson. 2002. The global biogeography of roots. *Ecological Monographs* 72:311–328.

Schmidt-Nielsen, K. 1998. *Animal Physiology: Adaptations and Environment,* 5th ed. Cambridge University Press, London and New York.

Somero, G. N. 1986. From dogfish to dogs: Trimethylamines protect proteins from urea. *News in Physiological Sciences* 1:9–12.

Thomas, D. N., and G. S. Dieckmann. 2002. Antarctic sea ice—a habitat for extremophiles. *Science* 295:641–644.

Tracy, R. L., and G. E. Walsberg. 2002. Kangaroo rats revisited: Re-evaluating a classic case of desert survival. *Oecologia* 133:449–457.

Tyree, M. 1997. The cohesion–tension theory of sap ascent: Current controversies. *Journal of Experimental Botany* 48(315):1753–1765.

Vogel, S. 1981. *Life in Moving Fluids: The Physical Biology of Flow.* Princeton University Press, Princeton, N.J.

Vogel, S. 1988. *Life's Devices.* Princeton University Press, Princeton, N.J.

Wijesinghe, D. K., et al. 2001. Root system size and precision in nutrient foraging: Responses to spatial pattern of nutrient supply in six herbaceous species. *Journal of Ecology* 89:972–983.

CHAPTER 3

Adaptations to the Physical Environment: Light, Energy, and Heat

A mong the mammals, kangaroo rats and camels are well suited to life in nearly waterless deserts. When air temperatures approach the maximum tolerable body temperature, animals can dissipate heat only by evaporating water from their skin and respiratory surfaces. In hot deserts, however, water is scarce, and evaporative cooling is costly. Instead, animals become less active, seek cool microclimates, and sometimes undertake seasonal migrations to cooler regions.

Kangaroo rats avoid the desert's greatest heat by venturing out only at night (Figure 3.1); during the blistering heat of the day, they remain comfortably below ground in their cool, humid burrows. Ground squirrels take a different approach. They remain active during the day, and as you would expect, their body temperatures rise as they forage above ground, exposed to the hot sun. However, before their body temperatures become dangerously high, they return to their cool burrows, where they can lose heat without losing water. When their body temperatures have dropped enough, they are back out on the surface foraging. By shuttling back and forth between their burrows and the surface, ground squirrels can extend their activity into the heat of the day and pay a relatively small price in terms of water loss.

Camels are famous desert animals. To conserve water, they too allow their body temperatures to rise in the heat of the day—by as much as 6°C. The camel's large body size gives it a distinct advantage, however. With increasing size, the surface area of an animal, across which it absorbs heat and intercepts solar radiation, increases less rapidly than the animal's volume, which is the bulk that heats up. Consequently, the camel heats up so slowly that it can remain in the sun most of the day. It dumps excess heat at night to the cooler desert surroundings.

(a)   (b)

FIGURE 3.1 Kangaroo rats and ground squirrels are adapted to the desert heat. (a) Kangaroo rats hide from the heat in their burrows during the day, then forage during the cooler night. (b) Ground squirrels forage during the day, retreating to their burrows periodically to cool down. Photo (a) by Mary MacDonald/Nature Picture Library; photo (b) by Peter Chadwick/Photo Researchers.

Clouds 0.8

Faced with the same problem of surviving in the intense heat of the desert, the kangaroo rat, ground squirrel, and camel take different approaches to avoiding excessive heat loads. Each in its own way makes use of spatial and temporal variation in the environment to lose excess heat without having to use the most limiting resource in the desert—water.

CHAPTER CONCEPTS

- Light is the primary source of energy for the biosphere
- Plants capture the energy of sunlight by photosynthesis
- Plants modify photosynthesis in environments with high water stress
- Diffusion limits uptake of dissolved gases from water
- Temperature limits the occurrence of life

- Each organism functions best under a restricted range of temperatures
- The thermal environment includes several avenues of heat gain and loss
- Homeothermy increases metabolic rate and efficiency

The ability to counteract external physical forces distinguishes the living from the nonliving. A bird in flight, expending energy to maintain itself aloft against the pull of gravity, supremely expresses this quality. Like internal combustion engines, organisms transform energy to perform work. An automobile engine burns gasoline chemically, and it transmits power from the cylinder to the tires mechanically. When a bird metabolizes carbohydrates to provide the energy to flap its wings, it follows related chemical and mechanical principles.

The ultimate source of energy for most life processes is light from the sun. Plants harness this energy by photosynthesis, which produces the high-energy bonds of the organic molecules that form the basis of the food chain in ecological systems. Sunlight is also the ultimate source of the thermal energy that creates suitable conditions for life. It imposes an excessive heat load in some environments, but organisms in cool environments can use it to warm themselves and speed their life processes.

In this chapter, we shall explore some of the challenges that plants and animals face, and some of the mechanisms they employ, to harness the energy of sunlight and manage gains and losses of heat. We shall begin by considering photosynthesis, which converts the energy of sunlight into the chemical energy that powers organism activities and the functioning of ecological systems.

Light is the primary source of energy for the biosphere

Solar radiation is essential for the existence of life on earth. Plants, algae, and some bacteria absorb sunlight and assimilate its energy by photosynthesis. Not all the sunlight

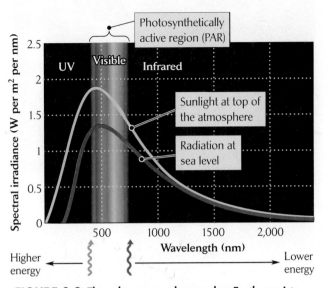

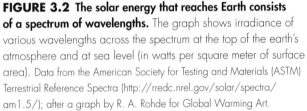

FIGURE 3.2 The solar energy that reaches Earth consists of a spectrum of wavelengths. The graph shows irradiance of various wavelengths across the spectrum at the top of the earth's atmosphere and at sea level (in watts per square meter of surface area). Data from the American Society for Testing and Materials (ASTM) Terrestrial Reference Spectra (http://rredc.nrel.gov/solar/spectra/am1.5/); after a graph by R. A. Rohde for Global Warming Art.

meter. The irradiance at the top of the earth's atmosphere is diminished by nighttime periods without light, reflection of light by clouds, and absorption of light by the atmosphere before the light even reaches the surface of the earth. At the earth's surface, still more light is reflected back into space by the oceans, snow and ice, and other surfaces. The proportion of light that is reflected by a particular surface is the **albedo** of that surface. Fresh snow and clouds have the highest recorded albedos, up to 80%–90%. Sand, dry soils, and deserts have albedos in the 20%–30% range; savannas, meadows, and most crops close to 20%, and forests and water surfaces 10% or less. The average albedo of the earth is about 30%, primarily because of reflection from clouds. All of this reflected light represents potential light energy lost to the earth.

The light absorption spectra of plants

The visible portion of the solar spectrum harnessed by photosynthetic organisms is also that portion of the solar spectrum with the greatest irradiance at the earth's surface. Leaves contain several kinds of pigments, particularly chlorophyll and carotenoids, that absorb this light and harness its energy (Figure 3.3). Chlorophyll, which is primarily responsible for capturing light energy in the

striking the earth's surface can be used in this way, however. As rainbows and prisms show, that light consists of a spectrum of wavelengths that we perceive as different colors (Figure 3.2). Visible light represents only a small part of the spectrum of electromagnetic radiation, which extends from gamma rays (the shortest wavelengths) to radio waves (the longest). Wavelengths are usually expressed in nanometers (nm; one-billionth of a meter). The visible portion of the spectrum, which corresponds to the wavelengths of light that are suitable for photosynthesis, ranges between about 400 nm (violet) and 700 nm (red). This range is called the **photosynthetically active region (PAR)** of the spectrum. Light of wavelengths shorter than 400 nm makes up the ultraviolet (UV) part of the spectrum. Light of wavelengths longer than 700 nm, called infrared (IR) radiation, is perceived by us primarily as heat.

Sunlight is packaged in small particle-like units of energy called photons. The energy intensity of individual photons varies inversely with their wavelength: the photons making up shorter-wavelength blue light vibrate more rapidly and have a higher energy level, or light intensity, than the photons that make up longer-wavelength red light.

Only a small proportion of the solar radiation that reaches the earth is converted into biological production through photosynthesis. The intensity of the light of all wavelengths impinging on a surface is referred to as **irradiance,** which can be quantified as watts per square

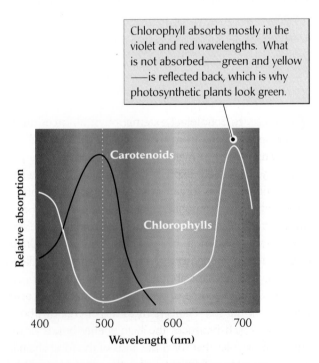

FIGURE 3.3 Two groups of photosynthetic pigments—chlorophylls and carotenoids—absorb different wavelengths of light. After R. Emerson and C. M. Lewis, *J. Gen. Physiol.* 25:579–595 (1942).

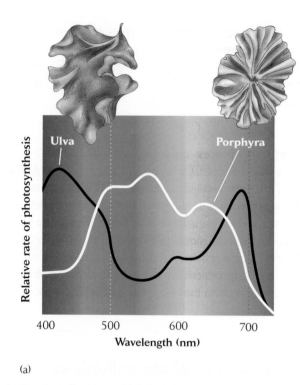

(a)

(b)

FIGURE 3.4 The photosynthetic pigments of aquatic algae are adapted to the available wavelengths of light. (a) Relative rates of photosynthesis by the green alga *Ulva* and the red alga *Porphyra* differ as a function of the color of light. (b) *Porphyra* appears red in this photograph because its photosynthetic pigments absorb light most strongly in the green portion of the spectrum and reflect red when photographed in artificial light resembling the spectrum at the surface. After F. T. Haxo and L. R. Blinks, *J. Gen. Physiol.* 33:389–422 (1950). Photo by Larry Jon Friesen/ Saturdaze.

light reactions of photosynthesis, absorbs red and violet light while reflecting green and blue light. Hence, leaves are predominantly green in color. Other pigments found in plant chloroplasts include two major classes of carotenoids: carotenes and xanthophylls. They are referred to as accessory pigments because they pass on the light energy they capture to chlorophyll to begin the sequence of reactions in photosynthesis. Carotenes, which give carrots their orange color, absorb primarily blue and green light and reflect light in the yellow and orange wavelengths of the spectrum. Thus, they complement the absorption spectrum of chlorophyll.

Water absorbs light in the visible region of the spectrum only weakly; therefore, a glass of water appears colorless. The transparency of a glass of water is deceptive, however. Although it appears colorless in small quantities, water absorbs or scatters enough light to limit the depth of the sunlit zone of the sea (referred to as the photic zone). In pure seawater, the intensity of light in the visible part of the spectrum diminishes to 50% of the surface value at a depth of 10 m, and it drops to less than 7% within 100 m. Moreover, water absorbs longer (red) wavelengths more strongly than shorter ones; most infrared radiation disappears within the topmost meter of water. The shortest visible wavelengths (violet and blue) tend to scatter when they strike water molecules, so they too fail to penetrate deeply. Because of the absorption and scattering of these wavelengths by water, green light predominates with increasing depth.

The photosynthetic pigments of aquatic algae parallel this spectral shift with depth. Algae that live near the surface of the oceans, such as the green sea lettuce *Ulva*, which grows in shallow water along rocky coasts, have pigments resembling those of terrestrial plants that absorb blue and red light and reflect green light. The deep-water red alga *Porphyra* has additional pigments that enable it to use green light more effectively (Figure 3.4).

Plants capture the energy of sunlight by photosynthesis

During photosynthesis, photons of light interact with pigments such as chlorophyll, to which the energy of light is transferred. Photosynthetic organisms then convert this energy into chemical energy stored in the high-energy bonds of organic compounds. They create these compounds by reducing an atom of carbon—the basic building block of organic compounds—from carbon dioxide (CO_2). The process of photosynthesis is often represented by a single equation describing the overall balance of reactants and products:

$$6 \, CO_2 + 6 \, H_2O + photons \rightarrow C_6H_{12}O_6 + 6 \, O_2$$

In fact, this simple equation summarizes a long chain of complex chemical reactions.

The light reactions

The first step in photosynthesis is the capture of light energy by photosynthetic pigments. When chlorophyll molecules in a chloroplast absorb photons, they release electrons, which are then passed along a chain of reactions to produce the high-energy compounds adenosine triphosphate (ATP) and NADPH. The cell then uses the energy in these compounds to reduce carbon and make glucose ($C_6H_{12}O_6$). The events from the absorption of light to the production of high-energy compounds are referred to collectively as the "light reactions" because of their dependence on light energy. Incidentally, chlorophyll molecules regain the electrons they lose in the light reactions by taking single electrons from water (H_2O) molecules, producing molecular oxygen (O_2) as a waste product.

C₃ photosynthesis

For most plants, the first stage in photosynthesis is the conversion of CO_2 into an atom of reduced carbon in a three-carbon sugar. In the process, a single molecule of CO_2, obtained from the atmosphere or surrounding water, is combined with a five-carbon sugar (ribulose bisphosphate, or RuBP) to eventually produce two molecules of glyceraldehyde 3-phosphate (G3P). This stage is one part of the light reactions. We can represent it as

$$CO_2 \; + \; RuBP \; \rightarrow \; 2 \, G3P$$
$$\text{1 carbon} \quad \text{5 carbons} \qquad \text{3 carbons}$$

Because the product of this stage is a three-carbon compound, biologists call this pathway **C₃ photosynthesis.**

The two molecules of G3P then enter what is known as the Calvin–Benson cycle, which regenerates one molecule of RuBP while making one reduced carbon atom available to synthesize glucose and other organic compounds. In most plants, these processes occur in the mesophyll cells of the leaves.

The enzyme responsible for the assimilation of carbon, **RuBP carboxylase-oxidase,** or **Rubisco,** has a low affinity for CO_2. Consequently, at the low concentrations of CO_2 found in mesophyll cells, plants assimilate carbon inefficiently. To achieve high rates of carbon assimilation, plants must pack their mesophyll cells with large amounts of Rubisco, which constitutes up to 30% of the dry weight of leaf tissue in some species.

Rubisco binds oxygen as well as carbon dioxide, particularly under high O_2 and low CO_2 concentrations, and especially at elevated leaf temperatures. When Rubisco binds O_2 instead of CO_2, it initiates a series of reactions that reverse the light reactions:

$$2 \, G3P \rightarrow CO_2 + RuBP$$

The overall process resembles respiration in that it uses O_2 and produces CO_2. Because it also requires ATP and NADPH from the light reactions, it is referred to as **photorespiration.** The tendency of Rubisco to undergo this reaction, which partially undoes what the enzyme accomplishes when it assimilates carbon, makes photosynthesis inefficient and self-limiting. Photorespiration is a wasteful and counterproductive process, and carbon assimilation therefore tends to inhibit itself as levels of CO_2 decline in the leaf tissue.

Plants modify photosynthesis in environments with high water stress

Because of the self-limiting nature of C₃ photosynthesis as CO_2 levels in leaves decrease, plants face serious limitations on their rate of photosynthesis and, therefore, on their growth and reproduction. Their solution to this problem is to maintain high CO_2 levels in their leaf cells. Plants can accomplish this to some extent by keeping the stomates of their leaves open to the surrounding atmosphere to allow free gas exchange.

Keeping the stomates open works as long as plants can replace the water they lose through the stomates by transpiration. But that may not be possible in hot, dry environments. Carbon dioxide has an extremely low concentration in the atmosphere (about 0.038% by volume at present). It enters plant cells because its concentration in the atmosphere is higher than its concentration in the cells, where it is continually used up by photosynthesis. However, the atmosphere-to-plant difference in the concentration of CO_2 is much, much less than the plant-to-atmosphere difference in the concentration of water vapor, which drives water out of plant cells into the surrounding air. This imbalance makes water conservation a problem for terrestrial plants, especially in hot, arid environments. Even the most drought-adapted plants evaporate a hundred or more grams of water from their leaves for every gram of carbon they assimilate (Figure 3.5).

C₄ photosynthesis

To address the problem of photorespiration, many herbaceous plants, particularly grasses growing in hot climates,

the Calvin–Benson cycle, just as it does in C_3 plants. The pyruvate is converted back to PEP, and the PEP moves back into the mesophyll cells to complete the C_4 carbon assimilation cycle.

This strategy solves the problem of photorespiration by allowing CO_2 to reach much higher concentrations within the bundle sheath cells than it could by diffusion from the atmosphere. At this higher CO_2 concentration, the Calvin–Benson cycle operates more efficiently. In addition, because the enzyme PEP carboxylase has a high affinity for CO_2, it can bind CO_2 at a lower concentration in the cell, thereby allowing the stomates to remain closed longer and reduce water loss. C_4 photosynthesis has two disadvantages that reduce its efficiency, however: less leaf tissue is devoted to photosynthesis, and some of the energy produced by the light reactions is used up in the C_4 carbon assimilation reactions. Because of their greater efficiency, C_3 plants are favored in cooler climates with abundant soil water. Nonetheless, many of our most important crop plants, such as corn (maize), sorghum, and sugarcane, are C_4 plants that are highly productive during hot growing seasons.

Carbon assimilation in CAM plants

Certain succulent plants that inhabit water-stressed environments, such as cacti and pineapple plants, use the same biochemical pathways as C_4 plants, but segregate CO_2 assimilation and the Calvin–Benson cycle between night and day. The discovery of this arrangement in plants of the family Crassulaceae (the stonecrop family; sedum is one example), and their initial assimilation and storage of CO_2 as four-carbon organic acids (malic acid and OAA), led botanists to call this photosynthetic pathway **crassulacean acid metabolism,** or **CAM.**

CAM plants open their stomates for gas exchange during the cool desert night, when transpiration is minimal. CAM plants initially assimilate CO_2 into four-carbon OAA, which is converted to malic acid and stored at high concentrations in vacuoles within the mesophyll cells of the leaf (Figure 3.6c). During the day, the stomates close, and the stored organic acids are gradually broken down to release CO_2 to the Calvin–Benson cycle. The enzyme responsible for the assimilation of CO_2 works best at the cool temperatures that occur at night, when the stomates are open. A different enzyme with a higher temperature optimum, geared to promote daytime photosynthesis, regulates the regeneration of PEP from pyruvate following the release of CO_2. Thus, CAM photosynthesis results in extremely high water use efficiencies and enables some types of plants to exist in habitats too hot and dry for other, more conventional species.

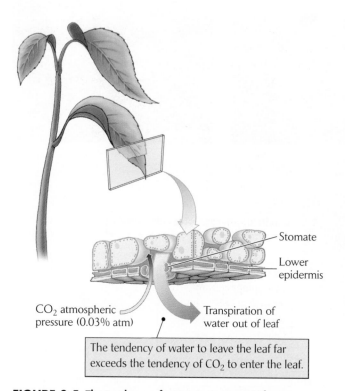

FIGURE 3.5 The tendency of water to evaporate from a leaf exceeds the tendency of CO_2 to enter a leaf. The surface of a leaf is relatively impermeable to water, so gas exchange occurs primarily through stomates. Because the plant uses CO_2 in photosynthesis, the concentration of that gas remains lower in the leaf than in the surrounding air, so CO_2 diffuses into the leaf. The movement of water vapor out of the leaf, however, is much more rapid than the diffusion of CO_2 in.

have modified the usual C_3 photosynthetic process (Figure 3.6a) by adding a step to the initial assimilation of CO_2. Biologists call this modification **C_4 photosynthesis** because CO_2 is first joined with a three-carbon molecule, phosphoenol pyruvate (PEP), to produce a four-carbon molecule, oxaloacetic acid (OAA):

$$CO_2 + PEP \rightarrow OAA$$

This reaction is catalyzed by the enzyme PEP carboxylase, which, unlike Rubisco, has a high affinity for CO_2. This preliminary assimilation step occurs in the mesophyll cells of the leaf. In most C_4 plants, actual photosynthesis (including the Calvin–Benson cycle) takes place in the bundle sheath cells surrounding the leaf veins (Figure 3.6b). To get carbon from the mesophyll into the bundle sheath cells, the plant converts oxaloacetic acid into malic acid, which then diffuses into the bundle sheath cells, where another enzyme breaks it down to produce CO_2 and pyruvate, a three-carbon compound. The CO_2 is then used in the light reactions to make G3P, which enters

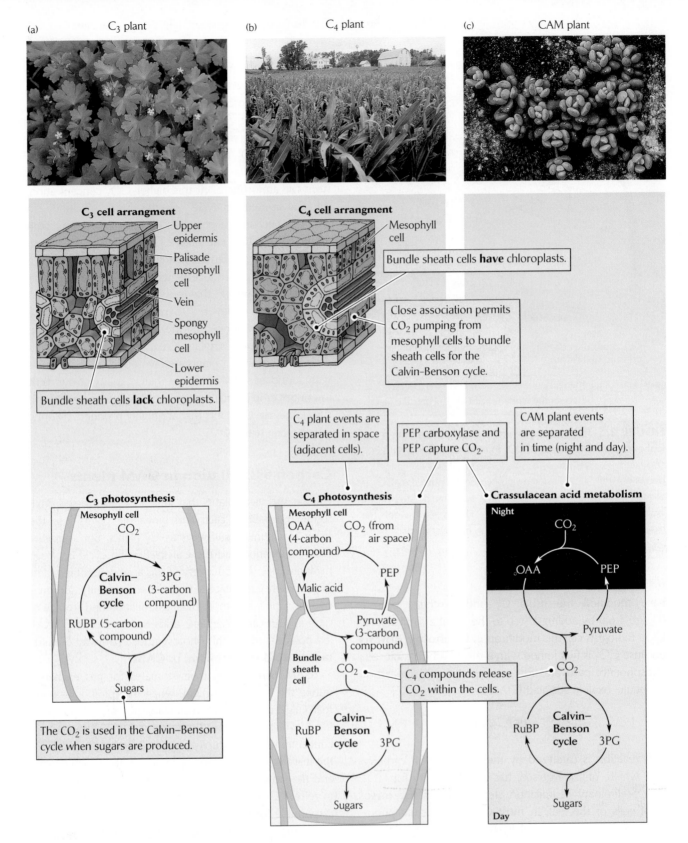

(a) C₃ plant   **(b)** C₄ plant   **(c)** CAM plant

C₃ cell arrangment

- Upper epidermis
- Palisade mesophyll cell
- Vein
- Spongy mesophyll cell
- Lower epidermis

Bundle sheath cells **lack** chloroplasts.

C₄ cell arrangment

- Mesophyll cell

Bundle sheath cells **have** chloroplasts.

Close association permits CO_2 pumping from mesophyll cells to bundle sheath cells for the Calvin–Benson cycle.

C₄ plant events are separated in space (adjacent cells).

PEP carboxylase and PEP capture CO_2.

CAM plant events are separated in time (night and day).

C₃ photosynthesis

Mesophyll cell

CO_2

Calvin–Benson cycle

3PG (3-carbon compound)

RUBP (5-carbon compound)

Sugars

The CO_2 is used in the Calvin–Benson cycle when sugars are produced.

C₄ photosynthesis

Mesophyll cell

OAA (4-carbon compound)

CO_2 (from air space)

Malic acid

PEP

Pyruvate (3-carbon compound)

Bundle sheath cell

CO_2

C₄ compounds release CO_2 within the cells.

Calvin–Benson cycle

RuBP

3PG

Sugars

Crassulacean acid metabolism

Night

CO_2

OAA

PEP

Pyruvate

CO_2

Calvin–Benson cycle

RuBP

3PG

Sugars

Day

FIGURE 3.6 The process of photosynthesis is modified in plants in water-stressed habitats. (a) A C₃ plant, the wild dovefoot geranium (*Geranium molle*). (b) A C₄ plant, cultivated sorghum (*Sorghum vulgare*). (c) A CAM plant, the Sierra sedum (*Sedum obtusatum*). Below the photos are idealized cross sections of a C₃ and a C₄ leaf, illustrating the arrangement of cells and the locations of chloroplasts (small dark green dots). At the bottom, the major steps of the Calvin–Benson cycle are shown for each plant type. Photo (a) by Bert Kragas/Visuals Unlimited; photo (b) by John Spragens, Jr.; photo (c) by John Gerlach/DRK Photo.

(a)

(b)

FIGURE 3.7 Spines and hairs help plants adapt to heat and drought. (a) Cross section and (b) surface view of the leaf of the desert perennial herb *Enceliopsis argophylla*, which uses this strategy. Courtesy of J. R. Ehleringer. From H. R. Ehleringer, in E. Rodrigues, P. Healy, and I. Mehta (eds.), *Biology and Chemistry of Plant Trichomes*, Plenum Press, New York (1984), pp. 113–132.

Structural adaptations to control water loss

In addition to these biochemical modifications of photosynthesis, heat- and drought-adapted plants have anatomic and physiological modifications that reduce transpiration across their surfaces, reduce heat loads, and enable the plants to tolerate high temperatures. When plants absorb sunlight, they heat up, and as their temperatures increase, they lose water more rapidly. Plants can minimize overheating by protecting their surfaces from direct sunlight with dense hairs and spines (Figure 3.7).

Spines and hairs also produce a still **boundary layer** of air that traps moisture and reduces evaporation. However, because thick boundary layers retard heat loss as well, hair-covered surfaces are prevalent in arid environments that are cool, but less so in hot deserts. Insulating boundary layers of still air also form on the flat surfaces of leaves, but those layers are broken up by air turbulence at leaf edges. Accordingly, many plants in hot deserts reduce their heat loads by producing finely subdivided leaves with a large ratio of edge to surface area (Figure 3.8). Some desert plants have no leaves at all. Many cacti rely entirely on their stems for

(a)

(b)

(c)

Mesquite (*Prosopis*) leaves are subdivided into leaflets that facilitate dissipation of heat.

Paloverde (*Cercidium*) its leaflets are tiny and the thick stems, which contain chlorophyll, are responsible for much of the plant's photosynthesis.

Limberbush (*Jatropha*) has broad, succulent leaves for a few weeks during the summer rainy season.

FIGURE 3.8 Leaves of desert plants have adaptations that increase heat dissipation. These three species from the Sonoran Desert in Arizona all have adaptations that help them cope with hot, dry conditions. (a) Leaves subdivided into numerous small leaflets facilitate the dissipation of heat because the leaf edges break down boundary layers of still air on the leaf surface. (b) Leaves of the paloverde (*Cercidium*) are tiny and the thick

stems, which contain chlorophyll, are responsible for much of the plant's photosynthesis (hence the name *paloverde*, which is Spanish for "green stick"). (c) Unlike most desert plants, limberbush (*Jatropha*) has broad succulent leaves, but it produces them for only a few weeks during the summer rainy season, then drops them. Photographs by R. E. Ricklefs.

FIGURE 3.9 Oleander plants reduce water loss by placing the stomates in their leaves in hair-filled pits. (a) Cross section of a leaf, showing a pit on the leaf's undersurface. (b) A pit in detail, magnified about 400 times. The hairs reduce water loss by slowing air movement and trapping water. Photos by Jack M. Bostrack/Visuals Unlimited.

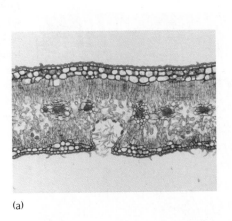

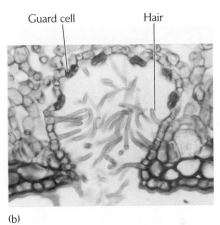

Guard cell   Hair

(a)   (b)

photosynthesis; their leaves are modified into thorns for protection.

Plants may further reduce transpiration by covering their surfaces with a thick, waxy cuticle that is impervious to water or by recessing the stomates in deep pits, often themselves filled with hairs (Figure 3.9).

Diffusion limits uptake of dissolved gases from water

Carbon dioxide

Getting enough carbon for photosynthesis is a particular challenge for aquatic plants and algae. The solubility of CO_2 in fresh water is about 0.0003 liters of gas per liter of water, which is 0.03% by volume, or about the same as its concentration in the atmosphere. When CO_2 dissolves in water, however, most of the molecules form carbonic acid (H_2CO_3). Depending on the acidity of the water, carbonic acid molecules release hydrogen ions (H^+) to form **bicarbonate ions** (HCO_3^-) or carbonate ions (CO_3^{2-}). Within the range of acidity that is typical of most fresh and salt water (pH values between 6 and 9), the more common form is bicarbonate, which dissolves readily in water. As bicarbonate forms, CO_2 is removed from solution, and more of the gas can then enter into solution from the atmosphere:

$$CO_2 + H_2O \rightarrow H_2CO_3 \rightarrow H^+ + HCO_3^-$$

This process continues until the concentration of bicarbonate ions is equivalent to 0.03–0.06 liters of CO_2 gas per liter of water (3%–6%), more than 100 times the concentration of CO_2 in air (Figure 3.10). Thus, bicarbonate ions provide a large reservoir of inorganic carbon in aquatic systems.

Inorganic carbon is abundant in water, to be sure, but rate of supply is the key, and because carbon moves so slowly through water, plants do not have free access to that supply. Carbon dioxide diffuses through unstirred water about 10,000 times more slowly than it does through air, and the larger bicarbonate ions diffuse even more slowly. Every surface of an aquatic plant, alga, or microbe is surrounded by a boundary layer of unstirred water through which carbon must diffuse. The thickness of this boundary layer may range from as little as 10 micrometers (μm) for single-celled algae in turbulent waters to 500 μm (0.5 mm) for a large aquatic plant in stagnant water (Figure 3.11). Thus, despite the high concentration of bicarbonate ions in the water surrounding these organisms, photosynthesis may be limited by carbon availability.

Both CO_2 and bicarbonate ions enter the cells of aquatic plants. Once inside the cells, bicarbonate ions can be used directly as a source of carbon for photosynthesis,

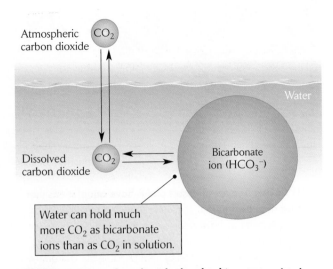

Atmospheric carbon dioxide CO_2

Water

Dissolved carbon dioxide CO_2 Bicarbonate ion (HCO_3^-)

Water can hold much more CO_2 as bicarbonate ions than as CO_2 in solution.

FIGURE 3.10 Carbon dioxide dissolved in water exists in equilibrium with a larger concentration of bicarbonate ions.

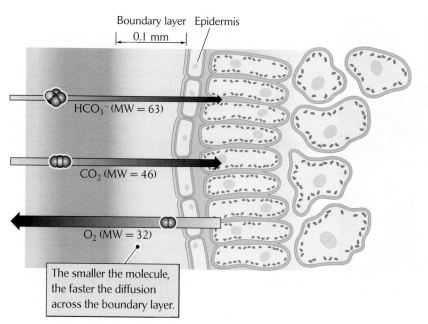

Boundary layer Epidermis

|← 0.1 mm →|

HCO_3^- (MW = 63)

CO_2 (MW = 46)

O_2 (MW = 32)

The smaller the molecule, the faster the diffusion across the boundary layer.

FIGURE 3.11 The boundary layer at the surface of an aquatic plant retards the exchange of gases between its leaves and the surrounding water. (MW = molecular weight.) After H. B. A. Prins and J. T. M. Elzenga, *Aquatic Botany* 34:59–83 (1989).

although not as efficiently as CO_2, which is the primary carbon source. As CO_2 itself is taken up from the water during photosynthesis and thereby depleted, bicarbonate ions associate once more with hydrogen ions to produce more CO_2 (Figure 3.12).:

$$H^+ + HCO_3^- \rightarrow CO_2 + H_2O$$

Thus, bicarbonate ions and CO_2 exist in a chemical equilibrium, which represents the balance achieved between H^+ and HCO_3^-, on one hand, and CO_2 and H_2O on the other.

Oxygen

Oxygen is abundant in the atmosphere, but much less so in water. The low solubility of oxygen in water often limits the metabolism of animals in aquatic habitats. Compared with its present-day concentration of 0.21 liters per liter

of air (21% by volume) in the atmosphere, the solubility of oxygen in water reaches a maximum (at 0°C in fresh water) of 0.01 liters per liter (1%). This limitation is compounded by the vastly lower rate of diffusion of oxygen in water than in air. Furthermore, below the photic zone in deep bodies of water and in waterlogged sediments and soils, no oxygen is produced by photosynthesis. Therefore, as animals and microbes living in deep water in lakes and in the mucky sediments of marshes use oxygen to metabolize organic materials, these habitats may become severely depleted of dissolved oxygen. Habitats that are devoid of oxygen are referred to as **anaerobic** or **anoxic.** Such conditions pose problems for terrestrial plants, whose roots need oxygen for respiration. Many plants that live in waterlogged habitats, such as bald cypress trees and many mangroves, have special vascular tissues extending from the roots that conduct air directly from the atmosphere (Figure 3.13).

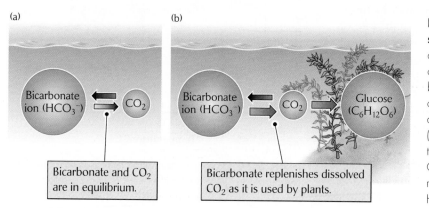

(a)

Bicarbonate ion (HCO_3^-) CO_2

Bicarbonate and CO_2 are in equilibrium.

(b)

Bicarbonate ion (HCO_3^-) CO_2 Glucose ($C_6H_{12}O_6$)

Bicarbonate replenishes dissolved CO_2 as it is used by plants.

FIGURE 3.12 Bicarbonate ions are a source of CO_2 in aquatic systems. When aquatic plants and algae deplete the supply of CO_2, it is replenished from the pool of bicarbonate ions in their immediate vicinity or within their cells. (a) Bicarbonate and dissolved CO_2 reach an equilibrium in water. (b) When plants and algae remove CO_2 from the water during photosynthesis, the reduced CO_2 concentration causes bicarbonate to release additional CO_2 into solution ($H^+ + HCO_3^- \rightarrow H_2O + CO_2$).

FIGURE 3.13 The knees of bald cypress trees conduct air from the atmosphere to their roots. This adaptation provides the roots with oxygen when the swamp where they grow is flooded and the waterlogged sediments contain little or no dissolved oxygen. Photo by David Muench/CORBIS.

Temperature limits the occurrence of life

All life depends on the energy of the sun, not just for the food supply created by photosynthesis, but also for the temperature conditions required by life. Most physiological processes occur only within the range of temperatures at which water is liquid: 0°–100°C at the earth's surface. Relatively few plants and animals can survive body temperatures above 45°C, which defines the upper limit of the physiological range for most eukaryotic organisms.

Heat and biological molecules

Much of the influence of temperature on physiological processes results from the way in which heat affects organic molecules. Heat imparts a high kinetic energy to living systems, causing biological molecules to move and change their shapes rapidly. By increasing the rate of movement of molecules, heat also accelerates chemical reactions. The rates of most biological processes increase between 2 and 4 times for each 10°C rise in temperature throughout the physiological range (Figure 3.14). The ratio of the rate of a physiological process at one temperature to its rate at a temperature 10°C cooler is referred to as the Q_{10} of that process.

Higher temperatures mean that organisms can develop more rapidly; swim, run, and fly faster; and digest and assimilate more food. Thus, increasing temperature has a

positive effect on biological productivity. In fact, because of the physics of kinetic energy, biological productivity within the physiological range is almost directly proportional to the temperature in degrees Celsius. The influence of temperature is so ubiquitous that some ecologists, including J. H. Brown and his colleagues at the University of New Mexico, have proposed a **metabolic theory of ecology,** which states that temperature has consistent effects on a range of processes important to ecology and evolution. These processes range from rates of metabolism and development of individuals to the productivity of ecosystems to rates of genetic mutation, evolutionary change, and species formation.

There is no question that heat energy accelerates life processes. Moreover, increasing global temperatures will undoubtedly speed these processes even more, with as yet unforeseen consequences. High temperatures can also have a depressing effect on life processes, however. In particular, proteins and other biological molecules become less stable at higher temperatures and may not function properly or retain their structure. The molecular motion caused by heat tends to open up, or denature, those molecules.

To exist at high temperatures, proteins must be bound by strong forces of attraction within and between molecules to resist being literally shaken apart. The proteins of **thermophilic** ("heat-loving") bacteria have higher proportions of amino acids that form strong bonds between one another than do the proteins of other, heat-intolerant organisms. Consequently, some photosynthetic bacteria can tolerate temperatures as high as 75°C, and some

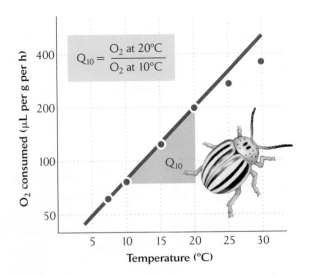

$$Q_{10} = \frac{O_2 \text{ at } 20°C}{O_2 \text{ at } 10°C}$$

FIGURE 3.14 Oxygen consumption increases as a function of temperature. These data are for the Colorado potato beetle. After M. Marzusch, *Zeitschr. Vergl. Physiol.* 34:75–92 (1952).

archaebacteria can live in hot springs at temperatures up to 110°C, the temperature of boiling water! No wonder that such organisms are referred to as *extremophiles*. Temperature affects other biological compounds as well. The physical properties of fats and oils, which are major components of cell membranes and constitute the energy reserves of animals, depend on temperature. When cold, fats become stiff (picture in your mind the fat on a piece of meat taken from the refrigerator); when warm, they become fluid.

Cold temperatures and freezing

Temperatures on the earth's surface rarely exceed 50°C, except in hot springs and at the soil surface in hot deserts. However, temperatures below the freezing point of water are common, particularly on the land and in small ponds, which may become solid ice during winter. When living cells freeze, the crystal structure of ice disrupts most life processes and may damage delicate cell structures, eventually causing death. Many organisms successfully cope with freezing temperatures, either by maintaining their body temperatures above the freezing point of water or by activating chemical mechanisms that enable them to resist freezing or tolerate its effects.

It might surprise you to learn that marine vertebrates are susceptible to freezing in cold seawater. You might wonder how blood and body tissues could freeze solid in liquid water. The answer is that dissolved substances depress the freezing temperature of water and other liquids. While pure water freezes at 0°C, seawater, which contains about 3.5% dissolved salts, freezes at −1.9°C, or almost 2°C colder. The blood and body tissues of most vertebrates contain less than half the salt content of seawater, and thus freeze at a higher temperature than seawater.

Two questions come to mind. First, why don't polar fish have high salt levels in their blood and tissues? Second, how can these fish survive at such low temperatures? Fish do not use salts to keep their body fluids from freezing because high salt concentrations would interfere with many biochemical processes. Instead, some Antarctic fish have circumvented their susceptibility to freezing by raising their blood and tissue concentrations of compounds such as glycerol—a three-carbon alcohol, common drugstore glycerin. A 10% glycerol solution lowers the freezing point of water by about 2.3°C without severely disrupting biochemical processes. Glycoproteins, the class of proteins that contain one or more carbohydrates, also lower the freezing temperature of water. Such antifreeze-like compounds in their tissues allow fish in Antarctic regions to

remain active in seawater that is colder than the freezing point of the blood of fish inhabiting temperate or tropical seas (Figure 3.15). Some terrestrial invertebrates also use the antifreeze approach; their body fluids may contain up to 30% glycerol, in extreme cases, as winter approaches.

Supercooling provides a second physical solution to the problem of freezing. Under certain circumstances, liquids can cool below the freezing point without ice crystals developing. Ice generally forms around some object, called a *seed*, which can be a small ice crystal or other particle. In the absence of seeds, pure water may cool more than 20°C below its freezing point without freezing. Such supercooling has been recorded to −8°C in reptiles and to −18°C in invertebrates. Glycoproteins in the blood of these cold-adapted animals impede ice formation by coating developing crystals, which would otherwise act as seeds.

FIGURE 3.15 Glycoproteins act as biological antifreeze in the Antarctic cod. The fish's blood and tissues are prevented from freezing by the accumulation of high concentrations of glycoproteins, which lower its freezing point to below the minimum temperature of seawater (−1.8°C) and prevent ice crystal formation. This fish is being pulled through a hole in the ice near McMurdo Station, Antarctica. Note the bright red color of its gills, which indicates a rich blood supply. Photo by John Bortniak, courtesy of NOAA.

Each organism functions best under a restricted range of temperatures

Each organism generally has a narrow range of environmental conditions to which it is best suited, which define its **optimum.** The optimum is determined by the properties of its enzymes and lipids, the structures of its cells and tissues, the form of its body, and other characteristics that influence the ability of the organism to function well under the particular conditions of its environment.

Temperature is a good example of a condition that must remain within a narrow range for an organism to function properly. Returning to the example of fish in the frigid oceans surrounding Antarctica, many species swim actively and consume oxygen at a rate comparable to fish living among tropical coral reefs. Put a tropical fish in cold water, however, and it becomes sluggish and soon dies; conversely, Antarctic fish cannot tolerate temperatures warmer than 5°–10°C.

How can fish from cold environments swim as actively as fish from the tropics? Swimming depends on a series of biochemical reactions, most of which are catalyzed by enzymes. Because most of these reactions proceed more rapidly at high temperatures than at low temperatures, cold-adapted organisms must have more of the substrate for a biochemical reaction, more of the enzyme that catalyzes the reaction, or a qualitative difference in the enzyme itself. A particular enzyme obtained from a variety of organisms that live under different conditions may exhibit different catalytic properties when tested over ranges of temperature, pH, salt concentration, and substrate abundance.

Organisms sometimes accommodate predictable changes in environmental conditions by having more than one form of an enzyme or structural molecule, each of which functions best within a different range of conditions. The rainbow trout, for example, experiences low temperatures in its native habitat during winter, when water temperatures may drop close to the freezing point, and much higher temperatures in summer. These seasonal changes in temperature are predictable, and the trout responds by producing different forms of many enzymes in winter and in summer.

One of these enzymes is acetylcholinesterase, which plays an important role in ensuring proper functioning of the nervous system by degrading neurotransmitters. The affinity of this enzyme for its substrate, the neurotransmitter acetylcholine, is a good measure of enzyme function. Substrate affinity in the winter form of the enzyme is high between 0°C and 10°C, but drops rapidly at higher temperatures. Substrate affinity in the summer form of the

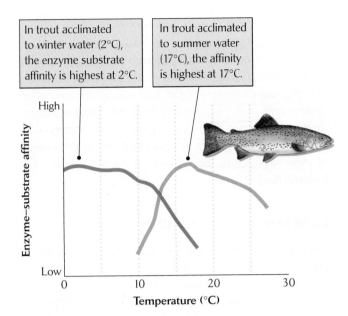

FIGURE 3.16 Some organisms can acclimatize to changing environmental conditions. Trout raised at winter and at summer temperatures produce different forms of the enzyme acetylcholinesterase. Data from J. Baldwin and P. W. Hochachka, *Biochemical Journal* 116:883–887 (1970).

enzyme is low at 10°C, rises to a peak between 15°C and 20°C, and drops slowly at higher temperatures (Figure 3.16). The form of the enzyme that a trout produces depends directly on the temperature of the water it lives in. When trout are maintained at 2°C, they produce the winter form; at 17°C, they produce only the summer form.

The thermal environment includes several avenues of heat gain and loss

Because body temperature influences physiological function so strongly, organisms must manage heat gain and heat loss carefully. The temperature of a substance reflects its heat content, and the ultimate source of heat at the surface of the earth is sunlight. Most of the solar radiation reaching the surface of the earth is absorbed by water, soil, plants, and animals and is converted to heat. Each object and each organism on earth continually exchanges heat with its surroundings (Figure 3.17). When the temperature of the environment exceeds that of an organism, the organism gains heat and becomes warmer. When the environment is cooler than the organism, the organism loses heat to the environment and cools.

Radiation is the emission of electromagnetic energy by a warm surface, which may then be absorbed by any

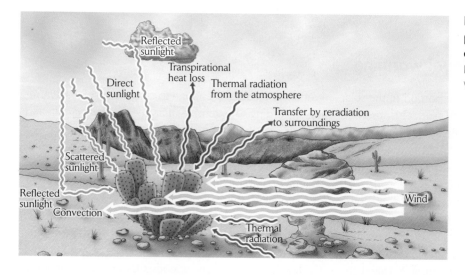

FIGURE 3.17 There are many pathways of heat exchange between an organism and its environment. After D. M. Gates, *Biophysical Ecology,* Springer-Verlag, New York (1980).

cooler surface. Sources of radiation in the environment include the sun, the sky (scattered light), and the landscape (which radiates heat it has absorbed from the sun). Lizards basking on rocks gain heat directly by radiation from the sun. How rapidly an object loses energy by radiation depends on the temperature of the radiating surface. The relationship is nonintuitive in that radiation increases with the *fourth* power of absolute temperature (K). (Absolute zero is 0 kelvin, 0 K, and is equal to $-273°C$.) Accordingly, a small mammal with a skin temperature of 37°C (310 K) radiates heat 30% more rapidly than a lizard of similar size with a skin temperature of 17°C (290 K). At night, objects that have warmed in the sunlight radiate their stored heat to colder parts of the environment (Figure 3.18).

Conduction is the transfer of the kinetic energy of heat between substances in contact with one another. Thus, a vacuum, which lacks all substance, conducts no heat. Water, because it is so much denser than air, conducts heat more than 20 times faster than air. The rate at which heat passes between an organism and its surroundings depends on the insulating value of the organism's surface (its resistance to heat transfer), its surface area, and the temperature difference between the organism and its surroundings. An organism can either gain or lose heat by conduction, depending on its temperature relative to that of the environment. That is why lizards often lie flat on hot rocks, warming their bodies by conduction from below as well as by radiation from the sun above.

Convection is the transfer of heat by the movement of liquids and gases: molecules of air or water next to a warm surface gain energy and move away from the surface. As we have seen, air conducts heat poorly. In still air, a boundary layer of air forms over a surface. A warm organism tends to warm its boundary layer to the temperature of its own body, effectively insulating itself against heat loss. A current of air flowing past a surface tends to disrupt the boundary layer so that heat can be conveyed away from the body by convection. This convection of heat away from the body surface is the basis of the "wind chill factor" we hear about in winter on the evening weather report. On a cold day, air movement makes you feel as cold as you would on an even colder windless day. For example, a wind blowing 32 km per hour at an air temperature of $-7°C$ has the cooling power of still air at $-23°C$.

Evaporation removes heat from a surface. The evaporation of 1 g of water from the body surface removes 2.43 kilojoules (kJ) of heat when the temperature of the

(continued on page 54)

FIGURE 3.18 Organisms lose heat to the environment by radiation. This thermal image of Canada geese on a cold day shows that the rate of heat loss is greatest through the head, neck, and legs. Because the image records infrared radiation, lighter areas indicate warmer temperatures. The birds are walking across a road onto a lawn. Courtesy of R. Boonstra, from R. Boonstra et al., *J. Field Ornithol.* 66:192–198 (1995).

GLOBAL CHANGE

Carbon dioxide and global warming

Most of the energy in the visible portion of the solar spectrum that reaches the earth's surface is absorbed by vegetation, soil, and surface waters and converted to heat energy. That heat is then radiated from the warmed surface of the earth back toward space as lower-intensity infrared radiation. Much of this radiation is absorbed by gases in the atmosphere, such as carbon dioxide, water vapor, and methane. The atmosphere thereby acts as a blanket covering the earth and keeping its surface warm. Because this warming effect resembles the manner in which glass keeps a greenhouse warm, it is called the **greenhouse effect** (Figure 1). Eventually, the absorbed energy reaches the upper levels of the atmosphere and is lost to space, but at a much slower rate than it would be in the absence of the infrared-absorbing components of the atmosphere—the so-called greenhouse gases.

Overall, the greenhouse effect greatly benefits life by maintaining temperatures on earth within a range that is favorable for life. However, the addition of CO_2 to the atmosphere by human activities, such as the clearing of forests and the burning of fossil fuels, has intensified the greenhouse effect—enough to cause rapid global warming, according to extensive analyses of climatic data and models of global carbon and energy dynamics.

In the late 1950s, Charles Keeling began recording atmospheric CO_2 concentrations atop 3,400-m-high Mauna Loa on the island of Hawaii. From the standpoint of air quality, this volcanic mountaintop is one of the most pristine places on earth. Keeling wanted to determine whether anthropogenic emissions were increasing the concentration of CO_2 in the atmosphere. At the time Keeling began his study, scientists had no accurate long-term measurements of atmospheric CO_2 concentrations. At the start of his observations in 1958, the CO_2 concentration was about 316 parts per million (ppm; 316 CO_2 molecules per million molecules of air, mostly nitrogen [N_2] and oxygen [O_2]).

Scientists have now developed ways of measuring CO_2 concentrations in atmospheric gases trapped in the Greenland and Antarctic ice caps. Those measurements show that CO_2 levels during the past 0.5 million years have varied with glacial cycles, from about 200 ppm during the peaks of glacial periods to 300 ppm during warm interglacial periods—including the recent epoch up to the beginning of the Industrial Revolution in the 1800s. In the decades following the beginning of Keeling's study, his measurements showed that atmospheric CO_2 concentrations were increasing dramatically, to 352 ppm by 1990 and 384 ppm by 2007, with no sign of leveling off (Figure 2). As demand for energy and agricultural land increases, the rate of emissions of CO_2 to the atmosphere is likely to increase even further.

Carbon dioxide is a potent greenhouse gas. Predictions about how much the surface temperature of the earth will warm as a result of increased atmospheric CO_2 vary considerably, however. Current climate models are consistent in predicting an increase of between 1.1°C and 6.4°C over 1990 temperatures by 2100. During the twentieth century, the earth's average surface temperature increased by 0.74°C (Figure 3). These temperature increases will not be distributed uniformly over the surface of the earth, however. It is likely that temperatures within the humid tropics will remain relatively stable, and the most dramatic increases will occur at northern high latitudes, including much of the area currently covered by boreal forest and tundra. The effects of temperature

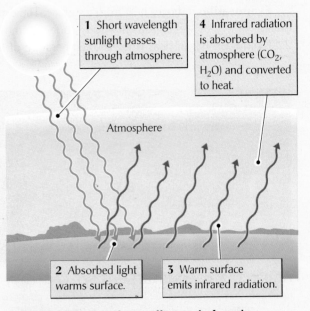

1 Short wavelength sunlight passes through atmosphere.

4 Infrared radiation is absorbed by atmosphere (CO_2, H_2O) and converted to heat.

Atmosphere

2 Absorbed light warms surface.

3 Warm surface emits infrared radiation.

FIGURE 1 The greenhouse effect results from the absorption of infrared radiation by CO_2 and other "greenhouse" gases in the atmosphere.

(a)

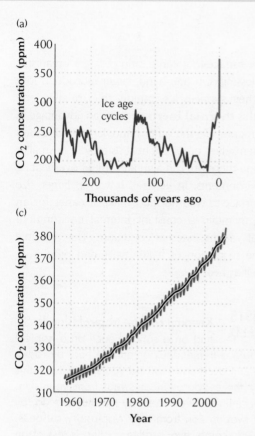

(b)

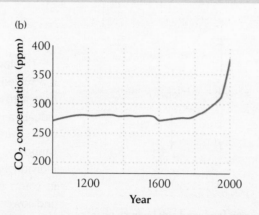

(c)

FIGURE 2 CO$_2$ concentrations in the atmosphere have changed over time. (a) Changes in atmospheric CO$_2$ concentrations estimated from gases trapped in the Antarctic ice cap during the past 250,000 years, including the last two major glacial cycles of the Pleistocene epoch. (b) Concentrations of CO$_2$ from ice cores dated over the past thousand years through the beginning of the Industrial Revolution. The combustion of fossil fuels accelerated in the early 1800s. (c) Direct measurements of atmospheric CO$_2$ concentrations at Mauna Loa, Hawaii. The curve oscillates because the CO$_2$ concentration in the Northern Hemisphere is lower during summer, when plant photosynthesis removes carbon from the atmosphere, and higher during winter, when respiration exceeds plant production. (a, b) Data from H. Fischer et al., *Science* 283:1712–1714 (1999); (c) data from NOAA (http://www.esrl.noaa.gov/gmd/ccgg/trends/co2_mm_mlo.dat), after a graph by R. A. Rohde for Global Warming Art.

increases in these regions are likely to accelerate global warming. Reduced snow and ice cover will reduce the albedo of the earth's surface, allowing it to absorb more solar radiation. Increased soil temperatures and thawing of permafrost will enhance the respiration rates of soil organisms and their release of CO$_2$ to the atmosphere.

Organisms are already beginning to respond to the climate changes of the past century. Long-term data for a wide range of primarily northern temperate zone species show that distributional limits are moving poleward at a rate of about 6 km per decade. The timing of spring events, such as leaf bud break, flowering time, and arrival of birds on northward migrations, is advancing by an average of 2.3 days per decade. The earth has been warm in the past, and it has recently gone through several glacial cycles, in which climatic conditions have changed from warm to cold and back again. The difference now is in the rate at which these changes are taking place and the disruptions to both natural systems and human populations that are likely to result from them. The timing and nature of these disruptions are difficult to predict, but it is certain that disruptions will occur.

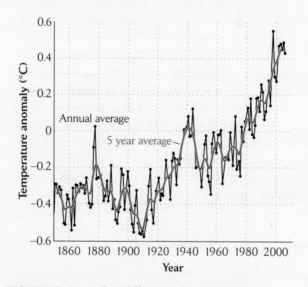

FIGURE 3 Anomalies (differences from the average temperature during the period 1961–1990) show the increase in the global mean surface temperature of the earth since 1850. Data compiled by the United Kingdom Meteorological Office Hadley Centre (dataset HadCRUT3); see P. Brohan et al., *Journal of Geophysical Research* 111:D12106, DOI:10.1029/2005JD006548 (2006).

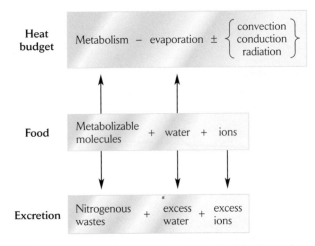

FIGURE 3.19 The heat, water, food, and salt budgets of animals are coupled by diet, evaporative water loss, and excretion.

surface is 30°C. As plants transpire and animals breathe, water evaporates from their exposed gas exchange surfaces, especially at higher temperatures. In dry air, the rate of evaporation nearly doubles with each 10°C increase in temperature.

All of the gains and losses of heat by an organism constitute its **heat budget,** which relates the rate of change in its heat content to gains and losses through radiation, conduction, convection, and evaporation, plus the internal heat it generates by metabolizing foods. When gains and losses are perfectly balanced, the change in heat content is zero. Because evaporation and metabolism influence heat content, the heat budget is connected to the organism's water, food, and salt budgets, as illustrated in Figure 3.19. Food is the source of metabolically produced heat, and it also contains water and salts. Evaporative heat loss is always accompanied by the loss of water, which can be replenished by drinking (where freestanding water is available). Water is also produced by the metabolism of organic compounds.

Body size and thermal inertia

Most exchanges of energy and materials between an organism and its environment occur across body surfaces. Larger organisms have less surface area compared with the bulk of their tissues than smaller organisms do, so exchanges between the organism and its environment become more difficult as body size increases. When organisms differ only in size, and not shape, surface area (S) tends to increase as the square of length (L), whereas volume (V) tends to increase as the cube of length. Accord-

ingly, the surface-to-volume ratio actually decreases in proportion to length:

$$\frac{S}{V} = \frac{L^2}{L^3} = \frac{1}{L}$$

The lower surface-to-volume ratio of larger organisms is a mixed blessing. On one hand, larger individuals lose heat across their surfaces less rapidly than smaller individuals, and this **thermal inertia** can be an advantage in cold environments. On the other hand, larger individuals cannot rid themselves of excess heat as rapidly as smaller individuals, and therefore run a greater risk of overheating in warm environments. In general, however, larger size and lower surface-to-volume ratio make it easier for an organism to maintain a constant internal environment in the face of varying external conditions. This principle applies to the regulation of water, salts, and other substances as well as heat.

ECOLOGISTS IN THE FIELD **Keeping cool on tropical islands.** Sitting on a sandy beach on a tropical island, you gain a tremendous amount of heat by radiation from the sun overhead. You rid yourself of much this heat load by the evaporation of sweat from your skin. Although few animals sweat the way that humans do, all lose heat by evaporation from their respiratory surfaces. Where water is scarce, evaporative cooling is less of an option, and animals tend to reduce their heat loads by staying out of the sun. Why, then, do several species of seabirds, such as sooty terns (Figure 3.20), nest in full sun on bare sand on small coral atolls in the tropics? Sooty terns are exposed to punishing levels of solar radiation during the middle of the day, including light reflected from

FIGURE 3.20 Sooty terns can tolerate a hot nesting environment. This sooty tern (*Sterna fuscata*) is sitting on its egg in the hot sun on Christmas Island, located on the equator in the central Pacific Ocean. Photo by R. E. Ricklefs.

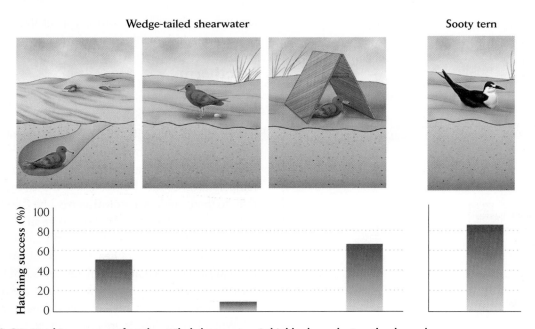

Wedge-tailed shearwater   Sooty tern

FIGURE 3.21 Hatching success of wedge-tailed shearwaters is highly dependent on the thermal environment. Hatching success is measured as the percentage of eggs laid that hatch. Individuals protected from the sun in burrows or provided with artificial shade have higher success than do these nesting in the open. Data courtesy of Paul Sievert.

the sand, while another species of similar size and coloration, the wedge-tailed shearwater, builds its nests in deep burrows beneath the surface of the sand.

Seabird biologist Paul Sievert wondered why the two species place their nests so differently. The conventional wisdom had been that shearwaters nest in burrows to avoid predators such as frigatebirds, which, ever watchful, swoop down to snatch unattended eggs and chicks. By chance, however, the density of shearwaters on Tern Island, in the northwestern Hawaiian Islands, is so great, and the sand so hard to dig through, that many shearwaters nest on the surface out of desperation. These birds were found to have very low nesting success because they were forced to abandon their eggs under the intense solar radiation. If the eggs weren't taken by frigatebirds, they heated up in the sun, and the developing embryos died. However, Sievert found that if he shaded surface nests with plywood A-frames, the shearwaters were able to reproduce successfully because the adults could remain on their eggs throughout the middle of the day (Figure 3.21).

This simple experiment demonstrated the importance of the thermal environment for shearwaters, but it did not explain how sooty terns can nest on the surface in full sun in the same environment. The key to this puzzle lies in the diets and foraging strategies of the two species. Sooty terns feed on fish and squid in areas close to their nesting sites. The male and female sooty terns alternate incubation duty, and neither one stays at the nest for more than a day or two at a time. Shearwaters have a diet similar to that of terns, but they feed hundreds of kilometers from their nest-

ing sites. They digest most of what they eat while foraging at sea, and convert the surplus energy to fat, which they metabolize during their weeklong spells of incubation. In contrast, sooty terns come back to their nests from the sea with a stomach full of water-laden food, which provides a reservoir of free water to compensate for evaporative heat loss. Remember that fish are hypo-osmotic with respect to seawater and thus provide a relatively inexpensive supply of free water. Shearwaters have plenty of fat to supply them with energy through a prolonged fast, but fat contains much less water than fresh fish, and even the water produced by fat metabolism is insufficient to dissipate the heat load they absorb under full sunlight. So why don't shearwaters drink the abundant seawater all about them? Seawater contains so much salt that they would have to use as much water as they consumed to excrete the salt through their salt glands. As Coleridge put it, "Water, water, every where, nor any drop to drink!"

Homeothermy increases metabolic rate and efficiency

Maintaining a constant internal body temperature is beneficial to an organism because its biochemical reactions can be adjusted to work most efficiently at that temperature. **Homeostasis** is an organism's ability to maintain constant internal conditions in the face of a varying external environment. All organisms exhibit homeostasis to

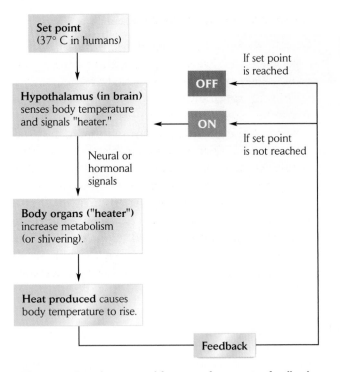

FIGURE 3.22 The essential features of a negative feedback system include sensors and switches. The hypothalamus, like a thermostat, compares body temperature with a desired set point. When the two differ, it signals the body organs to bring the temperature back into line with the set point.

some degree, as we have seen in the case of water and salt balance, although the occurrence and effectiveness of homeostatic mechanisms vary. Regardless of how organisms regulate their internal environments, all homeostatic systems exhibit **negative feedback,** meaning that when the system deviates from its desired state, or set point, internal response mechanisms act to restore that state (Figure 3.22). The thermostat used to regulate room temperature in your home works by the same principle. When the house is cold, a temperature-sensitive switch turns on a heater, which restores the temperature to its desired setting.

Most mammals and birds maintain their body temperatures between 36°C and 41°C, even though the temperature of their surroundings may vary from −50°C to +50°C. Such temperature regulation, which is referred to as **homeothermy** (the Greek root *homos* means "same"), creates constant temperature (homeothermic) conditions within the cells, under which biochemical processes can proceed efficiently. In contrast, the body temperatures of **poikilothermic** organisms, such as frogs and grasshoppers, conform to the external temperature (the Greek root *poikilo* means "varying"). Thus, frogs cannot function at either high or low temperature extremes, and so they

are active only within a narrow part of the temperature range over which mammals and birds thrive.

Ectotherms

Many organisms, including reptiles, insects, and plants, adjust their heat balance behaviorally, simply by moving into or out of shade, by changing their orientation with respect to the sun, or by adjusting their contact with warm substrates. Because the heat they use to elevate their body temperatures comes from outside the body, biologists refer to these animals as **ectotherms** (the Greek root *ecto* means "outside"). Ectotherms tend to be small (insects) or have low metabolic rates (reptiles and amphibians) that are not sufficient to offset heat loss in most environments. In contrast, animals that can generate sufficient heat metabolically to raise their body temperatures are referred to as **endotherms** (the Greek root *endo* means "inside").

When horned lizards are cold, they lie flat against the ground and gain heat by conduction from the sun-warmed surface. When they are hot, they decrease their exposure to the surface by standing erect on their legs. Basking behavior is widespread among reptiles and insects, which can use it effectively to regulate their body temperatures within a narrow range. Indeed, their temperatures may rise considerably above that of surrounding air, well into the range of birds and mammals.

Endothermic homeotherms

Organisms with high internal body temperatures, such as birds and mammals, gain the added benefit of accelerated biological activity, which makes them better able to forage, escape predators, and compete with other individuals. However, sustaining internal conditions that differ significantly from conditions in the external environment requires work and energy. Consider the costs to birds and mammals of maintaining constant high body temperatures in cold environments. As air temperature decreases, the gradient (difference) between internal and external environments increases. Heat is lost across body surfaces in direct proportion to this gradient. An animal that maintains its body temperature at 40°C loses heat twice as fast at an ambient (surrounding) temperature of 20°C (a gradient of 20°C) as at an ambient temperature of 30°C (a gradient of only 10°C). To maintain a constant body temperature, endothermic organisms must replace heat lost to their environment by generating heat metabolically. Thus, the rate of metabolism required to maintain body temperature increases in direct proportion to the difference between body and ambient temperatures, all other things being equal.

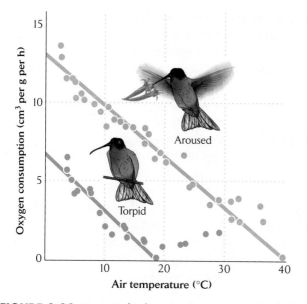

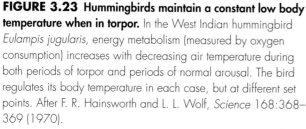

FIGURE 3.23 Hummingbirds maintain a constant low body temperature when in torpor. In the West Indian hummingbird *Eulampis jugularis*, energy metabolism (measured by oxygen consumption) increases with decreasing air temperature during both periods of torpor and periods of normal arousal. The bird regulates its body temperature in each case, but at different set points. After F. R. Hainsworth and L. L. Wolf, *Science* 168:368–369 (1970).

An organism's ability to maintain a high body temperature while exposed to low ambient temperatures is limited over the short term by its physiological capacity to generate heat and over the long term by its ability to gather food. At extremely low temperatures, animals may starve to death, rather than freeze to death, if they metabolize energy to maintain body temperature more rapidly than they can obtain that energy in food.

Because they are so small, hummingbirds have a large surface-to-volume ratio, and consequently lose heat rapidly relative to the amount of tissue that is available to produce heat. As a result, hummingbirds must sustain very high metabolic rates to maintain their resting body temperatures near 40°C. Species living in cool climates would starve overnight if they did not enter **torpor,** a voluntary, reversible condition of low body temperature and inactivity. For example, the West Indian hummingbird *Eulampis jugularis* drops its body temperature to 18°–20°C when resting at night. It does not cease to regulate its body temperature; it merely changes the set point on its thermostat to reduce the difference between ambient and body temperature, and thereby reduces the energy expenditure needed to maintain its temperature at the set point (Figure 3.23).

Countercurrent heat exchange

As we have seen, heat is conducted from warmer to cooler substances. Eventually, temperatures equalize, and net movement of heat comes to a standstill. Thus, conduction of heat, particularly from exposed extremities, works against the maintenance of a constant warm body temperature. Nature has devised many solutions to this problem, among the simplest and most effective of which is an arrangement of blood vessels in the extremities called **countercurrent circulation.** In a countercurrent circulation system, blood flowing from the body toward the extremities continuously encounters blood returning to the body. Because the legs and feet of most birds do not have feathers, they would be major avenues of heat loss in cold regions were they not held at a lower temperature than the rest of the body (Figure 3.24). Gulls standing

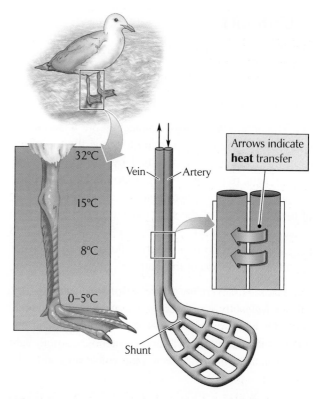

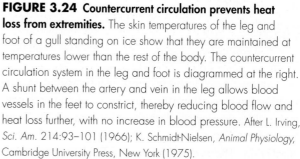

FIGURE 3.24 Countercurrent circulation prevents heat loss from extremities. The skin temperatures of the leg and foot of a gull standing on ice show that they are maintained at temperatures lower than the rest of the body. The countercurrent circulation system in the leg and foot is diagrammed at the right. A shunt between the artery and vein in the leg allows blood vessels in the feet to constrict, thereby reducing blood flow and heat loss further, with no increase in blood pressure. After L. Irving, *Sci. Am.* 214:93–101 (1966); K. Schmidt-Nielsen, *Animal Physiology,* Cambridge University Press, New York (1975).

on ice or swimming with their feet in frigid water conserve heat by means of countercurrent circulation in their legs. Warm blood in arteries leading to the feet cools as it passes close to veins that return cold blood to the body. In this way, heat is transferred from arterial to venous blood and transported back into the body rather than being lost to the environment. The feet themselves are kept only slightly above freezing, which minimizes heat transfer to the environment. The muscles used in swimming and walking are in the upper part of the leg, insulated by feathers and kept close to the core body temperature.

The countercurrent circulation principle appears frequently in adaptations that increase the flux of heat or materials between fluids. Tuna use the same principle to retain heat in the active swimming muscles close to the body core, a strategy that allows them to swim rapidly and capture smaller fish even in cold oceans. The gills of fish are designed so that blood and water flow in opposite directions to maximize the exchange of dissolved gases. Among terrestrial organisms, birds have a unique lung structure, which, unlike that of mammals, results in a one-way flow of air opposite to the flow of blood. This adaptation allows birds, with lungs whose weight and volume are small, to achieve the high rates of oxygen delivery required by their active lifestyles.

Many attributes of the physical environment, including the availability of water, nutrients and mineral ions, light, and heat, determine the abundance and productivity of life, drive evolutionary adaptations, and influence the distribution of animals and plants over the surface of the earth. As we shall see in the next chapter, the physical environment varies in predictable ways that shape the character of ecological systems.

SUMMARY

1. Most of the energy for life ultimately comes from sunlight. Solar radiation varies over a spectrum of wavelengths. The visible portion of the spectrum ranges between about 400 nm (violet) and 700 nm (red).

2. The intensity of light impinging on a surface is referred to as irradiance. The irradiance at the top of the earth's atmosphere is diminished by nighttime periods without light, reflection of light by clouds, and absorption of light by the atmosphere before it reaches the surface of the earth. Most of the solar radiation striking the earth is reflected back into space by oceans, snow, ice, and bare soil. The proportion of light reflected by a surface is known as the albedo of that surface.

3. Plants extract energy primarily from the high-intensity, short-wavelength portion of the spectrum, which roughly coincides with visible light. Different photosynthetic pigments, such as chlorophylls and carotenoids, absorb light of particular wavelengths within the visible portion of the spectrum.

4. Photosynthetic pigments absorb photons of light and convert their energy into high-energy compounds such as NADPH and ATP. These compounds can then be used as energy sources for other biochemical reactions.

5. During photosynthesis, most plants assimilate carbon through a reaction (the C_3 photosynthetic pathway) catalyzed by the enzyme Rubisco. This enzyme has a low affinity for CO_2, resulting in a low efficiency of carbon assimilation. The binding affinity of Rubisco for oxygen brings about photorespiration at low CO_2 concentrations and high temperatures.

6. Plants can increase the concentration of CO_2 in their leaves by opening their stomates. However, because CO_2 is scarce in the atmosphere (0.038%), it diffuses into leaves much more slowly than water moves out of them by transpiration, particularly in hot environments. Thus, a plant's need to avoid water loss can restrict its access to atmospheric CO_2 for photosynthesis.

7. Some plants adapted to high temperatures add a step to the carbon assimilation process that produces a four-carbon compound. This reaction, which occurs in the leaf mesophyll, is catalyzed by an enzyme, PEP carboxylase, that has a high affinity for CO_2. The assimilated carbon is then moved to the bundle sheath cells, where the light reactions and the Calvin–Benson cycle take place.

8. Many succulent desert plants, including cacti, use crassulacean acid metabolism (CAM), a pathway that is similar to C_4 photosynthesis, except that carbon assimilation takes place at night, when transpiration is minimal, and the assimilated carbon is released internally to the Calvin–Benson cycle during the day.

9. Heat- and drought-adapted plants have various adaptations to cut down on transpiration and reduce heat loads, including hairs on leaf surfaces that establish boundary layers of humid air, finely subdivided leaves that dissipate heat, waterproof leaf surfaces with waxy cuticles, and stomates recessed in hair-filled pits.

10. Although CO_2 is scarce in the atmosphere, it is more abundant in aquatic systems, where it dissolves to form bicarbonate ions. The availability of carbon in aquatic systems is limited, however, by the rate of diffusion of CO_2 gas and bicarbonate ions through water, especially through the boundary layers of still water that form at the surfaces of plants and algae.

11. Oxygen is abundant in the atmosphere, but is relatively scarce in water, where its solubility and rate of diffusion are low. Oxygen may be depleted by respiration (producing anoxic conditions), especially in environments where it cannot be replenished by photosynthesis.

12. The rates of most physiological processes increase by a factor of 2 to 4 for each 10°C increase in temperature within the physiological range. This factor is known as the Q_{10} of a process. The generality of this temperature effect has been captured in the metabolic theory of ecology.

13. Higher temperatures generally accelerate physiological processes, but can also cause proteins and other biological molecules to unfold and lose their structure and function. Some extremophiles can tolerate very high temperatures because their proteins are chemically designed to generate strong forces of attraction to hold molecules together.

14. Organisms in cold environments withstand freezing temperatures by lowering the freezing point of their body fluids with glycerol or glycoproteins, or by supercooling their body fluids.

15. Most organisms function best within a narrow range of environmental conditions. This optimum is determined by characteristics, such as the structure, function, and quantity of its enzymes, that influence the organism's ability to function under various conditions.

16. The temperature of an organism is closely tied to its thermal environment, which influences gains and losses of heat through radiation, conduction, convection, and evaporation. Together with metabolically produced heat, these factors make up the heat budget of the organism.

17. Maintenance of constant internal conditions, called homeostasis, depends on negative feedback mechanisms. Organisms sense changes in their internal environment and respond in such a manner as to return those conditions to a set point.

18. Homeostasis requires energy when a gradient between internal and external conditions must be maintained. For example, endotherms must generate heat metabolically to balance loss of heat to their cooler surroundings.

19. Organisms employ a variety of mechanisms to control heat loss. One of the most effective of these is countercurrent circulation. In the extremities of birds and mammals, countercurrent circulation transfers heat from arterial to venous blood, and the extremities are kept cooler than the rest of the body.

REVIEW QUESTIONS

1. Explain how light serves as the ultimate source of energy for a meat-eating animal.

2. What wavelengths of light should algae living in deep water use for photosynthesis?

3. Why is C_3 photosynthesis inefficient when the concentration of CO_2 in a leaf is low?

4. Describe the costs and benefits of a plant opening its stomates to increase the concentration of CO_2 in its leaves.

5. How does C_4 photosynthesis solve the problem of low CO_2 concentrations in a leaf?

6. How do CAM plants solve the problem of obtaining CO_2 for photosynthesis while minimizing water loss?

7. Explain how plants use structural adaptations to reduce water loss.

8. How do boundary layers that surround aquatic plants impede a plant's ability to obtain CO_2 for photosynthesis?

9. If oxygen is very abundant in air, why is the metabolism of aquatic organisms often limited by oxygen?

10. Describe the different adaptations that animals have evolved to survive freezing temperatures.

11. How does having different forms of an enzyme allow organisms to live across a wide range of temperatures?

12. If a snake is lying on a rock in the desert sun, how is the snake's body temperature affected by radiation, conduction, convection, and evaporation?

13. Why is torpor a particularly good adaptation for small-bodied endotherms?

SUGGESTED READINGS

Angilletta, M. J., Jr., P. H. Niewiarowski, and C. A. Navas. 2002. The evolution of thermal physiology in ectotherms. *Journal of Thermal Biology* 27:249–268.

Bennett, A. F., and J. A. Ruben. 1979. Endothermy and activity in vertebrates. *Science* 206:649–654.

Black, C. C., and C. B. Osmond. 2003. Crassulacean acid metabolism photosynthesis: "working the night shift." *Photosynthesis Research* 76:329–341.

Brock, T. D. 1985. Life at high temperatures. *Science* 230:132–138.

Brown, J. H., et al. 2004. Toward a metabolic theory of ecology. *Ecology* 85:1771–1789.

Ehleringer, J. R., et al. 1991. Climate change and the evolution of C$_4$ photosynthesis. *Trends in Ecology and Evolution* 6:95–99.

Fenchel, T., and B. J. Finlay. 1994. The evolution of life without oxygen. *American Scientist* 82:22–29.

Gates, D. M. 1965. Energy, plants, and ecology. *Ecology* 46:1–13.

Huey, R. B., and R. D. Stevenson. 1979. Integrating thermal physiology and ecology of ectotherms: A discussion of approaches. *American Zoologist* 19:357–366.

Karov, A. 1991. Chemical cryoprotection of metazoan cells. *BioScience* 41:155–160.

Keeley, J. E. 1990. Photosynthetic pathways in freshwater aquatic plants. *Trends in Ecology and Evolution* 5(10):330–333.

Lee, R. E., Jr. 1989. Insect cold-hardiness: To freeze or not to freeze. *BioScience* 39:308–313.

Munns, R. 2002. Comparative physiology of salt and water stress. *Plant, Cell & Environment* 25:239–250.

Sage, R. F. 2004. The evolution of C$_4$ photosynthesis. *New Phytologist* 161:341–370.

Watanabe, M. E. 2005. Generating heat: New twists in the evolution of endothermy. *BioScience* 55:470–475.

Whorf, T. P., and C. D. Keeling. 1998. Rising carbon. *New Scientist* 157(2124):54.

CHAPTER 4

Variation in the Environment: Climate, Water, and Soil

Few people make important decisions based on the evening weather report. Weather is notoriously irregular and unpredictable. On a global scale, among the most dramatic influences on weather patterns are so-called El Niño events, which are associated with periodic changes in air pressure patterns over the central and western Pacific Ocean. The cause of these changes is poorly understood, but their effects have been experienced, for better or worse, by most of the human population. For example, the 1991–1992 El Niño event, one of the strongest on record, was accompanied by the worst drought of the twentieth century in Africa, causing poor crop production and widespread starvation. The event brought extreme dryness to many areas of tropical South America and Australasia as well. Heat and drought in Australia reduced populations of red kangaroos to less than half their pre–El Niño levels. Outside the tropics and subtropics, El Niño events tend to increase, rather than decrease, precipitation, boosting the production of natural and agricultural systems, but also causing flooding. The El Niño event of 1997–1998 was blamed for 23,000 deaths—mostly from famine—and $33 billion in damages to crops and property worldwide.

Changes in climate—whether local or affecting most of the globe, whether lasting weeks or centuries—can be traced to changes in solar radiation, patterns of ocean circulation, the albedo of the earth's surface, or, over longer time scales, the shapes and positions of the earth's ocean basins, continents, and mountain ranges. On top of these variations, physical and biological processes can establish new patterns of variation as the outcome of unpredictable interactions among their components. Ecologists strive to understand both the origin of climatic variation and its consequences for ecological systems. Their efforts are becoming all the more important as human activities increasingly alter the earth's environments.

The physical environment varies widely over the surface of the earth. Differences in temperature, light, substratum, moisture, salinity, soil nutrients, and other factors have shaped the distributions and adaptations of organisms. The earth has many distinct climate zones whose extents are broadly determined by patterns of solar radiation and by the redistribution of heat and moisture by wind and water currents. Within climate zones, such geologic factors as topography and the composition of bedrock further differentiate the environment on a finer spatial scale. This chapter explores some important patterns of variation in the physical environment that underlie diversity in the biological components of ecosystems.

The surface of the earth, its waters, and the atmosphere above it make up a giant heat-transforming machine. Climatic patterns originate with differences in the intensity of sunlight falling on different parts of the earth's surface. Because its surface varies from bare rock to forested soil, open ocean, and frozen lake, its ability to absorb sunlight varies as well, thus creating differential heating and cooling. The heat energy absorbed by the earth eventually radiates back into space, after further transformations that perform the work of evaporating water and driving the circulation of the atmosphere and oceans. All these factors have created a great variety of physical conditions that, in turn, have fostered the diversification of ecosystems.

Global patterns in temperature and precipitation are established by solar radiation

In spite of its many variations, climate—the characteristic meteorological conditions that prevail at a particular place—exhibits some broadly defined patterns. The earth's climate tends to be cold and dry toward the poles and hot and wet toward the equator. On a global scale, this pattern originates in the greater intensity of sunlight at the equator than at higher latitudes. The sun warms the atmosphere, oceans, and land most when it lies directly overhead (Figure 4.1). A beam of sunlight spreads over a greater area when the sun approaches the horizon, and it also travels a longer path through the atmosphere, where much of its energy either is reflected or is absorbed and reradiated back into space as heat. The sun's highest position each day (its **zenith**) varies from directly overhead in the tropics to near the horizon in polar regions; thus, the warming effect of the sun diminishes from the equator to the poles.

Winds and ocean currents, mountain ranges, and even the positions of the continents create finer-scale climatic patterns. Changes over time follow astronomical cycles.

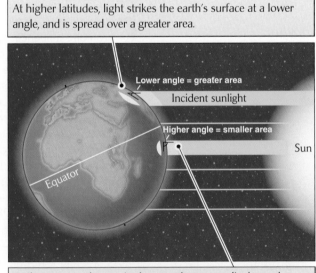

At higher latitudes, light strikes the earth's surface at a lower angle, and is spread over a greater area.

Lower angle = greater area

Incident sunlight

Higher angle = smaller area

Sun

Equator

At the equator, the sun is closer to the perpendicular and shines directly down on the earth's surface.

FIGURE 4.1 The warming effect of the sun is greatest at the equator. The sun's position in the middle of the day varies from directly overhead in the tropics to near the horizon in polar regions.

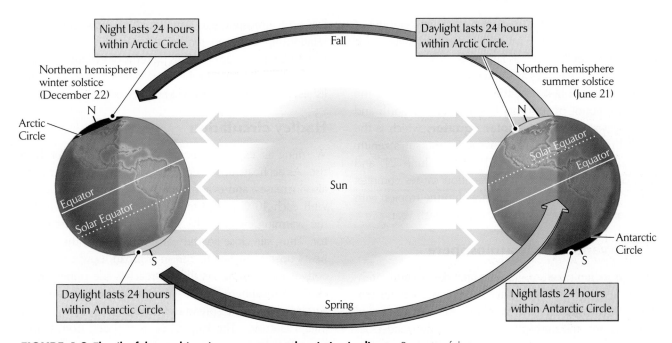

FIGURE 4.2 The tilt of the earth's axis causes seasonal variation in climate. Because of that tilt, the orientation of the earth's axis relative to the sun, and thus the incident solar radiation at each latitude, changes as the earth orbits the sun. The position of the solar equator also changes with the seasons.

The rotation of the earth on its axis causes daily cycles of light and dark, and of temperature; the revolution of the moon around the earth creates lunar cycles of 28 days in the amplitude of the tides; and the revolution of the earth around the sun brings seasonal change.

The distribution of solar energy with respect to latitude

The equator is tilted 23½° with respect to the path the earth follows in its orbit around the sun. Therefore, the Northern Hemisphere receives more solar energy than the Southern Hemisphere during the northern summer and less during the northern winter (Figure 4.2). Seasonal variation in temperature increases with distance from the equator, especially in the Northern Hemisphere, where there is less area of ocean to moderate temperature changes (Figure 4.3). At high latitudes in the Northern Hemisphere, mean monthly temperatures vary by an average of 30°C over the year, and extremes vary by more than 50°C annually. For example, at 60°N, the average coldest month is −12°C and the average warmest month

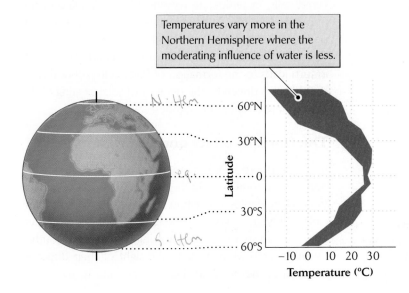

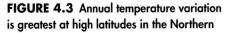

FIGURE 4.3 Annual temperature variation is greatest at high latitudes in the Northern Hemisphere. Mean monthly temperatures (red area) vary more over the year in the Northern Hemisphere because the moderating influence of water is less there.

is 16°C, a difference of 28°C. The average temperatures of the warmest and coldest months in the tropics are much higher, and they differ by as little as 2°–3°C.

The tilt of the earth's axis also results in a seasonal shift in the latitudinal belt near the equator that receives the greatest amount of sunlight. This area moves north and south seasonally with the **solar equator,** which is the parallel of latitude lying directly under the sun's zenith. The solar equator reaches 23½°N on June 21 and 23½°S on December 21. This variation causes complex seasonal patterns of precipitation within the tropics, with none, one, or two peaks of precipitation each year.

Water vapor in the atmosphere

At a given temperature, liquid water has a certain tendency to evaporate, and water vapor has a certain tendency to condense back to a liquid state. The amount of water vapor in the atmosphere when these two tendencies are balanced is referred to as the **equilibrium water vapor pressure.** Water vapor pressure is measured as the contribution of water vapor to the total pressure of the atmosphere, which is approximately 100 kilopascals (kPa), or 10^5 Pa, at sea level. The equilibrium vapor pressure of water increases with temperature, as shown in Figure 4.4. Thus, warm air can hold more water vapor than cold air.

Any air mass can contain less than the equilibrium water vapor pressure, in which case water will continue to evaporate from wet surfaces in contact with air. If the water vapor pressure exceeds the equilibrium value—for example, when the temperature of the air decreases rapidly—the excess water vapor (gas) will condense and

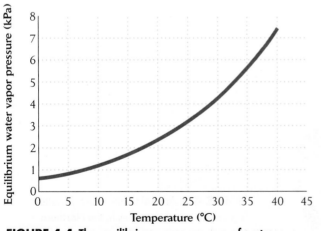

FIGURE 4.4 The equilibrium vapor pressure of water increases with temperature. From data in R. J. List, *Smithsonian Meteorological Tables,* 6th ed., Smithsonian Institution, Washington, D.C. (1966).

leave the atmosphere as rain (liquid) or snow (solid). This relationship between temperature and the equilibrium water vapor pressure controls patterns of evaporation and precipitation and, in combination with air currents, establishes the distributions of wet and dry environments.

Hadley circulation

Warming air expands, becomes less dense, and tends to rise. As air heats up, its equilibrium water vapor pressure also increases, and evaporation quickens, nearly doubling with each 10°C rise in temperature. We have seen that the warming effect of the sun is greatest near the equator. Thus, air close to the earth's surface in the tropics heats up and begins to rise in a broad, upward-moving convection current. As this air reaches the upper layers of the atmosphere, 10–15 kilometers above the earth's surface, it begins to spread to the north and south toward higher latitudes. This tropical air is replaced from below by surface-level air moving in from subtropical latitudes, which forms the trade winds.

The rising tropical air mass cools as it expands under the lower pressure of the upper atmosphere and radiates heat into space. By the time this air has extended to about 30° north and south of the equator, it has become dense enough to sink back to the earth's surface and spread out to the north and south, thus completing a cycle within the atmosphere (Figure 4.5). This type of circulation pattern is called **Hadley circulation,** and the closed cycle of rising and falling air within the tropics is referred to as a **Hadley cell.**

One Hadley cell forms immediately to the north of the equator and another to the south, like a pair of giant waistbands girdling the earth. The sinking air of the tropical Hadley cells drives less distinct secondary cells, called Ferrel cells, in temperate regions, which circulate in the opposite direction. The circulation of Ferrel cells in temperate latitudes (roughly 30°–60° north and south of the equator) causes air to rise at about 60°N and 60°S, which in turn leads to the formation of Polar cells. All of this air circulation is driven by the differential solar heating of the atmosphere at different latitudes.

The Coriolis effect and jet streams

In the Northern Hemisphere, the trade winds blow from northeast to southwest. In the early 1700s, George Hadley (for whom Hadley circulation is named) applied the principle that we now know as the **Coriolis effect** to explain why they do so, rather than flowing due north to south.

In general, winds veer to the right of their direction of travel in the Northern Hemisphere and to the left in

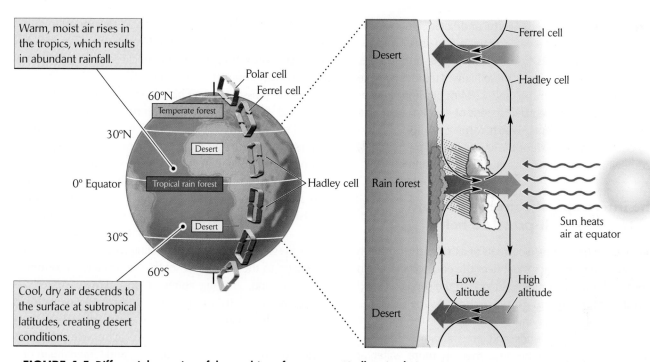

Warm, moist air rises in the tropics, which results in abundant rainfall.

Cool, dry air descends to the surface at subtropical latitudes, creating desert conditions.

FIGURE 4.5 Differential warming of the earth's surface creates Hadley circulation. Warm, moist air rises in the tropics, and cool, dry air moves toward the tropics from subtropical latitudes to replace it, forming Hadley cells. This circulation pattern drives the secondary Ferrel cells and Polar cells at higher latitudes.

the Southern Hemisphere. As Hadley realized, this is a direct consequence of the rotation of the earth and the conservation of momentum. As the earth rotates, a point on the surface at the equator is traveling from west to east at a rate of 1,670 km per hour relative to a fixed point— say, directly under the sun. This is also the velocity of the atmosphere at the earth's surface (fortunately, the ground underneath is moving at the same speed!). At 30°N, however, the circumference of the earth is smaller, and a point on the surface is traveling west to east at only 1,447 km per hour. Thus, the air that is rising at the equator is traveling more than 200 km per hour faster to the east than the air descending to the surface at 30°N. Accordingly, although this air inevitably loses some of its momentum to friction and turbulence, it gets far ahead to the east relative to the earth's surface as it moves north.

The opposite happens on the southward journey of the trade winds at the earth's surface. As they move to the south, they fall behind the rotation of the earth and therefore tend to veer to the west (Figure 4.6). Similarly, surface winds moving northward from about 30°N in the Ferrel cell veer to the east, becoming westerly winds. Thus, weather in temperate latitudes tends to move from west to east.

As the warm tropical air mass moving away from the equator in the upper atmosphere converges with cooler air

moving toward the equator from higher latitudes, it tends to form a rapidly moving west-to-east air current, the subtropical **jet stream,** about 10 km above the earth's surface. Although the formation of the jet stream is not fully understood, it is associated with the high-altitude junction of the Hadley and Ferrel cells. A similar, more powerful

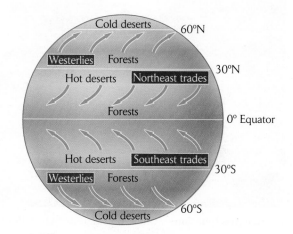

FIGURE 4.6 The Coriolis effect causes air currents to veer to the right in the Northern Hemisphere and to the left in the Southern Hemisphere. It results in the prevailing surface wind patterns known as the trade winds and the westerlies.

jet stream forms where the Ferrel and Polar cells meet, as surface air with west-to-east momentum gained at lower latitudes (the westerlies) rises into the upper atmosphere. The jet streams that form at these higher latitudes average 55 km per hour in summer and 120 km per hour in winter, with maximum recorded speeds of 400 km per hour. These rapid air currents, which form and dissipate and may course to the north or the south, have a tremendous, if somewhat unpredictable, influence on weather.

The intertropical convergence and subtropical high-pressure belt

The region where surface currents of air from the northern and southern subtropics meet near the equator and begin to rise under the warming influence of the sun is referred to as the **intertropical convergence.** As the moisture-laden tropical air rises and begins to cool, the moisture condenses to form clouds and precipitation. Thus, the tropics are humid not because there is more water at tropical latitudes than elsewhere, but because water cycles more rapidly through the tropical atmosphere. The

heating effect of the sun causes water to evaporate and warmed air masses to rise; the cooling of the air as it rises and expands causes precipitation because colder air has a lower equilibrium water vapor pressure.

The air masses moving high in the atmosphere to the north and south, away from the intertropical convergence, have already lost much of their water to precipitation in the tropics. Because this air has cooled, it becomes denser and begins to sink. This descending mass of heavy air creates a high atmospheric pressure, and so these regions north and south of the equator are known as the **subtropical high-pressure belts.** As the air sinks and begins to warm again at subtropical latitudes, its equilibrium water vapor pressure increases. Descending to ground level and spreading to the north and south, the air draws moisture from the land, creating zones of arid climate centered at approximately 30° north and south of the equator (Figure 4.7). The great deserts of the world—the Arabian, Sahara, Kalahari, and Namib of Africa; the Atacama of South America; the Mojave, Sonoran, and Chihuahuan of North America; and the Australian—all fall within subtropical high-pressure belts.

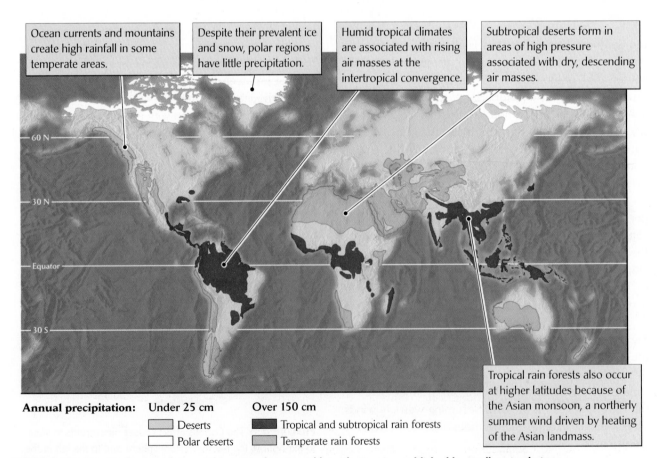

Ocean currents and mountains create high rainfall in some temperate areas.

Despite their prevalent ice and snow, polar regions have little precipitation.

Humid tropical climates are associated with rising air masses at the intertropical convergence.

Subtropical deserts form in areas of high pressure associated with dry, descending air masses.

Tropical rain forests also occur at higher latitudes because of the Asian monsoon, a northerly summer wind driven by heating of the Asian landmass.

Annual precipitation: Under 25 cm   Over 150 cm

Deserts   Tropical and subtropical rain forests
Polar deserts   Temperate rain forests

FIGURE 4.7 The distribution of the earth's major deserts and humid areas is established by Hadley circulation.

Ocean currents redistribute heat

Physical conditions in the oceans, like those in the atmosphere, are complex. Variation in marine conditions is caused partly by winds, which propel the major surface currents of the oceans, and partly by the topography of the ocean basins. In addition, deep-water currents are established by differences in the density of ocean water caused by variations in temperature and salinity. In large ocean basins, cold surface water circulates toward the tropics along the western coasts of continents, and warm surface water circulates poleward along the eastern coasts of continents (Figure 4.8). The direction of ocean circulation is another manifestation of the Coriolis effect: ocean currents tend to veer to the right (clockwise) in the Northern Hemisphere and to the left (counterclockwise) in the Southern Hemisphere.

Surface currents have profound effects on the climate of nearby continental landmasses. For example, the cold Peru Current of the eastern Pacific Ocean, which moves northward from the Southern Ocean along the coasts of Chile and Peru, creates cool, dry environments along the west coast of South America all the way to the equator. As a result, the coasts of northern Chile and Peru have some of the driest deserts on earth. Conversely, the warm Gulf Stream, emanating from the Gulf of Mexico, carries a mild climate far to the north into western Europe and the British Isles (see Figure 1.3).

Any upward movement of ocean water is referred to as **upwelling.** Upwelling occurs wherever surface currents diverge, as in the western tropical Pacific Ocean. As surface currents move apart, they tend to draw water upward from deeper layers. Strong upwelling zones are also established on the western coasts of continents where surface currents move toward the equator and then veer from the continental margins. As surface water moves away from the continents, it is replaced by water rising from greater depths. Because deep water tends to be rich in nutrients, upwelling zones are often regions of high biological productivity. The most famous of these support the

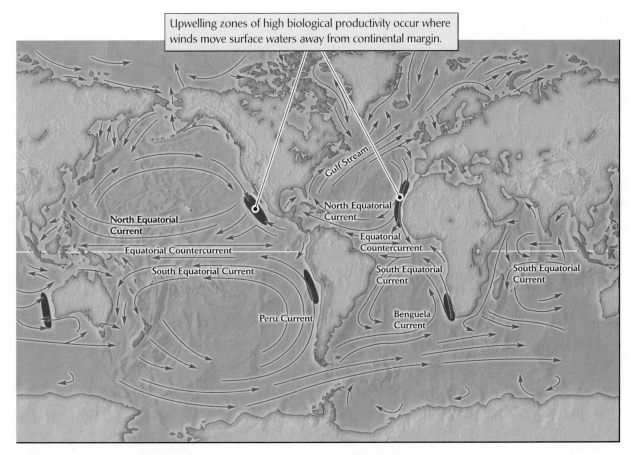

Upwelling zones of high biological productivity occur where winds move surface waters away from continental margin.

FIGURE 4.8 The major ocean surface currents are driven by winds and the earth's rotation. After A. C. Duxbury, *The Earth and Its Oceans,* Addison-Wesley, Reading, Mass. (1971).

FIGURE 4.9 Upwelling currents often support high biological productivity. The Benguela Current off the western coast of South Africa has a zone of upwelling and supports an important fishery. The Cape gannets in this dense nesting colony feed on small fish in the adjacent cold, nutrient-rich waters. Accumulated guano is occasionally scraped off the rocks during the nonbreeding season and used for fertilizer. Photo by R. E. Ricklefs.

rich fisheries of the Benguela Current along the western coast of southern Africa (Figure 4.9) and the Peru Current along the western coast of South America.

Thermohaline circulation

Both surface and deep-water currents are also driven by changes in the density of water caused by variations in temperature and salinity. This **thermohaline circulation** is responsible for the global movement of great masses of water between the major ocean basins. As wind-driven surface currents, such as the Gulf Stream, move toward higher latitudes, the water cools and becomes more dense. In the far north, toward Iceland and Greenland, the surface of the ocean freezes in winter. Because salts are excluded from the forming sea ice, the salt concentration of the underlying water rises. This cold water becomes even more dense and begins to sink, forming a current known as the North Atlantic Deep Water. Similar sinking currents are formed around the edges of Antarctica in the Southern Ocean. These dense waters then flow through the abyssal depths of the ocean basins back into equatorial regions, eventually surfacing as upwelling currents in distant corners of the globe. By one estimate, some of the North Atlantic Deep Water makes it into the North Pacific Ocean, by way of southern Africa and the Indian Ocean, after a journey of more than a millennium.

Thermohaline circulation causes extensive mixing of the oceans and, more importantly, distributes heat energy from the tropics to higher latitudes. The southward movement of the North Atlantic Deep Water toward the tropics is also crucial to the northward movement of the Gulf Stream on the surface. That is why oceanographers some-times refer to the global thermohaline circulation pattern as the ocean conveyor belt.

Shutdown of thermohaline circulation and the Younger Dryas

One of the concerns of scientists who study climate change, and the current warming of the earth's climate in particular, is that accelerated melting of the Greenland ice sheet and the sea ice of the Arctic Ocean will flood the North Atlantic with low-salinity surface waters and prevent the formation of the North Atlantic Deep Water. The disappearance of this deep-water current would effectively shut down the Gulf Stream as a conveyor belt for heat from the tropics. The effect on the climate of Europe would be devastating.

There is some evidence that such an event happened at the end of the most recent glacial period, about 12,700 years ago. As temperatures warmed and the glaciers covering much of northern Europe and North America began to melt, vast quantities of fresh water flowed out to sea and probably cut off the North Atlantic thermohaline circulation. The resulting disruption of the Gulf Stream precipitated a period of cold weather in the region—the Younger Dryas period—lasting about 1,300 years, even as the overall climate of the earth was leaving the glacial period behind. Because the Younger Dryas cold spell coincided in time with the origin of agriculture in what is now the Middle East, some authors have speculated that the development of agriculture was an inevitable consequence of this climate change. Colder climates would have made hunting so unproductive for the growing human population at this time that humans turned to raising crops and domesticating livestock, and thereafter to the establishment of permanent settlements.

Regardless of the forces that shaped the early development of human civilization, it is clear that climate is subject to variation over many time scales. Global climatic patterns can slowly shift over periods much longer than even those of the glacial cycles. Over millions and tens of millions of years, for example, they have been influenced by continental drift, which opens or closes connections between ocean basins and alters the flow of ocean currents, changing the distribution of heat over the earth's surface. At the other extreme, on the much shorter time scales experienced by individuals during their lifetimes, climate is influenced by more predictable factors, particularly over the course of the seasons each year.

Latitudinal shifting of the sun's zenith causes seasonal variation in climate

Within the tropics, the seasonal northward and southward movement of the solar equator determines when the rainy season comes. The intertropical convergence follows the solar equator, producing a moving belt of rainfall. Therefore, wet and dry seasons are most pronounced in broad latitudinal belts lying about 20° north and south of the equator.

Mérida, located on Mexico's Yucatán Peninsula, lies about 20° north of the equator. The intertropical convergence reaches Mérida only during the Northern Hemisphere summer, which is the rainy season for that region (Figure 4.10). During winter, the intertropical convergence lies far to the south of Mérida, and the local climate comes under the influence of the subtropical high-pressure belt. Rio de Janeiro, at the same latitude as Mérida, but to the south of the equator, has its rainy season during the Northern Hemisphere winter, roughly six months after Mérida. Close to the equator, at Bogotá, Colombia, the intertropical convergence passes overhead twice each year, at the time of the equinoxes, resulting in two rainy seasons, with peak rainfall in April and October. Thus, as the seasons change, tropical regions alternately come under the influence of the intertropical convergence, which brings heavy rains, and subtropical high-pressure belts, which bring clear skies.

Farther to the north, outside the tropics, climates come under the influence of the westerly winds that blow at middle latitudes. Here, temperatures, as well as rainfall, vary between winter and summer. The difference in climate between tropical and subtropical regions can be illustrated by temperature and rainfall graphs of three locations in northern Mexico and the southwestern United States (Figure 4.11). At 25°N, in the Chihuahuan Desert of central Mexico, rainfall comes only during the summer, when the intertropical convergence reaches its

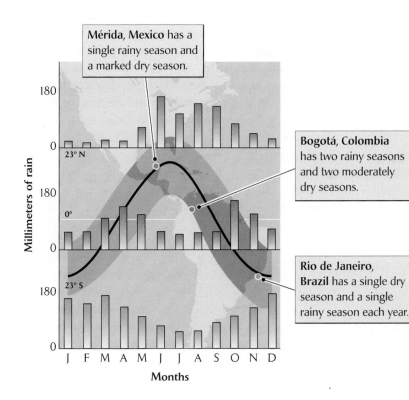

Mérida, Mexico has a single rainy season and a marked dry season.

Bogotá, Colombia has two rainy seasons and two moderately dry seasons.

Rio de Janeiro, Brazil has a single dry season and a single rainy season each year.

FIGURE 4.10 The movement of the intertropical convergence affects precipitation patterns. The seasonal latitudinal movement of the solar equator (see Figure 4.2) results in two seasons of heavy precipitation at the equator and a single rainy season alternating with a pronounced dry season at the edges of the tropics.

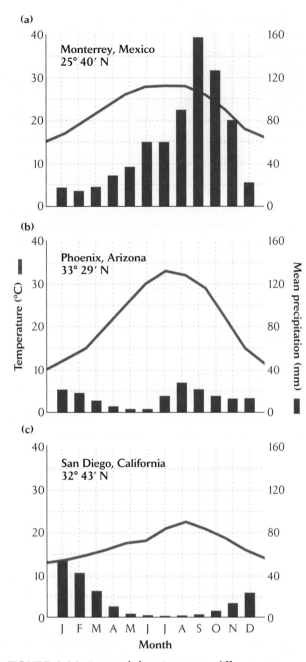

(a)

Monterrey, Mexico
25° 40′ N

(b)

Phoenix, Arizona
33° 29′ N

Temperature (°C)

Mean precipitation (mm)

(c)

San Diego, California
32° 43′ N

J F M A M J J A S O N D
Month

FIGURE 4.11 Seasonal climatic patterns differ among subtropical localities. (a) The Chihuahuan Desert in central Mexico has a summer rainy season. (b) The Sonoran Desert has a combined climatic pattern, with rainfall in summer and winter. (c) San Diego on the Pacific coast and the Mojave Desert have a winter rain and summer drought (Mediterranean) climatic pattern.

northward limit. During the rest of the year, this region falls within the subtropical high-pressure belt. Summer rainfall extends north into the Sonoran Desert of southern Arizona and New Mexico, at 32°N. This area also receives moisture during the winter from the Pacific Ocean, carried by the southwesterly winds emanating from the sub-

tropical high-pressure belt farther south. Thus, the Sonoran Desert experiences both a winter and a summer peak of rainfall. Southern California, at the same latitude, lies to the west of the summer rainfall belt and has a winter-rainfall, summer-drought climate, often referred to as a **Mediterranean climate.** Named for the Mediterranean region of Europe, which has the same seasonal pattern of temperature and rainfall, Mediterranean climates are also found in western South Africa, Chile, and Western Australia—all regions lying along the western sides of continents at about the same latitude north or south of the equator.

Temperature-induced changes in water density drive seasonal cycles in temperate lakes

As we have seen, water gains and loses heat slowly. This property of water tends to reduce temperature fluctuations in large bodies of water, such as oceans and large lakes, as well as in terrestrial environments lying near them. In contrast, small midcontinental lakes in the temperate zone respond quickly to the changing seasons (Figure 4.12). In such lakes, changes in temperature drive changes in water density, which determine the pattern of mixing of the lake water.

Where winters are cold and summers are warm, a lake undergoes two periods of vertical mixing and two periods when the water column is layered, with little vertical mixing. During the winter, such a lake exhibits an inverted **temperature profile;** that is, the coldest water (0°C) lies at the surface, just beneath the ice. Because the density of water increases between the freezing point and 4°C, the warmer water within this range sinks, and the temperature increases to as much as 4°C toward the bottom of the lake.

In early spring, the sun warms the lake surface gradually. But until the surface temperature exceeds 4°C, the sun-warmed surface water tends to sink into the cooler layers immediately below. This vertical mixing distributes heat throughout the water column from the surface to the bottom, resulting in a uniform temperature profile. At the same time, winds drive surface currents that can cause deep water to rise, in a manner similar to upwelling currents in the oceans. This **spring overturn** brings nutrients from the bottom sediments to the surface and oxygen from the surface to the depths.

Later in spring and in early summer, as the sun rises higher each day and the air above the lake warms,

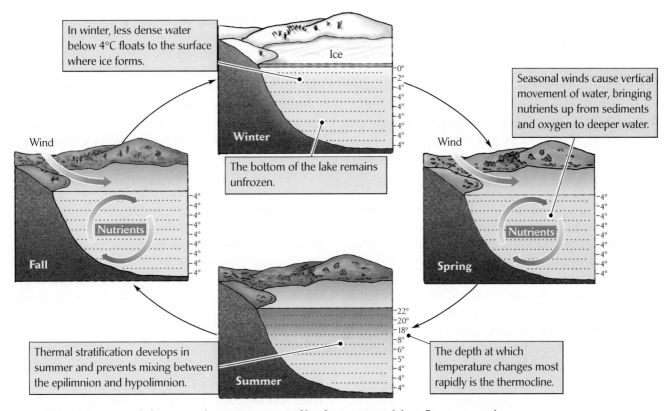

FIGURE 4.12 Seasonal changes in the temperature profile of a temperate lake influence vertical mixing of water layers. Vertical mixing is aided by wind-driven currents when the temperature of water is uniform from the surface to the bottom of the lake.

surface layers of water gain heat faster than deeper layers, creating a zone of abrupt temperature change at intermediate depth, called the **thermocline.** Once the thermocline is well established, water does not move across it because the warmer, less dense surface water literally floats on the cooler, denser water below. This condition is known as **stratification.** The depth of the thermocline varies with local winds and with the depth and turbidity of the lake. It may be located anywhere between 5 and 20 m below the surface; lakes less than 5 m deep usually lack stratification.

The upper layer of warm water above the thermocline is called the **epilimnion,** and the deeper layer of cold water below it is called the **hypolimnion.** Most of the production of a lake occurs in the epilimnion, where sunlight is most intense. Oxygen produced by photosynthesis supplements oxygen entering the lake at its surface, keeping the epilimnion well aerated and thus suitable for animal life. However, plants and algae often deplete the supply of dissolved mineral nutrients in the epilimnion. In doing so, they curtail their own production. The thermocline isolates the hypolimnion from the surface of the lake, so animals and bacteria that remain below the ther-

mocline, where there is little or no photosynthesis, may deplete the water of oxygen, creating anaerobic conditions. Oxygen is in particularly short supply deep in productive lakes that generate abundant organic matter in the epilimnion. Bacteria deep in the lake use up any available oxygen while decomposing the organic matter drifting down from the surface. During late summer, the productivity of temperate lakes may become severely depressed, as nutrients needed to support plant growth are depleted in surface waters and oxygen needed to support animal life is depleted in the depths.

During the fall, the surface layers of the lake cool more rapidly than the deeper layers, become denser than the underlying water, and begin to sink. This vertical mixing, called **fall overturn,** persists into late fall, until the temperature at the lake surface drops below 4°C and winter stratification ensues. Fall overturn speeds the movement of oxygen to deep waters and pushes nutrients to the surface. In lakes where the hypolimnion becomes warm by midsummer, deep vertical mixing may take place in late summer, while temperatures remain favorable for plant growth. The resulting infusion of nutrients into surface waters may cause an explosion in the population of phytoplankton—the **fall (autumn) bloom.** In deep, cold lakes, vertical mixing does not penetrate to all depths until late fall or early winter, when water temperatures are too cold to support phytoplankton growth.

The seasonality of vertical mixing is much less dramatic in lakes that are not exposed to continental climates. In tropical and subtropical lakes (and those in temperate climates closer to the oceans), lake water temperatures do not fall below 4°C. Such lakes do not stratify in the cool season, and many have only one mixing event each year, which follows summer stratification.

In some tropical lakes, a uniform temperature profile makes it possible for surface winds to foster deep vertical mixing. For example, the basins of Lake Tanganyika, a large tropical lake in East Africa, are more than 1,000 m deep, yet the temperature of the water in these basins varies by less than 1°C from a depth of 100 m to near the bottom. At both depths, the temperature is about 23°C. Deep vertical mixing in such lakes brings oxygenated water to the depths and mineral nutrients to the surface, supporting high overall productivity. In temperate zones, deep lakes are often permanently stratified and can be very unproductive. A concern in tropical regions is that climatic warming will increase the temperature of lake surface waters and create a thermocline at shallow depths, impairing vertical mixing and reducing lake production. There is already evidence that this is happening in Lake Tanganyika.

Climate and weather undergo irregular and often unpredictable changes

Everyone knows that weather is difficult to forecast far in advance. We often remark that a certain year was particularly dry or cold compared with others. Recent intense hurricanes along the Gulf Coast of the United States, flooding in Europe and southern Asia, drought in Africa—all drive home the capriciousness of nature. Such extremes occur infrequently, but they affect ecological systems disproportionately.

The rich Peruvian fishing industry thrives on the abundant fish in the cold, nutrient-rich waters of the Peru Current. The Peru Current flows north along the west coast of South America and finally veers offshore at Ecuador, heading west toward the Galápagos archipelago. North of this point, warm, tropical inshore waters prevail along the coast. Each year, a warm countercurrent known as **El Niño** ("little boy" in Spanish, a name referring to the Christ child because this countercurrent appears around Christmastime) moves down the coast toward Peru. In some years, it flows strongly enough and far enough south to force the cold Peru Current well offshore and shut down the local fishing industry.

During "normal" years between El Niño "events," the cool waters of the Peru Current warm up as they move westward across the equatorial Pacific Ocean. The temperature at the sea surface thus increases from east to west. This temperature difference creates a steady surface wind blowing across the equatorial central Pacific Ocean in the same direction, from an area of high atmospheric pressure and descending air in the east to an area with a warmer sea surface temperature, lower atmospheric pressure, and ascending air centered in the west (Figure 4.13a). The difference in atmospheric pressure over this gradient has traditionally been measured between Tahiti and Darwin, Australia. Typically, conditions are cooler and drier in the eastern equatorial Pacific, closer to the coast of South America, and warmer and wetter in the west.

An El Niño event appears to be triggered by a reversal of these pressure areas (the so-called **Southern Oscillation**) and of the winds that flow between them. As a result, the westward-flowing equatorial currents stop or even reverse, upwelling off the coast of South America weakens or ceases, and warm water—the El Niño current—piles up along the coast of South America (Figure 4.13b). Historical records of atmospheric pressure at Tahiti and Darwin, and of sea surface temperatures on the Peruvian coast, reveal pronounced El Niño–Southern Oscillation (ENSO) events at irregular intervals of 2 to 10 years (Figure 4.13c).

The climatic and oceanographic effects of an ENSO event extend over much of the world, affecting ecosystems in such distant areas as India, South Africa, Brazil, and western Canada. A strong ENSO event in 1982–1983 disrupted fisheries and destroyed kelp beds in California, caused reproductive failure of seabirds in the central Pacific Ocean, and killed off widespread areas of coral in Panama. Precipitation was also dramatically affected in many terrestrial ecosystems. The deserts of northern Chile, normally the driest place on earth, received their first recorded rainfall in over a century.

The 1982–1983 ENSO event drew the world's attention to the far-reaching effects of oceanographic and atmospheric changes in many parts of the world. For example, data from Zimbabwe for the period 1970–1993 show striking variation in yields of corn (maize). As one might expect, these variations in yield were correlated with variations in rainfall, but more surprisingly, they were also correlated with sea surface temperatures in the eastern tropical Pacific Ocean (Figure 4.14). One can see the

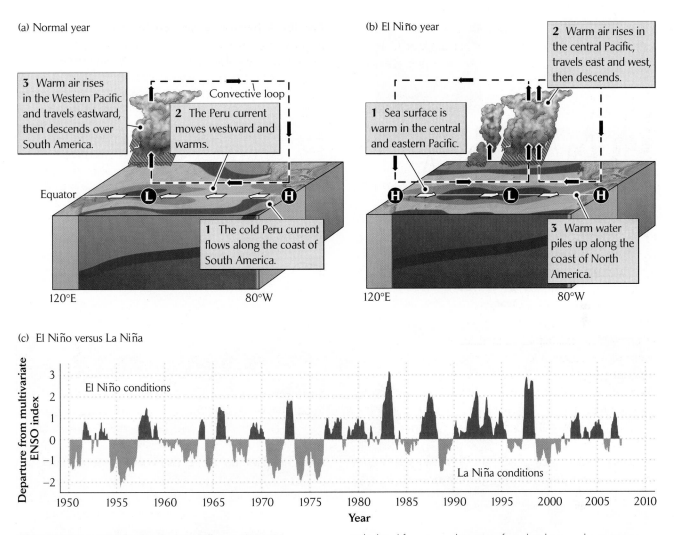

(a) Normal year

3 Warm air rises in the Western Pacific and travels eastward, then descends over South America.

Convective loop

2 The Peru current moves westward and warms.

Equator

1 The cold Peru current flows along the coast of South America.

120°E   80°W

(b) El Niño year

2 Warm air rises in the central Pacific, travels east and west, then descends.

1 Sea surface is warm in the central and eastern Pacific.

3 Warm water piles up along the coast of North America.

120°E   80°W

(c) El Niño versus La Niña

El Niño conditions

La Niña conditions

Departure from multivariate ENSO index

Year

FIGURE 4.13 El Niño–Southern Oscillation (ENSO) events result in dramatic climate changes. (a) Sea surface temperatures, ocean thermocline, and wind patterns during normal conditions in the Pacific, when warm surface waters are driven to the east. (b) Conditions during an ENSO event, when the trade winds weaken and warm water approaches the coast of South America. (c) ENSO events are marked by large positive anomalies in the multivariate ENSO index, which is calculated from a combination of sea-level atmospheric pressure, wind velocity, sea surface and surface air temperatures, and cloudiness fraction of the sky measured at various localities in the Pacific. (a, b) Courtesy of NOAA/Pacific Marine Environmental Laboratory/Tropical Atmosphere Ocean (TAO) project; (c) courtesy of NOAA/ESRL/Physical Science Division (http://www.cdc.noaa.gov/ENSO/enso.mei_index.html).

far-reaching effects of the 1982–1983 and 1991–1992 El Niño events in these data.

El Niño events also have predictable consequences for the climate of North America. The warm tropical waters that dominate the eastern Pacific Ocean during El Niño events create strong Hadley cell circulation, resulting in a persistent subtropical jet stream that brings cooler, wetter, often stormy weather to the southern United States and northern Mexico. The polar jet stream weakens, and warm, dry conditions settle in to the northern states and southern Canada and Alaska.

El Niño–Southern Oscillation events are often followed by **La Niña,** a period of strong trade winds that accentuate normal ocean surface and upwelling currents and bring extreme weather of a different sort than ENSO to much of the world. La Niña is characterized by heavy rainfall in many regions of the tropics, drought in north-temperate regions, and an increase in hurricane activity in the North Atlantic Ocean. Colder waters in the eastern Pacific weaken the subtropical jet stream and strengthen the polar jet stream.

(a)

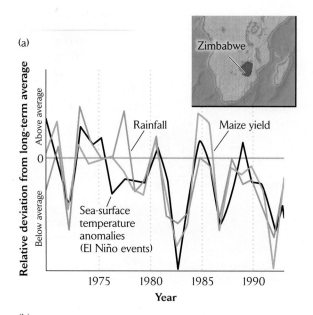

FIGURE 4.14 ENSO events have far-reaching effects.
(a) Deviations from the long-term average in rainfall and corn production in Zimbabwe are correlated with sea surface temperature anomalies in the eastern equatorial Pacific Ocean. In this graph, El Niño conditions are indicated by below-average values. (b) Areas affected by ENSO events from December through February in a typical ENSO year. Zimbabwe is located in the yellow area in southern Africa. (a) From M. A. Cane, G. Eshel, and R. W. Buckland, *Nature* 370:204–205 (1994); (b) from NOAA Climate Prediction Center.

(b)

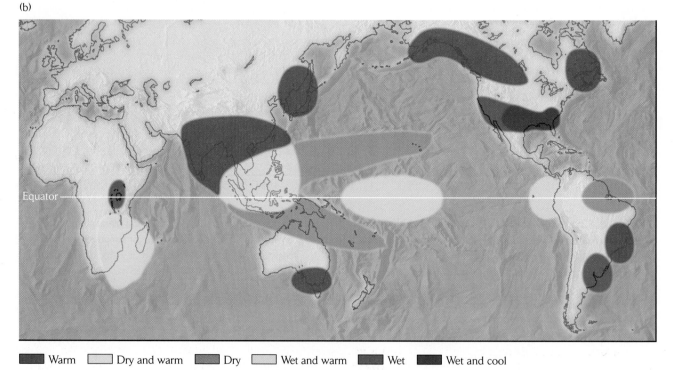

■ Warm   □ Dry and warm   ■ Dry   □ Wet and warm   ■ Wet   ■ Wet and cool

ECOLOGISTS IN THE FIELD **A half-million-year climatic record.** Humans have kept records of climate systematically for about 200 years and sporadically for several hundred years before that. Variation in the thickness of the growth rings in trees extends records of climate in some regions—at least from a tree's point of view—back to several thousand years ago. Whether a climatic record encompasses decades, centuries, or millennia, one sees both regular climatic cycles and irregular fluctuations. But what about longer periods? We know from geologic evidence that the Northern Hemisphere has undergone multiple glacial cycles during the past million years, and that these cycles reflect broader patterns of global climate change that influence the distributions and abundances of organisms and their evolutionary responses to environmental conditions. Scientists are now turning to isotope studies to obtain a direct picture of long-term climate change in our dynamic world. These studies are based on sensitive measurements of the proportions of stable isotopes of oxygen, carbon, and other elements in ocean sediments, ice cores, coral reefs, stalactites in caves, and other datable formations.

Sediments that accumulate in layers at the bottom of a lake or ocean preserve a record of the local conditions over time. The sediments of deep-sea basins consist largely of the calcium carbonate shells of small protists known as foraminifera (Figure 4.15). The shells of these long-dead creatures act as tiny permanent thermometers that provide a long-term record of temperature fluctuations. The foraminifera provide this record because they incorporate oxygen, in the form of carbonate, into their shells. Most of the oxygen in the biosphere has an atomic weight of 16, and is referred to as the form, or *isotope*, ^{16}O. Oxygen also occurs as an isotope with two additional neutrons, which has an atomic weight of 18. Oxygen-18, or ^{18}O, is relatively rare, making up only 0.2% of the oxygen in the biosphere. The heavier ^{18}O atom is incorporated less readily into calcium carbonate shells than is ^{16}O. This difference is represented as a delta ^{18}O value,

$$\delta^{18}O = 1000 \times \frac{(^{18}O/^{16}O_{sample} - {}^{18}O/^{16}O_{water})}{^{18}O/^{16}O_{water}}$$

which is the proportional difference in isotope concentration, expressed in parts per thousand, where "sample" refers to shell carbonate and "water" refers to Standard Mean Ocean Water (SMOW), a measure used as an international reference. Because the proportion of ^{18}O in the shells of foraminifera is less than the proportion dissolved in seawater, $\delta^{18}O$ values are negative in these analyses. More importantly for our purposes here, the proportion of ^{18}O incorporated into shells increases with temperature by approximately one part per thousand (that is, one unit of $\delta^{18}O$) for each 4°C increase in temperature.

Jerry McManus and his colleagues at the Woods Hole Oceanographic Institution analyzed a 65 m sediment core taken from the bottom of the North Atlantic Ocean northwest of Ireland. The record of $\delta^{18}O$ values from the sediment core is shown in Figure 4.15. As one might expect, temperatures indicated by the shells of the surface-dwelling foraminiferan *Neogloboquadrina pachyderma* are several degrees higher than those indicated by the bottom-dwelling *Cibicidoides wuellerstorfi*. (Sorry, they don't have common names.) The shells of both species, however, exhibit 100,000-year cycles of temperature, corresponding to glacial and interglacial climatic cycles. Temperature changes at the bottom of the ocean parallel those at the surface, confirming that no place on earth escapes variations in climate. Superimposed on the long-term temperature cycles are numerous variations of shorter duration. These variations correspond to a wide range of global climatic patterns resulting from periodic variations in the shape of the earth's orbit, which bring the earth slightly closer to or farther from the sun. ▮

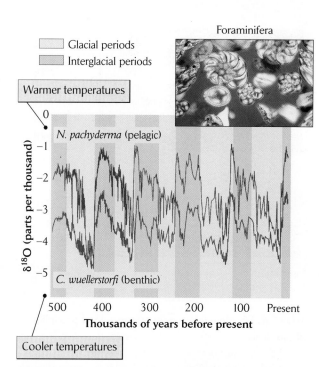

FIGURE 4.15 Variations in marine temperatures are recorded by foraminifera in deep-sea sediments. Variations in the proportions of oxygen isotopes incorporated into the shells of foraminifera in sediments from the North Atlantic Ocean during the last 500,000 years. The $\delta^{18}O$ value becomes more negative as the temperature of the water in which the foraminifera lived decreases. The record clearly shows five warm interglacial periods separated by cold glacial periods. From J. F. McManus, D. W. Oppo, and J. L. Cullen, *Science* 283:971–975 (1999). *Inset:* Shells of several foraminiferan species. Photo by Charles Gellis/Photo Researchers.

Topographic features cause local variation in climate

As we have seen, the primary global patterns in the earth's climate result from the unequal solar heating of the earth's surface from the equator to the poles. However, the positions of continental landmasses exert important secondary effects on temperature and precipitation. For example, at any given latitude, rain falls more plentifully in the Southern Hemisphere because oceans and lakes cover a greater proportion of its surface (81%, compared with 61% of the Northern Hemisphere). Water evaporates more readily from exposed surfaces of water bodies than from soil and vegetation. For the same reason, the interior of a continent usually experiences less precipitation than its coasts, simply because it lies farther from the major site of water evaporation, the surface of the ocean. Furthermore, coastal (maritime) climates vary less than interior

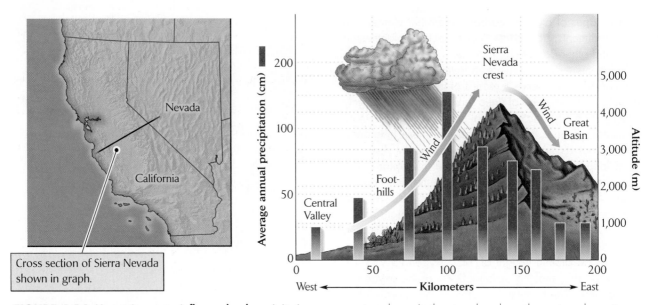

FIGURE 4.16 Mountain ranges influence local precipitation patterns. In the Sierra Nevada of California, the prevailing wind comes from the west across the Central Valley of California. As moisture-laden air is deflected upward by the mountains, it cools, and its moisture condenses, resulting in heavy precipitation on the western slope. As the air rushes down the eastern slope, it warms and begins to pick up moisture, creating arid conditions in the Great Basin. After E. R. Pianka, *Evolutionary Ecology*, 4th ed., Harper & Row, New York (1988).

(continental) climates because the heat storage capacity of ocean water reduces temperature fluctuations near the coast. For example, the hottest and coldest mean monthly temperatures near the Pacific coast of North America at Portland, Oregon, differ by only 16°C. Farther inland, this range increases to 18°C at Spokane, Washington; 26°C at Helena, Montana; and 33°C at Bismarck, North Dakota.

Surface winds and rain shadows

Global wind patterns interact with other features of the landscape to create precipitation. Mountains force air upward, causing it to cool and lose its moisture as precipitation on the windward side. As the dry air descends the leeward slope and travels across the lowlands beyond, it picks up moisture and creates arid environments called **rain shadows** (Figure 4.16). The Great Basin deserts of the western United States and the Gobi Desert of Asia are in the rain shadows of extensive mountain ranges.

Panama lies at 10°N, and like other areas in the northern part of the tropics, it experiences a dry and windy winter under the influence of the trade winds and a humid, rainy summer under the influence of the intertropical convergence. Because the trade winds come from the north and east, Panama's climate is wetter on the northern (Caribbean) side of the isthmus than on the southern (Pacific) side. Mountains intercept moisture coming from the Caribbean coast and produce a rain shadow

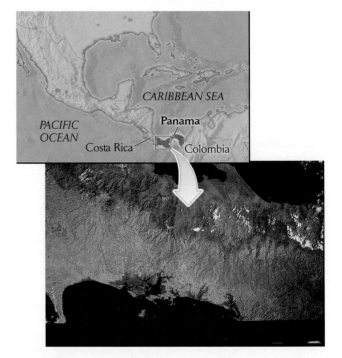

FIGURE 4.17 Trade winds create a rain shadow in Central America. This false-color satellite image of western Panama during the dry season shows heavy forest (brown) to the north of the continental divide, where the prevailing winds blow humid air from the Caribbean Sea. To the south of the continental divide, on the Pacific side of the isthmus, the green color indicates pasture and dry forest. Courtesy of Marcos A. Guerra, Smithsonian Tropical Research Institute.

FIGURE 4.18 Many trees shed their leaves during the dry season. These trees are growing in the rain shadow on the Pacific slope of Panama. Photo by R. E. Ricklefs.

(Figure 4.17). Indeed, the Pacific lowlands are so dry during the winter months that most trees lose their leaves to avoid the water stress (Figure 4.18).

Topographic influences on climate

Topography and geology can modify the environment on a local scale within regions of otherwise uniform climate. In hilly areas, the slope of the land and its exposure to the sun influence the temperature and moisture content of the soil. Soils on steep slopes may drain well, causing drought stress for plants on the hillside at the same time water saturates the soils of nearby lowlands. In arid regions, stream bottomlands and seasonally dry riverbeds may support well-developed **riparian** forests, which accentuate the contrasting bleakness of the surrounding desert. In the Northern Hemisphere, south-facing slopes receive more sunlight, and its warmth and drying power limit vegetation to shrubby, drought-resistant **xeric** forms. The adjacent north-facing slopes remain relatively cool and wet and harbor moisture-requiring mesic vegetation (Figure 4.19).

Air temperature decreases by about 6°–10°C for each 1,000 m increase in elevation, depending on the region. This decrease in temperature, which is caused by the expansion of air in the lower atmospheric pressures at higher altitudes, is referred to as **adiabatic cooling.** Climb high enough, even in the tropics, and you will encounter freezing temperatures and perpetual snow. In regions where the temperature at sea level averages 30°C, freezing temperatures are reached at about 5,000 m, the approximate elevation of the snow line on tropical mountains.

In north-temperate latitudes, a 6°C drop in temperature with each 1,000 m of elevation corresponds to the temperature change encountered over an 800 km increase in latitude. In many respects, the climate and vegetation of high-elevation locations resemble those of sea-level localities at higher latitudes. But despite these similarities, montane environments usually vary less from season to season than their low-elevation counterparts at higher latitudes. Temperatures in tropical montane environments vary less seasonally than those of montane environments at higher latitudes (although they can vary quite markedly from day to night), and some of these areas remain frost-free over the year, which makes it possible for many tropical plants and animals to live in the cool environments found there.

In the mountains of the southwestern United States, changes in plant communities with elevation result in more or less distinct belts of vegetation, which the nineteenth-century naturalist C. Hart Merriam referred to as **life zones.** Merriam's scheme of classification included

↳ don't need to know

FIGURE 4.19 Topography can modify the environment on a local scale. Exposure influences the vegetation growing on slopes of the San Gabriel Mountains, near Los Angeles, California. The cooler north-facing slope (left) supports pine–oak forest, while shrubby, drought-resistant chaparral vegetation grows on the south-facing slope (right). Photo by R. E. Ricklefs.

Lower Sonoran zone

Upper Sonoran zone

Upper Sonoran zone, upper edge

Transition zone

Canadian zone

Alpine zone

FIGURE 4.20 Vegetation changes with increasing elevation in the mountains of Arizona. At the lowest elevations (top photos), the Lower Sonoran zone supports mostly saguaro cactus, small desert trees such as paloverde and mesquite, numerous annual and perennial herbs, and small succulent cacti. Agave and grasses are conspicuous elements of the Upper Sonoran zone, and oaks appear toward its upper edge. At higher elevations, large trees predominate: ponderosa pine in the Transition zone, spruce and fir in the Canadian zone. These trees gradually give way to bushes, willows, herbs, and lichens in the Alpine zone above the tree line. Photos by Tom Bean/DRK Photo.

five broad zones, which he named, from low to high elevation (or from south to north), the Lower Sonoran, Upper Sonoran, Transition, Canadian (or Hudsonian), and Alpine (or Arctic–Alpine) (Figure 4.20). At low elevations, one encounters a cactus and desert shrub association characteristic of the Sonoran Desert of northern Mexico and southern Arizona. In the riparian forests along streambeds, the plants and animals have a distinctly tropical fla-

vor. Many hummingbirds and flycatchers, ring-tailed cats, jaguars, and peccaries make their only temperate zone appearances in this area. In the Alpine zone, 2,600 m higher, one finds a landscape resembling the tundra of northern Canada and Alaska. Thus, by climbing 2,600 m, one experiences changes in climate and vegetation that would occur in the course of a journey to the north of 2,000 km or more at sea level.

(a)   (b)

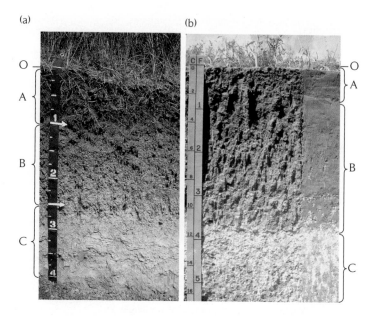

FIGURE 4.21 Soil profiles may show distinct layers, or horizons. (a) This prairie soil from Nebraska is weathered to a depth of about 3 feet (0.9 m), where the subsoil contacts the parent material, which consists of loosely aggregated, calcium-rich, wind-deposited sediments (loess). The B horizon (between the arrows) contains less organic material than the layers above it. Rainfall in Nebraska is not abundant, but it is sufficient to leach readily soluble ions completely from the soil; hence there is no redeposition of these ions in the B horizon. The C horizon is light-colored and has been leached of some of its calcium. (b) In this prairie soil from Texas the A horizon is only about 15 cm thick. The B horizon extends down to the bottom of the dark layer, which represents organic material redeposited from the A horizon. Considerable calcium has been redeposited at the base of the B horizon and in the C horizon below it. Because these soils have formed in dry climates, neither profile has a well-marked E horizon. Courtesy of the U.S. Department of Agriculture, Soil Conservation Service.

Climate and the underlying bedrock interact to diversify soils

Climate affects the distributions of plants and animals indirectly through its influence on the development of soil, which provides the substratum within which plant roots grow and many animals burrow. The characteristics of soil determine its ability to hold water and to make available the minerals required for plant growth. Thus, its variation provides a key to understanding the distributions of plant species and the productivity of biological communities.

Soil defies simple definition, but we may describe it as the layer of chemically and biologically altered material that overlies rock or other unaltered material at the surface of the earth. It includes minerals derived from the parent material, modified minerals formed anew within the soil, organic material contributed by plants, air and water within the pores of the soil, living roots of plants, microorganisms, and the larger worms and arthropods that make the soil their home.

Where a recent road cut or excavation exposes soil in cross section, one often notices distinct layers, which are called **horizons** (Figure 4.21). A generalized, and somewhat simplified, soil profile has several divisions, which, from the surface downward, are referred to as the O, A, E, B, C, and R horizons (Table 4.1). Five factors determine the characteristics of soils: climate, parent material, vegetation, local topography, and, to some extent, age. Soil horizons reveal the decreasing influence of climatic and biotic factors with increasing depth.

TABLE 4.1	Characteristics of the major soil horizons
Soil horizon	**Characteristics**
O	Primarily dead organic litter. Most soil organisms inhabit this layer.
A	A layer rich in humus, consisting of partly decomposed organic material mixed with mineral soil.
E	A region of leaching of minerals from the soil. Because minerals are dissolved by water—that is to say, mobilized—in this layer, plant roots are often concentrated here. Eluviation (hence the "E" horizon) refers to the downward movement of dissolved or suspended material within the soil by leaching.
B	A region of little organic material, whose chemical composition resembles that of the underlying rock. Clay minerals and oxides of aluminum and iron leached out of the overlying E horizon are sometimes deposited here (illuviation).
C	Primarily weakly altered material, similar to the parent material. Calcium and magnesium carbonates accumulate in this layer, especially in dry regions, sometimes forming hard, impenetrable layers or "pans."
R	Unaltered parent material.

| TABLE 4.2 | Soil types, their charcteristics, and their distribution |

Alfisols –	Moist, moderately weathered mineral soils	
Aridosols –	Dry mineral soils with little leaching and accumulations of calcium carbonate	
Entisols –	Recent mineral soils lacking development of soil horizons	
Histosols –	Organic soils of peat bogs; mucks	
Inceptisols –	Young, weakly weathered soils	
Mollisols –	Well-developed soils high in organic matter and calcium; very productive	
Oxisols/ Andisols –	Deeply weathered, lateritic soils of moist tropics (**not represented in continental United States**)	
Spodosols –	Acid, podsolized soils of moist, often cool climates with shallow leached horizon and a deeper layer of deposition	
Ultisols –	Highly weathered soils of warm, moist climates with abundant iron oxides	
Vertisols –	High content of swelling-type clays developing deep cracks in dry seasons	

Soils exist in a dynamic state, changing as they develop on newly exposed rock. Even after soils achieve stable properties, they remain in a constant state of flux. Groundwater removes some substances; other materials enter the soil from vegetation, in precipitation, as dust from above, and from the parent rock below. Where little rain falls, the parent material breaks down slowly, and plant production adds little organic detritus to the soil. Thus, arid regions typically have shallow soils, with bedrock lying close to the surface. Soils may not form at all where decomposed bedrock and detritus erode as rapidly as they form. Soil development also stops short on alluvial deposits, where fresh layers of silt deposited each year by floodwaters bury older material. At the other extreme, soil formation proceeds rapidly in parts of the humid tropics, where chemical alteration of parent material may extend to depths of 100 m. Most soils of temperate zones are intermediate in depth, extending to an average of about 1 m. The variety of soil types, their characteristics, and their distributions are presented in Table 4.2.

Weathering

Weathering—the physical and chemical alteration of rock material near the earth's surface—occurs wherever surface water penetrates. The repeated freezing and thawing of water in crevices physically breaks rock into smaller pieces and exposes a greater surface area to chemical action. Initial chemical alteration of the rock occurs when water dissolves some of its more soluble minerals, especially sodium chloride (NaCl) and calcium sulfate ($CaSO_4$). Other materials, such as the oxides of titanium, aluminum, iron, and silicon, dissolve less readily.

The weathering of granite illustrates some basic processes of soil formation. The minerals responsible for the grainy texture of granite—feldspar, mica, and quartz—consist of various combinations of oxides of aluminum, iron, silicon, magnesium, calcium, and potassium, along with other, less abundant compounds. The key aspect of the process of weathering is the displacement of certain elements in these minerals—notably calcium, magnesium, sodium, and potassium—by hydrogen ions, followed by the reorganization of the remaining oxides into new minerals. This chemical process provides the basic mineral structure of soil. Quartz, a type of silica (SiO_2), is relatively insoluble under cool, temperate conditions and remains little altered as grains of sand in soil derived from granite parent material.

Feldspar and mica grains consist of aluminosilicates of potassium, magnesium, and iron. Hydrogen ions percolating through granite displace potassium and magnesium ions, and the remaining iron, aluminum, and silicon form new, insoluble materials, particularly clay particles. These particles are important to the water-holding and nutrient-holding capacity of soils. When magnesium (Mg^{2+}) is replaced by hydrogen (H^+), a clay particle gains a negative charge; when aluminum (Al^{3+}) is replaced by iron (Fe^{2+}) or magnesium, the clay particle gains another negative charge. These negative charges accumulate on the outside surface of the clay particle, where they hold basic cations—positively charged ions, such as calcium (Ca^{2+}), magnesium (Mg^{2+}), potassium (K^+), and sodium (Na^+). The ability of a soil to retain these cations, called its **cation exchange capacity,** provides an index to the fertility of that soil. Young soils have relatively few clay particles and little added organic material, so the soil profile is poorly developed and the fertility of the soil is relatively low. Soil fertility improves with time, up to a point. Eventually, however, weathering breaks down clay particles, cation exchange capacity decreases, and soil fertility drops.

Where do the hydrogen ions involved in weathering come from? They are derived from two sources. One of these is the carbonic acid that forms when carbon dioxide dissolves in rainwater (see Chapter 2). In regions not affected by acidic pollution, concentrations of hydrogen ions in rainwater produce a pH of about 5. The other source of hydrogen ions is the oxidation of organic material in the soil itself. The metabolism of carbohydrates, for example, produces carbon dioxide, and dissociation of the resulting carbonic acid generates additional hydrogen ions. In the Hubbard Brook Experimental Forest of New Hampshire (see Chapter 24), these internal processes account for about 30% of the hydrogen ions needed for the weathering of bedrock; the remainder come from precipitation.

The changes in chemical composition as granite weathers from rock to soil in different climatic regions show that weathering is most severe under tropical conditions of high temperature and rainfall (Figure 4.22). Highly weathered tropical soils tend to have low cation exchange capacities and little natural fertility. The high productivity of some tropical rainforests depends more on the rapid cycling of nutrients close to the surface of the soil than on the nutrient content of the soil itself.

Podsolization

Under mild temperatures and moderate precipitation in the temperate zone, sand grains and clay particles resist weathering and form stable components of the soil. In acidic soils in cool, moist regions of the temperate zone, however, clay particles break down in the E horizon, and their soluble ions are transported downward and deposited in the lower B horizon. This process, known as **podsolization,** reduces the fertility of the upper layers of the soil.

Acidic soils occur primarily in cool regions where needle-leaved trees dominate the forests. The slow decomposition of the leaf litter shed by spruce and fir trees produces organic acids, which promote high concentrations of hydrogen ions. In addition, rainfall usually exceeds evaporation in regions of podsolization. Under these moist conditions, because water continuously moves downward through the soil profile, little clay-forming material is transported upward from the weathered bedrock below.

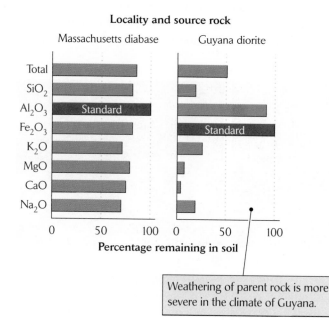

FIGURE 4.22 Weathering is more severe in tropical than in temperate climates. Differential weathering results in differential removal of minerals from granitic parent material in Massachusetts (42°N) and in Guyana (6°N). The bars show the amount of each mineral remaining in the soil as a percentage of the amount of the mineral (aluminum oxide or iron oxide) assumed to be the most stable component of the soil in its region (labeled Standard). After E. W. Russell, *Soil Conditions and Plant Growth,* 9th ed., Wiley, New York (1961).

FIGURE 4.23 Podsolized soils have reduced fertility. This 1-meter deep profile of a podsolized soil in northern Michigan shows strong leaching of the A horizon. The light-colored E horizon and the dark-colored B horizon immediately below it form distinct bands. Compare the general absence of roots in the heavily eluviated E horizon with their presence in the illuviated B horizon below it. Photo by R. E. Ricklefs.

In North America, podsolization advances farthest under spruce and fir forests in New England and the Great Lakes region and across a wide belt of southern and western Canada. A typical profile of a highly podsolized soil (Figure 4.23) reveals striking bands corresponding to the regions of leaching (eluviation) and redeposition (illuviation). The A horizon is dark and rich in organic matter. It is underlain by a light-colored E horizon that has been leached of most of its clay content. As a result, the E horizon consists mainly of sandy material that holds neither water nor nutrients well. One usually finds a dark band immediately below the E horizon. This is the uppermost layer of the B horizon, where iron and aluminum oxides are redeposited. Other, more mobile minerals may accumulate to some extent in lower parts of the B horizon, which then grades almost imperceptibly into a C horizon and the parent material (R horizon).

Laterization

Soils weather to great depths in the warm, humid climates of many tropical and subtropical regions. One of the most conspicuous features of weathering under these conditions is the breakdown of clay particles, which results in the leaching of silicon from the soil, leaving oxides of iron and aluminum to predominate in the soil profile. This process is called **laterization,** and the iron and aluminum oxides give lateritic soils their characteristic reddish coloration (Figure 4.24). Even though the rapid decomposition of organic material in tropical soils contributes an abundance of hydrogen ions, bases formed by the breakdown of clay particles neutralize them. Consequently, lateritic soils usually are not acidic, even though they may be deeply weathered. Laterization is enhanced in certain soils that develop on parent material that is deficient in quartz (SiO_2), but rich in iron and magnesium (basalt, for example); these soils contain little clay to begin with because they lack silicon. Regardless of the parent material, weathering reaches deepest, and laterization proceeds farthest,

(a)   (b)   (c)

FIGURE 4.24 Lateritic soils have little clay and hold few nutrients. (a) A fresh road cut in the Amazon basin in Ecuador shows a typical lateritic soil profile. (b) Notice the roots at the top of the B horizon in a layer of illuviated organic matter.

(c) Highly oxidized and deeply weathered soils are also found in the southeastern United States, as in this eroded area in western Tennessee. Photos (a) and (b) by R. E. Ricklefs; photo (c) courtesy of the U.S. Department of Agriculture, Soil Conservation Service.

on low-lying soils, such as those of the Amazon basin, where highly weathered surface layers are not eroded away and the soil profiles are very old.

One of the consequences of laterization is that many tropical soils have a low cation exchange capacity. In the absence of clay and organic matter, mineral nutrients are readily leached from the soil. Where soils are deeply weathered, new minerals formed by the decomposition of the parent material are simply too far from the surface to contribute to soil fertility. Besides, heavy rainfall keeps water moving down through the soil profile, preventing the upward movement of nutrients. In general, the deeper the ultimate sources of nutrients in the unaltered bedrock, the poorer the surface layers. Rich soils do, however, develop in many tropical regions, particularly in mountainous areas where erosion continually removes nutrient-depleted surface layers of soil, and in volcanic areas where the parent material of ash and lava is often rich in nutrients such as potassium.

Soil formation emphasizes the role of the physical environment—particularly climate, geology, and landforms—in creating the tremendous variety of environments for life that exist at the surface of the earth and in its waters. In the next chapter, we shall see how this variety affects the distribution of life forms and the appearance of biological communities.

ECOLOGISTS IN THE FIELD   **Which came first, the soil or the forest?** When the most recent glaciers retreated from most of Europe and North America, beginning about 18,000 years before the present (BP), dramatic changes in vegetation and soils moved across the landscape. In central Europe, cold, dry steppes were replaced by coniferous forests and then by the deciduous forests that occur throughout the region today. At about the same time as the coniferous–deciduous forest transition, there was a change from strongly podsolized soils (spodosols) to richer brown forest soils (alfisols). But, as British ecologist Kathy Willis and her colleagues at Cambridge University asked, "Which changed first? Did climatic warming result in a transformation from one soil type to another, which in turn resulted in a change in forest composition, or did the vegetation change first and subsequently alter the soil?"

The answer, at least for one area in northeastern Hungary, comes from a sediment core removed from small, shallow Lake Kis-Mohos Tó. Pollen grains (Figure 4.25) become trapped in lake sediments, as do minerals carried in water draining from soils surrounding the lake. The pollen and minerals in the sediments tell the story of changes in vegetation and soils over time.

What does the sediment core from Lake Kis-Mohos Tó reveal? First, the pollen record tells us that the local forest changed from coniferous to deciduous in a few centuries. You can see in Figure 4.26 that spruce, pine, and birch—trees typical of boreal forests—abruptly disappeared from the region about 9,500 years BP, and were just as quickly replaced by an oak–hornbeam deciduous forest. Up until this transition, most of the sediments in the lake were inorganic, suggesting that the area was cold and unproductive. Abundant aluminum, potassium, and magnesium in the sediment core suggested rapid breakdown and leaching of clay particles in the surrounding soils, typical of a heavily podsolized area. The first indication of change was a release of large amounts of strontium and barium into the lake. Spruce trees preferentially take up these elements from the soil instead of calcium. The strontium and barium are then deposited in the spruce needles that accumulate as a thick layer of litter on the forest floor. Willis and her colleagues interpreted the release of

(a)   (b)   (c)

FIGURE 4.25 Pollen grains from different types of plants have distinctive surface patterns that allow them to be identified. These scanning electron micrographs (×500) depict pollen grains of three subtropical plants from North America.

(a) *Callirhoe involucrata* (geranium poppy mallow), (b) *Ceanothus americanus* (New Jersey tea), and (c) *Polygonella americana* (small southern jointweed). Photos (a) and (b) by T. Nutall, J. Torrey, and A. Gray; photo (c) by F. von Fischer and C. von Meyer.

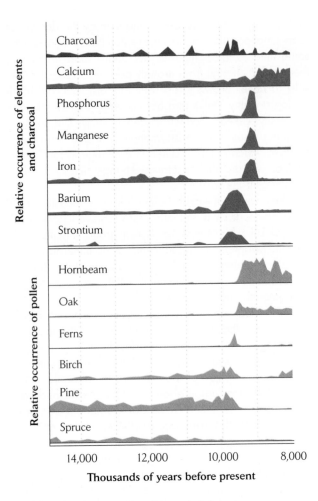

FIGURE 4.26 Layers of sediments in lakes preserve the history of environmental change in the surrounding watershed. The content of a sediment core from Lake Kis-Mohos Tó in Hungary shows the replacement of needle-leaved forest by broad-leaved deciduous forest and accompanying changes in soils about 10,000 years ago. From K. J. Willis et al., *Ecology* 78(3):740–750 (1997).

these elements into ground and surface water flowing into Lake Kis-Mohos Tó as resulting from a rapid breakdown of the spruce litter.

What triggered this rapid breakdown? It is difficult to know with certainty, but again, the sediment core provides a clue in the form of a contemporaneous increase in charcoal particles entering the lake. Climate modelers suggest that central Europe experienced a warm, dry period between 10,000 and 9,000 BP, after the end of the Younger Dryas. This climate may have promoted natural fires that burned away the litter layers of the coniferous for-

ests. The appearance of charcoal in the sediment core is also associated with a spike in fern spores, which is a sure sign of frequent fires. Ferns colonize burned areas quickly and produce luxuriant growth for a few years after a fire has swept through a forest (Figure 4.27). The fires mark the transition from coniferous to deciduous forest because pines disappear and are replaced by oaks at this time.

Once broad-leaved deciduous trees became established, large quantities of iron, magnesium, and phosphorus were released into the lake during another brief period. This release represents a period of leaching of these elements under the still acidic conditions of the forest soils, probably accompanied by a transient reduction in soil fertility. The final phase of the transition is marked by an increase in calcium in the sediment core. Calcium is not particularly abundant in the rock underlying the region, but deciduous trees such as oaks preferentially take up calcium from the soil and begin to enrich the calcium content of the top layers of soil through annual leaf fall.

So, which changed first, the soil or the forest? Clearly, the soil retained its acidic, podsolized nature until well after the establishment of deciduous vegetation, so apparently the vegetation change caused the soil change in this case, illustrating the contribution of vegetation to the development of soil. The change in the vegetation itself was evidently sparked, so to speak, by warmer and drier climates, which were less favorable for spruce and fostered fires that created openings in the pine forests. These openings allowed oak and other broad-leaved species to invade. ▮

FIGURE 4.27 Ferns grow abundantly in recently burned areas. The floor of this recently burned aspen forest in northern Michigan is covered with ferns. Photo by R. E. Ricklefs.

SUMMARY

1. Global climatic patterns result from differential input of solar radiation at different latitudes and from the redistribution of heat energy by winds and ocean currents.

2. Periodic climatic cycles follow astronomical cycles, including the rotation of the earth on its axis (daily), revolution of the moon around the earth (roughly monthly), and revolution of the earth around the sun (annually). Variations in atmospheric and ocean circulation occur at longer periods of tens to many thousands of years.

3. Solar radiation and winds are responsible for the evaporation and circulation of water vapor in the atmosphere and thus for global and seasonal patterns of precipitation. The equilibrium vapor pressure of water increases with temperature.

4. Air is warmed and rises at the equator, where solar radiation is most intense, then cools and sinks at about 30° north and south, forming Hadley cells over the tropics. The sinking air of the Hadley cells drives secondary cells, called Ferrel cells, over the temperate zones, which in turn drive Polar cells at higher latitudes. This global pattern is known as Hadley circulation.

5. Variation in marine conditions is determined on a global scale by wind-driven ocean currents. These currents redistribute heat over the surface of the earth and greatly affect climates on land. Upwelling currents, caused by winds, ocean basin topography, and variations in water density related to temperature and salinity, bring cold, nutrient-rich water to the surface in some areas.

6. Thermohaline circulation, caused by differences in the density of water masses, moves water masses at great depth between ocean basins. This circulation pattern can be interrupted by climate changes that melt glacial or sea ice, changing the salinity of surface waters.

7. Seasonality in terrestrial environments is caused by the tilt of the earth's axis of rotation in relation to the sun. In the tropics, the northward and southward movement of the intertropical convergence, which follows the move-

ment of the solar equator, results in pronounced rainy and dry seasons. At higher latitudes, the seasons are expressed primarily as annual cycles of temperature.

8. Seasonal warming and cooling influence the characteristics of lakes in the temperate zone that experience surface freezing in winter. During summer, such lakes are stratified, with a warm surface layer (epilimnion) separated from a cold bottom layer (hypolimnion) by a sharp thermocline. In spring and fall, the temperature profile becomes more uniform, allowing vertical mixing.

9. Irregular and unpredictable variations in climate, such as El Niño–Southern Oscillation events, may cause major changes in temperature and precipitation and disrupt biological communities on a global scale.

10. Topography and geology superimpose local variation in environmental conditions on more general climatic patterns. Mountains intercept rainfall, creating arid rain shadows on their leeward sides. At high latitudes, north- and south-facing slopes receive differing amounts of sunlight. Because temperature decreases about 6°C for every 1,000 m of elevation, conditions at higher elevations resemble conditions at higher latitudes.

11. The characteristics of soil reflect the influences of the parent material from which it forms as well as climate and vegetation. Weathering of bedrock results in the breakdown of some of its minerals and their incorporation into clay particles, which mix with organic detritus entering the soil from the surface. These processes usually result in distinct soil horizons.

12. Clay particles have negative charges on their surfaces, which hold cations. The cation exchange capacity of a soil determines its fertility.

13. In acidic (podsolized) soils of cool, moist regions of the temperate zone, and in deeply weathered (lateritic) tropical soils, clay particles break down and the fertility of the soil is much reduced.

REVIEW QUESTIONS

1. Why is solar energy input greater near the equator than near the poles?

2. Explain the factors that drive the movement of air in Hadley cells, Ferrel cells, and Polar cells.

3. Given that the position of the solar equator moves throughout the year, what does its varying position suggest about the location of the intertropical convergence throughout the year?

4. Based on our knowledge of the ocean conveyor belt, how might melting of the ice in the Arctic Ocean affect the climate of Europe?

5. What processes cause spring and fall overturn in temperate zone lakes?

6. If upwelling zones are important for marine fish production, what would you predict about the effect of an El Niño event on fish populations off the coast of Peru?

7. Why do many mountain ranges have high precipitation on one side and low precipitation on the other side?

8. Why might you expect to find similar plants living on mountains at low latitudes and in lowlands at high latitudes?

9. Compare and contrast the soil weathering processes of podsolization and laterization.

SUGGESTED READINGS

Barber, R. T., and F. P. Chavez. 1983. Biological consequences of El Niño. *Science* 222:1203–1210.

Buchdahl, J. 1999. *ARIC Global Climate Change Student Guide.* Atmosphere, Climate & Environment Information Programme, Manchester Metropolitan University. http://www.ace.mmu.ac.uk/resources/gcc/contents.html.

Cairns, S. C., and G. C. Grigg. 1993. Population dynamics of red kangaroos (*Macropus rufus*) in relation to rainfall in the South Australian pastoral zone. *Journal of Applied Ecology* 30:444–458.

Cane, M. A., G. Eshel, and R. W. Buckland. 1994. Forecasting Zimbabwean maize yield using eastern equatorial Pacific sea surface temperature. *Nature* 370:204–205.

Graedel, T. E., and P. J. Crutzen. 1995. *Atmosphere, Climate, and Change.* Scientific American Library, New York.

Hays, J. D., J. Imbrie, and N. J. Shackleton. 1976. Variations in the earth's orbit: pacemaker of the Ice Ages. *Science* 194:1121–1132.

Inouye, D. W., et al. 2000. Climate change is affecting altitudinal migrants and hibernating species. *Proceedings of the National Academy of Sciences USA* 97:1630–1633.

Jenny, H. 1980. *The Soil Resource: Origin and Behavior.* Springer-Verlag, New York.

McManus, J. F., D. W. Oppo, and J. L. Cullen. 1999. A 0.5-million-year record of millennial-scale climate variability in the North Atlantic. *Science* 283:971–975.

Muller, R. A., and G. J. MacDonald. 1997. Glacial cycles and astronomical forcing. *Science* 277:215–218.

Philander, G. 1989. El Niño and La Niña. *American Scientist* 77:451–459.

Rasmussen, E. M. 1985. El Niño and variations in climate. *American Scientist* 73:168–177.

Shelford, V. E. 1963. *The Ecology of North America.* University of Illinois Press, Urbana.

Sherman, K., L. M. Alexander, and B. D. Gold (eds.). 1990. *Large Marine Ecosystems: Patterns, Processes, and Yields.* American Association for the Advancement of Science, Washington, D.C.

Suplee, C. 1999. El Niño, La Niña. *National Geographic* 195(3):73–95.

Verburg, P., R. E. Hecky, and H. Kling. 2003. Ecological consequences of a century of warming in Lake Tanganyika. *Science* 301:505–507.

Willis, K. J., et al. 1997. Does soil change cause vegetation change or vice versa? A temporal perspective from Hungary. *Ecology* 78(3):740–750.

Wunsch, C. 2004. Quantitative estimate of the Milankovitch-forced contribution to observed Quaternary climate change. *Quaternary Science Reviews* 23:1001–1012.

Wurtsbaugh, W. A. 1992. Food-web modification by an invertebrate predator in the Great Salt Lake (USA). *Oecologia* 89:168–175.

CHAPTER 5

The Biome Concept in Ecology

Imagine that you are on safari on an East African savanna and one of your group shouts, "Look over there, a cactus tree!" With your training in botany, you know immediately that this can't be so, because the cactus family (Cactaceae) is restricted to the Western Hemisphere. Yet the plant looks just like cacti you have seen in similar environments in Mexico (Figure 5.1). Closer inspection of the flowers shows that the African plant is a cactus look-alike, a member of the spurge family (Euphorbiaceae).

Your friend was fooled by a common phenomenon in biology, that of convergence. **Convergence** is the process by which unrelated organisms evolve a resemblance to each other in response to similar environmental conditions. The leafless, thick, fleshy branches of both the cactus and the cactuslike euphorb evolved as adaptations to reduce water loss in semiarid environments. The two plants look alike because they evolved under the same conditions, although they descended from unrelated, different-looking ancestors. Natural selection and evolution are oblivious to the ancestry of a particular organism as long as it is capable of an adaptive response to a particular condition of the environment.

Convergence explains why we can recognize an association between the forms of organisms and their particular environments anywhere in the world. Tropical rain forest trees have the same general appearance no matter where they are found or to which evolutionary lineage they belong. The same can be said of shrubs inhabiting seasonally dry environments, which produce small, deciduous leaves and often arm their stems with spines to dissuade herbivores. The podocarp trees (Podocarpaceae) that grow in temperate forests of New Zealand resemble the broad-leaved trees of the Northern Hemisphere even though they are gymnosperms, more closely related to pines and firs than to oaks and maples.

FIGURE 5.1 Unrelated organisms can evolve similar forms in response to common environmental conditions. (a) A tree-forming cactus near Oaxaca, Mexico, and (b) an East African euphorb tree have converged in response to arid conditions. Photos by R. E. Ricklefs.

(a)   (b)

CHAPTER CONCEPTS

- Climate is the major determinant of plant growth form and distribution

- Climate defines the boundaries of terrestrial biomes

- Walter climate diagrams distinguish the major terrestrial biomes

- Temperate climate zones have average annual temperatures between 5°C and 20°C

- Boreal and polar climate zones have average temperatures below 5°C

- Climate zones within tropical latitudes have average temperatures exceeding 20°C

- The biome concept must be modified for freshwater aquatic systems

- Marine aquatic systems are classified principally by water depth

Climate, topography, and soil—and parallel influences in aquatic environments—determine the changing character of plant and animal life, as well as ecosystem functioning, over the surface of the earth. Although no two locations harbor exactly the same assemblage of species, we can group biological communities and ecosystems into categories based on climate and dominant plant form, which give them their overall character. These categories are referred to as **biomes.** Ecosystems belonging to the same biome type in different parts of the world develop a similar vegetation structure and similar ecosystem functioning, including productivity and rates of nutrient cycling, under similar environmental conditions. Thus, biomes provide convenient reference points for comparing ecological processes on a global scale. Ecosystems of the woodland/shrubland biome characteristic of Mediterranean climates (cool, wet winters and hot, dry summers),

for example, look similar and function similarly whether in southern California, southern France, Chile, South Africa, or Australia.

The important terrestrial biomes of the United States and Canada are tundra, boreal forest, temperate seasonal forest, temperate rain forest, shrubland, grassland, and subtropical desert. As one would expect, the geographic distributions of these biomes correspond closely to the major climate zones of North America. To the south in Mexico and Central America, tropical rain forest, tropical deciduous forest, and tropical savanna are important biomes. Although each biome is immediately recognizable by its distinctive vegetation, it is important to realize that different systems of classification make coarser or finer distinctions, and that the characteristics of one biome usually intergrade with those of the next. The biome concept is nonetheless a useful tool that enables ecologists throughout the world to work together toward understanding the structure and functioning of large ecological systems.

That biomes can be distinguished at all reflects the simple fact that no single type of plant can endure the whole range of conditions at the surface of the earth. If plants had such broad tolerance of physical conditions, the earth would be covered by a single biome. To the contrary, trees, for example, cannot grow under the dry conditions that shrubs and grasses can tolerate, simply because the physical structure, or **growth form,** of trees creates a high demand for water. The grassland biome exists because grasses and other herbs (called forbs) can survive the cold winters typical of the Great Plains of the United States, the steppes of Russia, and the pampas of Argentina.

This matching of growth form and environment allows us to understand the global distributions of vegetation types and the extents of biomes. If that were the whole of it, however, the study of ecology could simply focus on the relationships of individual organisms to their physical environments, and everything else in ecology would emanate from that point. However, we must remind ourselves that life is not so simple. In addition to physical conditions, two other kinds of factors influence the distributions of species and growth forms. The first of these is the myriad interactions between species—such as competition, predation, and mutualism—that determine whether a species or growth form can persist in a particular place. For example, grasses can grow perfectly well in eastern North America, as we see along roadsides and on abandoned agricultural lands, but trees predominate in that environment, and in the absence of disturbance, they exclude grasses, which cannot grow and reproduce under their deep shade.

The second kind of factor is that of chance and history. The present biomes have developed over long periods, during which the distributions of landmasses, ocean basins, and climate zones have changed continuously. Most species fail to occupy many suitable environments simply because they have not been able to disperse to all ends of the earth. This fact is amply illustrated by the successful introduction by humans of such species as European starlings (*Sturnus vulgaris*) and Monterey pines (*Pinus radiata*) to parts of the world that have suitable environmental conditions but which were far outside the restricted natural distributions of these species.

In addition, evolution has proceeded along independent lines in different parts of the world, leading in some cases to unique biomes. Australia has been isolated from other continents for the past 40 million to 50 million years, which accounts both for its unusual flora and fauna and for the absence of many of the kinds of plants and animals familiar to outsiders. Because of its unique history, areas of Australia with a climate that would support scrubland or oak savanna in California are clothed instead with tall eucalyptus woodland. Similarities between chaparral—as scrublands are called in California—and eucalyptus woodland include drought and fire resistance, but the predominant plant growth forms differ, primarily because of historical accident. We shall consider these biological and historical factors later in this book. As we shall see in the present chapter, the physical environment ultimately defines the character and distribution of the major biomes.

Climate is the major determinant of plant growth form and distribution

We can classify ecosystems into biomes because climate, along with other influences, determines the plant growth form best suited to an area, and because plants with particular growth forms are restricted to particular climates. These principles establish the close relationship between climate and vegetation. Keep in mind, however, that there are other, less conspicuous similarities among areas of the same biome type, including biological productivity, nutrient regeneration in soils, and the structure of animal communities.

One cannot understand the adaptations of an organism independently of the environment in which it lives. Different physical conditions characterize each biome, and its inhabitants are adapted to live under those conditions. The leaves of deciduous forest trees growing in the temperate seasonal forest biome are typically broad and thin, providing a large surface area for light absorption but little protection from desiccation or frost. In contrast, the leaves of many desert species are small and finely divided to dissipate heat (see Figure 3.8), and some desert species have no leaves at all.

Because of these adaptations, the vegetation of the temperate seasonal forest and subtropical desert biomes differs dramatically. These differences extend to the spacing of plants as well as their form. In temperate forests, trees form closed canopies, and the entire surface of the ground is shaded. In drier environments, including deserts, woodlands, and savannas, trees and shrubs are more widely spaced owing to competition among their root systems for limited water, and this spacing allows drought-resistant grasses to grow in the gaps between trees. In the most extreme deserts, much of the soil surface is bare because the scarce water cannot support an uninterrupted expanse of vegetation.

Given that organisms are adapted to the physical conditions of their biome, it is not surprising that the ranges of many species are limited by those same physical conditions. In terrestrial environments, temperature and moisture are the most important variables, particularly for plants. The distributions of several species of maples in eastern North America show how these factors operate. The sugar maple (*Acer saccharum*), a common forest tree in the northeastern United States and southern Canada, is limited by cold winter temperatures to the north, by hot summer temperatures to the south, and by summer drought to the west. Thus, the sugar maple is confined to roughly the northern portion of the temperate seasonal forest biome in North America (Figure 5.2). Attempts to grow sugar maples outside their normal range fail because these trees cannot tolerate average monthly summer temperatures above about 24°C or winter temperatures below about −18°C. The western limit of the sugar maple, determined by dryness, coincides with the western limit of forest in eastern North America. Because temperature and rainfall interact to control the availability of moisture, sugar maples require less annual precipitation at the northern edge of their range (about 50 cm) than at the southern edge (about 100 cm). To the east, the range of the sugar maple stops abruptly at the Atlantic Ocean.

The distributions of the sugar maple and other tree-sized maple species—black, red, and silver—reflect differences in the range of conditions within which each species can survive (Figure 5.3). Where their geographic ranges overlap, the maples exhibit distinct preferences for certain local environmental conditions created by differences in

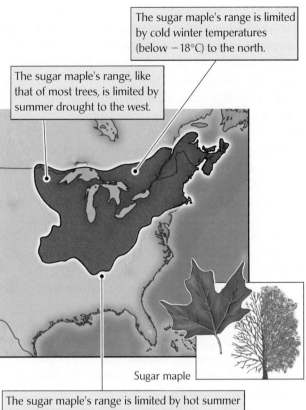

The sugar maple's range is limited by cold winter temperatures (below −18°C) to the north.

The sugar maple's range, like that of most trees, is limited by summer drought to the west.

The sugar maple's range is limited by hot summer temperatures (above 24°C) to the south.

Sugar maple

FIGURE 5.2 The distributions of species are limited by physical conditions of the environment. The red area shows the range of sugar maple in eastern North America. After H. A. Fowells, *Silvics of Forest Trees of the United States*, U.S. Department of Agriculture, Washington, D.C. (1965).

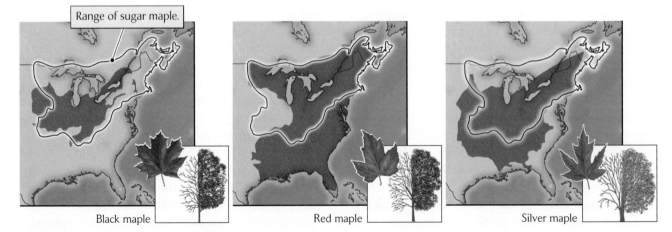

Range of sugar maple.

Black maple

Red maple

Silver maple

FIGURE 5.3 Related species may differ in their ecological tolerances. The red areas show the ranges of black, red, and silver maples in eastern North America. The range of the sugar maple is outlined on each map to show the area of overlap. After H. A. Fowells, *Silvics of Forest Trees of the United States*, U.S. Department of Agriculture, Washington, D.C. (1965).

soil and topography. Black maple (*A. nigrum*) frequently occurs in the same areas as the closely related sugar maple, but usually on drier, better-drained soils higher in calcium content (and therefore less acidic). Silver maple (*A. saccharinum*) occurs widely in the eastern United States, but especially on the moist, well-drained soils of the Ohio and Mississippi river basins. Red maple (*A. rubrum*) grows best either under wet, swampy conditions or on dry, poorly developed soils—that is, under extreme conditions that limit the growth of the other species. Nonetheless, these trees all have a similar growth form and naturally occur within—and partially define—the temperate seasonal forest biome.

Climate defines the boundaries of terrestrial biomes

One of the most widely adopted climate classification schemes is the **climate zone** system developed by the German ecologist Heinrich Walter. This system, which has nine major divisions, is based on the annual cycle of temperature and precipitation. The important attributes of climate and characteristics of vegetation in each of these zones are set out in Figure 5.4. The values of temperature and precipitation used to define climate zones correspond to the conditions of moisture and cold stress that are particularly important determinants of plant form. For example, within tropical latitudes, the tropical climate zone is distinguished from the equatorial climate zone by the occurrence of water stress during a pronounced dry season. The subtropical climate zone, which occurs at somewhat higher latitudes, is perpetually water-stressed. The typical vegetation types in these three climate zones are evergreen rain forest (equatorial), seasonal forest or savanna (tropical), and desert scrub (subtropical), respectively. We will look at Walter's climate zones in more detail below.

Many classification schemes for biomes exist. Walter's is based first on climate, with boundaries between climate zones drawn to match changes between major vegetation types. Cornell University ecologist Robert H. Whittaker defined biomes first by their vegetation type, and then devised a simple climate diagram on which he plotted

Biome name	Climate zone			Vegetation
Tropical rain forest	**I**	Equatorial:	Always moist and lacking temperature seasonality	Evergreen tropical rain forest
Tropical seasonal forest/ savanna	**II**	Tropical:	Summer rainy season and "winter" dry season	Seasonal forest, scrub, or savanna
Subtropical desert	**III**	Subtropical (hot deserts):	Highly seasonal, arid climate	Desert vegetation with considerable exposed surface
Woodland/shrubland	**IV**	Mediterranean:	Winter rainy season and summer drought	Sclerophyllous (drought-adapted), frost-sensitive shrublands and woodlands
Temperate rain forest	**V**	Warm temperate:	Occasional frost, often with summer rainfall maximum	Temperate evergreen forest, somewhat frost-sensitive
Temperate seasonal forest	**VI**	Nemoral:	Moderate climate with winter freezing	Frost-resistant, deciduous, temperate forest
Temperate grassland/ desert	**VII**	Continental (cold deserts):	Arid, with warm or hot summers and cold winters	Grasslands and temperate deserts
Boreal forest	**VIII**	Boreal:	Cold temperate with cool summers and long winters	Evergreen, frost-hardy needle-leaved forest (taiga)
Tundra	**IX**	Polar:	Very short, cool summers and long, very cold winters	Low, evergreen vegetation, without trees, growing over permanently frozen soils

FIGURE 5.4 Heinrich Walter classified the climate zones of the world according to the annual cycle of temperature and precipitation. The biome names given to these zones under Whittaker's classification scheme are shown in the left-hand column.

FIGURE 5.5 Whittaker's biomes are delineated according to average temperature and precipitation. Whittaker plotted the boundaries of observed vegetation types with respect to average temperature and precipitation. In climates intermediate between those of forest and desert biomes, climatic seasonality, fire, and soils determine whether woodland, grassland, or shrubland develops. *Inset:* Average annual temperature and precipitation for a sample of localities more or less evenly distributed over the land area of the earth. Most of the points fall within a triangular region that includes almost the full range of climates. Only the climates of high mountains do not fall within the triangle. From R. H. Whittaker, *Communities and Ecosystems,* 2nd ed., Macmillan, New York (1975).

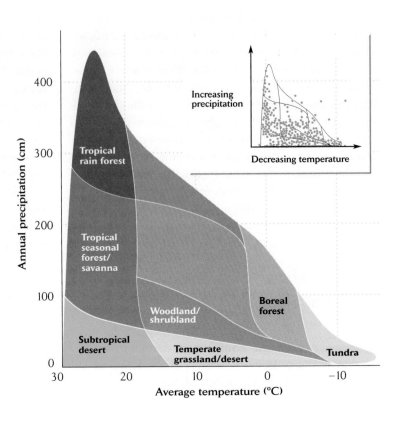

the approximate boundaries of his biomes with respect to average temperature and precipitation (Figure 5.5). The result is similar to Walter's scheme, as one would expect, and their nine biome types correspond directly. When plotted on Whittaker's diagram, most locations on earth fall within a triangular area whose three corners represent warm moist, warm dry, and cool dry climates. (Cold regions with high rainfall are rare because water does not evaporate rapidly at low temperatures and because the atmosphere in cold regions holds little water vapor.)

At tropical and subtropical latitudes, where average temperatures range between 20°C and 30°C, vegetation ranges from rain forest, which is wet throughout the year and generally receives more than 250 cm (about 100 inches) of rain annually (Walter's equatorial climate zone), to desert, which generally receives less than 50 cm of rain (Walter's subtropical climate zone). Intermediate climates support seasonal forests (150–250 cm rainfall), in which some or all trees lose their leaves during the dry season, or scrub and savannas (50–150 cm rainfall).

Plant communities at temperate latitudes follow the pattern of tropical communities with respect to rainfall, falling conveniently into four vegetation types: temperate rain forest (as in the Pacific Northwest of North America), temperate seasonal forest, woodland/shrubland, and temperate grassland/desert. At higher latitudes, precipitation varies so little from one locality to another that vegetation

types are poorly differentiated by climate. Where average temperature falls between 0°C and −5°C, boreal forest predominates. Where average annual temperatures are below −5°C, all plant communities may be lumped into one type: tundra.

Toward the drier end of the precipitation spectrum within each temperature range, fire plays a distinct role in shaping plant communities. The influence of fire is greatest where moisture availability is intermediate and highly seasonal. Deserts and moist forests burn infrequently because deserts rarely accumulate enough plant debris to fuel a fire and moist forests rarely dry out enough to become highly flammable. Grassland and shrubland have the combination of abundant fuel and seasonal drought that make fire a frequent visitor. In these biomes, fire is a dominating factor to which all community members must be adapted and, indeed, for which many are specialized. Some species require fire for germination of their seeds and growth of their seedlings. Toward the moister edges of African savannas and North American prairies, frequent fires kill the seedlings of trees and prevent the encroachment of forests, which could be sustained by the local precipitation if it were not for fire. Burning favors perennial grasses and forbs with extensive root systems and meristems (growth centers) that can survive underground. (Grasses tolerate grazing for the same reason.) After an area has burned, grass and forb roots sprout fresh shoots and quickly establish new vegetation above the surface of the soil. In the absence of frequent fires, tree seedlings become established and eventually shade out savanna and prairie vegetation.

As in all classification systems, exceptions appear, and boundaries between biomes are fuzzy. Moreover, not all plant growth forms correspond to climate in the same way; as mentioned earlier, Australian eucalyptus trees form forests under climatic conditions that support only scrubland or grassland on other continents. Finally, plant communities reflect factors other than temperature and rainfall. Topography, soils, fire, seasonal variations in climate, and herbivory all leave their mark. The overview of the major terrestrial biomes in this chapter emphasizes the distinguishing features of the physical environment and how these features are reflected in the form of the dominant plants.

 Biomes and Animal Forms. Why are biome definitions based on the predominant life forms of plants, rather than referring to their animal inhabitants?

 Characterizing Climate. Integrated descriptions of climate emphasize the interaction of temperature and availability of water.

Walter climate diagrams distinguish the major terrestrial biomes

Temperature and precipitation interact to determine the conditions and resources available for plant growth. It is not surprising, then, that the distributions of the major biomes of the earth follow patterns of temperature and precipitation. Because of this close relationship, it is important to describe climate in a manner that reflects the availability of water, taking into consideration changes in temperature and precipitation through the year.

Heinrich Walter developed a climate diagram that illustrates seasonal periods of water deficit and abundance and therefore permits ecologically meaningful comparisons of climates between localities (Figure 5.6). The Walter climate diagram portrays average monthly temperature and precipitation throughout the course of a year. The vertical scales of temperature and precipitation are adjusted so that when precipitation is higher than temperature on the diagram, water is plentiful and plant production is limited primarily by temperature. Conversely, when temperature is higher than precipitation, plant production is limited by availability of water. Walter's scales equate 20 mm of

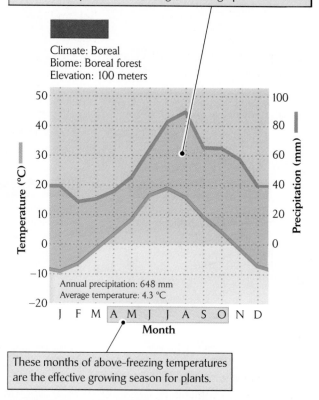

As a rule of thumb, about 20 mm of monthly precipitation for each 10°C in temperature provides sufficient moisture for plant growth. This occurs wherever the precipitation line (blue) is above the temperature line (orange) on this graph.

Climate: Boreal
Biome: Boreal forest
Elevation: 100 meters

Annual precipitation: 648 mm
Average temperature: 4.3 °C

These months of above-freezing temperatures are the effective growing season for plants.

FIGURE 5.6 Walter climate diagrams allow ecologically meaningful comparisons between localities. These diagrams, such as the one illustrated here for a hypothetical locality in the boreal forest biome, portray the annual progression of monthly average temperature (left-hand scale) and precipitation (right-hand scale).

monthly precipitation with 10°C in temperature. Thus, as a rule of thumb, at an average temperature of 20°C, 40 mm of monthly precipitation provides sufficient moisture for plant growth. We'll use Walter climate diagrams to compare the biomes characterized below.

Climate diagrams for locations in each of Walter's climate zones are shown in Figure 5.7. The seasonal distributions of wet and dry periods differ among the climate zones at lower latitudes. Equatorial climates (climate zone I) like that at Andagoya, Colombia, are aseasonal; that is, they are warm and wet throughout the year. Subtropical climates (III), such as that of Chiclayo, Peru, are warm and dry throughout the year. Summer rains and winter drought characterize tropical climates (II, Brasília, Brazil). Mediterranean climates (IV, Lisbon, Portugal) experience winter rains and summer drought. The climate of Sitka,

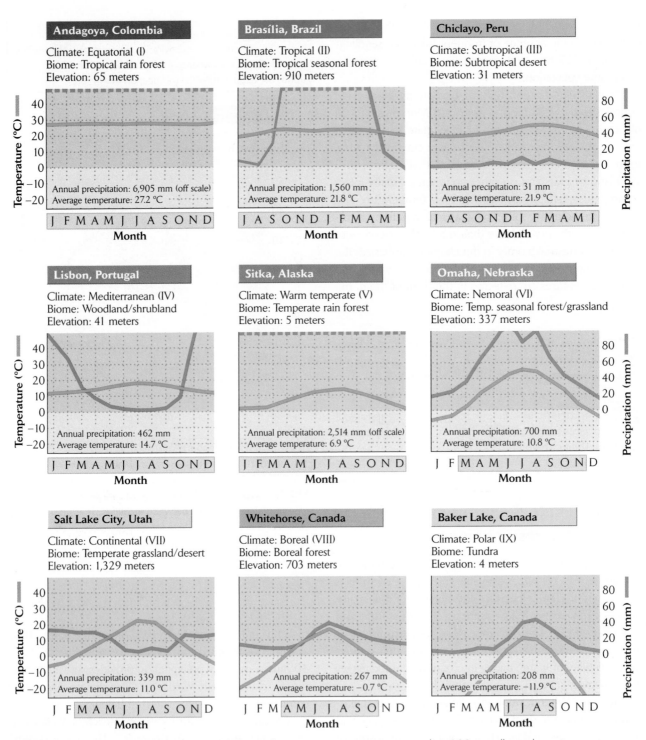

FIGURE 5.7 Each climate zone has a typical seasonal pattern of temperature and precipitation. Walter climate diagrams for representative locations in each of the nine major terrestrial climate zones are shown. The dashed blue line at the top of the graphs for climate zones I, II, and V indicates monthly precipitation exceeding 100 mm all year long. From H. Walter and S.-W. Breckle, *Ecological Systems of the Geobiosphere, I, Ecological Principles in Global Perspective*, Springer-Verlag, Berlin (1985).

Alaska (warm temperate, V), is wet and mild throughout the year and supports evergreen forest vegetation.

Seasonality of temperature is a major factor in climate zones VI–IX, which occur at middle and high latitudes. Precipitation is typically low, but because of the low temperatures, moisture is generally not limiting during the short summer growing season. Continental climates (VII, Salt Lake City, Utah) are typically dry throughout the year

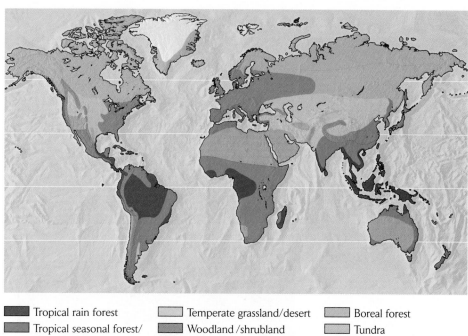

FIGURE 5.8 Global distribution of the major biomes.

Tropical rain forest

Tropical seasonal forest/ savanna

Subtropical desert

Temperate grassland/desert

Woodland /shrubland

Temperate seasonal forest

Temperate rain forest

Boreal forest

Tundra

Alpine

Polar ice cap

and become warm enough in summer to develop significant water stress. Such areas, which include much of the Great Basin of the western United States, support shrubby desert vegetation.

The same climate zones can be recognized where they occur around the world. For example, the tropical climates of Brasília (Brazil), Harare (Zimbabwe), and Darwin (Australia) all share the even year-round warm temperatures and summer rainfall typical of climate zone II. And each of these areas supports deciduous forest vegetation grading into savanna where precipitation is particularly low. Indeed, each climate zone supports characteristic vegetation that defines the biome type and makes it easy for us to recognize the general attributes of these ecosystems in any region.

Walter's climate zones are one of several systems of biome classification. While these systems differ in the number of biomes recognized, and some emphasize biological characteristics more than the physical environment, they all present essentially the same picture of ecosystem variation over the surface of the earth. For example, the World Wildlife Fund recognizes fourteen biomes, rather than Walter's nine, adding (i) temperate and (ii) tropical coniferous forests, both characterized by seasonal climates but tending to be drier and existing on poorer soils than biomes with broad-leaved trees; (iii) montane grasslands and shrublands, including the puna and páramo zones of the high Andes; (iv) seasonally flooded grasslands and savannas in both tropical and temperate regions, and

(v) mangrove wetlands, which comprise a specialized vegetation type within the marine intertidal zone. Its system is designed to identify the major ecological regions of the earth whose conservation would preserve the largest part of the diversity of the earth's ecosystems.

The worldwide distribution of biome types arranged by any system follows the same general patterns of temperature and precipitation over the earth (Figure 5.8). We shall consider the biomes and general ecological characteristics of each of the major Walter climate zones in the series of vignettes that follow. Because most readers of this book live within temperate latitudes, this is a good place to start.

Temperate climate zones have average annual temperatures between 5°C and 20°C

The climates within temperate latitudes are characterized by average annual temperatures in the range of 5°–20°C at low elevations. Such climates are distributed approximately between 30°N and 45°N in North America and Asia and between 40°N and 60°N in Europe, which is warmed by the Gulf Stream and by westerly winds. Frost is an important factor throughout the temperate latitudes, perhaps even defining their general character. Within those latitudes, biomes are distinguished primarily

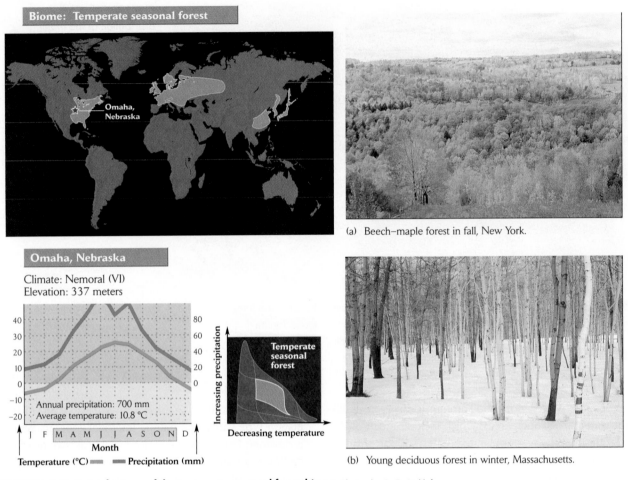

Biome: Temperate seasonal forest

Omaha, Nebraska

Omaha, Nebraska

Climate: Nemoral (VI)
Elevation: 337 meters

Annual precipitation: 700 mm
Average temperature: 10.8 °C

Month

Temperature (°C)   Precipitation (mm)

Increasing precipitation

Temperate seasonal forest

Decreasing temperature

(a) Beech–maple forest in fall, New York.

(b) Young deciduous forest in winter, Massachusetts.

FIGURE 5.9 Major features of the temperate seasonal forest biome. Photos by R. E. Ricklefs.

by total amounts and seasonal patterns of precipitation. The length of the frost-free season, which is referred to as the **growing season,** and the severity of frost are also important.

Temperate seasonal forest biome (climate zone VI)

Often referred to as deciduous forest, the temperate seasonal forest biome occurs under moderate conditions with winter freezing. In North America, this biome is found principally in the eastern United States and southeastern Canada; it is also widely distributed in Europe and eastern Asia (Figure 5.9). This biome is poorly developed in the Southern Hemisphere because the larger ratio of ocean surface to land moderates winter temperatures and prevents frost. In the Northern Hemisphere, the length of the growing season in this biome varies from 130 days at higher latitudes to 180 days at lower latitudes. Precipitation usually exceeds evaporation and transpiration; conse-

quently, water tends to move downward through soils and to drain from the landscape as groundwater and as surface streams and rivers. Soils are often podsolized, tend to be slightly acidic and moderately leached, and are brown in color owing to abundant organic matter. Deciduous trees are the dominant plant growth form. The vegetation often includes a layer of smaller trees and shrubs beneath the dominant trees as well as herbaceous plants on the forest floor. Many of these herbaceous plants complete their growth and flower early in spring, before the trees have fully leafed out.

Warmer and drier parts of the temperate seasonal forest biome, especially where soils are sandy and nutrient poor, tend to develop needle-leaved forests dominated by pines. The most important of these ecosystems in North America are the pine forests of the coastal plains of the Atlantic and Gulf states of the United States; pine forests also exist at higher elevations in the western United States. Because of the warm climate in the southeastern United States, soils there are usually lateritic and nutrient poor.

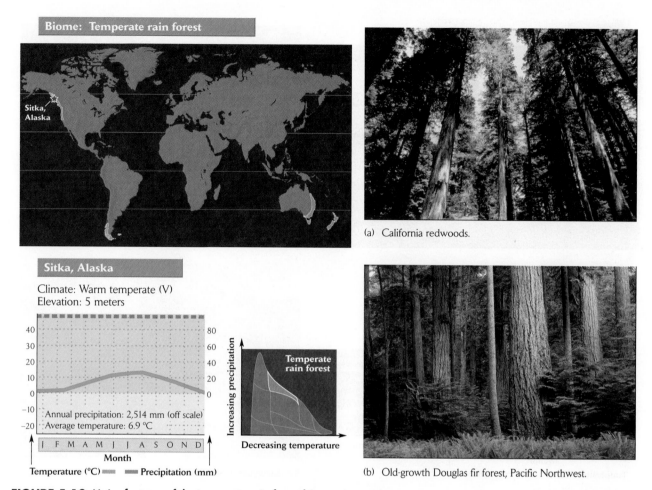

Biome: Temperate rain forest

Sitka, Alaska

(a) California redwoods.

Sitka, Alaska

Climate: Warm temperate (V)
Elevation: 5 meters

Annual precipitation: 2,514 mm (off scale)
Average temperature: 6.9 °C

Month

Temperature (°C) ▬▬ Precipitation (mm)

Increasing precipitation

Temperate rain forest

Decreasing temperature

(b) Old-growth Douglas fir forest, Pacific Northwest.

FIGURE 5.10 Major features of the temperate rain forest biome. Photo (a) by PhotoSphere Images/PictureQuest; photo (b) by Tom and Pat Leeson/Photo Researchers.

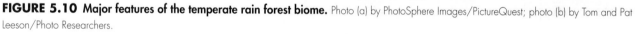

The low availability of nutrients and water favors evergreen, needle-leaved trees, which resist desiccation and give up nutrients slowly because they retain their needles for several years. Because soils tend to be dry, fires are frequent, and most species are able to resist fire damage.

Temperate rain forest biome (climate zone V)

In warm temperate climates near the Pacific coast in northwestern North America, and in southern Chile, New Zealand, and Tasmania, mild winters, heavy winter rains, and summer fog create conditions that support extremely tall evergreen forests (Figure 5.10). In North America, these forests are dominated toward the south by coast redwood (*Sequoia sempervirens*) and toward the north by Douglas-fir (*Pseudotsuga* spp.). These trees are typically 60–70 m tall and may grow to over 100 m. Ecologists do not understand why these sites are dominated by needle-leaved trees, but the fossil record shows that these plant

communities are very old and that they are remnants of forests that were vastly more extensive during the Mesozoic era, as recently as 70 million years ago. In contrast to rain forests in the tropics, temperate rain forests typically support few species.

Temperate grassland/desert biome (climate zone VII)

In North America, grasslands develop within continental climate zones where rainfall ranges between 30 and 85 cm per year and winters are cold (Figure 5.11). The growing season increases from north to south from about 120 to 300 days. These grasslands are often called **prairies.** Extensive grasslands are also found in central Asia, where they are called **steppes.** Precipitation is infrequent, so organic detritus does not decompose rapidly, and the soils are rich in organic matter. Because of their low acidity, prairie soils, which belong to the mollisol group, are not heavily leached and tend to be rich in nutrients. The

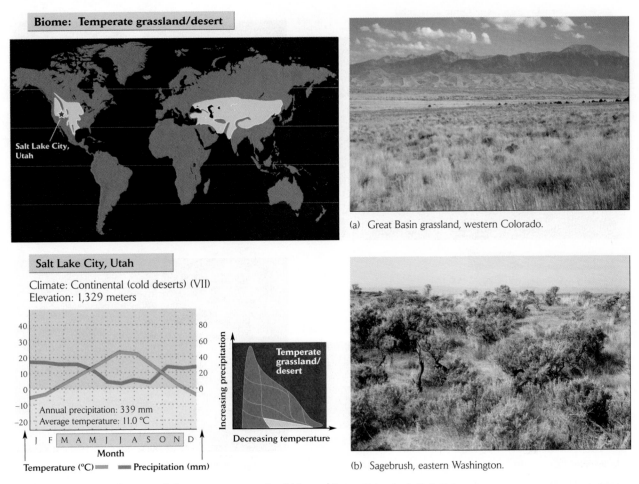

FIGURE 5.11 Major features of the temperate grassland/desert biome. Photos by R. E. Ricklefs.

vegetation is dominated by grasses, which grow to heights over 2 m in the moister parts of these grasslands and to less than 0.2 m in more arid regions. Forbs are also abundant. Fire is a dominant influence in these grasslands, particularly where the habitat dries out during the late summer. Most grassland species have fire-resistant underground stems, or **rhizomes,** from which shoots resprout, or they have fire-resistant seeds.

Where precipitation ranges between 25 and 50 cm per year, and winters are cold and summers are hot, grasslands grade into deserts. The temperate desert biome covers most of the Great Basin of the western United States. In the northern part of the region, sagebrush (*Artemisia*) is the dominant plant, whereas toward the south and on somewhat moister soils, widely spaced juniper and piñon trees predominate, forming open woodlands less than 10 m in stature with sparse coverings of grass. In these temperate deserts, evaporation and transpiration exceed precipitation during most of the year, so soils are dry and little water percolates through them to form streams and

rivers. Calcium carbonate leached from the surface layers of the soil tends to accumulate at the depths to which water usually penetrates. Fires are infrequent in temperate deserts because the habitat produces little fuel. However, because of the low productivity of the plant community, grazing can exert strong pressure on the vegetation and may even favor the persistence of shrubs, which are not good forage. Indeed, many dry grasslands in the western United States and elsewhere in the world have been converted into deserts by overgrazing.

Woodland/shrubland biome (climate zone IV)

The Mediterranean climate zone is found at 30°–40° north and south of the equator—and at somewhat higher latitudes in Europe—on the western sides of continental landmasses, where cold ocean currents and winds blowing from the continents dominate the climate. Mediterranean climates are found in southern Europe and

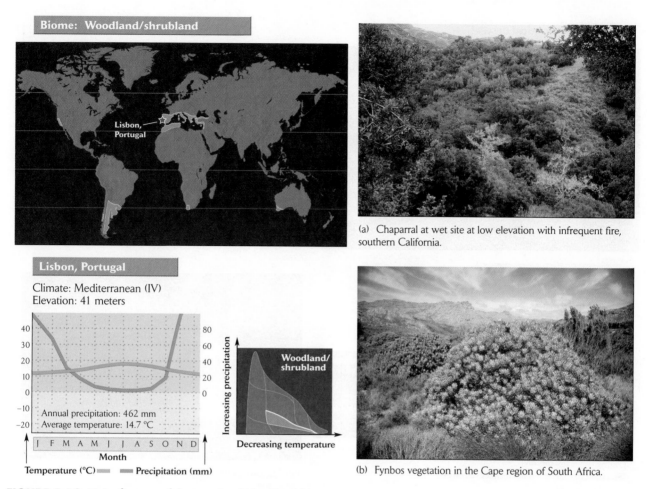

FIGURE 5.12 Major features of the woodland/shrubland biome. Photo (a) by Earl Scott/Photo Researchers; photo (b) by Fletcher & Baylis/Photo Researchers

southern California in the Northern Hemisphere and in central Chile, the Cape region of South Africa, and southwestern Australia in the Southern Hemisphere. Mediterranean climates are characterized by mild winter temperatures, winter rain, and summer drought. These climates support thick, evergreen, shrubby vegetation 1–3 m in height, with deep roots and drought-resistant foliage (Figure 5.12). The small, durable leaves of typical Mediterranean-climate plants have earned them the label of **sclerophyllous** ("hard-leaved") vegetation. Fires are frequent in the woodland/shrubland biome, and most plants have either fire-resistant seeds or root crowns that resprout soon after a fire.

Subtropical desert biome (climate zone III)

What people call "desert" varies tremendously. Many people refer to the dry areas of the Great Basin and of central Asia as deserts—the Gobi Desert is a name famil-

iar to most of us. But the climates of those "deserts" fall within Walter's continental climate zone, characterized by low precipitation and cold winters. These areas are referred to as cold deserts. In contrast, subtropical deserts (Figure 5.13), often called hot deserts, develop at latitudes 20°–30° north and south of the equator, in areas with high atmospheric pressure associated with the descending air of the Hadley cells (Chapter 4). Subtropical deserts have very sparse rainfall (less than 25 cm), high temperatures, and generally long growing seasons. Because of the low rainfall, the soils of subtropical deserts (aridosols) are shallow, virtually devoid of organic matter, and neutral in pH. Impermeable hardpans of calcium carbonate often develop at the limits of water penetration—at depths of a meter or less. Whereas sagebrush dominates the cold deserts of the Great Basin, creosote bush (*Larrea tridentata*) takes its place in the subtropical deserts of the Americas. Moister sites within this biome support a profusion of succulent cacti, shrubs, and small trees, such as mesquite (*Prosopis*) and paloverde (*Cercidium microphyllum*). Most

Biome: Subtropical desert

Chiclayo,
Peru

(a) Cholla cactus in northern Sonora, Mexico.

Chiclayo, Peru

Climate: Subtropical (hot deserts) (III)
Elevation: 31 meters

Annual precipitation: 31 mm
Average temperature: 21.9 °C

Month

Temperature (°C) ▬▬   ▬▬ Precipitation (mm)

Subtropical desert

Increasing precipitation

Decreasing temperature

(b) Saguaro cactus in southern Arizona.

FIGURE 5.13 Major features of the subtropical desert biome. Photos by R. E. Ricklefs.

subtropical deserts receive summer rainfall. After summer rains, many herbaceous plants sprout from dormant seeds, grow quickly, and reproduce before the soils dry out again. Many of the plants in subtropical deserts are not frost-tolerant. Species diversity is usually much higher than it is in temperate arid lands.

Boreal and polar climate zones have average temperatures below 5°C

At high latitudes, cold temperatures predominate. Precipitation is often very sparse because water evaporates slowly into the atmosphere at low temperatures, but soils are often saturated, and water availability is not an important limitation in high-latitude climate zones. Biological productivity during the short summer growing seasons is generally low, and cold temperatures slow the decomposition of organic matter and the release of nutrients in

the soil. As a result, plants retain their foliage for many years, and the vegetation tends to be evergreen and highly adapted to cold winter temperatures.

Boreal forest biome (climate zone VIII)

Stretching in a broad belt centered at about 50°N in North America and about 60°N in Europe and Asia lies the boreal forest biome, often called **taiga** (Figure 5.14). The average annual temperature is below 5°C, and winters are severe. Annual precipitation generally ranges between 40 and 100 cm, and because evaporation is low, soils are moist throughout most of the growing season. The vegetation consists of dense, seemingly endless stands of 10–20 m tall evergreen needle-leaved trees, mostly spruces and firs. Because of the low temperatures, plant litter decomposes very slowly and accumulates at the soil surface, forming one of the largest reservoirs of organic carbon on earth. The needle litter produces high levels of

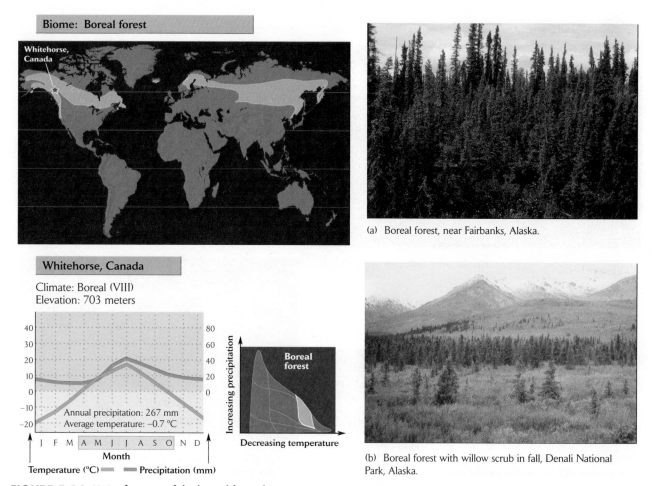

FIGURE 5.14 Major features of the boreal forest biome. Photos by R. E. Ricklefs.

organic acids, so the soils are acidic, strongly podsolized, and generally of low fertility. Growing seasons are rarely as much as 100 days, and often half that. The vegetation is extremely frost-tolerant, as temperatures may reach −60°C during the winter. Species diversity is very low.

Tundra biome (climate zone IX)

To the north of the boreal forest, in the polar climate zone, lies the Arctic tundra, a treeless expanse underlain by permanently frozen soil, or **permafrost** (Figure 5.15). The soils thaw to a depth of 0.5–1 m during the brief summer growing season. Precipitation is generally less, and often much less, than 60 cm, but in low-lying areas where drainage is prevented by permafrost, soils may remain saturated with water throughout most of the growing season. Soils tend to be acidic because of their high organic matter content, and they contain few nutrients. In this nutrient-poor environment, plants hold their foliage for years. Most

plants are dwarf, prostrate woody shrubs, which grow low to the ground to gain protection under the winter blanket of snow and ice. Anything protruding above the surface of the snow is sheared off by blowing ice crystals. For most of the year, the tundra is an exceedingly harsh environment, but during the 24-hour-long summer days, the rush of biological activity in the tundra testifies to the remarkable adaptability of life.

At high elevations within temperate latitudes, and even within the tropics, one finds vegetation resembling that of the Arctic tundra and even including some of the same species, or their close relatives. These areas of *alpine tundra* above the tree line occur most broadly in the Rocky Mountains of North America, the Alps of Europe, and especially on the Plateau of Tibet in central Asia. In spite of their similarities, alpine and Arctic tundra have important points of dissimilarity as well. Areas of alpine tundra generally have warmer and longer growing seasons, higher precipitation, less severe winters, greater

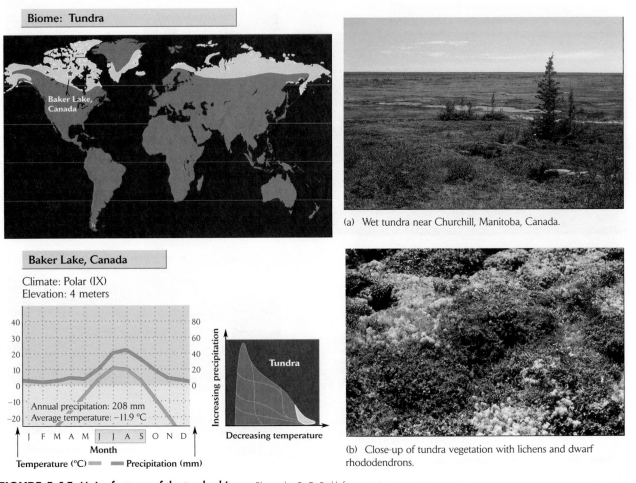

FIGURE 5.15 Major features of the tundra biome. Photos by R. E. Ricklefs.

(a) Wet tundra near Churchill, Manitoba, Canada.

(b) Close-up of tundra vegetation with lichens and dwarf rhododendrons.

productivity, better-drained soils, and higher species diversity than Arctic tundra. Still, harsh winter conditions ultimately limit the growth of trees.

Climate zones within tropical latitudes have average temperatures exceeding 20°C

Within 20° of latitude from the equator, the temperature varies more throughout the day than average monthly temperatures vary throughout the year. Average temperatures at sea level generally exceed 20°C. Climates within tropical latitudes are distinguished by differences in the seasonal pattern of rainfall. These differences create a continuous gradient of vegetation from wet, aseasonal rain forest through seasonal forest, scrub, savanna, and desert. Frost is not a factor in tropical biomes, even at high

elevations, and tropical plants and animals generally cannot tolerate freezing.

Tropical rain forest biome (climate zone I)

Climates where tropical rain forests develop (in Walter's equatorial climate zone) are always warm and receive at least 200 cm of precipitation throughout the year, with no less than 10 cm during any single month. These conditions prevail in three important regions within the tropics (Figure 5.16). First, the Amazon and Orinoco basins of South America, along with additional areas in Central America and along the Atlantic coast of Brazil, constitute the Neotropical rain forest. Second, the area from southernmost West Africa and extending eastward through the Congo River basin makes up the African rain forest (with an added area on the eastern side of the island of Madagascar). Third, the Indo-Malayan rain forest covers parts of Southeast Asia (Vietnam, Thailand, and the Malay

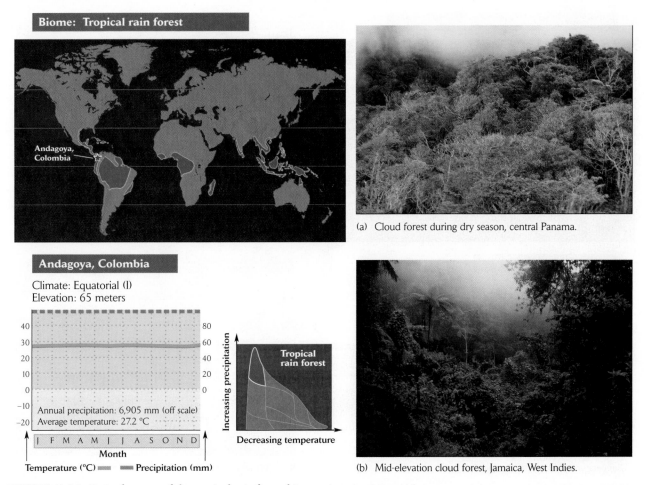

Biome: Tropical rain forest

Andagoya,
Colombia

(a) Cloud forest during dry season, central Panama.

Andagoya, Colombia

Climate: Equatorial (I)
Elevation: 65 meters

Annual precipitation: 6,905 mm (off scale)
Average temperature: 27.2 °C

J F M A M J J A S O N D
Month

Temperature (°C) ▬▬ ▬▬ Precipitation (mm)

Increasing precipitation

**Tropical
rain forest**

Decreasing temperature

(b) Mid-elevation cloud forest, Jamaica, West Indies.

FIGURE 5.16 Major features of the tropical rain forest biome. Photos by R. E. Ricklefs.

Peninsula); the islands between Asia and Australia, including the Philippines, Borneo, and New Guinea; and the Queensland coast of Australia.

The tropical rain forest climate often exhibits two peaks of rainfall centered on the equinoxes, corresponding to the periods when the intertropical convergence lies over the equator (see Chapter 4). Rain forest soils are typically old and deeply weathered oxisols. Because they are relatively devoid of humus and clay, they take on the reddish color of aluminum and iron oxides and retain nutrients poorly. In spite of the low nutrient status of the soils, rain forest vegetation is dominated by a continuous **canopy** of tall evergreen trees rising to 30–40 m. Occasional **emergent** trees rise above the canopy to heights of 55 m or so. Because water stress on emergent trees is great due to their height and exposure, they are often deciduous, even in a mostly evergreen rain forest. Tropical rain forests typically have several **understory** layers beneath the canopy, containing smaller trees, shrubs, and herbs, but these are usually quite sparse because so little

light penetrates the canopy. Climbing **lianas,** or woody vines, and **epiphytes,** plants that grow on the branches of other plants and are not rooted in soil (also called air plants; see Figure 1.5), are prominent in the forest canopy itself. Species diversity is higher than anywhere else on earth.

Per unit of area, the biological productivity of tropical rain forests exceeds that of any other terrestrial biome, and their standing biomass exceeds that of all other biomes except temperate rain forests. Because of the continuously high temperatures and abundant moisture, plant litter decomposes quickly, and the vegetation immediately takes up the released nutrients. This rapid nutrient cycling supports the high productivity of the rain forest, but it also makes the rain forest ecosystem extremely vulnerable to disturbance. When tropical rain forests are cut and burned, many of the nutrients are carted off in logs or go up in smoke. The vulnerable soils erode rapidly and fill the streams with silt. In many cases, the environment degrades rapidly and the landscape becomes unproductive.

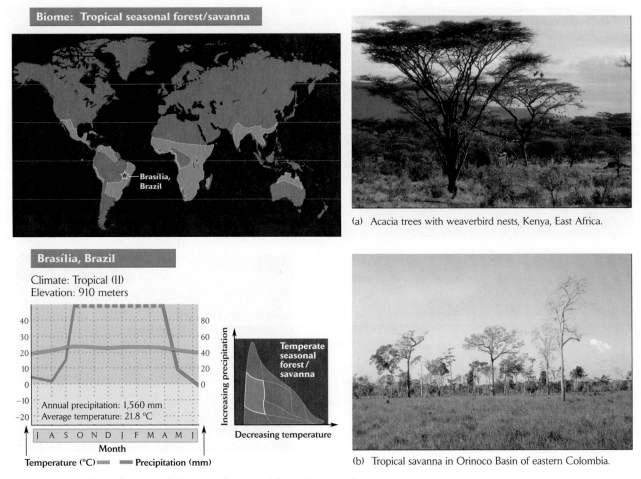

FIGURE 5.17 Major features of the tropical seasonal forest/savanna biome. Photos by R. E. Ricklefs.

Tropical seasonal forest/savanna biome (climate zone II)

Within the tropics, but beyond 10° from the equator (in Walter's tropical climate zone), there is typically a pronounced dry season, corresponding to winter at higher latitudes. Seasonal forests in this climate zone have a preponderance of deciduous trees that shed their leaves during the season of water stress (Figure 5.17). Where the dry season is longer and more severe, the vegetation becomes shorter, and thorns develop to protect leaves from grazing. With progressive aridity, the vegetation grades from dry forest into thorn forest and finally into true desert in the rain shadows of mountain ranges or along coasts with cold ocean currents running alongside. As in more humid tropical environments, the soils tend to be strongly lateritic and nutrient poor.

Savannas are grasslands with scattered trees. They are spread over large areas of the dry tropics, especially at moderate elevations in East Africa. Rainfall is typically 90–150 cm per year, but the driest three or four months bring less than 5 cm each. Fire and grazing undoubtedly play important roles in maintaining the character of the savanna biome, particularly in wetter regions, as grasses can persist better than other forms of vegetation under both influences. When grazing and fire are controlled within a savanna habitat, dry forest often begins to develop. Vast areas of African savanna owe their character to the influence of human activities, including burning, over many millennia.

The biome concept must be modified for freshwater aquatic systems

Terrestrial and aquatic ecologists have generated concepts and descriptive terms for ecological systems independently. The biome concept was developed for terrestrial ecosystems, where the growth form of the dominant vegetation reflects climatic conditions. In aquatic systems, however, depth, water temperature, flow rate, and oxygen and nutrient concentrations are the dominant physical factors, and the structural attributes of aquatic organisms do not differ much in relation to these factors. As a consequence, aquatic "biomes" do not exist in the sense in which the term is applied to terrestrial ecosystems. Indeed, defining aquatic biomes according to vegetation would be impossible, because the producers in many aquatic systems are single-celled algae, which do not form "vegetation" with a characteristic structure. As a result, aquatic systems have been classified primarily by such physical characteristics as salinity, water movement, and depth. The major kinds of aquatic environments are streams and rivers, lakes, wetlands, estuaries, and oceans, and each of these can be subdivided further with respect to many factors.

Flowing water: Streams and rivers

Streams form wherever precipitation exceeds evaporation and excess water drains from the land. Streams grow with distance as they join together to form rivers. Stream and river systems are often referred to as **lotic** systems, a term generally applied to flowing fresh waters. The continuous change in environments and ecosystems from the small streams at the headwaters of a river system to the mouth of the river is the basis for the **river continuum** concept. As one moves downstream, water flows more slowly and becomes warmer and richer in nutrients; ecosystems become more complex and generally more productive.

Within small streams, ecologists distinguish areas of **riffles,** where water runs rapidly over a rocky substratum, and **pools,** which are deeper stretches of more slowly moving water (Figure 5.18). Water is well oxygenated in riffles, whereas pools tend to accumulate silt and organic matter. Both areas tend to be unproductive because the nutrients needed for life are washed away in riffles, whereas the oxygen and sunlight needed for life are lacking in pools.

In general, streams lack the richness and diversity of life seen in other aquatic systems. Toward the headwaters of rivers, where small streams are often shaded and nutrient poor, the productivity of algae and other photosynthetic organisms tends to be low. Streams are usually bordered by a **riparian zone** of terrestrial vegetation that is influenced by seasonal flooding and elevated water tables. Much of the food web of headwater ecosystems depends on leaves and other organic matter that falls or washes into streams from this surrounding vegetation.

FIGURE 5.18 Within a stream, conditions differ between pools and riffles. Photo by Ed Reschke/Peter Arnold.

Such organic material that enters the aquatic system from the outside is termed **allochthonous.**

The larger a river, the more of its organic material is homegrown, or **autochthonous.** As one moves down the river continuum, rivers become wider, slower moving, more heavily nutrient laden, and more exposed to direct sunlight (Figure 5.19). The nutrients and sunlight support the growth of algae and plants within the river itself. However, rivers also become more heavily laden with sediments washed into them from the land and carried downstream. The high turbidity caused by suspended sediments in the lower reaches of silt-laden rivers can block light and reduce production. **Fluvial** systems, as rivers are sometimes called, are also distinguished by the fact that currents continuously move material, including animals, plants, and nutrients, downstream. To maintain a fluvial system in a steady state, this downstream drift must be balanced by the upstream movement of animals, production in the upstream portions of the system, and input of allochthonous materials.

All aquatic ecosystems interact with the terrestrial biomes that surround them. We have seen that streams receive runoff, groundwater, and organic matter from the surrounding land. A variety of organisms live their lives in both aquatic and terrestrial environments. Many frogs and salamanders, for example, have aquatic larval stages and terrestrial adult stages. Some terrestrial animals feed on organisms that grow in streams and lakes, effectively moving nutrients from aquatic to nearby terrestrial systems.

FIGURE 5.19 Nutrient-laden large rivers are highly productive. This river is a tributary in the vast wetland area of the lower Amazon River floodplain in Pará State, Brazil. Photo by Jacques Jangoux/Peter Arnold.

Conversely, many organisms with aquatic larval stages, such as mosquitoes, feed on terrestrial organisms. Thus, while aquatic and terrestrial biomes have recognizable borders, organisms readily cross these borders, and the borders themselves move, extending onto and retreating from floodplains as rivers rise and fall.

Lotic systems are extremely sensitive to any modification of their water flow. Tens of thousands of dams of all sizes interrupt stream flow in the United States alone. These dams are built for flood control, to provide water for irrigation, or to generate electricity. Dams alter rates of flow, water temperature, and sedimentation patterns. Typically, water behind dams becomes warmer, and bottom habitats become choked with silt, destroying habitat for fish and other aquatic organisms. Large dams used for hydroelectric power often release water downstream that has low concentrations of dissolved oxygen. Using dams for flood control changes the seasonal cycles of flooding necessary for maintaining many kinds of riparian habitats on floodplains. Dams also disrupt the natural movement of aquatic organisms upstream and downstream, fragmenting river systems and isolating populations. Thus, lotic systems are among the most vulnerable of all biomes to habitat modification.

Standing water: Lakes and ponds

Lakes and **ponds,** referred to as **lentic** systems, are distinguished by nonflowing water. Lakes and ponds can form in any kind of depression. They range in size from small, temporary rainwater pools a few centimeters deep to Lake Baikal, in Russia, which has a maximum depth of

1,740 m (about a mile) and contains about one-fifth of all the fresh water at the surface of the earth. Many lakes and ponds are formed by the retreat of glaciers, which leave behind gouged-out basins and blocks of ice buried in glacial deposits, which eventually melt. The Great Lakes of North America formed in glacial basins, overlain until 10,000 years ago by thick ice. Lakes are also formed in geologically active regions, such as the Great Rift Valley of Africa, where vertical shifting of blocks of the earth's crust creates basins within which water accumulates. Broad river valleys, such as those of the Mississippi and Amazon rivers, have oxbow lakes, which are broad bends of the former river cut off by shifts in the main channel.

An entire lake could be considered a biome, but it is usually subdivided into several ecological zones, each of which has distinct physical conditions (Figure 5.20). The **littoral zone** is the shallow zone around the edge of a lake or pond within which one finds rooted vegetation, such as water lilies and pickerelweed. The open water beyond the littoral zone is the **limnetic** (or **pelagic) zone,** where the producers are floating single-celled algae, or *phytoplankton*. Lakes may also be subdivided vertically on the basis of light penetration and the formation of thermally stratified layers of water (the epilimnion toward the surface and the hypolimnion at depth; see Figure 4.12). The sediments at the bottoms of lakes and ponds constitute the **benthic zone,** which provides habitat for burrowing animals and microorganisms.

Lakes and ponds are not permanent. Small temporary ponds can dry out each year, often multiple times during a season. Most small temperate lakes that formed when glaciers retreated will gradually fill in with sediment until

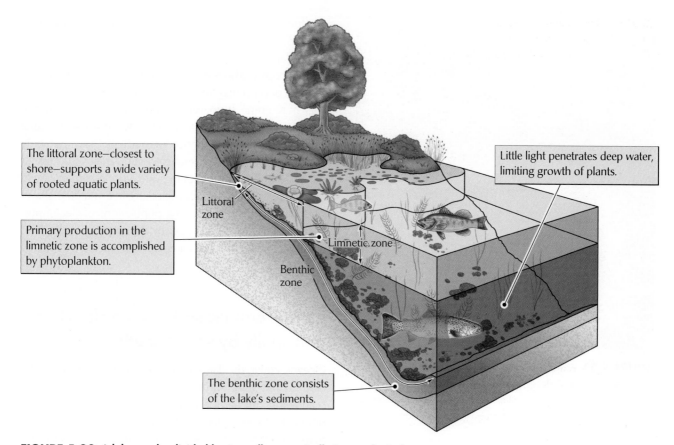

The littoral zone–closest to shore–supports a wide variety of rooted aquatic plants.

Primary production in the limnetic zone is accomplished by phytoplankton.

Little light penetrates deep water, limiting growth of plants.

Littoral zone

Limnetic zone

Benthic zone

The benthic zone consists of the lake's sediments.

FIGURE 5.20 A lake can be divided horizontally or vertically into ecological zones.

there is no open water. The formerly aquatic ecosystem will gradually change into a terrestrial ecosystem, first a wet meadow and later the natural terrestrial biome of the region.

Wetlands

Aquatic and terrestrial communities often come together in **wetlands,** which are areas of land consisting of soil that is saturated with water and supports vegetation specifically adapted to such conditions. Wetlands include swamps, marshes, and bogs when they derive from freshwater, and salt marshes and mangrove wetlands when they are associated with marine environments. Wetlands range in size from vernal pools formed in the aftermath of spring rains to vast areas of river deltas, such as the Okavango Swamp of Botswana, the Everglades of southern Florida, and the Pantanal of Brazil, Bolivia, and Paraguay—at 195,000 km^2, the world's largest wetland. Most of the plants that grow in wetlands can tolerate low oxygen concentrations in the soil; indeed, many are specialized for these anoxic conditions and grow nowhere else. Wetlands also provide important habitat for a wide variety of animals, notably waterfowl and the larval stages of many species of fish

and invertebrates characteristic of open waters. Wetlands protect coastal areas from the ravages of hurricanes and other storms. Wetland sediments immobilize potentially toxic or polluting substances dissolved in water and are thus natural water purifying plants.

Unfortunately, wetlands also occupy space, and they have been cut, drained, and filled to obtain wood products, to develop new agricultural lands, and for ever-increasing urban and suburban sprawl. Since the 1970s, increasing awareness of the natural values of wetland habitats, and legislation, such as the U.S. Clean Water Act (1977), have helped to conserve large areas of wetlands and restore them as closely as possible to their natural state.

Estuaries

Estuaries are found at the mouths of rivers, especially where the outflow is partially enclosed by landforms or barrier islands (Figure 5.21). Estuaries are unique because of their mix of fresh and salt water. In addition, they are abundantly supplied with nutrients and sediments carried downstream by rivers. The rapid exchange of nutrients between the sediments and the surface in the shallow waters of the estuary supports extremely high biological

FIGURE 5.21 Estuaries are extremely productive ecosystems. Estuaries develop at the mouths of rivers and are often bordered by extensive salt marshes, as in this view along the Georgia coast. Photo by S. J. Krasemann/Peter Arnold.

productivity. Because estuaries tend to be areas of sediment deposition, they are often edged by extensive tidal marshes at temperate latitudes and by mangrove wetlands in the tropics. Tidal marshes are among the most productive habitats on earth, owing to a combination of high nutrient levels and freedom from water stress. They contribute organic matter to estuarine ecosystems, which in turn support abundant populations of oysters, crabs, fish, and the animals that feed on them.

Human inputs into freshwater biomes

Freshwater biomes of all kinds are subject to a variety of inputs produced by human activities that can dramatically change their quality and ecological functioning. The most important of these are acid rain and eutrophication, which we shall discuss in more detail in later chapters. These inputs and their effects further demonstrate the intimate connections between terrestrial and aquatic biomes.

Acid rain forms when various gases produced by the combustion of fossil fuels, particularly sulfur dioxide and nitrogen oxides, dissolve in atmospheric moisture to form sulfuric and nitric acids. This acidified precipitation enters lakes and streams, where it can reduce the pH to as low as 4, well beyond the tolerance limits of many organisms. Acidified waters lose plant life and algae, and the low pH

disrupts the normal reproduction of fish and other aquatic animals. In the most extreme case, the entire ecosystem can collapse.

Eutrophication is the addition of limiting nutrients, such as phosphorus, to aquatic ecosystems. These nutrients may come from runoff carrying sewage, industrial wastes, or fertilizers or animal wastes from agricultural lands. A sudden abundance of nutrients may not only increase production dramatically, but may also disrupt normal ecosystem functioning by favoring certain organisms over others. The abundant organic material stimulates the growth of exploding populations of decomposing bacteria, but the process of decomposition depletes waters of oxygen that other organisms need.

Marine aquatic systems are classified principally by water depth

Oceans cover the largest portion of the surface of the earth. Beneath the surface of the ocean lies an immensely complex realm harboring a great variety of physical conditions and ecological systems (Figure 5.22). Variation in marine environments comes from differences in temperature, salinity, depth (which influences light and pressure), currents, substrata, and at the edge of the oceans, tides.

Many marine ecologists categorize marine ecological zones according to depth. The **littoral zone** (also called the *intertidal zone*) extends between the highest and lowest tidal water levels, and thus is exposed periodically to air (Figure 5.23). Ecological conditions within the littoral zone change rapidly as the tide flows in or out. A frequent consequence is the sharp **zonation** of organisms according to their ability to tolerate the stresses of terrestrial conditions, to which they are exposed to a varying extent depending on their position within the intertidal range. Beyond the range of the lowest tidal level, the **neritic zone** extends to depths of about 200 m, which correspond to the edge of the continental shelf. The neritic zone is generally a region of high productivity because the sunlit surface layers of water are close enough to the nutrients in the sediments below that strong waves can move them to the surface. Beyond the neritic zone, the seafloor drops rapidly to the great depths of the **oceanic zone.** Here, nutrients are sparse, and production is strictly limited. The seafloor beneath the oceanic zone constitutes the **benthic zone.** Both the neritic and the oceanic zones may be subdivided vertically into a superficial **photic zone,** in which there is sufficient light for photosynthesis, and an **aphotic zone** without light. Organisms in the

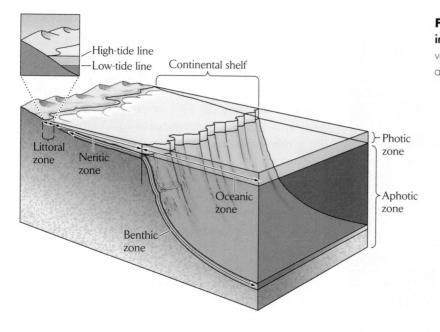

FIGURE 5.22 The oceans can be divided into several major ecological zones. This variation results from differences in factors such as temperature, depth, and tidal immersion.

aphotic zone depend mostly on organic material raining down from above.

Other systems of marine biome classification divide the oceans into biomes in different ways. One example is provided by the World Wildlife Fund's global list of 200 habitat types that are priorities for conservation. The World Wildlife Fund has singled out the following marine biomes as among the most productive and diverse on earth: polar, temperate shelves and seas, temperate upwelling, tropical upwelling, and tropical coral reefs. These biomes have traditionally provided most of the marine resources exploited by humans. Polar regions, which contain large areas of shallow seas, and continental shelves at temperate latitudes are highly productive because nutrients in seafloor sediments are not far below the surface waters, as indicated above. Upwelling zones are also highly productive because upwelling currents carry nutrients from the ocean depths to the sunlit surface waters.

Whereas the open ocean has been compared to a desert because of its low productivity, **coral reefs** are like tropical rain forests, both in the richness of their biological production and the diversity of their inhabitants (Figure 5.24). Reef-building corals are found in shallow waters of warm oceans, usually where water temperatures remain above 20°C year-round. Coral reefs often surround volcanic islands, where they are fed by nutrients eroding from the rich volcanic soil and by deep-water currents forced upward by the profile of the island. Corals are doubly productive because photosynthetic algae within their tissues generate the carbohydrate energy that fuels the corals' phenomenal rates of growth. Moreover, the complexity of the structure built by the corals over

time provides a wide variety of substrata and hiding places for algae and animals, making coral reefs among the most diverse biomes on earth. Unfortunately, rising sea surface temperatures in the tropics are killing the algal symbionts

FIGURE 5.23 The littoral zone is exposed to terrestrial conditions twice each day. Nonetheless, it may support prolific growth of algae and a variety of marine animals, as in this area of the New Brunswick coast in Canada. Photos by R. E. Ricklefs.

FIGURE 5.24 Coral reefs are productive and diverse ecosystems. In contrast to the open ocean, where productivity is low, coral reef ecosystems provide abundant food for a diverse biological community. This photo was taken in the Red Sea, near Egypt. Photo by Eric Hanauer.

of corals over large areas—a phenomenon known as coral bleaching. The stability of these biomes is now at risk.

Other marine biomes have physical conditions that foster unique forms of life and distinctive ecosystem properties. For example, the kelp forests that develop in shallow, fertile waters along continental coasts provide habitat for a rich variety of marine life (see Figure 1.16). Large areas of shallow polar seas are covered with pack ice that seals the air–water interface and increases the salinity of water as salts are excluded from ice. The result is a dim, salty environment without any wave disturbance. Hydrothermal vents are deep-sea environments dominated by the input of hot water laden with hydrogen sulfide, which provides the reducing power used by chemosynthetic bacteria to fuel high productivity in the otherwise sterile abyssal environment.

The physical qualities that characterize each terrestrial and aquatic biome constitute the environments to which their inhabitants are adapted in form and function. The close association between organisms and their environments over evolutionary time is the basis for ecological specialization and the resulting limits of the distributions of organisms and populations. Adaptations, however, reflect not only these physical factors in the environment, but also the many interactions of organisms with individuals of their own and other species. In the next part of this book, we shall examine the process of evolutionary adaptation and see how it has created the tremendous diversity of life on earth.

SUMMARY

1. The geographic distributions of plants are determined primarily by climate. Each climatic region has characteristic types of vegetation that differ in growth form.

2. Because plant growth form is directly related to climate, the major types of vegetation match temperature and precipitation closely. Major vegetation types can be used to classify ecosystems into categories called biomes.

3. Two ways of classifying biomes are represented by the climate zone approach of Walter and the vegetation approach exemplified by Whittaker. The first classifies regions on the basis of climate, within which a characteristic type of vegetation normally develops. The second classifies regions according to vegetation type, which generally reflects the local climate.

4. Climate zones and biomes are grouped within tropical, temperate, boreal, and polar latitudes. The adaptations of plants to different temperature ranges distinguish the

vegetation types of each of these latitudinal bands. Within each of these latitudinal bands, annual precipitation, the seasonality of precipitation, and additional factors such as fire further differentiate terrestrial biomes.

5. Within temperate latitudes, the major biomes are temperate seasonal forest, temperate rain forest, and temperate grassland/desert. The woodland/shrubland biome is found at lower temperate latitudes in areas with a Mediterranean climate. Subtropical deserts lie between temperate and tropical latitudes.

6. At high latitudes, one encounters boreal forest, usually consisting of needle-leaved trees with evergreen foliage on nutrient-poor, acidic soils, and tundra, a treeless biome that develops on permanently frozen soils, or permafrost.

7. Tropical latitudes are dominated by tropical rain forest and tropical seasonal forest, which grades from deciduous forest to thorn forest as aridity increases, and sometimes to savanna, which is grassland with scattered trees maintained by fire and grazing pressure.

8. Aquatic systems are not usually classified as biomes because they lack the equivalent of terrestrial vegetation with a characteristic structure. One may, however, distinguish streams (lotic systems), lakes (lentic systems), wetlands, estuaries, and oceans, and each of these systems can be further subdivided on the basis of other factors, such as depth of water.

9. The river continuum concept describes changes in the character of lotic systems from their headwaters to their mouths. Headwater streams tend to be shaded by surrounding vegetation, have low nutrient levels, and receive most of their organic material from outside (allochthonous) sources.

10. The lower reaches of rivers are wide and slow moving, carry heavy nutrient and sediment loads, receive plentiful sunlight, and grade into surrounding terrestrial habitats on floodplains.

11. Lakes and ponds vary tremendously in size, but are all distinguished by containing nonflowing water. Large

lakes can be differentiated into ecological zones that differ in water depth, temperature, nutrients, oxygen, and light.

12. Wetlands are areas in which the soil is saturated with water, including marshes, swamps, bogs, and mangrove wetlands. Wetlands support unique plants and animals and also serve important ecosystem functions, such as removing pollutants from water.

13. Estuaries, which occur at the mouths of rivers where fresh water mixes with seawater, support high levels of productivity. Because they are areas of sediment deposition, many estuaries are edged by extensive tidal marshes at temperate latitudes and by mangrove wetlands in the tropics.

14. Marine ecosystems are categorized mainly by depth. They include the littoral zone, on the shoreline between high and low tide levels; the neritic zone, made up of open waters to a depth of about 200 m; and the deep waters of the oceanic zone. Light penetration divides the oceans into a superficial photic zone and a deep aphotic zone, lacking all light. Many specialized types of marine ecosystems are found in connection with upwelling currents and tropical coral reefs.

REVIEW QUESTIONS

1. Why do unrelated plants often assume the same growth form in different parts of the world?

2. Which types of environmental conditions limit the distributions of plants?

3. What climatic conditions are used to define biomes?

4. What types of plants are found in each of the four biomes at temperate latitudes, and what environmental condition differs among these biomes?

5. Why is the boreal forest biome found on several different continents, including North America, Europe, and Asia?

6. Explain why tropical rain forests experience two peaks of rainfall.

7. How can fire shift an area from one biome type to another?

8. Compare and contrast the factors used to categorize terrestrial biomes with those used to categorize aquatic biomes.

9. How do headwater streams and larger rivers differ in their major source of organic material?

10. How do the littoral zone and the limnetic zone of a lake differ in their source of production?

11. What conditions allow coral reefs to be highly productive?

SUGGESTED READINGS

Allan, J. D. 1995. *Stream Ecology: Structure and Function of Running Waters.* Chapman & Hall, London.

Allen, C. D., and D. D. Breshears. 1998. Drought-induced shift of a forest–woodland ecotone: Rapid landscape response to climate variation. *Proceedings of the National Academy of Sciences USA* 95:14839–14842.

Barnes, R. S. K., and R. N. Hughes. 1999. *An Introduction to Marine Ecology.* 3rd ed. Blackwell Scientific Publications, Oxford.

Cushing, C. E., and J. D. Allan. 2001. *Streams: Their Ecology and Life.* Academic Press, San Diego.

Dodson, S. 2004. *Introduction to Limnology.* McGraw-Hill, New York.

Giller, P. S., and B. Malqvist. 1998. *The Biology of Streams and Rivers.* Oxford University Press, Oxford and New York.

Jeffree, E. P., and C. E. Jeffree. 1994. Temperature and biogeographical distributions of species. *Functional Ecology* 8:640–650.

McLusky, D. S. 1989. *The Estuarine Ecosystem.* 2nd ed. Chapman & Hall, New York.

Nybakken, J. W. 2005. *Marine Biology: An Ecological Approach.* 6th ed. Benjamin Cummings, San Francisco.

Olson, D. M., and E. Dinerstein. 2002. The global 200: Priority ecoregions for global conservation. *Annals of the Missouri Botanical Garden* 89:199–224.

Olson, D. M., et al. 2001. Terrestrial ecoregions of the world: A new map of life on earth. *BioScience* 51(11):933–938.

Prentice, I. C., et al. 1992. A global biome model based on plant physiology and dominance, soil properties and climate. *Journal of Biogeography* 19:117–134.

Primack, R., and R. Corlett. 2005. *Tropical Rain Forests: An Ecological and Biogeographical Comparison.* Blackwell Publishers, Malden, Mass.

Smith, T. M., H. H. Shugart, and F. I. Woodward. 1997. *Plant Functional Types: Their Relevance to Ecosystem Properties and Global Change.* Cambridge University Press, Cambridge.

Teal, J., and M. Teal. 1969. *Life and Death of a Salt Marsh.* Little Brown, Boston.

Vannote, R. L., et al. 1980. The river continuum concept. *Canadian Journal of Fisheries and Aquatic Sciences* 37(1):130–137.

Woodward, F. I. 1987. *Climate and Plant Distribution.* Cambridge University Press, Cambridge.

CHAPTER 6

Evolution and Adaptation

The Galápagos archipelago, lying a thousand kilometers off the Pacific coast of Ecuador, was a source of inspiration to Charles Darwin 175 years ago on his famous globe-circling voyage on H.M.S. *Beagle*. Darwin noticed that several of the organisms living in the archipelago had different forms on different islands. He surmised that these differences must have arisen by independent modification of the descendants of the original colonists, which came from the mainland of South America. This idea helped Darwin to develop his theory of evolution by natural selection. Ever since Darwin's time, the Galápagos archipelago has held a special fascination for evolutionary biologists, and many have returned there to pursue evolutionary studies. Peter and Rosemary Grant, of Princeton University, have observed populations of Darwin's finches for many years. Among their many findings was that the reproductive success and survival of individuals with different-sized beaks differed between El Niño and La Niña years.

The Galápagos archipelago normally comes under the influence of the cold Peru Current and is relatively dry. During El Niño years, however, prolonged warm sea surface temperatures greatly increase rainfall, and the resulting luxuriant growth of vegetation produces abundant insects and seeds for the populations of birds and reptiles that rely on these foods (Figure 6.1). La Niña years can bring periods of unremitting drought and food scarcity.

The medium ground finch (*Geospiza fortis*) subsists primarily on seeds, which it cracks open with its beak. During one La Niña drought period in the mid-1970s, seeds became scarce as the vegetation shriveled and died. The seeds that did remain were generally the hardest to crack, and were thus avoided by the finches during times of plenty. However, with few seeds available, the finches had no choice other than to tackle the hard ones. Consequently, the population of the medium ground finch on the tiny island of Daphne Major dropped from about 1,400 individuals in 1975 to about 200 by the end of 1977.

(a)   (b)

FIGURE 6.1 Heavy rains during El Niño events support lush plant growth in the Galápagos archipelago. These photos show a hillside on Tower Island (Isla Genovesa) at the end of a normal dry season in January 1982 (a) and in the middle of a strong El Niño event in March 1983 (b). The most important difference is the dramatic increase in understory shrubs and vines. The larger *Bursera* trees are not affected by the exceptionally wet conditions. Courtesy of Robert L. Curry, Villanova University.

Finches do not survive or die at random. Because the average hardness of seeds increased as the drought intensified and the softest seeds were consumed, birds with larger beaks that could generate the forces needed to crack hard seeds survived better than individuals with smaller beaks. Thus, the average beak size of surviving individuals, and their progeny, increased significantly between 1976 and 1978 (Figure 6.2). Here was a case of evolution in action! During the exceedingly wet El Niño year of 1983, small seeds were produced in abundance. Birds with smaller beaks handled the smaller seeds more efficiently, so more of them survived, and they produced more offspring than individuals with larger beaks. Consequently, the average beak size of the population returned to a lower value. Although these evolutionary responses were small, they nonetheless illustrate the capacity of a population to respond to changes in the environment.

FIGURE 6.2 Darwin's finches show evolutionary responses to changes in food resources associated with climate changes. (a) Changes in the abundance of seeds and in the population size of the medium ground finch (*Geospiza fortis*) on Daphne Major, in the Galápagos archipelago, during the drought period of 1975–78. (b) Changes in relative seed hardness and in the average beak size in the medium ground finch population during the same period. From P. R. Grant, *Ecology and Evolution of Darwin's Finches*, Princeton University Press, Princeton, N.J. (1986).

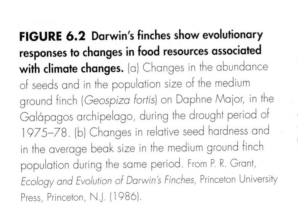
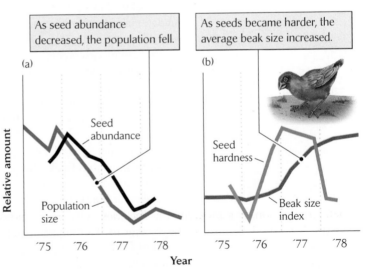

- The phenotype is the outward expression of an individual's genotype

- Adaptations result from natural selection on heritable variation in traits that affect evolutionary fitness

- Evolutionary changes in allele frequencies have been documented in natural populations

- Individuals can respond to their environments and increase their fitness

- Phenotypic plasticity allows individuals to adapt to environmental change

Evolution provides a straightforward explanation for the way in which the form and functioning of organisms is shaped by the properties of their environments—including physical conditions, food resources, and interactions with competitors, predators and pathogens, and other individuals of the same species. Through the process of evolution, traits of individuals within populations are continually adjusted to changes in the environment, just as beak size in the medium ground finch population responded to the availability of hard and soft seeds in the example at the opening of this chapter. The results of evolution cannot be understood without considering the environment, so ecology establishes a context for evolution. In this chapter, we shall examine the many ways in which populations and individuals respond to changes in their surroundings.

We are largely products of our genes, and all of us differ in our genetic makeup. Individuals with a particular genetic makeup may produce a greater number of offspring over their lifetimes than other individuals. In this way, some variations of genes will be passed on to future generations more frequently than others and will come to predominate in the population. These changes appear slowly in populations as parental generations are replaced by their progeny, but organisms can also respond individually to changes in the environment, whether they occur in an instant or over most of an individual's life. Population evolution and individual responses enable a species to adapt to the conditions of its environment and produce the close correspondence that we observe between form, function, and environment.

tions, and a phenotype is the rendering, or expression, of the genotype in the form of an organism. Of course, the environment also influences this rendering, but the genetic makeup of an individual sets limits on phenotypic expression. Obviously, an animal with the genotype of a mouse cannot develop into a rat, but it might become a large or a small mouse depending on its particular genetic makeup and the nature of its environment. To put it another way, the genotype is to the phenotype as blueprints are to the structure of a building. In this analogy, the effects of environmental influences are like details in a blueprint that are left up to the discretion of the building contractor, which may hinge, for example, on unpredictable changes in the availability of certain construction materials. However, when presented with the blueprints for a house, the contractor will not build a warehouse.

While all phenotypic traits have a genetic basis, they are also influenced by variations in the environment, either through the effects of environmental conditions on individuals (such as the influence of temperature or food supply on growth and development) or through the behavioral and physiological responses of individuals to variation in their environments. The capacity of an individual to exhibit different responses to its environment, called *phenotypic plasticity,* may itself be an evolved trait. That is, the way in which the individual responds to environmental variation is also subject to evolution by natural selection. We shall look at phenotypic plasticity in more detail later in this chapter, but we should keep in mind the difference between these plastic responses of individuals and the evolutionary responses of populations as we consider the relationship of organisms to their environments.

The phenotype is the outward expression of an individual's genotype

Each individual in a population is endowed with a unique genetic constitution, or **genotype,** that includes all of the individual's genes. The outward expression of the genotype in the individual's structure and function is called the **phenotype.** Thus, a genotype is a set of genetic instruc-

Genetic variation

Different forms of a particular gene are referred to as **alleles.** In many cases, alleles create perceptible and measurable differences in an organism's phenotype. The ABO blood groups in humans, for example, are determined by the inheritance of one of three alleles (A, B, and O) from each parent. The gene involved is responsible for production of the antigens A and B (the O allele does not produce an antigen), which are molecules on the surfaces of our red blood cells that interact with the immune system. Individuals with blood type A have AA or AO genotypes; individuals with blood type B have BB or BO genotypes; and the remaining individuals have either AB or OO genotypes. In this case, the link between the genotype and the phenotype is direct, and the pattern of inheritance is straightforward. Children of an AA father and a BB mother will all have the AB genotype, for example. Some other traits, such as eye color in humans, reflect several genes acting together. Eye color reflects the influence of at least three genes that control pigments in different parts of the iris of the eye. In this case, patterns of phenotype inheritance can be quite complex, as they depend on interactions among the genotypes for each gene.

Every individual has two copies of each gene, one inherited from its mother and one from its father (exceptions include sex-linked genes and organisms that reproduce without the sexual union of gametes). An individual that has two different alleles of a particular gene is said to be **heterozygous** for that gene, as in the case of a person with the AB blood type. When both copies of a gene are the same, the individual is **homozygous** (AA, for example). When an individual is heterozygous, the two different alleles may produce an intermediate phenotype, or one allele may mask the expression of the other. In the latter case, one allele is said to be **dominant** and the other **recessive.** When heterozygotes have an intermediate phenotype—as in the case of a person with the AB blood type, who expresses both alleles—the alleles are said to be **codominant.** Most harmful alleles are recessive, and the normal gene product of the dominant allele masks the defective function of the gene product of the recessive mutant allele in heterozygotes.

All the alleles of all the genes of every individual in a population constitute that population's **gene pool.** The gene pools of most sexually reproducing populations contain substantial genetic variation. With respect to the ABO blood type gene, the human population of the United States includes 61.3% O alleles, 30.0% A alleles, and 8.7% B alleles. The proportions of these alleles vary among populations. For example, Asians tend to have higher frequencies of B alleles, and the Irish more of the O alleles. Of course, population biologists are very much interested how this variation arises and how it is maintained. The answers to these questions require an understanding of the genetic mechanisms of inheritance and the influence of the alleles in each genotype on the evolutionary fitness of individuals.

Sources of genetic variation

How does genetic variation arise? The human genome contains 20,000–25,000 protein-coding genes, which represent only a small fraction of the genome's total DNA. A part of the rest produces many types of RNA, some of which are used to control gene expression. However, the function of much of our DNA remains a mystery. Genes that encode proteins are responsible for producing the structural molecules and enzymes that make up our bodies and carry out the biochemical transformations of life. Associated with these genes are nearby regions of the DNA that control the expression of genes, switching them on and off in response to variation in the physical and chemical environment of the nucleus of the cell. Changes in the DNA may cause changes in either the structure of the protein product itself or in the amount of the protein produced.

One important way in which such changes to proteins occur is through **mutation.** Mutations result from any change in the sequence of the nucleotides (adenine, thymine, cytosine, guanine, represented by the letters A, T, C, and G in the genetic code) that make up a gene or in regions of the DNA that control the expression of a gene. Such changes include substitution of one nucleotide for another, deletions or insertions of nucleotides, rearrangements of the DNA molecules involving inverted regions, duplicated regions, or exchanges of DNA sequences between chromosomes. Some of these events cause drastic, often lethal, changes in the phenotype. However, other events, particularly simple nucleotide substitutions, may alter the appearance, physiology, or behavior of the individual. Many human genetic disorders, such as sickle-cell anemia, Tay-Sachs disease, cystic fibrosis, and albinism, as well as tendencies to develop certain cancers and Alzheimer's disease, are caused by single-nucleotide mutations of individual genes. Still other nucleotide substitutions have no detectable effect and are referred to as silent, or synonymous, mutations. Occasionally, mutations produce new phenotypes that are better suited to the local environment, and these phenotypes will tend to increase in the population.

Many mutations have multiple effects. For example, a mutation of a gene called *daf*-2 lengthens the life span of the roundworm *Caenorhabditis elegans*—a common animal model in laboratory studies of genetics and development. Not surprisingly, *daf*-2 and related genes in other organisms, including humans, are of great interest to biologists. However, the action of the mutant *daf*-2 allele is quite complex because the gene influences the expression of other genes and gene regulators, resulting in a cascade of effects with different influences on life span and other physiological processes. Such effects of a single gene on multiple traits are referred to as **pleiotropy.** Indeed, worms with the *daf*-2 mutation overexpress dozens of gene products compared with "wild-type" (nonmutant) individuals and underexpress many others. With the many thousands of genes in the genome of even a simple worm, the relationship between the genotype and the phenotype is bound to be complex.

The genetic basis of continuously varying phenotypic traits

Up to this point, we have discussed phenotypic traits mostly in terms of simple genetic determination, in which a single gene with two or more alternative alleles gives rise to discrete, readily distinguishable differences in form or function, such as the A and B antigens produced by the gene for the ABO blood types. Many genes act in this fashion. However, many phenotypic traits with ecological relevance vary continuously over a range of values (Figure 6.3). Body size is a good example. In most populations, individuals exhibit a normal, or bell-shaped, distribution of body sizes, with a concentration of individuals toward the middle of the distribution and fewer toward the large and small extremes.

A part of this continuous variation might be due to differences in the environments individuals experience during development, but much of it can be attributed to the actions of many genes, each with a relatively small influence on the value of the trait. Thus, if each of several genes has an allele that tends to increase body size, while the alternative allele tends to decrease body size, the actual size of an individual will depend on the mix of alleles for all of these genes in its genome. The tendency of individuals to be concentrated toward the center of the distribution reflects the relative improbability of an individual inheriting mostly alleles causing large size, or mostly

alleles causing small size. Think of this in terms of flipping pennies. The chance of getting ten tails in a row (about one in a thousand) is much more remote than the chance of getting a more even mix of tails and heads. Exactly five heads and five tails will come up about one time in four.

Adaptations result from natural selection on heritable variation in traits that affect evolutionary fitness

Genetic variation among individuals within a population is pervasive and has many consequences. The most important of these consequences for the study of ecology is evolution by natural selection. The term **evolution** pertains to any change in a population's gene pool. When genetic factors cause differences among individuals in survival and reproductive success, evolutionary change comes about through natural selection. Individuals whose traits enable them to achieve higher rates of reproduction leave more descendants, and therefore the alleles responsible for those traits increase in the gene pool of the population. The traits themselves are referred to as *adaptations,* or *evolutionary adaptations,* and the process itself is often referred to as *adaptation.* The process of evolution by natural selection has three main ingredients:

1. Variation among individuals

2. Inheritance of that variation

3. Differences in survival and reproductive success, or *fitness,* related to that variation

Variation refers to the differences in a particular trait among individuals in a population, and thus to the variation in that trait contained in the gene pool of the population as a whole. Bird beaks, for example, come in a variety of sizes under the control of genetic factors; different individuals have different-sized beaks.

The *inheritance* of variation is the genetic basis of evolution. Heritable traits remain stable as they are passed from parent to offspring. Thus, a trait, such as the size of a bird's beak, has an existence of its own in a population independently of the particular individual to which it belongs. An individual does not possess a trait except for the brief period of its lifetime. The individual merely "borrows" a trait, or rather the genetic factors that influence it, from the population by way of its ancestors and then passes it on to its descendants. Thus, the trait—or rather, its heritable underlying genetic influences—is a part of the makeup of the population as a whole.

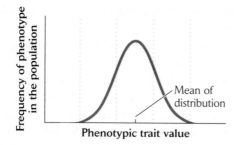

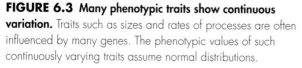

FIGURE 6.3 Many phenotypic traits show continuous variation. Traits such as sizes and rates of processes are often influenced by many genes. The phenotypic values of such continuously varying traits assume normal distributions.

Fitness refers to the production of descendants over an individual's lifetime. The more offspring an individual produces, the greater the fitness of that individual, and the greater the contribution of its "borrowed" traits to the makeup of future generations. Heritable traits that promote reproductive success are passed on at a high rate and eventually replace those traits conferring lower fitness. The resulting change in the genetic makeup of the population is called evolution.

Consider how these principles apply in the following example of evolutionary change in a California citrus pest. Early in the twentieth century, certain species of scale insects were serious pests in citrus orchards in southern California. Growers discovered that they could control scale populations effectively by fumigating orchards with cyanide gas. However, after several years of such treatment, the gas killed fewer of the insects, and before long the scale regained its pest status. Researchers determined that scale insects had evolved a genetically based resistance to cyanide poisoning (Figure 6.4). Furthermore, when they surveyed orchards in areas that had never been fumigated, they found that small numbers of individuals possessed an innate resistance to cyanide. In other words, cyanide resistance had a genetic basis, and the alleles involved were present at low frequencies in scale populations. The presence of a few individuals with these alleles endowing cyanide resistance was the initial genetic variation required for evolution.

Cyanide fumigation was initially very effective, killing off all nonresistant individuals. As a result, only those individuals with alleles for cyanide resistance reproduced and left offspring. Over time, the proportion of resistant individuals in the remaining population increased, until all the individuals were resistant and fumigation was no longer effective. Thus, despite their initial successes, the fumigation programs had, in the end, changed the insects' environment so as to favor reproduction by cyanide-resistant individuals, whose progeny then increased to epidemic proportions.

Any phenotypic variation can cause variation in fitness among individuals. The traits of those individuals that leave the most offspring are said to be *selected*, and the differential survival and reproduction of individuals having different traits is referred to as **natural selection.** Biologists use "natural" to distinguish what happens in nature from **artificial selection,** the result of conscious decisions made by humans concerning desirable qualities of domesticated or laboratory animals and crops.

Sometimes the environment can influence the phenotype directly. Variation in the availability of food to individuals, for example, can produce variation in body size

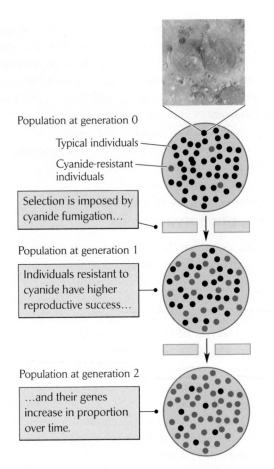

Population at generation 0

Typical individuals

Cyanide-resistant individuals

Selection is imposed by cyanide fumigation…

Population at generation 1

Individuals resistant to cyanide have higher reproductive success…

Population at generation 2

…and their genes increase in proportion over time.

FIGURE 6.4 Evolutionary change in a population may result from a change in the environment. Alleles that confer resistance to cyanide are present at low frequencies in populations of scale insects that have never been exposed to cyanide, simply because of recurrent mutations. In the absence of cyanide, the trait may actually be mildly harmful. When populations are fumigated on a regular basis, however, the gene for cyanide resistance confers high fitness, and its frequency in populations rapidly rises. Photo by Jack Kelly Clark, courtesy of the University of California Statewide IPM Project.

and other attributes. Such phenotypic changes in an individual sometimes improve that individual's relationship to its environment (think of the suntan that reduces damage from solar radiation to the underlying skin). Although such changes can beneficial, they are not adaptations in the evolutionary sense, but are referred to as *acclimatization*. Such environmental influences create phenotypic variation in the population, which may result in differential survival and reproduction (that is, differences in fitness). Without an underlying genetic basis for the variation, however, the gene pool of the population remains unchanged by such selection, and individuals in successive generations will exhibit the same range of traits. It is important to realize that an evolutionary response across generations requires a genetic basis for variation in the phenotype, for it is only genes that are passed on from generation to generation and which can therefore cause heritable change in the phenotypes of a population. We'll have more to say about acclimatization later in this chapter.

Most evolutionary biologists believe that the diversification of living beings over the long history of life has been guided primarily by natural selection. It is important to understand, however, that natural selection is not an external force that urges organisms toward some predetermined goal, in the sense that humans artificially "select" cows to achieve a higher rate of milk production. Quite the opposite. Natural selection occurs because of differences in reproductive success among individuals endowed with different form or function in a particular environment. The process that creates selection is ecological—the interaction of individuals with their environments, including physical conditions, food resources, predators, other individuals of the same species, and so on. A cold winter wind doesn't care whether a bird is well insulated by its plumage. Whether a rabbit runs fast or not is irrelevant to evolution. All that matters is whether fast rabbits leave more offspring, perhaps because they are more likely to escape foxes. One presumes that foxes would prefer to chase slow rabbits, but, alas, by catching the slow ones, they end up favoring reproduction by faster ones.

ECOLOGISTS IN THE FIELD   **Rapid evolution in response to an introduced parasitoid.** All attributes of morphology, physiology, and behavior are subject to evolutionary modification. For example, the males of many animal species display structures or behaviors that attract females for the purpose of mating. These traits are under strong selection by choosy females that find certain male traits desirable because they signal some attribute of the male that can contribute to the success of their offspring. Female choice is an important component of sexual selection, which is responsible for many of the differences between the sexes that one observes in natural populations (see Chapter 8). However, attractiveness to females can be a double-edged sword.

Consider the recent history of the field cricket *Teleogryllus oceanicus* in Hawaii. This insect, which is native to Australia and islands of the South Pacific Ocean, was introduced to Hawaii in the 1800s. There it coexists with another more recently introduced insect, from North America in this case: the parasitoid fly *Ormia ochracea*. This fly locates potential hosts (such as field crickets) by homing in on their mating calls. Female flies lay their eggs in the hapless victims, which are consumed from the inside by the developing fly larvae. In the course of studies on sexual

selection on the mating call of the field cricket on Kauai, Marlene Zuk, of the University of California at Riverside, noticed that males had become mostly silent during the past two decades. One can imagine the crickets becoming mute with fear in the presence of the parasitoid fly. In this case, however, the silence of the crickets is caused by a mutant allele that has increased in the cricket gene pool. Male field crickets make their mating calls by rubbing together structures on the wing that work like a file and scraper to produce the sound (Figure 6.5). Females lack these structures. The mutant allele produces male wings that look like female wings and are incapable of making sound.

Clearly, the *Ormia* fly's use of sound to find hosts would apply strong selection favoring any modification of the male cricket wing that reduced sound production—a case of predator behavior having a selective effect on the prey that makes life more difficult for the predator. The mutant wings solved one problem for the crickets, but what about the problem of attracting mates? If the mutant males had failed to reproduce, any advantage they had gained by avoiding parasitism would have been cancelled out. In fact, a few calling males remain, and the silent males tend to gather around these individuals and attempt to mate with females attracted by their calls. Indeed, when the investigators played male calls in the presence of other males, those with silent wings approached the speakers more closely than the normal-winged males. In populations with all calling males (in locations lacking the parasitoid flies, of course), individuals space themselves to avoid competition for the mates they attract. Evidently, strong selection favoring silent wings in males has also favored behaviors that draw males close to callers so that they can compete for the attracted females.

This example illustrates the basic features of adaptation in response to a change in the environment (the presence of a parasitoid fly). Wing morphology is inherited and varies by genetic mutation; under new conditions, a different morphology confers more fitness and increases in frequency. In this case, because selection was very strong (as in the case of cyanide fumigation and the scale insect), the adaptive response was nearly completed in a few tens of generations. Notice also that the advantages of a particular adaptation (parasite avoidance) may be partly balanced by disadvantages conferred by some other selective agent (in this case, choosy females). Therefore, most adaptations represent the best compromise between alternatives. ▮

Stabilizing, directional, and disruptive selection

Selection can influence the distribution of traits in a population in three different ways, depending on how the environment varies over time and space. **Stabilizing**

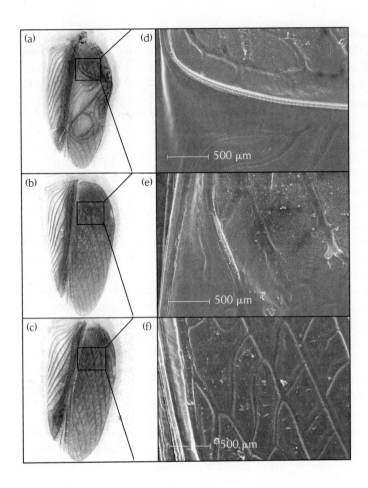

(g)

FIGURE 6.5 Parasitoid flies have caused evolutionary changes in the wings of male field crickets. Underside of the right forewing from (a) a normal-winged male, (b) a silent-winged male, and (c) a female *Teleogryllus oceanicus*. (d–f) SEM micrographs of these wings. (d) The many evenly spaced teeth and the scraper of the normal-winged male are used to produce the mating call. (e) These structures are much reduced in size and relocated on the wing in the silent-winged male. (f) The sound-producing apparatus is absent in the female. (g) A male *Teleogryllus oceanicus*. (a–f) Courtesy of R.M. Tinghitella, from M. Zuk, J. T. Rotenberry, and R. M. Tinghitella, *Biol. Lett.* 2:521–524 (2006); (g) Robin M. Tinghitella.

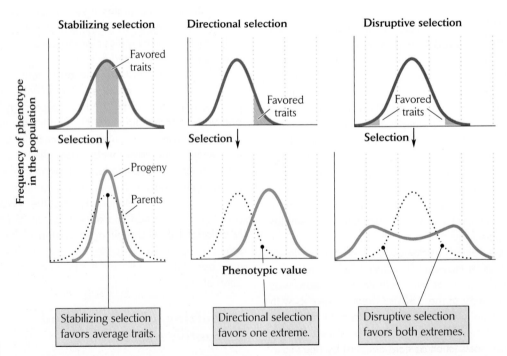

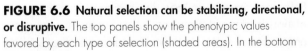

FIGURE 6.6 Natural selection can be stabilizing, directional, or disruptive. The top panels show the phenotypic values favored by each type of selection (shaded areas). In the bottom panels, the distributions of phenotypes after selection (in the progeny generation; solid lines) are compared with those before selection (in the parental generation; dotted lines).

selection occurs when individuals with intermediate, or average, phenotypes—generally the commonest phenotypes in a population—have higher reproductive success than those with extreme phenotypes (Figure 6.6). Stabilizing selection tends to draw the distribution of phenotypes within a population toward an intermediate, optimum point, and thus counteracts the tendency of phenotypic variation to increase through mutation. In the case of discrete traits, stabilizing selection tends to maintain a single fittest phenotype. Stabilizing selection performs genetic housekeeping for a population, sweeping away harmful genetic variation. When the environment of a population is relatively unchanging, stabilizing selection is the dominant mode, and little evolutionary change takes place.

Directional selection occurs when the fittest individuals have a more extreme phenotype than the average of the population. In this case, individuals whose phenotypes fall at one end of the population distribution produce the most progeny, and the distribution of phe-notypes in succeeding generations shifts toward a new optimum. When that new optimum is reached, selection becomes stabilizing. In the case of discrete traits, directional selection results in the replacement of one trait by another over time, as we shall see below in the case of the peppered moth.

Individuals with extreme phenotypes at either end of the population distribution might, under some circumstances, have higher fitness than individuals with intermediate phenotypes. This situation leads to **disruptive selection,** which tends to increase genetic and phenotypic variation within a population and, in the extreme case, creates a bimodal distribution of phenotypes with peaks toward both ends of the original distribution. Disruptive selection is thought to be relatively uncommon. It might occur, however, when individuals can specialize on one of a small number of food resources that differ in size or some other attribute (Figure 6.7). Strong competition among individuals for a preferred resource might also increase the fitness of individuals that avoid that competition by specializing on one of several alternative resources.

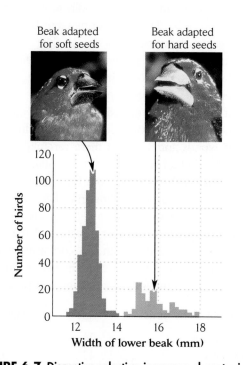

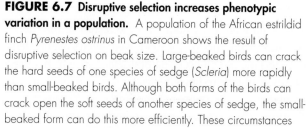

FIGURE 6.7 Disruptive selection increases phenotypic variation in a population. A population of the African estrildid finch *Pyrenestes ostrinus* in Cameroon shows the result of disruptive selection on beak size. Large-beaked birds can crack the hard seeds of one species of sedge (*Scleria*) more rapidly than small-beaked birds. Although both forms of the birds can crack open the soft seeds of another species of sedge, the small-beaked form can do this more efficiently. These circumstances have led to a bimodal distribution of beak size. After T. B. Smith, *Nature* 329:717–719 (1987).

Evolutionary changes in allele frequencies have been documented in natural populations

Cyanide resistance in scale insects was one of the first documented cases of a population that responded genetically to a change in its environment. Similar cases of pesticide and herbicide resistance among agricultural pests and disease vectors, as well as increasing antibiotic resistance among pathogenic bacteria, are further examples of how the gene pools of populations can respond to changes wrought by humankind in their environments. In each case, preexisting variation in the gene pool allowed the population to respond rapidly to the changed conditions.

Selection and change in the frequency of melanistic moths

One of the most striking demonstrations of evolution in action is the case of melanism in the peppered moth (*Biston betularia*), a species widely distributed in woodlands (Figure 6.8). Early in the nineteenth century, occasional dark, or *melanistic,* specimens of the peppered moth were collected in England. Over the next hundred years, this dark form became increasingly common in forests near heavily industrialized regions, which is why the phenomenon is often referred to as **industrial melanism.** In the

FIGURE 6.8 Industrial melanism in the peppered moth demonstrates genetic change in response to selective factors in the environment. Melanistic (*left*) and typical forms of the peppered moth at rest on a soot-darkened tree trunk. Photo by Stephen Dalton/Photo Researchers.

absence of factories and other heavy industry, the pale, or typical, form of the moth still prevailed.

This phenomenon aroused considerable interest among geneticists, who showed by mating pale and dark forms that melanism in the peppered moth is an inherited trait determined by a single dominant allele. Because melanism is inherited, its increase and spread must have resulted from genetic changes (evolution) in the population. It seemed reasonable to suppose that natural selection had led to the replacement of typical pale individuals with melanistic individuals. That is, where melanistic individuals had become common, the environment must somehow have been altered to give dark forms a survival advantage over pale forms.

Working during the 1950s, the English biologist H. B. D. Kettlewell tested this hypothesis by measuring the relative fitnesses of the two forms. He had to do this independently of the changes in their frequencies. That is, Kettlewell could not infer the relative fitnesses of typical and melanistic moths from the fact that one had decreased and one had increased in frequency, as that would be circular reasoning. To determine whether the melanistic form had greater fitness than the typical form in regions where melanism occurred, Kettlewell chose the mark–recapture method. He marked adult male moths of both forms with a dot of cellulose paint and then released them. The mark was placed on the underside of the wing so that it would not attract the attention of predators to a moth resting on a tree trunk. Kettlewell recaptured moths by attracting them to a mercury vapor lamp in the center of the woods or to caged virgin females at the edge of the woods. (Only males could be used in the study because females are attracted neither to lights nor to virgin females.)

In one experiment, Kettlewell marked and released 201 typical and 601 melanistic individuals in a wooded area near industrial Birmingham. The results were as follows:

	Typical	Melanistic
Number of moths released	201	601
Number of moths recaptured	34	205
Percentage recaptured	16	34

These figures indicated that a higher percentage of the dark individuals had survived over the course of the experiment. A similar experiment in a nonindustrial area revealed higher survival by pale individuals.

The specific agent of selection was easily identified. Peppered moths rest on branches of trees during the day. Kettlewell reasoned that in industrial areas, air pollution had darkened the branches so much that typical moths stood out against them and were readily found by predators (see Figure 6.8). The dark forms were better camouflaged against darkened branches, so their coloration increased their chances of survival. Eventually, differential survival of dark and pale forms had led to changes in their relative frequencies in those populations.

To test this idea, Kettlewell placed equal numbers of pale and dark moths on tree trunks in polluted and unpolluted woods and watched them carefully at a distance from behind a blind. (A *blind* is a tentlike structure intended to conceal observers from their subjects, more appropriately called a *hide* in England.) He quickly discovered that several species of birds regularly searched tree trunks for moths and other insects, and that these birds more readily found a moth that contrasted with its background than one that resembled the bark it clung to. Kettlewell tabulated the following instances of predation:

Individuals taken by birds	Typical	Melanistic
Unpolluted woods	26	164
Polluted woods	43	15

These data were consistent with the results of the mark–recapture experiments. Together, Kettlewell's results demonstrated the operation of natural selection, which, over a century, had resulted in genetic changes—evolution—in populations of the peppered moth in polluted areas.

Although Kettlewell was praised for providing the first demonstration that natural selection had caused evolutionary changes in a natural system—truly a textbook example—he also had his detractors. They complained that the moths he had used for the mark—recapture experiments were reared in the laboratory, which might have affected their behavior—such as their choice of resting locations—and that he had released the experimental moths at unnaturally high densities, which might have affected the behavior of predators so as to bias the results. However, recent detailed studies by Michael Majerus at Oxford University and others have fully confirmed Kettlewell's original results.

One of the most gratifying aspects of the peppered moth story is that, with the advent of pollution control programs and the return of forests to a cleaner state, frequencies of melanistic moths have decreased, as evolutionary theory predicted they should. In the area around the industrial center of Kirby in northwestern England, for example, the melanistic form decreased from more than 90% of the population to about 30% over a period of 20 years (Figure 6.9).

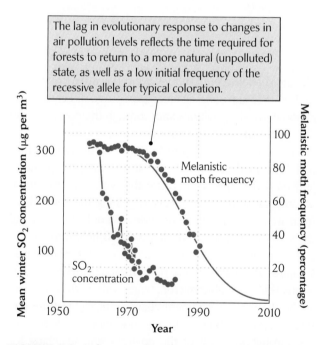

The lag in evolutionary response to changes in air pollution levels reflects the time required for forests to return to a more natural (unpolluted) state, as well as a low initial frequency of the recessive allele for typical coloration.

FIGURE 6.9 A change in the environment can result in a change in the frequency of a phenotype. The frequency of the melanistic form of the peppered moth has decreased since the beginning of pollution control programs in England in the 1950s. The index to pollution is the winter level of sulfur dioxide (SO_2), which directly affects lichens growing on tree trunks, against which moths rest by day. After C. A. Clarke, G. S. Mani, and G. Wynne, *Biol. J. Linn. Soc.* 26:189–199 (1985); G. S. Mani and M. E. N. Majerus, *Biol. J. Linn. Soc.* 48:157–165 (1993).

Population genetics and the prediction of evolutionary change

The dynamics of natural selection and genetic change in populations are a focus of the branch of ecology known as population genetics. The study of **population genetics** has shown that populations are continually engaged in dynamic evolutionary relationships with their environments that shape ecological interactions. A primary goal of population genetics since the 1920s has been to develop methods for predicting changes in gene frequencies in response to selection. As we have just seen, such changes are the essence of evolution, and the ability to predict them is important because it can tell us whether the genetic changes we observe in nature are consistent with our understanding of how evolution works. The theory of population genetics has led to mathematical models that allow us to predict such changes as well as the direction of evolution under a particular type of selection. The topic is too extensive to take up in detail here, but additional information on the Web explains some of the basic features of population genetic analyses.

MORE ON THE WEB *Rates of Evolution in Populations.* Population genetic models can predict how rapidly genes will be replaced within populations.

MORE ON THE WEB *Selection on Traits that Exhibit Continuous Variation.* Most attributes of organisms are controlled by many genes with small effects, which must be analyzed by the methods of quantitative genetics.

MORE ON THE WEB *Modeling Selection Against a Deleterious Recessive Gene.* The rate of change in allele frequency depends on how the allele is expressed in the phenotype and the strength of selection.

Population genetics has a number of important messages for ecologists. First, every population harbors some genetic variation that influences fitness. This means that the potential for evolution exists in all populations.

Second, changes in the environment will almost always be met by an evolutionary response that shifts the frequencies of genotypes within the population. These shifts are adaptive in that over time, a larger proportion of individuals will inherit structures and mechanisms that increase their fitness under the new environmental conditions. The magnitude of the evolutionary response is not always predictable and depends on the genetic variation present in the population at a given time. Most continuously varying traits, such as size, have enough genetic variation to respond to selection, but the range and extent of a response may be limited by the negative fitness consequences of other changes influenced by the same genetic factors. Given enough time, populations may reach some sort of evolutionary optimum and become stabilized, at least until the environment changes again.

Third, rapid environmental changes brought about by the appearance of new adaptations in populations of enemies, or by human-caused changes in the environment, such as habitat destruction or the introduction of predators or pathogens, can exceed the capacity of a population to respond by evolution. Under these circumstances, the decline of a population toward extinction is a distinct possibility.

Although much remains to be learned, it is clear that populations are engaged in evolutionarily dynamic relationships with their environments, particularly with the biological components of their environments (competitors, predators, and pathogens), which are also evolving in response to other kinds of organisms. These interactions among different species can exert powerful effects on populations, influencing the evolutionary and population dynamics of species and determining whether species can coexist with one another.

FIGURE 6.10 The cactus wren is adapted to its desert environment. The cactus wren (*Campylorhynchus brunneicapillus*) is a conspicuous resident of deserts in the southwestern United States and northern Mexico. Photo by Craig K. Lorenz/Photo Researchers.

Individuals can respond to their environments and increase their fitness

Evolution occurs through the replacement of less fit individuals by the progeny of more fit individuals in a population over time. Although evolutionary responses help to match the adaptations of organisms to their environments, the individual itself does not benefit from evolution. It is the gene pool of the population that evolves, not the individual. Nevertheless, individuals can undergo changes that help them cope with variation in their environments during their lifetimes. The *capacity* to respond to environmental variation, which is called **phenotypic plasticity,** can itself be an adaptation that enhances the individual's fitness. Environments change over time and space, and individuals adjust to these changes by altering their behavior, their physiology, and even their structure.

Animals are free to move about their environment, and at any given time they may choose that part of the environment that best enhances their survival or reproduction because of its temperature, moisture, salinity, or other conditions. Parts of the environment that can be distinguished by their conditions are referred to as **microhabitats** or **microenvironments.** In deserts, for exam-

ple, the shaded ground under a shrub is often cooler and moister than surrounding areas exposed to direct sunlight, although these conditions vary through the course of the daily cycle and with the seasons.

The cactus wren (Figure 6.10) is an insectivorous bird that lives in deserts of the southwestern United States and northern Mexico. Because the wren has no source of drinking water, it must avoid gaining too much heat from its environment. Otherwise, it would have to use the water in its body to dissipate excess heat through the cooling effect of evaporation from its respiratory tract (see Chapter 3). In the desert near Tucson, Arizona, cactus wrens seek favorable microhabitats within which to feed as the thermal environment changes throughout the day (Figure 6.11). During cool early mornings, the wrens forage throughout most of the environment, searching for food among foliage and on the ground. As the day brings warmer temperatures, the wrens select cooler parts of their habitat, particularly in the shade of small trees and large shrubs, always managing to avoid being where the temperature exceeds 35°C. When the minimum temperature in the environment rises above 35°C, at which point the birds must use evaporative cooling to maintain their body temperatures even when inactive, the wrens stop feeding and perch quietly in deep shade.

provide a suitable environment day and night, in hot and cool weather.

During the long breeding period (March through September) in southern Arizona, cactus wrens usually rear several broods of young. Early in spring, most individuals build their nests so that the entrances face away from the direction of the cold winds; during the hot summer months, they orient their nests to face prevailing afternoon breezes, which circulate air through the nest chamber and facilitate heat loss (Figure 6.12). This strategy makes a difference! Nests oriented properly for the season are consistently more successful (82% produce viable offspring) than nests facing in the wrong direction (only 45% are successful). The behavioral flexibility of the cactus wren in choosing where to forage and how to orient its nest is a good example of the more general ability of the phenotype to respond to variation in the environment.

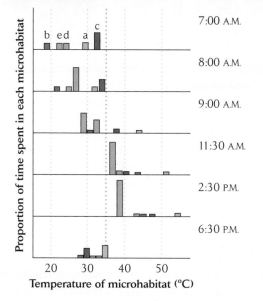

FIGURE 6.11 Temperature affects microhabitat use by cactus wrens. The microhabitats available to the wrens, from exposed ground (a) to the deep shade of trees (e), vary in their degree of thermal stress. The bars show the proportion of their time wrens spent in each microhabitat type as temperatures changed over the course of a day in late spring. From R. E. Ricklefs and F. R. Hainsworth, *Ecology* 49:227–233 (1968). Photo by R. E. Ricklefs.

Although an adult cactus wren can move without restraint to any part of its habitat, its nest is fixed in place, and wren chicks cannot move among microhabitats until they are old enough to leave the nest. The microenvironment of the nest must therefore be within the tolerance range of chicks at all times. Cactus wrens appear to achieve this both by choosing particular nest sites and by orienting their nests in particular directions. Cactus wrens build untidy, enclosed nests—bulky, somewhat haphazardly constructed balls of grass—with side entrances. Of course, once a nest is built, its position and orientation cannot be changed. For a month and a half, from the time the first egg is laid until the young fly off, the nest must

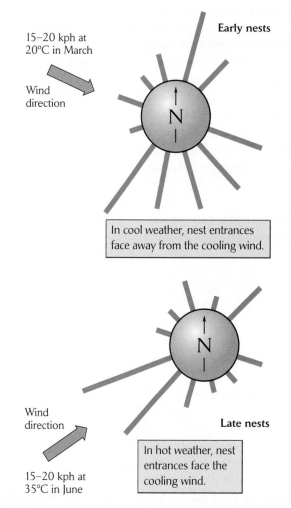

FIGURE 6.12 The orientation of cactus wren nest entrances changes over the breeding season. The lengths of the bars represent relative numbers of nests with each orientation. After R. E. Ricklefs and F. R. Hainsworth, *Condor* 71:32–37 (1969).

Phenotypic plasticity allows individuals to adapt to environmental change

Virtually all attributes of an individual are affected by environmental conditions and by the response of the individual to those conditions. The observed relationship between the phenotype of an individual and the environment is referred to as a **reaction norm** (Figure 6.13).

Some reaction norms are a simple consequence of the influence of the physical environment on life processes. The same genotype may result in different phenotypes under different conditions. Heat energy, for example, accelerates most life processes. This response to heat explains why caterpillars of the swallowtail butterfly *Papilio canadensis* grow faster at higher temperatures. The relationship between temperature and an individual's growth rate describes the reaction norm of growth rate with respect to temperature for that individual. However, individuals of the same butterfly species from Michigan and from Alaska exhibit different relationships between growth rate and temperature (Figure 6.14). In one experiment, larvae from Alaskan populations grew more rapidly than those from Michigan populations at low temperatures, and larvae from Michigan grew more rapidly than those from Alaskan populations at high temperatures, as one might have predicted from the temperatures typically found in their environments during the growing season. This finding indicates that reaction norms may be modified by evolution to improve performance under the particular conditions experienced by a population, as shown diagrammatically in Figure 6.15.

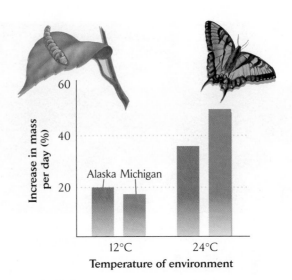

FIGURE 6.14 The reaction norms of populations adapted to different environments may differ. Caterpillars of the swallowtail butterfly *Papilio canadensis* were obtained from populations in Alaska and in Michigan and reared on balsam poplar leaves in the laboratory at temperatures of 12°C and 24°C. Although caterpillars from both environments grew more rapidly at the higher temperature, caterpillars from Alaska grew more rapidly than those from Michigan at the lower temperature. After M. P. Ayres and J. M. Scriber, *Ecol. Monogr.* 64:465–482 (1994).

Acclimatization

Growing thicker fur in winter, producing smaller leaves during the dry season, increasing the number of red cells in the blood at high elevations, producing enzymes with different temperature optima or lipids that remain fluid at different temperatures—all of these are forms of adaptive phenotypic plasticity referred to as **acclimatization.** Acclimatization may be thought of as a shift in an individual's range of physiological tolerances. The body's structure and metabolic machinery are actually modified to perform better under the new conditions of temperature or water availability or elevation. Such modifications sometimes require days to weeks. Thus, acclimatization is generally useful in response to seasonal and other persistent changes in conditions. Acclimatization is reversible, so it allows organisms to follow the ups and downs of their environments. As long as the environmental change is persistent, acclimatization is a good strategy. However, increased tolerance of one extreme often brings reduced tolerance of the other.

By producing differing forms of enzymes and other molecules with different temperature optima, animals and plants can adjust their metabolism in response to prevailing environmental conditions, as described in Chapter 3. *Larrea divaricata* (creosote bush) inhabits subtropical des-

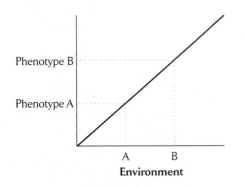

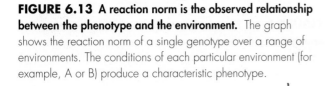

FIGURE 6.13 A reaction norm is the observed relationship between the phenotype and the environment. The graph shows the reaction norm of a single genotype over a range of environments. The conditions of each particular environment (for example, A or B) produce a characteristic phenotype.

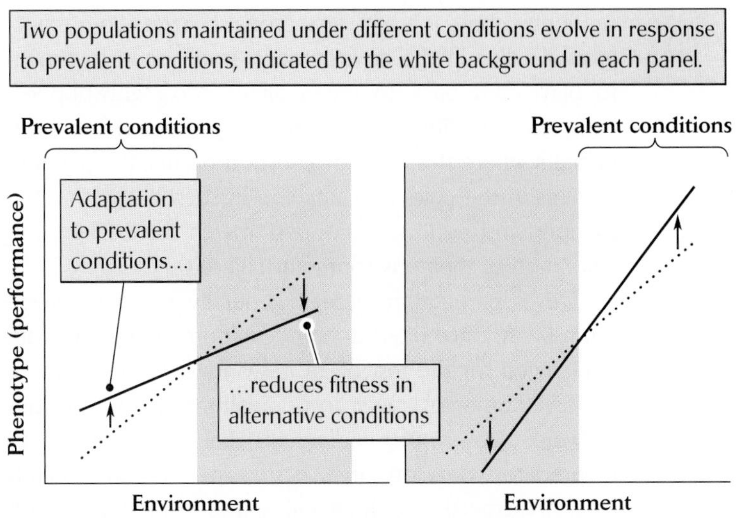

Two populations maintained under different conditions evolve in response to prevalent conditions, indicated by the white background in each panel.

Prevalent conditions

Prevalent conditions

Adaptation to prevalent conditions...

...reduces fitness in alternative conditions

Phenotype (performance)

Environment

Environment

FIGURE 6.15 Reaction norms may be modified by evolution. Reaction norms may diverge when two populations of the same species exist for long periods under different conditions. Very often an increase in performance under the prevalent conditions (indicated by the white background in each panel) is accompanied by a decrease in performance when individuals are exposed to conditions outside the population's normal range.

erts in western North America and maintains photosynthetic activity during the cool winters as well as the hot summers. In response to changing temperatures, these plants shift their temperature optima with the seasons (Figure 6.16). When individual plants are acclimatized to different temperatures in a greenhouse and then tested over a wide range of temperatures, the maximum photosynthetic rate reaches the same level in plants acclimatized to 20°C and to 45°C, but plants acclimatized to 20°C do not perform as well at 45°C as plants acclimatized to that temperature, and vice versa. In these plants, acclimatization seems to result from changes in the viscosity of the membranes directly involved in photosynthesis. These changes are brought about by shifts in the biosynthetic pathways for the fatty acids that form membrane lipids, reflecting increases or decreases in the expression of genes encoding the enzymes involved in these processes.

An organism's capacity for acclimatization often reflects the range of conditions experienced in its natural environment. Where the environment is relatively

constant, we would not expect organisms to evolve the capacity to respond strongly to environmental variation or to tolerate conditions that differ from the norm. Evolution favors economical designs, and the capacity to respond to environmental change imposes a cost on an organism. Some plants that experience narrow ranges of temperatures through the year have limited phenotypic plasticity. *Atriplex glabriuscula* is a species of saltbush native to cool coastal regions of California, where temperatures during the growing season rarely exceed 20°C. Unlike *Larrea*, if *Atriplex* is maintained at 40°C in a greenhouse, it does not increase its photosynthetic rate at high temperatures (see Figure 6.16), although it may respond to high temperatures in other ways. The physiological changes that occur during acclimatization to high temperatures also cause saltbush plants to perform less well at lower temperatures. In contrast, the thermophilic (heat-loving) species *Tidestromia oblongifolia* acclimatizes poorly to low temperatures. Prolonged maintenance of this plant at low temperatures reduces its photosynthetic rate over a wide range of leaf

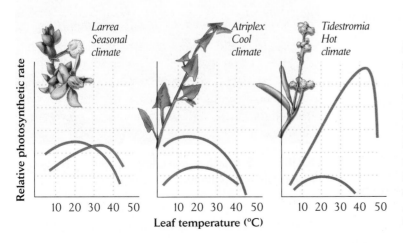

Larrea
Seasonal climate

Atriplex
Cool climate

Tidestromia
Hot climate

Relative photosynthetic rate

10 20 30 40 50   10 20 30 40 50   10 20 30 40 50

Leaf temperature (°C)

FIGURE 6.16 A species' capacity for acclimatization may reflect the range of conditions in its environment. Photosynthetic rate as a function of leaf temperature is shown for three species of plants (genera *Larrea*, *Atriplex*, and *Tidestromia*) maintained at 20°C (blue line) and at 45°C (red line). From P. W. Hochachka and G. N. Somero, *Biochemical Adaptation*, Princeton University Press, Princeton, N.J. (1984); after O. Bjorkman, M. R. Badger, and P. A. Arnold, in N. C. Turner and P. J. Kramer (eds.), *Adaptation of Plants to Water and High Temperature Stress*, Wiley, New York (1980), pp. 231–249.

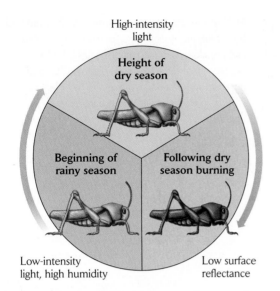

FIGURE 6.17 Irreversible developmental responses can match organisms to environmental conditions. The epidermal coloration of the grasshopper *Gastrimargus africanus* develops differently under differing seasonal conditions. The coloration of these grasshoppers responded to laboratory conditions designed to mimic light and humidity experienced during the wet season (green), the dry season (brown), and following burning (black). The color can change only during the molt between each growth stage, when a new epidermis is formed. After C. H. Fraser Rowell, *Anti-locust Bull.* 47:1–48 (1970).

temperatures. The responses of *Atriplex* to maintenance under high temperatures and of *Tidestromia* to maintenance under low temperatures appear to be generalized stress responses that allow individuals to survive under extreme conditions, rather than mechanisms that effectively broaden the range of conditions over which they can be active. It is also clear that plants living in different environments evolve different capacities to respond to environmental change.

Irreversible developmental responses

When conditions persist for long periods, the environment may influence individual development so as to modify the size or other attributes of the individual for long periods, even for its remaining lifetime. These changes are referred to as **developmental responses.** A striking example is the coloration of the African grasshopper *Gastrimargus africanus.* It is important that the color of these insects match the color of their environment if they are to avoid detection by predators. In tropical habitats with seasonal precipitation, the onset of the wet season stimulates the growth of lush, green vegetation. During the early part of the dry season, this vegetation browns and dies, often exposing reddish brown earth. As the seasonal drought intensifies, fires blacken the ground over vast areas. Con-

sequently, there is a regular seasonal progression of background color in the environment, from green to brown to black and back to green again. Most grasshoppers complete their life cycles within a single season, and so in habitats where this color progression occurs, the pigment systems in the epidermis develop in such a way that the nymphs and adult grasshoppers match the background coloration of their environment (Figure 6.17).

Developmental responses generally do not reverse themselves; once fixed during development, they remain unchanged for the rest of an individual's life (or particular developmental stage). In populations of some species of water fleas, young individuals that survive exposure to predators develop large, hard cases over their heads (called helmets) and long tail spines (Figure 6.18). Of course, these structures dissuade predators and contribute to survival, even though they are costly in terms of both the materials needed to make the helmet and the water flea's own feeding efficiency. This developmental type of phenotypic plasticity that produces a persistent structure is nonetheless useful because a particular pond is likely to contain predators, or not, for an entire growing season.

Genotype–environment interaction

The genetic makeup of the individual and the individual's environment interact to determine its phenotype. When the reaction norms of two genotypes cross for some

FIGURE 6.18 Immature water fleas exposed to predators develop defensive structures. The water flea (*Daphnia lumholtzi*) on the left has developed a protective helmet and a long tail spine. The individual on the right has not been exposed to predators and has not developed these structures. Photo by LaForsch/Tollrian.

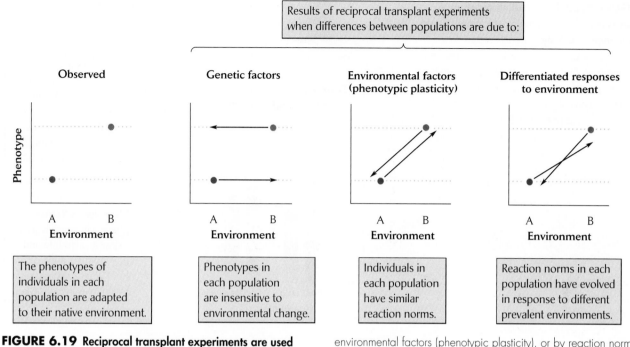

FIGURE 6.19 Reciprocal transplant experiments are used to investigate the cause of differences between populations. The results of reciprocal transplant experiments show different patterns depending on whether phenotypic differences between populations are determined solely by genetic factors, solely by environmental factors (phenotypic plasticity), or by reaction norms that have differentiated between the populations. The arrows indicate the transplantation of individuals from one environment to the other, as well as the change in the phenotype.

aspect of performance, then individuals with each genotype perform better in one environment and worse in the other, as in the case of the swallowtail butterfly (see Figure 6.14). Such a relationship is referred to as a **genotype–environment interaction** because each genotype responds differently to environmental variation.

Whether differences between populations are due to genetic differences, phenotypic plasticity, or genotype–environment interactions can often be revealed by **reciprocal transplant experiments.** Transplant studies compare the observed phenotypes of individuals kept in their native environment with those of individuals transplanted to a different environment (Figure 6.19). Reciprocal transplants involve the switching of individuals between two localities. When the phenotypic values of native and transplanted individuals are the same, we can conclude that the traits of interest are genetically determined—that is, that the trait values reflect the population of origin (genotype) rather than the local conditions (environment). When the trait values reflect where an individual is living (environment) rather than where it comes from (genotype), then the results of the experiment are consistent with phenotypic plasticity. When the individual's response to the environment depends on where it comes from, then the trait values reflect genotype—environment interactions.

ECOLOGISTS IN THE FIELD   **A reciprocal transplant experiment.** Peter Niewiarowski and Willem Roosenberg transplanted young eastern fence lizards (*Sceloporus undulatus*) between the nutrient-poor Pine Barrens of New Jersey and nutrient-rich tallgrass prairies in Nebraska. The effect of the switch on the lizards' growth rates revealed both genetic determination and phenotypic plasticity (Figure 6.20). Nebraska lizards grow about twice as fast as New Jersey lizards in their native environments. When Nebraska lizards were transplanted to New Jersey, their growth rates decreased by half—to the New Jersey level. In contrast, New Jersey lizards did not grow faster in Nebraska.

A simple interpretation of these results is that resources for growth are consistently scarcer in New Jersey than in Nebraska and that Nebraska lizards transplanted to New Jersey cannot gather resources fast enough to support their natural growth rates. Apparently, New Jersey lizards have a genetically regulated growth rate that is adapted to a low resource level. That is, they have lost the ability to modify their individual growth rates in response to higher resource levels—levels that they probably experience rarely, if ever.

It is fair to ask whether the slower growth of Nebraska lizards under New Jersey conditions is adaptive or is merely a consequence of reduced resources and more stressful conditions. If their slower growth was an example

FIGURE 6.20 The growth rates of fence lizards reveal both genetic determination and phenotypic plasticity. Juvenile eastern fence lizards (*Sceloporus undulatus*) from populations in Nebraska and New Jersey were exchanged in a reciprocal transplant experiment. Arrows indicate the growth responses of the transplanted populations. From data in P. H. Niewiarowski and W. Roosenberg, *Ecology* 74:1992–2002 (1993).

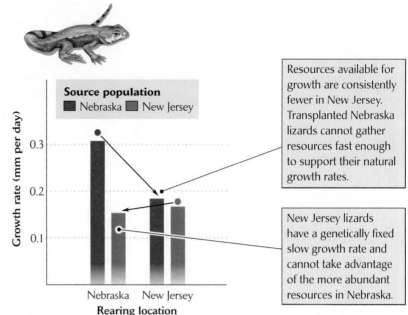

Resources available for growth are consistently fewer in New Jersey. Transplanted Nebraska lizards cannot gather resources fast enough to support their natural growth rates.

New Jersey lizards have a genetically fixed slow growth rate and cannot take advantage of the more abundant resources in Nebraska.

of adaptive phenotypic plasticity, it would reduce the negative effects of environmental change on their fitness. That is, we would expect an adaptation to compensate in some beneficial way for a change in environmental conditions. This remains an open question in the case of fence lizards in New Jersey. The study shows, however, that organisms may have little control over their rate of growth under poor conditions. If resources are not available, then individuals cannot grow rapidly and achieve large size. Nonetheless, other aspects of their lives could be modified in response to growth performance to offset their small size. ▌

MORE ON THE WEB *Ecotypes and Reaction Norms.* The response of yarrow plants to variation in growing conditions has been studied in a reciprocal transplant experiment.

MORE ON THE WEB *Phenotypic Plasticity and Contrasting Mechanisms of Growth and Reproduction in Animals and Plants.* The modular organization of plants allows them to respond to the challenge of herbivory and to changes in light and nutrients by differential growth of root and shoot tips.

MORE ON THE WEB *Rate of Phenotypic Response.* The mechanisms that organisms use to respond to their environment, as shown by the example of wing length polymorphism in water striders, must match the pattern of environmental change.

SUMMARY

1. Through the process of evolution, the traits of populations are gradually adjusted to changes in the environment. Organisms can also respond individually to changes in the environment.

2. The genotype includes all the genetic factors that determine the structure and functioning of an individual. The phenotype is the outward expression of the genotype. The phenotype can also respond directly to variation in the environment.

3. Phenotypic traits with continuous variation result from the expression of many genes influencing a single characteristic, such as body size.

4. Evolution by natural selection occurs when genetic factors influence survival and reproductive success. Evolution is the direct consequence of the existence of heritable variation in fitness in a population. The genetic characteristics of those individuals that achieve the highest reproductive success increase in frequency from generation to generation.

5. When selection is strong, an evolutionary response can occur within a few generations, provided that genes that convey higher fitness are present in the population.

6. Natural selection can be stabilizing, directional, or disruptive. Stabilizing selection maintains average or optimal phenotypes in a population. Directional selection shifts the phenotype toward a new optimum.

7. When individuals with extreme phenotypes have the highest fitness, selection can be disruptive. Disruptive selection increases genetic and phenotypic variation and can create a bimodal distribution of phenotypes in a population.

8. The evolution of melanism in populations of the peppered moth (*Biston betularia*) illustrates natural selection and the evolutionary response over a short period.

9. The branch of ecology known as population genetics has produced mathematical models that allow us to predict changes in gene frequencies.

10. Individual organisms can respond to changes in their environments by altering their behavior, physiology, or morphology. For example, animals select microhabitats whose physical conditions fall within a suitable range, thereby optimizing their relationship to the environment.

11. The capacity of the phenotype to respond to the environment is referred to as phenotypic plasticity. The observed relationship between the phenotype of an individual and the environment is called a reaction norm.

12. Acclimatization is a shift in an individual's range of physiological tolerances in response to changes in the environment. It involves reversible changes in structure or metabolic pathways. Such changes require longer periods (usually days or weeks) than behavioral changes. Acclimatization plays a prominent role in responses of long-lived organisms to seasonal change.

13. Different environmental conditions during development may lead to different characteristic, irreversible structures and appearances. Such developmental responses result from the interaction between an organism and its environment during its growth.

14. Different reaction norms in populations from different environments result from genotype–environment interactions, which show that phenotypic plasticity is subject to evolutionary change to benefit individuals in their particular environments.

REVIEW QUESTIONS

1. Why is it essential that traits be inherited for evolution to occur?

2. The insecticide DDT has been widely used to control the mosquitoes that carry malaria. How would you explain the fact that many mosquito populations are now resistant to DDT?

3. What are the sources of genetic variation, and why is genetic variation important to evolution?

4. Compare and contrast evolution by artificial selection with evolution by natural selection.

5. How should stabilizing selection and disruptive selection affect the magnitude of phenotypic variation from one generation to the next?

6. Suppose that rapid environmental change—for example, change caused by human activities—exerts very strong selection on a population. Contrast the changes that one might observe in allele frequencies and in population size.

7. Individual cactus wrens orient their nests in one direction during the cool spring months, but in a different direction during the hot summer months. What does this observation suggest about the kinds of traits that selection has acted on during the evolution of this behavior?

8. Why are organisms living in variable environments more likely to evolve phenotypically plastic traits than organisms living in relatively constant environments?

SUGGESTED READINGS

Dodson, S. 1989. Predator-induced reaction norms. *BioScience* 39:447–452.

Majerus, M. E. N. 1998. *Melanism: Evolution in Action.* Oxford University Press, Oxford, New York.

Palumbi, S. R. 2001. *The Evolution Explosion: How Humans Cause Rapid Evolutionary Change.* Norton, New York and London.

Quayle, H. J. 1938. The development of resistance to hydrocyanic acid in certain scale insects. *Hilgardia* 11:183–210.

Ricklefs, R. E., and F. R. Hainsworth. 1968. Temperature dependent behavior of the cactus wren. *Ecology* 49:227–233.

Schlichting, C. D., and M. Pigliucci. 1998. *Phenotypic Evolution: A Reaction Norm Perspective.* Sinauer Associates, Sunderland, MA.

Schluter, D. 1993. Adaptive radiation in sticklebacks: Size, shape and habitat use efficiency. *Ecology* 74:699–709.

Via, S., et al. 1995. Adaptive phenotypic plasticity: Consensus and controversy. *Trends in Ecology and Evolution* 10:212–217.

Zuk, M., J. T. Rotenberry, and R. M. Tinghitella. 2006. Silent night: Adaptive disappearance of a sexual signal in a parasitized population of field crickets. *Biology Letters* 2:521–524.

CHAPTER 7

Life Histories and Evolutionary Fitness

A remarkable fact about reproductive success is that the end result is always nearly the same. That is, each individual, on average, produces one offspring that lives to reproduce. This must be so, for otherwise populations would either dwindle rapidly to extinction because individuals failed to replace themselves, or they would grow out of all bounds.

Nonetheless, how organisms grow and produce offspring varies in all imaginable ways. A female sockeye salmon, after swimming up to 5,000 km from her Pacific Ocean feeding ground to the mouth of a coastal river in British Columbia, faces another 1,000 km upriver journey to her spawning ground. There she lays thousands of eggs, and then promptly dies, her body wasted from the exertion. A female African elephant produces a single offspring at a time at intervals of several years, lavishing intense care on her baby until it is old enough and large enough to fend for itself in the world of elephants (Figure 7.1). Thrushes start to reproduce when they are 1 year old, and may produce several broods of three or four chicks each year, but rarely live beyond 3 or 4 years. Storm petrels, which are seabirds the size of thrushes, do not begin to reproduce until they are 4 or 5 years old, and rear at most a single chick each year, but may live for 30 or 40 years.

The schedule of an individual's life—age at maturity, number of offspring, life span—makes up what ecologists call the **life history** of the individual. In effect, life histories include all the behavioral and physiological adaptations of organisms, and all the individual responses of organisms to their environments, because all of these ultimately contribute to reproductive success. Life histories are complex phenomena influenced by physical conditions, food supply, predators, and other aspects of the environment, and constrained by the general body plan and lifestyle of the individual.

FIGURE 7.1 Organisms differ in their degree of parental investment. Elephant mothers produce one offspring at a time and care for it for many years. Photo by R. E. Ricklefs.

CHAPTER CONCEPTS

- Trade-offs in the allocation of resources provide a basis for understanding life histories

- Life histories vary along a slow–fast continuum

- Life histories balance trade-offs between current and future reproduction

- Semelparous organisms breed once and then die

- Senescence is a decline in physiological function with increasing age

- Life histories respond to variation in the environment

- Individual life histories are sensitive to environmental influences

- Animals forage in a manner that maximizes their fitness

As we have seen, organisms are generally well suited to the conditions of their environments. Their form and function are influenced by physical and biological factors, both through the evolutionary responses of populations and through the behavioral and physiological responses of individuals to temporal and spatial variation in their environments. Whether modifications of form and function have evolved or result from individual responses, we presume that they are adaptive and that they increase the reproductive success of individuals. The close color matching of peppered moths and grasshoppers to their backgrounds makes sense when one understands that these insects are eaten by visually hunting predators. Life histories, too, are shaped by natural selection. An organism's life history represents a solution to the problem of allocating limited time and resources so as to achieve maximum reproductive success.

For decades, ecologists have used observation, mathematical modeling, and experimentation to explore why life histories differ so much among species. Songbirds in the tropics, for example, lay fewer eggs at a time (two or three, on average) than their counterparts that breed

at higher latitudes (generally four to ten, depending on the species) (Figure 7.2). In 1947, David Lack of Oxford University was the first person to recognize that these differences in reproductive strategy had evolved in response to differences between tropical and temperate environments. Lack recognized that birds could increase their overall reproductive success, and hence their evolutionary fitness, by increasing the size of their broods, unless reduced survival of offspring in larger broods offset this advantage. He hypothesized that the ability of adults to gather food for their young was limited and, accordingly, that chicks in large broods, where too many hungry mouths clamored for the food brought by the parents, would be undernourished and survive poorly. Lack further noted that at temperate and Arctic latitudes, birds had longer days in which to gather food during summer, when they produce their young. Therefore, it made sense that birds at high latitudes could rear more offspring than birds breeding in the tropics, where day length remains close to 12 hours year-round.

Lack made three important points. First, he stated that because life history traits, such as the number of eggs in a

(a)

(b)

FIGURE 7.2 Organisms differ in the number of offspring they produce. Birds breeding at high latitudes typically lay larger clutches of eggs than tropical species. (a) This five-egg clutch was laid by a snow bunting (*Plectrophenax nivalis*) in Alaska. (b) This two-egg clutch was laid by a red-capped manakin (*Pipra mentalis*) in Panama. Photos by R. E. Ricklefs.

clutch (the set of eggs laid together in a nest), contribute to reproductive success, they also influence evolutionary fitness. Second, he demonstrated that life histories vary consistently with respect to factors in the environment, such as the length of time available for feeding young. This observation suggested that life history traits are molded by natural selection. Third, he proposed a hypothesis that could be subjected to experimental testing. In the case of clutch size, Lack suggested that the number of offspring that parents can rear is limited by food supply. To test this idea, one could add eggs to nests to create enlarged clutches and broods. According to Lack's hypothesis, parents should be unable to rear added chicks because they cannot gather the additional food required by a larger brood.

This experiment has been conducted many times over the last several decades, usually with the result predicted by Lack. For example, the Swedish ecologist Gören Hogstedt manipulated the clutch sizes of European magpies by moving eggs between nests, enlarging some clutches and reducing the sizes of others. His results showed that magpies lay a clutch that corresponds to the maximum number of offspring that a pair can rear to the point at which they leave the nest, or *fledge*. Either adding or subtracting eggs resulted in fewer offspring fledged (Figure 7.3), just as Lack had predicted.

Because evolution tends to increase the average fitness of individuals within a population, you might wonder why populations do not continually increase in numbers, at an ever faster rate, as evolution improves the match of the gene pool to the environment. The answer is that selection favors individuals with the highest fitness *rela-tive to other individuals in the same population*. Evolution is a property of a single population, but other populations are also evolving to exploit their food resources or to avoid being caught by their predators. Consequently, relationships among species are kept more or less in balance by evolution taking place within each of the populations, and no species gains an advantage, on average, over the long run. We shall return to the regulation of population size later in this book. In this chapter, we shall focus on the factors that shape the life histories of organisms.

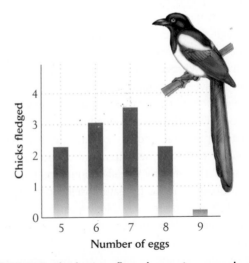

FIGURE 7.3 Clutch size reflects the maximum number of offspring the parents can rear. The average European magpie (*Pica pica*) clutch size of seven was manipulated by adding or removing eggs to make up clutches of five to nine eggs. The most productive clutch size was seven. After G. Hogstedt, *Science* 210:1148–1150 (1980).

Trade-offs in the allocation of resources provide a basis for understanding life histories

Organisms have limited time, energy, and nutrients at their disposal. Adaptive modifications of form and function either increase the resources available to individuals or allow them to use those resources to their best advantage. Many such modifications involve **trade-offs,** meaning that time, energy, or materials devoted to one body structure, physiological function, or behavior cannot be allotted to another. Therefore, the adaptations of an organism must solve the problem of **allocation:** given that time and resources are limited, how can the organism best use them to achieve its maximum possible fitness?

Practical solutions to the allocation problem depend on how a change in any given structure or function affects the fitness of the individual. When the modification of a trait influences several components of survival and reproduction, as is often the case, the evolution of that trait can be understood only by considering the entire life history strategy. For example, an increase in the number of seeds produced by an oak tree may contribute to fitness by increasing the number of offspring. But such a modi-

fication may also reduce the survival of seedlings (if seed size is reduced to make more of them), the survival of adult trees (if resources are shifted from root growth to support increased seed production), or subsequent seed production (if seed production in one year reduces adult tree growth, and therefore size, or nutrient reserves in subsequent years).

From an evolutionary point of view, individuals exist to produce successful progeny—as many as possible. Reproduction involves many allocation problems: when to begin to breed, how many offspring to have at one time, how much care to bestow upon them. How the resolution of these problems influences an individual's survival and reproduction at each age governs the evolution of the life history. Each life history has many components, the most important of which are age at **maturity**—that is, first reproduction; **parity,** or number of episodes of reproduction; **fecundity,** or number of offspring produced per reproductive episode; and **longevity** (Figure 7.4).

MORE ON THE WEB *Metabolic Ceilings.* Can organisms increase overall performance without trading one function off against another?

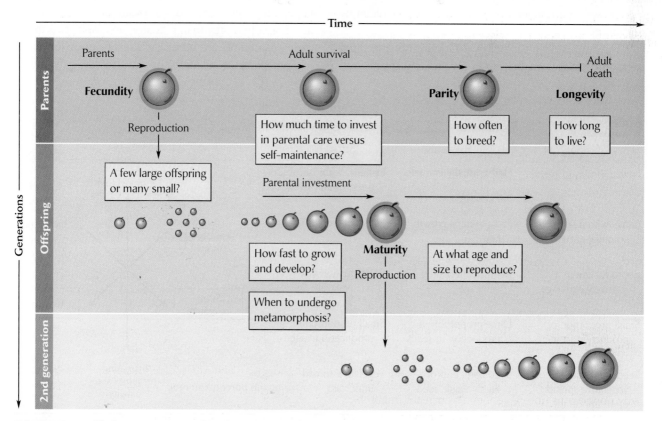

FIGURE 7.4 A life history is governed by the resolution of allocation problems influencing survival and reproduction.

Life histories vary along a slow–fast continuum

Life history traits vary widely among different species and even among different populations of the same species. Two points can be made about this variation. First, life history traits often vary consistently with respect to habitat or conditions in the environment. Seed size, for example, is generally larger among plants of the forest than among plants of grasslands. Second, variation in one life history trait is often correlated with variation in other traits. For example, the number of independent offspring produced each year is positively correlated with adult annual mortality rate. As a result, variations in many life history characteristics are organized together along a single continuum of values.

We can refer to one extreme as the "slow" end of the spectrum. At this extreme, organisms such as elephants, albatrosses, giant tortoises, and oak trees exhibit long life, slow development, delayed maturity, low fecundity, and high parental investment. At the fast end of the spectrum are mice, fruit flies, and weedy plants, which exhibit the opposite life history characteristics. The correlation between mortality and fecundity across species must in part reflect the fact that if populations are to persist, births and deaths, on average, must balance. In addition, however, these life history traits can be modified by evolution.

The English plant ecologist J. P. Grime emphasized the relationship between the life history traits of plants and certain environmental conditions. He envisioned variation in life history traits as lying between three extreme apexes, like points of a triangle, and called plants with life histories at these extremes "stress tolerators," "ruderals" (weeds), and "competitors" (Table 7.1). As their name implies, stress tolerators live under extreme environmental conditions. They grow slowly and conserve resources. Because seedling establishment is difficult in stressful environments, these plants rely heavily on vegetative spread. Where conditions for plant growth are more favorable, ruderals and competitors occupy opposite ends of a spectrum of habitat disturbance. Ruderals, which colonize disturbed patches of habitat, exhibit rapid growth, early maturation, high reproductive rates, and easily dispersed seeds. These traits enable them to reproduce quickly and disperse their progeny to other disturbed sites—"growing like weeds"—before being overgrown by superior competitors. Competitors tend to grow large, mature at large sizes, and exhibit long life spans. The competitor life history therefore requires a more constant and predictable environment for its success.

More generally, larger organisms tend to have longer life spans and lower reproductive rates than smaller organisms. This pattern is partly a function of the physical and physiological consequences of body size and partly a result of the different environmental factors that affect large and small organisms. For example, small organisms are more often prey than predator, and rapid growth and early maturation are generally advantageous for escaping predation. In addition, because so few offspring of

---

TABLE 7.1   Typical life histories of plants in environments with different selective factors

Competitors	Ruderals	Stress tolerators
Herbs, shrubs, or trees	Herbs, usually annuals	Lichens, herbs, shrubs, or trees; usually evergreen
Large, with a fast potential growth rate	Fast potential growth rate	Potential growth rate slow
Reproduction at a relatively early age	Reproduction at an early age	Reproduction at a relatively late age
Small proportion of production to seeds	Large proportion of production to seeds	Small proportion of production to seeds
Seed bank sometimes, vegetative spread often important	Seed bank and/or highly vagile seeds	Vegetative spread important

Competitors

Increasing disturbance   Increasing stress

Stress tolerators   Ruderals

Increasing resources and stability

Source: J. P. Grime, *Plant Strategies and Vegetation Processes*, Wiley, Chichester (1979).

prey organisms survive, selection favors the production of many, small young rather than a small number of large young.

MORE ON THE WEB *Allometry and the Consequences of Body Size for Life Histories.* Metabolic rate, fecundity, and mortality tend to decrease with increasing body mass.

Life histories balance trade-offs between current and future reproduction

Life history traits to some extent result from the physiological, behavioral, and developmental responses of individuals, but these responses are not without constraint. As we have seen, a life history represents the best resolution of conflicting demands on the organism. For example, watching more carefully for predators takes time away from feeding, and so while this tactic may increase survival, the watchful individual may produce fewer young. Each of these responses affects other aspects of an individual's life. Because breeding takes time and resources from other activities and entails risks, investment in offspring generally diminishes the survival of parents. In many cases, rearing offspring drains a parent's resources so much that it produces fewer offspring later.

An optimized life history is one that resolves conflicts between the competing demands of survival and reproduction to the best advantage of the individual in terms of its fitness. A critical effort in the study of life histories has been to understand the fitness consequences of changing the allocation of limited time and resources to such competing functions.

ECOLOGISTS IN THE FIELD **The cost of parental investment in the European kestrel.** Although it is widely believed that trade-offs between functions constrain life histories, demonstrating such trade-offs has proved difficult. In some cases, experimentally manipulating individual components of the phenotype will expose such trade-offs. As we saw above, by adding eggs to the nests of birds, investigators can increase competition among the young for food brought by the parents. The chicks with more competing siblings grow more slowly, and fewer survive to reach adulthood. Consequently, production of offspring is often greatest from broods of intermediate size. Sometimes, however, having more mouths to feed stimulates parents to hunt harder for food for their chicks. In this case, an artificially enlarged brood might result in higher

reproductive success, but impose a cost on parents in the form of smaller future broods or decreased survival.

Cor Dijkstra, Serge Daan, and their colleagues at the University of Groningen in the Netherlands have conducted extensive studies of the ecology and reproduction of European kestrels (*Falco tinnunculus*). These small falcons search for voles and shrews in open fields, often by hovering high overhead. Thus, kestrel foraging requires a high rate of energy expenditure, but small mammals are so abundant that kestrel pairs normally can catch enough prey to feed their brood in a few hours each day. Kestrels lay an average of five eggs. When the broods in a sample of nests were about a week old, the investigators either removed two chicks or added two chicks, creating artificially reduced and enlarged broods. They also left some brood sizes unchanged to control for disturbance caused by the manipulations themselves. Parents that were provided extra chicks worked harder to feed their enlarged broods, increasing their foraging time and energy expenditure.

The fruits of the kestrels' increased parental investment were an increase in the average number of chicks successfully fledged per brood. However, in spite of the increased efforts of their parents, chicks in the enlarged broods were somewhat undernourished, and only 81% survived to fledging, compared with 98% in control and reduced broods. Consequently, the extra parental investment netted the harder-working parents only an extra 0.8 chick per nesting attempt, and this gain may have been diminished by the later deaths of some of the underweight fledglings. A more telling effect of the increased parental effort of adults with enlarged broods was seen in the lower survival of these adults to the next breeding season (Figure 7.5). Clearly, at some level of parental investment, the law of diminishing returns sets in, and further parental effort reduces the possibility of future reproductive success more than it increases the success of the present brood. ▌

Most allocation problems concerning life histories can be phrased in terms of three questions: When should an individual begin to produce offspring? How often should it breed? How many offspring should it attempt to produce in each breeding episode? The variation among species in these life history traits illustrates the different ways of resolving the fundamental trade-off between fecundity and adult growth and survival—that is, between present and future reproduction.

Age at first reproduction

When should an animal or plant begin to breed? Long-lived organisms typically begin to reproduce at an older age than short-lived ones. For example, adults of most albatrosses have very high annual survival rates, up to 97%

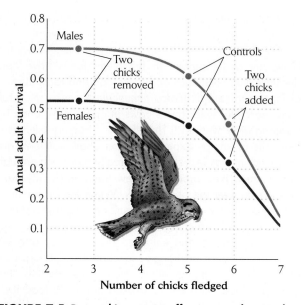

FIGURE 7.5 Parental investment affects parental survival.
The annual survival rate of male and female European kestrels (*Falco tinnunculus*) was affected by the number of chicks they reared. From data in C. Dijkstra et al., *J. Anim. Ecol.* 59:269–285 (1990).

in some species, but they often do not become sexually mature until 10 years of age. Many small songbirds, with survival rates of only 50% each year, breed for the first time at 1 year of age, or less. Why should this be so? At every age, an individual may either reproduce or abstain from reproduction. These life history alternatives weigh the benefits and costs of breeding at a particular age. The benefits appear as an increase in fecundity at that age. The costs may appear as reduced survival to older ages or reduced fecundity at older ages. Natural selection favors the age at maturity that results in the greatest number of offspring over the lifetime of the individual.

Consider the following hypothetical example. A type of lizard continues to grow only until it reaches sexual maturity (a pattern referred to as **determinate growth**). Its fecundity varies in direct proportion to its body size at maturity. Suppose that the number of eggs that a female lays per year increases by 10 for each year that she delays reproduction. Thus, individuals that begin to breed in their first year produce 10 eggs that year and the same number each year thereafter; individuals first breeding in their second year produce 20 eggs per year; individuals maturing in their third year produce 30 eggs; and so on. Now, when we add up the total number of eggs that a female lays during her lifetime, we find that the optimum age at maturity—that which maximizes lifetime reproduction—varies in direct proportion to the life span (Table 7.2). For example, for a lizard with a life span of 3 years, maturing at 2 years results in the greatest lifetime reproduction. When the life span is 7 years, 4 years is the best age to mature.

Evolutionary optimization of life histories must resolve the trade-off between reproducing now and surviving to reproduce later. Individuals that delay breeding avoid the risks and resource requirements of preparations for reproduction, such as courtship, nest building, and migration to breeding areas. Presumably, life experience gained with age also reduces the risks associated with breeding. It may also increase foraging efficiency and hence the number of offspring that parents can provision. All these factors favor delayed reproduction. Humans are extreme in this regard, as we require many years to acquire the physical and social skills, and the knowledge about our environments (imagine that you are a hunter–gather now), required to be successful adults. Tending to offset the advantages of delayed reproduction are factors that reduce the expectation of future reproduction. These factors include high

TABLE 7.2	Total eggs produced by individuals in a hypothetical population as a function of life span and age at first reproduction							
Age at first reproduction (years)	Life span (years)							
	1	2	3	4	5	6	7	8
1	**10***	**20**	30	40	50	60	70	80
2	0	**20**	**40**	**60**	80	100	120	140
3	0	0	30	**60**	**90**	**120**	150	180
4	0	0	0	40	80	**120**	**160**	**200**
5	0	0	0	0	50	100	150	**200**
6	0	0	0	0	0	60	120	180

*Bold type indicates most productive ages at first reproduction for a given life span.

predation rates, encroaching senescence at old age, and, for organisms with a life span of a year or less that live in seasonal environments, the end of the reproductive season.

MORE ON THE WEB *Annual and Perennial Life Histories.* Why should some plants grow, reproduce, and die within one season while others persist from year to year?

The trade-off between fecundity and survival

We have just seen that increasing the number of offspring produced today can reduce the number produced tomorrow. Natural selection should optimize the trade-off between present and future reproduction. But what factors influence the resolution of this conflict? Our intuition tells us that high mortality rates for adults should tip the balance in favor of reproducing early. Conversely, when the life span is potentially long, adults should reduce their investment in today's young so as not to jeopardize future reproduction. We can demonstrate the truth of this idea with a simple proof.

We assume that trade-offs obey the law of diminishing returns, which means that an adult can increase its fecundity with added reproductive investment, but the benefits decrease, and the costs increase, with each additional increment of effort. The effect of diminishing returns is shown graphically in Figure 7.6a, in which fecundity and adult survival are each plotted as a function of reproductive investment. By plotting the values for adult survival

as a function of fecundity for each level of reproductive investment (Figure 7.6b), we obtain a trade-off curve for survival and fecundity as a function of reproductive investment. Notice that this curve is outwardly concave because of the law of diminishing returns. Then compare the curve to the relationship between adult survival and fecundity for the European kestrel shown in Figure 7.5. Both adult survival and fecundity contribute to fitness, so different points on the trade-off curve correspond to different fitness values. The question is, which point on the trade-off curve represents the greatest fitness, and hence the optimal resolution of the survival–fecundity conflict?

The genotype representing the point of highest fitness on the trade-off curve will be the one that makes the greatest contribution to the gene pool of the population in the future. An individual's contribution to the gene pool in the following breeding period is the sum of its probability of personal survival (S) over that interval and the number of its offspring that are recruited into the breeding population. For simplicity, let's assume that individuals mature at the end of their first year and breed once each year. Thus, an adult's contribution to next year's population through its production of offspring is the product of the number of offspring produced (B) and their survival to one year of age (S_0). The adult's fitness (F) can be expressed as

$$F = S + S_0B$$

We now partition adult survival into two components, one directly related to reproduction (S_R) and the other independent of reproduction (S_N). The portion related

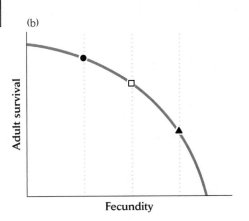

FIGURE 7.6 The effects of reproductive investment on fecundity and adult survival produce a convex trade-off curve. (a) Both adult survival of reproductive risk and fecundity show diminishing returns with increasing reproductive investment.

(b) When adult survival is plotted as a function of fecundity, the resulting trade-off curve is outwardly convex. Combinations of adult survival and fecundity at different levels of reproductive investment are portrayed along the trade-off curve.

to reproduction includes the chance of mortality due to risks such as the stress of breeding, greater susceptibility to parasites and disease because of a depressed immune system, and increased exposure to predation as a parent forages to feed offspring. The portion that is independent of reproduction includes the chance of mortality due to risks such as predation and starvation during the nonbreeding period of the year. Now, an individual's fitness may be expressed as

$$F = S_N S_R + S_0 B$$

[contribution to future population] =
     [adult survival] + [recruitment of offspring]

This expression can be rearranged to

$$S_N S_R = F - S_0 B$$

and then to

$$S_R = \frac{F}{S_N} - \left(\frac{S_0}{S_N}\right)B$$

which describes the relationship between values of S_R and B that have equal fitness F. This relationship, which is called the fitness function, is a straight line with a slope of $-S_0/S_N$; lines farther from the origin of the graph ($S_0 = 0$ and $S_N = 0$) represent lines of increasing fitness (Figure 7.7a). S_N and S_0 are determined by attributes of each species and its environment that are unrelated to breeding.

We are now ready to find the point of highest fitness on the trade-off curve relating S_R to B. Imagine superimposing the trade-off curve over Figure 7.7a. The point of highest fitness on the trade-off curve is the point just touched by the fitness function having the highest value of F. That tells us what the best trade-off is for species having a particular ratio of adult to first-year survival (S_N and S_0). But our goal is to find out how adult life span influences the point of highest fitness, so we need to consider what happens when S_N and S_0 vary. Now, when we compare two species with the same S_R–B trade-off, but different values of S_N and S_0, the optimum position on the trade-off curve depends on the ratio of S_0 to S_N. As you can see in Figure 7.7b, as adult survival increases relative to offspring survival, the slope of the fitness function becomes shallower, and the optimum point shifts around to the left (lower fecundity, but higher survival) on the trade-off curve. When adult survival is low compared with survival of the young, the slope of the fitness function is steep, and the optimum shifts to a higher value of B, which implies a higher level of parental investment.

Thus, when adults have a low probability of survival from one year to the next and offspring survival is relatively good, the best strategy is to invest heavily in offspring. When offspring survival is relatively poor, a parent should produce fewer of them each year and increase its own reproductive life span. This prediction is supported by the strong inverse relationship between the average number of offspring produced and the average life span of adults observed across a wide range of organisms.

The trade-off between growth and fecundity

Many plants and invertebrates, as well as some fishes, reptiles, and amphibians, do not have a characteristic adult

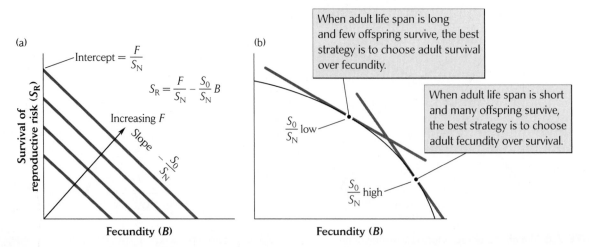

(a)

Survival of reproductive risk (S_R)

Intercept $= \dfrac{F}{S_N}$

$$S_R = \frac{F}{S_N} - \frac{S_0}{S_N} B$$

Increasing F

Slope $-\dfrac{S_0}{S_N}$

Fecundity (B)

(b)

When adult life span is long and few offspring survive, the best strategy is to choose adult survival over fecundity.

When adult life span is short and many offspring survive, the best strategy is to choose adult fecundity over survival.

$\dfrac{S_0}{S_N}$ low

$\dfrac{S_0}{S_N}$ high

Fecundity (B)

FIGURE 7.7 The optimum trade-off between adult survival and fecundity is influenced by the survival rates of adults and young. (a) The parallel lines represent combinations of survival of reproductive risk (S_R) and fecundity (B) with equal fitness (F) given the survival rate of juveniles (S_0) and the survival of nonreproductive risk by adults (S_N). (b) When superimposed on a trade-off curve (see Figure 7.6b), these fitness functions determine the optimum levels of reproductive investment.

size. They grow, although often at a continually decreasing rate, throughout their adult lives (a pattern referred to as **indeterminate growth**). Fecundity is directly related to body size in most species with indeterminate growth: the larger a female grows, the more eggs she can produce. Because egg production and growth draw on the same resources of assimilated energy and nutrients, however, increased fecundity during one year must be weighed against reduced growth, and thus reduced fecundity, in subsequent years. Accordingly, organisms with a long life expectancy should favor growth over fecundity during each year. For organisms with less chance of living to reproduce in future years, allocating limited resources to growth rather than eggs wastes potential fecundity at a young age.

Consider two hypothetical female fish, both of which weigh 10 g at sexual maturity, but which allocate resources to growth and reproduction differently thereafter. Both gather enough food each year to reproduce their weight in new tissue or eggs. Fish A allocates 20% of its production to growth and 80% to eggs, whereas fish B allocates half of its production to growth and half to eggs. In this model, fecundity is directly proportional to body size. Calculations of growth, fecundity, and cumulative fecundity (Table 7.3) show that for fish living 4 or fewer years, on average, high fecundity and slow growth result in greater cumulative weight of eggs produced, whereas for fish living longer than 4 years, low fecundity and rapid growth are more productive. Adult longevity, therefore, determines the optimal allocation of resources between growth and reproduction.

Many studies have compared populations in high- and low-mortality environments to demonstrate how mortality rates affect the evolution of life histories. One of the most striking of these studies was conducted by David Reznick, of the University of California at Riverside, and his colleagues on the guppy *Poecilia reticulata,* which is common in streams in Trinidad. In the lower reaches of these streams, the guppies live with a number of predatory fish species, such as the pike cichlid *Crenocichla alta,* which preys on adult guppies, and the smaller killifish *Rivulus hartii,* which preys primarily on juvenile guppies. At higher elevations in the mountains of northern Trinidad, one encounters numerous small waterfalls that the guppies have been able to ascend, but which block the upstream movement of their predators. Thus, above the waterfalls, the guppies live in a relatively predator-free environment.

Reznick and his colleagues found that below the waterfalls, where the guppies are exposed to intense predation, males mature at a smaller size, and females allot more of their body mass to reproduction, than in populations above the falls (Figure 7.8). Because the offspring are also subject to high predation rates, females below the waterfalls also produce more, smaller offspring. These differences are consistent with predictions from the life history hypotheses we have discussed.

What is unusual about the guppy story is that it was possible to transplant predators from the lower reaches of the streams to the areas above the waterfalls to determine whether the imposition of a strong selective force would shift the life history strategies of the local populations.

TABLE 7.3	Numerical comparisons of the strategies of slow growth/high fecundity and rapid growth/low fecundity in two hypothetical fish					
	Years					
Characteristic	1	2	3	4	5	6
Slow growth/high fecundity						
Body weight	10	12	14.4	17.3	20.8	25.0
Growth increment	2	2.4	2.9	3.5	4.2	5.0
Weight of eggs	8	9.6	11.5	13.8	16.6	20.0
Cumulative weight of eggs	8	17.6	29.1	42.9	59.5	79.5
Rapid growth/low fecundity						
Body weight	10	15	22.5	33.8	50.7	76.1
Growth increment	5	7.5	11.3	16.9	25.4	38.1
Weight of eggs	5	7.5	11.3	16.9	25.4	38.1
Cumulative weight of eggs	5	12.5	23.8	40.7	66.1	104.2

Note: All weights in grams. Body weight + growth increment = next year's body weight. Cumulative weight of eggs to last year + weight of eggs = cumulative weight of eggs to this year. Growth increment and weight of eggs in each year are equal to the body weight.

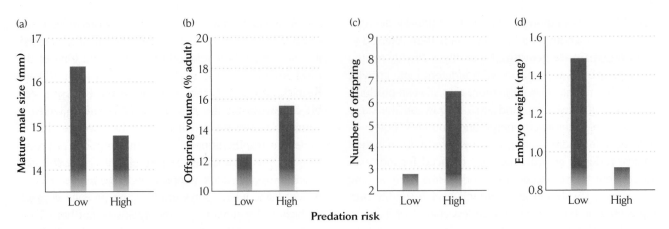

FIGURE 7.8 Predation on guppies selects for shifts in life histories. Populations exposed to strong predation in streams below waterfalls mature at an earlier age and smaller size (a), allocate more of their resources to reproduction (b), and produce more (c) and smaller (d) offspring than populations that are free from predation. After D. N. Reznick et al., *Am. Nat.* 147:319–338 (1996).

Indeed, within a few generations, the life histories of the populations above the waterfalls experimentally subjected to predation came to resemble those of the populations below. This finding not only confirmed the basic ideas about the optimization of life history patterns, but also demonstrated the strength of predation as a selective force in evolution. Humans, of course, exert just such selection on most of the populations we exploit for food. Fisheries tend to take the oldest and largest individuals from a population first, selecting for early maturation at a small size. Trophy hunters apply strong selection for reduced antler size in deer and elk and reduced tusk size in elephants. Heavy mortality of any kind in adults reduces the value of future reproduction and selects for increased investment in present reproduction, as we have seen in populations of Trinidadian guppies.

Semelparous organisms breed once and then die

Unlike most fish, which breed repeatedly, some species of salmon grow rapidly for several years, then undertake a single episode of breeding. During this one burst of reproduction, females convert a large portion of their body mass into eggs, males waste away fighting and courting females, and both sexes die shortly after spawning. Because salmon make such a great effort to migrate upriver to reach their spawning grounds, it might be to their advantage to make the trip just once, rather than assuming this cost repeatedly. Then, after arriving at their breeding areas, because they will have no chance to reproduce again, they should produce as many eggs as possible, even if this supreme reproductive effort results in the wastage of most body tissues and ensures death. This pattern is called **pro-grammed death** because death is a direct consequence of adaptation to maximize reproductive success.

Ecologists call the salmon life history pattern **semelparity.** This term comes from the Latin *semel* ("once") and *pario* ("to beget"). The opposite of semelparity is **iteroparity,** from *itero* ("to repeat"). Semelparity is not the same as an annual life history, which refers to completion of the life cycle within a single year. For one thing, annuals can undertake more than one episode of reproduction, or even prolonged continuous reproduction, within a season; for another, like perennial species (those that potentially live for many years), semelparous individuals must survive at least one nonbreeding season—and usually many—before maturing, reproducing, and then dying. Semelparity is rare among long-lived animals and plants.

The best-known cases of semelparous reproduction in plants occur in bamboos and agaves, two distinctly different groups, although this life history pattern has been reported even for some tropical forest trees. Most bamboos are plants of tropical or warm temperate climates that form dense stands in disturbed habitats. Reproduction in bamboos does not appear to require substantial preparation or resources, as in the case of salmon. But bamboos probably have few opportunities for successful seed germination. Once established, a bamboo plant increases by asexual reproduction (vegetative growth), continually sending up new stalks until the habitat in which it germinated is fairly packed with bamboo. Only then, when vegetative growth becomes severely limited, does the plant benefit from producing seeds, which can colonize newly disturbed sites. In many species of bamboo, breeding is highly synchronous over large areas, after which the future of the entire population rests with the crop of seeds. Synchronous breeding may facilitate fertil-

roots descend deep into the desert soil to tap persistent sources of groundwater; agaves have shallow, fibrous roots that catch water percolating through the surface layers of soil after rain showers, but are left high and dry during drought periods. Thus, the semelparous agaves may experience greater variation in moisture availability from year to year than the iteroparous yuccas.

Several explanations have been proposed for the occurrence of semelparous and iteroparous reproduction in plants. First, variable environments might favor iteroparity, which would reduce variation in lifetime reproductive success by spreading reproduction over both good and bad years. This tactic is referred to as **bet hedging.** However, this hypothesis can be rejected in the case of semelparous plants that live in more variable (usually drier) environments than their iteroparous relatives. Alternatively, variable environments might favor semelparity if a plant can time its reproduction to occur during a very favorable year. Storing resources and then using them for the big event makes sense. *Carpe diem:* seize the day. Semelparity should be particularly favored when the interval between good years is long. Under such conditions, plants that take advantage of good conditions when they do occur might have higher fitness. Finally, attraction of pollinators by massive floral displays might favor plants that put all their effort into one reproductive episode. For example, in the semelparous rosette plant *Lobelia telekii,* which grows high on the slopes of Mount Kenya in Africa (Figure 7.10), a doubling of inflorescence size was seen to result in a fourfold increase in seed production.

FIGURE 7.9 Agaves are semelparous plants. The Parry agave (*Agave parryi*) of Arizona grows as a rosette of thick, fleshy leaves for many years. Then it rapidly sends up a huge flowering stalk and sets fruit, after which the rosette dies. Photo by Tom Bean/DRK Photo.

ization in this wind-pollinated plant group, or perhaps it overwhelms seed predators, which cannot consume such a large crop of seeds.

The environments and habits of agaves occupy the opposite end of the spectrum from those of bamboos. Most species of agaves live in arid climates with sparse and erratic rainfall. Each agave plant grows vegetatively as a rosette of leaves produced from a single meristem over several years (the number of years varies from species to species). The plant then sends up a gigantic flowering stalk. The growth of the stalk is too rapid to be fully supported by photosynthesis or uptake of water by the roots. As a consequence, the nutrients and water necessary for stalk growth are drawn from the leaves, which die soon after the seeds are produced (Figure 7.9).

Agaves frequently live side by side with yuccas, a closely related group of plants that have a similar growth form but which are iteroparous. Yuccas typically are branched and have multiple rosettes of leaves. The root systems of agaves and yuccas also differ markedly. Yucca

FIGURE 7.10 The semelparous plant *Lobelia telekii* is found on the slopes of Mount Kenya. The giant inflorescences in the foreground are *L. telekii;* the stalked rosette plants in the background are *L. keniensis.* Courtesy of Truman P. Young.

TABLE 7.4	Ecological, life history, demographic, and reproductive traits of *Lobelia telekii* and *Lobelia keniensis* on Mount Kenya	
Trait	*Lobelia telekii*	*Lobelia keniensis*
Life history	Semelparous	Iteroparous
Habitat	Dry rocky slopes	Moist valley bottoms
Growth form	Unbranched	Branched
Reproductive output	Larger inflorescences, more seeds	Smaller inflorescences, fewer seeds
Variation in inflorescence size	Highly variable, increases with soil moisture	Relatively invariable, independent of soil moisture
Demography	Virtually no adult survival	Populations in drier sites have lower adult survival and less frequent reproduction
Variation in number of seeds per pod	Strongly positively correlated with inflorescence size	Independent of inflorescence size, positively correlated with number of rosettes
Effects of pollinators	Increased seed quality, but not seed quantity	Increased seed quality, but not seed quantity

Source: T. P. Young, *Evol. Ecol.* 4:157–171 (1990).

Comparison of *Lobelia telekii* with its iteroparous relative *L. keniensis* (Table 7.4), like the comparison between agaves and yuccas, suggests that semelparity is associated with dry habitats that are highly variable in both space and time. Presumably, infrequent conditions that support seedling establishment trigger the massive flowering episodes in these plants. In summary, semelparity appears to arise either when preparation for reproduction is extremely costly, as it is for species that undertake long migrations to breeding grounds, or when the payoff for reproduction is highly variable but favorable conditions are predictable from environmental cues.

One of the most remarkable cases of semelparity in animals is the life cycle of periodical cicadas (Figure 7.11). Many cicadas have annual life cycles, and their mating calls in the trees grace summer evenings in many parts of the Northern Hemisphere. Unlike their annual cousins, some species remain in the soil as larvae, taking nutrients from the xylem tissue of plant roots, for 13 or 17 years before emerging above ground as adults to mate. Periodical cicada emergences are marked by nearly deafening noise as the males call to attract females during their brief mating period. Not only are the life cycles of these animals very long, they are also completely synchronous within a given area, so that an entire brood emerges at the same time. The long life cycle gives the larvae time to grow to adulthood on a diet of low nutritional quality. The synchrony is probably a mechanism to overwhelm potential predators. Most of the occasional individuals that

FIGURE 7.11 Periodical cicadas (*Magicicada* sp.) during an emergence in the midwestern United States. Courtesy of Claus Holzapfel.

emerge a year late or a year early are grabbed by predators attracted to their loud mating calls. Many observers have noted that the common periods of 13 and 17 years found in periodical cicada species are prime numbers, which makes it difficult for pathogens or predators to synchronize their emergences to coincide with the cicadas.

Perhaps you are wondering how cicadas can count to 17. Do they count the years by the warming and cooling of the soil, or by their hosts' physiological cycles? Richard Karban and his colleagues at the University of California at Davis performed a clever experiment by growing 17-year periodical cicadas on peach trees that had been artificially selected to produce two fruit crops annually, dropping their leaves and flowering twice each year. The cicadas emerged after 17 fruiting seasons had passed, rather than 17 years, demonstrating that they are sensitive to the reproductive cycles of their hosts rather than annual physical changes in their environments. How they count to 17 is still a mystery.

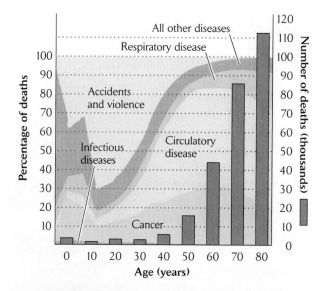

FIGURE 7.12 The incidence of death from cancer and circulatory disease rises with age. Ages at death and causes of death for males in the English population in the early 1980s are shown. From N. Coni, W. Davison, and S. Webster, *Ageing: The Facts* (2nd ed.), Oxford University Press, Oxford (1992).

Senescence is a decline in physiological function with increasing age

Although few long-lived organisms exhibit programmed death associated with reproduction, most experience a gradual increase in mortality and a decline in fecundity as physiological function deteriorates over time. This phenomenon is known as **senescence.** Humans are no exception to the general pattern seen in virtually all animals. Reproductive decline and death in old age do not result from abrupt physiological changes. Instead, they follow upon a gradual decrease in physiological function. Most physiological functions in humans decrease in a roughly linear fashion between the ages of 30 and 85 years; for example, rates of nerve conduction and basal metabolism decrease by 15%–20%, volume of blood circulated through the kidneys by 55%–60%, and maximum breathing capacity by 60%–65%. The function of the immune system and other repair mechanisms declines, and with their decline, the incidence of death from tumors and cardiovascular disease rises (Figure 7.12). Birth defects in offspring and infertility generally occur with increasing prevalence in women after 30 years of age, and fertility decreases dramatically in men after 60 years.

Why does senescence exist, when survival and reproduction would presumably increase an individual's fitness at any age? Perhaps physiological decline is just a fact of life, and natural selection can do nothing about it. Senescence might simply reflect the accumulation of molecular defects that fail to be repaired, just as an automobile eventually wears out and has to be junked. Ultraviolet radiation and highly reactive forms of oxygen break chemical bonds, macromolecules become inactivated, and DNA accumulates mutations. This wear and tear cannot be the entire explanation for patterns of aging, however, because maximum longevity varies widely even among species of similar size and physiology. For example, many small insectivorous bats achieve ages of 10–20 years in captivity, whereas mice of similar size rarely live beyond 3–5 years. Furthermore, cellular mechanisms for reducing the production of reactive forms of oxygen and for repairing damaged DNA and protein molecules appear to be better developed in long-lived animal species than in their short-lived relatives.

These observations suggest that, while senescence may be inevitable, rates of senescence are under the influence of natural selection. Mechanisms for preventing and repairing molecular damage might require allocation of resources and might thus be costly to the organism. If such trade-offs exist, then lengthening the maximum potential life span by postponing senescence could have costs, such as reduced reproduction at younger ages. If survival were so poor because of predation and other extrinsic mortality factors that an individual had little chance of living to old age, it might be better, in terms of fitness, for the

GLOBAL CHANGE

Global warming and flowering time

You may know of Henry David Thoreau as a nineteenth-century writer who spent a year in a small cabin at Walden Pond, in Concord, Massachusetts, and wrote numerous essays about the natural world (Figure 1). You may not have known, however, that during the years he lived in Concord, he also collected data on flowering plants that would provide important insights into the effects of global warming more than a century later.

From 1852 to 1858, Thoreau took notes several days each week on the dates when more than 500 plant species first began flowering around the town. Thoreau planned on publishing this work, but he died at the relatively young age of 44 from tuberculosis before realizing this goal.

Thoreau was not alone in his interest in flowering times around the Concord area. A local shopkeeper, Alfred Hosmer, continued Thoreau's work by observing the first flowering times of more than 700 plants species in 1878 and between 1888 and 1902. Sixty years later, another resident of Concord, Pennie Logemann, began keeping records of the first flowering times of more than 250 species of plants on her property, continuing her observations from 1963 to 1993.

Two ecologists from Boston University, Abraham Miller-Rushing and Richard Primack, learned about the existence of these data and realized their tremendous

FIGURE 1 Henry David Thoreau (1817–1862). © Bettmann/Corbis.

potential for addressing whether long-term changes in global temperatures might be associated with changes in the initial flowering times of plants. Because flowering time is sensitive to temperature as well as to photoperiod, they predicted that warmer global temperatures would cause plants to flower earlier today than in Thoreau's time. The researchers first had to determine whether the local temperatures in Concord had increased over the past century and a half. Fortunately, a nearby observa-

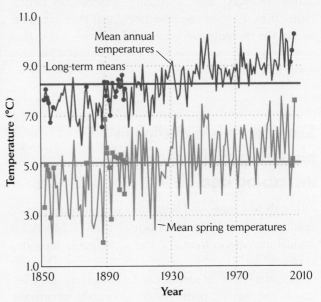

FIGURE 2 Temperatures at Blue Hill Meteorological Observatory, 1852–2006. The lower graph shows the mean "spring" temperature, which is the average of mean monthly temperatures for January, April, and May; the temperatures in these months are highly correlated with flowering times for many species. The horizontal lines show the long-term means (over the entire period shown) for each graph. The symbols indicate years in which flowering data were recorded. After A. J. Miller-Rushing and R. B. Primack, *Ecology* 89:332–341 (2008).

individual to allocate resources to reproduction early in life and let the body fall apart with age.

Even in the absence of senescence, extrinsic causes of death result in progressively fewer individuals remaining alive at older ages. Consequently, gradual physiological deterioration and loss of reproductive capacity have

little effect on average lifetime reproduction because most adults in most natural populations never achieve old age. By this reasoning, selection for changes in survival or fecundity in old age is weak. Thus, selection should tend to favor improvements in reproductive success at young ages over those later in life. Because this effect is greater

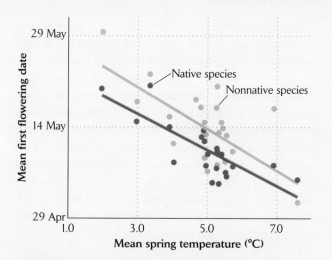

FIGURE 3 Response of first flowering date to mean spring temperature near Concord, Massachusetts. After A. J. Miller-Rushing and R. B. Primack, *Ecology* 89:332–341 (2008).

tory had kept monthly temperature records going back to 1852, and those records indicated that mean annual temperatures had increased by 2.4°C from 1852 to 2006 (Figure 2).

Miller-Rushing and Primack updated the historical data on flowering times by conducting their own survey in the vicinity of Concord from 2003 to 2006. When they analyzed the complete dataset, encompassing the data from Thoreau's time to their own, they found that for the 43 most common species of plants, flowering time today is an average of 7 days earlier than it was 154 years ago, in Thoreau's time (Figure 3). Interestingly, not all plants responded to the temperature change in the same way. In some species, initial flowering time has remained unchanged, perhaps because the influence of day length (which has not changed!) predominates over temperature. Other species, such as highbush blueberry and yellow wood sorrel, flower 3 to 4 weeks earlier now

than they did in 1852 (Figure 4). Surprisingly, even closely related species of plants exhibit contrasting responses to the warming temperatures.

These unique data collected over a century and a half indicate that a seemingly small change in average annual temperature has been associated with dramatic changes in initial flowering time. Because species have responded differently to the warming conditions, the assemblage of plant species in flower that is available to pollinators on a given day has changed from Thoreau's time. No doubt the seasonal progression of pollinator species has also changed. Although Henry David Thoreau did not live to see his work on flowering times published, one wonders how he might have felt knowing that his data provided valuable insights int o the global effects of human activities on the environment 150 years later.

FIGURE 4 Highbush blueberry bush.
Copyright © Richard Shiell/Animals Animals/ Earth Scenes

in populations with lower adult survival rates (Figure 7.13), senescence should appear earlier and progress faster in populations with high rates of mortality due to extrinsic causes (accidents, predators, weather).

This prediction is consistent with observations of natural and human populations. For example, because they

can fly to escape predators, bats and birds lead safer lives than similar-sized terrestrial mammals, and their maximum potential life spans are correspondingly longer. Birds and bats simply age slowly. A small storm petrel, which weighs only 40 grams, can live more than 40 years in spite of having a metabolic rate similar to that of a rodent

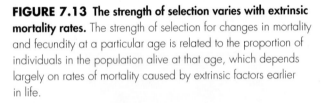

In populations with lower survival rates, few individuals survive to old age, and there is little selection to delay senescence.

Populations with higher survival rates have an older age structure, and hence they have a greater chance of selection for changes in survival or fecundity.

FIGURE 7.13 The strength of selection varies with extrinsic mortality rates. The strength of selection for changes in mortality and fecundity at a particular age is related to the proportion of individuals in the population alive at that age, which depends largely on rates of mortality caused by extrinsic factors earlier in life.

of the same size. Thus, variation in rates of aging is consistent with the predictions of evolutionary theory based on trade-offs between reproductive success early and late in life, in combination with changes in the strength of selection as a function of age. Because modern technology has made life so much safer for humans, and because a larger proportion of the human population is reaching old age and suffering the consequences of aging-related decline, it is possible that increased selection to prolong reproductive life span even further will result in greater longevity in future generations.

Studies of aging in a variety of animals demonstrate that senescence is an inevitable consequence of natural wear and tear. It is impossible to build a body that will not wear out eventually. It is also clear, however, that the rate of wear can be modified by a variety of physiological mechanisms that either prevent or repair damage. These mechanisms are under genetic control, and therefore, like most life history traits, they can be modified by evolution. Mechanisms of prevention and repair require investments of time, energy, nutrients, and tissues, and therefore the allocation of resources to these mechanisms depends on the expected life span of the individual. In this case, as in so many others we have seen in this chapter, consideration of life history patterns demonstrates the power of evolutionary theory to explain variation in biological systems.

Life histories respond to variation in the environment

Under constant environmental conditions, a single life history strategy can provide an optimal solution to conflicts resulting from competing demands on time and resources. When the environment varies, however, no single pattern of time and resource allocation is optimal under all conditions, and the life history must ensure reasonable fitness under a range of conditions. This requirement can often be satisfied when the individual stores resources during good times and uses those reserves to sustain itself during lean times. Furthermore, the individual must reproduce when resources are plentiful and conditions are good for offspring survival. In this way, variation in the environment can influence the optimum life history.

Storage of food and buildup of reserves

Where environmental changes plunge organisms from feast into famine, storing resources acquired during periods of abundance for use in times of scarcity may be a way to cope. During infrequent rainy periods, desert cacti swell with water stored in their succulent stems. Plants growing on infertile soils absorb more nutrients than they require in times of abundance and use them when soil nutrients are depleted. In habitats that frequently burn—such as the chaparral of southern California—perennial plants store food reserves in fire-resistant root crowns, which sprout and send up new shoots shortly after a fire (Figure 7.14). Of course, by allocating resources to storage against future contingencies, organisms reduce their short-term reproductive success.

Many temperate and Arctic animals accumulate fat during mild weather in winter as a reserve of energy for periods when snow and ice make food sources inaccessible. Heavier animals are often slower and less agile, however, and therefore are more likely to be caught by predators. One way to avoid this particular cost of food storage is to store food before consuming it. Some winter-active mammals (such as beavers, squirrels, and pikas) and birds (such as acorn woodpeckers and jays) cache food supplies underground or under the bark of trees for later retrieval. Many of these hoards are immense and can sustain individuals for long periods.

Organisms do not individually anticipate the need to keep stored reserves. Rather, in an environment where food shortages are frequent, those individuals that accumulate reserves have higher fitness than those that do not, and the trait evolves in the population in response to variation in the environment. Where the food supply varies little over time, storage of reserves would needlessly take

FIGURE 7.14 Chaparral plants store food reserves in fire-resistant root crowns. In chaparral habitat in southern California, chamise (*Adenostoma fasciculatum*) and other species resprout from root crowns following a fire. These photos were taken September 12, 1979, and April 20, 1980, following a fire near Los Angeles, California. Photos by Tom McHugh/Photo Researchers.

away resources that could be used to increase reproductive success, and storage would not evolve.

Dormancy

Environments sometimes become so cold, dry, or nutrient-depleted that animals and plants can no longer function normally. In such circumstances, some species that are not capable of migrating elsewhere can enter physiologically inactive states, collectively referred to as **dormancy,** which then become a characteristic part of their life histories. For example, many tropical and subtropical trees shed their leaves during seasonal periods of drought, and many temperate and Arctic trees shed theirs in the fall before the onset of winter frost and long nights. Ground squirrels **hibernate** (spend winter in a dormant state) because they cannot find food in winter, not because they are physiologically unable to cope with the harsh physical environment.

In most species, environmental conditions requiring dormancy are anticipated by a series of physiological changes (for example, production of antifreezes, dehydration, or fat storage) that prepare the individual for a partial or complete shutdown of activity. Before winter, some insects enter a resting state known as **diapause,** in which water is chemically bound or reduced in quantity to prevent freezing, and metabolism drops so low that it is barely detectable. Drought-resistant insects that enter diapause in summer dehydrate themselves and tolerate the desiccated condition of their bodies, or secrete an impermeable outer covering to prevent drying. Plant seeds and spores of bacteria and fungi exhibit similar dormancy mechanisms. Indeed, there are many cases of seeds stored in ancient burial chambers or recovered in

other archeological settings that have sprouted after hundreds of years of dormancy. By whatever mechanism it occurs, dormancy reduces exchanges between organisms and their environments, enabling the organisms to "ride out" unfavorable conditions.

An obvious cost of dormancy is that dormant individuals cannot gather resources to support additional reproduction. Thus, dormancy represents one resolution of a survival–reproduction trade-off. Because dormancy is costly, individuals should minimize the time spent in a dormant state. Doing so, however, requires that organisms be able to sense appropriate cues in the environment that predict conditions that are unfavorable for activity.

Stimuli for change

What stimulus indicates to birds wintering in the tropics that spring is approaching in northern forests? What urges salmon to leave the oceans and migrate upstream to their spawning grounds? How do aquatic invertebrates in the Arctic sense that if they delay entering diapause, a quick freeze may catch them unprepared for winter? Many events in the life history of an organism are timed to match predictable changes in the environment. With the return of warm temperatures and increasing ecosystem productivity in spring, most organisms shift from a winter maintenance state to a state of growth and reproduction. Getting the timing right so that behavior and physiology match the environment is essential. Many adjustments of life history traits at this time are direct responses to changes in food supplies or perceived predation risk. Other adjustments require time and must be started before the environmental change takes place. To make these adjustments, organisms must rely on various indirect cues in the environment. In 1938, J. R. Baker made an important distinction between two kinds of cues that trigger life history changes. **Proximate factors** are cues, such as day length, by which an organism can assess the state of the environment but which do not directly affect its fitness. **Ultimate factors** are features of the environment, such as food supplies, that bear directly on the fitness of the organism.

Virtually all organisms sense **photoperiod** (the length of daylight) as a proximate factor that indicates season, and many can distinguish periods of lengthening and shortening days. Different populations of a single species living in different locations may differ strikingly in their responses to photoperiod, reflecting different relationships of environmental changes to day length. Southern populations (at 30°N) of side oats grama grass (*Bouteloua curtipendula*) flower in autumn, when the photoperiod is 13 hours, whereas more northerly populations (at 47°N) flower in summer, only when the photoperiod exceeds 16 hours each day. In Michigan, at 45°N, water fleas (*Daphnia*) enter diapause in mid-September, at photoperiods of 12 hours or less. In Alaska, at 71°N, related species enter diapause when the photoperiod decreases to fewer than 20 hours, which happens in mid-August. Thus, it is not day length per se that is important to these organisms, but rather the particular changes in physical conditions and food supplies that day length portends at a particular place. Clearly, the sensitivity of individuals to proximate cues has been adjusted by natural selection to match their information content in a particular environment. Water fleas never see 20-hour days in Michigan, but Alaskan water fleas would perish if they waited for 12-hour days before entering diapause. Currently, global climate change is altering environmental conditions without adjusting the proximate cues, such as day length, that organisms have evolved to use. This mismatch between cues and conditions is already beginning to cause inappropriate timing of behavioral and physiological shifts in many organisms.

Individual life histories are sensitive to environmental influences

Although the life histories of individuals in a population can be modified over time by evolutionary response to environmental conditions, individuals also can adjust their life histories in response to variations in the environment. The optimum life history for an individual depends on the particular conditions under which it lives. As we saw in Chapter 6, when an individual's environment changes, it can respond by altering its behavior, its physiology, or

even its development. Individuals can also alter their life history traits by changing their feeding and growth rates, shifting between life history stages at different times or at different sizes, or adjusting their behavior with respect to perceived risk in the environment. The capacity to respond to variation in the environment is itself an aspect of the life history that is subject to natural selection.

Many types of organisms undergo dramatic changes during the course of their development. Metamorphosis from larval to adult forms and sexual maturation are the most prominent of these changes. The best time to undergo such transitions depends on the presence of resources and natural enemies in the environment, and their timing is made more complicated by variations in the rate of growth due to food supply, temperature, and other environmental factors.

Imagine two growth curves resulting from two levels of food supply (Figure 7.15). An individual with access to abundant food matures at a given mass and age. A poorly nourished individual clearly cannot reach the same mass at a given age, and therefore must mature at a different point with respect to size, age, or both. The response of age and size at maturation to different levels of food availability is the reaction norm for maturation. Faced with such environmental variation, an individual can follow one of two pathways, or some intermediate course between them. First, the individual may mature when it achieves a certain mass, however long that takes. With poor nourishment, it will take longer to achieve that mass,

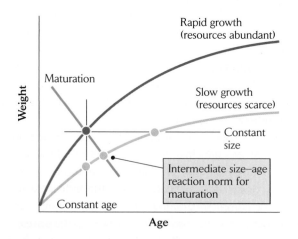

FIGURE 7.15 Relationships between age and size at maturation may differ when growth rates differ. Individuals may mature at a constant age, at a constant size, or at some point intermediate between the two. The response of metamorphosis to different levels of food availability is the reaction norm of metamorphosis. From S. C. Stearns, *Am. Zool.* 23:65–76 (1983), and S. C. Stearns and J. Koella, *Evolution* 40:893–913 (1986).

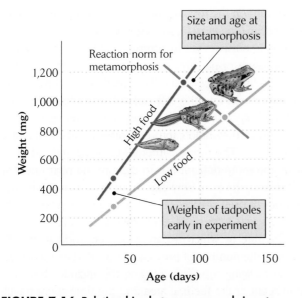

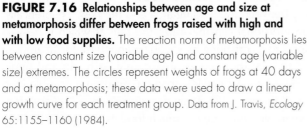

FIGURE 7.16 Relationships between age and size at metamorphosis differ between frogs raised with high and with low food supplies. The reaction norm of metamorphosis lies between constant size (variable age) and constant age (variable size) extremes. The circles represent weights of frogs at 40 days and at metamorphosis; these data were used to draw a linear growth curve for each treatment group. Data from J. Travis, *Ecology* 65:1155–1160 (1984).

and maturation will be delayed. Consequently, the individual will be exposed to a longer period of risk prior to reproduction. Alternatively, the individual may mature at a predetermined age. With poor nourishment, this strategy will result in a smaller size at maturity, and perhaps a reduced reproductive rate as an adult. The optimum solution, which is usually somewhere in between, depends in part on the risk of death as a juvenile (high risk favors earlier maturation at a smaller size) and the slope of the relationship between fecundity and size at maturity (higher values favor later maturation at a larger size because the fecundity payoff is greater).

When frog tadpoles are raised under experimental conditions of high and low food availability, they exhibit different growth rates, as expected. In one experiment, tadpoles given a poor diet metamorphosed into adult frogs at a smaller size, but a later age, than tadpoles reared with abundant food (Figure 7.16). This finding supports the prediction that the timing of metamorphosis should take into account both age and size: poor nutrition slows the developmental program in frogs, but does not stop it altogether. Predation risk is also an important factor. Studies on the size at metamorphosis of mayflies in high-elevation streams in western Colorado showed that in streams with trout, which are important predators on mayfly larvae, the mayflies matured at smaller sizes and left the streams

earlier than in comparable streams lacking trout. Growth rates in the two types of streams were similar, so the difference was due entirely to the risk of predation.

As the mayfly example shows, the risk of predation or other causes of mortality can influence the life history in ways that minimize that risk. In another example, in wooded areas of Connecticut, juvenile individuals in some populations of the spotted salamander (*Ambystoma maculatum*) are preyed on by the larger marbled salamander (*Ambystoma opacum*). However, marbled salamanders cannot eat spotted salamanders after those prey have grown above a certain size, simply because their mouths are not big enough. This pattern is referred to as "gape-limited predation." In *A. maculatum* populations at risk of predation by *A. opacum*, the young spotted salamanders forage for their own food much more actively than where marbled salamanders are absent. Although foraging is risky business because the salamanders are vulnerable while they are looking for food, fueling their growth above the critical size to escape predation altogether is strongly favored in these populations.

Risks of all sorts depend on size, and those risks influence the allocation of resources between functions that support growth and those that support maintenance and survival. For example, in the Kalahari sand vegetation of Zimbabwe, plants with larger stems are able to recover more rapidly from fires and have a higher chance of surviving them (Figure 7.17). Where fires are frequent, there

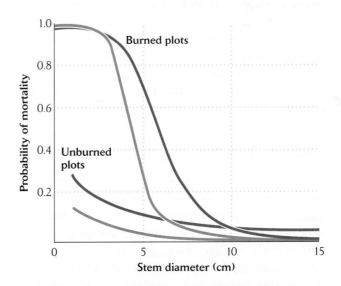

FIGURE 7.17 The probability of mortality from fire decreases with increasing stem diameter. These data for several species in the plant family Combretaceae were gathered in the Kalahari sand vegetation of western Zimbabwe. The effect of fire is particularly strong when the stems have also been damaged by frost. After R. M. Holdo, *Plant Ecol.* 180:77–86 (2005).

is strong selection for rapid growth of the stem at the expense of developing the root system.

In addition to being capable of altering their growth and development, as well as other aspects of their physiology, most animals can respond to rapid change in their environments by altering their behavior. Appropriate behavioral responses presumably increase the individual's fitness compared with alternative behaviors. Because of the close connection between behavior and fitness, various types of behavior can be considered a part of the life history of the organism. To illustrate this principle, let's consider how individuals allocate their time as they forage to satisfy their food requirements.

Animals forage in a manner that maximizes their fitness

Every detail of an individual's life involves the resolution of conflicting demands on its time and resources. Therefore, we can apply the principles of optimal allocation that we have just applied to life history stages to individual behaviors, such as foraging. Food supplies vary spatially, temporally, and with respect to the quality of food items. Furthermore, foraging is dangerous because it exposes the individual to predation, yet individuals must eat to live and reproduce. Foraging individuals must resolve conflicts among these time, reward, and risk factors by deciding where to forage, how long to feed in a certain patch of habitat, which types of foods to eat, and so on. Evolutionary theories of **optimal foraging** seek to explain these behavioral responses in terms of the likely costs and benefits of each possible alternative behavior. Let's see how this works in the case of parent birds provisioning their offspring.

Central place foraging

When birds feed their offspring in a nest, the chicks are tied to a single location, while the parents are free to search for food at some distance from the nest. This situation is referred to as **central place foraging.** As the parents' foraging distance increases, more food is potentially available to them. However, greater travel distances also increase the time, energy, and risk costs of foraging. Is there some best distance from the nest at which a parent bird should forage, and how much food should the parent bring to its brood with each trip? That is, how much time should the parent spend gathering food before it returns to its nest?

Studies on the foraging behavior of European starlings allowed investigators to address these questions from an economic perspective. In summer, starlings typically forage on lawns or pastures for leatherjackets, which are the larvae of tipulid flies (crane flies). Starlings feed by thrusting their beaks into the soft turf and spreading the mandibles to expose prey. When they are gathering food for their young, they hold captured leatherjackets at the base of the beak. The more leatherjackets a starling has in its beak, the more difficult it is to capture the next one. For this reason, the time between captures increases with the number of prey already caught, until a starling with eight leatherjackets in its beak can hardly feed at all.

The rate at which a parent delivers food to its offspring is the number of prey caught divided by the length of the foraging trip. The foraging trip includes both the time spent at the feeding area and the time spent traveling between the feeding area and the nest. Theoretically, an individual can maximize the rate at which it delivers food to its offspring by spending an intermediate amount of time at the feeding area during each trip and bringing back something less than the maximum possible food load (Figure 7.18). Imagine yourself in a grocery store where you must buy as much food as you can in an hour and you have to carry your items to the checkout counter by hand. How frequently would you take your items to the cashier? Carrying one item at a time is clearly a bad choice, particularly if there is a long line waiting to be checked out (analogous to a long foraging trip). Trying to carry more food than you can handle well, and having to spend time picking dropped items up from the floor, also seems unproductive. As in the case of the starling with a beak full of leatherjackets, the "law of diminishing returns" sets in: the longer the feeding period, the lower the average rate of prey capture. The best strategy is somewhere in the middle. The optimum load varies in direct relation to the traveling time, or more generally, to any fixed cost per foraging trip.

DATA ANALYSIS MODULE 1 *Spatially partitioned foraging by oceanic seabirds.* Oceanic seabirds may intersperse long and short foraging trips to gather food alternately for themselves and their chicks. You will find this module on page 156.

ECOLOGISTS IN THE FIELD   **Optimal foraging by starlings.** To what extent do organisms actually forage optimally? Figure 7.18 is theory. In reality, we have all seen some inefficient shoppers in our local grocery stores. Are starlings better economists than we are?

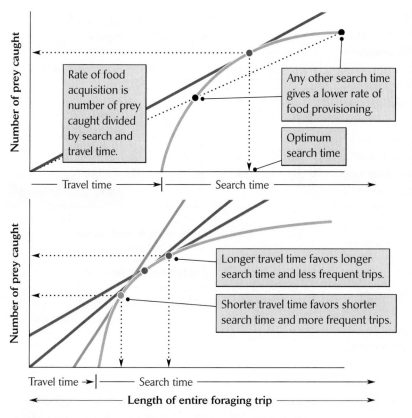

FIGURE 7.18 Optimal foraging models can be used to predict behavior. For a given prey accumulation curve (orange line; the increase in captured prey as a function of time), the slope of the line passing through the origin of the graph (the beginning of the foraging trip) and tangent to the prey accumulation curve indicates the maximum average rate of food acquisition (number of prey caught per unit of time). As shown in the lower graph, the optimal search time on an individual foraging trip increases as the length of the travel time decreases.

This question was addressed in a clever experiment by behavioral ecologist Alex Kacelnik of Oxford University. Instead of letting starlings feed on their natural prey, he trained them to visit feeding tables at which mealworms could be provided through a plastic tube at precisely timed intervals. A starling would arrive at the table, "capture" the first mealworm, and then wait for the next one to be delivered. To mimic the longer intervals at which a starling would catch leatherjackets as its beak became full, Kacelnik adjusted the timing so that each successive mealworm would arrive at a longer interval. He then placed the feeding tables at different distances from nests and observed how many mealworms a starling would wait for at different travel times. As expected, starlings increased their load size as travel time increased (Figure 7.19). Kacelnik concluded that starlings are good economists, at least when it comes to gathering food. ▐

Risk-sensitive foraging

The value of a feeding area depends not only on the rate at which an individual can gather food, but also on its relative safety. Every activity, including foraging, carries a risk of mortality. For many animals, predation is the most significant mortality risk, and the presence of a predator, or even the perceived threat of predation, can reduce the value of an otherwise good feeding area. The extra food is simply not worth the increased risk of becoming food.

The predation factor has been incorporated into optimal foraging theory in studies of **risk-sensitive foraging.**

James F. Gilliam and Douglas F. Fraser elegantly demonstrated the principle of risk-sensitive foraging in a simple experiment with fish. They constructed cages with

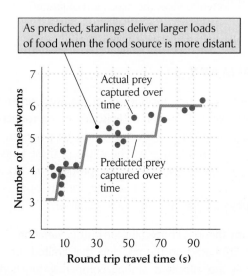

FIGURE 7.19 Food loads increase with travel times. The number of mealworms brought by starlings to their broods increased with the total length of the foraging trip. From A. Kacelnik, *J. Anim. Ecol.* 53:283–299 (1984).

two compartments and placed them in an experimental stream. The subjects in the experimental system were small minnows (juvenile creek chubs), and the predators were adult creek chubs. The minnows were provided with tubifex worms buried in mud placed in small trays in the compartments. A refuge area that permitted passage of the minnows, but not the adult chubs, connected the two compartments.

In the experiment, minnows were presented with a low density of food (0.17 worms per cm^2) and only one predator in the first compartment, and a higher density of food but two or three predators in the second compartment. The researchers gradually increased the amount of food in the more dangerous compartment to determine at what point the minnows would expose themselves to greater risk in order to obtain more food. The minnows were very sensitive to predation risk. When the more dangerous compartment contained two adult chubs, minnows switched to foraging there only after prey density was increased to more than 0.33 worms per cm^2, or twice the density in the less risky compartment (Figure 7.20). In the presence of three predators, the food level had to be more than 4 times that in the safer compartment to entice the minnows to switch.

MORE ON THE WEB *Variable Food Supplies and Risk-Sensitive Foraging.* Would you choose a predictable supply of a lower-quality food or a more variable food supply with a higher average reward?

In this chapter, we have examined ways in which natural selection influences the allocation of limited time and

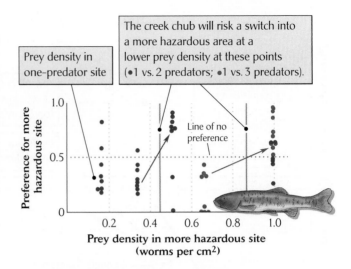

FIGURE 7.20 Foraging fish are risk-sensitive. Minnows switched to a more hazardous feeding site only when the prey density in that site exceeded a certain critical level. That critical density increased with the relative risk of foraging. After J. F. Gilliam and D. F. Fraser, *Ecology* 68:1856–1862 (1987).

resources by individuals to determine their life histories—and indeed, all aspects of their morphology, physiology, and behavior. We have also seen that the life histories of organisms are sensitive to environmental conditions and to variation in those conditions. A part of the environment of every individual includes other individuals of the same species with which it interacts. In the next two chapters, we shall consider how interactions with mates, family, and society members shape many features of the life histories of organisms.

SUMMARY

1. Life history traits include age at maturity (first reproduction), parity (number of episodes of reproduction), fecundity (number of offspring produced per reproductive episode), and longevity. The values of these traits can be interpreted as solutions to the problem of allocating limited time and resources among various body structures, physiological functions, and behaviors.

2. Many theories concerning life history variation among species are based on the principle that limited time and resources are allocated among competing functions in such a way as to maximize lifetime fitness.

3. Life history traits often vary consistently with respect to the environment. Variation in one life history trait is

often correlated with variation in others. Many life history characteristics are organized together along a "slow–fast" continuum of values.

4. J. P. Grime recognized three clusters of life history attributes among plants associated with ruderal, stressful, and highly competitive environments.

5. An optimized life history is one that resolves conflicts between the competing demands of survival and reproduction to the best advantage of the individual in terms of its fitness.

6. Delayed reproduction is favored when the life span is relatively long and when increased growth or accumulation of experience results in greater fecundity later in life.

7. High adult mortality rates favor increased investment in present reproduction at the expense of growth, adult survival, and future reproduction.

8. When reproduction requires costly preparation, selection may favor a single all-consuming reproductive event followed by death, as in salmon. This pattern of reproduction, called semelparity, is the converse of iteroparity, or repeated reproduction.

9. Senescence, the progressive deterioration of physiological function with age, causes declines in fecundity and probability of survival. Senescence is caused by the natural wear and tear of living and is inevitable. Rates of senescence, however, are subject to natural selection.

10. Owing to extrinsic causes of mortality, few individuals in most natural populations survive to old age. As a result, the strength of selection on traits expressed at progressively later ages diminishes. Individuals in populations subjected to higher extrinsic mortality rates age faster, as predicted by evolutionary theory.

11. When environmental conditions result in food shortages or exceed the range of physiological tolerance, organisms can rely on resources stored during periods of abundance or enter inactive states, such as diapause or hibernation.

12. In many cases, the individual must anticipate environmental changes to respond to them successfully. Organisms rely on proximate cues, such as photoperiod, to predict changes in ultimate factors, such as food supply, that directly affect their well-being.

13. Individuals can adjust certain life history traits, such as age or size at metamorphosis and age at reproductive maturity, in response to variation in the environment, particularly variation in the risk of mortality.

14. Food supplies vary spatially, temporally, and with respect to quality. Thus, foraging individuals must resolve conflicts about when, where, and how to feed so as to maximize fitness.

15. Central place foragers deliver food to a fixed place, such as a nest with young. They must balance the costs and risks of travel against the size of the area within which they can forage.

16. The quality of a feeding area is affected by the risk of predation on a foraging individual. Many animals avoid feeding in high-risk areas even though food may be plentiful there. This strategy is referred to as risk-sensitive foraging.

REVIEW QUESTIONS

1. Why are trade-offs among life history traits so commonly observed?

2. In J. P. Grime's categorization of plant life history traits, why might ruderals (weeds) spread via easily dispersed seeds whereas stress tolerators spread vegetatively?

3. Why should organisms with low annual survival rates begin to reproduce at an early age?

4. Explain why organisms face a fundamental trade-off between growth and fecundity.

5. Compare and contrast semelparous and iteroparous life history strategies.

6. Why might natural selection act more strongly on traits that improve reproductive success early in life rather than later in life?

7. If we think of humans as central place foragers, what would you predict about the amount of groceries brought back by individuals living close to a grocery store compared with those far from a grocery store?

8. Under what conditions is dormancy an effective life history strategy?

9. The snowshoe hare is a relative of the rabbit that lives in Canada and has fur that is brown in the summer but turns white in the winter. Propose a proximate and an ultimate cause for this change in color.

SUGGESTED READINGS

Bazzaz, F. A., et al. 1987. Allocating resources to reproduction and defense. *BioScience* 37:58–67.

Bielby, J., et al. 2007. The fast–slow continuum in mammalian life history: An empirical reevaluation. *American Naturalist* 169:748–757.

Borchert, R., et al. 2005. Photoperiodic induction of synchronous flowering near the equator. *Nature* 433:627–629.

Dijkstra, C., et al. 1990. Brood size manipulations in the kestrel (*Falco tinnunculus*): Effects on offspring and parental survival. *Journal of Animal Ecology* 59:269–286.

Fleming, I. A., and M. R. Gross. 1989. Evolution of adult female life history and morphology in a Pacific salmon (Coho: *Oncorhynchus kisutch*). *Evolution* 43:141–157.

Fontaine, J. J., and T. E. Martin. 2006. Parent birds assess nest predation risk and adjust their reproductive strategies. *Ecology Letters* 9:428–434.

Gilliam, J. F., and D. F. Fraser. 1987. Habitat selection under predation hazard: Test of a model with foraging minnows. *Ecology* 68:1856–1862.

Grime, J. P. 1979. *Plant Strategies and Vegetation Processes*. Wiley, Chichester.

Gross, M. R. 1996. Alternative reproductive strategies and tactics: Diversity within sexes. *Trends in Ecology and Evolution* 11:92–98.

Hamilton, S. L., et al. 2007. Size-selective harvesting alters life histories of a temperate sex-changing fish. *Ecological Applications* 17:2268–2280.

Janzen, D. H. 1976. Why bamboos wait so long to flower. *Annual Review of Ecology and Systematics* 7:347–391.

Kacelnik, A. 1984. Central place foraging in starlings (*Sturnus vulgaris*). I. Patch residence time. *Journal of Animal Ecology* 53:283–299.

Karban, R., C. A. Black, and S. A. Weinbaum. 2000. How 17-year cicadas keep track of time. *Ecology Letters* 3:253–256.

Miller-Rushing, A. J., and R. B. Primack. 2008. Global warming and flowering times in Thoreau's Concord: A community perspective. *Ecology* 89:332–341.

Peckarsky, B. L., et al. 2001. Variation in mayfly size at metamorphosis as a developmental response to risk of predation. *Ecology* 82:740–757.

Reznick, D. 1985. Costs of reproduction: An evaluation of the empirical evidence. *Oikos* 44:257–267.

Reznick, D. N., H. Bryga, and J. A. Endler. 1990. Experimentally induced life-history evolution in a natural population. *Nature* 346:357–359.

Ricklefs, R. E., and C. E. Finch. 1995. *Aging: A Natural History*. Scientific American Library, New York.

Ricklefs, R. E., and M. Wikelski. 2002. The physiology–life history nexus. *Trends in Ecology and Evolution* 17(10):462–468.

Schlichting, C. D. 1989. Phenotypic integration and environmental change. *BioScience* 39:460–464.

Sibly, R. M., and J. H. Brown. 2007. Effects of body size and lifestyle on evolution of mammal life histories. *Proceedings of the National Academy of Sciences USA* 104:17707–17712.

Stearns, S. C. 1992. *The Evolution of Life Histories*. Oxford University Press, Oxford.

Strathmann, R. R. 1990. Why life histories evolve differently in the sea. *American Zoologist* 30:197–207.

Urban, M. C. 2007. Risky prey behavior evolves in risky habitats. *Proceedings of the National Academy of Sciences USA* 104:14377–14382.

Williams, G. C. 1966. Natural selection, the costs of reproduction, and a refinement of Lack's principle. *American Naturalist* 100:687–690.

Young, T. P., and C. K. Augspurger. 1991. Ecology and evolution of long-lived semelparous plants. *Trends in Ecology and Evolution* 6:285–289.

DATA ANALYSIS MODULE 1

Spatially Partitioned Foraging by Oceanic Seabirds

Most of the birds encountered in forests, fields, and gardens confine themselves to small areas of habitat during the breeding season. In contrast, many seabirds range over vast expanses of ocean to gather food for their young. The abundances of fish, squid, and other prey vary tremendously depending on the productivity of local waters, which is influenced by depth, currents, runoff from coastal areas, and vertical mixing between bottom and surface waters, among other factors. Furthermore, the best places to feed are not always the best places to breed. Like other birds, seabirds must lay their eggs and rear their young on land (DA Figure 1.1). Suitable nesting sites are sometimes far from abundant supplies of food. Because of this conflict between breeding and feeding requirements, oceanic seabirds must make trade-offs in order to allocate their time in a way that maximizes their fitness. The nature of these trade-offs has been the focus of several studies.

Obtaining information about the behavior of a bird at sea presents many logistical difficulties, but much can be learned indirectly by observing the comings and goings of parent birds at their nest sites. Two French biologists, Thierry Chaurand and Henri Weimerskirch, documented an unusual pattern of food delivery to chicks by parent blue petrels nesting on Kerguelen Island in the South

Atlantic Ocean. We shall take a close look at their data and perform some analyses to better understand the problems seabirds face and how they solve them.

Step 1: Create a bar graph to determine whether foraging trips exhibit a unimodal or a bimodal distribution of lengths.

DA Table 1.1 shows the lengths of completed foraging trips (that is, for which both the beginning and the end were recorded) taken by each adult at 10 nests between

DA FIGURE 1.1 An oceanic seabird, the gray-headed albatross (*Thalassacrche chrysostoma*), and its chick. R. E. Ricklefs.

DA TABLE 1.1		Lengths of completed foraging trips					
Nest	Sex	Lengths of successive foraging trips (days)					
28	M	6	2	6	1	7	1
	F	2	6	2	8	2	
95	M	2	7	5	1		
	F	3	6	2	6	3	
240	M	2	7	2	8		
	F	8	2	7			
532	M	2	7	6	2		
	F	2	7	2	9		
370	M	7					
	F	7	2				
257	M	8	2	7	1		
	F	2	8	1	8		
231	M						
	F	7	8				
232	M	10	2	6	3		
	F	1	8	2	7		
491	M	8	1	8			
	F	1	7	2	8		
278	M	2	7	2			
	F	1	7	1	7		

January 13 and February 7, 1990. You can use the data for males and females without regard to sex.

Chaurand and Weimerskirch (1994) suggested that foraging trips fell into two distinct classes: short trips and long trips. Alternatively, this pattern might be the result of random variation in the small sample of foraging trips. The bar graph looks like a bimodal distribution, which is to say that it has two peaks, or modes. Nonetheless, small samples of observations drawn from an underlying unimodal (single-peaked) distribution of trip lengths sometimes appear to be bimodal. Thus, we should first consider this statistical question: do trip durations really fall into two discrete classes, or could the apparent pattern have occurred just by chance?

There are several ways to determine the statistical significance of the bimodality in this sample. They all involve calculating the probability that one might obtain the observed distribution by chance. Chaurand and Weimerskirch found that the durations of the 70 feeding trips were between 1 and 10 days. Suppose the underlying distribution were uniform, with an average of 10% (proportion 0.10) of all trips occurring at each of 1, 2, 3, . . . , 10 days. This uniform distribution is not realistic, because

biological patterns most often exhibit normal distributions with single peaks, but it demonstrates the principle of testing an observed distribution against some expectation.

• With a hypothetical uniform distribution, how many trips of each length would you expect? Note that intermediate trip lengths (3–5 days) are underrepresented, and particularly, that the investigators observed no trips of 4 days. Is this a significant deviation from what you would expect of a uniform distribution? Let's test a simple prediction based on that distribution.

The probability of any one trip being exactly 4 days long is 0.10, so the probability that a trip is not 4 days long is 0.90 (90%). The probability that neither of two trips drawn at random from a uniform distribution is 4 days long is simply the product of the probability that each trip is not 4 days long, which is 0.90^2. For three trips, the probability becomes 0.90^3, and so on.

• What is the probability that none of 70 trips is 4 days long? Do you consider it unlikely that no trip would be 4 days long?

Now suppose that the observed trip lengths were drawn from an underlying normal (unimodal) distribution, with the intermediate lengths the most common. Would the expected number of 4-day trips be higher or lower under the unimodal distribution than under the uniform distribution? Does this mean that a statistical test based on a uniform distribution would be relatively conservative—that is, would it offer a higher probability than a normal distribution of observing no 4-day trips?

Regardless of the statistical test we use, finding so few trips of intermediate length (3–5 days) is highly improbable. Confident that we have a bimodal distribution consisting of distinct classes of short and long trips, we can ask further questions about that pattern.

• Do blue petrels embark on short and long trips at random—that is, without regard to the length of the previous trip?

Step 2: Determine whether there is a temporal pattern to the duration of foraging trips.

First, divide the foraging trip durations into long trips and short trips, choosing a convenient duration as a cutoff point. What proportion of the trips are long and what proportion of the trips are short? From the trip records in DA Table 1.1, tabulate the number of times a short trip was followed by a long trip, and vice versa. The table could be structured as shown in DA Table 1.2.

We can test for a significant statistical association between long and short trips (positive or negative) with a

DA TABLE 1.2	Sample table of recorded trip lengths	

		Following trip	
		Short	Long
Preceding trip	Short	Number of short trips followed by short trips	Number of short trips followed by long trips
	Long	Number of long trips followed by short trips	Number of long trips followed by long trips

χ^2 test, which compares observed values with the values expected according to the "null" hypothesis of no association. Calculate the proportion of short and long following trips in the whole sample. Then, for the short preceding trips, calculate the expected number of short and long following trips. This number is simply the number of following trips multiplied by the proportion of short and long following trips in the whole sample. Do the same calculation for the long preceding trips. You now have the number of trips expected for each of the combinations of short–short, short–long, long–short, and long–long trips. You calculate the χ^2 value as the sum, over these four combinations, of (observed − expected)^2/expected. That is,

$$\chi^2 = \sum \frac{(O - E)^2}{E}$$

The value of χ^2 has a well-known distribution, and the probability of obtaining a value greater than 4.0 in a comparison of this type is less than 5%, the usual probability level for statistical significance.

• After calculating the χ^2 value, do you think there is significant heterogeneity, or nonrandomness, in the data? Do trip lengths tend to alternate (negative association) or cluster together (positive association)?

Step 3: Interpret the data: Relate the rate of food delivery to the chick and the effect of long and short trips on adult condition.

Adults appeared to alternate long foraging trips averaging 7 days with short foraging trips averaging 2 days. Another interesting result was that, regardless of the length of the foraging trip, the amount of food delivered to the single chick in each nest was the same: on average, about 60 grams. (This is a huge amount, considering that an adult blue petrel weighs only 170 grams on average. Why do you think meals are so large and infrequent?) Furthermore, adults lost an average of about 12 grams of body mass during short trips and gained almost 10 grams during long trips.

• How might you explain these observations? How might spatial variation in the food supply in the surrounding ocean produce a best strategy of alternating long and short foraging trips? What cues should the parent use to make the choice between traveling a long or a short distance? How should it balance the consequences of its decision for its chick? For itself?

The amount of food brought to the chick is the same whether the foraging trip is long or short, but of course, the rate of food delivery (grams per day) is much higher for short trips. Why don't parents make only short trips? Consider the fact that adults tend to lose weight during short trips. Does this observation imply that parents give most of the food they catch during short trips to their chick and digest and store little for themselves? Remember, flight requires a lot of energy. If that is the case, then why do the parents take long trips? Does the fact that they gain weight during long trips suggest that they use this time to replenish their own depleted reserves? Chaurand and Weimerskirch argued that long trips represented flights to distant areas with exceptionally abundant food supplies.

• Is this hypothesis better than the alternative, in which parents on long trips feed in the same areas they use for short trips, but keep for themselves most of the prey they catch?

As a follow-up, subsequent studies of other oceanic seabirds using satellite tracking have shown that many species alternate long and short foraging trips, and that long trips target distant feeding areas of exceptionally high productivity.

Literature Cited

Chaurand, T., and H. Weimerskirch. 1994. The regular alternation of short and long foraging trips in the blue petrel *Halobaena caerulea*: A previously undescribed strategy of food provisioning in a pelagic seabird. *Journal of Animal Ecology* 63:275–282.

See also: Weimerskirch, H., et al. 1994. Alternate long and short foraging trips in pelagic seabird parents. *Animal Behavior* 47:472–476.

CHAPTER 8

Sex and Evolution

Nature is full of bizarre creatures, and few are more bizarre-looking than the stalk-eyed flies (*Teleopsis*) of Malaysia, whose eyes are widely separated at the ends of long projections emanating from the head. Both males and females have these stalks, but in some species they are up to twice as long in males as they are in females (Figure 8.1). Stalk-eyed flies aggregate at night to mate, and field biologists have observed that the mating success of males increases in direct relation to the distance between their eyes. Evidently, the sex difference in eye span results from selection by females on the expression of this trait in males: females prefer to mate with males with large eye spans. Ecologists refer to this mechanism of evolution as *sexual selection*.

Why does this difference between the sexes exist? If a wide eye span improved visual detection of food or predators, one would expect male and female eye spans to be similar. Indeed, in some *Teleopsis* species, male and female eye spans do not differ. How, then, can we explain the larger eye span of males in other species? Perhaps eye span in these species provides information about some aspect of male quality that is important to females.

Two sexually dimorphic *Teleopsis* species (*T. dalmanni* and *T. whitei*) have biased sex ratios in nature. Their populations contain only about one-third male individuals, whereas in most flies, including monomorphic species of *Teleopsis,* the sex ratio is close to half and half. Genetic analyses have revealed the cause of this biased sex ratio: in many males, sperm cells bearing the Y chromosome are defective, so that most of their progeny are female (XX) rather than male (XY).

While most males of *T. dalmanni* and *T. whitei* produce few Y-bearing sperm, some males—those with wider eye spans than average—produce practically normal

FIGURE 8.1 Sexual dimorphism results from sexual selection. Two large *Teleopsis whitei* males on a rootlet approach each other to compare eye spans at dusk in peninsular Malaysia. Courtesy of Gerald S. Wilkinson, University of Maryland.

sperm counts, and thus father roughly equal numbers of male and female offspring. This was demonstrated dramatically in artificial selection experiments carried out by Gerald Wilkinson, at the University of Maryland, and his colleagues Daven Presgraves and Lili Crymes. They set out to duplicate the effects of sexual selection by mating only male flies with exceptionally wide eye spans. After 22 generations of such mating, the average eye span of males in the experimental population had increased about 1 mm, or 10%. Moreover, the percentage of males among the progeny of these flies had increased to 50% or more. These results show a genetic link between factors in the males that corrected the deficiency in sperm production and factors that increased eye span. Consequently, females that choose males with a wider eye span are also choosing males that will father more male offspring.

Why should a female want to produce more male offspring than the population average? The answer is relatively simple: when a population contains a smaller proportion of individuals of one sex, it is to the advantage of an individual to produce more of that sex among its progeny. Each individual receives one set of genes from its mother and one set from its father. Therefore, each generation has equal amounts of genetic material contributed by males and by females. When there are many females in a population, it is to a mother's advantage to produce male offspring, because each of her sons will, on average, contribute more sets of genes (her genes) to future generations than will each of her daughters. Thus, when females choose males that are likely to produce more sons than the population average, they increase their own evolutionary fitness.

CHAPTER CONCEPTS

- Sexual reproduction mixes the genetic material of two individuals

- Sexual reproduction is costly

- Sex is maintained by the advantages of producing genetically varied offspring

- Individuals may have female function, male function, or both

- The sex ratio of offspring is modified by natural selection

- Mating systems describe the pattern of pairing of males and females within a population

- Sexual selection can result in sexual dimorphism

Sex is a basic component of the life histories of all species of animals and plants. Many aspects of sex, however—such as the proportions of males and females in a population, the allocation of resources among male and female sexual function, and even the presence of sexual reproduction itself—vary greatly from species to species.

Reproduction, as we have seen, is the ultimate goal of an individual's life history. In most species of multicellular organisms, most reproduction is sexual, which means that new individuals come from the union of female and male gametes. A few species have given up sex altogether, and as we shall see, these exceptions can help us to understand the prevalence of sexual reproduction in nature. Some of the most important and fascinating attributes of life concern sexual function. Among these are sex differences, sex ratios, and the various devices and behaviors used to enhance the success of an individual's gametes.

FIGURE 8.2 Much of what we observe in nature has evolved to improve an organism's reproductive success. A male peacock spreads his elaborate tail to attract females. Photo by Norbert Rosing/Animals Animals.

The peacock's glorious tail, whose purpose is to make its bearer more attractive to females, is one of nature's most fantastic productions (Figure 8.2). Indeed, sex underlies much of what we see in nature. In this chapter, we shall consider how sexual function influences the evolutionary modification of organisms and many of their behaviors as individuals. A good place to start is with sex itself.

Sexual reproduction mixes the genetic material of two individuals

In most animals and plants, reproductive function is divided between two sexes, and reproduction is accomplished through the production of **gametes.** A male gamete (the sperm) and a female gamete (the egg or ovule) join together in an act of **fertilization** to form a single cell, called a **zygote,** from which a new individual develops. This sequence of events is referred to as **sexual reproduction.** The mixing of genetic material from two parents results in new combinations of genes in the offspring. Because of this mixing, siblings can differ from one another genetically. Thus, in a variable environment, at least some offspring of a sexual union are likely to have a genetic constitution that enables them to survive and reproduce, regardless of the particular conditions. Sexual reproduction may also produce new combinations of genes previously absent from a population. The expression of any one gene can be influenced by other genes, and so new combinations of old genes may provide new variation for natural selection to work on. Indeed, many biologists believe that sexual reproduction evolved as a means of generating the genetic diversity necessary to respond through evolution to a varied and changing environment.

The gametes themselves are formed by **meiosis,** a special type of cell division that takes place in the germ cells within the primary sexual organs, or **gonads.** The cell products of meiosis are **haploid**—that is, they contain only one member of each of the chromosome pairs present in the individual's other, **diploid,** cells. Each of these haploid cells contains a single full set of chromosomes, but whether a particular chromosome was inherited from the father or the mother is, in most instances, random. These haploid cells are the ones that eventually develop into gametes. As a consequence of meiosis, the genetic makeup of each zygote is a unique, random combination of the genetic material of each of the individual's four grandparents.

In contrast to sexual reproduction, progeny produced by **asexual reproduction** are generally identical to one another and to their single parent, and thus none of them is likely to be well adapted to novel conditions. Asexual reproduction is most common in plants, most of whose cells retain the ability to produce an entire new individual. For example, new shoots that sprout from roots or rhizomes (underground shoots), or even from the margins of leaves, may give rise through so-called **vegetative reproduction** to separate individuals with genotypes identical to that of the "parent" plant (Figure 8.3). The individuals that descend asexually from the same parent and bear the same genotype are referred to collectively as a **clone.** Similarly, many simple animals, such as hydras, corals, and their relatives, produce buds in the body wall that develop into new individuals. When these offspring remain attached to the parent, a colony develops, as in the case of hydroids, corals, bryozoans, and many other aquatic animals.

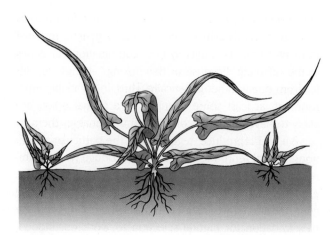

FIGURE 8.3 Many plant species reproduce asexually. The walking fern sprouts a fully formed plant from the tip of a leaf. After V. A Greulach and J. E. Adams, *Plants: An Introduction to Modern Botany,* Wiley, New York (1962).

Some animals reproduce asexually by forming diploid eggs. This type of reproduction, referred to as **parthenogenesis,** has cropped up in all-female populations of fishes, lizards, and some insects, to name a few. In some of these animals, germ cells develop directly into egg cells without going through meiosis, and all of an individual's eggs are therefore genetically identical. In other parthenogenetic species, meiosis proceeds through the first meiotic division, but suppression of the second meiotic division results in diploid egg cells. Although a sexual union is not involved, these eggs differ from one another genetically because of recombination (the exchange of genes between homologous chromosomes) and independent assortment of chromosomes in the first meiotic division. In another variation of parthenogenesis, meiosis proceeds to completion, but female gamete-forming cells then fuse to form diploid eggs. This process is a type of self-fertilization, and its products vary genetically, but not as much as when two parents are involved in producing them. Finally, individuals that have both male and female sexual organs may form both male and female gametes and then fertilize themselves. This method of sexual reproduction is most frequently encountered as **selfing** in plants. It is sexual in that both types of gametes are made and fertilization takes place, but it resembles asexual reproduction in that offspring have only a single parent.

Sexual reproduction is costly

Sexual and asexual reproduction are both viable life history strategies. Asexual reproduction is most common among plants and is found sporadically in all major groups of animals, with the exception of birds and mammals. Perhaps we should be surprised that sex occurs at all, considering its costs to the organism. Gonads are expensive organs that confer little direct benefit to the individual beyond procreation, and they require resources that could be devoted to other purposes. Mating itself is a major production for animals and plants, involving floral displays to attract pollinators and elaborate courtship rituals to please mates. These activities require time and resources, and in many cases they elevate risks of predation and parasitism.

For organisms in which the sexes are separate—that is, in which individuals are either male or female—sexual reproduction has a much higher cost. This cost is a consequence of the fact that only half the genetic material of each individual offspring comes from each parent. Compared with asexually produced offspring, which contain only the genes of their single parent, the offspring of a sexual union contribute only half as much to the evolutionary fitness of either parent. This 50% genetic cost of sexual reproduction to the individual parent is sometimes referred to as the **twofold cost of meiosis.**

Natural selection favors those traits that reproduce the greatest number of copies of the genes that encode them. Genes for asexual reproduction propagate themselves much faster than genes for sexual reproduction, at least initially. A female can produce only a limited number of eggs; thus, from the standpoint of a hypothetical individual female, producing offspring asexually would result in twice as many copies of her genes in the next generation as producing the same number of offspring sexually (Figure 8.4). Under this scenario, male offspring would not only be superfluous, but mating with a male would reduce the female's genetic contribution to her offspring by 50%.

The twofold cost of meiosis does not necessarily apply to individuals having both male and female sexual function, as in the case of most plants and many invertebrates. When all of its offspring result from sexual unions (outcrossing), such an individual contributes one set of its genes to each of its offspring produced through female function and an equivalent number of sets, on average, to offspring produced through male function. Even in this situation, however, an individual that developed some of its eggs parthenogenetically would have higher fitness because it would allocate fewer resources to male function and pass on two copies of its genome rather than one. The twofold cost of meiosis also does not apply when the sexes are separate but males contribute, by means of parental care, as much as females to the number of offspring produced. When male parental investment doubles the number of

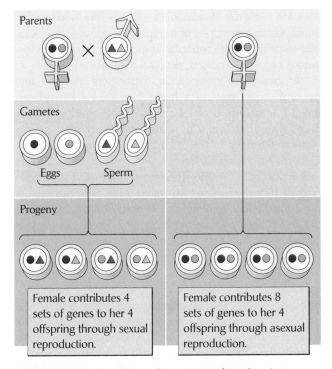

Parents

Gametes

Eggs   Sperm

Progeny

Female contributes 4 sets of genes to her 4 offspring through sexual reproduction.

Female contributes 8 sets of genes to her 4 offspring through asexual reproduction.

FIGURE 8.4 Sexual reproduction is costly. A female contributes only half as many sets of her genes to her progeny when she reproduces sexually as when she reproduces asexually.

offspring that a female could rear on her own, the cost of sex to the female is canceled.

Sex is maintained by the advantages of producing genetically varied offspring

If sex is so costly, then why does it exist? The high fitness cost of sexual reproduction is presumably offset by the advantage of producing genetically varied offspring when the environment itself varies over time or space. Clearly, a parent that survives to reproduce is well adapted to the conditions of its environment. Genetic variation among its offspring increases the chance that at least some of them will be well adapted to conditions that differ from those of the parent's environment. Another factor that might favor sex is the purging of mutations that would otherwise accumulate in clones of asexually produced individuals. Mutations crop up in each generation, and most are deleterious. Recombination during meiosis makes the removal of mutations possible. But are these advantages enough to overcome the twofold cost of meiosis?

A partial answer to this question comes from the sporadic distribution of asexual reproduction among com-

plex animals. For example, most vertebrate species that reproduce asexually belong to genera, such as *Ambystoma* (salamanders), *Poeciliopsis* (fishes), and *Cnemidophorus* (lizards), in which other species are sexual. This observation suggests that purely asexual species typically do not have long evolutionary histories; if they did, we would expect to see older and larger taxonomic groups of species sharing this trait. Thus, the long-term evolutionary potential of asexual populations appears to be low, possibly because of their greatly reduced genetic variation.

An important exception to this pattern is the bdelloid rotifers, a group of simple freshwater and soil organisms comprising more than 300 species. Males are unknown among the bdelloids, which apparently have reproduced asexually since their origin tens of millions of years ago. Recently, researchers in Alan Tunnacliffe's laboratory at Cambridge University found something unexpected in one of the genes of one bdelloid species: the gene has been duplicated within the bdelloid's genome, and each of the two copies produces a slightly different protein with a different function. Both proteins provide resistance to drying stress, but they function in different ways in different locations within the cells. The researchers believe that this functional divergence of the copies of a duplicated gene could be a mechanism for generating diversity in the gene pool of an asexual species. Indeed, because these rotifers have no sex to break up favorable gene combinations, asexual reproduction might even be favorable to creating genetic diversity in this way.

Regardless of such rare cases of parthenogenesis, sexual reproduction is the rule among multicellular organisms. Because the twofold cost of meiosis is so great, ecologists believe that, if we are to understand how evolution maintains sex, it is important to find a substantial short-term advantage of sexual reproduction. Otherwise, asexual reproduction might evolve in most populations. Most theoretical models based on temporal and spatial variation in the physical environment simply do not produce a great enough advantage to offset the twofold cost of meiosis. One promising alternative explanation is that sex provides the genetic variation necessary to respond to *biological* change in the environment—particularly changes in pathogens.

Parasites, especially microbes, that cause disease in their hosts are called **pathogens.** These organisms can evolve very rapidly because their population sizes are large and their generation times short compared with those of their hosts. The ability of pathogens to evolve responses to their hosts' defenses places a premium on rapid evolutionary responses by host populations, which might otherwise be driven to low numbers, and perhaps to extinction, by increasingly virulent pathogens. Any parent should benefit by producing offspring genetically different from itself, with unique combinations of genes for defenses to which the parent's pathogens are not well adapted. In this way, sex and genetic recombination could provide a moving target for the evolving pathogens and keep them from getting the upper hand. This idea is called the **Red Queen hypothesis,** after the famous passage in Lewis Carroll's *Through the Looking Glass and What Alice Found There,* in which the Red Queen tells Alice, "Now, here, you see, it takes all the running you can do, to keep in the same place." For this model to work, pathogens must have the potential to severely reduce the fitness of their hosts, and their effects must be strongly dependent on the genotypes of their hosts.

ECOLOGISTS IN THE FIELD

Parasites and sex in freshwater snails. One of the most compelling tests of the Red Queen hypothesis has been conducted by Curt Lively and his coworkers at Indiana University. Their test focuses on the freshwater snail *Potamopyrgus antipodarum,* a common inhabitant of lakes and streams in New Zealand. Most of the snails are asexual, all-female clones, but populations in some localities have about 13% males—enough to maintain some genetic diversity. Trematode worms of the genus *Microphallus* commonly infect the snails and sterilize them. The definitive hosts in the com-

plex life cycle of *Microphallus*—that is, the hosts within which the sexual stages of the parasite occur—are ducks. Not surprisingly, *Microphallus* is most abundant in shallow waters of lakes, where ducks feed (Figure 8.5).

A laboratory competition experiment showed that asexual snails reproduce faster than sexual individuals. Indeed, in nature, asexual clones tend to predominate in areas where *Microphallus* is absent or uncommon, particularly in the deep water of large lakes. Where the prevalence of *Microphallus* infection is high, however, sexual individuals are common. This finding suggests that in spite of their higher reproductive rates, asexual clones cannot persist in the face of high rates of parasitism. According to the Red Queen hypothesis, because the snails in asexual clones are genetically uniform, *Microphallus* can evolve rapidly enough to overcome their defenses.

Lively and his coworkers were able to test this idea by taking snails from three different depths of Lake Alexandrina, New Zealand, and exposing them to parasites obtained from each group of snails. They reasoned that if the parasites had evolved to specialize on local (depth-specific) populations of snails, then they should have the greatest success in infecting the populations they evolved with. That is exactly what happened, as shown in Figure 8.6: snails taken from shallow water were infected most readily by parasites taken from shallow water, and so on. Furthermore, infection rates were relatively low in deep-water snails because the parasites had had little opportunity to specialize on those snail populations. Remember that the definitive hosts—ducks—feed mostly in shallow

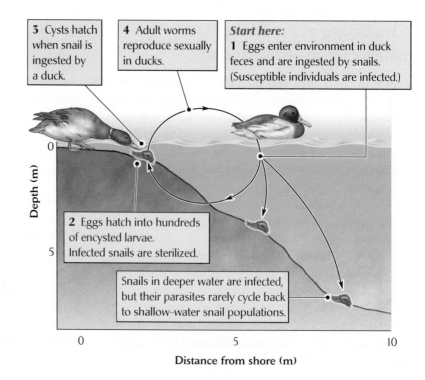

FIGURE 8.5 The trematode worm *Microphallus* is a parasite with a complex life cycle. Adult worms reproduce sexually in ducks. Larval stages reproduce asexually in snails, rendering infected snails sterile. After C. M. Lively and J. Jokela, *Proc. R. Soc. Lond.* B 263:891–897 (1996).

3 Cysts hatch when snail is ingested by a duck.

4 Adult worms reproduce sexually in ducks.

Start here:
1 Eggs enter environment in duck feces and are ingested by snails. (Susceptible individuals are infected.)

2 Eggs hatch into hundreds of encysted larvae. Infected snails are sterilized.

Snails in deeper water are infected, but their parasites rarely cycle back to shallow-water snail populations.

Depth (m)

Distance from shore (m)

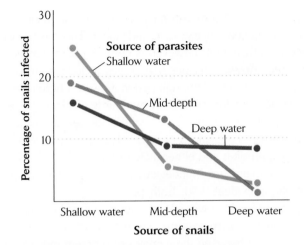

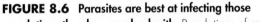

FIGURE 8.6 Parasites are best at infecting those populations they have evolved with. Populations of snails (*Potamopyrgus antipodarum*) obtained from each of three different depths in Lake Alexandrina were exposed in the laboratory to parasites (*Microphallus*) obtained from snails at each of the three different depths. Data from C. M. Lively and J. Jokela, *Proc. R. Soc. Lond.* B 263:891–897 (1996).

water, and so only the shallow-water parasite populations cycled regularly through snail host populations. In Lake Alexandrina, snails from deeper water were infected mostly by parasites from shallow-water populations when ducks roosted in deep water and left parasites behind in fecal matter. Because deep water provided a partial refuge from parasites, sexual lineages of snails did not compete well against asexual lineages there, and the prevalence of male snails was low.

If parasites evolved higher probabilities of infecting individuals from more common clones of snails over time, then one would expect parasites to be less prevalent in rare clones because natural selection does not favor specialization on rare resources. However, rare clones that experience low rates of parasitism should tend to become more abundant over time because they remain fertile and can replace highly infected clones in a population. Then, as a clone becomes common, parasites should evolve to specialize on it, and should eventually reduce its abundance. Accordingly, we should see cycling in the relative abundances of clones in an asexual population over time. Mark Dybdahl and Curt Lively found just such a cycle when they surveyed clones of *Potamopyrgus* along the shoreline of Lake Poerua, New Zealand, over 5 years, or about fifteen snail generations. The data showed marked variation in the abundances of four common clones, as well as a marked increase in the rate of infection by *Microphallus* in years following increases in the abundance of a clone (Figure 8.7). These patterns are consistent with the predictions of the Red Queen hypothesis. ▌

Several recent studies have supported the Red Queen hypothesis. For example, investigators have found higher rates of outcrossing—that is, mating with other individuals rather than self-pollinating—in species of plants attacked by a greater variety of fungal pathogens. In a study of the planarian flatworm *Schmidtea polychroa* in a lake in northern Italy, investigators found that parthenogenetic individuals were more frequently infected with various protozoan parasites than were sexual individuals. Purging of mutations seemed to be an additional benefit

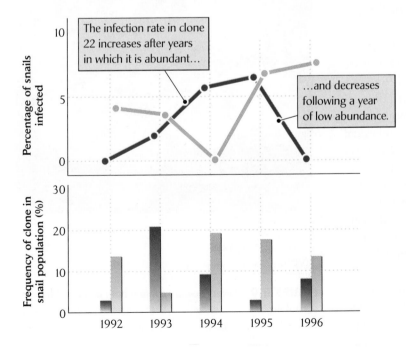

FIGURE 8.7 Cycles of parasite prevalence follow cycles of abundance in asexual clones. The frequencies of two different clones in a population of *Potamopyrgus* snails from Lake Poerua and rates of infection of each clone by *Microphallus* are shown over 5 years. After M. F. Dybdahl and C. M. Lively, *Evolution* 52:1057–1066 (1998).

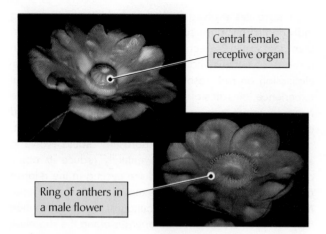

FIGURE 8.8 Dioecious plants have two separate sexes. The dioecious tree *Clusia grandiflora* has sexually dimorphic flowers, female (above) and male (below). Photos by Volker Bittrich.

of sexual reproduction in these planaria: embryo mortality was higher in the parthenogenetic clones, a sign that the clones had accumulated deleterious mutations over time. Moreover, rates of parasitism and of embryo mortality were correlated among parthenogenetic clones, suggesting that parasitism and mutation accumulation work together to favor sexual reproduction.

Despite the success of these research programs, sex remains one of the most challenging questions that face biologists. At this point, we shall accept that sex is with us, and we shall turn to exploring some of the consequences of sexual reproduction in the lives of organisms.

Individuals may have female function, male function, or both

We humans are used to thinking in terms of two sexes, female and male. But female and male sexual functions may be combined in the same individual, or an individual may change its sex during its lifetime. When both sexual functions occur in the same individual, that individual is a **hermaphrodite** (named after the mythological Hermaphroditus, son of Hermes and Aphrodite, who while bathing became joined in one body with a not-so-shy nymph). Male and female functions in hermaphrodites may be **simultaneous,** as in the case of many snails and most worms, or they may be **sequential:** male first in some mollusks, echinoderms, and plants; female first in some fishes.

Plants that exhibit separate sexes in different individuals are called **dioecious,** from the Greek *di-* ("two") and *oikos* ("dwelling;" also the root of the word "ecology")

(Figure 8.8). **Monoecious** plants bear distinct male and female flowers on the same individual. The most common configuration, however, is seen in plants that bear **perfect flowers** (Figure 8.9), which include both male and female parts. Although perfect-flowered hermaphrodites account for more than two-thirds of plant species, nearly all imaginable combinations of sexual patterns are known. Populations of some plant species have both hermaphrodites and either male or female individuals, or male, female, and monoecious individuals, or hermaphroditic individuals with both perfect flowers and either male or female flowers. Most populations of hermaphrodites are fully **outcrossing,** which means that fertilization takes place between the gametes of different individuals. The rarer case of self-fertilization will be discussed in Chapter 13.

The sexual pattern that occurs in a given sexual, outcrossing population depends on the relative fitness costs and benefits to an individual of having either or both sexual functions. One can measure the fitness contributions of male and female sexual function by the number of sets of genes transmitted to offspring through either male or

FIGURE 8.9 Perfect flowers contain both male and female sexual organs. The perfect flower of *Miconia mirabilis* possesses both anthers and carpels. Photo by R. E. Ricklefs.

female gametes. When females can achieve added fitness through some amount of male function by giving up a smaller amount of their female function, selection favors individuals that shift some resources to male function. Similarly, males that can add female function and not cut deeply into the fitness they achieve through male function are also favored by selection (Figure 8.10). It would seem that both male and female flowers can add the other sexual function with little cost. After all, the basic flower structure and the floral display necessary to attract pollinators are already in place in single-sexed flowers. Under these circumstances, we would expect hermaphroditism to arise frequently, as it has among plants and most simple forms of animal life.

Many hermaphroditic plants have mechanisms to prevent self-fertilization and ensure outcrossing. Selfing in many such species is prevented by self-incompatibility (SI) genes. Individuals with the same SI genotypes (including an individual mating with itself) cannot produce offspring. In many cases, pollen that lands on a stigma with the same SI allele does not germinate and grow. In species with SI systems, selection favors new alleles that avoid incompatibility with existing alleles in the population, so allelic diversity is often very high. Thus, the SI system provides some of the advantages of separate sexes while retaining both male and female sexual function in the same individual.

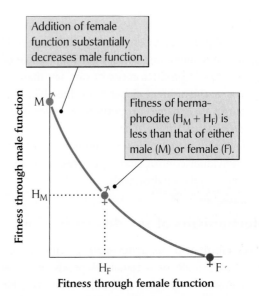

FIGURE 8.11 **Natural selection sometimes favors separate sexes.** When male and female functions interfere with each other in the same individual, hermaphrodites are less successful than males and females and are excluded from a population.

Separate sexes are favored by selection when gains in fitness from adding one sexual function bring about even greater losses in the other sexual function (Figure 8.11). This may be the case when establishing a new sexual function entails a substantial fixed cost before any gametes can be produced. Sexual function in complex animals requires gonads, ducts, and other structures for transmitting gametes. Moreover, in many animals, maleness requires specializations for mate attraction and combat with other males, and femaleness requires specializations for egg production or brood care. Such fixed costs may put hermaphroditism at a disadvantage compared with sexual specialization. In fact, hermaphroditism occurs rarely among animal species that actively seek mates and that engage in brood care. It is much more common among sedentary aquatic animals that simply shed their gametes into the water.

MORE ON THE WEB *Sequential Hermaphroditism.* Some organisms are male first, and then become female later in their lives, or vice versa.

The sex ratio of offspring is modified by natural selection

Male and female individuals differ with respect to their ecological requirements and their social interactions within populations. Consequently, the relative fitnesses of males

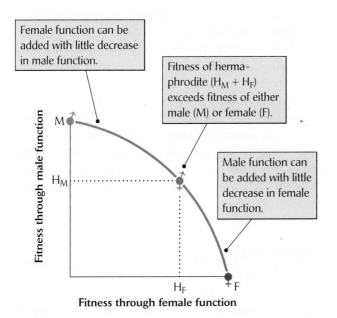

FIGURE 8.10 **Natural selection sometimes favors hermaphrodites.** When male or female function can be added with little loss of fitness achieved through the opposite sexual function, hermaphrodites can exclude males and females from a population.

and females can change depending on the availability of resources, population density, and other factors. Depending on the relative fitnesses of male and female offspring, parents should produce more of one sex than the other. Of course, this strategy depends on the ability of parents to control the relative proportion of males and females (the **sex ratio**) among their offspring. In fact, many studies have shown that the sex ratio is under both genetic and individual control, and that it responds to selective influences in the environment.

Mechanisms of sex determination

In organisms with separate sexes, whether an individual becomes a male or a female depends on a variety of mechanisms. In humans and other mammals, birds, and many other organisms, sex is determined by inheritance of sex-specific chromosomes. Female mammals have two X chromosomes, and thus have an XX genotype, whereas males have one X chromosome and one Y chromosome, and thus have an XY genotype. The Y chromosome is specific to males, and a single copy of the Y chromosome inherited from the father determines that a zygote will develop into a male. Males produce X-bearing and Y-bearing sperm in approximately equal numbers. Females produce only X-bearing gametes. Thus, on average, half the progeny in a population will be female and half male.

If this were the end of the story, there would be little more to say about sex ratios in populations, but in fact it is just the beginning. Many factors can alter the sex ratio in an XY sex determination system, including competition among X- and Y-bearing sperm to fertilize eggs or selective abortion of male or female embryos. In birds and butterflies, the female is the sex with XY sex chromosomes, so females can adjust the sex ratio of their offspring during egg formation by controlling whether an X or a Y chromosome gets into the egg. When a single haploid egg is formed from a diploid germ cell, only one of the four sets of haploid nuclei resulting from meiosis is passed to the egg. As a result, the female reproductive system has some control over whether X-bearing or Y-bearing chromosome sets are partitioned into the egg. This control is so precise that in many species of birds, the chance of an offspring being male changes predictably from the first to the last laid egg in the clutch, possibly as a way of controlling competitive interactions between male and female siblings.

In some species, sex is determined by the physical environment. In several species of turtles, lizards, and alligators, the sex of an individual is determined by the temperature at which it develops in the egg. In turtles, embryos that

FIGURE 8.12 The blue-headed wrasse is a sequential hermaphrodite. Young individuals of this species develop into males (like the individual shown here) or females depending on their social context. © Zigmund Leszczynski/Animals Animals; all rights reserved.

develop at lower temperatures produce males, and those that develop at higher temperatures produce females; the reverse is true in alligators and lizards.

MORE ON THE WEB *Environmental Sex Determination.* In many reptiles, sex is determined by the temperature at which the embryo grows.

In some fish species, whether a young individual develops into a male or a female depends strongly on the social environment. The blue-headed wrasse (*Thalassoma bifasciatum*), a common coral reef species (Figure 8.12), is a sequential hermaphrodite. When individual wrasses are raised in isolation, they invariably develop into females. When they are raised in small groups, however, at least one of the individuals develops initially into a male without passing through a female phase. Females may become males later in life, when they grow large enough to compete for territories on the reef, but the primary males never change their sex.

ECOLOGISTS IN THE FIELD **Effects of fishing on sex switching.** The coast of southern California and Baja California is home to a fascinating fish called the California sheephead. This fish is a sequential hermaphrodite: once it achieves sexual maturity, it breeds first as a female, but then, after growing to a larger size, changes its sex and breeds as a male. Like most life history traits, the timing of sex switching is related to reproductive success.

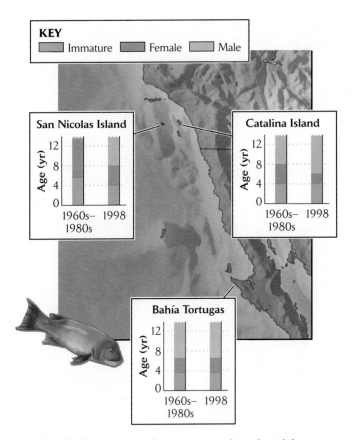

FIGURE 8.13 Heavy fishing pressure has altered the ages of maturity and sex switching in the California sheephead. Ages at maturity and at switching from female to male are shown for populations of the California sheephead (*Semicossyphus pulcher*) in the 1960s–1980s and in 1998 at Bahía Tortugas (little fishing), Catalina Island (moderate recreational fishing), and San Nicolas Island (heavy commercial fishing). After S. L. Hamilton et al., *Ecological Applications* 17:2268–2280 (2007).

Males are territorial, and small individuals cannot compete with large males to secure mating sites. Thus, small individuals enjoy higher reproductive success as females. Larger individuals, however, can obtain more matings and produce more offspring by breeding as males.

The California sheephead has been the target of recent commercial and recreational fishing. According to Scott Hamilton and his colleagues from the University of California at Santa Barbara, some sheephead populations have experienced relatively little fishing pressure, whereas other populations have been recreationally fished since the mid-1960s or commercially fished since the late 1980s. Because the biggest fish are preferred in both fisheries, the researchers asked whether sheephead populations experiencing continuous removal of the largest individuals would undergo changes in their life histories.

By comparing historical data on fish collected from the late 1960s through the early 1980s with data on fish that they collected in 1998, Hamilton and colleagues were able to determine whether historical and current life histories differed. In one population that had experienced relatively little fishing (Bahía Tortugas), they found that neither age at maturity nor age at switching from female to male had changed. In a population that had experienced recreational fishing (Catalina Island), age at maturity had not changed, but fish in this population now switch sex at a younger age than they did only a couple of decades before. A population under heavy pressure from commercial fishing (San Nicolas Island) both matures at an earlier age and changes sex at an earlier age (Figure 8.13).

In heavily fished populations, as mentioned in Chapter 7, an earlier age at maturity is favored because few individuals live to old age, and any individual that can breed at a younger age will be favored. An earlier age of sex switching is favored because removal of large fish translates to removal of male fish. Hence, any individual that can change from female to male at a younger age will have a larger number of potential females with which it can mate and little competition from larger males. In this way, human activities can influence the life histories of organisms over relatively short periods. ▮

Evolution of the sex ratio

When the sexes are separate, one can define the sex ratio among the progeny of an individual, or within a population, as the number of males relative to the number of females. Because females and males occur at a ratio of approximately 1 : 1 in the human population, and in the populations of most other species as well, we consider the 1 : 1 sex ratio the usual condition and regard deviations from this ratio as special cases. Yet there are many such deviations.

As we have seen, an XY sex determination system tends to produce a 1 : 1 sex ratio. However, rather than using this tendency to explain the roughly equal numbers of males and females observed, we should ask why this particular mechanism of sex determination is so common. That is, why is a mechanism that produces a 1 : 1 sex ratio seemingly favored by natural selection?

We can explain the predominant 1 : 1 sex ratio by the following simple reasoning outlined at the opening of this chapter: Every offspring of a sexual union has exactly one mother and one father. Consequently, if the sex ratio of a population is not 1 : 1, individuals of the rarer sex will enjoy greater reproductive success because they will compete for matings with fewer others of the same sex (Figure 8.14). For example, if a population of 2 males and 5 females produced 10 offspring, each male

FIGURE 8.14 The rare sex advantage leads to a 1:1 sex ratio.
Because each offspring has equal genetic contributions from its mother and its father, individuals of the rarer sex in a population contribute, on average, more sets of their genes to the next generation. This fact explains the nearly equal proportions of male and female individuals observed in most populations.

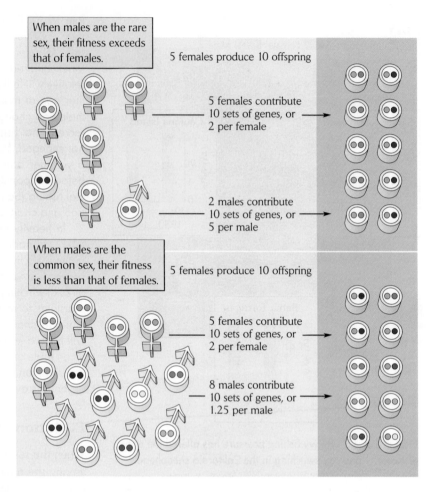

When males are the rare sex, their fitness exceeds that of females.

5 females produce 10 offspring

5 females contribute 10 sets of genes, or 2 per female

2 males contribute 10 sets of genes, or 5 per male

When males are the common sex, their fitness is less than that of females.

5 females produce 10 offspring

5 females contribute 10 sets of genes, or 2 per female

8 males contribute 10 sets of genes, or 1.25 per male

would contribute 5 sets of genes to those offspring, but each female would contribute only 2 sets of genes. Consequently, individuals of the rarer sex would contribute more sets of their genes to subsequent generations than the more common sex would. Thus, when a population has more females than males, natural selection will favor any genetic tendency on the part of a parent to produce a larger proportion of male offspring. This will increase the frequency of males in the population and bring the sex ratio closer to 1:1. Similarly, when females are the rarer sex, genotypes that increase the proportion of female progeny will be favored, and the frequency of females in the population will increase. When males and females are equally numerous, individuals of both sexes contribute equally to future generations on average, and different frequencies of males and females produced at random among the progeny of one individual are of no consequence to its relative long-term reproductive success. Because the fitness of genes affecting the sex ratio depends on the frequencies of males and females in a population, the evolution of the sex ratio is said to be the product of **frequency-dependent selection.**

MORE ON THE WEB *Female Condition and Offspring Sex Ratio.* Female mammals should produce male offspring only when they are in excellent condition and can raise sons that will be superior competitors.

This explanation for the 1:1 sex ratio depends, however, on individuals having the opportunity to mate with unrelated individuals within a large population. When individuals do not disperse far from where they are born, or when they mate prior to dispersal, mating often takes place among close relatives (a situation known as **inbreeding**). In the extreme case, mating may occur among the progeny of an individual parent. In this situation, known as **local mate competition,** competition among males for mates takes place among brothers. From the standpoint of the parent of these siblings, one son would serve just as well as many to fertilize his sisters and propagate the parent's genes. In this situation, the number of copies of her genes that a mother passes on to her grandoffspring depends only on the number of daughters she produces, because each son's genetic contribution to

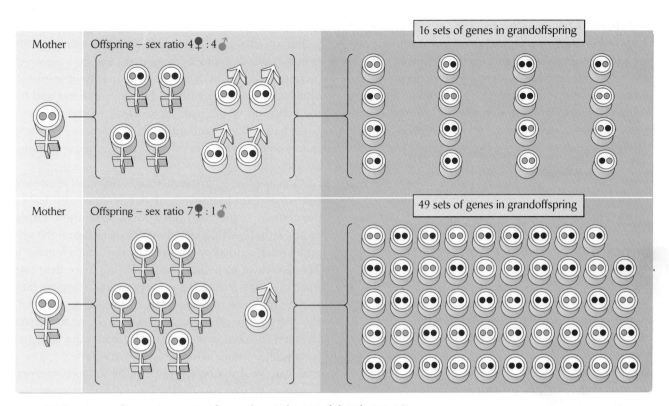

FIGURE 8.15 Local mate competition favors the production of daughters. When mating opportunities are restricted to siblings, and total fecundity (number of female plus male offspring) is limited, females should produce a high proportion of daughters in their progeny.

those daughters' offspring will also come from the mother. Thus, females that produce daughters at the expense of sons will have more grandoffspring and greater evolutionary fitness (Figure 8.15).

Brother–sister (sib) mating occurs commonly in certain wasps that parasitize other insects or lay their eggs and complete their larval development within the fruits of certain plants (Figure 8.16). For many of these species, hosts are so scarce, and mates are so difficult to find, that females mate where they hatch before they disperse to find new hosts on which to lay their own eggs. These wasps can determine the sex ratio among their progeny, and that ratio can be predicted by the degree of inbreeding their offspring will experience.

When a single female wasp parasitizes a host, her female offspring will be limited to mating with their brothers. Under these circumstances, male offspring will contribute little to the reproductive success of their mother, as described above. Therefore, these wasps skew the sex ratio of their progeny greatly in favor of females—to the point of producing only one male per brood in some species. Males of many of these species lack wings, and in extreme cases, fertilize their sisters as larvae within the host. When two or more females lay their eggs in the same host, however, their male offspring can mate either with their sisters or with the daughters of the other females. Just as one would expect, the proportion of males in a brood increases as the possibility that sons might inseminate the female offspring of another wasp increases.

FIGURE 8.16 Many parasitic wasps mate with siblings in their host. A female braconid wasp (*Nasonia vitripennis*) deposits her eggs in a fly pupa. Her daughters will mature and mate with their brothers in the pupa before they disperse to lay their own eggs. Courtesy of John H. Werren.

How do wasps control the sex of their offspring? Hymenopterans (bees, ants, and wasps) determine the sex of their offspring by an unusual mechanism: fertilized eggs produce females and unfertilized eggs produce males. Consequently, females are diploid and males are haploid, a condition known as **haplodiploidy.** Reproductive females can control the sex ratio of their offspring simply by storing sperm when they mate and using it—or not—to fertilize their eggs.

Mating systems describe the pattern of pairing of males and females within a population

The **mating system** of a population describes the pattern of matings between males and females—for example, the number of simultaneous or sequential mates each individual has and the permanence of the pair bond between them. Like the sex ratio, the mating system of a population is subject to natural selection and evolutionary modification. Consequently, mating systems can usually be explained by the ecological relationships of individuals.

Mating systems reflect variation in male and female reproductive success

It is a basic asymmetry of life that female and male functions contribute differently to an individual's evolutionary fitness. A female's reproductive success depends on her ability to make eggs and otherwise provide for her offspring. Each large female gamete requires far more resources than each tiny male gamete, so a female's ability to gather resources to make eggs determines her fecundity. A male's reproductive success usually depends on the number of eggs he can fertilize.

Males that mate with as many females as they can locate often provide their offspring with nothing more than sets of genes. Such males are said to be promiscuous. **Promiscuity** usually precludes a lasting pair bond. Among animal taxa as a whole, promiscuity is by far the commonest mating system, and it is universal among outcrossing plants. Promiscuity is associated with a high degree of variation in male mating success: some individual promiscuous males may obtain dozens of matings, while others get none.

When eggs and sperm are released directly into the water, or pollen is shed into the wind, much of the variation in male mating success is simply random. Whether a particular sperm is the first to find an egg is largely a matter of chance. When males attract or compete for mates, however, reproductive success can be influenced by factors such as body size and the quality of courtship displays, which are controlled by genetic factors and by the condition of the male, as we shall see below. Even when fertilization is random, males that produce the most sperm or pollen are bound to father the most offspring on average.

A mating system in which a single individual of one sex forms long-term bonds with more than one individual of the opposite sex is called **polygamy.** Most often, a male mates with more than one female, in which case the system is referred to as **polygyny** (literally, "many females"). The rare cases of a single female having more than one male mate are referred to as **polyandry.** Polygyny may require the male to defend several females against mating attempts by other males (Figure 8.17), or to defend territories or nesting sites to which females are attracted to raise their young. Thus, polygyny may arise because a male can prevent access by other males to more than one female, in which case his contribution to his progeny may be primarily genetic, or because he can control or provide resources that females need for reproduction.

Monogamy is the formation of a pair bond between one male and one female that persists through the period that is required for them to rear their offspring, and which may endure even until one of the pair dies. Monogamy is favored primarily when males can contribute substantially

FIGURE 8.17 Elephant seals are polygynous. Successful males attract many females and defend them against the sexual advances of other males. Frans Lanting/Minden Pictures.

to the number and survival of their offspring by providing parental care. Hence, it is most common in species with dependent offspring that can be cared for equally well by either sex. Monogamy is not common in mammals because males neither carry the developing embryo nor produce milk. But it is common among birds, especially those in which parents feed their offspring. Male and female birds can incubate eggs and feed the young equally well.

Recent genetic surveys of monogamous bird populations have revealed that males other than a female's mate may father some of her offspring as a result of so-called **extra-pair copulations,** or **EPCs.** A third or more of the broods produced by some monogamous species contain one or more offspring sired by a different male. Most EPCs are matings with males on neighboring territories. This behavior surely increases, at relatively little cost, the fitness of those neighboring males. It is not known whether EPCs increase the fitness of a mated female, but they could do so if the neighboring males have better genotypes than her mate or if her reproductive success is improved by greater genetic variation among her offspring. The constant threat of EPCs has also selected strongly for **mate guarding** behaviors on the part of males during their mates' periods of fertility.

Plants can also have very complex mating systems, even though most species consist of hermaphroditic individuals. In one variant system, called *gynodioecy,* hermaphrodites and females coexist in the same population. Gynodioecy occurs when mutations arise among hermaphroditic plants that lead to male sterility, thereby creating plants that are, for all intents and purposes, females. In many cases, these genes are present in the chloroplast and are transmitted by cytoplasmic inheritance from the female parent. Such is the case for common thyme (*Thymus vulgaris*), a native of the Mediterranean region that is best known as an herb used to flavor food.

Among flowering plants, cytoplasmic genes are transmitted only through female gametes—the ovules. The male gametes consist almost entirely of nuclei and pass on very little cytoplasm to the zygote. Thus, it is the female parent that transmits chloroplasts, and the genes they contain, to the zygote along with the cytoplasm in the ovule. When male function (i.e., male flower and pollen production) competes with female function (ovule production) for a plant's resources, cytoplasmic genes that reduce male function or even cause male sterility are strongly favored because male function does not contribute to the fitness of cytoplasmic genes. However, selection pressure on cytoplasmic genes is at odds with that on nuclear genes, which are transmitted equally through male and female function. Should male sterility genes increase in

frequency, populations may not have enough hermaphrodites with male function to provide pollen for fertilization, and female reproductive success may decline. Under these circumstances, nuclear restorer genes, which block the action of the cytoplasmic sterility genes, are strongly favored to restore male sexual function.

These opposing selection pressures on cytoplasmic and nuclear genes set up a constant conflict that drives changes in the sex ratios of thyme populations. The actual sex ratio of a population depends on the availability of male sterility and restorer genes. Each of these types of genes arises sporadically and spontaneously by mutation. Over time, the sex ratio for thyme is rarely at equilibrium—it is constantly changing back and forth between relatively high and low levels of male sterility.

This brief overview of mating systems has barely scratched the surface of this fascinating and complex topic. You might wish to search for information on bluegill sunfish, in which males come in two varieties: large, dominant individuals that defend breeding territories and small "sneaker" males that steal copulations from females attracted to the dominant males' territories. Indeed, fishes may have the widest range of reproductive tactics of any animal group. Some parthenogenetic species have done away with males, but the females must nonetheless mate with males of a different species to initiate the development of their eggs. Conflict between the sexes, particularly in promiscuous species, is a common theme in the animal world. Females can be badly injured by the repeated copulation attempts of males, and in many species females have strategies to prevent this. Much of this amazing diversity of behavior has been related to the different ecological circumstances of species, as we shall see next.

MORE ON THE WEB   *Alternative Male Reproductive Strategies.* Males of different species take different approaches to winning a female's favor.

The polygyny threshold model: Relating mating systems to ecology

In cattail marshes throughout North America, male red-winged blackbirds (*Agelaius phoeniceus*) establish territories in early spring (Figure 8.18). Marsh habitat is heterogeneous with respect to vegetation cover and water depth, which affect food supply and the safety of nests from predators, and therefore these territories vary greatly in their intrinsic quality. Females return to the breeding grounds after the males, by which time the males have established territories. What do these females look for in a mate?

FIGURE 8.18 The male red-winged blackbird has a conspicuous display. Males establishing territories in spring use these displays to attract females and defend their territories against other males. Photo by Richard Day/Animals Animals.

Female red-winged blackbirds appear to assess the quality of male territories, and the first females to arrive pair monogamously with the best males—that is, those holding the best territories. A latecomer is faced with the choice between pairing monogamously with a male holding a lower-quality territory and pairing polygynously with a male holding a higher-quality territory, but sharing its resources with one or more other females. Polygyny is favored in this system by the large variation in resources in the marsh environment. In contrast to blackbirds, many forest birds live in habitats that are more homogeneous than marshes. Because bird territories in forests vary less in quality, most species of forest birds are primarily monogamous; few territories rise above the polygyny threshold.

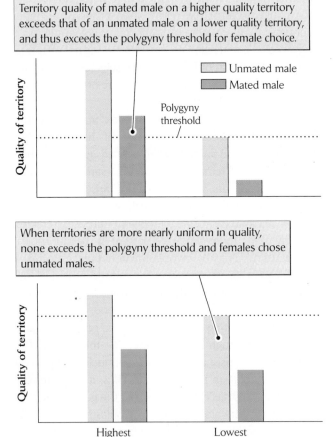

FIGURE 8.19 The polygyny threshold model predicts the variation in habitat quality at which polygyny will occur. When habitat quality varies enough, females may enjoy greater reproductive success by mating polygynously with a male holding a high-quality territory than by choosing an unmated male on a low-quality territory.

As we have seen, a female increases her fecundity by choosing a territory or a mate of high quality. A male gains fitness by increasing the number of his mates, as long as his territory holds sufficient resources. Thus, polygyny arises when a female can obtain greater reproductive success by sharing a male with one or more other females than she can by forming a monogamous relationship.

Suppose that the quality of two males' territories differs so much that a female could rear as many offspring on the better territory, while sharing it with other females and having little or no help from her mate, as she could on the poorer territory with help from a monogamous mate. The point at which the reproductive success of a polygynous female on a better territory equals that of a monogamous female on a poorer territory is referred to as the **polygyny threshold** (Figure 8.19). According to the polygyny threshold model, polygyny should occur only when the quality of male territories varies so much that some females will have higher reproductive success when mated to a polygynous male on a high-quality territory than they would when mated to a monogamous male on a low-quality territory.

Sexual selection can result in sexual dimorphism

Regardless of the mating system, the initial stages of reproduction in many species involve choosing mates. In polygynous and promiscuous mating systems, males gain by mating with as many females as they can, but the choice of mates is usually the prerogative of the female. How should a female choose among the males that court her attention? If males differed in obvious features that could affect a female's reproductive success, and if her offspring could inherit those features, she should choose to mate with the male of highest quality. Of course, males should do everything in their power to advertise their quality—they should strut their stuff. This intense competition among males for mates has resulted in the evolution of male attributes for use in combat with other males or in attracting females. Such selection by one sex for specific characteristics in individuals of the opposite sex is referred to as **sexual selection.**

A common result of sexual selection is **sexual dimorphism,** meaning a difference in the outward appearance of male and female individuals of the same species. Sexual selection tends to produce sexual dimorphism in body size, ornamentation, coloration, and courtship behavior. Such traits that distinguish the sexes, over and above the sexual organs themselves, are known as **secondary sexual characteristics.** Charles Darwin, in his book *The Descent of Man and Selection in Relation to Sex,* published in 1871, was the first to propose that sexual dimorphism could be explained by selection applied uniquely to one sex.

Sexual dimorphism can arise in three ways. First, the dissimilar sexual functions of males and females lead to different considerations in the evolution of their life histories and ecological relationships. For example, because females produce large gametes, their fecundity often increases in direct relation to their body size; this may explain why females are larger than males in many species, such as orb-weaving spiders (Figure 8.20). A size difference is particularly likely to evolve when fertilization is internal and producing large numbers of sperm is not a major consideration for males.

Second, sexual dimorphism may result from contests between males, which may favor the evolution of elaborate weapons for combat, such as the antlers of elk and the horns of mountain sheep (see Figure 11.11). Males that win such contests are more likely to gain access to females. When large size confers an advantage in these contests, males may be larger than females (see Figure 8.17).

FIGURE 8.20 Females of many spider species are larger than males. These male and female golden silk spiders (*Nephila clavipes*) were photographed in a web in northeastern Florida. Photo by Millard H. Sharp.

Third, sexual dimorphism may arise through the direct exercise of mate choice. With few exceptions, females do the choosing, and males attempt to persuade them with magnificent courtship displays. That females choose, and males compete among themselves for the opportunity to mate, is a consequence of the asymmetry of reproductive investment that defines the male and female functions. As we saw above, males enhance their fecundity in direct proportion to the number of matings they obtain. The fecundity of females is limited by the number of eggs they can produce, but they stand to improve the fitness of their offspring by choosing to mate with males that have superior genotypes.

Female choice

Most males experience female choice at some level. One of the first demonstrations of female choice in nature came from an experimental study of tail length in male

FIGURE 8.21 Sexual selection may favor elaborate courtship displays. The tail of the male long-tailed widowbird (*Euplectes progne*) is a handicap in flight, but is attractive to females. Photo by Gregory G. Dimijian, M. D./Photo Researchers.

long-tailed widowbirds (*Euplectes progne*). This polygynous species inhabits open grasslands of central Africa. The females, which are about the size of a sparrow, are mottled brown, short-tailed, and altogether ordinary in appearance. During the breeding season, the males are jet black, with a red shoulder patch, and they sport a half-meter-long tail that is conspicuously displayed in courtship flights (Figure 8.21). The most successful males may attract up to a half dozen females to nest in their territories, but they provide no care for their offspring. The tremendous variation in male reproductive success in this species provides classic conditions for sexual selection.

In a simple yet elegant experiment, researchers cut the tail feathers of some males to shorten them and glued the clipped feather ends onto the feathers of other males' tails to lengthen them. Tail length had no effect on a male's ability to maintain a territory, but males with experimentally elongated tails attracted significantly more females than those with shortened or unaltered tails (Figure 8.22). This result strongly suggests that females choose mates on the basis of tail length. Many subsequent studies have demonstrated that females choose their mates on the basis of such conspicuous differences among males.

MORE ON THE WEB   *The Origin of Female Choice.* Many issues regarding female choice remain unresolved: Which came first, female choice or male traits that indicate fitness? How

are the various ornaments of males related to their fitness? Why don't low-quality males cheat by taking on a high-quality appearance?

Runaway sexual selection

Once female choice is established in a population, it exaggerates fitness differences among males and may create what is known as **runaway sexual selection.** Female widowbirds might have intrinsically preferred males with longer tails; alternatively, tail length might have indicated male fitness, and females might therefore have evolved a preference for long tails. In either case, the females' mating preference would give long-tailed males a fitness advantage. If females choose mates by comparing among males, rather than by comparing males to some idealized standard of beauty, then their mating preferences will continually select for further elaboration of male traits. In other words, if longer tails in males are what females prefer, then longer tails will evolve. The peacock's tail, as well as the other outlandish (to our eyes) sexual ornaments and behaviors liberally spread throughout the animal kingdom, provide convincing evidence that some sort of runaway process must be at work. This process can be brought to a halt only when genetic variation for further elaboration of a trait is exhausted or the costs of producing the trait balance its reproductive benefits.

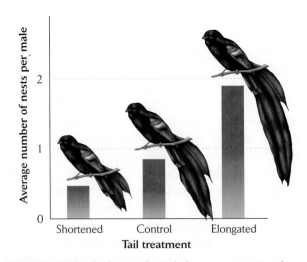

FIGURE 8.22 The longer the tail, the more attractive the male. Male long-tailed widowbirds with artificially elongated tails attracted more females to nest in their territories than did control males or males with shortened tails. The tail feathers of the control males were cut and reattached to simulate the experimental treatment without affecting the length of the tail. After M. Andersson, *Nature* 299:818–820 (1982).

If sexually selected traits indicate—at least initially, before runaway sexual selection takes hold—intrinsic attributes of male quality, we are then faced with a paradox. Presumably, such outlandish traits as the tail of the long-tailed widowbird burden males by making them more conspicuous to predators and by requiring energy and resources to maintain. How, then, can such traits indicate, let alone contribute to, male quality?

The handicap principle

One intriguing possibility, suggested by the Israeli biologist Amotz Zahavi, is that elaborate male secondary sexual characteristics act as handicaps. That a male can survive while bearing such a handicap indicates to a female that he has an otherwise superior genotype. This idea is known as the **handicap principle.** It may sound crazy, but if you wanted to demonstrate your strength to someone, you might make your point by carrying around a large set of weights. A weaker individual couldn't do it, and thus could not falsely advertise strength. Accordingly, the greater the handicap an individual bears, the greater its ability to offset that handicap with other virtues—and to pass genes for those virtues on to its offspring. One small European songbird, the wheatear, takes the iron-pumping analogy literally and festoons its nesting ledge with up to 2 kilograms of small stones carried from a distance in its beak.

One virtue that males might possess, and which might be demonstrated by producing a showy plumage, is resistance to parasites and pathogens. William D. Hamilton and Marlene Zuk first proposed this idea in 1982. They suggested that only individuals with genetic factors that allow them to resist parasite infection could produce or maintain a bright and showy plumage. Thus, a well-maintained courtship display of elaborate plumage may provide a convincing demonstration of high male fitness, even when the display itself is an encumbrance. The importance of parasites to this theory is that they evolve rapidly and thereby continually apply selection for genetic resistance factors. We have already discussed similar reasoning for the evolutionary maintenance of sex itself.

The Hamilton–Zuk hypothesis, along with its subsequent modifications, comes under the general heading of **parasite-mediated sexual selection.** Its general assumptions—that parasites reduce host fitness, that parasites alter male showiness, that parasite resistance is inherited, and that females choose less parasitized males—are generally supported by experiments and field observations. For example, feather lice produce obvious damage by eating the downy portions of feathers and the barbules of feather vanes (Figure 8.23). In feral rock doves, highly

FIGURE 8.23 Feather lice produce substantial damage. (a) Scanning electron microscopic view of a louse on a host's feather. The louse is about 1 mm long, and is seen from a dorsal view. (b) Average (*center*) and heavy (*right*) damage to abdominal contour feathers by feather lice. A normal feather is at the left. *Courtesy of D. H. Clayton, from D. H. Clayton, Am. Zool. 30:251–262 (1990).*

infested males had higher metabolic requirements in cold weather because their damaged feathers reduced the insulation value of their plumage, and they were lighter in body mass. Female rock doves preferred clean to lousy males by a ratio of three to one.

A particularly elegant set of studies on ring-necked pheasants, performed by Torbjorn von Schantz and his colleagues at the University of Lund in Sweden, showed that females prefer males with long spurs (a spikelike projection from the back of a male pheasant's lower leg), and that long spurs are linked genetically to major histocompatibility complex (MHC) genes that influence susceptibility to disease. Males with longer spurs had MHC alleles that were linked to longer life spans. Therefore, females that choose to mate with long-spurred males should tend to produce offspring with a higher chance of surviving to reproduce as adults.

Sexual selection remains an active area of research, and much has yet to be learned. Studies of sexual displays show quite clearly, however, the power of natural selection to modify structures and behaviors, and the ways in which these changes can be directed by the asymmetry of sexual function in males and females.

SUMMARY

1. In most multicellular species, reproductive function is divided between two sexes. Sexual reproduction involves the production of male and female gametes with haploid chromosome numbers. Haploid gametes are formed by meiosis, in which chromosome number is halved and maternal and paternal sets of genes are mixed. Male and female gametes unite to form the zygotes that start a new generation.

2. Species with separate sexes incur the twofold cost of meiosis: sexual females pass on only half as many copies of their genes to their progeny as asexually reproducing individuals. This high fitness cost is offset by the advantage of producing genetically varied offspring, which increases the probability that at least some of those offspring may be well suited to varied conditions, and by the possibility of purging deleterious mutations.

3. An alternative explanation for the maintenance of sex is the Red Queen hypothesis, which states that the production of genetically varied offspring slows the evolution of virulence in parasites and pathogens.

4. Most plants and some animals are hermaphrodites, meaning that they have both male and female sexual functions. Separation of the sexual functions between individuals occurs infrequently among plants, but commonly among animals. Hermaphroditism is favored when one sex can add the sexual function of the other at little cost. Separate sexes are favored when either sexual function imposes large fixed costs.

5. The sex ratio in a population balances the genetic contributions of males and females to future generations. In general, because the rarer sex is favored, most populations have equal numbers of males and females.

6. In some parasitic wasps, males compete with siblings for matings, and mothers shift the sex ration of their offspring in favor of females. In wasps and other hymenopterans, the sex of offspring is determined by whether an egg is fertilized or not, and thus is under direct control of the mother.

7. Mating systems may be promiscuous (individuals mate at large within the population, without lasting pair bonds), polygamous (one individual, usually male, has more than one mate), or monogamous (a lasting pair bond is formed between one male and one female).

8. Promiscuity may arise when males contribute little, other than their genes, to the number or survival of their offspring; this is the common condition in all plants and most animals.

9. Monogamy usually occurs in species in which males can increase their fitness by caring for offspring. In birds, monogamy is most frequent in species in which offspring are fed by their parents.

10. Polygyny arises when males can monopolize either resources or mates. According to the polygyny threshold model, polygyny occurs when some females can gain greater fitness by mating with an already mated male that holds a high-quality territory than by mating monogamously with a male that holds a low-quality territory.

11. When males attract or compete for mates, females can choose among them. Female choice leads to sexual selection of male traits that indicate fitness. Eventually, female choice itself confers fitness on males with favored traits. When females chose males by comparing their traits, males can evolve extreme traits through runaway sexual selection.

12. Sexually selected structures may function as "handicaps" that only the more fit males in a population can bear without encumbrance.

13. Because parasites can evolve rapidly, and because they may directly affect the appearance or survival of males with elaborate ornaments or displays, females that choose mates on the basis of these displays may be choosing males with genetic factors for resistance to parasites. This idea is referred to as parasite-mediated sexual selection.

REVIEW QUESTIONS

1. In what three ways can organisms exhibit parthenogenesis, and how does each affect genetic variation among the resulting offspring?

2. Describe the costs and benefits associated with sexual reproduction.

3. How does the Red Queen hypothesis help us understand the fitness benefits of sexual reproduction?

4. When the fitness increment of increased male function results in a larger cost in fitness through female function, why should a population evolve separate sexes rather than hermaphrodites?

5. When a population is composed of two sexes, why does the rarer sex have a fitness advantage?

6. How does local mate competition favor the production of female-biased sex ratios in offspring?

7. Compare and contrast monogamy, polygyny, and polyandry.

8. According to the polygyny threshold model of mating systems, how does resource availability affect a female's mating decision?

9. Why might exaggerated secondary sexual characteristics in males demonstrate a superior genotype to females?

SUGGESTED READINGS

Almond, D., and L. Edlund. 2007. Trivers–Willard at birth and one year: Evidence from US natality data 1983–2001. *Proceedings of the Royal Society of London* B 274:2491–2496.

Andersson, M., and Y. Iwasa. 1996. Sexual selection. *Trends in Ecology and Evolution* 11:53–58.

Barrett, S. C. H., and L. D. Harder. 1996. Ecology and evolution of plant mating. *Trends in Ecology and Evolution* 11:73–79.

Borgia, G. 1995. Why do bowerbirds build bowers? *American Scientist* 83:542–547.

Bruvo, R., et al. 2007. Synergism between mutational meltdown and Red Queen in parthenogenetic biotypes of the freshwater planarian *Schmidtea polychroa*. *Oikos* 116:313–323.

Busch, J. W., M. Neiman, and J. M. Koslow. 2004. Evidence for maintenance of sex by pathogens in plants. *Evolution* 58:2584–2590.

Cameron, E. Z., and W. L. Linklater. 2007. Extreme sex ratio variation in relation to change in condition around conception. *Biology Letters* 3:395–397.

Chapman, T. 2006. Evolutionary conflicts of interest between males and females. *Current Biology* 16:R744–R754.

Chapman, T., A. Pomiankowski, and K. Fowler. 2005. Stalk-eyed flies. *Current Biology* 15:R533–R535.

Crews, D. 2003. Sex determination: Where environment and genetics meet. *Evolution and Development* 5:50–55.

Dybdahl, M. F., and C. M. Lively. 1998. Host–parasite coevolution: Evidence for rare advantage and time-lagged selection in a natural population. *Evolution* 52:1057–1066.

Ebert, D., and W. D. Hamilton. 1996. Sex against virulence: The coevolution of parasitic diseases. *Trends in Ecology and Evolution* 11:79–82.

Foerster, K., et al. 2003. Females increase offspring heterozygosity and fitness through extra-pair matings. *Nature* 425:714–717.

Godfray, H. C. J., and J. H. Werren. 1996. Recent developments in sex ratio studies. *Trends in Ecology and Evolution* 11:59–63.

Hamilton, S. L., et al. 2007. Size-selective harvesting alters life histories of a temperate sex-changing fish. *Ecological Applications* 17:2268–2280.

Howard, R. S., and C. M. Lively. 1994. Parasitism, mutation accumulation and the maintenance of sex. *Nature* 367:554–557.

Klinkhamer, G. L., T. J. de Jong, and H. Metz. 1997. Sex and size in cosexual plants. *Trends in Ecology and Evolution* 12(7):260–265.

Kruger, O., et al. 2005. Successful sons or superior daughters: Sex-ratio variation in springbok. *Proceedings of the Royal Society of London* B 272:375–381.

Lively, C. M. 1996. Host–parasite coevolution and sex. *BioScience* 46:107–114.

Lively, C. M., and J. Jokela. 1996. Clonal variation for local adaptation in a host–parasite interaction. *Proceedings of the Royal Society of London* B 263:891–897.

Majerus, M. E. N. 2003. *Sex wars: Genes, bacteria, and biased sex ratios.* Princeton, N.J.: Princeton University Press.

Manicacci, D., et al. 1998. Gynodioecy and reproductive trait variation in three *Thymus* species. *International Journal of Plant Sciences* 159:948–957.

Mays, H. L., and G. E. Hill. 2004. Choosing mates: Good genes versus genes that are a good fit. *Trends in Ecology and Evolution* 19:554–559.

Munday, P. L., P. M. Buston, and R. R. Warner. 2006. Diversity and flexibility of sex-change strategies in animals. *Trends in Ecology and Evolution* 21:89–95.

Neff, B. D., and T. E. Pitcher. 2005. Genetic quality and sexual selection: An integrated framework for good genes and compatible genes. *Molecular Ecology* 14:19–38.

Reynolds, J. D. 1996. Animal breeding systems. *Trends in Ecology and Evolution* 11:68–72.

Ross, K. G., and L. Keller. 1998. Genetic control of social organization in an ant. *Proceedings of the National Academy of Sciences USA* 95:14232–14237.

Shine, R. 1999. Why is sex determined by nest temperature in many reptiles. *Trends in Ecology and Evolution* 14:186–189.

Soler, M., et al. 1996. The functional significance of sexual display: Stone carrying in the black wheatear. *Animal Behavior* 51:247–254.

Vollrath, F. 1998. Dwarf males. *Trends in Ecology and Evolution* 13(4):159–163.

von Schantz, T., et al. 1996. MHC genotype and male ornamentation: Genetic evidence for the Hamilton–Zuk model. *Proceedings of the Royal Society of London* B 263:265–271.

Werren, J. H. 1987. Labile sex ratios in wasps and bees. *BioScience* 37:498–506.

Wilkinson, G. S., D. C. Presgraves, and L. Crymes. 1998. Male eye span in stalk-eyed flies indicates genetic quality by meiotic drive suppression. *Nature* 391:276–279.

Family, Society, and Evolution

ale lizards interact with one another through a variety of social behaviors. They show off their size and coloration in displays intended to intimidate one another. They chase and fight with one another when these displays fail to achieve the desired result. Barry Sinervo, now at the University of California at Santa Cruz, and Curt Lively, at Indiana University (whom we met through his study of sex and parasitism in freshwater snails, described in Chapter 8), analyzed the peculiar social organization in a population of the side-blotched lizard (*Uta stansburiana*) in northern California. Males in this population come in three varieties, or *morphs* (Figure 9.1). Their differences in appearance and behavior are genetically determined. Orange, or O morph, lizards are large, aggressive, short-lived, and dominant over blue (B morph) lizards. B lizards, which are smaller than O lizards, are vigilant, and they are dominant over yellow (Y morph) lizards, which mimic females in size, coloration, and behavior. Y lizards use their resemblance to females to sneak into other males' territories and mate with the females there.

Although all three morphs coexist in the same population, this coexistence is not stable, and the frequencies of the morphs vary over time. Consider how this works. When O males are numerous, numbers of B males are depressed by aggression from O males, but Y males can sneak onto other males' territories because the O males are busy chasing B males, and they don't discriminate Y males from females. Thus, when O is the commonest morph, the frequency of Y males increases among the progeny produced in each generation. However, when the Y morph becomes numerous, the vigilant B males are not fooled by their female appearance, and they chase them out of their territories. Thus, B males increase when Y males are common. When B males are numerous, O males can dominate B males, and the proportion of O males increases. These fitness relationships might sound like a late-night sitcom, but, as you can see, they lead to cycling in the frequencies of the three male morphs in the population (Figure 9.2).

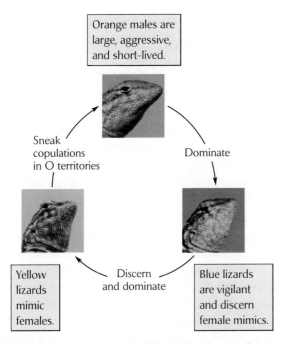

Orange males are large, aggressive, and short-lived.

Sneak copulations in O territories

Dominate

Yellow lizards mimic females.

Discern and dominate

Blue lizards are vigilant and discern female mimics.

FIGURE 9.1 The side-blotched lizard has three male morphs. The three morphs of *Uta stansburiana* differ in their behavior as well as in their coloration. Photos courtesy of Barry Sinervo.

The story is further complicated by female choice. For example, when O males are most common, females should benefit by producing a high proportion of Y males among their offspring, and therefore should chose Y males as their mates. Because the frequency of Y males increases when O males are common, Y males have the higher fitness under this condition, and are indeed preferred by females.

Sinervo and Lively noted that the fitness relationships among the three male morphs resemble the rock–paper–scissors game many of us played when we were children. Rock can be covered by paper, which can be cut by scissors, which can be broken by rock. Thus, the outcome of any one choice depends on whether the opponent plays rock, paper, or scissors. In the same way, the fitness of each male morph of the side-blotched lizard depends on the frequency of the other morphs in the population—an example of frequency-dependent selection. The most common morph is always being replaced by a less common morph, leading to a cycling of frequencies.

Clearly, the social and family environment of an individual, along with its relationship to members of the opposite sex, applies strong selection on behavior and, indirectly, on life histories and ecological relationships.

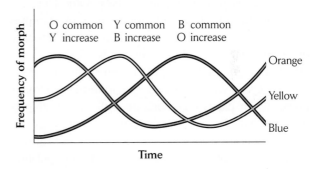

FIGURE 9.2 The fitness of each lizard morph varies with its frequency in the population. Each male morph of the side-blotched lizard is outcompeted by another when it becomes common, setting up a continual cycling in the frequency of the morphs.

CHAPTER CONCEPTS

- Territoriality and dominance hierarchies organize social interactions within populations

- Individuals gain advantages and suffer disadvantages from living in groups

- Natural selection balances the costs and benefits of social behaviors

- Kin selection favors altruistic behaviors toward related individuals

- Cooperation among individuals in extended families implies the operation of kin selection

- Game theory analyses illustrate the difficulties for cooperation among unrelated individuals

- Parents and offspring may come into conflict over levels of parental investment

- Insect societies arise out of sibling altruism and parental dominance

During the course of its life, each individual interacts with many others of the same species: mates, offspring, other relatives, and unrelated members of its social group. Each interaction requires that the individual perceive the behavior of others and make appropriate responses. Most behaviors have a genetic component, which means that natural selection can shape behaviors and responses in a population. Social interactions are a special kind of behavior because they involve members of the same population, and hence the same gene pool, and because they often involve individuals that are related and share many of the same genes. This has resulted in the evolution of behaviors that lend cohesiveness to family and social groups and constrain antagonistic interactions within populations.

When individuals behave in a more friendly and supportive manner toward close relatives than they do toward unrelated individuals, it may be because relatives share genes inherited from a common ancestor and therefore have a common evolutionary interest. Mates, too, must cooperate if they are to raise offspring successfully. Nonetheless, social behaviors emphasize that all interactions between members of the same species delicately balance conflicting tendencies of cooperation and competition, altruism and selfishness.

Although we tend to think of social behavior in the forms most familiar to us, such as conspicuous visual or vocal displays, most organisms have some kind of a social life. Even bacteria and protists can sense the presence of others of the same species and react in "friendly" or "aggressive" ways, often through chemical secretions. Free-living slime molds respond to others during parts of their life cycles when they aggregate to form large fruiting bodies. Even plants communicate with one another by means of volatile compounds to signal damage by herbivores. Social interactions pervade nature and form an important part of the environment to which populations adapt.

Humans are the most social of all animals. Our societies are sustained by role specialization among society members, the interdependence attendant upon that specialization, and the cooperation that interdependence requires. Yet humans are also competitive, to the point of violence, within this mutually supportive structure. Our social life balances contrasting tendencies toward mutual help and conflict. Some animal populations exhibit much of the complexity of human societies. The social insects—ants, bees, wasps, and termites—are remarkable for their division of labor and behavioral coordination among individuals within the hive or nest. Similar subtleties of social interaction, including role specialization and altruistic behavior, are being discovered increasingly in other animals, especially mammals and birds.

Social behavior includes all kinds of interactions between individuals of the same species, from cooperation to antagonism. In this chapter, we shall explore some of the consequences for individuals of interactions within social and family groups, and we shall describe various ways in which social relationships are managed by the behaviors of individuals toward others in the population.

Territoriality and dominance hierarchies organize social interactions within populations

Any area defended by one or more individuals against the intrusion of others may be regarded as a **territory.** Territories can be transient or more or less permanent, depending on the stability of the resources they contain and an individual's need for those resources. Because territories require active defense, most territorial animals are highly mobile (Figure 9.3). Many migratory species establish territories on both their breeding and wintering grounds; shorebirds, for example, defend feeding areas for a few hours or days at stopover points on their long migrations. Hummingbirds and other nectar feeders defend individual flowering bushes and abandon them when they cease producing flowers. Male ruffs and grouse defend a few square meters of space on a communal display ground, to which they attract females to mate. As long as a resource is defensible and the rewards of its defense outweigh the costs, animals are likely to maintain exclusive territories.

In some situations, territoriality is impractical because of the pressures of high population density, the transience of critical resources, or the overriding benefits of living in a group. In such circumstances, when conflicts occur, social rank, rather than space, may be the winner's prize. Once individuals order themselves into a **dominance hierarchy,** subsequent contests between them are resolved quickly in favor of higher-ranking individuals. When a dominance hierarchy is linearly ordered among individuals in a group, the first-ranked member dominates all others, the second-ranked dominates all but the first-ranked,

FIGURE 9.3 Territoriality is often most conspicuous in highly mobile animals. Researchers mark male damselflies with painted numbers to follow their movements and behavioral interactions. Photo by Ola Fincke.

and so on down the line to the last-ranked individual, who dominates none.

Occupation of space and social rank are opposite sides of the same coin, and they are often directly related. The position of an individual in a dominance hierarchy is sometimes reflected by its spatial position within its social group. In large foraging flocks of European wood pigeons, for example, low-ranking individuals tend to be at the periphery, where they are more vulnerable to predators than the high-ranking individuals at the flock's center. These peripheral birds appear nervous and, because they spend much of their time looking up from feeding, they are often undernourished. Birds in the center of the flock remain calmer and feed more because they are protected from surprise attack by the vigilance of the birds at the periphery.

Whether an individual lives within a territorial system or in a group setting, its social rank is determined by its ability to win contests. The outcomes of these contests are all-important to the individual because they determine the quality and extent of the space it can defend, which in turn determines its access to food and mates. Each contest between two individuals can be resolved only through behavioral decisions made by each participant. One spider confronting another over a particularly good place to build a web assesses the situation and decides either to back down or to escalate the contest. Sometimes the projected outcome of a physical contest is obvious, and the smaller individual retreats. When the outcome is more difficult to judge in advance, the two spiders may engage in a series of elaborate displays that help them to weigh each other's fighting abilities, each hoping (although not consciously, as far as I know) that the other will be duly impressed and back down. If the match appears to be close and the outcome uncertain, the contest may then escalate to actual fighting, with the risk of serious injury or death to one or both participants.

Optimal behavior in a contest depends on each contestant's assessment of the likely outcome and the payoffs of winning or losing. What actually happens—that is, how the contest plays out—also depends on the decisions made by each contestant. Each individual should behave in a way that brings it the greatest benefit, but the outcome of the contest also depends on the behavior of the other participant, over which the first individual has little control. Humans are faced with such behavioral decisions all the time, not only in social behavior but also in business, war, and other competitive and cooperative enterprises. Optimal behaviors in these situations are the subject of **game theory,** a method of analyzing the outcomes of behavioral decisions when those outcomes depend on the behavior of other players.

Game theory analysis is based on the *payoffs,* or fitness consequences, of behaviors. Consider the spider's decision whether to escalate a contest or not. If the other contestant backs down, the payoff to the first spider—the contested territory—is large, and the cost is small. If the other contestant meets the challenge, then the payoff depends on the chance of winning the contest, and the cost (win or lose) is much higher. Even without making a quantitative analysis, you can see that an individual's behavior should depend on its best estimate of the other contestant's response and on the reward for winning. When the first spider is much larger than the second, it is likely that the second will back down from any confrontation, so escalation carries little risk of harmful conflict—an easy win, so to speak. When the two are evenly matched, both the response of the second spider and the outcome of the conflict are more difficult to predict, and the probability of getting hurt is higher. Under such circumstances, both escalation and meeting the challenge are likely to occur only when the potential reward for winning a contest is large. It is no surprise, then, that spiders are observed to fight only over the best web sites, and only when the two contestants are similar in size.

MORE ON THE WEB   *Ritualized Antagonistic Behavior Reduces the Incidence of Fighting.* Certain appearances or behaviors signal high social status and discourage aggression by subordinate individuals.

Individuals gain advantages and suffer disadvantages from living in groups

Animals get together for a variety of reasons. Sometimes they are independently attracted to the same habitat or resource and form aggregations, such as vultures around a carcass or dung flies on a cowpat. Within such groups, individuals may interact, usually to compete for space, resources, or mates. Sometimes offspring remain with their parents to form family groups, and aggregation results from their failure to disperse. True social groups, however, arise through the attraction of unrelated individuals to one another—that is, through a purposeful joining together.

Animals form groups to increase their chances of surviving, their rate of feeding, or their success in finding mates. When they are in groups, individuals tend to spend more time feeding and less time looking out for predators.

(a)

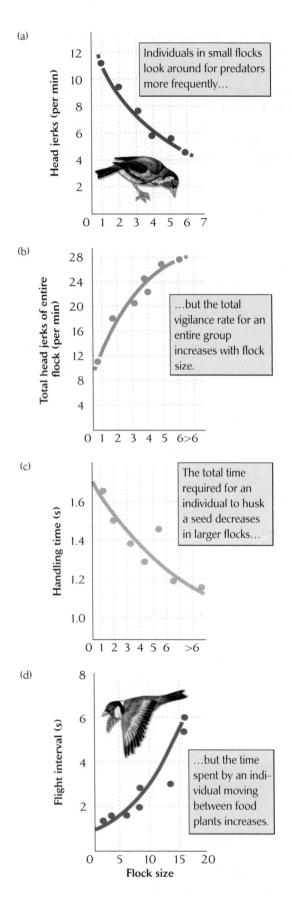

Individuals in small flocks look around for predators more frequently…

(b)

…but the total vigilance rate for an entire group increases with flock size.

(c)

The total time required for an individual to husk a seed decreases in larger flocks…

(d)

…but the time spent by an individual moving between food plants increases.

Flock size

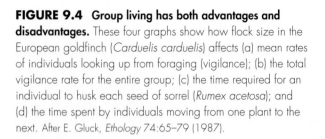

FIGURE 9.4 Group living has both advantages and disadvantages. These four graphs show how flock size in the European goldfinch (*Carduelis carduelis*) affects (a) mean rates of individuals looking up from foraging (vigilance); (b) the total vigilance rate for the entire group; (c) the time required for an individual to husk each seed of sorrel (*Rumex acetosa*); and (d) the time spent by individuals moving from one plant to the next. After E. Gluck, *Ethology* 74:65–79 (1987).

Consider the data presented in Figure 9.4 for the European goldfinch, which feeds on the seed heads of plants in open fields and hedgerows. Two factors control the optimal flock size in these birds. As flock size increases, each individual spends less time looking out for predators. If you watch closely as birds feed, you will notice that they raise their heads and look around from time to time. In a larger flock, an individual goldfinch can spend less time looking around and more time going about the business of eating, and can gather and husk seeds more rapidly, because the total vigilance of the flock is higher. Balancing this advantage of reduced individual vigilance time, a larger flock depresses a local food supply faster, so individuals are forced to fly farther between suitable plants, using valuable feeding time and energy and perhaps increasing their vulnerability to predators. Thus, joining a flock is a good choice for an individual as long as the flock is not too large.

 Social Groups as Information Centers. Watching your neighbors can provide valuable information about food resources and habitat quality.

Natural selection balances the costs and benefits of social behaviors

Most social interactions can be dissected into a series of behavioral acts by one individual, the **donor** of the behavior, directed toward another, the **recipient** of the behavior. One individual delivers food, the other receives it; one threatens, the other is threatened. When one individual attacks another, the attacker may be thought of as the donor of a behavior. The attacked individual (the recipient in this case) may respond by standing its ground or by fleeing; in either case, it becomes the donor of a subsequent behavior. The donor–recipient distinction is useful because each act has the potential to affect the fitness of both the donor and the recipient. These increments of fitness may be positive or negative, depending on the interaction.

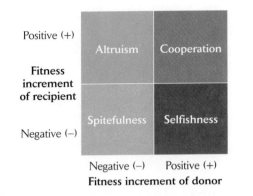

FIGURE 9.5 Social interactions can be organized into four categories. Behaviors can be classified according to their effects on the fitness of donors and recipients.

Four combinations of cost and benefit to donor and recipient can be used to organize social behaviors into four categories (Figure 9.5). **Cooperation** and **selfishness** both benefit the donor and therefore should be favored by natural selection. **Spitefulness**—behavior that reduces the fitness of both donor and recipient—cannot be favored by natural selection under any circumstance, and presumably does not occur in natural populations (even though we have a word for this type of behavior in human populations). The fourth type of behavior, **altruism,** benefits the recipient at a cost to the donor.

Altruism presents a difficult problem for evolutionary theory because it requires the evolution of behaviors that reduce the fitness of the individuals performing them. We would expect selfishness to prevail to the exclusion of altruism because selfishness increases the fitness of the donor.

However, altruism appears to have arisen in colonies of social insects, in which workers forgo personal reproduction (they are sterile in most species!) to rear the offspring of the queen, their mother. Although you might reject the comparison with ants, we humans also like to think that we are not only capable of altruistic behavior, but that such interactions hold together the fabric of our society. But how could such behaviors ever have evolved?

Kin selection favors altruistic behaviors toward related individuals

The evolutionary problem posed by the apparent altruism of social insects is resolved when one recognizes that their colonies are discrete family units, containing mostly the offspring of a single female (the queen). Therefore, social interactions within an ant colony or beehive occur between close relatives—in this case, between siblings. When an individual directs a behavior toward a sibling or other close relative, it influences the fitness of an individual with which it shares more genes than it does with an individual drawn at random from the population. This special outcome of social behavior among close relatives is referred to as **kin selection.**

Close relatives have a certain probability of inheriting copies of the same gene from a particular ancestor. The likelihood that two individuals share copies of any particular gene is their probability of **identity by descent,** the value of which varies with the degree of genealogical relationship between the individuals (Figure 9.6). For

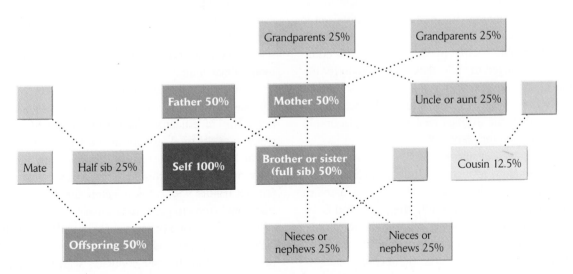

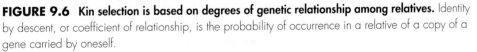

FIGURE 9.6 Kin selection is based on degrees of genetic relationship among relatives. Identity by descent, or coefficient of relationship, is the probability of occurrence in a relative of a copy of a gene carried by oneself.

example, two siblings have a 50% probability of inheriting copies of the same gene from one parent. This probability is also called their **coefficient of relationship.** Two cousins, for example, have a probability of one in eight (12.5%) of inheriting copies of the same gene from one of their grandparents, which are their closest shared ancestors.

When an individual behaves in a particular way toward a close relative, that act influences not only its own personal fitness, but also the fitness of an individual that shares a portion of its genes. Suppose that an altruistic act is directed toward a sibling. The probability that the recipient of the behavior (the sibling) will have a copy of any particular one of the donor's genes is 50%. Therefore, if a tendency to perform a particular behavior is inherited, the fitness of a gene influencing that behavior will be determined both by its influence on the fitness of the donor and by its influence on the fitness of the recipient, weighted by their coefficient of relationship.

Biologists refer to the total fitness of a gene responsible for a particular behavior as its **inclusive fitness.** Inclusive fitness is the contribution of the gene to the fitness of the donor resulting from its own behavior, plus the change in fitness of the recipient discounted by its coefficient of relationship to the donor (that is, the probability that the recipient carries a copy of the same gene) (Figure 9.7). Therefore, the inclusive fitness of a gene for an altruistic behavior would exceed that of its selfish alternative as long as the cost to the altruist was less than the benefit to the recipient multiplied by the average genetic relationship of the recipient to the donor.

Algebraically, a gene promoting altruistic behavior will have a positive inclusive fitness, and will increase in the population, when the cost (C) of a single altruistic act is less than the benefit (B) to the recipient times their coefficient of relationship (r); that is, when $C < Br$. This equation can be rearranged to show that the condition for the evolution of altruism is $C/B < r$; that is, the cost–benefit ratio, which is a measure of how altruistic the behavior is, must be less than the average coefficient of relationship between donor and recipient.

While kin selection makes possible the evolution of altruism among close relatives, it also constrains the evolution of selfish behavior toward relatives. For a selfish behavior, B represents the benefit to the donor and C the cost to the recipient. Accordingly, selfish behavior among close relatives can evolve only when $B > Cr$, or $C/B < 1/r$. The cost–benefit ratio (C/B) is, in this case, a measure of the selfishness of the behavior. A higher coefficient of relationship (r) between donor and recipient reduces the level of selfishness that can evolve (Figure 9.8).

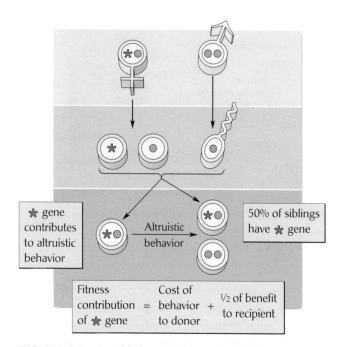

FIGURE 9.7 Social behaviors directed toward relatives have consequences for inclusive fitness. The inclusive fitness of a gene controlling altruistic behavior toward relatives is the cost of the behavior to the donor plus the benefit to the recipient multiplied by their coefficient of relationship.

The maintenance of altruistic behavior by kin selection requires that such behaviors have a low cost to the donor and be restricted to close relatives. Individuals of many species tend to associate in family groups, and limited dispersal often keeps close relatives together. Moreover, individuals of many species can sense their degree of relationship to others by chemical or behavioral cues, even when they have had no family experience. Thus, the opportunity for altruistic behavior to evolve by kin selection is real, and such behavior is perhaps inevitable in many social animals.

A study of wild turkeys in California has shown how altruistic behavior can be maintained through kin selection. The investigator, Alan Krakauer, performed the study as a doctoral student at the University of California at Berkeley. Male turkeys, like the ruffs and grouse mentioned earlier in this chapter, display at leks to attract females (Figure 9.9). A **lek** is a gathering of males within a traditional arena to perform courtship displays. In the case of wild turkeys, two or more males may form a tight association within a lek and display together. Usually, only one of the males in the pair copulates with the females the pair attracts, although both males display. Using genetic data, Krakauer determined that the males associated in such "coalitions" were more closely related than two

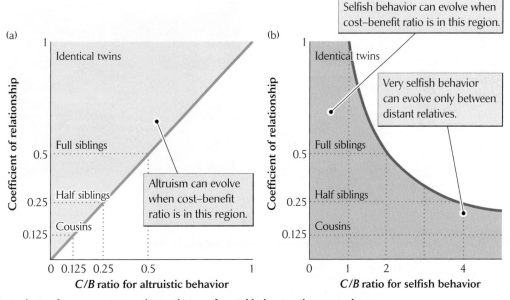

FIGURE 9.8 Inclusive fitness constrains the evolution of social behaviors between close relatives. (a) The level of altruism that can evolve in response to kin selection increases with the degree of genetic relationship between the interacting individuals. (b) The level of selfishness that can evolve is increasingly constrained as the degree of genetic relationship increases.

males drawn at random from the population would be. Indeed, their average coefficient of relationship ($r = 0.42$) was consistent with the coalitions being a mixture of full and half sibs.

Krakauer then determined the average number of offspring sired by dominant coalition males (6.1) compared with the average number sired by solo males (0.9). The subordinate coalition male (the one that does not mate) produces no offspring of his own, but increases his inclusive fitness as the donor of an altruistic behavior to the dominant male. From his standpoint, the inclusive fitness of joining a coalition is calculated as the benefit to the recipient times the degree of their relationship minus the cost. The benefit is the number of offspring produced by the dominant male ($B = 6.1$). The cost to the subordinate male is the number of offspring that he could have produced on his own ($C = 0.9$). Thus, the inclusive fitness in this example is equal to $rB - C$, or $0.42 \times 6.1 - 0.9 = 1.66$—a strongly positive value, for which the criterion $C/B < r$ ($0.9/6.1 < 0.42$) is satisfied.

As well as fostering altruistic behavior, kin selection also limits selfish behavior among close relatives. David Pfennig of the University of North Carolina has demonstrated this principle for the tiger salamander *Ambystoma tigrinum*. When grown at high densities, the larvae of these salamanders can develop a predatory morphology that includes an enlarged mouth and impressive set of teeth (Figure 9.10). These predatory morphs often cannibalize smaller tiger salamander larvae—a most selfish behavior. Pfennig showed, however, that when these salamander larvae are grown with close relatives—either siblings or first cousins—the development of the predatory morphology is delayed. That delay reduces the level of cannibalism considerably.

FIGURE 9.9 Male wild turkeys team up in joint courtship displays. Only the dominant male in a coalition of wild turkeys (*Meleagris gallopavo*) mates with the females the pair attracts, but the subordinate male gains inclusive fitness. Photo by Rolf Nussbaumer.

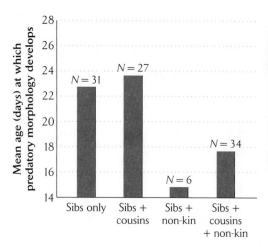

FIGURE 9.10 Close kinship delays development of predatory morphology in larval tiger salamanders. Salamander larvae raised only with siblings or with siblings and cousins delayed acquiring a predatory (and potentially cannibalistic) morphology by a week compared with larvae raised with unrelated salamanders. From D. W. Pfennig and J. P. Collins, *Nature* 362:836–838 (1993); photo courtesy of David W. Pfennig.

MORE ON THE WEB *Alarm Calls as Altruistic Behaviors.* Belding's ground squirrels give alarm calls warning of predators more often in the presence of close relatives.

ECOLOGISTS IN THE FIELD **Are cooperative acts always acts of altruism?** Not all behaviors that benefit a social group are altruistic. In the meerkat (*Suricata suricatta*), a group-living mongoose of southern Africa (Figure 9.11), individuals take up positions on raised structures, such as termite mounds or dead trees, and stand guard while others in the group forage. Timothy Clutton-Brock, of the University of Cambridge, and his coworkers have spent thousands of hours observing meerkats in the field in Kalahari Gemsbok Park, South Africa. One issue they have addressed is whether this guarding behavior is altruistic. That is, does a guarding individual suffer a decrease in personal fitness to increase the fitness of other members of its group?

FIGURE 9.11 Meerkats stand guard while others forage. Alerted by their sense of danger, all the members of this group in Kalahari Gemsbok Park, South Africa, have assumed a guarding stance. Typically, some individuals stand guard and warn foraging group members of approaching enemies. Photo by J & B Photographers/Animals Animals.

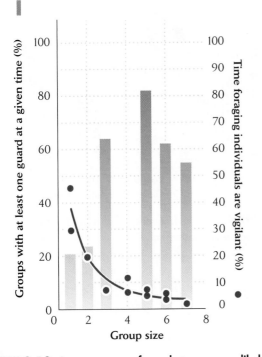

FIGURE 9.12 Larger groups of meerkats are more likely to be protected by guarding individuals. Foraging individuals in larger groups spend more time foraging and less time being vigilant; therefore, they satisfy their food needs in less time and are available to stand guard. From T. Clutton-Brock et al., *Science* 284:1640–1644 (1999).

In this situation, the answer appears to be no. An individual assumes a guard position only after it has filled its own stomach (typically by digging small invertebrates out of the soil). Thus, guarding does not detract from foraging. Furthermore, a guarding individual is free to keep watch from a safe site close to a burrow. Guards are usually the first to see approaching predators and, after emitting warning calls, they are the first to reach safety underground. So the cost of guarding is likely to be low. The larger a group of meerkats, the more likely it is to be guarded at any particular time (Figure 9.12). In a larger group, more individuals are potentially available to guard, and foraging individuals fill their stomachs more quickly because they spend less time in vigilant behavior.

A simple field experiment demonstrated the importance of satisfying food requirements to guarding behavior in meerkats. Ten individuals were each fed 25 grams of hard-boiled egg, and their guarding behavior was compared with that on previous days when they had not received food supplements. When fed, these individuals stood guard more often for longer periods and were more likely to go on guard before foraging in the morning. That meerkats bother to stand guard at all may reflect the fact that their groups are extended families with a single, domi-

nant breeding female. Therefore, most of the individuals in the group are close relatives. Regardless of the benefits of guarding to its relatives, the cost to the donor evidently is small.

Cooperation among individuals in extended families implies the operation of kin selection

Human extended families include the nuclear family (a mated pair and their young progeny) as well as, to varying degrees, grandparents, uncles and aunts, cousins, nephews and nieces, and sometimes individuals of uncertain relationship to the rest. These families are complex social units with a tremendous variety of social interactions, most of them cooperative, but many competitive enough to stress the bonds that hold a family unit together. Rarely do human extended families include more than one child-producing pair, and at least a portion of the behavior of non-nuclear members of the family is directed toward supporting the well-being and upbringing of the children.

Studies of the white-fronted bee-eater (*Merops bullockoides*) in East Africa by Stephen Emlen, Peter Wrege, and Natalia Demong of Cornell University have revealed complex extended families in this species as well (Figure 9.13). These families are typically multigenerational groups of three to seventeen individuals, often including two or three mated pairs plus assorted single birds—unpaired young and widowed older individuals. Careful observations of individually marked birds over several years have shown that these social groups are truly extended families, made up of related individuals and their mates, which normally come from other families. Although relationships within extended families tend to be cooperative, bee-eater family groups are hardly models of harmonious behavior; one sees the usual squabbling over food, nest sites, and mates. Remarkably, however, selfless and selfish acts appear to be directed toward other individuals very much in accordance with degree of relationship: brothers and sisters ($r = 0.50$) are treated better than half siblings and uncles ($r = 0.25$), for example, and cousins ($r = 0.125$) fare almost as badly as nonrelatives outside the family group (Figure 9.14).

Through their behavior, bee-eaters tell us that individuals know their relatives and can distinguish subtle differences in degree of relationship. We can also conclude from the distribution of helpful and harmful behaviors in this species that inclusive fitness is the appropriate measure

FIGURE 9.13 White-fronted bee-eaters live in extended family groups. Courtesy of Natalia Demong.

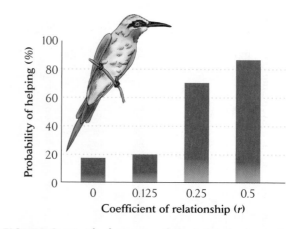

FIGURE 9.14 **The frequency of altruistic behavior varies with degree of relationship.** White-fronted bee-eaters engaged in more helping behaviors toward close relatives than toward more distant relatives. From S. T. Emlen, P. H. Wrege, and N. J. Demong, *Am. Sci.* 83:148–157 (1995).

of selection on social behavior. In other words, altruistic behaviors can indeed evolve among close relatives by kin selection. It is likely that much of human altruistic social behavior has evolved through this mechanism.

We will return to interactions between family members later in this chapter. Before doing so, however, let's consider whether altruistic behavior can evolve among nonrelatives. Clearly, social groups can form out of the self-interest of group members seeking protection from predators, or perhaps because of some efficiency gained by foraging or hunting with other individuals. Whether groups of unrelated individuals can take the next step toward true cooperation, in which each individual forfeits some personal fitness to benefit another, is a fundamental issue in the evolution of social behavior. We shall address this issue by means of a simple game theory analysis.

Game theory analyses illustrate the difficulties for cooperation among unrelated individuals

Self-interest rules behavior among unrelated individuals. A paradox of social behavior is that conflict can reduce the fitness of selfish individuals below that of cooperative individuals. A consequence of natural selection on phenotypes within a population is that the average fitness of individuals in a population increases as the frequency of favored phenotypes grows. One might think, therefore, that behaviors that raise the average fitness of cooperating individuals should also increase in a population. The problem with this reasoning is that when most of a group consists of cooperative members, a selfish individual can greatly increase its personal fitness by "cheating." Thus, selfish behavior would always be favored by natural selection, which would prevent groups from crossing the threshold of cooperative behavior to become true societies.

The logic of this somewhat pessimistic argument can be shown by a simple game theory analysis. The approach we will use is called the **hawk–dove game** (it is also known in a different context as "the prisoner's dilemma"). The term "hawk" refers generally to individuals that behave selfishly, and "dove" to individuals that behave cooperatively, regardless of the kind of organism involved. Let's assume that one type of individual—a hawk—always behaves selfishly in conflict situations, is always willing to fight over a contested resource, and takes all of the resource when it wins. In contrast, doves

TABLE 9.1	Payoff matrix for the hawk–dove game	
	Recipient of the behavior	
Donor of the behavior	Hawk	Dove
Hawk	½B − C	B
Dove	0	½B

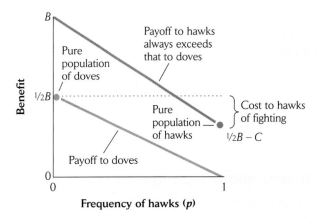

FIGURE 9.15 Hawk behavior is an evolutionarily stable strategy. The payoff to hawks in the hawk–dove game always exceeds the payoff to doves regardless of the frequency of hawks and doves in the population.

never compete over a resource, but share it evenly with other doves. Thus, a dove's behavior is altruistic in that it gives up potential resources. Each contest between two individuals has a potential reward, or benefit (B), and it may also have a cost (C) if the contest results in physical conflict. The payoff to an individual—either hawk or dove—depends on whether the second contestant is a hawk or a dove, as shown in the **payoff matrix** for the game (Table 9.1). For example, two hawks always fight, and on average, each has a 50% chance of winning the resource, so the payoff for hawk behavior toward another hawk is ½B − C, which is one-half of the average benefit minus the cost of fighting. When a hawk confronts a dove, the hawk gains the entire uncontested resource at no cost; thus, the payoff for the hawk is B, and the payoff for the dove is zero. When two doves come together at a resource, they share it, and incur no cost of fighting, so the payoff is ½B.

The average payoff (fitness increment) to hawks and doves depends on the relative proportions of the two types of individuals in a population. Let p be the proportion of hawks and $(1 − p)$ the proportion of doves in the population. Let's assume that interactions are randomly distributed among hawks and doves in proportion to the frequency of each type of individual in the population. The payoffs are now as follows: hawks receive $p(½B − C) + (1 − p)B$, and doves receive $½(1 − p)B$. A population consisting only of hawks ($p = 1$) has an average payoff of ½B − C, which is less than the average payoff of ½B in a population consisting only of doves ($p = 0$). Clearly, the dove strategy would be the best all around from a social point of view because resources would be distributed evenly without the cost of fighting.

The problem is that dove behavior is not an **evolutionarily stable strategy.** That is, it cannot resist invasion by an alternative strategy—namely, hawk behavior. A single hawk in a population of doves (p close to 0) receives twice the average payoff that doves do (B versus ½B) because it never encounters another hawk, and resources are never contested (Figure 9.15). Thus, in a world of doves, the hawk strategy increases rapidly.

Not only can hawk behavior invade a dove population, but a pure hawk population is also resistant to invasion by doves, except when the cost of conflict greatly outweighs the benefit. When p is close to 1 (a pure hawk population), the payoff to hawks is ½B − C, and that to doves is 0. Thus, hawk behavior is an evolutionarily stable strategy as long as $B > 2C$. Only when the benefit is less than twice the cost of conflict can doves invade the hawk population. In that case, doves can survive because the hawks incur such high costs by fighting among themselves. The eventual outcome is a mixed population of hawks and doves with the proportion of hawks (p) equal to ½B/C (Figure 9.16). The persistence of both hawks and doves is

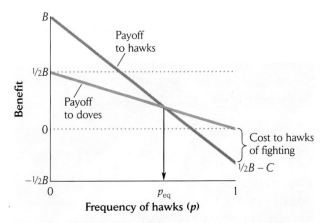

FIGURE 9.16 A stable mixed strategy in the hawk–dove game is possible under certain conditions. When the cost of fighting among hawks is more than one-half the benefit they gain by fighting, the hawk strategy is less fit than the dove strategy. As a result, doves can invade such a population and establish an evolutionarily stable equilibrium frequency (p_{eq}).

referred to as an evolutionarily stable **mixed strategy.** Either type of individual can increase in frequency when it is rare, thereby keeping both in the game.

MORE ON THE WEB *The Reciprocal Altruism Game.* A game that is more complex than the hawk–dove game illustrates a possibility for altruism among unrelated individuals.

Parents and offspring may come into conflict over levels of parental investment

Behavioral interactions between parents and offspring are more complex than interactions among individuals in the population as a whole. While offspring depend on their parent(s) for some part of their lives, the fitness of a parent is wrapped up in the survival and reproductive success of its offspring. In addition, offspring compete with one another as individuals, within limits set by their inclusive fitness. Therefore, the best interests of offspring and parents can diverge.

Rather than passively accepting whatever their parents offer, most offspring actively solicit care. Young animals beg for food and solicit brooding (Figure 9.17); eggs actively take up yolk from the ovarian tissues or bloodstream of the mother; developing seeds take up nutrients from the maternal tissues in the ovary of a flower. For the most part, the evolutionary interests of parent and offspring are compatible: when progeny thrive, so do their parents' genes. But when the selfish accumulation of resources by one offspring deprives its siblings and reduces the overall fecundity of its parents, the interests of parent and offspring can come into conflict. We may define **parent–offspring conflict** as a situation that arises when parent and offspring differ over the optimal level of parental investment.

Each act of parental care and each unit of parental investment benefits an individual offspring by enhancing its survival, but has costs to the parent. Resources allocated to one offspring cannot be delivered to others, prolonged parental care delays the birth of subsequent offspring, and the risks of caring for today's offspring decrease the probability that a parent will survive to rear tomorrow's offspring (see Chapter 7). Thus, for the parent, there is always a conflict between present and future reproductive success. Offspring try to resolve that conflict in favor of present reproductive success (that is, in favor of themselves); parents benefit from a more balanced distribution of their parental investment.

From the standpoint of a parent, its offspring are all genetically equivalent, and the parent should not bias its investment in any of them. From the standpoint of an individual offspring, however, the self has twice the genetic value of a sibling because a sibling shares only half of the individual's genes (Figure 9.18). Therefore, when an individual possesses a gene that increases the care it receives from its parents—perhaps by causing it to beg more persistently—that trait is favored as long as the cost to the parents, in terms of number of siblings raised, is less than twice the benefit to the individual. This is the limit to the evolution of selfish behavior under kin selection that we discussed above (see Figure 9.8).

As young individuals mature and become better able to care for themselves, the benefits of parental care dwindle. When the benefit–cost ratio of parental care drops below 1, a parent should cease to provide care to those offspring in favor of producing additional ones. Suppose, however, that an offspring has a gene that prolongs its solicitation of parental care. Because the inclusive fitness of the offspring includes only one-half the cost of not delivering care to its future siblings, the offspring "prefers" that parental care continue until the benefit–cost ratio is 0.5 when its parents' future offspring are full siblings, and even less when they are not. Thus, the time between the offspring age at which B/C is 1 and the age at which it is 0.5 is a period of conflict between parent and offspring.

Biologists believe that such conflicts can be seen in many species of mammals and birds that exhibit extensive postnatal care. Young animals, often fully capable of taking care of themselves, sometimes hound their parents mercilessly for food. You would think that parents would have the upper hand in any such conflict with their offspring, but remember that parents are adapted to respond positively to solicitations by their offspring when they are

FIGURE 9.17 Offspring actively solicit care from their parents. Nestling cedar waxwings beg for food from their parent. Photo by Ralph Reinhold/Animals Animals.

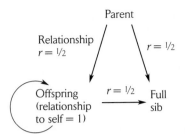

FIGURE 9.18 The evolutionary interests of parents and their offspring may conflict. The asymmetry in evolutionary interest between self and siblings, contrasting with the symmetrical interest of parents in each of their offspring, creates conflict between parents and offspring over the optimal allocation of parental investment.

growing and not yet independent. By prolonging their juvenile appearance and dependent behavior, offspring may be able to take advantage of their parents' responsiveness and prolong parental care.

Insect societies arise out of sibling altruism and parental dominance

The most extreme manifestation of family living is seen in the social insects. Most individuals in these insect species forgo sexual maturation and reproduction to stay with their parents and help them to rear siblings. The origins of these insect societies are still actively debated—particularly whether they evolved because they enhanced the individual fitness of despotic parents, the inclusive fitness of the altruistic sterile offspring (workers), or the fitness of unrelated individuals cooperating in social groups.

These complex societies present a formidable challenge to evolutionary theory, primarily because of the existence of nonreproductive individuals. How can natural selection produce individuals with no reproductive output—that is to say, with no individual fitness? Before considering the evolutionary issues raised by insect societies, let's have a quick look at their natural history.

There are several grades of sociality in the animal world, the highest of which is **eusociality.** This grade is distinguished by several characteristics:

1. Several adults living together in a group

2. Overlapping generations—that is, parents and offspring living together in the same group

3. Cooperation in nest building and brood care

4. Reproductive dominance by one or a few individuals, including the presence of sterile **castes**

Thus defined, eusociality is limited among the insects to the termites (Isoptera, which are actually social cockroaches) and the ants, bees, and wasps (Hymenoptera). Elements of eusociality are present in at least one mammal, the naked mole-rat of Africa. The social insects not only are of evolutionary interest, but also are major players in ecosystem processes. Social insects are pollinators of plants, consumers of plant and animal material on a large scale, and recyclers of wood and organic detritus. Their numerical and functional dominance in the world is due in large part to the immense success of eusociality (Figure 9.19).

From its distribution across taxonomic groups, it is clear that eusociality has evolved independently many times in the hymenopterans. The evolutionary steps that lead to eusociality are less clear. The most widely accepted sequence of evolutionary events includes a lengthened period of parental care for the developing brood, with parents either guarding their nests or continuously provisioning their larvae in a manner similar to birds feeding their young. If such parents survived and continued to produce eggs after their first progeny emerged as adults, then their offspring would be in a position to help raise subsequent broods consisting of their younger siblings. Once progeny remain with their mother after they attain adulthood, the way is open to their relinquishing their own reproductive function solely to support hers.

Organization of insect societies

Social insect societies are dominated by one or a few egg-laying females, which are referred to as **queens.** The queens in colonies of ants, bees, and wasps mate only

FIGURE 9.19 Fungus-growing ants form some of the largest colonies of any social insects. Workers of the ant *Acromyrmex echinatior* tend fungus gardens deep in their underground nest. Courtesy of David Nash.

once during their lives and store enough sperm to produce all their offspring—up to a million or more over 10–15 years in some army ants. The nonreproductive progeny of a queen gather food and care for developing brothers and sisters, some of which become sexually mature, leave the colony to mate, and establish new colonies.

Bee societies are organized simply: the offspring of a queen are divided among a sterile worker caste, which are all genetically female, and a reproductive caste, produced seasonally, that consists of males and females. Whether an individual becomes a sterile worker or a fertile reproductive is controlled by the quality of nutrition it receives as a developing larva. Substances produced by a queen and fed to her larvae can inhibit the development of reproductive organs. In bees, the worker caste represents an arrested stage in the development of reproductive females, stopped short of sexual maturity.

Unlike ant, bee, and wasp societies, termite colonies are headed by a mated pair—the king and queen—which produce all the workers by sexual reproduction. Termite workers are both male and female, but none of these workers mature sexually unless either the king or the queen dies.

Coefficients of relationship in hymenopteran societies

Hymenopterans have a *haplodiploid* sex determination mechanism (see Chapter 8). Females—both workers and queens—are produced from fertilized eggs (Figure 9.20). Males, which develop from unfertilized eggs, appear in colonies only as reproductive individuals (drones) that leave to seek mates. Haplodiploidy creates strong asymmetries in coefficients of relationship within these insect societies (Table 9.2). In particular, a female worker's coefficient of relationship to a female sibling is 0.75, whereas to a male sibling it is 0.25. The queen herself has the same genetic relationship to sons and to daughters (0.50), so she can be relatively ambivalent about the sex of her reproductive offspring (the drones and new queens that leave the nest), especially when the sex ratio among reproductive individuals in the population as a whole is near equality. The skewed genetic relationship among siblings means that cooperation is likely to be greater among all-female castes than among male castes or, especially, among mixed castes. This may explain why workers in hymenopteran societies are all female, and why broods of reproductive individuals usually favor females, by about 3 : 1 on a weight basis, in spite of the queen's ambivalence about the sex ratio. Furthermore, when a female worker

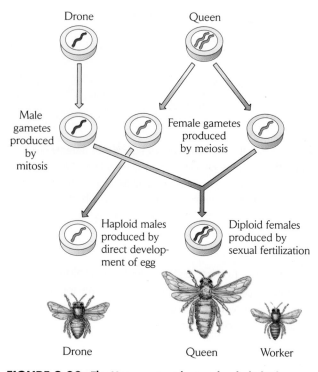

FIGURE 9.20   The Hymenoptera have a haplodiploid sex determination system. A queen can determine the sex of her offspring by using stored sperm to fertilize eggs, producing diploid females (workers or queens), or by not fertilizing them, producing haploid males (drones).

can help to rear more female than male reproductive individuals, her own inclusive fitness may actually be higher than it would be if she raised a brood of her own consisting of an equal number of males and females. Under these bizarre circumstances, it is not surprising that sterile castes might have evolved. Nevertheless, considering that termite societies have a more typical mix of diploid males and females, other factors must come into play.

TABLE 9.2	Coefficients of relationship between male and female hymenopterans and their relatives	
Coefficient of relationship to	Male	Female
Mother	0.50	0.50
Father	0.00	0.50
Brother	0.50	0.25
Sister	0.25	0.75
Son	0.00	0.50
Daughter	1.00	0.50

Multiple queens in ant colonies

In many species of social insects, colonies have more than one queen, and often those queens are not close relatives. In such situations, the degree of genetic relationship among workers is, on average, greatly reduced. Moreover, kin selection for altruistic behavior may not be strong enough to maintain peace within the colony. Indeed, the fact that such multi-queen social colonies exist at all suggests that the queens exercise control over the development of their eggs into various caste types. Otherwise, their female progeny could increase their inclusive fitness by maturing sexually and reproducing.

The fire ant (*Solenopsis invicta*) is an introduced species in the southern United States that forms two types of colonies. One colony type has a single queen, while the other type can have hundreds of queens, each of which lays relatively few eggs. These colonies produce smaller workers than in the single-queen colonies and a number of sexually mature females, or new queens. New colonies are established by several unrelated queens, and new queens may even be adopted from outside the colony. The difference in behavior between these two types of colonies appears to be controlled by a single gene, named *Gp-9*, involved in the sensing of chemical pheromone signals. Different forms of the gene probably influence the ability of workers to recognize queens and limit their numbers.

Recent research on the Argentine ant shows how genetic relationship can affect aggressiveness and invasiveness. This species was introduced to the United States from South America about 100 years ago and has displaced several native species of ants. It appears that the introduced population experienced a genetic bottleneck that reduced the species' genetic variation: Argentine ants in the southern United States have only about half as much genetic variation at seven gene loci as ant populations back home in Argentina. Ants usually act aggressively toward unrelated individuals, but tolerate related individuals. Because Argentine ants in the United States are relatively uniform genetically, individuals from different colonies fail to recognize colony differences and thus cooperate with one another. Thus, Argentine ants may form networks of connected colonies. This may help explain why they are so effective at displacing native ant species.

Behavioral relationships among the social insects represent one extreme along a continuum of social organization, from animals that live alone except to breed to those that aggregate in large groups organized by complex behavior. Regardless of their complexity, all social behaviors balance costs and benefits to the individual and to close relatives affected by its behavior. Like morphology and physiology, behavior is strongly influenced by genetic factors and thus is subject to evolutionary modification by natural selection. The evolution of behavior becomes complicated when individuals interact within a social setting, and the interests of individuals within a population may either coincide or conflict. Understanding the evolutionary resolution of social conflict in animal societies continues to be one of the most challenging and important concerns of biology.

SUMMARY

1. Selection imposed by interactions with family members and with unrelated individuals within a population provides the basis for evolutionary modification of social behavior.

2. Territoriality is the defense of an area from intrusion by other individuals. Animals are more likely to maintain territories when the resources they gain by doing so are rewarding and defensible.

3. Dominance hierarchies order individuals within social groups by rank, which is often established by direct confrontation. Because rank is generally respected, dominance relationships may reduce conflict within the group.

4. Living in social groups may benefit individuals by enabling them to better detect and defend against predators or to obtain food more efficiently. Groups form to the extent that such benefits outweigh the costs of group living, such as competition among group members.

5. Isolated acts of social behavior involve a donor and a recipient. When both benefit from their interaction, the behavior is termed cooperation; when the donor benefits at a cost to the recipient, the behavior is selfish; when the recipient benefits at a cost to the donor, the behavior is altruistic.

6. Altruistic behavior has been explained in terms of kin selection. When an individual interacts with a relative, it affects the fitness of the genes it shares with that relative through inheritance from a common ancestor.

7. Inclusive fitness expresses the cost (or benefit) of a behavior to the donor plus the benefit (or cost) to the recipient adjusted by the coefficient of their relationship.

8. In general, the distribution of cooperation and altruism within social groups is sensitive to degree of genetic relationship between individuals.

9. Game theory analyses, such as the hawk–dove game, indicate that cooperative behavior is unlikely to evolve among nonrelatives, even though the average benefit to individuals in a purely cooperative social group would exceed that gained by individuals through confrontation and conflict. The reason is that cooperative behavior is not an evolutionarily stable strategy, but can be invaded by selfish cheaters.

10. Conflict may arise between parents and offspring over the optimal level of parental investment. All siblings have an equal coefficient of relationship to their parents, which therefore have no preference among them. Siblings have a coefficient of relationship among themselves, however, of only 0.50. Therefore, individual offspring should prefer unequal parental investment in themselves at a cost to their siblings, even when parental fitness is reduced as a result.

11. Social insects (termites, ants, wasps, and bees) live in extended family groups in which most offspring are retained in a colony as sterile workers, increasing their mother's fitness by rearing reproductive siblings.

12. The haplodiploid sex determination mechanism of the Hymenoptera results in females having a coefficient of relationship of 0.75 to sisters, but only 0.25 to brothers. This asymmetry has probably contributed to the sterile workers in ant, bee, and wasp colonies all being female and to the production of more female reproductives than males.

REVIEW QUESTIONS

1. Why might individuals give up defending territories if the density of their population increases?

2. Explain the costs and benefits that influence the optimal flock size in birds.

3. Why should natural selection not favor spiteful behavior?

4. How could helping a relative improve the helper's fitness?

5. In the kin selection explanation for the evolution of altruism, why is the benefit to the recipient weighted by the coefficient of relationship to the donor?

6. Why are selfish behaviors less favored when the donor and recipient are related to each other?

7. In the hawk–dove game, why is a population comprising individuals using the "dove" strategy susceptible to invasion by individuals using the "hawk" strategy?

8. If a parent bird has two chicks to feed, what is the conflict between the parent's selfish interest in distributing food between the chicks and either of the chicks' self-interest in obtaining food?

9. Compare the coefficient of relationship between brothers and sisters in diploid organisms and in haplodiploid organisms.

10. How does a haplodiploid genetic system favor the evolution of eusociality?

SUGGESTED READINGS

Bernhard, H., U. Fischbacher, and E. Fehr. 2006. Parochial altruism in humans. *Nature* 442:912–915.

Brommer, J. E. 2000. The evolution of fitness in life-history theory. *Biological Reviews* 75:377–404.

Clutton-Brock, T. H., et al. 1999. Selfish sentinels in cooperative mammals. *Science* 284:1640–1644.

Elgar, M. A. 1989. Predator vigilance and group size in mammals and birds: A critical review of the empirical evidence. *Biological Reviews* 64:13–33.

Emlen, S. T., P. H. Wrege, and N. J. Demong. 1995. Making decisions in the family: An evolutionary perspective. *American Scientist* 83:148–157.

Fletcher, J. A., and M. Zwick. 2006. Unifying the theories of inclusive fitness and reciprocal altruism. *American Naturalist* 168: 252–262.

Foster, K. R., T. Wenseleers, and F. L. W. Ratnieks. 2006. Kin selection is the key to altruism. *Trends in Ecology and Evolution* 21:57–60.

Gordon, D. M. 1995. The development of organization in an ant colony. *American Scientist* 83:50–57.

Griffin, A. S., and S. A. West. 2003. Kin discrimination and the benefit of helping in cooperatively breeding vertebrates. *Science* 302:634–636.

Heinrich, B., and J. Marzluff. 1995. Why ravens share. *American Scientist* 83:342–349.

Hoffman, E. A., and D. W. Pfennig. 1999. Proximate causes of cannibalistic polyphenism in larval tiger salamanders. *Ecology* 80: 1076–1080.

Honeycutt, R. L. 1992. Naked mole-rats. *American Scientist* 80: 43–53.

Hunt, J. H., and G. V. Amdam. 2005. Bivoltinism as an antecedent to eusociality in the paper wasp genus *Polistes. Science* 308: 264–267.

Krause, J. 1994. Differential fitness returns in relation to spatial position in groups. *Biological Reviews* 69:187–206.

Krieger, M. J. B., and K. G. Ross. 2002. Identification of a major gene regulating complex social behavior. *Science* 295:328–332.

Mock, D. W., and G. A. Parker. 1997. *The Evolution of Sibling Rivalry.* Oxford University Press, Oxford.

Pfennig, D. W., and J. P. Collins. 1993. Kinship affects morphogenesis in cannibalistic salamanders. *Nature* 362:836–838.

Queller, D. C., and J. E. Strassmann. 1998. Kin selection and social insects. *BioScience* 48(3):165–175.

Ratnieks, F. L. W., K. R. Foster, and T. Wenseleers. 2006. Conflict resolution in insect societies. *Annual Review of Entomology* 51: 581–608.

Sharp, S. P., et al. 2005. Learned kin recognition cues in a social bird. *Nature* 434, 1127–1130.

Sherman, P. W., J. U. M. Jarvis, and S. H. Braude. 1992. Naked mole-rats. *Scientific American* 267:72–78.

Sinervo, B., and C. M. Lively. 1996. The rock–paper–scissors game and the evolution of alternative male strategies. *Nature* 380:240–243.

Strassmann, J. E., Y. Zhu, and D. C. Queller. 2000. Altruism and social cheating in the social amoeba *Dictyostelium discoideum. Nature* 408:965–967.

Toth, A. L., et al. 2007. Wasp gene expression supports an evolutionary link between maternal behavior and eusociality. *Science* 318:441–444.

Trivers, R. L. 1974. Parent–offspring conflict. *American Zoologist* 14: 249–264.

Trivers, R. L. 1985. *Social Evolution.* Benjamin Cummings, Menlo Park, Calif.

Tsutsui, N. D., et al. 2000. Reduced genetic variation and the success of an invasive species. *Proceedings of the National Academy of Sciences USA* 97:5948–5953.

West, S. A., A. S. Griffin, and A. Gardner. 2007. Evolutionary explanations for cooperation. *Current Biology* 17:R661–R672.

Wilson, E. O. 1975. *Sociobiology.* Harvard University Press, Cambridge, Mass.

Wilson, E. O., and B. Hölldobler. 2005. Eusociality: Origin and consequences. *Proceedings of the National Academy of Sciences USA* 102:13367–13371.

The Distribution and Spatial Structure of Populations

A primary threat to populations of many species is the fragmentation of their habitat. As forests are cleared, roads are built, and rivers are channeled, suitable habitat for many organisms is broken up into small patches, greatly restricting their movements. Organisms can use a particular habitat patch only if they can gain access to it by moving through less favorable surrounding habitats. Human land uses—particularly those related to agriculture, forestry, housing, and transportation—have not only reduced the area of many types of forest, grassland, and wetland habitat, but have also reduced access to livable patches of those habitats for many species. Small, isolated subpopulations in small habitat fragments can die out because they lose genetic diversity and cannot adapt to changed conditions, or because they cannot escape such disasters as fire, drought, or disease. With global climatic warming causing environments to shift across the landscape, access to dispersal routes has become even more critical for the maintenance of natural communities of plants and animals. Conversely, species that can use disturbed habitats have been able to spread more widely. Habitats altered by human activities have become highways for expanding populations of some introduced species.

Fragmentation also reduces habitat quality. The smaller a habitat patch, the closer any particular point in that patch is to a habitat edge. Increases in edge can have unexpected consequences. For example, in tropical rain forests, trees within 100 meters of the edge of a clear-cut are exposed to higher winds and can die from excessive water loss from their leaves. Such edge effects have caused losses of up to 15 tons of tree biomass per hectare each year in one study area in the central Amazon basin (Figure 10.1).

North Americans need not travel so far to feel the effects of habitat fragmentation. For example, throughout much of the eastern and midwestern United States, forest

FIGURE 10.1 Habitat fragmentation places organisms closer to the edges of habitat patches. This patch of rain forest habitat near Manaus, Brazil, was created when the surrounding forest was converted to pasture. The mortality rate of trees at the edges of such patches is several times higher than that of trees in the middle of intact forest, owing to drying and wind damage. Courtesy of Eduardo M. Venticinque, National Institute for Research in the Amazon.

fragmentation has brought populations of forest birds into contact with the parasitic brown-headed cowbird, which lays its eggs in the nests of other songbirds, reducing the reproductive success of its hosts. Cowbirds prefer open farms and fields, but they enter the edges of woodlots to seek out host nests. Nest parasitism on the Kentucky warbler (a female is shown feeding a cowbird nestling in the chapter opening photo) exceeds 50% within 300 meters of forest edges in southern Illinois, and this edge effect is still discernible more than a kilometer into the forest. Similarly, various nest predators, including many small rodents that normally hunt in fields, typically do not venture far into the forest. However, habitat fragmentation has created so much forest edge that populations of some forest-dwelling songbirds have dropped precipitously in parts of eastern North America.

Habitat fragmentation is just one of many threats to population viability. The species with which we share this planet have persisted over many thousands or even millions of years. That so many are now threatened with extinction is cause for great concern. To understand why this has happened and what we can do about it, we must understand how the environment shapes the structure and dynamics of populations.

CHAPTER CONCEPTS

- Populations are limited to ecologically suitable habitats
- Ecological niche modeling predicts the distributions of species
- The dispersion of individuals reflects habitat heterogeneity and social interactions
- The spatial structure of populations parallels environmental variation

- Three types of models describe the spatial structure of populations
- Dispersal is essential to the integration of populations
- Macroecology addresses patterns of range size and population density

A **population** consists of the individuals of a species within a given area. However, a population is more than a collection of individuals. It has integrity as a unit of organization in ecology because individuals within a population come together to reproduce, thereby mixing the gene pool of the population and ensuring its continuity through time (Figure 10.2).

Each population lives primarily within areas of suitable habitat. The natural environment is a mosaic of different habitats: patches of woodland within savannas, wetlands threaded along rivers through prairies, serpentine barrens scattered through forests, dry south-facing slopes across from moister north-facing slopes in mountainous regions. The patchy distribution of suitable habitat means that

many populations are divided into smaller **subpopulations,** or *local populations,* between which individuals move less frequently than they would if the habitat were homogeneous.

Populations also have a characteristic extent and size. Population extent, or **distribution,** is the geographic area occupied by a population. The distribution of a population in space is also called its **geographic range. Population size** is the number of individuals in a population. That number may vary with food supplies, predation rates, nest site availability, and other ecological factors.

Population structure encompasses a number of attributes, including the density and spacing of individuals within suitable habitat and the proportions of individuals

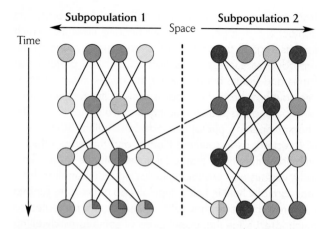

FIGURE 10.2 Interbreeding within a population maintains its integrity in space and its continuity through time. Lines connect offspring to their parents. The population is divided into two subpopulations, separated by a partial barrier to dispersal (dashed line), with different genetic makeup (indicated by different colors). Individuals within a subpopulation come together to reproduce, as shown by the lines indicating the parentage of each progeny generation. Individuals occasionally disperse between the two subpopulations, allowing some mixing of genes between the two subpopulations so that they remain parts of one overall population.

of each sex and in each age class. The part of population structure that encompasses the density and spacing of individuals is called **spatial structure.** It and population distribution together give a picture of the arrangement of populations in space. In this chapter, we shall consider factors that influence the distribution and spatial structure of populations. These factors operate mainly through the behavioral and physiological responses of individuals to variation in their environments as well as to other individuals. In the next chapter we shall examine the other components of population structure, especially the distribution of individuals among age classes.

The measures used to describe population structure provide a snapshot of a population, but populations, of course, are constantly changing—in size, distribution, age structure, and genetic composition. We shall consider this dynamic behavior of populations in Chapter 11 and subsequent chapters. Changes in populations result from variation in births, deaths, and movements of individuals, all of which are influenced by the interactions of individuals with one another and with their environment. Thus, an understanding of population dynamics also provides insights into community structure and ecosystem function.

Populations are limited to ecologically suitable habitats

The natural world varies from place to place. Uniform habitats extending over vast areas simply do not exist. Instead, the natural world varies over space in a mosaic of habitat patches. For any particular species, some of these patches are suitable and others are not. The range of physical conditions over which species can persist is referred to as the **fundamental niche** of the species. Within this range of conditions, predators, pathogens, and competitors can limit the distribution of a species to a smaller **realized niche.** The niche concept ties the distributions of populations to their environments. Environmental conditions influence population abundance and distribution by influencing birth, death, and dispersal.

Within the geographic range of a population, climate, topography, soils, vegetation structure, and other factors influence the distribution and abundance of individuals. Sugar maples, for example, cannot live in marshes, serpentine barrens, newly formed sand dunes, recently burned areas, and a variety of other habitats that simply lie outside their fundamental niche. Hence, the geographic range of the sugar maple is actually a patchwork of occupied and unoccupied areas.

The distribution of the perennial shrub *Clematis fremontii* in Missouri reflects a hierarchy of limiting factors (Figure 10.3). Climate, perhaps in combination with interactions with ecologically similar plants, restricts this species of *Clematis* to a small area of the midwestern United States. The distinctive variety of *Clematis fremontii* named *riehlii* occurs only in Jefferson County, Missouri. Within its geographic range, *Clematis fremontii* var. *riehlii* is restricted to dry, rocky soils on outcroppings of limestone. Small variations in relief and soil quality further confine these plants within each limestone glade to sites with suitable soil structure, moisture, and nutrients. Local aggregations occurring on each of these sites consist of many more or less evenly distributed individuals.

An experimental study of populations of two species of monkeyflowers (*Mimulus*) that live at different elevations in the Sierra Nevada of California shows how the limits of their distributions are influenced by the suitability of the environment. The study compared how well the two species survived and reproduced within and outside their natural elevational distributions. The investigators, Amy Angert and Doug Schemske of Michigan State University, transplanted individuals of the lower-elevation species, *M. cardinalis,* to experimental garden plots within and

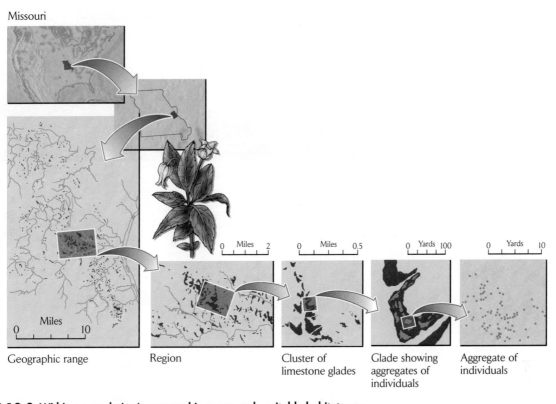

FIGURE 10.3 Within a population's geographic range, only suitable habitats are occupied. Different scales of mapping reveal a hierarchy of patterns in the distribution of *Clematis fremontii* var. *riehlii* in east central Missouri. After R. O. Erickson, *Ann. Mo. Bot. Gard.* 32:416–460 (1945).

above its natural elevational distribution, and transplanted the higher-elevation species, *M. lewisii,* to plots within and below its natural distribution. Survival, growth, and flower production were uniformly higher in plots within the normal elevational distribution of each species (Figure 10.4). Although population growth rates were not estimated in this study, it is evident that each species can sustain its population only within the narrow range of environmental conditions that make up the species' niche.

Dispersal limitation

The presence or absence of suitable habitat often determines a population's distribution, but other factors, including barriers to dispersal, also have an influence. For example, the natural distribution of the sugar maple in the United States and Canada corresponds primarily to its tolerance limits for stressful physical conditions: aridity to the west, cold winters to the north, and hot summers to the south (see Figure 5.2). Much suitable habitat for the species exists, however, in other parts of the world, especially in Europe and Asia, where one finds close relatives of the sugar maple in the genus *Acer.* Of course,

sugar maple seeds cannot disperse such great distances across the oceans to colonize these areas. The absence of a population from suitable habitat because of barriers to dispersal is called **dispersal limitation.**

The point that barriers to long-distance dispersal often limit geographic ranges is driven home when we see what happens when those barriers are overcome by human intervention. Humans have carried useful plants and animals with them since the beginning of their migrations. Aboriginal peoples brought dogs to Australia, and Polynesians distributed pigs (and rats) throughout the small islands of the Pacific. In more recent times, foresters have transplanted fast-growing eucalyptus trees from Australia and pines from California all over the world for timber and fuelwood. Other species have hitched rides on human conveyances, hidden among cargo, in ballast, and on the hulls of ships. Many of these introduced species have become established and thrived in their new lands and waters. Indeed, some introduced populations far exceed their natural source populations in number and distribution. Occasionally, individuals cross formidable barriers and disperse long distances under their own power. How else would plants and animals have populated remote

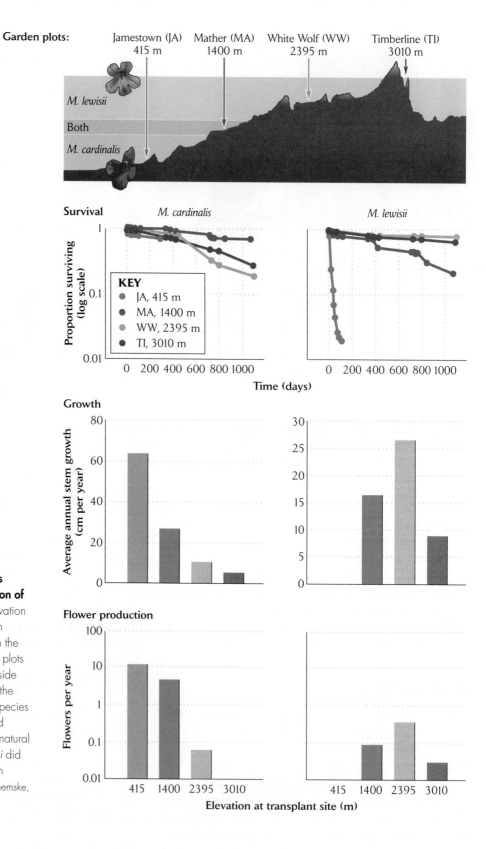

FIGURE 10.4 Individual fitness is highest within the natural distribution of a species. Individuals of the high-elevation *Mimulus lewisii* and the low-elevation *M. cardinalis* were transplanted from the center of their distributions to garden plots at different elevations within and outside of their natural elevational ranges in the Sierra Nevada of California. Each species survived the best, grew the most, and produced the most flowers within its natural elevational range; indeed, *M. lewisii* did not survive below its natural elevation range. After A. L. Angert and D. W. Schemske, *Evolution* 59:222–235 (2005).

islands, such as the Hawaiian Islands, before human colonization? We'll have more to say about such invasive organisms later in this book, but their success in many places outside their native ranges emphasizes the role of barriers to dispersal in limiting species' distributions.

Migration

It is important to remember that the geographic range of a population includes all of the areas its members occupy during their entire life history. Thus, for example, the distribution of sockeye salmon includes not only the rivers of western Canada that are their spawning grounds, but also vast areas of the North Pacific Ocean where individuals grow to maturity before making the long migration back to their birthplace. Arctic terns (*Sterna paradisea*) travel 30,000 kilometers per year from their North Atlantic breeding grounds to the Antarctic, where they spend the Southern Hemisphere summer, and back again. Satellite tracking of electronically tagged Atlantic bluefin tuna (*Thunnus thynnus*) has revealed remarkable migrations that span the entire breadth of the Atlantic Ocean (Figure 10.5). One individual tagged in 1999 off the coast of South Carolina stayed for a year in the western Atlantic before crossing, in the space of a few weeks, to waters off Europe, where it remained, making regular seasonal movements between the area north of Ireland to its spawning area in the western Mediterranean Sea, until it was recaptured in 2003.

Many animals, particularly those that fly or swim, undertake extensive migrations. Each fall, hundreds of species of land birds leave temperate and Arctic North America, Europe, and Asia for the south in anticipation of cold winter weather and dwindling supplies of their invertebrate foods. Populations of monarch butterflies migrate between their wintering areas in the southern United States and Mexico and their summer breeding areas far to the north into southern Canada. In East Africa, many large ungulates, such as wildebeests, migrate long distances, following the geographic pattern of seasonal rainfall and fresh vegetation (Figure 10.6).

Some migratory movements are a response to occasional failure or depletion of local food supplies, which forces individuals to move out of an area in search of new feeding places. Such movements are perhaps best known from outbreaks, or *irruptions,* of migratory locusts. These mass migrations occur when locusts leave areas of high local density where food has been depleted. The migrations can reach immense proportions and spread extensive crop damage over wide areas (Figure 10.7). Irruptive behavior in locusts is a developmental response to population density. When locusts occur in sparse populations, they become solitary and sedentary as adults. In dense populations, however, frequent contact with other locusts stimulates young individuals to develop gregarious, highly mobile behavior, which can develop into a mass migration. The phenomenon of locust migration emphasizes the dynamic nature of populations and their distributions.

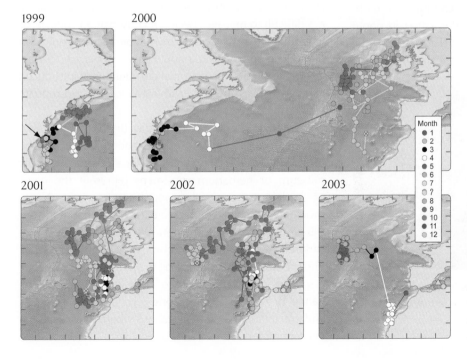

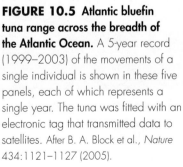

FIGURE 10.5 Atlantic bluefin tuna range across the breadth of the Atlantic Ocean. A 5-year record (1999–2003) of the movements of a single individual is shown in these five panels, each of which represents a single year. The tuna was fitted with an electronic tag that transmitted data to satellites. After B. A. Block et al., *Nature* 434:1121–1127 (2005).

FIGURE 10.6 The migration of wildebeests follows their food supply. The distribution of wildebeest populations of the Serengeti ecosystem of northern Tanzania and southern Kenya (shaded area) is shown for three times during the annual cycle over the years 1969–1972. The migrations follow the lush growth of grasses after seasonal rains. The size of each dot indicates the relative size of the population in that area. After L. Pennycuick, in A. R. E. Sinclair and M. Norton-Griffiths (eds.), *Serengeti: Dynamics of an Ecosystem,* University of Chicago Press, Chicago (1979), pp. 65–87; photos courtesy of A. R. E. Sinclair.

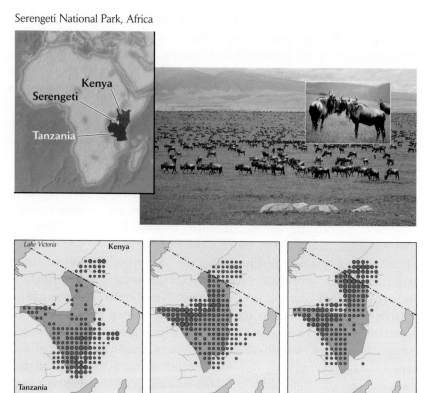

Serengeti National Park, Africa

December to April   May to July   August to November

Ecological niche modeling predicts the distributions of species

As we have seen, the distributions of species reflect the range of physical conditions that support the survival and reproduction of individuals. As a general rule, the more suitable the conditions for a species, the denser its population, and the greater its productivity. This fundamental relationship between distribution and environment allows ecologists to predict the actual or potential distributions of species. Consider some of the applications of this knowledge. In many parts of the world, especially in the tropics, the distributions of species are poorly known. Without accurate assessments of population sizes and distributions, it is difficult to manage species of conservation concern or to identify areas that would be suitable for reintroduction of reduced or extirpated populations. Conversely, managers able to predict the potential distributions of introduced invasive species are better able to develop appropriate countermeasures.

The basic problem is to predict the distribution of a species from limited information about the occurrence of individuals. If we assume that locations recorded for museum or herbarium specimens represent the range of conditions over which a population can exist, then it should be possible to extrapolate the remainder of the distribution from this information. We can do this by a procedure called **ecological niche modeling** (Figure 10.8). The modeler starts by mapping the known occurrences of a species in geographic space, then catalogs the combination of ecological conditions—generally temperature and precipitation—at the locations where the species has been recorded. The catalog of ecological conditions is a

FIGURE 10.7 The migration of locusts is a developmental response to high population density. A dense swarm of migratory locusts (*Locusta migratoria*) moves over Somalia, Africa, in 1962. Courtesy of the U.S. Department of Agriculture.

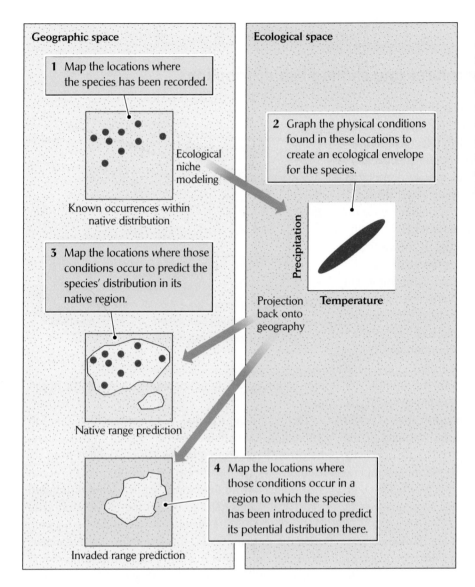

Geographic space

1 Map the locations where the species has been recorded.

Ecological niche modeling

Known occurrences within native distribution

3 Map the locations where those conditions occur to predict the species' distribution in its native region.

Native range prediction

4 Map the locations where those conditions occur in a region to which the species has been introduced to predict its potential distribution there.

Invaded range prediction

Ecological space

2 Graph the physical conditions found in these locations to create an ecological envelope for the species.

Precipitation

Projection back onto geography

Temperature

FIGURE 10.8 Ecological niche modeling can be used to predict the actual or potential distributions of species. The ecological envelope developed from a small number of recorded locations of individuals can be used to predict the distribution of a population, or to predict the distribution of an introduced species in a new region. After A. T. Peterson. *Quarterly Review of Biology* 78:419–433 (2003).

species' **ecological envelope.** Then, because geographic patterns of temperature and precipitation are well known, the modeler can map the geographic area that has the same combination of conditions to predict the broader occurrence of the species within a region.

Many statistical procedures have been developed to construct the ecological envelope of a species, but their details and the differences between them are not important here. The accuracy of the geographic predictions can be evaluated by dividing the data into a "training" set used to construct the ecological envelope and a "test" set used to determine the accuracy of the predicted distribution. The value of producing ecological envelopes can be shown in cases in which sampling of species distributions is particularly thorough—that is, the entire geographic range is known, allowing very accurate ecological niche modeling. For example, the distribution of three species of

Eucalyptus trees across 6,080 plots in southeastern New South Wales, Australia, shows a clear separation of the species with respect to climatic variables (Figure 10.9 on p. 208). Climate is not the whole story in this case, however, because each species also occurs predominately on soils derived from different types of underlying rock: volcanics in the case of *E. rossii; E. muellerana* on sedimentary rocks; and *E. pauciflora* on granites.

In Figure 10.10, the range of the Chinese bush clover (*Lespedeza cuneata*) in its native region in eastern Asia is projected from an ecological envelope produced on the basis of the 28 locations in eastern China and Japan where the plant has been collected. Note that the projected range includes most of temperate southern China, the Korea Peninsula, and Japan. On the basis of 30 known locations, another Asian species, the aquatic plant *Hydrilla verticillata,* is projected to have a more southerly range,

GLOBAL CHANGE

Changing ocean temperatures and shifting fish distributions

Ecologists commonly observe that the physical conditions of terrestrial and aquatic environments help determine where a species can live. Given the importance of conditions such as temperature, precipitation, and salinity, one would expect that a species' distribution would change as these conditions change over time.

Global warming has had a dominating influence on ecological change during recent decades. During the past century, the average temperature of the earth has increased by 0.6°C, and this warming has influenced the oceans as well as environments on land. Temperatures in the bottom waters of the shallow North Sea, for example, have increased 0.7°C per decade, a 2°C increase since the mid-1970s (Figure 1). Could such temperature changes cause species to alter their distributions?

J. G. Hiddink, from Bangor University in the United Kingdom, and R. ter Hofstede, from the Institute for Marine Resources and Ecosystem Studies in the Netherlands, investigated whether marine fish species might adjust their ranges in response to the warming of the North Sea. If many fish species from more southerly waters were to move north with the warming temperatures, the entire fish community of the North Sea could be substantially altered, with potential consequences for the fisheries of the region.

To understand how fish species are distributed relative to temperature, the researchers first needed temperature data for the North Sea region. Fortunately, ocean temperatures had been collected for every degree of latitude and longitude in the region from 1977 to 2003 by the International Council for the Exploration of the Sea (ICES). These data provided detailed geographic

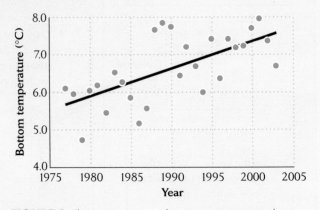

FIGURE 1 The average winter bottom temperature in the North Sea has increased over the past 30 years. After Hiddink and ter Hofstede, *Global Change Biology* 14:453–460 (2008).

information on where temperatures varied on the ocean floor.

During the same period, ICES compiled data on fish distributions by pulling large nets (called "trawls") along the ocean floor. From 1985 to 2006, scientists from six countries worked together to fish the bottom of the ocean at 300 locations in the North Sea that were distributed at a spatial scale similar to that of the temperature data (0.5° latitude × 1° longitude; or about 56 × 56 km). Using the resulting 7,000 trawl samples, the researchers could determine how many fish species were present at each location. From these data, they could deduce how fish species richness was related to temperature and how it had changed over the years.

The researchers found that fish species richness in the North Sea had increased steadily over 22 years, from

extending through Southeast Asia to New Guinea and tropical northern Australia. Nonetheless, the distributions of the two species should overlap extensively in southern and eastern China and in northern Laos and Vietnam.

Both *Hydrilla* and *Lespedeza* are highly invasive species in North America. Based on the ecological envelopes developed for their native distributions, we can predict that *Hydrilla* should have a more southerly distribution than *Lespedeza* in North America, and it does. One often finds exceptions to such predictions, however. Notice

that *Hydrilla* has become well established in central and southern California, which are not in the core area of its ecological envelope. Extensive irrigation for agriculture has created suitable conditions for *Hydrilla* in these areas even though the climate is generally too dry to support productive aquatic systems. Conversely, *Hydrilla* is absent from some areas that fall within the most favorable parts of its ecological envelope, particularly the Mississippi River drainage of Arkansas, Oklahoma, Missouri, Mississippi, and western Tennessee. These absences might be

FIGURE 2 Anchovies are one of the small-bodied fish species that are expanding their ranges northward. Tom McHugh/Photo Researchers.

about 60 species in the mid-1980s to nearly 90 species two decades later. This change reflected dozens of more southerly species expanding their ranges northward. Indeed, the ranges of 34 species, including anchovies (*Engraulis encrasicolus;* Figure 2) and red mullet (*Mullus surmuletus*), expanded, whereas the ranges of only 3 species that were already present in the area contracted.

Interestingly, most of the fish species that moved northward were relatively small-bodied.

The increase in species richness was positively correlated with the increase in bottom-water temperatures in the North Sea (Figure 3). This correlation suggests that warmer temperatures are more hospitable to a greater variety of species, and that the warming of the North Sea has allowed more southerly species to expand the northern edges of their ranges into the area. Hence, this is a case of global change increasing the diversity of species in a region.

The increase in the diversity of species in the North Sea is not only dramatically changing the fish community that is present there, but may also affect the important commercial fisheries that depend on this community. For example, the three species whose ranges have contracted (wolffish, spurdog, and ling) are all commercially important, whereas more than half of the species with expanded ranges have little or no commercial value. As a result, these shifts in species distributions with warming temperatures may increase fish diversity, but decrease the value of the commercial fisheries of the North Sea.

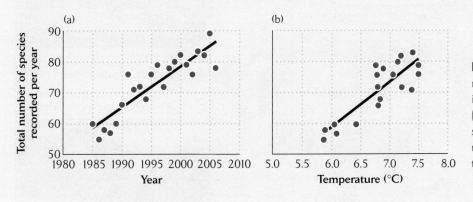

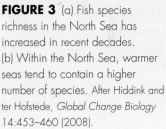

FIGURE 3 (a) Fish species richness in the North Sea has increased in recent decades. (b) Within the North Sea, warmer seas tend to contain a higher number of species. After Hiddink and ter Hofstede, *Global Change Biology* 14:453–460 (2008).

explained either by local ecological conditions that differ from those in the native range, but were unaccounted for in the ecological niche model, or by failure of the species to disperse to these areas. Indeed, absences of species from ecologically suitable areas have been used to study the role of dispersal limitation in the distributions of species. The world is constantly changing, and sometimes species have difficulty moving with shifting geographic areas of suitable conditions. This is a major concern given the current rapid warming of the earth's environment.

The dispersion of individuals reflects habitat heterogeneity and social interactions

Dispersion describes the spacing of individuals with respect to one another within the geographic range of a population. (Keep in mind the distinction between *dispersion* and *dispersal*, which refers to the movements of individuals.) Patterns of dispersion range from **clumped**

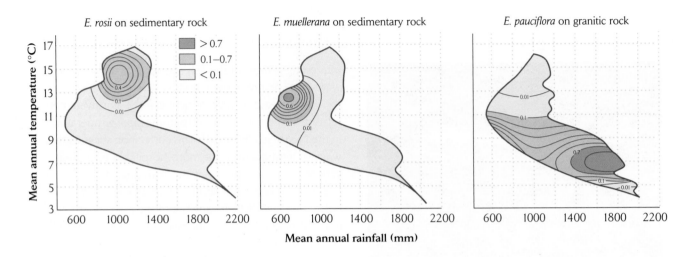

FIGURE 10.9 The value of ecological niche modeling can be tested using species with well-known distributions. Each of three species of *Eucalyptus* has a distinct ecological envelope. The curves show the probability of the species being found in areas of particular temperature and rainfall. Note that the ecological envelopes differ somewhat for locations on sedimentary and granitic rocks. From M. P. Austin et al., *Ecological Monographs* 60:161–177 (1990).

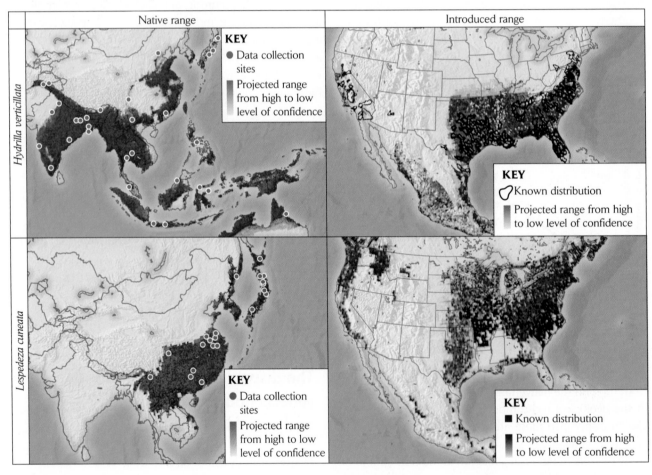

FIGURE 10.10 Ecological envelopes can be used to predict the distributions of invasive species. A small number of records of *Hydrilla verticillata* and *Lespedeza cuneata* were used to predict the distributions of these plants, both in their native region (eastern Asia) and in a new region where they are becoming invasive (North America). From A. T. Peterson et al., *Weed Science* 51:863–868 (2003).

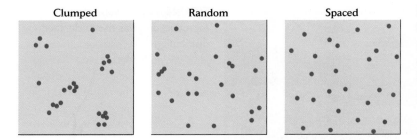

Clumped   Random   Spaced

FIGURE 10.11 Dispersion patterns describe the spacing of individuals. Different population processes produce clumped or evenly spaced distributions. In the absence of such processes, individuals are distributed without regard to the positions of others in a so-called random pattern.

distributions, in which individuals are found in discrete groups, to evenly **spaced** distributions, in which each individual maintains a minimum distance between itself and its neighbors (Figure 10.11).

Spaced and clumped distribution patterns derive from different processes. Even spacing most commonly arises from direct interactions between individuals. For example, plants positioned too close to larger neighbors often suffer from shading and root competition; as these individuals die, individuals become more evenly spaced (Figure 10.12). Clumped distributions may result from social predisposition to form groups, clustered resources, or tendencies of progeny to remain close to their parents. Birds in large flocks may find safety in numbers. Salamanders and sow bugs aggregate under logs because they are attracted to dark, moist places. Some species of trees, such as the quaking aspen, form clumps of stems—all parts of the same individual—by vegetative reproduction (Figure 10.13). Within such a clump, however, the stems tend to be regularly spaced.

Aggregation can also occur when trees scatter their seeds over limited distances. The distributions of tree species in a tropical rain forest at Pasoh, Malaysia, reflect their seed dispersal mechanisms (Figure 10.14). Animals carry

the seeds of *Baccaurea racemosa* fruits over relatively long distances, spreading them evenly through the environment. Seeds of *Shorea leprosula* have wings and gyrate to the ground like little helicopters. *Croton argyratus* disperses its small seeds by popping them out of drying capsules (so-called ballistic dispersal). Viewed on a scale of hundreds of meters, both of the latter species have clumped distributions that reflect their limited dispersal abilities. On smaller scales of meters, however, competition between individuals tends to create more regular spacing, just as in an aspen clone. Thus, the dispersion pattern that we perceive depends on the scale of our viewpoint because different processes act over different spatial scales.

Individuals are rarely free of the influence of others, whether that influence takes the form of competition or mutual attraction. If the individuals in a population were distributed in a completely uniform environment without regard to the positions of others in the population, we would say that they were distributed at **random.** Although a random distribution is unlikely, it is a null hypothesis against which alternative hypotheses invoking other processes must be tested. Even if we wished to know simply whether the dispersion pattern of a population is clumped or spaced, we would have to demonstrate statistically that its distribution differed significantly from random. Ecologists have devised several tests for this purpose.

DATA ANALYSIS MODULE *Parting the Dispersion of Individuals Against a Nonrandom Dispersion.* Patterns of dispersion—clumped, random, and spaced—may be distinguished by comparison with a Poisson distribution. You will find this module at http://www.whfreeman.com/ricklefs6e.

The spatial structure of populations parallels environmental variation

We have seen that species inhabit patches of suitable habitat and do not inhabit unsuitable environments. However, even suitable environments vary in quality. Some

FIGURE 10.12 Evenly spaced distributions result from interactions between individuals. The even spacing of these desert shrubs in Sonora, Mexico, results from competition for water in the soil. Photo by R. E. Ricklefs.

FIGURE 10.13 Vegetative reproduction gives rise to clumped distributions. This photograph, taken in Coconino National Forest, Arizona, shows many different clones of aspen trees, which can be distinguished from one another by timing of leaf fall: some clones are bare, while others still have their yellow autumn foliage. Within each clone, each individual stem ("tree") has grown from a common root system that developed from a single seedling. Photo by Tom Bean/DRK Photo.

patches, for example, may have the resources to support a large population, while others of similar size, with fewer resources or a different vegetation structure, may support a much smaller population. Thus, within the geographic range of a population, the number of individuals per unit of area, or population **density,** can vary markedly.

Variation in population density

Environmental conditions vary over broad gradients—for example, from warm to cold as one travels northward in the Northern Hemisphere or higher up a mountain. Each species is well adapted to a relatively narrow range of conditions. Areas with those conditions tend to support the highest population densities and are often found close to the center of a species' distribution. As one travels away from this point, conditions become less

favorable, and fewer individuals can be supported per unit of area.

Consider, for example, the occurrence of the dickcissel (*Spiza americana*), a small songbird of prairies and grasslands, throughout the midwestern United States (Figure 10.15). The species tends to be most numerous in the center of its range and to decrease in density toward the periphery. However, because environmental conditions do not vary smoothly, the pattern is highly irregular, mapping onto the underlying distribution of prairie and grassland habitat.

Ideal free distributions

If an individual has perfect knowledge of variation in the quality of habitat patches, it should choose among them so to maximize its access to resources. Of course, all indi-

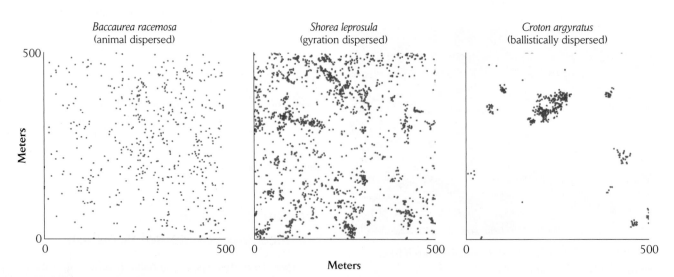

FIGURE 10.14 Dispersion patterns in rain forest trees reflect dispersal distances. Trees with animal-dispersed seeds, such as *Baccaurea racemosa*, are spread more evenly through a Malaysian forest than are trees whose seeds fall passively close to the parent, such as *Shorea leprosula* and *Croton argyratus*. After T. G. Seidler and J. B. Plotkin, *PLoS Biology* 4:e344 (2006).

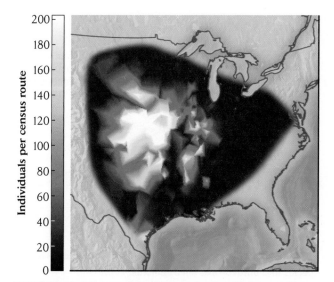

FIGURE 10.15 Population density is often highest at the center of a species' geographic range. The map shows the average number of individuals recorded per 24.5-mile census route (with stops every 0.5 miles) throughout the range of the dickcissel (*Spiza americana*) in North America. Note the vertical "gash" in the center of the range, which outlines unsuitable (for dickcissels) forested areas in Arkansas, southern Missouri, and southern Illinois. After B. McGill and C. Collins, *Evolutionary Ecology Research* 5:469–492 (2003).

viduals in a population have this same goal, which means that good patches tend to fill up rapidly. Thus, individuals should settle in the highest-quality patches first. As the populations in such patches build up, however, their resources become depleted, and conflicts over resources increase. Therefore, the quality of those patches decreases until inherently poor patches become equally good choices (Figure 10.16). In the end, individuals move from better to poorer patches until each patch has the same value for individual fitness, regardless of the intrinsic patch quality. This outcome is called an **ideal free distribution.**

Tendencies toward an ideal free distribution were first investigated in laboratory studies, in which the quality of patches could be controlled. For example, Manfred Milinski, of the Max Planck Institute for Limnology in Ploen, Germany, tested the patch-seeking behavior of stickleback fish (genus *Gasterosteus*) by providing food (water fleas) at different rates at opposite ends of an aquarium. Each end of the aquarium thus constituted a patch with its own food supply. The experimental system had the following properties: the two patches differed in quality; quality decreased as the number of fish using a patch increased (more mouths, less food); and fish were free to move between patches.

Milinski placed fish in the aquarium about 3 hours ahead of time to ensure they would be hungry when the experiment began. Before any food was added to the aquarium, the fish were distributed equally between the two halves. In one experiment, 30 water fleas were added per minute to one end of the aquarium and 6 per minute to the other—a ratio of 5 to 1. Within 5 minutes, the fish had distributed themselves between the two halves in the ratio predicted for an ideal free distribution—that is, 5 to 1. When the provisioning ratio was changed, or the high- and low-quality patches were switched between ends of the aquarium, the fish quickly followed suit. How they achieved this ideal free distribution was not determined, but they must have used the rate at which they encountered food items, and perhaps the number of other fish close by, as cues to patch quality. It seems remarkable that organisms are so sensitive to the conditions of their environments and able to alter their behavior appropriately.

Dispersal from high-density to low-density patches

Under an ideal free distribution, realized quality is evened out among patches, and the reproductive success of each individual is the same regardless of the intrinsic quality of the patch it occupies. However, populations rarely attain this ideal in nature. First, individuals rarely have perfect knowledge of patch quality and thus cannot make fully informed choices. Second, dominant individuals may force subordinates to leave high-quality patches. Whenever reproductive success has been measured in the field, it has been higher in preferred habitats, often because dominant individuals have monopolized the best

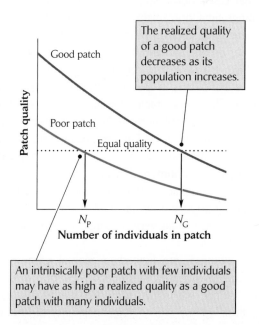

FIGURE 10.16 In an ideal free distribution, each individual exploits a patch of the same realized quality.

resources. As a result, populations in the best habitats tend to produce surplus individuals, whereas in poorer habitats deaths exceed births and populations cannot maintain their numbers. The result is a net movement of individuals from growing populations to shrinking populations. In this way, continuous immigration from crowded high-quality patches can maintain populations in lower-quality patches.

In southern Europe, a small songbird called the blue tit (*Parus caeruleus*) breeds in two kinds of forest habitat, one dominated by the deciduous downy oak (*Quercus pubescens*) and the other by the evergreen holm oak (*Quercus ilex*). Long-term studies by Jacques Blondel and his colleagues at the University of Montpellier, in southern France, have revealed many details about the populations of these birds. The deciduous oak habitat produces more caterpillars as food for the tits, and their population densities (90 pairs of birds per 100 ha in deciduous oak habitat versus 14 pairs in evergreen oak habitat) reflect this difference in resources. Parents produce about 50% more offspring in deciduous oak forest (5.8 per pair per year) than in evergreen oak forest (3.6 per pair per year). Annual survival of adults breeding in the two habitats is about the same (50%), so populations in the deciduous oak forest would increase at about 9% per year if individuals—mostly young birds—did not disperse to the evergreen oak habitat, where local populations would decrease at a rate of about 13% per year in the absence of immigration.

Given that populations can persist in suboptimal habitats, what limits the ecological and geographic ranges of a species? This question can be difficult to answer for particular cases, but a simple illustration will show how such range limitation might work. Imagine a population distributed over an environmental gradient with favorable conditions on one end and unfavorable conditions on the other (Figure 10.17). The population growth rate under favorable conditions is positive; that is, births exceed deaths. Under unfavorable conditions, deaths exceed births, and local populations decline when no individuals are added by immigration. Dispersal from more favorable habitats can offset these deaths and maintain a population in a marginal environment, but only to a limited extent. The capacity to replace population losses is limited by the ability of the population in favorable habitats to produce new members, so that very marginal habitats cannot be populated at all.

If distributions are limited in general because individuals are not adapted to conditions beyond the edges of their ranges, we might reasonably ask why marginal populations do not evolve greater tolerance of local conditions that would allow them to increase in numbers and

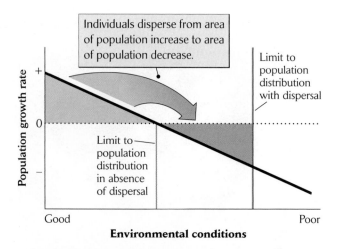

FIGURE 10.17 Dispersal may extend the distribution of a species into unfavorable environments. Individuals move away from habitats with positive population growth rates (blue area), which would become overcrowded in the absence of emigration, to less suitable habitats with negative population growth rates (red area), helping to maintain populations there. This movement does not affect the net population growth rate throughout the distribution. This capacity to replace population losses is limited by the productivity of the local populations in favorable regions, which ultimately constrains the population's distribution.

expand their ranges. In an important paper published in 1997, Mark Kirkpatrick and Nick Barton suggested that local adaptation at the periphery of a species range is prevented by immigration of dispersing individuals from the center of the range, which bring genomes adapted to conditions in the optimal habitat for the species. According to this scenario, dispersal hinders range expansion rather than promoting it, but biologists have not yet weighed the balance between these conflicting processes. In Chapter 13, we shall look more closely at the genetic consequences of individual movements within populations.

Three types of models describe the spatial structure of populations

The patchiness of the natural world and the movement of individuals between patches require a spatially complex concept of the population. Ecologists have developed this concept through three types of models—metapopulation, source–sink, and landscape—that include the spatial structure of the environment. **Metapopulation models** describe a set of subpopulations occupying patches of a particular habitat type between which individuals move occasionally. The intervening habitat, which is referred to as the **habitat matrix,** serves only as a barrier to

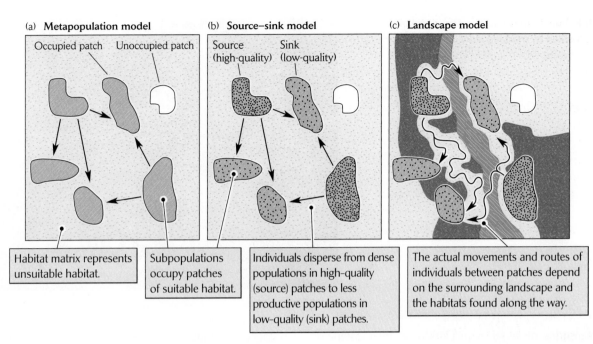

FIGURE 10.18 Models of spatial population structure differ with respect to variation in habitat patch quality and in the intervening matrix. Arrows represent movements of individuals between patches. (a) A metapopulation model with six patches of suitable habitat, five of which are occupied. Populations occupy habitat patches surrounded by a featureless matrix of unsuitable habitat. Thus, the quality of the matrix has no effect on interpatch movements, but the arrangement of patches and distances between patches may. (b) In a source–sink model, growing populations in high-quality habitat patches (source populations) produce excess individuals. These individuals disperse to less suitable habitat patches, where immigration maintains less productive sink populations. (c) In a landscape model, patches of suitable habitat are overlaid on a landscape composed of a mosaic of many different habitat types. The landscape affects the movements of individuals between patches. After J. A. Wiens, in I. A. Hanski and M. E. Gilpin (eds.), *Metapopulation Biology: Ecology, Genetics, and Evolution*, Academic Press, San Diego (1997), pp. 43–62.

the movement of individuals between subpopulations (Figure 10.18a). These models emphasize the contributions of colonization and extinction of subpopulations to changes in overall population size. We shall explore these dynamics in more detail in Chapter 12.

Source–sink models add differences in the quality of suitable habitat patches, as we have seen in the example of the blue tit in oak forests of southern Europe. Where resources are abundant, individuals produce more offspring than required to replace themselves, and the surplus offspring disperse to other patches. Such populations serve as **source populations.** In patches of poorer habitat, populations are maintained by immigration of individuals from elsewhere because too few offspring are produced locally to replace those that die. These populations are known as **sink populations** (Figure 10.18b).

The **landscape model** goes a step beyond the metapopulation model by considering the effects of differences in habitat quality within the habitat matrix (Figure 10.18c). Accordingly, the quality of a habitat patch can be altered by the nature of nearby habitats. It might be improved, for example, if other nearby patches in the landscape provide resources such as safe roosting sites, nesting materials, pollinators, or water. Other kinds of neighboring habitat patches might deliver serious negatives if they harbored predators and vectors of pathogens. The habitat matrix also influences the movement of individuals from one subpopulation to another. Clearly, some travel routes are more attractive than others because of the habitat types encountered along the way.

Dispersal is essential to the integration of populations

When individuals disperse throughout a population, they link different subpopulations together and make the whole population function and evolve as a single structure. When dispersal is limited, different parts of a population behave independently of one another. Measurements of dispersal are therefore important to ecologists who wish to understand population dynamics.

FIGURE 10.19 Movements within populations can be tracked by marking individuals. Here, a female shoveller duck (*Anas clypeata*) is fitted with an individually identifiable beak tag and a small radio transmitter. These devices do not harm the birds or impair their movement and feeding. Photo by R. E. Ricklefs.

Measuring dispersal, particularly over long distances, requires marking and recapturing individuals (Figure 10.19). Because individuals in some populations can disperse long distances in any direction, large areas must be searched for marked individuals to ensure accurate sampling of movements. Population biologists often resort to ingenious techniques to measure dispersal. For example, one of the first attempts to measure dispersal in natural populations involved measuring movements away from a release point by fruit flies (*Drosophila*) that could be distinguished by a visible mutation.

A convenient measure of movement within populations is the average **lifetime dispersal distance,** which indicates how far individuals move, on average, from their birthplace to where they reproduce. If we draw a circle with its origin at an individual's birthplace and a radius equal to the average lifetime dispersal distance for its species, it will circumscribe the individual's *lifetime dispersal area* (Figure 10.20). In a uniformly distributed population, an individual's lifetime dispersal area will include most of the other individuals that it could potentially interact with or mate with in its lifetime. The number of those individuals, which is estimated by the lifetime dispersal area times the population density, defines the **neighborhood size** for each member of a population.

To illustrate the application of the neighborhood size concept to natural populations, small songbirds of eight species were marked with leg bands as nestlings and recaptured as breeding adults. Lifetime dispersal distances for these species averaged between 344 and 1,681 meters, population densities varied between 16 and 480 individuals per square kilometer, and neighborhood sizes varied between 151 and 7,679 individuals. A lifetime dispersal

distance of about 1 kilometer per generation is thus not unusual for populations of songbirds. At this rate, the descendants of a particular individual might traverse an entire population, distributed across a reasonably sized continent, in a thousand generations or so.

This rate, however, is slow compared with the rate at which the edge of a population distribution can expand through an unoccupied region. In 1890 and 1891, 160 European starlings were released in the vicinity of New York City (by an individual who wished to introduce all the birds mentioned in Shakespeare's works to the

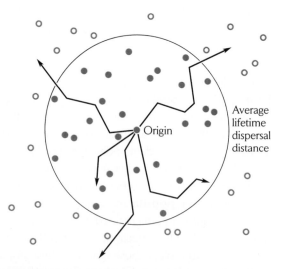

FIGURE 10.20 Neighborhood size is the number of individuals within a circle whose radius is the average lifetime dispersal distance. An individual's birthplace is indicated by the origin in the center of the circle. The radius of the circle represents the average distance that individuals travel between birth and reproduction.

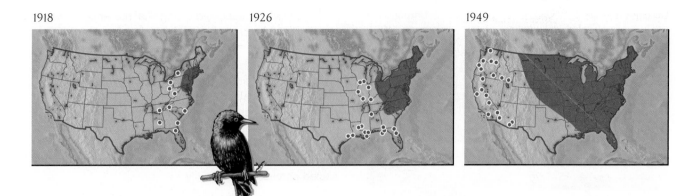

FIGURE 10.21 Introduced populations can spread rapidly outside their established range. Western expansion of the range of the European starling (*Sturnus vulgaris*) in the United States has occurred through long-distance dispersal. The shaded areas represent the breeding range; dots indicate records of birds—primarily young individuals—in preceding winters outside of the breeding range. Such long-distance movements of unmarked juveniles would probably not be detected within an established population. The starling population now covers more than 3 million square miles of North America from coast to coast. After B. Kessel, *Condor* 55:49–67 (1953).

New World). Within 60 years, the population had spread 4,000 kilometers from New York to California, at an average rate of about 67 kilometers, or 67,000 meters, per year (Figure 10.21). This rapid expansion resulted when a few individuals, primarily young ones, dispersed much longer distances than the average and established new populations beyond the species' range boundary. Although the few individuals that might move over such long distances are exceptions, they can have large effects on population distributions and on the movement of alleles within a population.

Average dispersal distance bears a surprisingly constant relationship to population density across many species. For example, in three populations of the land snail *Cepaea nemoralis,* which obviously holds no speed records, dispersal distances after 1 year varied between 5.5 and 10 meters. However, because snail populations are dense, neighborhood sizes were similar to those for the songbirds described above: 1,800 to 7,600 individuals. Mark–recapture data on a rusty lizard (*Sceloporus olivaceous*) population near Austin, Texas, revealed average lifetime dispersal distance to be 89 meters and neighborhood size to be between 225 and 270 individuals. Thus, for a variety of different animal populations, neighborhood sizes are rather more similar than one might expect from either dispersal distance or population density alone.

 **Effects of habitat corridors on dispersal and distributions in an Atlantic coastal plain pine forest.** On the Atlantic coastal plain

of the southeastern United States, loblolly and longleaf pines form open forests with a diverse assemblage of understory plants. Many of these understory species grow best in open patches created when storms topple trees or fire sweeps through an area. As a result, many species of shrubs and herbs have patchy distributions that are limited, to some extent, by their dispersal abilities.

Nick Haddad, of North Carolina State University, and many of his students and colleagues conducted a large experimental study at the Savannah River Ecology Laboratory in South Carolina to investigate one factor contributing to the distributions of animals and plants in this landscape: the effect of corridors connecting suitable habitats. The plan of the study was simple: The investigators created eight sets of five equal-area (1.375 ha) clear-cut habitat patches. Each set had a central patch that served as a source area. Of the four peripheral patches, one was connected to the central patch by a 25-meter-wide clear-cut corridor. The remaining three isolated patches had different shapes to control for the effect of the increased habitat edge within the corridor (Figure 10.22).

The investigators used a number of methods for measuring dispersal. They followed movements of the common buckeye butterfly (*Junonia coenia*) and the variegated fritillary butterfly (*Euptoieta claudia*) by marking individuals in each of the central patches and recapturing them in the peripheral patches. A similar technique was used to follow movements of insect-borne pollen and bird-dispersed fruits and seeds. To track pollen movement, the investigators planted three mature female holly plants (*Ilex verticillata*) in each peripheral patch and eight male plants in each central patch. Fertilized flowers producing fruits in the peripheral patches indicated pollen movement from the central patch. To track seed and fruit movements, they

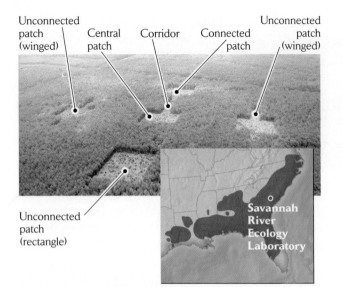

FIGURE 10.22 Experimental patches can be used to investigate the effects of habitat corridors. The Savannah River Ecology Laboratory is located in South Carolina, in the loblolly and longleaf pine forests of the Atlantic coastal plain (red area). Investigators created sets of clear-cut experimental patches in the forest habitat to investigate the effects of habitat connectivity and patch shape. The central patch served as a source area for dispersing animals and plants. After E. I. Damschen et al., *Science* 313:1284–1286 (2006); D. J. Levey et al., *Science* 309:146–148 (2005); photo courtesy of Ellen Damschen.

planted individuals of a local holly (*Ilex vomitoria*) and wax myrtle (*Myrica cerifera*) in the central patch, then collected samples of bird droppings in traps placed under 16 artificial perches in each peripheral patch. In some cases, the fruits in the central patch were dusted with colored fluorescent powder. The powder would fluoresce

under ultraviolet light, indicating the presence of seeds from fruits in the central patch in the fecal material.

The results of these experiments, some of which are shown in Figure 10.23, demonstrated that butterflies, insect pollinators, and birds all made use of habitat corridors. The connected peripheral patches received more traffic from the central patch than did the isolated peripheral patches. Indeed, the movement of fruits and seeds through the corridors was so frequent that the number of species of herbs and shrubs increased faster in the connected patches. ▌

Macroecology addresses patterns of range size and population density

Although it may be difficult to understand the factors that shape the distribution and population size of a particular species, the patterns revealed by large samples of species can provide useful insights. Analyzing and interpreting these patterns is one objective of the developing discipline of **macroecology.** In 1984, James H. Brown, at the University of New Mexico, suggested that the distribution and population size of a species reflects the distribution of conditions to which individuals of the species were well adapted. If these conditions are common and widespread, then the population should also be common and widespread. Furthermore, as we have seen, a species should be most abundant in the center of its distribution, where conditions are most favorable. As one travels away from this point, fewer individuals can be supported per unit of area, until a point is reached at which the population can no longer be supported by local reproduction and immigration from more productive areas.

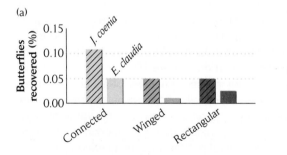

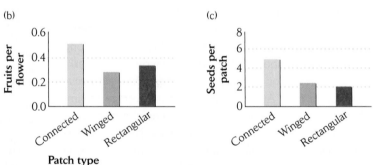

FIGURE 10.23 Insects and birds use corridors to move between patches of habitat. Results of experiments at the Savannah River Ecology Laboratory showed more frequent movement between connected patches. (a) Proportions of butterflies marked in the central patch that were recovered in the different types of peripheral patches. (b) Fruit set of female *Ilex* bushes in peripheral patches. Fruit set indicates pollen movement by bees and other insects from male plants in the central patch. (c) Number of seeds of *Ilex* and *Myrica* bushes recovered in samples of bird droppings in the peripheral patches. Recovery of seeds indicates movement of those seeds by birds from plants in the central patch. From J. J. Tewksbury et al., *Proceedings of the National Academy of Sciences USA* 99:12923–12926 (2002).

Range size and population density

Brown used the methods of macroecology to determine which factors explain differences in the densities and distributions of different species. According to his view, species that are adapted to a broad variety of resources and conditions are likely to be more abundant in the center of their distribution than are species that more specialized. Furthermore, broadly adapted organisms are likely to find their preferred resources distributed over a broader geographic area. Accordingly, Brown predicted that geographic range and population density in the center of the distribution should be positively correlated. This prediction can be tested statistically in a sample of species for which both geographic range and population density have been measured. One such test looked at populations of birds within North America (Figure 10.24), for which data are particularly good. The results illustrate two features of species distributions. First, the area occupied by a species and its maximum population density are positively correlated, as Brown predicted. However, the data exhibit considerable scatter around this relationship. In the North American bird sample, population density statistically explains only 13% of the variation in range size.

The source of this variation is of considerable interest to ecologists, as it must result from the special relationship that each species has with its environment. From the perspective of Brown's hypothesis, variation in the underlying distributions of resources is also important. For example, a bird species might specialize on particular features of productive riparian habitats. Such a species might densely populate the floodplains of rivers, but be only rarely present in surrounding areas. Although ecologists have considerable data on the distributions and abundances of species, they have much less information about the distribution of factors that limit populations.

Two additional considerations will help us to better understand these differing patterns of distribution and abundance among species. First, we often fail to find individuals of a particular species within areas of apparently suitable habitat, and we often observe that patterns of species presence and absence vary over time. Of course, habitat suitability may itself be changing with the rise and fall of prey populations, predators, and pathogens. In addition, births, deaths, and movements of individuals are, to some extent, random events that create unpredictable variation in populations. The dynamics of these processes are considered in the next chapter.

Second, many biologists have noticed that even closely related species can have very different ranges. For example, the breeding distribution of the Kirtland's warbler (*Dendroica kirtlandii*) is restricted to a small area in central Michigan, where the population numbered fewer than 1,500 pairs in 2002. In contrast, millions of individuals of the yellow-rumped warbler (*Dendroica coronata*) are spread across the breadth of North America in boreal forests. These differences in distribution cannot be caused by characteristics of morphology, physiology, and behavior, which the two species share because of their close evolutionary relationship. Rather, the differences reflect special attributes or circumstances that have influenced each of the species differently—possibly the historical distributions of the habitats they prefer or their relationships with specialized pathogens.

Body size, distribution, and abundance

Returning to more general patterns, macroecologists have paid considerable attention to the relationship between population density and body size—asking, for example, why mice are more abundant than elephants. In fact, body size is an important factor determining the density of individuals in a population. Over a wide range of sizes—say, from bacteria or diatoms to giraffes or whales—population density decreases with increasing organism size, as shown for herbivorous mammals in Figure 10.25. Part of this relationship is simply a matter of size relative to space. A square meter of soil harbors hundreds of thousands of small arthropods; that many elephants simply would not fit. However, even if they could be squeezed in, a square meter of soil would not produce enough food to sustain

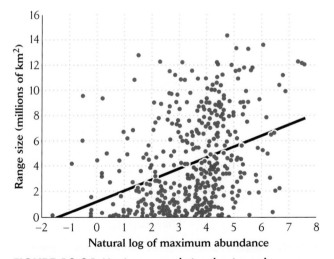

FIGURE 10.24 Maximum population density and geographic range are generally positively correlated. Among the 457 species of North American birds sampled, those species with larger ranges tend to have higher maximum abundances. After B. McGill and C. Collins, *Evolutionary Ecology Research* 5:469–492 (2003).

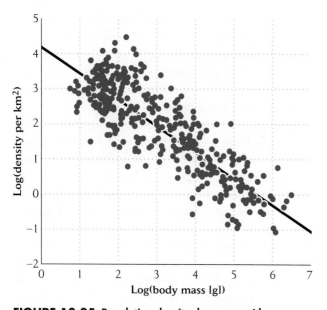

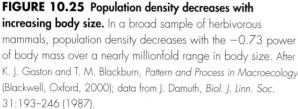

FIGURE 10.25 Population density decreases with increasing body size. In a broad sample of herbivorous mammals, population density decreases with the −0.73 power of body mass over a nearly millionfold range in body size. After K. J. Gaston and T. M. Blackburn, *Pattern and Process in Macroecology* (Blackwell, Oxford, 2000); data from J. Damuth, *Biol. J. Linn. Soc.* 31:193–246 (1987).

a large number of elephants. Larger individuals require more food and other resources than small ones, and the density of a population balances the requirements of individuals with the capacity of the environment to support them.

The metabolic rates of organisms, and therefore their food requirements, increase with body mass, but not in direct proportion to body mass. Rather, the increase in food requirement is equal to the increase in mass raised to a power less than 1, where 1 represents proportionality. In many analyses, this power is observed to be close to $3/4$ or $2/3$. When food requirement increases, for example, as the $2/3$ power of body mass, a tenfold larger organism requires only 4.6 times as much food, and a hundredfold larger organism requires only 22 times as much food.

It is probably more than a coincidence that population density decreases with body mass at the same rate that food requirement increases. The total food consumption of a population per unit of area is equal to the average consumption per individual multiplied by the local population density. Let's use the symbol M to represent body mass. The $3/4$ power of body mass is thus $M^{3/4}$. If the food requirement per individual increases by this amount, and population density decreases by the same amount (see Figure 10.25), then the local food consumption by populations of different-sized organisms can be compared

by the product $M^{3/4} \times M^{-3/4} = 1$. This result means that populations tend to consume the same amount of food per unit of area regardless of the size of individuals. In other words, elephants and mice would have about the same food requirement per hectare, and by implication, they would have similar effects on population and ecosystem processes. This principle is referred to as the **energy equivalence rule.**

The relationship between consumer populations and their resources has been particularly well documented for mammals of the order Carnivora, whose prey typically include other mammals. Carnivores range in size from the least weasel (*Mustela nivalis*, 0.14 kg) to the polar bear (*Ursus maritimus*, 310 kg). Carnivore density over this range of body size decreases with the −0.88 power of mass, which is steeper than one would expect on the basis of food requirements. Carnivore density per ton of prey individuals decreases as approximately the −1 power of mass, indicating that the ratio of predator mass to prey mass is constant over the range of predator size. When the slower reproductive rate of larger prey organisms is taken into account, however, the energy equivalence rule holds up well. Thus, compared with the rate of biomass production by their prey populations per unit of area, the density of carnivore populations decreases as the −0.66 power of body mass, which closely parallels their food requirements (Figure 10.26). It makes sense that food supplies limit populations of carnivores.

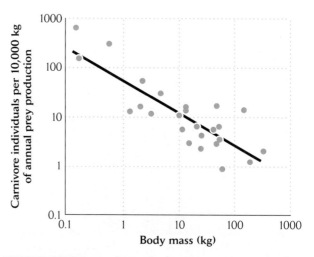

FIGURE 10.26 Population density in carnivores is closely related to their food supply. The number of carnivores relative to the annual biomass production of their prey populations decreases in relation to the food requirement of the carnivores per unit of body mass (−$2/3$ power) rather than their body mass alone (−1 power). After C. Carbone and J. L. Gittleman, *Science* 295:2273–2276 (2002).

Variation in populations over space and time

Just as populations vary spatially, they also vary over time. No population has a static structure; one's perception of a population depends on where and when one looks at it. Most of the theory and empirical and experimental work on population dynamics has focused on local populations so that the complexities of spatial variation do not need to be accounted for. However, we must not forget that populations vary simultaneously in both space and time.

Variation in population density or overall population size in response to changes in climate, resources, or predators and pathogens is often mirrored by changes in distribution, particularly in small organisms with short life spans. Long-term records of the population of the chinch bug (*Blissus leucopterus*) in Illinois illustrate this point (Figure 10.27). Because this insect damages cereal crops, the Office of the State Entomologist, and later the Illinois Natural History Survey, realized the importance of monitoring chinch bug populations, which they estimated from county reports of crop damage. These estimates were calibrated by local studies of the relationship between population size (determined by direct counts on small plots) and crop damage.

Consider the numbers involved. During 1873, when the bugs infested crops over most of the state, ballpark estimates of the population indicated an average density of 1,000 chinch bugs per square meter over an area of 300,000 km^2, or a total of 3×10^{14} pests (300 trillion, more or less). By contrast, farmers reported little damage in 1870 and 1875. These population fluctuations are revealed in Figure 10.27 in the waxing and waning of infestations across the state. We shall consider the processes that underlie such changes in population size in the next chapter, and then go on in the following chapter to address the dynamic behavior of populations.

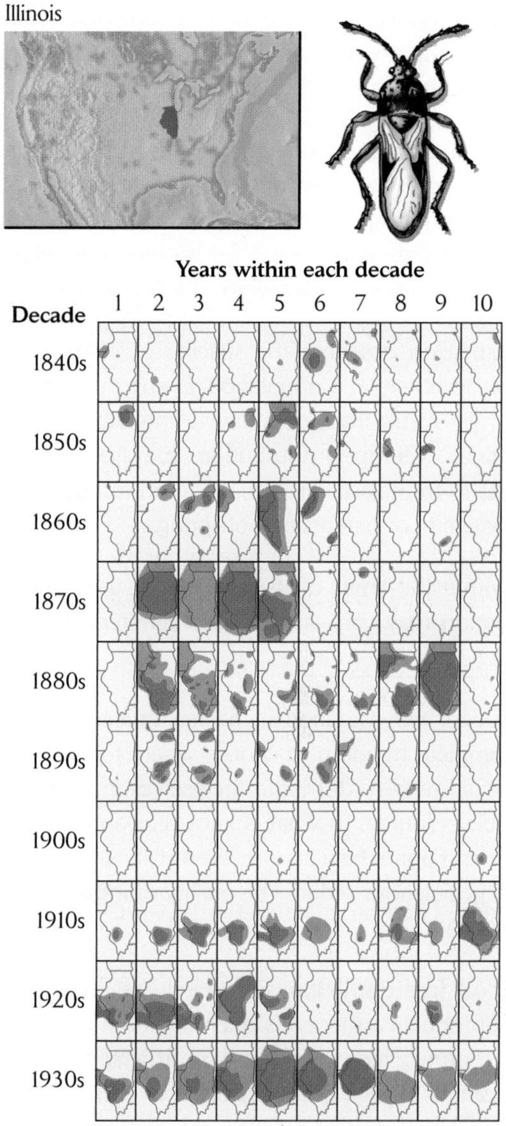

Illinois

FIGURE 10.27 Population density changes over time and space. The distribution of crop damage caused by chinch bugs (*Blissus leucopterus*) in Illinois varied dramatically over the period between 1840 and 1939. Yellow indicates low densities of chinch bugs and blue, high densities. From V. E. Shelford and W. P. Flint, *Ecology* 24:435–455 (1943).

SUMMARY

1. A population consists of the individuals of a species within a given area. Where suitable habitat is distributed in patches, populations may be subdivided into a number of subpopulations.

2. The distribution of a population in space is its geographic range, which is generally limited by the extent of suitable habitat and by barriers to dispersal. The range of physical conditions within which a species can persist is referred to as its fundamental niche. Within this ecological space, the distribution of a species may be further limited by predators and competitors to its realized niche.

3. Populations may be absent from potentially suitable habitats because barriers to dispersal prevent individuals from colonizing these areas. This phenomenon is called dispersal limitation.

4. Many mobile animals undertake long-distance migrations to remain within areas with suitable conditions as the environment changes over time.

5. Ecological niche modeling can be used to predict the actual or potential distribution of a population. Climatic variables at locations where a species has been recorded are used to determine the species' ecological envelope: the range of conditions under which the species can persist.

6. Dispersion describes the spacing of individuals with respect to others in a population. Evenly spaced distributions may result from competitive interactions between individuals. Clumped distributions may result from tendencies to form social groups, clustered resources, or spatial proximity of parents and offspring.

7. Within the limits of its distribution, the density of a population may vary according to differences in habitat quality.

8. Faced with variation in habitat quality, and assuming complete freedom to choose where to live, individuals should tend to distribute themselves in proportion to available resources in what is known as an ideal free distribution. Poorer habitats are eventually settled because dense populations reduce the quality of intrinsically superior habitats.

9. Ideal free distributions are rarely realized in nature, and reproductive success is often higher in some habitats than in others. Individuals tend to disperse from growing populations in high-quality habitats to shrinking populations in lower-quality habitats, thereby maintaining the populations in those habitats.

10. Three types of models—metapopulation, source–sink, and landscape—have been developed to represent the spatial complexity of populations and their environment. Metapopulation models describe a set of patches between which individuals move through a matrix of unsuitable habitat. Source–sink models add variation in habitat quality; individuals disperse from source populations in high-quality patches to sink populations in low-quality patches. Landscape models consider differences in the quality of the habitat matrix

11. Dispersal maintains the spatial integrity of populations and tends to link the dynamics of subpopulations. Ecologists characterize movement within populations by the average lifetime dispersal distances of individuals from their birthplaces. Neighborhood size is the number of individuals within a circle whose radius is equal to the average lifetime dispersal distance. Neighborhood sizes are surprisingly similar across a variety of animal species.

12. Macroecology is devoted to understanding patterns in the sizes of geographic ranges and in the densities and distribution of individuals within those ranges. Much of the variation in macroecological patterns can be related to the body mass and energy requirements of individuals.

13. Although macroecology has identified many general patterns, species are often absent from what appear to be suitable environments, and closely related species, which presumably have similar ecological requirements, often occupy distributions of strikingly different extent.

14. Populations often fluctuate dramatically in space and time, often as the result of biological interactions. Therefore, an understanding of the dynamics of populations is needed to interpret species distributions.

REVIEW QUESTIONS

1. What is the difference between population distribution and population structure?

2. Why is the realized niche considered a subset of the fundamental niche?

3. The American bullfrog is native to eastern North America, but, following its movement by humans, now thrives in western North America. What does this suggest about the cause of the bullfrog's historical range limit?

4. How can ecological niche modeling be used to predict the spread of introduced species?

5. What mechanisms could cause evenly spaced distributions of individuals within populations? What mechanisms could cause clustered distributions?

6. Suppose that 100 cows were allowed to graze in either of two pastures. If the grass was 3 times more productive in pasture A than in pasture B, how many cows would be in each pasture if they followed an ideal free distribution? What might prevent this distribution of cows from happening?

7. What realities of nature do source–sink models and landscape models include that metapopulation models do not include?

8. How would lifetime dispersal distance and neighborhood size differ for a plant whose seeds are carried by the wind and a plant whose seeds drop near the parent plant? How might these plants' dispersal mechanisms influence their population structures?

9. What is the basis for the hypothesis that species with larger ranges should exhibit higher abundances in the center of their range compared with species with smaller ranges?

10. What is the logic underlying the energy equivalence rule?

SUGGESTED READINGS

Angert, A. L., and D. W. Schemske. 2005. The evolution of species' distributions: Reciprocal transplants across the elevation ranges of *Mimulus cardinalis* and *M. lewisii*. *Evolution* 59:222–235.

Block, B. A., et al. 2005. Electronic tagging and population structure of Atlantic bluefin tuna. *Nature* 434:1121–1127.

Blondel, J., et al. 1999. Selection-based biodiversity at a small spatial scale in a low-dispersing insular bird. *Science* 285:1399–1402.

Broennimann, O., et al. 2007. Evidence of climatic niche shift during biological invasion. *Ecology Letters* 10:701–709.

Brown, J. H. 1995. *Macroecology*. University of Chicago Press, Chicago.

Brown, J. H., et al. 2004. Toward a metabolic theory of ecology. *Ecology* 85:1771–,1789.

Carbone, C., and J. L. Gittleman. 2002. A common rule for the scaling of carnivore density. *Science* 295:2273–2276.

Damschen, E. I., et al. 2006. Corridors increase plant species richness at large scales. *Science* 313:1284–1286.

Damuth, J. 2007. A macroevolutionary explanation for energy equivalence in the scaling of body size and population density. *American Naturalist* 169:621–631.

Dias, P. C., and J. Blondel. 1996. Local specialization and maladaptation in the Mediterranean blue tit (*Parus caeruleus*). *Oecologia* 107:79–86.

Dunning, J. B., B. J. Danielson, and H. R. Pulliam. 1992. Ecological processes that affect populations in complex landscapes. *Oikos* 65:169–175.

Gaston, K. J. 2003. *The Structure and Dynamics of Geographic Ranges*. Oxford University Press, Oxford.

Gaston, K. J., and T. M. Blackburn. 2000. *Pattern and Process in Macroecology*. Blackwell Science, Oxford.

Guisan, A., et al. 2006. Using niche-based models to improve the sampling of rare species. *Conservation Biology* 20:501–511.

Kirkpatrick, M., and N. H. Barton. 1997. Evolution of a species' range. *American Naturalist* 150:1–23.

Laurance, W. F., et al. 1997. Biomass collapse in Amazonian forest fragments. *Science* 278:1117–1118.

Marquet, P. A., S. A. Naverrete, and J. C. Castilla. 1995. Body size, population density, and the Energetic Equivalency Rule. *Journal of Animal Ecology* 64:325–332.

McGill, B., and C. Collins. 2003. A unified theory for macroecology based on spatial patterns of abundance. *Evolutionary Ecology Research* 5:469–492.

Milinski, M. 1979. An evolutionarily stable feeding strategy in sticklebacks. *Zeitschrift für Tierpsychologie* 51:36–40.

Morse, S. F., and S. K. Robinson. 1999. Nesting success of a Neotropical migrant in a multiple-use, forested landscape. *Conservation Biology* 13:327–337.

Neal, D. 2004. *Introduction to Population Biology*. Cambridge University Press, Cambridge.

Pearson, R. G., and T. P. Dawson. 2003. Predicting the impacts of climate change on the distribution of species: Are bioclimatic envelope models useful? *Global Ecology and Biogeography* 12:361–371.

Peterson, A. T. 2003. Predicting the geography of species' invasions via ecological niche modeling. *Quarterly Review of Biology* 78:419–433.

Peterson, A. T., J. Soberón, and B. Sanchez-Cordero. 1999. Conservatism of ecological niches in evolutionary time. *Science* 285:1265–1267.

Pulliam, H. R. 2000. On the relationship between niche and distribution. *Ecology Letters* 3:349–361.

Seidler, T. G., and J. B. Plotkin. 2006. Seed dispersal and spatial pattern in tropical trees. *PLoS Biology* 4:2132–2137.

Tewksbury, J. J., et al. 2002. Corridors affect plants, animals, and their interactions in fragmented landscapes. *Proceedings of the National Academy of Sciences of the United States of America* 99:12923–12926.

Wiens, J. A., et al. 1993. Ecological mechanisms and landscape ecology. *Oikos* 66:369–380.

Wiens, J. J., and C. H. Graham. 2005. Niche conservatism: Integrating evolution, ecology, and conservation biology. *Annual Review of Ecology, Evolution, and Systematics* 36:519–539.

CHAPTER 11

Population Growth and Regulation

The size of the human population passed the 6 billion mark on October 12, 1999, as far as United Nations demographers could tell. Since then, the population has increased by more than half a billion. Many fewer humans existed when you were born, and there were only one-fifth as many in 1850. The growth of the human population during the last 10,000 years, since the advent of agriculture, has been one of the most significant ecological developments in the history of the earth. It ranks with the massive displacements caused by glaciations during the past million years and the wholesale global extinctions caused by an asteroid impact near Mexico's present-day Yucatán Peninsula 65 million years ago.

One of the most remarkable aspects of human population growth is that its rate has continued to increase even as the population becomes more crowded. Estimates of the size of the human population in ancient times are understandably crude, but it is likely that a million years ago, our ancestors numbered about a million individuals. The population increased slowly until the development of agriculture about 10,000 years ago, by which time the global human population might have reached 3–5 million. As crops became increasingly abundant, food supply no longer limited survival and reproduction, and the rate of human population growth picked up, leading to the development of the first large cities. By the beginning of the eighteenth century, our numbers had swelled a hundredfold since the invention of agriculture (Figure 11.1), even with the occasional setbacks of war, famine, and disease. A hundredfold increase over nearly 10,000 years is equivalent to an average exponential growth rate of about 2% per century. The Industrial Revolution, which began about 1700, provided another impetus to human population growth, particularly with improvements in public health and medicine and increasing material wealth. Over the past 300 years of industrialization, humans have increased in number from perhaps 300 million to 6 billion, roughly a twentyfold increase, or an average exponential growth rate of close to 100% per century (1% per year). The most recent doubling of the

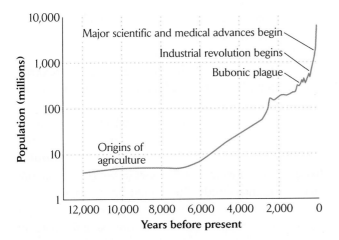

FIGURE 11.1 The human population has increased rapidly with the development of technology. The growth of the human population increased with the development of agriculture, and it accelerated rapidly at the beginning of the Industrial Revolution. Major disease epidemics, such as the bubonic plague, had little lasting effect on human population growth.

human population, from 3 billion to 6 billion, required only 40 years (1.7% per year).

The earth is becoming a very crowded place (Figure 11.2). Many believe that the human population has long since exceeded the ability of the earth to support it and that we are depleting the earth's resources rapidly. How the earth and its inhabitants, including humans, will tolerate these effects in the future is uncertain. What is certain is that continued human population growth will further stress the biosphere and will lead to the continued degradation of many natural environments.

When, and at what level, will human population growth cease? Predicting the future is difficult because there are so many unknown possibilities, including changes in technology, emergence of epidemic diseases of humans or their crops and livestock, and changes in material wealth,

education, and culture. At present, the growth rate of the human population is decreasing, and some estimates indicate that it will reach a plateau at about 9 billion or 10 billion by 2050.

FIGURE 11.2 Human populations have become extremely crowded in both rich and poor countries. © Tom Uhlman/Alamy.

CHAPTER CONCEPTS

- Populations grow by multiplication rather than addition
- Age structure influences population growth rate
- A life table summarizes age-specific schedules of survival and fecundity

- The intrinsic rate of increase can be estimated from the life table
- Population size is regulated by density-dependent factors

Ever since humankind began to understand the rapid increase in its own numbers, human population growth has been cause for concern. This concern has led to the development of mathematical techniques to predict the growth of populations—one aspect of the discipline of **demography,** or the study of populations—and to

intensive study of natural and laboratory populations to learn about mechanisms of population regulation. The origins of modern demography can be traced to a seventeenth-century London cloth merchant, John Graunt, who developed a variety of population statistics, particularly the probability of death at different ages (which provided a foundation for the life insurance industry) and the rate of growth of the human population. Toward the end of the eighteenth century, the British economist Thomas Malthus calculated that the human population would soon outstrip its food supply, an insight that inspired Charles Darwin to develop his theory of evolution by natural selection. Subsequent work to the present day has given us a general understanding of the dynamics of populations. In this chapter, we shall explore the nature of population growth and examine the factors that limit population size, showing how their effects on birth and death rates change with increasing population density in such a way as to bring population growth under control.

Populations grow by multiplication rather than addition

A population increases in proportion to its size, just as a bank account earns interest on its principal. Because population growth depends on the reproduction and deaths of individuals, biologists conveniently describe the rate of growth on a per-individual, or **per capita,** basis. Thus, a population growing at a constant per capita rate expands ever faster as the number of individuals increases. For example, a 10% annual rate of increase adds 10 individuals in 1 year to a population of 100, but the same rate of increase adds 100 individuals to a population of 1,000. Allowed to increase at this rate, the population would rapidly climb toward a very large number. As Charles Darwin wrote in *On the Origin of Species,* "There is no exception to the rule that every organic being naturally increases at so high a rate, that, if not destroyed, the earth would soon be covered by the progeny of a single pair." To make his case as forcefully as possible, Darwin offered a conservative example:

> The elephant is reckoned the slowest breeder of all known animals, and I have taken some pains to estimate its probable minimum rate of natural increase; it will be safest to assume that it begins breeding when thirty years old, and goes on breeding till ninety years old, bringing forth six young in the interval, and surviving till one hundred years old; if this be so, after a period of from 740 to 750 years there would be nearly nineteen million elephants alive, descended from the first pair.

Because baby elephants grow up, mature, and themselves have babies, the elephant population grows by multiplication. Darwin's estimates were a little off because he did

not have the benefit of modern demographic methods, but his logic was sound.

Population growth, to be useful for comparisons, must be expressed as a rate, such as the number of new individuals per unit of time. It would make little sense to say that one population increased by 10% and another by 50% without knowing the intervals over which those changes occurred. Time, of course, flows continuously, and change can occur at every instant. Biologists can use a **continuous-time** approach to model the way in which populations change instantaneously. They often find it more convenient, however, to work with time intervals—days or years, for example—that match the natural cycles of activities in populations and the ways in which ecologists sample populations. This method is referred to as a **discrete-time** approach.

For all practical purposes, the human population grows continuously because babies are born and added to the population throughout the year, and because the population is so large that the interval between one birth and the next, or one death and the next, is very brief. In fact, at present, the global human population is increasing by approximately 2.5 individuals per second. Such continuous increase is unusual in natural populations, in which reproduction is typically restricted to the time of year when resources are most abundant. Accordingly, many populations grow during a reproductive season, then decline between one reproductive season and the next. In the California quail population, for example, the number of individuals doubles or triples each summer as adults produce their broods of chicks, but then dwindles by nearly the same amount during autumn, winter, and spring (Figure 11.3). Within each year, the population growth rate follows these seasonal changes in the balance of births and deaths. If we wished to measure the long-term population growth rate, it would be pointless to compare numbers of quail in August, recently augmented by the chicks born that year, with numbers in May, after winter had taken its toll. One must count individuals at the same time each year, so that all counts are separated by the same cycle of birth and death processes. Such an increase (or decrease) over discrete intervals is referred to as **geometric growth.**

Calculating population growth rates

The rate of geometric growth is most conveniently expressed as the ratio of a population's size in one year to its size in the preceding year (or other time interval). Demographers have assigned the symbol λ (the lowercase Greek letter lambda) to this ratio, which expresses the fac-

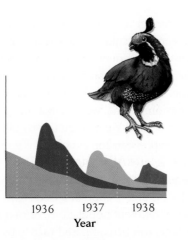

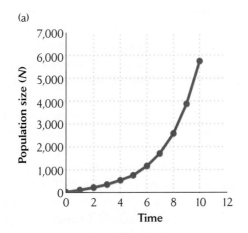

FIGURE 11.3 Populations with discrete reproductive seasons increase by geometric growth. In California quail, the number of individuals increases during the reproductive season, then declines. Each cohort of individuals born or hatched in a particular year is colored differently. After J. T. Emlen, Jr., *J. Wildl. Mgmt.* 4:2–99 (1940).

tor by which a population changes from one time interval to the next. Thus, if $N(t)$ is the size of a population at time t, then its size one time interval later would be

$$N(t + 1) = N(t)\lambda$$

Because there cannot be a negative number of individuals, the value of λ is always positive.

To show how geometric growth proceeds, let's suppose that the present is time $t = 0$, at which point a population has $N(0)$ individuals. After one unit of time, the population will have increased by the factor λ, so $N(1) = N(0)\lambda$. During the next time interval, the population will

grow by the same factor, so $N(2) = N(1)\lambda$. Because $N(1) = N(0)\lambda$, we can say that $N(2) = [N(0)\lambda]\lambda$, or $N(0)\lambda^2$. Similarly, $N(3) = N(0)\lambda^3$, and more generally,

$$N(t) = N(0)\lambda^t.$$

For example, when a population increases by 50% over one time interval—say, a year—the number of individuals increases by a factor of 1.5 ($\lambda = 1.50$). Accordingly, an initial population of 100 individuals would grow to $N(0)\lambda = 100 \times 1.50 = 150$ at the end of 1 year, $N(0)\lambda^2 = 225$ at the end of 2 years, and $N(0)\lambda^{10} = 5,767$ at the end of 10 years.

The same relationship expressed in continuous time is called **exponential growth.** A population growing continuously increases according to the equation

$$N(t) = N(0)e^{rt}$$

The variable r is the exponential growth rate. The constant e is the base of natural logarithms; it has a value of approximately 2.72.

Notice that the equation for exponential growth is identical to the equation for geometric growth, except that e^r takes the place of λ. Thus, geometric and exponential growth are related by

$$\lambda = e^r$$

and

$$\log_e \lambda = r$$

Because of their one-to-one relationship, the exponential and geometric growth equations describe the same data equally well (Figure 11.4). In addition, values of λ and r

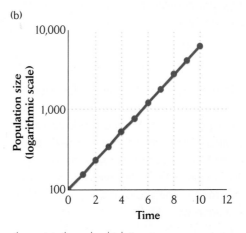

FIGURE 11.4 Geometric and exponential growth curves can be superimposed. The curves in each of these diagrams have equivalent rates of increase (geometric: $\lambda = 1.5$; exponential, $r = 0.47$). $N(0) = 100$ individuals. In each diagram, the exponential growth curve is represented by a continuous line; the points through which it passes represent geometric growth. (a) Change in population size plotted on an arithmetic scale. (b) Change in population size plotted on a logarithmic scale, on which geometric and exponential growth curve both form a straight line.

correspond directly to each other, as summarized in the following table:

	Geometric growth (λ)	Exponential growth (r)
Decreasing population	$0 < \lambda < 1$	$r < 0$
Constant population size	$\lambda = 1$	$r = 0$
Increasing population	$\lambda > 1$	$r > 0$

LIVING GRAPHS   To access interactive tutorials on exponential and geometric growth, go to http://www.whfreeman.com/ricklefs6e.

Calculating changes in population size

When a population undergoes geometric growth at a constant rate, the number of individuals added to or removed from the population varies with the size of the population. The change in population size from one time period to the next is referred to as ΔN, where Δ is the capital Greek letter delta, meaning in this context "a change in." The change over one interval is the difference between the population size at the beginning and at the end of the interval; hence $\Delta N = N(t + 1) - N(t)$, thus $\Delta N = N(t)\lambda - N(t)$, and $\Delta N = (\lambda - 1)N(t)$. As you can see from the last expression, when λ exceeds 1, ΔN is positive, and the population grows; when λ is less than 1, ΔN is negative, and the population declines.

Change in population size is more conveniently expressed for exponential growth. In this case, change is instantaneous, and the rate at which individuals are added to or removed from a population is the derivative of the exponential equation $N(t) = N(0)e^{rt}$, which is

$$\frac{dN}{dt} = rN$$

This equation is the slope of the curve illustrated in Figure 11.4a.

There are two points to make about this equation. First, the exponential growth rate (r) expresses population increase (or decrease) on a per capita basis. Second, the rate of increase of the population as a whole (dN/dt) varies in direct proportion to the size of the population (N). In words, this equation would read

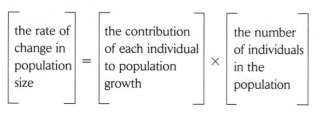

There is another reason to express population growth as an exponential equation. Whereas the change in absolute number of individuals depends on the population size, the change in the logarithm of population size is linear over time. That is, if we plot the natural logarithm (ln) of N as a function of time, we obtain a straight line whose slope is the value of r. We can see this by taking the logarithm of the expression for exponential growth, $N(t) = N(0)e^{rt}$, which is $\ln[N(t)] = \ln[N(0)] + rt$. As you can see, this is the equation for a straight line with an intercept at $t = 0$ of $\ln[N(0)]$ and a slope of r (see Figure 11.4b). The growth rates of different populations, or of a single population over time, can be compared readily by plotting the logarithm of population size over time.

Calculating population growth rates from birth and death rates

The individual contribution to population growth (r in the case of exponential growth) is the difference between birth rate (b) and death rate (d) calculated on a per capita basis; thus, $r = b - d$. In the case of geometric growth, the corresponding per capita growth rate per unit of time (R) is the difference between the per capita rates of birth (B) and death (D) per unit of time; thus, $R = B - D$, and $\lambda = 1 + R$. If we were considering a subpopulation that exchanged individuals with other subpopulations of the same species, as in a metapopulation model, we would also have to include gains through immigration (i) and losses through emigration (e), thus, $r = b - d + i - e$. For the remainder of this chapter, however, we shall take the simpler course of ignoring exchanges of individuals between populations, leaving this complication for later.

Rates of birth and death are averages for a population and do not pertain to individuals except as probabilities. In other words, an individual groundhog dies only once, so it cannot have a personal death rate. Similarly, babies are produced in discrete litters separated by the intervals of time needed for gestation and parental care, not at constant rates such as 0.05 young per day. But when births and deaths are averaged over a population, they take on meaning as rates of demographic events in the population. If 1,000 individuals produced 10,000 progeny in a year, the average per capita birth rate would be $B = 10$ per year, and we could assume that a population of 1 million would produce 10 million progeny—still 10 per capita—under the same conditions. Conversely, if only half the groundhogs alive on their big day in one particular year survived to February 2 of the next year, the population would have a death rate of 50%, or 0.50 per individual per year, although some of the groundhogs would have died "completely," and others would not have died at all. $D = 0.50$ represents the probability that any one groundhog will have died during the 1-year period.

In the case of geometric growth, $\Delta N = B - D$, so $\lambda = 1 + B - D$. In the case of exponential growth, $r = b - d$, and $dN/dt = (b - d)N$. The exponential expression is simpler and has been widely adopted in the development of population theory.

Age structure influences population growth rate

When birth rates and death rates have the same values for all members of a population, we can estimate the future size of the population from the total population size (N) at the present time. But when birth and death rates vary with the ages of individuals, the contributions of younger and older individuals to population growth must be calculated separately. The proportions of individuals in each age class make up the population's **age structure.** If two populations have identical birth and death rates, but different age structures, they will grow at different rates, at least for a while. A population composed wholly of prereproductive adolescents and postreproductive oldsters, for example, cannot increase until the young individuals reach reproductive age. This represents an extreme case, but smaller variations in age distribution can also profoundly influence population growth rates.

A little pencil-and-paper (or spreadsheet) figuring will demonstrate these effects. Let's start with a hypothetical population of 100 individuals having the age-specific survival and fecundity rates and the age distribution shown in Table 11.1. In this population, all the adult individuals reproduce in a single reproductive season each year, and

TABLE 11.1 Life table for a hypothetical population of 100 individuals

Age (x)	Survival (s_x)	Fecundity (b_x)	Number of individuals (n_x)
0	0.5	0	20
1	0.8	1	10
2	0.5	3	40
3	0.0	2	30

some individuals die between that reproductive season and the next. All 3-year-olds die after they reproduce (survival at age 3, $s_3 = 0.0$), so there are no 4-year-olds in the population. Newborn individuals have a fecundity of zero (fecundity at age 0, $b_0 = 0$), as seems biologically reasonable, and a 50% probability of surviving to age 1. We count our population just after the reproductive season, so our count includes newborn individuals (n_0). Alternatively, we could have counted individuals just before the reproductive season; the calculations in that case would be somewhat different, but the results would be the same.

Now, let's use the schedule of births and deaths in Table 11.1 to project our hypothetical population into the future (Table 11.2). The first step is to calculate the number of individuals surviving from one year to the next in each age group. Remember that in moving from one year to the next, each individual becomes a year older. You can see that the 20 newborn individuals at time 0 have a survival rate of 0.5 and become 10 1-year-olds in the following year; the 10 1-year-olds have a survival rate of 0.8 and become 8 2-year-olds in the following year; and so on.

TABLE 11.2 Steps in the projection of a population through one time period of survival and reproduction

Age	(1) Census of population just after reproduction ($t = 0$)	(2) Survival rate of individuals (s_x) (Table 11.1)	(3) Number surviving to next reproductive season (1) × (2)	(4) Number of offspring per adult (b_x) (Table 11.1)	(5) Total number of offspring produced (3) × (4)	(6) Census of population just after reproduction ($t = 1$) (3) and sum of (5)
0	20	0.5		0		74
1	10	0.8	10	1	10	10
2	40	0.5	8	3	24	8
3	30	0.0	20	2	40	20
4	0		0			0
Total	100		38		74	112

TABLE 11.3 Projection of age distribution and total size through time for the hypothetical population in Table 11.1

					Time interval					
	0	1	2	3	4	5	6	7	8	%
n_0	20	74	69	132	175	274	399	599	889	63.4
n_1	10	10	37	34	61	87	137	199	299	21.3
n_2	40	8	8	30	28	53	70	110	160	11.4
n_3	30	20	4	4	15	14	26	35	55	3.9
N	100	112	118	200	279	428	632	943	1,403	100
λ		1.12	1.05	1.69	1.40	1.53	1.48	1.49	1.49	

Note: We designate age by the symbol x and the number of individuals in each age class at time t as $n_x(t)$. The population was projected by multiplying the number of individuals in each age class by the survival to obtain the number in the next older age class in the next time period: $n_x(t) = n_{x-1}(t-1)s_x$. Then the number of individuals in each age class was multiplied by its fecundity to obtain the number of newborns: $n_0(t) = \Sigma n_x(t)b_x$.

With the first step done, we now know the number of breeders of each age that will produce next year's newborns. With this information, we can calculate the total number of newborn individuals as the sum of the number of breeding adults in each age class times their fecundity. In the example in Table 11.2, in year 1 ($t = 1$), the total number of newborn individuals is 10×1 (1-year-olds) + 8×3 (2-year-olds) + 20×2 (3-year-olds) = 74. Now we see that the total population after 1 year is the number of surviving adults (38) plus their offspring (74), or a total of 112 individuals.

The same exercise can be repeated for each year into the future. The results of such a population projection are shown in Table 11.3, along with the total number of individuals in the population after each reproductive season. With this information, we can calculate the population growth rate for each year. Because we are projecting growth over discrete time intervals, we should calculate the geometric rate of population growth, which is the

ratio of the population size after one year to that at the beginning of the year:

$$\lambda(t) = \frac{N(t + 1)}{N(t)}$$

Our hypothetical population at first grows erratically, with λ fluctuating between 1.05 and 1.69. However, provided that the age-specific rates of survival and fecundity remain unchanged, the population eventually assumes a **stable age distribution.** Under such conditions, each age class in a population grows or declines at the same rate, and therefore, so does the total size of the population. As you can see in Table 11.3, the population growth rate λ eventually settles down to a constant value of 1.49. At that point, the population has achieved a stable age distribution, which is to say that the percentages of individuals in each age class also remain constant (Figure 11.5). The stable age distribution and constant

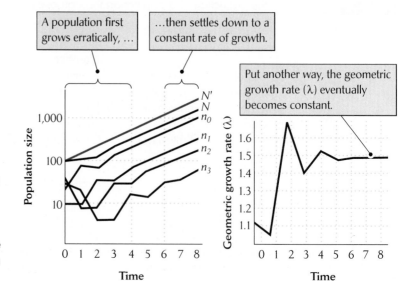

FIGURE 11.5 In a stable age distribution, each age class grows at the same rate. Notice that the growth rate of the population as a whole also stabilizes eventually. The data used to create this graph are taken from Table 11.3. N' represents the growth of a population of 100 individuals having a stable age distribution at time 0.

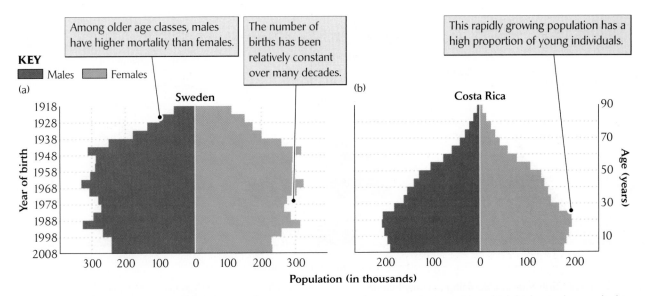

FIGURE 11.6 The age structures of human populations reflect their history of birth and survival rates. (a) Sweden, 2008. Because this population has grown slowly, it is weighted toward older age classes. (b) Costa Rica, 2008. Rapid

population growth, caused by a high birth rate, has resulted in a bottom-heavy age structure, but birth rates have decreased since the 1990s. After U.S. Census Bureau, International Data Base.

growth rate achieved by a particular population depend on the age-specific survival and fecundity rates of its individuals. Any change in those rates alters the stable age distribution and results in a new rate of population growth.

The effect of population growth rate on age structure stands out in comparisons of stable and growing human populations (Figure 11.6). The population of Sweden in 2008 (Figure 11.6a), for example, had been stable for many years, and the age structure of the population primarily reflects the survival of most individuals from infancy through old age. Thus, the tapering of the age distribution at the top of the figure is the result of rapidly decreasing survival after age 50, particularly among males. In contrast, the rapid growth of the population of Costa Rica (Figure 11.6b) has resulted in a bottom-heavy age structure, with large proportions of young individuals. A general rule of thumb is that age structure pyramids with broad bases reflect growing populations and pyramids with narrow bases reflect stable or declining populations.

In many industrialized countries today, the proportion of older individuals is increasing because declining birth rates are reducing the population growth rate. When the age structure of the German population in 1910 is compared with that in 2005 and a projection to 2025 (Figure 11.7), it shows a striking demographic transition to a declining population with a small number of school-age children and a larger proportion of individuals past retirement age.

DATA ANALYSIS MODULE 2 *Birth and Death Rates Influence Population Age Structure and Growth Rate.* Simulate the effects of changes in fecundity, survival,

and the initial age distribution of individuals on a population's growth rate and age structure. You will find this module on page 246.

A life table summarizes age-specific schedules of survival and fecundity

Life tables, of which Table 11.1 is an example, can be used to model the addition and removal of individuals in a population (in the absence of immigration and emigration). Because it is hard to ascertain paternity in many species, life tables are usually based on females. For some populations with highly skewed sex ratios or unusual mating systems, this can pose difficulties, but in most cases a female-based life table provides a workable population model.

As we have seen in Table 11.1, age is designated in a life table by the symbol x, and the subscript x indicates age-specific variables. Thus, n_x refers to the number of individuals of age x in a population. When reproduction occurs during a brief reproductive season each year, each age class is composed of a discrete group of individuals born at approximately the same time. When reproduction is continuous, as it is in the human population, each age class can be designated arbitrarily as comprising individuals between ages $x - \frac{1}{2}$ and $x + \frac{1}{2}$.

The **fecundity** of females is often expressed in terms of female offspring produced per reproductive season or age interval and is designated by b_x (think of b for "births"). Life tables portray mortality in several ways. The fundamental measure of mortality is the probability of **survival**

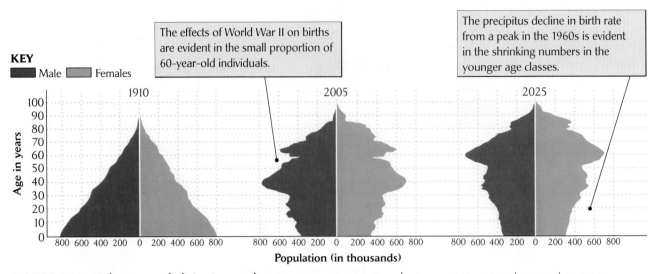

KEY
Male   Females

The effects of World War II on births are evident in the small proportion of 60-year-old individuals.

The precipitus decline in birth rate from a peak in the 1960s is evident in the shrinking numbers in the younger age classes.

FIGURE 11.7 Birth rates are declining in many human populations. The age structure of the human population of Germany in 1910, in 2005, and projected to 2025 shows the effects of population stabilization. Over this period, the proportion of retirement-age men and women has grown dramatically, while the proportion of children has decreased. The population is declining at the present time. After J. W. Vaupel and E. Loichinger, *Science* 312:1911–1913 (2006).

(s_x) between ages x and $x + 1$. Probabilities of survival over many age intervals are summarized by **survivorship** to age x, designated by l_x (think of l for "living"), which is the probability that a newborn individual will be alive at age x.

Ideally, a life table would contain statistics from age 0. However, it is possible to compile a life table from age 0 only for populations in which the fates of all individuals can be followed from birth (age = 0). These include laboratory populations, of course, as well as populations isolated on habitat islands where dispersal is not possible and populations in which young organisms can be tagged and followed wherever they might go. More commonly, organisms disperse widely before they mature, so biologists cannot keep track of young individuals. In these cases,

practical realities dictate that the life table begin at the age of first reproduction, after which movements are limited.

Table 11.4 describes the demography of female individuals of the pied flycatcher (*Ficedula hypoleuca*) at a study location in Sweden. These birds are attracted to nest boxes provided by the investigators, so they can be marked with numbered leg bands and followed individually. Adults are sedentary, nesting in the same area throughout their lives, but the young disperse widely during their first year and cannot be followed. Accordingly, the life table begins at age $x = 1$. The numbers of individuals in the column "Numbers alive" are recorded in the field, but the numbers in all the other columns can be calculated from that single column. The survival rate (s_x), for example, is cal-

TABLE 11.4   Life table of the pied flycatcher (*Ficedula hypoleuca*) in Sweden

Age (years)	Number alive	Survivorship (l_x)	Survival rate (s_x)	Mortality rate (m_x)	Exponential mortality rate (k_x)	Death rate (d_x)	$d_x \times$ age	Expectation of further life (e_x)
1	777	1.000	0.502	0.498	0.689	0.498	0.498	0.977
2	390	0.502	0.523	0.477	0.648	0.239	0.479	0.946
3	204	0.263	0.490	0.510	0.713	0.134	0.402	0.809
4	100	0.129	0.420	0.580	0.868	0.075	0.299	0.650
5	42	0.054	0.452	0.548	0.793	0.030	0.148	0.548
6	19	0.024	0.158	0.842	1.846	0.021	0.124	0.211
7	3	0.004	0.333	0.667	1.099	0.003	0.018	0.333
8	1	0.001	0.000	1.000		0.001	0.010	0.000
9	0	0.000				0.000	0.000	

Source: H. Sternberg, in I. Newton (ed.), *Lifetime Reproduction in Birds,* Academic Press, London (1989), pp. 55–74 (Table 4.1).

TABLE 11.5 Summary of life table variables

Symbol	Description
l_x	Survival of newborn individuals to age x, often known as survivorship; $l_x = n_x$ divided by n_0 or some other starting number of individuals.
d_x	Death rate; proportion of individuals dying during the interval x to $x + 1$; that is, $d_x = l_x - l_{x+1}$.
m_x	Mortality rate; proportion of individuals of age x dying by age $x + 1$; $m_x = d_x/l_x$.
s_x	Survival rate; proportion of individuals of age x surviving to age $x + 1$; $s_x = 1 - m_x$, and $s_x = l_{x+1}/l_x$.
k_x	Exponential mortality rate between age x and $x + 1$, $k_x = -\log_e s_x$.
e_x	Expectation of further life of individuals of age x; that is, the weighted average number of time intervals that individuals alive at age x will survive. It is calculated by the expression

$$e_x = \frac{1}{l_x} \sum_{i-x}^{\infty} (i - x)(l_i - l_{i+1}) = \left[\frac{1}{l_x} \sum_{i=x}^{\infty} i(l_i - l_{i+1}) \right] - x$$

culated as the number of individuals alive at age $x + 1$ divided by the number alive at age x; thus, $s_x = n_{x+1}/n_x$.

Survivorship (l_x) is the probability that a newborn individual will be alive at age x, but we can also use a higher age than 0 as a starting point for the calculation of l_x. For example, in Table 11.4, all individuals being tracked are alive at age 1, so $l_1 = 1$. The proportion of these individuals alive at age 2 is the probability of surviving from age 1 to age 2; hence, $l_2 = l_1 s_1$. Similarly, $l_3 = l_2 s_2$, or $l_3 = l_1 s_1 s_2$, and, by extension, $l_x = l_1 s_1 s_2 s_3 \ldots s_{x-1}$. Because $l_1 = 1$, it can be deleted from the expression for l_x. For a life table that begins with newborn individuals at age 0, $l_0 = 1$. Subsequently, proportion $l_1 = s_0$ of newborns survive from age 0 to age 1, proportion $l_2 = s_0 s_1$ survive to age 2, and, by extension, $l_x = s_0 s_1 s_2 \ldots s_{x+1}$.

Survivorship is the most important variable in the life table for calculating the growth rate of a population with a stable age structure. However, demographers use several other variables, all of which can be derived from one another, to describe life and death in a population in different ways (Table 11.5). For example, the death rate (d_x) describes the proportion of the population that perishes in each time interval, and hence the distribution of ages at death. The death rates for each age class must sum to 1 because death is the fate of all individuals. The mortality rate (m_x) describes the probability of death for individuals that are alive at each age, which can range from 0 to 1. The exponential mortality rate (k_x) expresses the same probability as an instantaneous rate, in which case its value can exceed 1.

For an individual of age x, the number of additional years that it can expect to live is the expectation of further life (e_x), which depends on the probability of survival through each subsequent age interval. The expectation of further life measures the probability that a particular individual will be a member of a population in the future. If

the survival rate (s_x) were the same for all adults regardless of age, then each individual would have the same expectation of *further* life, even though the number of individuals in the population would dwindle continuously with age; for example, individuals 2 and 3 years old would have the same life expectancy of 3 years. In most populations, however, the survival rate decreases with age due to senescence, so the expectation of further life becomes shorter in older individuals; in other words, the probability of dying increases. In the life table of the pied flycatcher, the survival rate of young adults is about 50% per year, but that rate decreases among older age classes. One can also see that the expectation of further life is about 1 year for young adults, but then decreases, on average, to 0.65 years at age 4 and 0.21 years at age 6. In other populations, expectation of further life can be low when survival rates of young are low, but then increase as individuals grow and mature.

Two types of life tables

Life tables can be constructed in two ways. The first is to construct a **cohort** (or dynamic) **life table,** which follows the fate of a group of individuals born at the same time from birth to the death of the last individual. This method is readily applied to populations of plants and sessile animals, in which marked individuals can be continually tracked over the course of their life spans. It works fine, as long as individuals do not live too long or are not too mobile to track easily. Another disadvantage of the cohort method is that time and age are confounded, so it is difficult to disentangle the effects of age from the effects of conditions in particular years. Suppose, for example, you begin a cohort life table study during a wet year, which is followed by a period of declining precipitation and finally drought. You might find that survival and fecundity decrease over time, but does their decline reflect age or the particular environmental conditions during the study?

A **static** (or time-specific) **life table** sidesteps this problem by considering the survival and fecundity of individuals of known age during a single time interval, and thus under the same conditions. Of course, to apply this technique, it is necessary to know the ages of individuals (which may be estimated by growth rings, tooth wear, or some other reliable index). However, predictions made from a static life table are accurate only under the conditions experienced by a population at a particular time. In a variable environment, a single static life table could mislead us about the long-term growth of a population. Thus, the best approach whenever possible is to construct static life tables for several time periods to sort out the contributions of age and environmental conditions to survival and reproductive rates. The following two examples illustrate some of the difficulties of constructing life tables for natural populations.

ECOLOGISTS IN THE FIELD   **Building life tables for natural populations.** Dedicated field ecologists have followed marked individuals of many species in their natural environments to understand the dynamics of their populations. One of the finest examples of such work is that of Peter and Rosemary Grant, of Princeton University, who have studied several species of ground finches of the genus *Geospiza* on the island of Daphne Major in the Galápagos archipelago. This small (40-hectare), uninhabited island lies right on the equator, about 1,000 km to the west of the coast of Ecuador. The Grants were able to capture all the birds on the island and mark them with

uniquely colored plastic leg bands. In addition, because Daphne is so isolated, few birds left the island or arrived from elsewhere.

The Grants followed finch populations on Daphne for over 15 years and were able to construct cohort life tables for many of the birds born early in the study. The fates of 210 cactus finch (*Geospiza scandens*) chicks fledged in 1978 are shown in Figure 11.8. Notice how much longer these finches live than pied flycatchers, undoubtedly due to the absence of predators and winter weather. As it is in many species, mortality was high (57%) during the first year of life. Mortality decreased in the second year, and was quite variable through 10 years of age, before increasing again among the oldest birds. Variation in survival from year to year reflects swings in precipitation on the island related to El Niño and La Niña climatic patterns (see Chapter 6). El Niño years are wet, and vegetation grows luxuriantly, producing abundant food for the finches and resulting in high survival. La Niña years are periods of drought and food scarcity.

The cactus finch data emphasize a disadvantage of cohort life tables: variation in survival with age may be obscured by variation in the environment. In Figure 11.8, for example, does the rise in mortality late in life reflect senescence, or could it have resulted from a prolonged drought? Static life tables can avoid this problem because survival of all age classes is considered under the same conditions.

In another classic demographic study, the mammalogist Olaus Murie used the distribution of ages at death to construct a static life table for Dall mountain sheep (*Ovis dalli*) in Denali (then called Mount McKinley) National Park, Alaska, during the 1930s. The size of the horns, which grow continuously during the lifetime of an

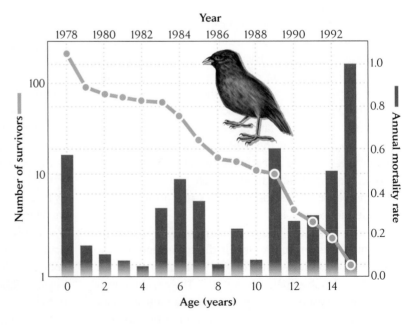

FIGURE 11.8 Survival rates may vary from year to year. The cohort of cactus finches fledged on Daphne Major in 1978 shows high variation in survival rates. Mortality of immature birds during the first year is high. Adult mortality rates generally increase with age, but also vary with climatic patterns. Data from P. R. Grant and B. R. Grant, *Ecology* 73:766–784 (1992).

FIGURE 11.9 Static life tables require a means of estimating the ages of organisms. The size of the horns of the Dall mountain sheep, which grow continuously during the animal's lifetime, provide a good estimate of an individual's age. Steven Kaslowski/Nature Picture Library.

individual sheep (Figure 11.9), provided an estimate of age at death. Of 608 skeletal remains that Murie recovered, he judged 121 to have been less than 1 year old at death, 7 between 1 and 2 years old, 8 between 2 and 3 years old, and so on, as shown in Table 11.6.

Murie constructed the life table by using the following reasoning: All 608 dead sheep must have been alive at birth; all but the 121 that died during the first year must have been alive at the age of 1 year (608 − 121 = 487), all but 128 (the 121 dying during the first year and the 7 dying during the second) must have been alive at the end of the second year (608 − 128 = 480), and so on, until the two oldest sheep died during their fourteenth year. Survivorship (l_x) was calculated by converting the number of sheep alive at the beginning of each age interval to a decimal fraction of those alive at birth. Thus, for example, the 390 sheep alive at the beginning of the seventh year is decimal fraction 0.640 (or 64.0%) of 608. By proceeding in this way, Murie was able to construct a life table for a population of long-lived organisms during a relatively short time interval. Such static life tables allow investigators to determine how changes in the environment affect population processes. ▌

| TABLE 11.6 | Life table for Dall mountain sheep constructed from the ages at death of 608 sheep in Denali National Park |

Age interval	Number dying during age interval	Number surviving at beginning of age interval	Number surviving as a fraction of newborns (l_x)	Survival rate (s_x)	Expectation of further life (e_x)
0–1	121	608	1.000	0.801	6.57
1–2	7	487	0.801	0.986	7.20
2–3	8	480	0.789	0.983	6.31
3–4	7	472	0.776	0.985	5.42
4–5	18	465	0.765	0.961	4.50
5–6	28	447	0.735	0.937	3.68
6–7	29	419	0.689	0.931	2.92
7–8	42	390	0.641	0.892	2.14
8–9	80	348	0.572	0.770	1.40
9–10	114	268	0.441	0.575	0.82
10–11	95	154	0.253	0.383	0.42
11–12	55	59	0.097	0.068	0.10
12–13	2	4	0.007	0.500	0.50
13–14	2	2	0.003	0.000	0.00
14–15	0	0	0.000		
Total	608				

Source: Based on data of O. Murie, *The Wolves of Mt. McKinley*, U.S. Department of the Interior, National Park Service, Fauna Series No. 5, Washington, D.C. (1944); quoted by E. S. Deevey, Jr., *Quarterly Review of Biology* 22:283–314 (1947).

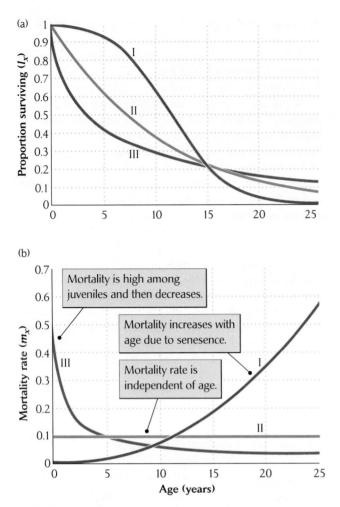

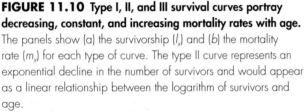

FIGURE 11.10 Type I, II, and III survival curves portray decreasing, constant, and increasing mortality rates with age. The panels show (a) the survivorship (l_x) and (b) the mortality rate (m_x) for each type of curve. The type II curve represents an exponential decline in the number of survivors and would appear as a linear relationship between the logarithm of survivors and age.

The shapes of survivorship curves

Demographers recognize that when survivorship data are plotted on a graph, the resulting curves assume three basic shapes (Figure 11.10):

- A type II survivorship curve is the result of a constant survival rate, or mortality rate (m), with age. Thus, survivorship declines exponentially with age according to $l_x = e^{-mx}$. Plotted on a logarithmic scale, $\ln(l_x) = -mx$ and therefore decreases linearly with age, with slope m.

- A type I survivorship curve is the result of a high initial survival rate, which then falls off abruptly with age as the age-specific mortality rate increases. Type I curves characterize populations, such as the human population,

in which mortality is low early in life and then increases rapidly later in life.

- A type III survivorship curve results from high mortality early in life. A type III curve typifies species, such as small invertebrates, plants, and many fishes and amphibians, in which young individuals are extremely vulnerable to predation and other risk factors, which they escape as they grow larger and mature.

In reality, most populations have survivorship curves that combine features of type I and type III curves, with both early vulnerability and later senescence, and with the survival rate reaching its highest point in early adulthood. You can see these characteristics in the Dall mountain sheep life table, in which mortality during the first year of life is very high (20%) in spite of intense parental care, drops to 1%–2% between ages 1 and 4, and then gradually rises to over 50% by age 10.

The intrinsic rate of increase can be estimated from the life table

The **intrinsic rate of increase** of a population, indicated by λ_m or r_m (often called the Malthusian parameter, after Thomas Malthus), is the geometric or exponential growth rate (λ or r) assumed by a population with a stable age distribution. In practice, populations rarely achieve stable age distributions and therefore rarely grow at their intrinsic rates of increase. Instead, the age structure of a population continuously readjusts to changes in environmental conditions that alter the schedule of birth and death rates in the life table. Thus, the actual growth rate of a population depends as much on past conditions, which determine its age structure, as on current life table values. Therefore, the intrinsic rate of increase shows how a population would grow if environmental conditions remained constant. In a varying world, λ_m or r_m cannot project long-term population growth accurately.

Each life table compiled under a particular set of environmental conditions has a single intrinsic rate of increase. Calculating that intrinsic rate of increase requires data for both survival and fecundity on each age class, including newborn individuals, which will contribute to future populations. To find the exact value of λ_m or r_m, one must solve a complicated equation. However, λ_m and r_m can be approximated by a simple formula based on life table values, which yields the values λ_a and r_a (the letter a indicates an approximation). Before we can calculate either of these values, we must add a few columns to our hypothetical

TABLE 11.7	Expansion of life table values used to calculate the exponential rate of increase for the hypothetical population in Table 11.1				
x	s_x	l_x	b_x	$l_x b_x$	$x l_x b_x$
0	0.5	1.0	0	0.0	0.0
1	0.8	0.5	1	0.5	0.5
2	0.5	0.4	3	1.2	2.4
3	0.0	0.2	2	0.4	1.2
Net reproductive rate (R_0)				2.1	
Expected number of births weighted by age					4.1

Note: The sums of the $l_x b_x$ column (net reproductive rate) and the $x l_x b_x$ column are used to estimate r_a as described in the text.

life table in Table 11.1 (Table 11.7). One of these is a column with the product of l_x and b_x, which represents the expected number of offspring that a newborn individual will produce at age x. Another is the product of x, l_x, and b_x, the expected number of births weighted by age, which we will use shortly to estimate the average age at which a female gives birth to her offspring. We then add the values in the $l_x b_x$ and $x l_x b_x$ columns to get the totals for each column.

 To access interactive tutorials on life table analysis, go to http://www.whfreeman.com/ricklefs6e.

The sum of the $l_x b_x$ terms is the **net reproductive rate** (R_0) of individuals in the population. One may think of R_0 as the expected total number of female offspring produced by an average female over the course of her life span. In Table 11.7, $R_0 = 2.1$. Because this rate exceeds the replacement rate of 1 offspring per individual, the population should grow rapidly as long as the life table values do not change.

The average age at which an individual gives birth to its offspring (T) is now calculated as

$$T = \frac{\Sigma x l_x b_x}{\Sigma l_x b_x}$$

T is sometimes referred to as the **generation time,** the average period between the birth of an individual and the birth of its offspring. In our hypothetical population, T is 4.1/2.1, or approximately 1.95 years.

Now we are prepared to estimate the intrinsic rate of increase. Remember that $N(t) = N(0)\lambda^t$, and therefore $N(t)/N(0) = \lambda^t$. If t represents the generation time T, then the ratio $N(T)/N(0)$ is the net reproductive rate; hence, $R_0 = \lambda^T$, and $\lambda = R_0^{1/T}$. For the hypothetical population

in Table 11.7, $\lambda_a = 2.1^{1/1.95} = 1.46$, which is close to the observed value of about 1.49 after the population achieves a stable age distribution (see Table 11.3). For exponential growth, $r_a = \log_e R_0 / T$, in which case $r_a = \log_e(2.1)/1.95$, or 0.38. Because $\lambda = e^r$, this value is equivalent to $\lambda_a = e^{0.38} = 1.46$.

You can see that a population's intrinsic rate of increase depends on both the net reproductive rate (R_0) and the generation time (estimated by T). A population grows ($\lambda > 1$, $r > 0$) when R_0 exceeds 1, which is the replacement level of reproduction for a population, and declines when $R_0 < 1$. The rate at which a population grows ($r > 0$) or shrinks ($r < 0$) increases as young are born to their mothers at younger ages; that is, as the generation time (T) decreases. The shorter the generation time, the faster population size can change.

The growth potential of populations

We can appreciate the capacity of a population for growth by observing the rapid increase of organisms introduced into a new region with a suitable environment. In 1937, 2 male and 6 female ring-necked pheasants were released on Protection Island, Washington. They increased to 1,325 adults within 5 years. This 166-fold increase represents a 178% annual rate of increase ($r = 1.02$, $\lambda = 2.78$). In other words, the population almost tripled, on average, each year. Even such an unlikely creature as the northern elephant seal (*Mirounga angustirostris*), whose population along the western coast of North America was all but obliterated by hunting during the nineteenth century, increased from about 100 individuals in 1900 to 150,000 in 2000, a 1,500-fold increase ($r = 0.073$, $\lambda = 1.076$). If you are unimpressed, consider that another century of unrestrained growth would find 225 million elephant seals crowding surfers and sunbathers off California beaches

FIGURE 11.11 Populations have the potential to increase rapidly. The northern elephant seal (*Mirounga angustirostris*), which was nearly extirpated in the nineteenth century, has rebounded from near extinction. These females and young males will spend several weeks on this beach while they are molting their fur. Photo by François Gohier/Photo Researchers.

(Figure 11.11). Before the end of the following century, the shorelines of the Western Hemisphere would give lodging to more than 30 trillion of the beasts.

Elephant seal populations do not hold any records for growth potential—quite the contrary. Life tables of populations maintained under optimal conditions in laboratories have exhibited potential annual growth rates (λ) as great as 24 for the field vole (a small mouselike mammal), 10 billion (10^{10}) for flour beetles, and 10^{30} for water fleas (*Daphnia*).

Another way of expressing the growth rate of a population is its **doubling time** (t_2), which produces figures a little more familiar to us. Doubling time can be calculated as

$$t_2 = \frac{\log_e 2}{\log_e \lambda}$$

or

$$t_2 = \frac{\log_e 2}{r}$$

The value of $\log_e 2$ is 0.69. Hence, for the field vole ($\lambda = 24$), $t_2 = 0.69/\log_e 24$, which is 0.22 years, or 79 days. The same calculation gives doubling times of 246 days for the ring-necked pheasant, 11 days for the flour beetle, and only 3.6 days for the water flea. The potential growth rates of populations of bacteria and viruses under ideal conditions are almost unimaginable.

Limits on population growth

It is not surprising that experimental laboratory populations kept under controlled conditions and supplied with abundant food can increase at such rapid rates. Even the most slowly reproducing species would cover the earth in a short time if its population growth were unrestrained. Yet most populations we observe in nature remain within reasonably confined limits.

Nearly two centuries ago, Thomas Malthus understood that this fact "implies a strong and constantly operating check on population from the difficulty of subsistence." In *An Essay on the Principle of Population* (1798), he wrote,

> Through the animal and vegetable kingdoms, nature has scattered the seeds of life abroad with the most profuse and liberal hand. She has been comparatively sparing in the room and the nourishment necessary to rear them. The germs of existence contained in this spot of earth, with ample food, and ample room to expand in, would fill millions of worlds in the course of a few thousand years. Necessity, that imperious all pervading law of nature, restrains them within the prescribed bounds. The race of plants, and the race of animals shrink under this great restrictive law.

Darwin echoed this view in *On the Origin of Species:*

> As more individuals are produced than can possibly survive, there must in every case be a struggle of existence, either one individual with another of the same species, or with the individuals of distinct species, or with the physical conditions of life. It is the doctrine of Malthus applied with manifold force to the whole animal and vegetable kingdoms; for in this case there can be no artificial increase of food, and no prudential

restraint from marriage. Although some species may be now increasing, more or less rapidly, in numbers, all cannot do so, for the world would not hold them.

This essentially modern view of the regulation of populations grew out of an awareness of the immense capacity of populations to increase. A population's growth potential and the relative constancy of its numbers cannot be logically reconciled otherwise. It makes sense that when a population increases, the resources available to each individual diminish, and either the birth rate declines or the death rate increases. Fewer resources mean that fewer offspring can be nourished, and those offspring survive less well. Crowded populations also aggravate social strife, promote the spread of disease, and attract the attention of predators, as we shall see in a later chapter. Many such factors may act together to slow, and finally halt, population growth. An important step in understanding how populations are regulated was the development of mathematical descriptions of population growth processes early in the twentieth century.

The logistic equation

In 1920, Raymond Pearl and L. J. Reed, at the Institute for Biological Research of Johns Hopkins University, published a paper in the *Proceedings of the National Academy of Sciences* titled "On the Rate of Growth of the Population of the United States since 1790 and Its Mathematical Representation." Thorough and accurate population data had been gathered even in colonial times. Indeed, the phenomenal population growth of the American colonies had impressed upon Malthus how rapidly humans could multiply; this was not as evident in the more crowded European countries of his time, even though John Gaunt had carefully noted the population increase in London in the 1600s.

Pearl and Reed wished to project the future growth of the U.S. population, which they supposed must eventually reach a limit. Data to 1910, the latest census then available, had revealed a decline in the exponential rate of growth (Figure 11.12). Pearl and Reed reasoned that if this decline followed a regular pattern that could be described mathematically, it would be possible to predict the future course of population growth, as long as the decline in the exponential growth rate continued. They also reasoned that the changes in the exponential growth rate must be related to the size of a population rather than to time, because time scales are arbitrary. And so, in place of a constant value of r in the differential equation for unrestrained population growth ($dN/dt = rN$), Pearl and

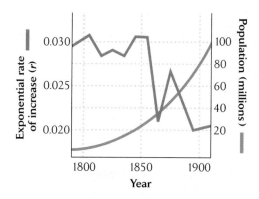

FIGURE 11.12 Population growth rates decrease as a population grows larger. The rate of growth slowed as the size of the human population of the United States increased between 1790 and 1910. When the exponential rate of increase (r) was calculated for each 10-year period, it showed a downward trend, indicating a decreasing rate of population growth. Data from R. Pearl and L. J. Reed, *Proc. Natl. Acad. Sci.* 6:275–288 (1920).

Reed suggested that r decreases as N increases, according to the relation

$$r = r_0 \left(1 - \frac{N}{K} \right)$$

In this expression, r_0 represents the intrinsic exponential growth rate of a population when its size is very small (that is, close to 0), and K—the **carrying capacity** of the environment—represents the number of individuals that the environment can support. The differential equation describing restricted population growth now became

$$\frac{dN}{dt} = r_0 N \left(1 - \frac{N}{K} \right)$$

In words, this equation may be expressed as

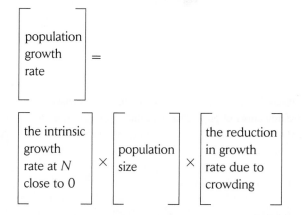

According to this equation, which is called the **logistic equation,** the exponential rate of increase decreases as a linear function of the size of a population. Such a decrease

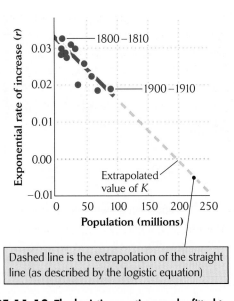

To access an interactive tutorial on the logistic equation, go to http://www.whfreeman.com/ricklefs6e.

As long as population size does not exceed the carrying capacity—that is, as long as N/K is less than 1—a population continues to increase, albeit at a slowing rate. When N exceeds K, however, the ratio N/K exceeds 1, the term in parentheses $(1 - N/K)$ becomes negative, and the population decreases. Because populations below K increase and those above K decrease, K is the eventual steady-state, or equilibrium, size of a population that is growing according to the logistic equation.

The influence of population size (N) on the per capita rate of increase (r) and the overall rate of increase (dN/dt) is shown in Figure 11.14. The overall growth rate is small when the population is small because there are so few individuals producing babies. The growth rate falls off as the population approaches the carrying capacity because declining resources limit reproduction. Thus, the curve for dN/dt has its maximum at an intermediate population size—specifically, at $N = K/2$, or half the carrying capacity. This **inflection point** (which occurs at time i) separates the early accelerating phase of population growth from the later decelerating phase.

The differential form of the logistic equation can be integrated to describe the population size at time t ($N(t)$) as a function of the initial population size at time 0 ($N(0)$) and the intrinsic rate of population growth (r_0):

$$N(t) = \frac{K}{1 + e^{-r_0(t-i)}}$$

This equation describes a sigmoid, or S-shaped, curve (Figure 11.15). In other words, the population grows slowly at first, then more rapidly as the number of individuals increases, and finally more slowly again as it approaches the carrying capacity, K.

FIGURE 11.13 The logistic equation can be fitted to patterns of human population growth. The exponential rate of increase in the United States during each decade between 1790 and 1910 (from Figure 11.12) is plotted here as a function of the population size (the geometric mean of the beginning and ending numbers) during that decade. The best fit to the data is a straight line, which can be extrapolated into the future. This extrapolation suggested that the U.S. population would level off ($r = 0$) at just under 200 million individuals.

reasonably approximated the data for the population of the United States through 1910 (Figure 11.13).

One of the curiosities of population biology is that the logistic equation had been derived almost a century earlier, in 1838, by the Belgian mathematician Pierre François Verhulst (1804–1849). Verhulst had read Thomas Malthus's essay and sought to formulate a natural law governing the growth of populations. He called his expression the *équation logistique,* but Pearl and Read did not discover Verhulst's work until after they had derived their own function.

FIGURE 11.14 The logistic equation incorporates the influences of population size and per capita growth rate. The overall rate of population growth (dN/dt) is the product of the per capita exponential rate of increase ($r = 1/N\, dN/dt$) and population size (N). The value of r declines as a linear function of population size (N), from r_0 at $N = 0$ to 0 when $N = K$. The overall growth rate of a population reaches a maximum at the inflection point, at which the population size is one-half the carrying capacity ($K/2$).

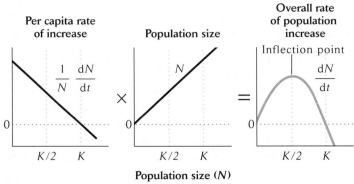

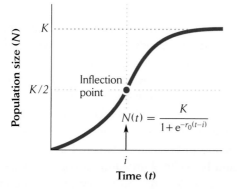

FIGURE 11.15 Logistic growth follows an S-shaped curve. The curve is symmetrical about the inflection point (K/2); that is, accelerating and decelerating phases of population growth have the same shape.

When Pearl and Reed applied the logistic equation to the growth of the U.S. population from 1790 to 1910 (Figure 11.16), they obtained the best fit with $K = 197,273,000$ individuals and $r_0 = 0.03134$. Although the population in 1910 was only 91,972,000, Pearl and Reed could extrapolate its future growth to twice the 1910 level based on its earlier growth. Such projections often prove incorrect, however, when circumstances change. The U.S. population reached 197 million between 1960

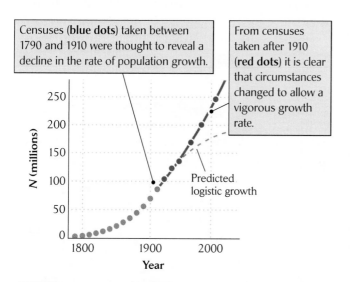

FIGURE 11.16 Population projections may prove incorrect if life table values change. When the logistic equation is fitted to population data for the United States between 1790 and 1910, it predicts a leveling off at about 197 million people. Subsequent censuses, however, have shown numbers well above the projected population curve. The current population of the United States exceeds 300 million individuals and continues to grow slowly.

and 1970, when it was still growing vigorously. Improved public health and medical treatment raised survival rates substantially, particularly for infants and children, between the 1920s and the 1970s. Moreover, the logistic equation did not incorporate the millions of immigrants to the United States. The present population of the United States is about 300 million and still growing slowly.

Population size is regulated by density-dependent factors

The logistic equation describes reasonably well the growth of many populations observed in the laboratory and in natural habitats. The fact that it does so suggests that extrinsic factors limiting growth exert stronger effects on mortality and fecundity as a population grows (see Figure 11.14). That is why the intrinsic growth rate of a population decreases as population size increases. But what are those factors, and how do they operate?

Many things influence rates of population growth, but only **density-dependent** factors, whose effect increases with crowding, can bring the size of a population under control. Of prime importance among these factors are food supplies and places to live, which are relatively fixed in amount and number. Additionally, the effects of predators, parasites, and diseases are felt more strongly in crowded populations than in sparse ones. Other factors, such as temperature, precipitation, and catastrophic events, alter birth and death rates largely without regard to the numbers of individuals in a population. Such **density-independent** factors can influence the growth rate of a population, often causing dramatic fluctuations in population size, but they do not regulate population size.

Density dependence in animals

Numerous experimental studies have revealed a variety of mechanisms of density dependence. For example, when a single pair of fruit flies is confined to a bottle with a fixed supply of food, its descendants increase in number rapidly at first, but soon reach a limit. When different numbers of pairs of flies were introduced into otherwise identical culture bottles, the number of progeny raised per pair varied inversely with the density of flies in the bottle (Figure 11.17). The competition among the larvae for food caused high mortality in dense cultures. Adult life span also declined, but only at densities well above the levels at which survival of larvae dropped off. Juvenile stages often suffer the adverse effects of density-dependent factors more than adults do.

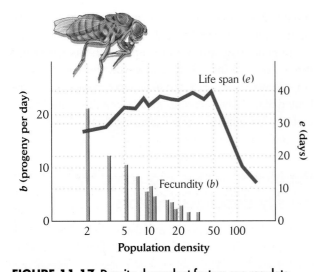

FIGURE 11.17 **Density-dependent factors can regulate population growth.** Fecundity and life span decrease as population density increases in laboratory populations of the fruit fly *Drosophila melanogaster*. After R. Pearl, *Q. Rev. Biol.* 2:532–548 (1927).

MORE ON THE WEB *Density Dependence in Laboratory Cultures of Water Fleas.* How does density influence life table values to determine population growth rate?

Most studies of density dependence have focused on laboratory populations, in which the factors that affect population growth can be controlled. The simplicity of such systems leaves some doubt about the relevance of laboratory findings to populations in more complex natural surroundings, where physical conditions change continually and the experimenter does not control factors such as food supply or predation. Nonetheless, many long-term studies of natural populations, particularly of those recovering from population declines brought on by hunting or other human interference, provide evidence of density-dependent regulation. A long-term study by Ian Nisbet described changes in the populations of common terns (*Sterna hirundo*) that nest on small islands in Buzzards Bay, Massachusetts (Figure 11.18). The factor limiting population growth on a particular island appears to be the availability of suitable nesting sites. The tern population began to expand into this area in the mid-1970s on Bird Island, where tern numbers increased from about 200 to 1,800 individuals by 1990. After that year, most of the suitable breeding sites on Bird Island were occupied, and numbers leveled off. Soon afterward, birds began to colonize Ram Island, where the population increased to just over 2,000 birds before leveling off by the year 2000. By this time, the overflow was beginning to nest on Penikese

Island. Clearly, the intrinsic growth rate of the tern population in Buzzards Bay is very high, but the bay provides a limited number of islands offering suitable nesting sites, which ultimately limits the population size.

A severe winter can sharply reduce the population of song sparrows (*Melospiza melodia*) on Mandarte Island, a 6-hectare speck of land off the coast of British Columbia. Variation in the severity of winter weather has caused the population to fluctuate between 4 and 72 breeding females and between 9 and 100 breeding males. These fluctuations provide a natural experiment on the effects of population density on reproductive success during the summer breeding season. During years with large summer populations, a higher percentage of males cannot obtain territories and so cannot breed. Due to competition for food, breeding females rear fewer offspring, on average, during those years, and a lower percentage of juveniles survive into autumn (Figure 11.19). These density-dependent responses undoubtedly contribute to the regulation of the song sparrow population on Mandarte Island

Although natural variation in population size provides a way of visualizing density dependence, ideally we would like to conduct the same experiment in the field as in the laboratory—that is, to alter the density of individuals in the population while keeping everything else constant. Sometimes such experiments happen serendipitously on a large scale—for example, when managers increase

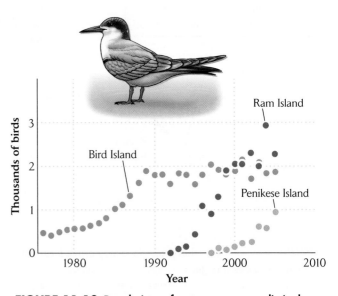

FIGURE 11.18 **Populations of common terns are limited by nesting space.** Populations of common terns (*Sterna hirundo*) on several islands in Buzzards Bay, Massachusetts, have grown rapidly, then leveled off as suitable nesting sites were occupied. Data courtesy of Ian C. T. Nisbet.

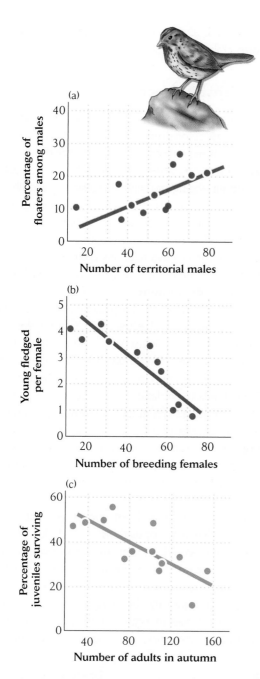

or decrease populations of game animals such as deer. Deer browse leaves, and they require large quantities of new growth, which has a high nutritional content, to grow rapidly and reproduce successfully. When hunting is restricted by regulation or by the inaccessibility of an area, and natural predators, such as wolves and mountain lions, are absent, deer can become so abundant that they can seriously deplete their food supply.

A survey of harvested white-tailed deer (*Odocoileus virginianus*) in five regions of New York State in the 1940s showed that the proportion of females that were pregnant and the average number of embryos per pregnant female were directly related to range quality (Figure 11.20). In addition, deer living on low-quality range showed a greater difference between the number of embryos and

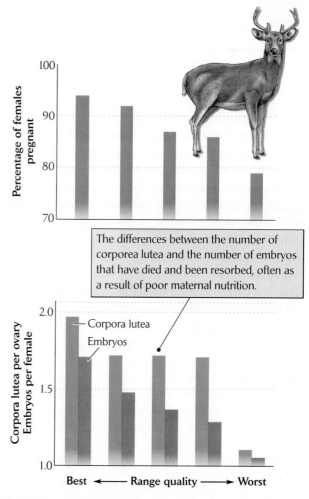

FIGURE 11.20 Game management practices may provide natural experiments in population dynamics. Reproductive parameters of white-tailed deer (*Odocoileus virginianus*) harvested in 1939–1949 reflected range quality in five regions of New York State. After E. L. Chaetum and C. W. Severinghaus, *Trans. N. Am. Wildl. Conf.* 15:170–189 (1950).

FIGURE 11.19 Density-dependent factors can control the size of natural populations. The responses of the song sparrow population on Mandarte Island to population fluctuations caused by environmental variation show that density-dependent factors are at work even in these circumstances. (a) The proportion of males prevented from acquiring territories ("floaters") increases with increased crowding on the small island. (b, c) The number of fledglings produced per female (b) and the survival of those offspring through autumn and winter (c) decrease with crowding. After P. Arcese and J. N. M. Smith, *J. Anim. Ecol. 57*:119–136 (1988); J. N. M. Smith, P. Arcese, and W. M. Hochachka, in C. M. Perrins, J.-D. Lebreton, and G. J. M. Hirons (eds.), *Bird Population Studies: Relevance to Conservation and Management*, Oxford University Press, Oxford (1991), pp. 148–167.

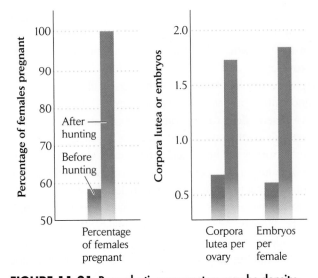

FIGURE 11.21 Reproductive parameters may be density-dependent. Measures of fecundity in the population of white-tailed deer in the DeBar Mountain area of the Adirondack Mountains of New York State increased after hunting had reduced population density. After E. L. Chaetum and C. W. Severinghaus, *Trans. N. Am. Wildl. Conf.* 15:170–189 (1950).

the number of corpora lutea than deer on high-quality range. The number of corpora lutea in each ovary indicates the number of eggs ovulated. When the number of corpora lutea exceeds the number of embryos, some of the embryos have died and been resorbed, often as a result of the pregnant female receiving poor nutrition from range of low quality. In the central Adirondack area, where habitat for deer was very poor, even ovulation was greatly reduced.

Selective hunting to thin dense deer populations can often reverse range deterioration caused by overgrazing. When DeBar Mountain, an area of very poor range, was opened to hunting, the deer population decreased, range

quality recovered, and reproduction improved dramatically (Figure 11.21).

Positive density dependence

Increasing population density usually depresses survival and birth rates; therefore, this type of population response is referred to as **negative density dependence.** In some cases, however, population growth rates actually increase with increasing population density, especially at low population densities. That is, as the population grows, its intrinsic rate of increase (r) goes up, instead of decreasing as it would if it followed the logistic growth equation. Because of the positive relationship between growth rate and density, this pattern is referred to as **positive** or **inverse density dependence.**

Several processes might cause positive density dependence. One is the so-called Allee effect (named for the pioneering population biologist W. C. Allee of the University of Chicago), in which individuals are better able to find mates as population density increases. Individuals might also suffer less predation in a larger population in which more individuals can detect the presence of predators and give warning signals, as we saw in Chapter 9. Larger populations also harbor greater genetic diversity, and harmful mutations are less frequently expressed because inbreeding is less common, as we shall see in a later chapter.

Fisheries managers have been concerned that overfishing could drive fish populations to such low levels that lack of mates would prevent fish stocks from recovering. A direct measure of population growth potential is **recruitment:** the number of new offspring per individual breeder that join the breeding population in the future. As you can see in Figure 11.22, one population of herring (the Downs stock, which spawns in the North

FIGURE 11.22 Some fish stocks exhibit positive density dependence at low population sizes. (a) The ratio of recruits (new offspring that survive to breeding age) to spawners declines with decreasing population size in the spring-breeding Iceland herring stock, showing a classic Allee effect. (b) The Downs herring stock exhibits negative density dependence, which is the more common pattern. To make variations at low population sizes more apparent, the population size is plotted on a square root scale. From R. A. Myers et al., *Science* 269:1106–1108 (1995).

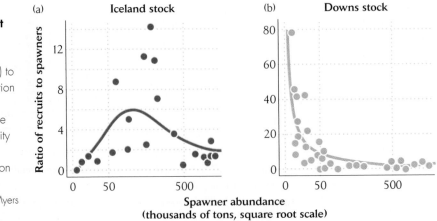

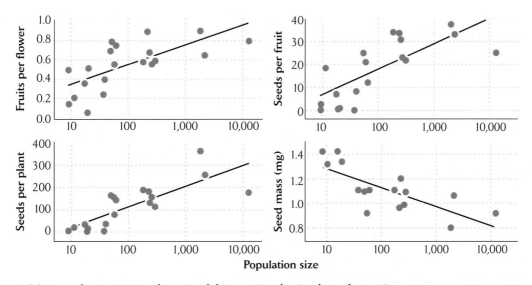

FIGURE 11.23 Reproduction in *Primula veris* **exhibits positive density dependence.** Plants in small populations produce fewer fruits and seeds than those in large populations. Seed mass is the average mass of an individual seed. After M. Kéry et al., *J. Ecol.* 88:17–30 (2000).

Sea) shows an increase in the ratio of recruits to spawners at low population densities, whereas another stock of spring-spawning herring near Iceland exhibits a classic Allee effect, with extremely low recruitment ratios at the lowest stock sizes. Fortunately, negative density dependence, as exhibited by the Downs stock, appears to be the more common pattern, suggesting that most stocks of fish can recover readily when released from excessive fishing pressure.

Allee effects are not limited to animals. Many plant species have self-incompatibility mechanisms to avoid inbreeding, so their flowers must be pollinated by other individuals in the population. When populations are small and individuals are widely spread, pollination, whether by wind or by animals, becomes less likely. *Primula veris* is a small herbaceous plant that can be found on nutrient-poor grasslands in Europe. Agricultural practices are fragmenting the plant's habitat, and local populations are markedly declining. To determine how these smaller populations are faring, Marc Kéry and his colleagues at the University of Zürich looked at the relationship between reproductive success and the sizes of local populations (Figure 11.23). They found that in populations of fewer than 100 individuals, both the number of fruits produced per flower and the number of seeds per fruit were substantially reduced, evidently because too little pollen was available. It is likely that such small numbers of plants cannot attract bees and other pollinators, which concentrate their foraging on more abundant species.

Density dependence in plants

Plants experience increased mortality and reduced fecundity at high densities, just as animals do. A common developmental response of plants to intense competition for resources is slowed growth, which can reduce survival and fecundity. The sizes of flax (*Linum*) plants grown to maturity at different densities reveal this flexibility (Figure 11.24). When seeds were sown sparsely at a density of 60 per square meter, the average dry weight of individuals was between 0.5 and 1 g, and many plants exceeded 1.5 g by the end of the experiment. When seeds were sown at densities of 1,440 and 3,600 per square meter, most of the individuals weighed less than 0.5 g, and few grew to large sizes. Plants grow larger, on average, at low densities because more resources are available to each individual. The variation in average plant size over the three treatments is a reaction norm with respect to resource availability. Variation in the size of individuals within each planting results from chance factors early in the seedling stage, particularly the date of germination and the quality of the particular microhabitat in which an individual seedling has grown. Early germination in a favorable spot gives a plant an initial growth advantage over others, which increases as larger plants grow and crowd their smaller neighbors.

Despite the phenotypic plasticity of plant growth, mortality can be high in crowded situations. When horseweed (*Erigeron canadensis*) seeds were sown at a density of

FIGURE 11.24 Plants may respond to competition for resources by slowing their growth. The average sizes of individual flax plants were smaller when seeds were sown at higher densities. After J. L. Harper, *J. Ecol.* 55:247–270 (1967).

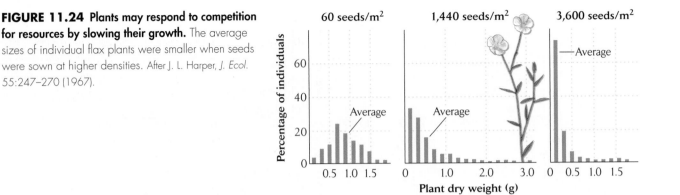

100,000 per square meter (equivalent to about 10 seeds in the area of your thumbnail), young plants competed vigorously. As the seedlings grew, many died, decreasing the population density for the surviving seedlings (Figure 11.25). At the same time, the growth rates of the surviving individual plants exceeded the rate of decline of the population, and the total weight of the planting increased. Over the entire growing season, a thousandfold increase in the average weight of the surviving individuals more than balanced the hundredfold decrease in population density.

In Figure 11.25b, in which plant weight and plant density are plotted on logarithmic scales, the data recorded over the growing season fall on a line with a slope of approximately −3/2. Plant ecologists call this relationship between average plant weight and density a **self-thinning curve.** Such is the regularity of this relationship that many have referred to it as the **−3/2 power law.** These density-dependent responses in plant populations are important considerations for spacing plants in agricultural fields and forest plantations to maximize individual plant size and yields.

Density-dependent factors tend to bring populations under control and maintain their size close to the carrying capacity set by the availability of resources and conditions of the environment. Changes in these resources and conditions continually establish new equilibrium values toward which populations grow or decline. Furthermore, density-independent changes in the environment, brought about by events such as a sudden freeze, a violent storm, or a shift in an ocean current, often reduce populations far below their carrying capacities and initiate periods of rapid population growth when favorable conditions return. Thus, although density-dependent factors regulate all populations, variations in the environment also cause populations to fluctuate about their equilibrium sizes. We shall explore population dynamics in more detail in the next chapter.

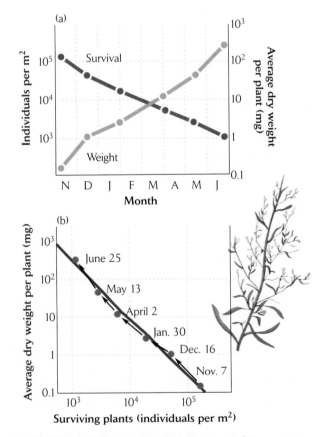

FIGURE 11.25 Plant populations increase in biomass even as numbers of individuals decrease. (a) Progressive change in average plant weight and population density in an experimental planting of horseweed (*Erigeron canadensis*) sown at a density of 100,000 seeds per square meter. (b) Relationship between plant density and average plant weight as the season progressed. The straight line, which has a slope of −3/2, illustrates the −3/2 power law. After J. L. Harper, *J. Ecol.* 55:247–270 (1967).

SUMMARY

1. Demography—the study of populations—has a long history in Western science, stretching back to John Gaunt and Thomas Malthus in seventeenth- and eighteenth-century England.

2. Populations grow by multiplication because newborn individuals grow up to reproduce themselves. The factor by which a population increases in one unit of time is the geometric growth rate of the population (λ). Populations with discrete reproductive seasons grow geometrically by periodic increments according to the relation $N(t + 1) = N(t)\lambda$.

3. Population growth can also be described in continuous time by the exponential growth rate (r) in the expression $N(t) = N(0)e^{rt}$. In population growth rate equations, λ and e^r are interchangeable.

4. The instantaneous rate of increase of an exponentially growing population is $dN/dt = rN$. In other words, the exponential growth rate of a population depends on both its size and its per capita birth and death rates.

5. The population growth rate in a population in which there is no immigration or emigration is the difference between the per capita birth rate (b) and the per capita death rate (d) ($r = b - d$).

6. When birth rates and death rates vary according to the ages of individuals, one must know the proportion of individuals in each age class to calculate the population growth rate. A population in which these rates do not change over time assumes a stable age distribution.

7. The life table of a population displays fecundities (b_x) and probabilities of survival (s_x) of individuals by age class (x). These variables can be used to calculate survival to a particular age and expectation of future survival of individuals at a particular age.

8. The fates of individuals born at the same time and followed throughout their lives may be used to produce a cohort life table. Survival of individuals of known age during a single period, the age structure of a population at a particular time, or the age distribution of individuals at death may be used to produce a static life table.

9. Survivorship curves tend to take on one of three basic shapes, depending on whether the mortality rate remains constant (type II), increases (type I), or declines with age (type III). Most populations exhibit a mixture of type I and type III curves.

10. In a population with a stable age distribution, numbers in each age class, as well as in the population as a whole, increase at the same geometric or exponential rate, known as the intrinsic rate of increase (λ_m or r_m).

11. The incongruity between the potential of all populations for rapid growth and the relative constancy of population sizes observed over long periods led naturally to the idea that extrinsic factors cause population growth to decrease as population size increases.

12. The growth of most populations can be described by the logistic equation, which follows a sigmoid curve as the population grows slowly at first, then more rapidly as the number of reproducing individuals increases, and finally more slowly again as the population approaches the carrying capacity.

13. The fit of the logistic equation to the growth of most populations suggests that density-dependent factors, such as dwindling supplies of food and increasing pressure from predators and disease, limit population growth.

14. Laboratory studies of animal and plant populations under controlled conditions have shown how density-dependent factors affect population growth rates. Similar studies of plants and animals in natural habitats have also revealed density-dependent effects.

15. In some species, per capita population growth rates decrease in smaller populations because individuals fail to find mates (the Allee effect) or because genetic diversity is lost.

16. Plants often respond to high population density by slowed growth and reduced reproduction, as well as reduced survival. The relationship between average plant weight and density in dense populations, referred to as a self-thinning curve, often assumes the form of a power relationship with a slope of $-\frac{3}{2}$.

REVIEW QUESTIONS

1. What is the difference in approach between the geometric growth equation and the exponential growth equation?

2. Given the relationship between λ and r in the geometric and exponential growth equations, can you demonstrate mathematically why λ must be 1 when r is 0?

3. Why might a model with age-specific demographic variables reflect the reality of population growth better than a model that lacks age structure?

4. What is the difference between a stable population and a stable age distribution?

5. If a life table projects a population size of 100 females and the sex ratio is 1:1, how large is the entire population?

6. In a life table, what is the fundamental difference between survival (s_x) and survivorship (l_x)?

7. Compare and contrast a cohort life table and a static life table.

8. What is the relationship between generation time and the rate of population growth?

9. What are the different causes of slow population growth at low population sizes versus high population sizes?

10. What evidence would you need to determine whether a population experiences negative density dependence or positive density dependence?

SUGGESTED READINGS

Chase, M. K., N. Nur, and G. R. Geupel. 2005. Effects of weather and population density on reproductive success and population dynamics in a song sparrow (*Melospiza melodia*) population: A long-term study. *Auk* 122:571–592.

Clutton-Brock, T. H., M. Major, and F. E. Guinness. 1985. Population regulation in male and female red deer. *Journal of Animal Ecology* 54:831–846.

Courchamp, F., T. Clutton-Brock, and B. Grenfell. 1999. Inverse density dependence and the Allee effect. *Trends in Ecology and Evolution* 14:405–410.

Gotelli, N. J. 1995. *A Primer of Ecology*. Sinauer Associates, Sunderland, Mass.

Harrison, S. and N. Cappuccino. 1995. Using density-manipulation experiments to study population regulation. In N. Cappuccino and P. W. Price (eds.), *Population Dynamics: New Approaches and Synthesis*, pp. 131–147. Academic Press, New York.

Kéry, M., D. Matthies, and H.-H. Spillmann. 2000. Reduced fecundity and offspring performance in small populations of the declining grassland plants *Primula veris* and *Gentiana lutea*. *Journal of Ecology* 88:17–30.

Kingsland, S. E. 1985. *Modeling Nature: Episodes in the History of Population Ecology*. University of Chicago Press, Chicago.

Lack, D. 1954. *The Natural Regulation of Animal Numbers*. Oxford University Press, London.

Murdoch, W. W. 1994. Population regulation in theory and practice. *Ecology* 75:271–287.

Myers, R. A., et al. 1995. Population dynamics of exploited fish stocks at low population levels. *Science* 269:1106–1108.

Pollard, E., K. H. Lakhani, and P. Rothery. 1987. The detection of density dependence from a series of annual censuses. *Ecology* 68:2046–2055.

Skogland, T. 1985. The effects of density-dependent resource limitations on the demography of wild reindeer. *Journal of Animal Ecology* 54:359–374.

Stephens, P. A., and W. J. Sutherland. 1999. Consequences of the Allee effect for behavior, ecology and conservation. *Trends in Ecology and Evolution* 14:401–405.

Turchin, P. 1995. Population regulation: Old arguments and new synthesis. In N. Cappuccino and P. W. Price (eds.), *Population Dynamics: New Approaches and Synthesis*, pp. 19–40. Academic Press, New York.

Weiner, J. 1988. Variation in the performance of individuals in plant populations. In A. J. Davy, M. J. Hutchings, and A. R. Watkinson (eds.), *Plant Population Ecology*, pp. 59–81. Blackwell Scientific Publications, Oxford.

Weller, D. E. 1987. A reevaluation of the $-3/2$ power rule of plant self-thinning. *Ecological Monographs* 57:23–43.

Wolff, J. O. 1997. Population regulation in mammals. *Journal of Animal Ecology* 66:1–13.

DATA ANALYSIS MODULE 2

Birth and Death Rates Influence Population Age Structure and Growth Rate

A population's growth rate reflects the balance between births and deaths. Clearly, when the birth rate increases or the death rate decreases, the growth rate of a population increases. If we wish to know by how much, we have to calculate the magnitude of the increase through the effect of these changes on the life table. There are two general principles involved in this calculation. First,

changes in birth or death rates at older ages tend to have smaller effects on population growth than changes at younger ages, simply because fewer individuals survive long enough to be affected. Second, changes in the life table that increase the population growth rate also tend to shift the age structure of the population to younger ages. We are seeing the opposite effect in many human populations, in which growth rates have been declining and the proportion of older individuals has been increasing.

Let's look at the effect of changing the life table values in Table 11.1 on the growth rate and age structure of a population. In this hypothetical population, the growth rate settled to a value of $\lambda = 1.49$ at a stable age distribution of 63.4% newborns, 21.3% age 1, 11.4% age 2, and 3.9% age 3 (see Table 11.3). This age structure might be typical of a bird such as the prairie chicken or a population of white-footed mice on a barrier island on the Gulf Coast of the United States. Because life table analysis shows the contributions of births and deaths at each age to the overall growth rate of the population, it is an important management tool for conservation as well as for the control of pest species.

Step 1: Create a population projection table in an Excel spreadsheet.

Use the template in the spreadsheet titled "module5data.xls" to perform the calculations in Table 11.2 (page 227), based on the life table in Table 11.1 (page 227). The template extends through 10 time intervals, but otherwise it should provide the same numbers (within rounding errors) as in Table 11.3 (page 228). When the spreadsheet is set up properly, you might try adding older age classes to make the population more complex.

Make changes in the life table and in the initial age distribution of individuals in the population to see how these changes affect population growth rate and age structure.

• Does the initial age distribution of individuals have any effect on the eventual age structure of the population? Does it influence how long the population takes to achieve a stable age distribution?

• Do changes in survival and fecundity have similar effects on the population growth rate? How do changes in survival and fecundity differ in their influence on the life table?

Step 3: Increase and decrease fecundity at ages 2 and 3 by the same amount.

• Do changes at younger or older ages have the greater effect? You can fill in DA Table 2.1 to obtain the data for this exercise.

DA TABLE 2.1	Data for the effect of changing fecundity on population growth rate		
Fecundity b_2	Population growth rate λ	Fecundity b_3	Population growth rate λ
1.5		1.5	
2		2	
2.5		2.5	
3		3	
3.5		3.5	

Step 4: Plot a graph of the relationship between the population growth rate at the stable age distribution and the fecundity at age 2 and at age 3, leaving all other variables unchanged.

These relationships show the sensitivity of population growth to changes in life table parameters. For example, the slopes of the lines in your graph are the sensitivities of population growth with respect to change in fecundity at ages 2 and 3 ($d\lambda/db_2$ and $d\lambda/db_3$).

• Are these strictly straight lines? If not, where would you evaluate the slopes to obtain the sensitivities?

• How might conservationists use this type of sensitivity analysis to identify critical periods in the lives of organisms in terms of their effects on population growth?

• The spreadsheet contains columns for calculating the survival of newborn individuals to each age and the product of fecundity at each age times survival to that age (see page 235). The sum of the $l_x b_x$ column is called the net reproductive rate (R_0). Express in words what this rate represents. Can you see any relationship between the value of R_0 and the growth rate of the population?

• If you were a manager concerned with an endangered population of prairie chickens in southwestern Kansas, how would you determine whether to invest in increasing the survival of eggs and chicks (thereby increasing annual fecundity) or in reducing mortality of adult birds during winter (perhaps by supplemental feeding during bad weather or creating shelter habitat)?

Temporal and Spatial Dynamics of Populations

The lore of many countries at high northern latitudes includes tales of dramatic fluctuations in local mammal populations. Lemmings, whose numbers alternate between scarcity and epidemic proportions, are the most famous of these. Lemming "plagues" occur approximately every 4 years, at which times their extremely high densities set off mass dispersal movements reminiscent of locust migrations. Some of these movements terminate at the edge of the sea, where the forward ranks are pushed into the water by those behind, inspiring the legend that lemmings commit suicide to reduce the density of their populations. Of course, their behavior is neither suicidal nor altruistic, but only an unfortunate outcome of severe overcrowding.

Many populations fluctuate, varying between high and low numbers more or less periodically. The ecologist Charles Elton first drew the attention of the scientific community to the regularity of some population cycles in 1924. For centuries, however, such cycles had been evident to naturalists, trappers, and others attentive to variations in economically important species.

From the late seventeenth century through the end of the eighteenth century, Danish royalty transported gyrfalcons from Danish territories in Iceland to Copenhagen and presented them as diplomatic gifts to the courts of Europe. Falconry was a popular pastime among European royalty at this time, and the gyrfalcon, the largest of the falcons and one of the most beautiful, was especially prized (Figure 12.1). Birds were usually caught by trapping them at their nests, perched on ledges on high cliffs. Records kept between 1731 and 1793 of the number of gyrfalcons exported from Denmark show 10-year cycles of striking regularity (Figure 12.2). After 1770, the numbers being exported declined, and the fluctuations were not so apparent, primarily because falconry had become less popular and the demand for birds could be satisfied even during years of population lows.

CHAPTER CONCEPTS

- Fluctuation is the rule for natural populations
- Temporal variation affects the age structure of populations
- Population cycles result from time delays in the response of populations to their own densities

- Metapopulations are discrete subpopulations linked by movements of individuals
- Chance events may cause small populations to go extinct

FIGURE 12.1 The gyrfalcon has been a favorite with falconers for centuries. Photo by J. Krimmel/Cornell Laboratory of Ornithology.

Under the influence of density-dependent factors, populations tend to increase or decrease toward an equilibrium size determined by the carrying capacity of the environment. However, we also know that popula-tion sizes can vary widely over time. Two kinds of factors cause populations to fluctuate. First, birth and death rates respond to changes in environmental conditions, such as temperature, moisture, salinity, acidity, and other physical factors. Environmental conditions may affect these rates directly, by influencing the performance of individuals, or indirectly, by altering the food supply. Second, variation in population size may result from intrinsic characteristics of the population itself. Some biological systems are inher-ently unstable and tend to develop oscillations.

Ecological conditions vary both in space and in time, creating differences in population dynamics from place to place and changes in population dynamics over time. When distance isolates subpopulations from one another, they can fluctuate independently. Changes in an entire population are the sum of changes in all of its subpop-ulations, but because the dynamics of large and small populations differ, subdivided populations possess unique properties.

In this chapter, we shall discuss the causes of variation in population size, explore how this variation affects small and large populations, and examine the consequences of migration among discrete subpopulations. The dynamics of small populations and metapopulations have become increasingly relevant as many species dwindle toward extinction and human activities fragment habitats into ever smaller, more isolated patches.

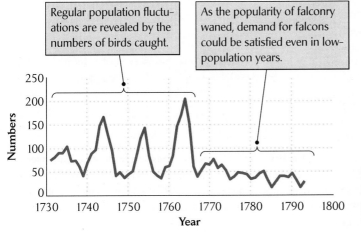

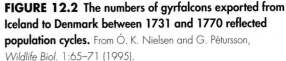

FIGURE 12.2 The numbers of gyrfalcons exported from Iceland to Denmark between 1731 and 1770 reflected population cycles. From Ó. K. Nielsen and G. Pétursson, *Wildlife Biol.* 1:65–71 (1995).

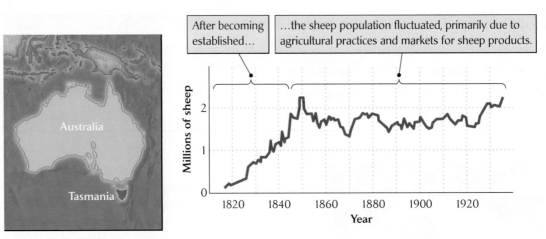

FIGURE 12.3 Domestic sheep on the island of Tasmania maintain a relatively stable population. Sheep were introduced to Tasmania in the early 1800s. After increasing rapidly to the carrying capacity in less than 30 years, their numbers varied by no more than a factor of 2. After J. Davidson, *Trans. R. Soc. S. Aust.* 62:342–346 (1938).

Fluctuation is the rule for natural populations

Variation in the density of a population depends on the two factors just mentioned: the magnitude of fluctuation in the environment and the intrinsic stability of the population. Some populations tend to remain relatively stable over long periods. Once domestic sheep became established on Tasmania, a large island off the coast of Australia, their population varied irregularly between 1,230,000 and 2,250,000—by less than a factor of 2—over nearly a century (Figure 12.3). We know this because sheep were, and still are, important to the economy of Tasmania, and their numbers were carefully recorded. Moreover, much of the variation in their numbers can be attributed to changes in grazing practices, markets for wool and meat, and pasture management.

In sharp contrast, populations of small, short-lived organisms may fluctuate wildly over many orders of magnitude within short periods. Populations of the green algae and diatoms that make up the phytoplankton may soar and crash over periods of a few days or weeks (Figure 12.4). These rapid fluctuations overlie changes with longer periods that occur, for example, with the change in seasons.

What accounts for the difference between these two populations? Sheep and algae differ in their sensitivity to environmental change and in the response times of their populations. Because sheep are larger, they have a greater capacity for homeostasis and can better resist the physiological effects of environmental change. Furthermore, because sheep live for several years, the population at any one time includes individuals born over a long period, which tends to even out the effects of short-term fluctuations in birth rate. Thus, sheep populations possess a high intrinsic stability. The lives of single-celled phytoplankton span only a few days, so their populations turn over rapidly; in other words, individuals have high mortality rates and are quickly replaced in the population by young individuals. As a consequence of this rapid population **turnover,** population size depends on continued reproduction, which is sensitive to food availability, predation, and physical conditions. Consequently, phytoplankton populations are intrinsically unstable and bear the full effect of a capricious environment.

Populations of similar species living in the same place often respond to different environmental factors. For example, the densities of four species of moths, whose

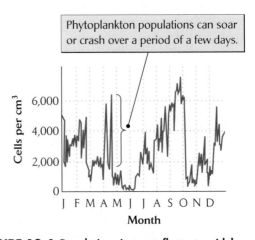

FIGURE 12.4 Population sizes can fluctuate widely over short periods. The density of phytoplankton in samples of water taken from Lake Erie during 1962 varied not only over the course of the year, but also over periods of a few days or weeks. After C. C. Davis, *Limnol. Oceanogr.* 9:275–283 (1964).

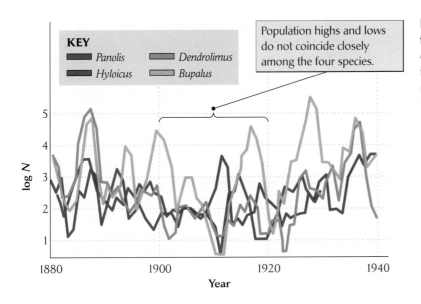

FIGURE 12.5 Populations of four moth species in the same habitat fluctuate independently. Numbers of pupae or hibernating larvae in a managed pine forest in Germany were recorded over 60 consecutive midwinter counts. After G. C. Varley, *J. Anim. Ecol.* 18:117–122 (1949).

larvae all feed on pine needles, were found to fluctuate more or less independently in a pine forest in Germany (Figure 12.5). The populations all varied over three to five orders of magnitude (a thousandfold to a hundred thousandfold) with irregular periods of a few years. Furthermore, the highs and lows of the four populations did not coincide closely. These observations suggested that, even though all four species feed on the same trees, their populations are governed independently by different factors, possibly specialized parasites or pathogens.

Some populations show **periodic cycles**—that is, the intervals between successive highs or lows are often remarkably regular. Among the most striking population phenomena in nature are the periodic cycles of abundance of certain mammals and birds at high latitudes, such as the lemmings and gyrfalcons described at the opening of this chapter. Annual counts of three species of grouse over a 20-year period in several regions of Finland revealed peaks of abundance separated by periods of 6 or 7 years (Figure 12.6). As in the 10-year cycle of gyrfalcon

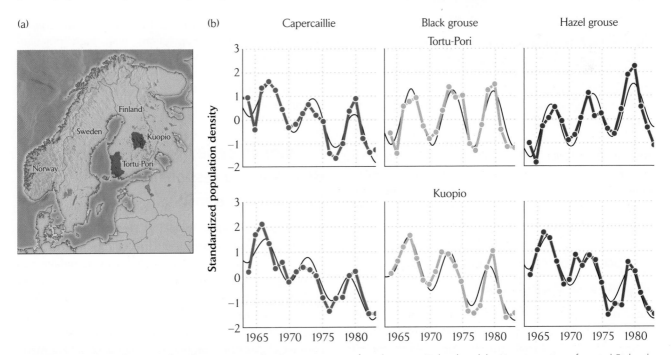

FIGURE 12.6 Population cycles of grouse in Finland are synchronized across species and areas. Cycles in the numbers of capercaillie (*Tetrao urogallus*), black grouse (*T. tetrix*), and hazel grouse (*Bonasa bonasia*) counted in the Turku-Pori region of southwestern Finland and the Kuopio region of central Finland from 1964 to 1983 can be superimposed almost exactly. From J. Lindström et al., *Oikos* 74:185–194 (1995).

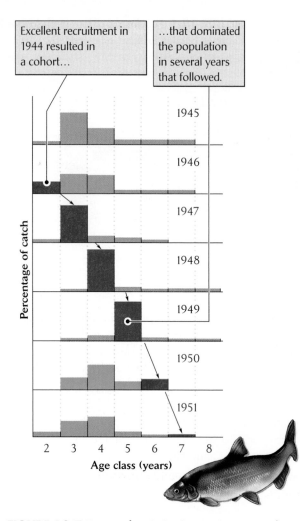

Excellent recruitment in 1944 resulted in a cohort...

...that dominated the population in several years that followed.

FIGURE 12.7 Temporal variation in recruitment may be evident in the age structure of a population. Samples from the commercial whitefish catch in Lake Erie between 1945 and 1951 show that the cohort of fish spawned in 1944 dominated the population for several years afterward. From G. H. Lawler, *J. Fish. Res. Bd. Can.* 22:1197–1227 (1965).

example, the age composition of samples from the Lake Erie commercial whitefish catch for the years 1945–1951 shows that during 1947, 1948, and 1949, most of the individuals caught belonged to the age class spawned in 1944 (Figure 12.7). Biologists estimated the ages of fish from growth rings on their scales. Their data showed that 1944 was an excellent year for spawning and recruitment. Thus, the 1944 cohort constituted an atypically large proportion of the total population over the next several years.

Similarly, the age structures of stands of trees often show wide variation in recruitment from year to year. Foresters estimate the ages of trees in seasonal environments by counting the growth rings in the woody tissue of the trunk; under normal circumstances, one ring is added each year. The age composition of a virgin stand of timber surveyed near Hearts Content, Pennsylvania, in 1928 shows that individuals of most species were recruited sporadically over the nearly 400-year span of the record (Figure 12.8). Many oaks and white pines became established between 1620 and 1710, undoubtedly following a major disturbance, possibly associated with the severe drought and fire year of 1644. Fire can open up a forest enough to allow the establishment of white pine seedlings, which do not tolerate deep shade. In contrast, beech— a species whose seedlings can grow under a closed canopy—exhibited a relatively even age distribution.

DATA ANALYSIS MODULE *Tracking Environmental Variation.* Populations with high potential growth rates keep up with environmental change more closely than do those with low potential growth rates. You will find this module at http://www.whfreeman.com/ricklefs6e.

Population cycles result from time delays in the response of populations to their own densities

Except for daily, lunar (tidal), and seasonal cycles, environmental variation tends to be irregular rather than periodic. As we saw in Chapter 4, years of abundant rain or drought, extreme heat or cold, and natural disasters such as fires and hurricanes occur unpredictably. Biological responses to these factors are similarly irregular. Why, then, do the sizes of many populations rise and fall in periodic cycles (for example, as in Figure 12.6)? For many years, ecologists believed that such cycles must be caused by extrinsic environmental factors that exhibit similar peri-

populations in Iceland, this regularity is distinctly nonrandom. Furthermore, the periodic cycles of the three species showed remarkable synchrony over large areas. The cause of such cycles, and of their synchrony, has been one of the most interesting and persistent questions in ecology.

Temporal variation affects the age structure of populations

Variation in population size over time often leaves its mark on the age structure of a population. As we have seen in Chapter 11, changes in age structure can affect the rate of population growth. The sizes of age classes also provide a history of population changes in the past. In one classic

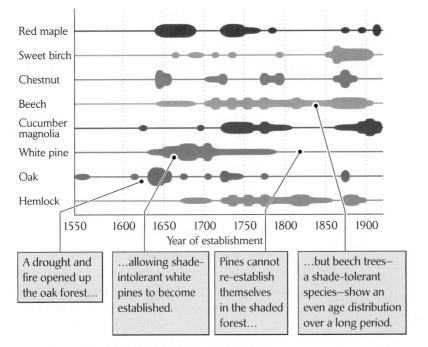

KEY

Percentage of
each species
established in
the year shown

0–5′
5–15′
15–40′

A drought and fire opened up the oak forest…

…allowing shade-intolerant white pines to become established.

Pines cannot re-establish themselves in the shaded forest…

…but beech trees—a shade-tolerant species—show an even age distribution over a long period.

FIGURE 12.8 Age distributions of forest trees show the effects of disturbances on seedling establishment. These data were collected near Hearts Content, Pennsylvania, in 1928. The thickness of each line represents the percentage of the white pines in the stand that were established in each time period. After A. F. Hough and R. D. Forbes, *Ecol. Monogr.* 13:299–320 (1943).

odic variation. A regular 11-year cycle in sunspot numbers was frequently cited to explain population cycles of snowshoe hares, but the sunspot cycle never matched the hare cycles well, and ecologists could find no mechanism connecting the two.

The source of regular population cycling was finally discovered through mathematical models developed in the 1920s and 1930s, which showed that the cycles reflect intrinsic dynamic qualities of biological systems. Some populations subjected to even minor, random environmental fluctuations will begin to **oscillate** back and forth across their equilibrium value. Populations have an intrinsic periodicity, just as a pendulum has an inherent frequency of swinging depending on its length. Momentum imparted to a pendulum by the acceleration of gravity carries it past the equilibrium point and causes it to swing back and forth in a regular pattern. Similarly, "momentum" imparted to a population by high birth rates at low densities can cause it to grow rapidly and overshoot its carrying capacity. Then, low survival rates at high densities cause the population to decrease below its carrying capacity. Such periodic cycling can result from **time delays** in the responses of birth and death rates to changes in the environment.

Time delays and oscillations in discrete-time models

Population models based on the logistic equation (see Chapter 11) have been used to study population cycles. Time delays that cause populations to oscillate are inherent in models of populations with discrete generations. In such discrete-time models, population growth occurs at intervals associated with breeding episodes. Thus, the response of a population to conditions at one time is not expressed as a change in the number of adults until the end of the next time interval. Because the population responds by increments, it cannot continuously readjust its growth rate as its size approaches the carrying capacity. Thus, the population may grow (or shrink) past its equilibrium as its size draws closer to the carrying capacity. Whether, and how much, such a population oscillates around the equilibrium depends on whether the growth increment exceeds the difference between the size of the population and the carrying capacity.

In a discrete-time model, the change in population size from one time interval to the next is $\Delta N(t) = N(t + 1) - N(t)$. We now set $\Delta N(t) = RN(t)$, where R is the proportional increase or decrease in N per unit of time. Now, let's make R density-dependent by a factor of $1 - N(t)/K$, in which case the increment of change in population size from one time interval to the next is $\Delta N(t) = RN(t)[1 - N(t)/K]$. This expression can be rearranged more conveniently to show how the intrinsic growth rate of the population can cause periodic cycles:

$$\Delta N(t) = \frac{RN(t)}{K}[K - N(t)]$$

Now, $K - N(t)$ is the difference between the size of the population and its carrying capacity at time t. When $\Delta N(t)$ exceeds this difference, the population will overshoot its carrying capacity. This occurs when $RN(t)/K$ is greater than 1.

The value of R can be used to predict whether population size will oscillate (Figure 12.9). Notice how, as $N(t)$ approaches the carrying capacity (K), the ratio $N(t)/K$ approaches 1 and $\Delta N(t)$ approaches R times the amount by which the population is below or above K. Thus, when R is less than 1, each growth increment brings the population closer to K, but it does not exceed the carrying capacity. Accordingly, the population will approach the carrying capacity (K) directly, without oscillation.

When R exceeds 1 but is less than 2, a population will tend to overshoot its equilibrium because $\Delta N(t)$ is greater than $K - N(t)$, but it will nonetheless end up closer to the equilibrium than it was before. Thus, the population will oscillate back and forth across the equilibrium value, get-

ting closer with each generation, until it eventually settles at K. This behavior is called **damped oscillation.**

When R exceeds 2, the population can end up farther from the equilibrium with each generation, and oscillations tend to increase. The population may nonetheless settle into a stable pattern of oscillations called a **limit cycle,** in which numbers bounce back and forth between high and low values. With increasing R above 2, these oscillations can take on very complex, eventually unpredictable forms referred to as **chaos.**

Time delays and oscillations in continuous-time models

Continuous-time population models have no built-in delays in the response of a population to its environment. Instead, time delays result from the developmental period that separates reproductive episodes between generations. These time delays can create population cycles when birth and death rates are a response to the population's density at some time in the past, rather than its density at the present. For example, if poor resource availability in the past caused few young to survive, by the time those survivors reach breeding age, the population may have declined, even if resources are plentiful, because of the low number of reproducing adults. The length of such a time delay is referred to by the symbol τ (the lowercase Greek letter tau).

In continuous-time models based on the logistic equation, time delays can be indicated by

$$\frac{dN(t)}{dt} = rN(t)[1 - N(t - \tau)/K]$$

Thus, the growth rate of the population at time t is a response to density-influenced birth and death rates at τ time units in the past, when the population might have been larger or smaller than at time t.

Whether population size oscillates, and how, depends on the product of the intrinsic rate of increase (r) and the time delay, or $r\tau$. High intrinsic rates of increase and long time delays increase the amplitude of population cycles. Oscillations are damped as long as the product $r\tau$ is less than $\pi/2$ (about 1.6). When $r\tau$ is less than $1/e$ (0.37), the population increases or decreases to the carrying capacity (K) without overshooting it, so the population size does not oscillate. When $r\tau$ is greater than $\pi/2$, the oscillations increase to form a limit cycle, whose amplitude grows as the time delay lengthens (Figure 12.10). For example, at $r\tau = 2$ the maximum population size is nearly 3 times K, and at $r\tau = 2.5$ it is nearly 5 times K. The periods of these limit cycles, measured from peak to peak, increase from about 4τ to more than 5τ with increasing $r\tau$. Thus,

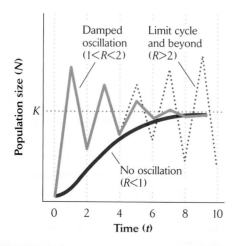

FIGURE 12.9 A population may adopt one of three oscillation patterns. In a discrete-time model based on the logistic equation, the oscillation pattern depends on the per capita growth rate (R). When $R < 1$, the population returns without fluctuation to the carrying capacity. When $1 < R < 2$, the population exhibits damped oscillations, which become limit cycles when $R > 2$.

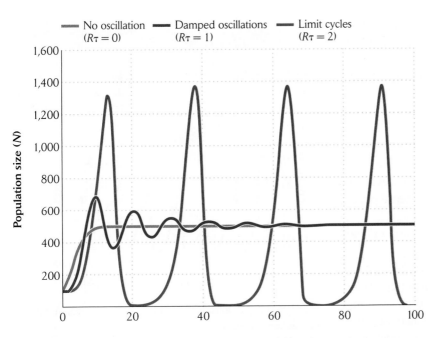

FIGURE 12.10 Time delays and density dependence create oscillations in population size. In population models based on the logistic equation, increasing time delays (τ) in the response of population growth to density result in oscillations of greater amplitude. In all these discrete-time models, $R = 0.50$ and $K = 500$.

a population cycle with a period of 10 years would imply a time delay of about 2 years.

Cycles in laboratory populations

Population cycles observed in many laboratory cultures of single species have provided clues to the mechanisms responsible for creating time delays. In one study, populations of the water flea *Daphnia magna* exhibited marked oscillations when cultured at 25°C, but these oscillations disappeared at 18°C (Figure 12.11). The period of the cycle at 25°C appeared to be just over 60 days, suggesting a time delay in the response of growth rate to density of about 12−15 days. The average age at which water fleas give birth at 25°C is about 12−15 days.

The time delay arose in the following manner. As population density increased, fecundity decreased, falling nearly to zero when the population exceeded 50 individuals. Survival was less sensitive to density even at the highest densities, and adults lived at least 10 days (a pattern similar to that shown for fruit flies in Figure 11.17). Thus, crowding at the peak of the cycle prevented births, and the population began to crash as adults died off. By the time the population density had dropped low enough to permit reproduction, most of the adults were senescent, nonreproducing individuals, and thus the population continued to decline. The beginning of a new phase of population increase had to await the accumulation of young, fecund individuals. The length of the time delay at high densities was approximately the average adult life span.

At the lower temperature, reproduction fell off quickly with increasing density, and the life span was longer than that seen at 25°C at all densities. Populations at the lower

temperature apparently lacked a time delay because deaths were more evenly distributed over all ages and some individuals gave birth even at high population densities. Consequently, generations overlapped more broadly.

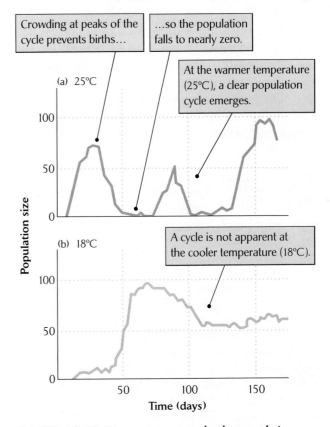

Crowding at peaks of the cycle prevents births…

…so the population falls to nearly zero.

At the warmer temperature (25°C), a clear population cycle emerges.

A cycle is not apparent at the cooler temperature (18°C).

(a) 25°C

(b) 18°C

FIGURE 12.11 Warm temperatures lead to population cycles in *Daphnia magna*. Populations were maintained at (a) 25°C and (b) 18°C. After D. M. Pratt, *Biol. Bull.* 85:116−140 (1944).

At the higher temperature, water fleas behaved according to a discrete-time model with its built-in time delay of one generation. At the lower temperature, they behaved according to a continuous-time model with little or no time delay. Thus, time delays in the response of *Daphnia magna* population growth rates to density are tied to both development time and life span.

Storage of food reserves by individuals can reduce the sensitivity of mortality to crowding and introduce time delays into death rates. Energy and nutrient reserves allow adults to continue to reproduce even after a dense population has overeaten the food supply. *Daphnia galeata,* a large water flea species, stores energy in the form of lipid droplets during periods of high food abundance (that is, at low population densities). It can then live on these stored reserves when food supplies dwindle at high population densities. Females also transfer energy to each offspring through lipid droplets in their eggs, thereby increasing the survival of young, prereproductive water fleas under poor feeding conditions. In contrast, the smaller water flea *Bosmina longirostris* stores fewer lipids, so mortality increases quickly in response to increases in population density. The consequences of this difference for population growth are predictable. In one study, *Daphnia* exhibited pronounced limit cycles with a period of 15 to 20 days, whereas *Bosmina* populations grew quickly to an equilibrium size. The rate of increase (r) in the *Daphnia* populations was about 0.3 per day. With a cycle period of 15 to 20 days, τ would have been about 4 to 5 days, and therefore $r\tau$ was about 1.2 – 1.5. Because the value of $r\tau$ was less than $\pi/2$, the cycles in the *Daphnia* population should have damped out eventually.

ECOLOGISTS IN THE FIELD **Time delays and oscillations in blowfly populations.** Slight differences in laboratory culture conditions or in the life histories of species can tip the balance between a population that does not oscillate and one that sustains a limit cycle. Australian entomologist A. J. Nicholson's pioneering experimental manipulations of time delays in cultures of the sheep blowfly (*Lucilia cuprina*) demonstrated dramatically the relationship of time delays to population cycles.

In one set of population cages, Nicholson provided blowfly larvae with 50 grams of ground liver per day, while giving adults unlimited food. The number of adults in the population cycled from a maximum of about 4,000 to a minimum of 0 (at which point all the individuals were either eggs or larvae) with a period of 30–40 days (Figure 12.12). These regular fluctuations were caused by a time delay in the responses of fecundity and mortality to the density of adults in the cages. At high population densities, adults laid many eggs, resulting in strong larval

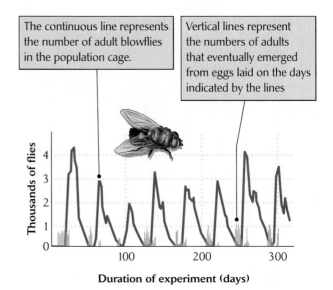

The continuous line represents the number of adult blowflies in the population cage.

Vertical lines represent the numbers of adults that eventually emerged from eggs laid on the days indicated by the lines

FIGURE 12.12 Introduction of time delays results in regular population cycles. Limiting the food supply available to larvae in a laboratory population of sheep blowflies (*Lucilia cuprina*) caused a time delay in density-dependent effects on population numbers and resulted in regular population cycles. Larvae were provided with 50 grams of liver per day; adults were given unlimited supplies of liver. After A. J. Nicholson, *Cold Spring Harbor Symp. Quant. Biol.* 22:153–173 (1958).

competition for the limited food supply. None of the larvae that hatched from eggs laid during adult population peaks survived, primarily because they did not grow large enough to pupate. Therefore, large adult populations gave rise to few adult progeny, and because adults lived less than 4 weeks, the population soon began to decline. Eventually, so few eggs were laid on any particular day that most of the larvae survived, and the size of the adult population began to increase again.

A time-delayed logistic model with $r\tau = 2.1$ provides a good fit to the observed oscillations in these blowfly populations. This value predicts that the ratio of the maximum population size to the carrying capacity ($N/K = e^{r\tau}$) should be 8.2 and that the cycle period should be 4.5τ. The experiment clearly reveals that density-dependent factors did not raise the mortality rates of adults immediately as the population increased, but were felt a week or so later when their progeny were larvae. Larval mortality did not express itself in the size of the adult population until those larvae emerged as adults about 2 weeks after eggs were laid. As in the *Daphnia* population maintained at a high temperature, crowding in the blowfly population created discrete, nonoverlapping generations with an intrinsic time delay (τ) equal to the larval development period, about 10 days.

The hypothesis that time delays in the response of growth rates to population densities cause population cycles can be tested directly by eliminating time delays—

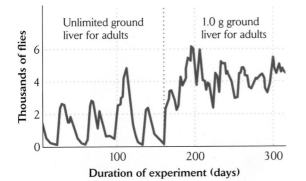

FIGURE 12.13 Elimination of time delays results in the elimination of population cycles. Limiting the food supply available to adults removed the time delay and eliminated fluctuations in a sheep blowfly population. This experiment was similar in all other respects to that depicted in Figure 12.12. After A. J. Nicholson, *Cold Spring Harbor Symp. Quant. Biol.* 22:153–173 (1958).

that is, by making the deleterious effects of high densities felt immediately. Nicholson did this by adjusting the amount of food provided to his flies so that food availability limited adults as severely as it did larvae. Adult flies require protein to produce eggs. By restricting the amount of liver available to adults to 1 gram per day, Nicholson curtailed egg production to a level determined by the availability of liver. Under these conditions, recruitment of new individuals into the population was determined at the egg-laying stage rather than at the later larval stage, and most of the larvae survived. Consequently, fluctuations in the population subsided (Figure 12.13). ▌

We have seen that both development time and storage of nutrients can put off deaths to a later point in the life cycle or to a later time, creating time delays in response to density. In contrast, fecundity can respond to changes in population density with little delay when adults produce eggs quickly from resources accumulated over a short time. Populations able to adjust birth rates quickly should not exhibit marked oscillations.

A population at its equilibrium point will remain there until perturbed by some outside influence, whether a change in the carrying capacity (K) or a catastrophic change in population size (N). Once displaced from its equilibrium, a population may move toward a stable limit cycle, depending on the nature of the time delay and the response time. Alternatively, it may return to the equilibrium directly or through damped oscillations. Population cycles can be reinforced through interactions with other species—prey, predators, parasites, even competitors—that have similar rates of response to changes in population density, as we shall see in later chapters.

Metapopulations are discrete subpopulations linked by movements of individuals

Areas of habitat with the necessary resources and conditions for a population to persist are called **habitat patches,** or simply **patches.** The individuals of a species that live in a habitat patch constitute a *subpopulation*. As we saw in Chapter 10, areas of unsuitable habitat often separate habitat patches; individuals may move through these areas on occasion, but cannot persist in them. A set of discrete subpopulations connected by occasional movement of individuals between them is referred to as a **metapopulation** (Figure 12.14). The metapopulation concept has become one of ecology's most important tools for understanding the dynamics of species living in fragmented habitats. Thus, as forest clearing, road building, and other human activities create patchworks of different types of habitats, metapopulation models help us manage and conserve populations that cannot move readily through a fragmented landscape.

Two sets of processes contribute to the dynamics of metapopulations. The first is the growth and regulation of subpopulations within patches—processes that we have already discussed in Chapter 11. The second is the colonization of empty patches by migrating individuals to form new subpopulations and the extinction of established subpopulations.

Because subpopulations are typically much smaller than the metapopulation as a whole, local catastrophes and chance fluctuations in numbers of individuals have greater effects on their population dynamics. Indeed, the smaller the subpopulation, the higher its probability of extinction during a particular time interval. When individuals move frequently between subpopulations, however, such fluctuations are damped out, and changes in subpopulation size mirror those in the metapopulation. Thus, a high rate of migration transforms metapopulation dynamics into the dynamics of a single large population.

At the other extreme, when no individuals move between subpopulations, the subpopulations in each patch behave independently. When those subpopulations are small, they have high probabilities of extinction, as we shall see below, so the total population may gradually go extinct as one subpopulation after another dies out.

At intermediate levels of migration, some patches left unoccupied by subpopulation extinction will be colonized again. Under such circumstances, the entire metapopulation exists as a shifting mosaic of occupied and unoccupied patches. This mosaic has its own dynamics and

FIGURE 12.14 A metapopulation is a set of discrete subpopulations having partially independent dynamics. Southern California spotted owls are distributed as a metapopulation over patches of suitable old-growth forest habitat in the mountains of southern California. At any one time, some patches are occupied, while others are not. From W. S. Lahaye, R. J. Gutiérrez, and H. R. Akçakaya, *J. Anim. Ecol.* 63:775–785 (1994).

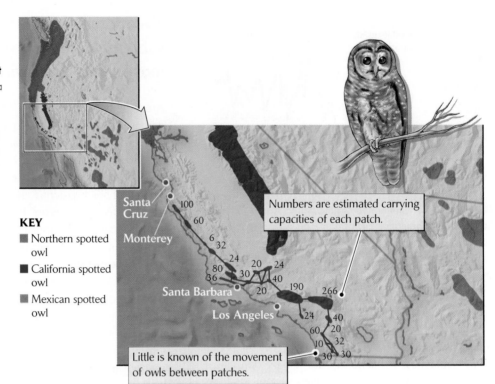

KEY
- ■ Northern spotted owl
- ■ California spotted owl
- ■ Mexican spotted owl

Numbers are estimated carrying capacities of each patch.

Little is known of the movement of owls between patches.

equilibrium properties, which can be understood in terms of a simple metapopulation model.

The basic model of metapopulation dynamics

Consider a population divided into discrete subpopulations. We assume that within a given time interval, each subpopulation has a probability of going extinct, which we shall refer to as *e*. Therefore, if *p* is the fraction of suitable habitat patches occupied by subpopulations, then subpopulations go extinct at the rate *ep*. The rate of colonization of empty patches depends on the fraction of patches that are empty $(1 - p)$ and the fraction of patches sending out potential colonists (p). Thus, we may express the rate of colonization within the metapopulation as a whole as a single rate constant c times the product $p(1 - p)$. The rate of change in patch occupancy under this model is therefore

$$\frac{dp}{dt} = cp(1 - p) - ep$$

[colonization]   [extinction]

A metapopulation attains equilibrium size when colonization equals extinction; that is, when $cp(1 - p) = ep$.

This expression may be rearranged to express the proportion of occupied patches at metapopulation equilibrium,

$$\hat{p} = 1 - \frac{e}{c}$$

The equilibrium proportion of occupied patches is indicated by the little hat (^) over the *p*. The equilibrium is stable because when *p* is below the equilibrium point, colonization exceeds extinction, and vice versa.

This simple model shows the critical importance of the relative rates of extinction and colonization (*e*/*c*). When $e = 0$, $\hat{p} = 1$, and all patches are occupied because all are eventually colonized and none of the subpopulations disappears. (This does not mean that the populations in the patches cease to vary, only that they are large enough, or otherwise stable enough, not to suffer extinction.) When $e = c$, $\hat{p} = 0$, and the metapopulation heads toward extinction. Intermediate values of *e*—that is, greater than 0 but less than *c*—result in a shifting mosaic of occupied and unoccupied patches. Thus, when the rate of colonization exceeds the rate of extinction, the fraction of occupied patches attains an equilibrium between 0 and 1. When extinction exceeds colonization, the fraction of occupied patches declines to zero, and the entire metapopulation goes extinct. This pattern makes clear the importance of keeping habitat patches from becoming too isolated or,

(a)

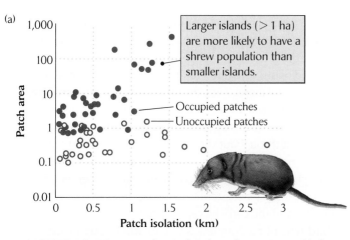

(b)

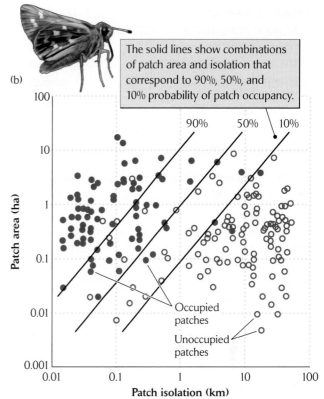

FIGURE 12.15 Larger, less isolated patches are more likely to be occupied. (a) In a metapopulation of the shrew *Sorex araneus* on islands in two lakes in Finland, islands larger than 1 ha were more likely to be occupied than smaller islands. Patch isolation appears to be relatively unimportant in this metapopulation. (b) In a metapopulation of the butterfly *Hesperia comma* on patches of calcareous grassland in England, both patch size and patch isolation appeared to be important. (a) from I. Hanski, *Biol. J. Linn. Soc.* 42:17–38 (1991); (b) after C. D. Thomas and T. M. Jones, *J. Anim Ecol.* 62:472–481 (1993).

alternatively, maintaining migration corridors between patches in a managed landscape.

The model outlined above portrays a highly simplified metapopulation in which (1) all patches are equal, (2) rates of extinction and colonization for each patch are the same, (3) each occupied patch contributes equally to dispersal, (4) colonization and extinction in each patch occur independently of other patches, and (5) the colonization rate is proportional to the fraction of occupied patches. More realistically, patches vary in size, habitat quality, and degree of isolation from other patches. Larger patches can support larger subpopulations, which have lower probabilities of extinction. Smaller, more isolated patches are less likely to be occupied.

Figure 12.15 shows the occupation of small islands in two lakes in Finland by subpopulations of the shrew *Sorex araneus* and the occupation of patches of calcareous grassland in England by subpopulations of the butterfly *Hesperia comma*. The Finnish islands varied in size from about 0.1 to 1,000 ha, and in distance from other islands or the shore of the lake from less than 0.1 to more than 2 km. The grassland patches varied from about 0.01 to 10 ha in area, and some were as far removed from other patches as 10–100 km. Subpopulations of shrews were sensitive to patch area, as few occupied islands smaller

than 1 ha. Patch isolation did not exert a marked effect in this case; shrews appeared to colonize more distant islands as readily as close ones. For the butterflies, patch area and distance both influenced patch occupancy, but the distances between patches were much greater than in the shrew study.

One of the most extensive studies of metapopulations has been conducted by the Finnish ecologist Illka Hanski on the Glanville fritillary butterfly (*Melitaea cinxia*) on the Åland Islands of Finland. Hanski observed that the butterflies occupied patches of dry meadows on the islands, but that of about 1,600 suitable patches, only 30% were occupied at any given time. Over the study area as a whole, patch occupancy was highest in areas that have larger and more numerous patches, as one would expect. Such a snapshot of patch occupancy, however, does not reveal the dynamics of the metapopulation: the extinction of subpopulations and the recolonization of empty patches.

In one experiment, Hanski and his colleagues introduced populations of the Glanville fritillary to 10 of 20 suitable habitat patches on the small, isolated island of Sottungia, which previously lacked this butterfly species, in August 1991. Over the next 10 years, the number of extinctions varied between 0 and 12 per year, and the number of colonizations between 0 and 9. The number of subpopulations started at 10, dropped to as few as 2, and increased to as many as 14, ending the decade at 11. Although the turnover rate of subpopulations was high, and although none of the original 10 subpopulations survived the decade, the metapopulation as a whole persisted.

The rescue effect

From a metapopulation standpoint, a patch is either occupied or it is not. However, the probability that a subpopulation will become extinct depends to some extent on its size. Immigration from large, productive subpopulations can keep declining subpopulations from dwindling to small numbers and eventually becoming extinct. This phenomenon is known as the **rescue effect.** Clearly, dispersal is critical not just for colonizing empty patches, but also for maintaining established populations.

The rescue effect can be incorporated into metapopulation models by specifying that the rate of extinction (e) decreases as the fraction of occupied patches (p) increases (that is, with more numerous sources of migrants, or rescuers). In one version of a such a metapopulation model, the rate of change in patch occupancy due to extinction becomes $dN/dt = -ep(1 - p)$. This expression predicts that the proportion of occupied patches ($\hat{p}$) will either increase to 1 or decrease to 0, depending on the relative rates of colonization (c) and extinction (e). The rescue effect further highlights the importance of patch connectivity to the persistence of metapopulations.

Chance events may cause small populations to go extinct

Analyses of metapopulations have shown that small subpopulations are more likely to die out than large ones. Patch size itself may be a factor, as smaller areas—which support fewer individuals—are ecologically less diverse and more likely to suffer declines in habitat quality as environmental factors vary. However, small populations are also vulnerable to extinction just by chance or bad luck, regardless of changes in environmental conditions. The population models we have considered thus far assume large population sizes and use average values of birth and death rates as if they applied to every individual without variation. Such models, whose outcomes we can predict with certainty, are called **deterministic** models. In the real world, however, random variations—in deaths over a particular interval, or in the number and sex of offspring, for example—can influence the course of population growth.

Ecologists recognize three types of randomness that affect populations. First, an unpredictable **catastrophe,** such as the appearance of a predator or infectious disease or an intense fire, may cause reproductive failure or a high death toll throughout the population. The second type of randomness is environmental variation. Randomly occurring changes in physical conditions and other environmental factors persistently influence rates of population growth and the carrying capacity of the environment. The third type of randomness is due to **stochastic** processes, which can cause population size to vary even in a constant environment. The death of an individual, for example, is a chance event that has some probability of occurring during a particular interval. The number of deaths within a population, however, has a probability distribution. That is to say, a stochastic death process allowed to run repeatedly in the same population model would produce a range of outcomes from few to many deaths, just by chance. The average value of this distribution would be the number of individuals in the population times the probability of death. The actual number of deaths observed in a particular population would vary above or below this value at random.

Coin tossing is another example of a stochastic process. Suppose you repeatedly toss a set of 10 pennies. Although the probability of a head turning up on each toss is one-half, any one trial might turn up 6 heads, or it might produce only 3. When the test is repeated frequently enough, the average of the outcomes settles down to 5 heads, but many trials turn up 4 or 6 heads, somewhat fewer yield 3 or 7 heads, and runs with all heads occur once in 1,024 trials, on average.

Chance events exert their influence more forcefully in small populations than in large ones. If you repeatedly toss a set of 5 pennies, the probability of obtaining 5 heads in a trial is 1 in 32, compared with the smaller chance of 1 in 1,024 of obtaining 10 heads in a trial. Thus, if we visualize each individual in a population as a coin, and heads means death, a population of 5 individuals clearly has a higher probability of extinction, just by chance, than a population of 10 individuals.

A simple discrete-time population model will illustrate this point. Suppose that adults have a probability of 0.5 per year of successfully rearing a single offspring. A population of 10 individuals should therefore produce 5 offspring per year on average, but if birth is a stochastic process, the actual number in a particular year is likely to

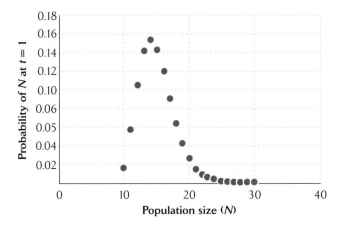

FIGURE 12.16 Stochastic population processes produce a probability distribution of population size. Plotted here are the probabilities of different population sizes (N) after one time interval in a population undergoing a pure birth process with initial size $N(0) = 10$, $B = 0.5$, and $\lambda = 1.5$. In a discrete-time model, the maximum value of N would be 20 individuals: the 10 present at time 0 plus 10 produced during the first time interval. This graph is based on a continuous-time model, in which young can also give birth and thus larger numbers are possible. After E. C. Pielou, *Mathematical Ecology*, Wiley, New York (1977).

diverge from that value. What will be the effect on population growth? Consider a simple birth process (no deaths) in which a population grows geometrically according to $N(t) = N(0)\lambda^t$. Now suppose that during each year, adults produce a single offspring with a probability of 0.5. Thus, $N(0) = 10$, $B = 0.5$, and $\lambda = 1.5$ (the 10 adults present at time 0 remain alive, plus each adult produces a single offspring with a probability of 0.5). Accordingly, a population of 10 individuals would increase to 15 individuals, on average, after one year, but could also remain at 10 (if no births occurred) or increase to as many as 20 (if all adults gave birth), just by chance (Figure 12.16).

Stochastic extinction of small populations

A population subject to stochastic birth and death processes is said to take a **random walk,** meaning that its numbers may increase or decrease strictly by chance. When the size of such a population does not respond to changes in density (a point we shall return to below), its ultimate fate is extinction, regardless of how its size might increase in the meantime.

Every population eventually faces a series of unlucky intervals with few births and many deaths. Theorists have

derived mathematical expressions for the probability that a population will die out within a time interval t. For mathematical convenience, these models use a continuous-time approach and are therefore based on the exponential growth rate ($r = b - d$). As you would expect, large populations persist longer than small ones, on average. Furthermore, extinction arrives sooner in populations with higher turnover rates.

In the simplest model of stochastic extinction, b and d are equal—that is, births balance deaths, and the average change in population size is zero. In this case, the probability that a population will become extinct by time t [$p_0(t)$] decreases with increasing population size (N) and increases with larger b (and d), which indicate more rapid population turnover. The probability of extinction also increases with time (t).

The relationship of the probability of extinction within time interval t to population size N is shown in Figure 12.17 for a population in which $b = d = 0.5$. These are reasonable values for adult mortality and recruitment in a population of terrestrial vertebrates. We see, for example, that for a population with 10 individuals, the probability of extinction is 0.16 within 10 years and 0.82 within 100 years; extinction becomes virtually certain (0.98) within 1,000 years. For a population with an initial size of

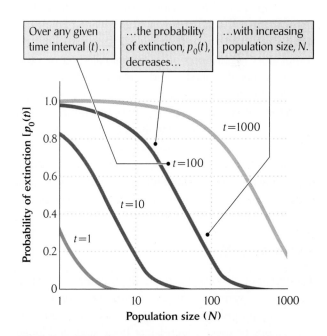

FIGURE 12.17 The probability of stochastic extinction increases over time (t), but decreases as a function of initial population size (N). In this example, birth rate and death rate equal 0.5. The functions are described by $p_0(t) = [bt/(1 + bt)]^N$.

1,000, the probability of extinction is about 0.13 within a millennium; it becomes virtually certain (0.999, not shown) within a million years. As a rule of thumb, the median time to extinction (when half of a set of populations with these characteristics would have died out) is $1.4N/b$.

Stochastic extinction with density dependence

Most stochastic extinction models do not include density-dependent changes in birth and death rates. In those that do, extinction becomes exceedingly rare except in the smallest populations because, as a population drops below its carrying capacity, birth rates typically increase and death rates decrease. This response to lower density greatly improves the probability that a declining small population will start to increase rather than decrease further. Accordingly, we should consider whether density-independent stochastic models are relevant to natural populations. The answer is that they are, for several reasons.

First, human land use patterns and habitat fragmentation are such that many species now exist as collections of exceedingly small subpopulations, often so isolated that their eventual demise cannot be prevented by immigration from other populations. Second, changing environmental conditions are likely to reduce the fecundity of populations trapped in isolated habitat patches and bring them closer to the abyss of extinction. Third, when endangered species compete for resources with other species, the competitors might appropriate additional food or other resources that the endangered species would otherwise gain because of their low density. In this case, high population densities of one species could depress small populations of other species, causing them to behave as though they were density *independent*. Finally, small populations sometimes exhibit positive rather than negative density dependence owing to inbreeding and difficulty locating mates (the Allee effect), so their numbers may decline even more rapidly.

DATA ANALYSIS MODULE 3 *Stochastic Extinction with Variable Population Growth Rates.* Variation in population size reduces the expected time to extinction of a population. You will find this module on page 264.

Size and extinction of natural populations

When populations dwindle to a small size, they become more susceptible to extinction, particularly on small islands. Populations on small islands are restricted geographically and are rarely augmented by immigration. In fact, extinction occurs so often on small islands that we can determine its probability from historical records. These data have confirmed the theoretical predictions of models of stochastic extinction. For example, species lists compiled in 1917 and in 1968 for birds on the Channel Islands off the coast of southern California revealed several extinctions of island populations during the 51-year interval between censuses. Seven of 10 species disappeared from Santa Barbara Island (3 km² in area), but only 6 of 36 species disappeared from the larger Santa Cruz Island (249 km²). (New colonists of different species replaced some of the species that went extinct on each island.) On an annual basis, these extinction figures can be expressed as 1.7% and 0.1% of the avifauna per year, respectively. In this case, extinction rate and island size were clearly inversely related.

Disappearances of populations from isolated islands dramatize the role of extinction of subpopulations in the dynamics of metapopulations. The extinction rate of subpopulations, which influences the number of patches occupied at equilibrium, depends on the number of individuals in a subpopulation and hence on the size of the patch it occupies. These considerations emphasize the interaction of spatial and temporal dynamics in population processes and remind us that we must understand the spatial structure of populations if we are to manage them intelligently.

SUMMARY

1. Most populations fluctuate in numbers, either in response to variations in the environment or because they have intrinsic properties that make them unstable. Populations of species with larger body sizes and longer life spans tend to respond less rapidly to changes in their environments.

2. The age structure of a population often indicates variations in recruitment over time. For example, seedlings of certain tree species tend to become established in forests primarily following a major disturbance.

3. Discrete-time models of density-dependent populations show that the sizes of some populations tend to oscillate when perturbed. The behavior of the population depends on the proportional rate of increase (R) per unit of time. For R between 0 and 1, the size of a population approaches equilibrium (K) without oscillation. When R

falls between 1 and 2, population size undergoes damped oscillations and eventually settles down to K. When R exceeds 2, oscillations in population size increase in amplitude until either a stable limit cycle is achieved or the population fluctuates irregularly (chaos).

4. Continuous-time models predict population cycles when there are time delays in density-dependent responses to changes in the environment. Such populations exhibit no oscillation when the product of the intrinsic rate of increase (r) and the time delay (τ) lies between 0 and $1/e$ (0.37), damped oscillations when $r\tau$ is less than $\pi/2$ (1.6), and limit cycles with a period of 4τ or more when $r\tau$ exceeds $\pi/2$.

5. Many laboratory populations exhibit oscillations that arise from time delays in the responses of individuals to population density. These time delays are related to the period of development and may be enhanced by storage of nutrients. In laboratory populations of sheep blowflies, A. J. Nicholson experimentally circumvented a time delay and was thereby able to eliminate population cycles.

6. Populations that are subdivided into discrete subpopulations occupying patches of suitable habitat are referred to as metapopulations. The dynamics of metapopulations depend not only on population growth and regulation processes within patches, but also on migration of individuals between patches. When the rate of extinction of subpopulations is small relative to the rate of colonization of unoccupied patches, a metapopulation exists as a changing mosaic of an equilibrium number of occupied patches.

7. Migration of individuals dispersing from large, productive subpopulations can rescue declining subpopulations from extinction. This rescue effect demonstrates the importance of maintaining connectivity between habitat patches in metapopulations.

8. The dynamics of small populations, such as a subpopulation in an individual habitat patch, depend to a large degree on chance events. Stochastic models demonstrate that the probability of extinction due to random fluctuations in population size is greater in smaller populations. The median time to extinction for small density-independent populations when birth rate (b) equals death rate is approximately $1.4N/b$, where N is the population size. Even small density-dependent populations, however, are vulnerable to extinction.

REVIEW QUESTIONS

1. What two characteristics of a species help determine the magnitude of fluctuations in population size?

2. In discrete-time models, what is the relationship between r and the stability of the population?

3. Using a continuous-time approach, contrast the stability of populations with a low intrinsic rate of growth and a short time delay and populations with a high intrinsic rate of growth and a long time delay.

4. What is the relationship between the amount of movement among subpopulations in a metapopulation and the synchrony of fluctuations of the subpopulation?

5. If you were trying to save an endangered species that lived in a metapopulation, how might you try to increase the proportion of occupied patches?

6. In the basic model of metapopulation dynamics, how might the rescue effect alter both the probability of colonization and the probability of extinction?

7. Why does the probability of extinction due to stochastic processes decline as population size increases? Why does it decline with low birth and death rates?

8. In models of stochastic extinction, why is extinction rarer in models that include density dependence than in models that exclude density dependence?

SUGGESTED READINGS

Beckerman, A., et al. 2002. Population dynamic consequences of delayed life-history effects. *Trends in Ecology and Evolution* 17:263–269.

Belovsky, G. 1987. Extinction models and mammalian persistence. In M. Soulé (ed.), *Viable Populations for Conservation*, pp. 35–57. Cambridge University Press, Cambridge and New York.

Berryman, A. 1996. What causes population cycles of forest Lepidoptera? *Trends in Ecology and Evolution* 11:28–32.

Berryman, A., and P. Turchin. 2001. Identifying the density-dependent structure underlying ecological time series. *Oikos* 92:265–270.

Brown, P. M., and R. Wu. 2005. Climate and disturbance forcing of episodic tree recruitment in a southwestern ponderosa pine landscape. *Ecology* 86:3030–3038.

Daniel, C. J., and J. H. Myers. 1995. Climate and outbreaks of the forest tent caterpillar. *Ecography* 18:353–362.

Ehrlich, P. R., and I. Hanski (eds.). 2004. *On the Wings of Check-erspots: A Model System for Population Biology.* Oxford University Press, Oxford.

Eriksson, O. 1996. Regional dynamics of plants: A review of evidence for remnant, source–sink and metapopulations. *Oikos* 77:248–258.

Gilbert, F., A. Gonzalez, and I. Evans-Freke. 1998. Corridors maintain species richness in the fragmented landscapes of a microecosystem. *Proceedings of the Royal Society of London* B 265:577–582.

Gotelli, N. J. 2008. *A Primer of Ecology,* 4th ed. Sinauer Associates, Sunderland, Mass.

Goulden, C. E., and L. L. Hornig. 1980. Population oscillations and energy reserves in planktonic Cladocera and their consequences to competition. *Proceedings of the National Academy of Sciences USA* 77:1716–1720.

Gutiérrez, R. J., and S. Harrison. 1996. Applying metapopulation theory to spotted owl management: A history and critique. In D. R. McCullough (ed.), *Metapopulations and Wildlife Conservation,* pp. 167–185. Island Press, Washington, D.C.

Hanski, I., and M. E. Gilpin (eds.). 1997. *Metapopulation Biology: Ecology, Genetics, and Evolution.* Academic Press, San Diego.

Hassell, M. P., J. H. Lawton, and R. M. May. 1976. Patterns of dynamical behaviour in single-species populations. *Journal of Animal Ecology* 45:471–486.

Husband, B. C. and S. C. H. Barrett. 1996. A metapopulation perspective in plant population biology. *Journal of Ecology* 84:461–469.

Keith, L. B. 1990. Dynamics of snowshoe hare populations. *Current Mammalogy* 2:119–195.

Kendall, B. E., et al. 1999. Why do populations cycle? A synthesis of statistical and mechanistic modeling approaches. *Ecology* 80:1789–1805.

Lindström, J., et al. 1995. The clockwork of Finnish tetraonid population dynamics. *Oikos* 74:185–194.

Lundberg, P., et al. 2000. Population variability in space and time. *Trends in Ecology and Evolution* 15:460–464.

Mennechez, G., N. Schtickzelle, and M. Baguette. 2003. Metapopulation dynamics of the bog fritillary butterfly: Comparison of demographic parameters and dispersal between a continuous and a highly fragmented landscape. *Landscape Ecology* 18:279–291.

Myers, J. H. 1993. Population outbreaks in forest Lepidoptera. *American Scientist* 81:240–251.

Nielsen, Ó. K., and G. Pétursson. 1995. Population fluctuations of gyrfalcon and rock ptarmigan: Analysis of export figures from Iceland. *Wildlife Biology* 1:65–71.

Pimm, S. L., H. L. Jones, and J. M. Diamond. 1988. On the risk of extinction. *American Naturalist* 132:757–785.

Pulliam, H. R. 1988. Sources, sinks, and population regulation. *American Naturalist* 132:652–661.

Ranta, E., V. Kaitala, and P. Lundberg. 1997. The spatial dimension in population fluctuations. *Science* 278:1621–1623.

Schtickzelle, N., et al. 2005. Metapopulation dynamics and conservation of the marsh fritillary butterfly: Population viability analysis and management options for a critically endangered species in Western Europe. *Biological Conservation* 126:569–581.

Schtickzelle, N., G. Mennechez, and M. Baguette. 2006. Dispersal depression with habitat fragmentation in the bog fritillary butterfly. *Ecology* 87:1057–1065.

Villard, M.-A., G. Merriam, and B. A. Maurer. 1995. Dynamics in subdivided populations of Neotropical migratory birds in a fragmented temperate forest. *Ecology* 76:27–40.

DATA ANALYSIS MODULE 3

Stochastic Extinction with Variable Population Growth Rates

The random nature of births, the number and sex of offspring, and particularly deaths can lead to variation in population size even in a constant environment. Such stochastic variation generally is not a problem for large populations because these chance events average out over many individuals. However, small populations can suffer from random variations in births and deaths, which can lead to random variation in population size and even to extinction.

The kakapo (*Strigops habroptilus*) is a large, flightless parrot found only in New Zealand. Because it is flightless, it is vulnerable to introduced predators such as cats, opossums, and weasels. By 1976, only 14 kakapos were known to be alive on New Zealand's South Island; sadly, all of them were males. If males and females were hatched with equal frequency on average, what is the probability that 14 individuals would include no females? If each individual is considered a trial, and being male is considered a success with probability (p) = 0.5, then the probability that n trials will all be successes is (p^n). If you

think this is unlikely, why might the New Zealand Wildlife Service have failed to locate any female kakapos? Fortunately, additional kakapos were later discovered on Stewart Island, at the southern tip of South Island, and kakapos were then introduced onto two small islands from which all predators had been removed. (For more on this fascinating bird, see http://en.wikipedia.org/wiki/Kakapo; http://animaldiversity.ummz.umich.edu/site/accounts/information/Strigops_habroptila.html.)

When the environment varies, all individuals in a population can be affected in the same way, and dramatic changes, even in large populations, can result. All members of a population feel a prolonged drought or a cold snap. Conversely, a period of exceptionally favorable conditions may increase the fecundity of all individuals or increase their probability of survival. The kakapo, for example, breeds primarily in years when the rimu tree, an endemic conifer of the podocarp group, produces good fruit crops. Variations in environmental conditions may be random and essentially unpredictable—what is referred to as *stochastic environmental variation*—or they may occur with some regularity. Understanding the connection between changes in the environment and changes in population

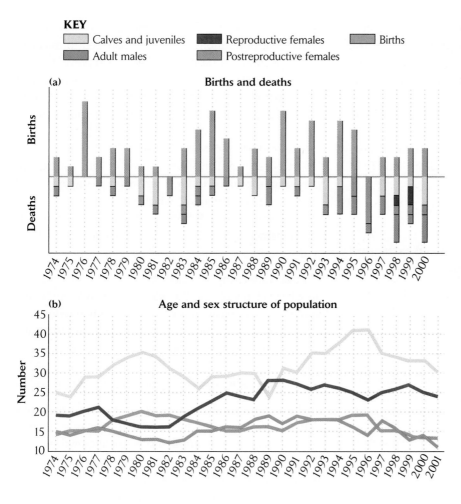

KEY

- ☐ Calves and juveniles
- ☐ Adult males
- ■ Reproductive females
- ☐ Postreproductive females
- ☐ Births

(a) Births and deaths

(b) Age and sex structure of population

DA FIGURE 3.1 Stochastic variation in births and deaths of a population of killer whales. Data from Taylor and Plater, 2001.

size can suggest interventions, such as supplemental feeding during critical periods of limited food supply, that can reduce the chance of population decline and extinction.

DA Figure 3.1 illustrates stochastic variation in births and deaths in a population of killer whales (orcas) (*Orcinus orca*) that reside off the coast of British Columbia, Canada. These data are reproduced in the Excel spreadsheet that accompanies this module as well. You can see that although the number of reproductive females in the population varied between 16 and 28 individuals, the number of young born each year varied between 0 and 8. Annual deaths varied between 0 and 7.

Step 1: Calculate the average rate of population change and its standard deviation.

Using the numbers of births and deaths and the initial population size of 73 individuals in 1974, calculate the changes in population size in the Excel spreadsheet through 2001.

The average exponential rate of population growth (*r*) during year *i* is calculated from the number of individuals at the beginning and at the end of the year as

$$r_i = \ln\left(\frac{N_{i+1}}{N_i}\right)$$

Remember that the exponential growth rate (*r*) is equal to the natural logarithm of the geometric growth rate λ (see page 225).

• Did the population increase or decrease over this period? Was the average exponential growth rate positive or negative?

To explore how stochastic environmental variation affects the size of a population and the probability of its extinction, we need to develop a model based on random changes in the factors that determine changes in population size. Such models often include an upper limit to the size of the population to incorporate the effect of density dependence. However, in the simplest case, the exponential rate of increase of a population is independent of its size (density-independent) and has a mean of *r* and a standard deviation of *S*. Thus, on average, the population grows at an exponential rate of *r*, but during some periods the growth rate is above this level, and during other periods it is below this level. If a population experiences many periods of below-average growth, it runs the risk of extinction. The critical parameter that influences the average time to extinction under random environmental variation is the variance in r.

The variance is the square of the standard deviation, or S^2. In the special case in which population size is, on average, balanced ($r = 0$; that is, births equal deaths under average conditions), the average time to extinction (T) of a population of size N is

$$T(N) = \frac{2}{S^2}\ln(1 + S^2N) + 1$$

Step 2: Estimate the average time until extinction for a population of killer whales.

Assume that for our killer whale population, the average growth rate (r) is 0.

• Starting with the population size in 2001, what is your estimate of the time to extinction?

• How does $T(N)$ change with the size of the initial population and with the variance in the rate of change in population size?

• Fill in the expected times to extinction for the range of population sizes (N) and standard deviations of the population growth rate (S) in DA Table 3.1.

Assuming that the time units are years, these values suggest that small populations in particular have relatively short life expectancies.

• What would $T(N)$ be for the killer whale population at its largest and smallest sizes? If a population grows just by chance, does this mean that its prospects for long-term persistence improve? Assume that the sample standard deviation of r in the spreadsheet accurately estimates the underlying value of S.

DA TABLE 3.1	Calculating time to extinction			
	Initial population size			
S	10	100	1,000	10,000
0.05				
0.1				
0.2				
0.5				

The implication of this model for conservation is clearly that populations should be managed to maintain as large a population size as possible and to prevent strong depression in the population growth rate during periods of poor environmental conditions. The latter strategy might involve supplemental feeding or predator and pathogen control programs at critical times.

The model described here lacks density dependence. Normally, ecologists believe that the growth potential of populations reduced to low numbers is greatly increased during normal conditions, which should allow them to increase and draw back from the brink of extinction. This is a fundamental message of the logistic equation and one of the most basic foundations of ecology.

• If this were always the case, why should we be worried about small populations? Under what conditions might you expect a population not to increase when reduced to low numbers? This certainly has been the case for many endangered species that have become extinct or now teeter on the brink of extinction. Do some populations simply not "have what it takes" to maintain healthy numbers?

Let's consider the effects of adding normal density dependence to models incorporating stochastic environmental variation the intrinsic exponential growth rate, r_0. According to the logistic equation, the average growth rate of populations reduced below their usual carrying capacity always exceeds $r = 0$, and such populations tend to recover quickly. But a long series of unfavorable periods might still be enough to drive population size below 1 individual and cause extinction. The important parameters for predicting the average time to extinction in models with density dependence are, first, the product of the average value of r_0 and K, and second, the ratio of the standard deviation of r_0 to its mean; in other words, S/r_0. The equations for time to extinction under density dependence are messy, but as you would expect, the addition of negative density dependence greatly increases the expected time to extinction. For example, when $N = 100$, $r_0 = 0.1$, and $S = 0.22$; $T(100)$ is equal to nearly 26,000 time units, rather than the value of about 81 (see DA Table 3.1) in the absence of density dependence.

Literature Cited

Taylor, M., and B. Plater. 2001. Population viability analysis for the southern resident population of the killer whale (*Orcinus orca*). The Center for Biological Diversity, Tucson, Arizona (http://www.biologicaldiversity.org/swcbd/species/orca/pva.pdf).

CHAPTER 13

Population Genetics

Wolves have been persecuted by humans everywhere, and the last individuals on the Scandinavian Peninsula in northern Europe had disappeared by the 1960s. In 1983, a single new breeding wolf pack was discovered in the region over 900 km from the nearest known packs in Finland and Russia. The new Scandinavian population had only one breeding pair at any one time and did not exceed 10 individuals until 1991, when the number of wolves began to grow exponentially, reaching a size of 9–10 packs with 90–100 individuals by 2001. The sudden increase was unrelated to changes in climate, habitat, prey abundance, or human protection. What happened to the population?

Genetic studies have shown that the population was founded by a single pair of unrelated wolves that migrated to the Scandinavian Peninsula from an unknown region to the east. All the individuals in the original pack were descended from this one pair. Because the pack consisted entirely of close relatives, its genetic diversity was very low. Beginning in 1991, however, pups were born that carried new genetic variation, signaling the addition of a new breeding male from outside the pack. Reproductive success soared, and the population started its steady climb. Evidently, the infusion of new genes into the pack by a single immigrant male created new gene combinations among the offspring and greatly increased the number successfully reared.

CHAPTER CONCEPTS

- The ultimate source of genetic variation is mutation

- Genetic markers can be used to study population processes

- Genetic variation is maintained by mutation, migration, and environmental variation

- The Hardy–Weinberg law describes the frequencies of alleles and genotypes in ideal populations

- Inbreeding reduces the frequency of heterozygotes in a population

- Genetic drift in small populations causes loss of genetic variation

- Population growth and decline leave different genetic traces

- Loss of variation by genetic drift is balanced by mutation and migration

- Selection in spatially variable environments can differentiate populations genetically

With growing human populations and increasing fragmentation of natural habitats, stories like that of the disappearance of the Scandinavian Peninsula wolf population are becoming all too common. Against this somewhat gloomy backdrop, the subsequent natural recovery of the wolf population casts a ray of hope. The story also emphasizes the importance of genetic factors to the overall health and productivity of natural populations, and it demonstrates the use of genetic markers in reconstructing the history of the population and providing a plausible explanation for its sudden good fortune. Without the ability to analyze the genetic variation harbored by individuals and genetic differences between individuals, it would have been difficult for researchers to identify inbreeding depression as a potential limit to the new Scandinavian wolf population or to identify the arrival of new genetic variation from outside the population as an agent of rescue. Indeed, genetic analyses are finding new applications to understanding the geographic structure of natural populations, mating relationships, movements of individuals within populations, and histories of change in population size. In this chapter, we will explore some basic aspects of genetic variation in natural populations and see how this variation is being increasingly used in ecological studies.

The ultimate source of genetic variation is mutation

Ever since biologists figured out the nature of genetic inheritance, they have recognized that all populations contain variations in the genetic material. No two individuals are alike genetically unless, like identical twins, they are derived from the first cell division of a zygote, or unless they are members of an asexually produced clone. Many obvious differences among individuals, such as eye color and blood type in humans, result from genetic factors. Such visible differences in the phenotypes of individuals were the first such variations identified by biologists. Increasingly sophisticated techniques have uncovered much more genetic variation. Protein electrophoresis, a technique that separates proteins having different electrical charges, can reveal amino acid differences among proteins, particularly enzymes. When this technique was first applied to natural populations of *Drosophila* fruit flies in the 1960s, geneticists were astonished to find that the amino acid sequences of one-third of the many metabolic enzymes in a *Drosophila* cell varied within a population— that is, the enzymes were **polymorphic.** Surveys of the entire human genome, made possible recently by DNA sequencing technology, have shown that variation is the

rule in the DNA of our genes as well. Whereas evolutionary biologists had long considered natural selection to be a purifying process that removed deleterious genetic variation from a population, they are now confronted with explaining the abundant genetic diversity in most populations.

The structure of DNA

You learned in your introductory biology course that genetic information is contained in the molecule **deoxyribonucleic acid, or DNA** for short, and that genetic variation is caused by changes in the DNA molecule. DNA has four kinds of subunits, which are called nucleotides: adenine (A), thymine (T), cytosine (C), and guanine (G). Genetic information is encoded in the particular order of the different nucleotides, just as the order of letters in a word conveys information.

A DNA strand serves as a template from which a cell manufactures other nucleic acids and proteins. Proteins are unbranched chains composed of up to twenty different amino acids. Each amino acid is encoded by one or more specific sequences of three nucleotides called **codons.** All the codons that specify the amino acid sequence for a single protein, along with any sequences that regulate its expression, make up the gene for that protein.

Mutations

The sequences of nucleotides in DNA, as we saw in Chapter 6, are subject to errors. The most common of these errors are nucleotide substitutions, but deletions, additions, and rearrangements of nucleotides also occur. These mistakes can result from random copying errors when the genetic material replicates during cell division, or they can be caused by certain highly reactive chemical agents or ionizing radiation. Our genomes also contain unusual genetic entities such as transposons, which are segments of DNA that can replicate themselves and insert copies into new positions within the genome of their host. The origins of transposons, and whether they serve any purpose for the host organism, are unknown, but they can disrupt normal genetic function.

A substitution of one of the nucleotides in a DNA codon can change the amino acid that it specifies. Consider this sequence of DNA nucleotides and the corresponding amino acids:

DNA:	GAA	TGG	CGA	GAA	ATA	GGG
Amino acid:	Leucine	Serine	Alanine	Leucine	Tyrosine	Proline

If the guanine occupying the eighth position were changed to thymine, the third codon would be altered from CGA to CTA, and it would now encode the amino acid aspartine instead of alanine:

DNA:	GAA	TGG	CTA	GAA	ATA	GGG
Amino acid:	Leucine	Serine	Aspartine	Leucine	Tyrosine	Proline

Such changes are called **point mutations.**

Several codons can specify the same amino acid: for example, both GAA and GAG code for leucine. Because of this redundancy in the genetic code, some nucleotide substitutions leave the amino acid specified by a codon unchanged, and thus have no effect on the phenotype. If the first codon in our example were changed from GAA to GAG, it would still specify leucine. Such changes are called **silent mutations,** because they are not "heard from," or *neutral mutations,* because they have no consequence for fitness, or *synonymous mutations,* because they have the same meaning for the protein.

When a mutation results in an amino acid substitution, the new protein produced by the mutant gene may display different properties, which might be beneficial to the individual, but are much more likely to be harmful. Mutations of the gene encoding the beta-chain of hemoglobin—the principal oxygen-binding molecule in the bloodstream of vertebrates, including ourselves—are a case in point. Most mutations in this gene produce hemoglobins with impaired ability to bind oxygen, and the affected individual suffers from anemia, or lack of oxygen. The disease sickle-cell anemia, for example, is caused by a mutation that changes the sixth amino acid in beta-hemoglobin from glutamic acid to valine. As a result, the structure of hemoglobin molecules changes such that when they release oxygen from the red cells in the bloodstream, they become stacked close together in long, fibrous helices, which gives the red blood cells a peculiar sickle-like shape—hence the name of the disease (Figure 13.1). The severely distorted red blood cells of individuals with two copies of the mutant allele block blood flow in the capillaries, starving tissues of oxygen and causing severe and debilitating anemia.

Alleles

Alternative forms of the same gene, such as the two forms of the beta-hemoglobin gene, are known as alleles. By alternative forms, we mean DNA sequences that differ by one or more nucleotide substitutions. Alleles may repre-

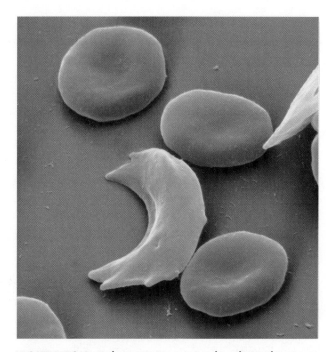

FIGURE 13.1 A change in just one nucleotide can have dramatic phenotypic effects. A single-nucleotide mutation in the gene encoding the human beta-hemoglobin molecule changes the structure of that molecule, causing red blood cells to assume a sickle-like shape. Photo by Meckes/Ottawa/Photo Researchers, in A. J. Griffiths et al., *Introduction to Genetic Analysis,* 7th ed., W. H. Freeman and Company, New York (2000).

sent neutral variation, recognizable only at the nucleotide sequence level, or they may be expressed in the phenotype of the individual. Blood types in humans, for example, result from different alleles of a gene that encodes molecules on the surfaces of red blood cells, as described in Chapter 6.

Each gene can have many alleles. Because each diploid individual has two copies of each gene, one inherited from its mother and one from its father, an individual can be homozygous (have two copies of the same allele) or heterozygous (have two different alleles) for any particular gene. When a population's gene pool includes more than two alleles of a particular gene, the individual carries only a portion of the total variation in the population.

Genetic markers can be used to study population processes

Only a small portion of the DNA in an individual's genome consists of protein-coding genes. Some DNA sequences encode RNA, and others serve as sites for the attachment of molecules that regulate the expression of a gene. Much of the DNA in the genome, however, has no known purpose. Strikingly, much of this DNA consists of identical sequences of nucleotides repeated over and over. One type of repeat, called a **microsatellite,** is a tandem repeat of sequences of two, three, or four nucleotides.

Mutations occur in this silent, "nongenetic" DNA as well. These mutations may not have any apparent effect on the organism, but the variation they create has turned out to be very useful in scientific studies, including studies of populations. One especially useful type of variation is present in repetitive sequences, such as microsatellites and the related **minisatellites:** these sequences may vary from individual to individual in the number of repeats they contain, creating a range of alleles of different sizes.

This wonderful genetic diversity provides a unique opportunity to study population processes that influence the pattern of genetic variation. Changes in population size, movements of individuals between populations, and mating between related individuals all leave characteristic traces in patterns of genetic variation. Researchers can use many different kinds of genetic variation, referred to as **genetic markers,** to study the dispersal of individuals within and between subpopulations in a metapopulation, mating patterns and pedigrees, and the history of changes in population size and distribution, as well as the results of evolution by natural selection.

Before the advent of modern methods for visualizing genetic markers, population processes could be studied only by tracking the movements of individuals and determining the pedigrees of individuals within populations. Such studies required marking and intensive observation of individuals. These methods, although still important, are now complemented by analyses of genetic markers.

Although any variation that has a genetic basis can serve as a genetic marker, most markers used in present-day research are variations in individual nucleotides at particular points in the DNA sequence. These markers may be located in protein-coding genes or elsewhere. In practice, most genetic markers are located in nongenetic DNA because mutations in that DNA are not removed from the gene pool by selection and can therefore become common in a population. A variety of techniques are used to detect these polymorphisms, from direct sequencing of the DNA to several indirect approaches, such as electrophoresis.

Direct sequencing of DNA allows researchers to compare the nucleotide sequences of different individuals in a population and identify polymorphisms. In most applications, millions of copies of a particular sequence of DNA are produced by the polymerase chain reaction (PCR) technique, and the copies are then sequenced by a process that makes use of a particular bacterium's DNA replication machinery. PCR is also used to copy other DNA markers that vary in length and can be characterized by electrophoresis without sequencing, such as the microsatellite repeats described above. Regardless of the type of genetic marker used, all of these approaches can provide useful characterizations of genetic diversity within populations.

Genetic variation is maintained by mutation, migration, and environmental variation

Natural selection cannot produce evolutionary change without genetic variation. Yet natural selection tends to reduce genetic variation by eliminating less fit individuals, and the alleles they carry, from the gene pool. How is genetic variation maintained in a population under these circumstances? Much genetic variation is beyond the reach of natural selection because it has no consequence for the fitness of the individual. The maintenance of such neutral variation in a population from generation to generation depends entirely on chance, as we shall see below. In addition, every population is supplied with new genetic variation by mutation and immigration. Furthermore, natural selection can itself maintain genetic variation when environmental conditions are variable or when individuals that are heterozygous for certain genes have superior fitness.

Mutation rates

For any particular nucleotide in a DNA sequence, the rate of mutation in eukaryotic organisms is extremely low, from roughly 1 in 100 million (10^{-8}) per generation in the mitochondrial DNA of birds to as little as 1 in 10 billion (10^{-10}) in nuclear DNA. These low rates, however, when multiplied by the hundreds or thousands of nucleotides in a gene, and by the trillion or so nucleotides in such complex organisms as vertebrates, mean that each individual is likely to sustain a new mutation in some part of its genome. Measured rates of mutation in expressed (visible) genes average about 1 in 100,000 (10^{-5}) to 1 in 1 million (10^{-6}) per gene per generation, with rates for some genes being much higher. For example, mutations with readily visible effects, such as changes in the color of kernels on an ear of corn (Figure 13.2) or defects in the wing structure of fruit flies, occur at rates of 1 in 10,000 (10^{-4}) to 1 in 100,000 gametes per generation. In plants,

FIGURE 13.2 Mutations are a source of genetic variation. Mutations with readily visible phenotypic effects, such as changes in the color of kernels of corn, have been used extensively for analyses of inheritance. Photo by Gregory G. Dimijian/Photo Researchers.

chlorophyll deficiencies, which are lethal because plants cannot assimilate carbon without photosynthesis, appear by mutation at rates of almost 1 in 100 (10^{-2}) to 1 in 10,000 of the many genes required for producing chlorophyll. In laboratory strains of fruit flies (*Drosophila*), water fleas (*Daphnia*), and mouse-ear cress (*Arabidopsis*), mutations with lethal effects arise in the genomes of about 2% of individuals each generation, and mutations with mildly detrimental effects on survival and reproduction arise at a rate of about one per individual genome. There is no question that mutation is actively churning out genetic variation.

Migration

Migration of individuals within a population or between populations can affect genetic variation in two ways. On one hand, high migration rates integrate populations into larger units, which tend to retain genetic variation just because of their size. On the other hand, movement of individuals between habitats with different environmental conditions can mix genes that have been selected under those different conditions and increase genetic variation within the population, both locally and as a whole. When different alleles of a gene are favored in different environments, both can be maintained locally and spread through the population with dispersing individuals, as we shall see below.

Environmental variation and frequency-dependent selection

Spatial and temporal variation in the environment can maintain genetic variation as well. When different alleles

are favored at different times because of environmental variation, heterozygotes may have greater fitness than any homozygous genotype. Suppose, for example, that the environment varies in such a way that individuals having one allele of the A gene, A_1, are favored in some years, and individuals with another allele, A_2, are favored in other years. Because heterozygotes have both alleles, their phenotype is likely to be superior to that of either homozygote in such a varying environment (Figure 13.3).

One of the most remarkable cases of such heterozygote superiority involves the sickle-cell allele (S) of the beta-hemoglobin molecule in humans. When homozygous (SS), this gene produces sickle-cell anemia, a debilitating disease (described above) that often leads to an early death. Yet, in some parts of tropical Africa, the frequency of the S allele may reach 20% of the gene pool. The reason for this is that in the heterozygous state (AS), the sickle-cell allele confers protection against malaria. Where malaria is prevalent and virulent, the fitness of heterozygous individuals may be 25% greater than the fitness of individuals who are homozygous for the normal allele (AA) and are susceptible to infection by malaria parasites. This heterozygous advantage is a case of **frequency-dependent selection** because the fitness of each allele depends on its frequency in the population. Another important set of genes in vertebrates for which genetic variation is maintained at high levels is the major histocompatibility complex (MHC), which controls aspects of cell and pathogen recognition and thus is an important part of the organism's defense against disease.

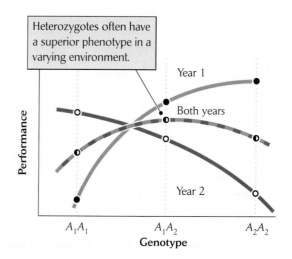

FIGURE 13.3 In environments that favor different alleles in different years, the heterozygote may be more fit than either homozygote. Such environmental variation maintains genetic variation within a population.

Most genetic variation within populations has less dramatic fitness consequences than these examples—in most cases, none at all. This variation is controlled less by natural selection than by random processes arising from variations in birth and death rates and in the gametes that are lucky enough to form future generations.

The Hardy–Weinberg law describes the frequencies of alleles and genotypes in ideal populations

The frequencies of alleles and genotypes in a population provide clues to the processes that have shaped genetic variation within that population. One of the most important tools in population genetics is the Hardy–Weinberg law, which describes expected allele and genotype frequencies in an ideal population (one in which genetic variation remains unchanged over time). The Hardy–Weinberg law is a great source of insight because, by comparing the ideal population it describes with real populations, researchers can find evidence of processes that have increased or decreased genetic variation.

We have already described some processes that act to increase genetic variation: mutation, migration, and even, in some circumstances, natural selection. Through the lens of the Hardy–Weinberg law, we will now examine some processes that decrease genetic variation. These processes include natural selection, of course, but they also include inbreeding and the loss of genetic variation through stochastic processes. Then we will see how processes that increase and decrease genetic variation may be balanced within the same population.

Allele frequencies and genotype frequencies

The **frequency** of an allele in a population is simply the number of copies of that allele divided by the total number of copies of that gene in the population. If individuals are diploid, then each individual has two copies of each nuclear gene, and there are $2N$ copies of the gene in a population, where N is the population size.

The **genotype** of an individual is the set of alleles that it bears. For diploid organisms, the genotype consists of the two alleles for each nuclear gene. In the case of haploid organisms or haploid portions of the genome (gametes, mitochondria, chloroplasts), the genotype is referred to as a **haplotype.** Returning to the diploid case, suppose that a gene for hair color has alleles A_1 and A_2. There are three possible genotypes: the A_1 homozygote (A_1A_1), the

A_2 homozygote (A_2A_2), and the heterozygote (A_1A_2). The frequency of a genotype in a population is simply the number of those genotypes divided by the total number of individuals in the population. Suppose that a population contains 48 individuals with A_1A_1 genotypes, 36 with A_1A_2 genotypes, and 16 with A_2A_2 genotypes. The total number of individuals in the population is $48 + 36 + 16 = 100$. The frequencies of the three genotypes are $48/100 = 0.48$, $36/100 = 0.36$, and $16/100 = 0.16$. Note that the genotype frequencies must add to 1.00.

Allele frequencies can be calculated from genotype frequencies. In the above population, the number of A_1 alleles is $48 \times 2 = 96$ in A_1A_1 homozygotes plus 36×1 in A_1A_2 heterozygotes $= 132$. The number of A_2 alleles is 16×2 in A_2A_2 homozygotes plus 36×1 in A_1A_2 heterozygotes $= 68$. The total number of alleles in the population is 100 individuals $\times$ 2 alleles per individual $= 200$. Therefore, the frequency of the A_1 allele is $132/200 = 0.66$, and the frequency of the A_2 allele is $66/200 = 0.34$.

The Hardy–Weinberg equilibrium

The **Hardy–Weinberg law** states that the frequencies of alleles and genotypes will remain constant from generation to generation in a population if that population has (1) a large (infinite) number of individuals, (2) random mating, (3) no natural selection, (4) no mutation, and (5) no migration between populations. In other words, the process of sexual reproduction, by itself, produces no evolutionary change: allele frequencies and genotype frequencies stay the same from generation to generation. This law, named for the two geneticists who independently described it in 1908, shows that changes in allele or genotype frequencies can result only from the action of additional processes on the gene pool of a population.

In an ideal population that exists in **Hardy–Weinberg equilibrium (HWE),** the proportions of homozygotes and heterozygotes take on equilibrium values, which we can calculate from the frequencies of each allele in a population. Consider again gene A, which has two alleles, A_1 and A_2, which occur in proportions p and q ($p + q = 1$, and therefore $q = 1 - p$). At Hardy–Weinberg equilibrium, the three genotypes that can result will occur in the following proportions:

Genotype:	A_1A_1	A_1A_2	A_2A_2
Frequency:	p^2	$2pq$	q^2

Notice that $p^2 + 2pq + q^2 = 1$: the proportions of all the genotypes in the population add to 1.

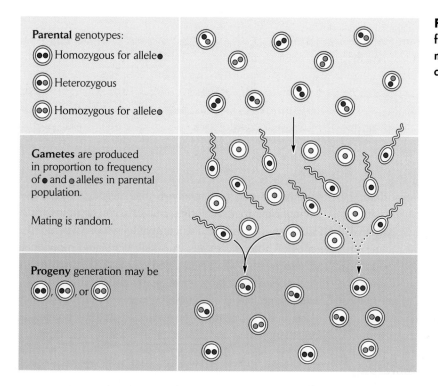

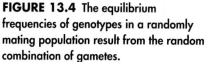

FIGURE 13.4 The equilibrium frequencies of genotypes in a randomly mating population result from the random combination of gametes.

These proportions come from the probabilities that each type of zygote will be formed from the random combination of any two gametes (Figure 13.4). To form an A_1A_1 homozygote, both gametes must have an A_1 allele. When the probability of drawing one A_1 allele at random is p, the probability of drawing two A_1 alleles together is simply the probability of each multiplied by the other, $p \times p$, or p^2. The proportion of heterozygotes is $2pq$ because an A_1A_2 heterozygote will result from an A_1 egg and an A_2 sperm, with probability $p \times q$, and from an A_2 egg and an A_1 sperm, with probability $p \times q$.

This reasoning applies to gametes drawn at random from a pool that reflects the allele frequencies in a population. Fair enough for sea urchins that broadcast their eggs and sperm directly into the water column to mix willy-nilly. What happens when fertilization takes place between the eggs and sperm of paired individuals, none of which individually represents the entire gene pool of the population? When fertilization is internal, each combination of genotypes in a mating pair will produce unique proportions of offspring genotypes. Nonetheless, as long as mating is random, the Hardy–Weinberg frequencies still pertain to the population as a whole. One of the remarkable accomplishments of Hardy and Weinberg was to show that random mating among individuals (as well as among gametes) produces the HWE values in a single generation.

We can calculate the numerical values of the proportions of genotypes under Hardy–Weinberg equilibrium as in the following example. Suppose one allele (A_1) occurs in a population with a frequency of 0.7, and the other (A_2) with a frequency of 0.3 (0.7 + 0.3 = 1). Accordingly, 49% ($0.7^2 = 0.49$) of the genotypes in the population will be A_1 homozygotes, 42% ($2 \times 0.7 \times 0.3 = 0.42$) will be heterozygotes, and 9% ($0.3^2 = 0.09$) will be A_2 homozygotes. Notice that 0.49 + 0.42 + 0.09 = 1.

LIVING GRAPHS To access an interactive tutorial on the Hardy–Weinberg equation, go to http://www.whfreeman.com/ricklefs6e.

Inbreeding reduces the frequency of heterozygotes in a population

Violation of any of the assumptions of the Hardy–Weinberg law can result in genotype frequencies that differ from the Hardy–Weinberg equilibrium. Not surprisingly, essentially all natural populations violate those assumptions. However, the Hardy–Weinberg equilibrium is relatively insensitive to departures from the assumptions of infinite population size, no mutation, no natural selection, and no migration. It is much more sensitive to departures from random mating.

Assortative mating occurs when individuals choose mates nonrandomly with respect to their own genotypes. Mating of like with like is referred to as **positive assortative mating.** Mating with unlike partners is referred

FIGURE 13.5 Inbreeding decreases the frequency of heterozygotes in a population. Selfing—the most extreme form of inbreeding—decreases the frequency of heterozygotes by one-half in each generation. Note, however, that inbreeding does not change the frequencies of alleles.

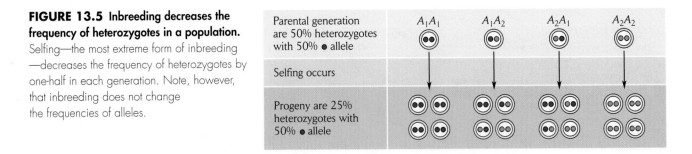

to as **negative assortative mating.** Assortative mating does not directly change the frequencies of alleles within populations, only the frequencies of genotypes. Negative assortative mating—for example, preferential mating of A_1A_1 homozygotes with A_2A_2 homozygotes, producing A_1A_2 offspring—increases the proportion of heterozygotes in a population at the expense of homozygotes, but the proportions of A_1 and A_2 alleles do not change. Positive assortative mating has the opposite effect of reducing the proportion of heterozygotes. So does mating with close relatives, which is referred to as **inbreeding** (Figure 13.5).

The most extreme form of inbreeding is selfing—mating with oneself or with an individual with an identical genotype. Selfing is found primarily among hermaphroditic plants as well as an odd assortment of simple animals. When selfing occurs, homozygous individuals produce only homozygous offspring, but the offspring of heterozygous individuals are half heterozygous and half homozygous. Thus, the frequency of heterozygotes decreases by one-half each generation. Under these circumstances, it does not take long for heterozygotes to disappear from a population. Mating between close relatives, such as siblings or cousins, has the same effect of reducing the frequency of heterozygotes with each generation—it just takes longer.

The high proportion of homozygotes produced by inbreeding can lead to the expression of deleterious recessive alleles, resulting in a reduction in offspring fitness known as **inbreeding depression.** Many plant species are self-compatible, so it is possible to demonstrate this effect by self-fertilizing entire plant populations experimentally to achieve a desired level of inbreeding. In one study, selfed populations of the monkey flower (*Mimulus guttatus*) showed a progressive reduction in ovule number and in male fertility over five generations (Figure 13.6). Presumably, an increasing number of deleterious recessive alleles were exposed as the proportion of homozygous genotypes increased with each generation.

Most species employ mechanisms to reduce the occurrence of inbreeding, including dispersal of progeny,

recognition of close relatives, and negative assortative mating. Mammals, including humans, can distinguish differences in the major histocompatibility complex (MHC) genes of potential mates by smell. Unrelated individuals are less likely to share MHC genes than are close relatives. Hermaphroditic plants have many mechanisms to prevent self-pollination, including genetic self-incompatibility (described in Chapter 8), temporal separation of male and female functions, and elaborate flower structures designed to make selfing difficult.

The inbreeding coefficient

One way to quantify the loss of heterozygosity resulting from inbreeding in a population is the **inbreeding coefficient (F),** which describes the departure of the observed frequency of heterozygotes (H_{obs}) from the Hardy–Weinberg equilibrium values (H_{HWE}). The inbreeding coefficient, which is directly related to homozygosity, is calculated by

$$F = \frac{H_{HWE} - H_{obs}}{H_{HWE}}$$

For example, when the frequency of the A_1 allele is $p = 0.6$ ($1 - p = 0.4$), the Hardy–Weinberg equilibrium frequency of heterozygotes is $2p \times (1 - p) = 0.48$. If the observed frequency of heterozygotes is 0.36, then $F = (0.48 - 0.36)/0.48 = 0.25$. When $H_{obs} = H_{HWE}$, $F = 0$. When a population lacks heterozygotes, $F = H_{HWE}/H_{HWE} = 1$. Because the proportion of heterozygotes in a population of selfing individuals decreases by one-half each generation, the inbreeding coefficient after a single generation of selfing would increase from $F = 0$ to $F = 0.5$. Eventually, F would increase to 1 as the proportion of heterozygotes dwindled until all individuals were homozygous.

In addition to quantifying the effect of inbreeding on the proportion of heterozygotes in a population, F is also the probability that the two copies of a gene in an individual's genotype are **identical by descent,** meaning that they were derived from the same ancestral copy. Some

(a)

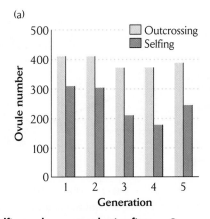

(b)

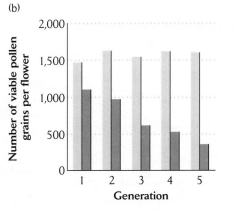

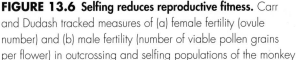

FIGURE 13.6 Selfing reduces reproductive fitness. Carr and Dudash tracked measures of (a) female fertility (ovule number) and (b) male fertility (number of viable pollen grains per flower) in outcrossing and selfing populations of the monkey flower (*Mimulus guttatus*) over five generations. Both measures decreased steadily in the selfing population, but not in the outcrossing population. After D. E. Carr and M. R. Dudash, *Evolution* 51:1797–1807 (1997).

convenient benchmarks for the inbreeding coefficients of individuals include the offspring of matings among siblings (*F* = 0.25) and of matings among first cousins (*F* = 0.125).

It can be shown that in the absence of mutation or migration, the value of *F* in a population increases continually through stochastic processes, even with completely random mating, until *F* = 1 (all individuals are homozygous for the same allele). When all alleles but one are lost from a population, the remaining allele is said to be **fixed.** Thus, *F* is also a measure of allele fixation in a population and is therefore called the **fixation index.**

Inbreeding and fitness in natural populations

In most populations, the inbreeding coefficient of an individual can be calculated from its *pedigree*—the history of its ancestry. In large populations in which outcrossing is the rule, the average *F* for an individual across all its genes tends to be close to zero. In small populations in which opportunities for mating are limited, the average *F* can increase to a significant level. For example, in the small population of song sparrows (*Melospiza melodia*) on Mandarte Island, British Columbia (described in Chapter 11), Lukas Keller calculated that the average *F* of the offspring of 671 mated pairs was 0.031, and that the offspring of 51 of these pairs had an average *F* exceeding 0.125, the level of first-cousin matings. Keller was able to show that the survival of young decreased dramatically with increasing *F*. Compared with completely outbred young (*F* = 0), offspring with *F* = 0.031 survived 8% less well, and offspring with *F* = 0.125 survived 29% less well.

A similar study focused on the population of gray wolves (*Canis lupus*) described at the opening of this chapter. The pedigree of every wolf in the growing population was reconstructed by genotype analysis, and the inbreeding coefficient of the pups produced in the first litter of each breeding pair was calculated from their pedigree. *F* varied from 0.00 to 0.41 (very nearly the level for selfing). The number of surviving pups per litter decreased dramatically with increasing *F* (Figure 13.7). From this relationship, the researchers concluded that the gene pool of that wolf population contained deleterious genetic variation equivalent to six lethal recessive genes per individual genome. Considering that populations contain so much potentially harmful genetic variation, selection for mechanisms to avoid inbreeding must be very strong. In small populations, however, choice of mates is limited, and some degree of inbreeding is inevitable.

ECOLOGISTS IN THE FIELD | **Inbreeding depression and selective abortion in plants.** *Banksia spinulosa* is an Australian shrub that is pollinated by small nectar-feeding birds. *Banksia* plants can self-pollinate, but they normally outcross. Each inflorescence has about 800 flowers, but normally produces fewer than 50 fruits, each containing 0, 1, or 2 seeds. Thus, most of the developing fruits in each inflorescence are aborted. The pollinated flowers compete for the resources needed to develop into a fruit, and there are too few resources to go around. Fruit production appears to be resource-limited rather than pollen-limited, because removal of one-third of the flowers from either the base or the top of an inflorescence does not significantly depress fruit production.

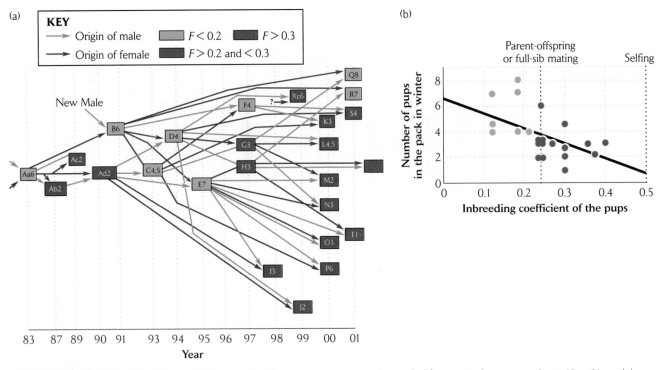

FIGURE 13.7 Inbreeding depresses reproductive fitness in a newly founded population of wolves. (a) This pedigree shows the high level of inbreeding in the small population of wolves on the Scandinavian Peninsula, which was founded by only three individuals. Each box indicates a breeding pair; the lines trace back to the pairs that each parent came from. Notice that in 1987 and 1989, all the individuals in the population were descended from a single outcrossed pair ($F = 0$), and that mating therefore occurred between siblings (pairs Ab2 and Ac2; $F = 0.25$). Pair Ad2 represents a mother–son mating ($F = 0.375$); pairs D4 and C4,5 represent first-cousin matings ($F = 0.125$); and so on. (b) The number of pups surviving to winter decreased dramatically as the inbreeding coefficient of a litter increased. After O. Liberg et al., *Biology Letters* 1:17–20 (2005).

To determine whether a *Banksia* plant can distinguish an ovule pollinated by its own pollen from an ovule pollinated by that of another plant, Australian botanists Glenda Vaughton and Susan Carthew hand-pollinated *Banksia* inflorescences with pollen obtained either from the same plant (selfed pollen) or from neighboring plants (outcrossed pollen). In some of the plants, they pollinated half the inflorescence with selfed pollen and half with outcrossed pollen (mixed pollination). After fruits had developed, they counted the fruits and seeds on each half of each inflorescence.

Compared with cross-pollination, self-pollination reduced the number of seeds produced by 38% (24 versus 39 seeds per half-inflorescence) and increased the proportion of developed fruits with aborted seeds from 8% to 16% (Figure 13.8). These results clearly indicate inbreeding depression. When one-half of an inflorescence was cross-pollinated and the other half was self-pollinated, the number of seeds produced per self-pollinated half dropped further to 14, and 28% of the fruits aborted their seeds. This experiment shows that self-pollinated ovules, which are likely to have inferior genotypes, do not fare well in competition with cross-pollinated ovules. Thus, plants are capable of making distinctions among developing seeds on the basis of their genotypes. ▌

Genetic drift in small populations causes loss of genetic variation

In a population of infinite size in which individual movements are unlimited, mating can be completely random. In reality, however, all populations are finite, and many are quite small. Populations of conservation concern may number in the tens or hundreds of individuals. In any finite population, mating cannot be completely random because mate choice is limited and all individuals are related to some degree, regardless of how remote the connection might be.

Coalescence

Here is a remarkable property of populations: because of the randomness of births and deaths, all the copies of a particular gene in a population will have descended, just by chance, from a single copy that existed at some time in the past, referred to as the **coalescence time**. Because of recombination and the independent assortment of chromosomes during meiosis, each gene has a somewhat independent history of descent through the generations

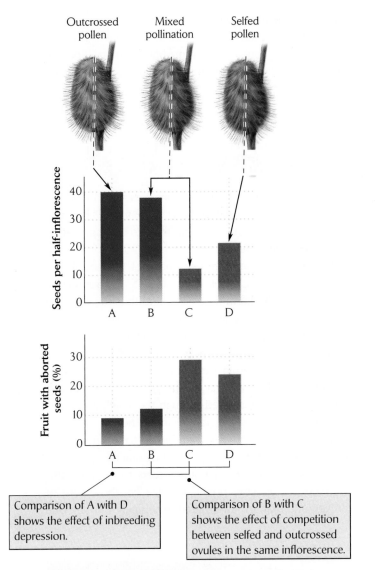

FIGURE 13.8 Plants discriminate between selfed pollen and outcrossed pollen. In a pollination experiment with the Australian shrub *Banksia spinulosa*, the two halves of individual inflorescences were fertilized with pollen either from the same plant or from neighboring plants. The results show both inbreeding depression in the selfed flowers (D compared with A) and the further discrimination against selfed ovules in competition with outcrossed ovules growing in the same inflorescence (C compared with B). Data from G. Vaughton and S. M. Carthew, *Biol. J. Linn. Soc.* 50:35–46 (1993).

Within figure:

Outcrossed pollen   Mixed pollination   Selfed pollen

Seeds per half-inflorescence

Fruit with aborted seeds (%)

Comparison of A with D shows the effect of inbreeding depression.

Comparison of B with C shows the effect of competition between selfed and outcrossed ovules in the same inflorescence.

of a population, and coalescence time therefore varies among genes, just by chance. For nuclear genes in diploid organisms, the average coalescence time is equivalent to $4N$ generations, where N is the size of the population. Thus, in the absence of the introduction of genetic variation by immigration or mutation, all the copies of a particular gene in a population become identical by descent from a single ancestral copy that existed, on average, $4N$ generations in the past. In this situation, all the individuals

in the population are homozygous for, and have identical copies of, that particular gene, and all matings between individuals are the equivalent of selfing with respect to that gene.

Coalescence happens throughout the genome. In the absence of mutation and immigration, a population will eventually become genetically uniform, with all individuals homozygous for all genes. This is exactly the result desired by researchers attempting to develop inbred laboratory populations of mice and fruit flies. These inbred populations provide a uniform genetic background for studies of the effects of introducing single mutations into the genome. However, a similar level of inbreeding is disastrous for natural populations.

Genetic drift

The process by which allele frequencies change and genetic variation is lost due to random variations in fecundity, mortality, and inheritance of gene copies through male and female gametes is called **genetic drift.** Genetic drift has its greatest effects on small populations, just as stochastic variation in birth and death rates causes greater variation in population size in small populations (see Chapter 12).

Let's think about how genetic drift works. Suppose a population contains 95 homozygotes with genotype A_1A_1 and 5 heterozygotes with genotype A_1A_2. If each individual has a 50% chance of surviving to reproduce, then the probability that all five of the A_1A_2 individuals will fail to reproduce, just by chance—in which case the A_2 allele will disappear from the population—is 1 in 32 (about 3%). Even if the heterozygotes produce offspring, half of their gametes will bear A_1 alleles, and their A_2 gene copies may not be transmitted to the next generation, again just by chance. If the A_2 allele were lost from the population in either way, the A_1 allele would become fixed. The rate of allele fixation is inversely related to the size of a population. Thus, genetic variation decreases more rapidly in small populations than in large ones.

A single episode of small population size, as might occur during the colonization of an island or a new habitat by a few individuals from a large parent population, can reduce genetic variation in the colonizing population. Such episodes are known as **founder events.** When founding populations consist of ten or fewer individuals, they typically contain only a small random sample of the total genetic variation of the parent population (Figure 13.9). Continued existence of the population at small sizes results in further loss of genetic variation due to genetic drift. This situation is often referred to as a **population bottleneck.** Such a situation appears to have

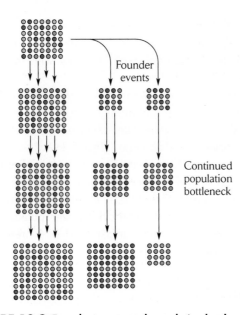

FIGURE 13.9 Founder events and population bottlenecks may result in reduced genetic variation. When a small group of individuals colonizes a new habitat, the colonists take with them a small random sample of the alleles in the parent population. Continued existence at small population sizes following colonization or any other form of population decline (a population bottleneck) results in further loss of genetic diversity through random sampling of each generation's gene pool in the following generation.

occurred in the recent past in the population of cheetahs in East Africa (Figure 13.10), which exhibit practically no genetic variation.

The fragmentation of natural populations into small subpopulations may subject them to genetic drift. The resulting losses of genetic variation could eventually restrict the evolutionary responsiveness of those subpopulations to the selection pressures of changing environments, making them more vulnerable to extinction. Furthermore, loss of genetic variation in the major histocompatibility complex (MHC) genes, which are involved in immune function, may reduce the natural resistance of individuals to disease.

Effective population size

In most of our calculations, we assume that populations remain at a constant size and that all individuals contribute equally to the production of the next generation. Of course, nature is not so ideal. Populations vary in numbers and individuals have different numbers of offspring—some are even excluded from breeding. To account for these variations, the population geneticist Sewall Wright devised the concept of **effective population size (N_e),** which can be thought of as the size of an ideal popula-

tion that undergoes genetic drift at the same rate as an observed population. Many factors influence effective population size, but variation in population size and the participation of individuals in reproduction are among the most important.

When a population varies in size from generation to generation, the effective population size is calculated by

$$\frac{1}{N_e} = \frac{1}{t}\left(\frac{1}{N_1} + \frac{1}{N_2} + \cdots + \frac{1}{N_t}\right) = \frac{1}{t}\sum_{i=1}^{t}\frac{1}{N_i}$$

where t is the number of generations over which the population has been measured. N_e is the *harmonic mean* of the population size. Suppose that the size of a population varies over five generations from 10 to 100, 50, 20, and 100 individuals. The arithmetic mean of the population size is $(10 + 100 + 50 + 20 + 100)/5 = 56$ individuals. However, the harmonic mean, and thus the effective population size, is the inverse of $(0.1 + 0.01 + 0.02 + 0.05 + 0.01)/5$, which is $5/0.19 = 26.3$ individuals. In this calculation, smaller population sizes contribute more to the effective population size than larger population sizes do, reflecting the fact that all the genes in the population have to be transmitted through the smallest number of individuals in any generation. Imbalance in the numbers of males and females contributing to future generations, as occurs in many promiscuous mating systems in which a few males monopolize access to females, also decreases the effective population size. For those parts of the genome that are transmitted exclusively through one

FIGURE 13.10 Cheetahs in East Africa exhibit virtually no genetic variation. This observation suggests that the cheetah population recently passed through a severe population bottleneck. Photo by R. E. Ricklefs.

sex (such as the mitochondrial genome and the Y chromosome), the effective size of the population is less than it is for biparentally inherited parts of the genome.

Population growth and decline leave different genetic traces

Population growth and population decline have different effects on genetic variation. A growing population that starts out small may have little genetic variation, as we saw in case of the Scandinavian wolf population. If the population grows rapidly, it may increase to a large size before additional genetic variation is introduced by mutation or immigration. Thus, a large population size with little genetic variation is a sign that a population has been small in the recent past, because of a founder event or because of a persistent bottleneck. In contrast, a population that decreases from a large size retains much of its genetic variation, as loss of alleles by genetic drift is a relatively slow process. Thus, a small population size

with substantial genetic variation is a good sign of a recent population decrease. Complex mathematical relationships allow researchers to estimate how rapidly a population has increased or decreased, but these are beyond the scope of this discussion. However, the following example will illustrate the principle involved.

A survey of mitochondrial diversity in several populations of giant tortoises (*Geochelone nigra*) on Isabela Island in the Galápagos archipelago revealed substantial differences in genetic variation. Remember that the haploid mitochondrial genome is inherited as a single unit without recombination and only through the female parent. Isabela is made up of several volcanoes that have merged, and the tortoise population on each of these volcanoes is genetically distinct. The tortoise populations on the southern volcanoes of Sierra Negra (100–300 individuals) and Cerro Azul (400–600 individuals) exhibited considerable variation, showing a large number of nucleotide substitutions between mitochondrial haplotypes (Figure 13.11). In the larger population (3,000–5,000 individuals) on Volcano Alcedo to the north, however, most of

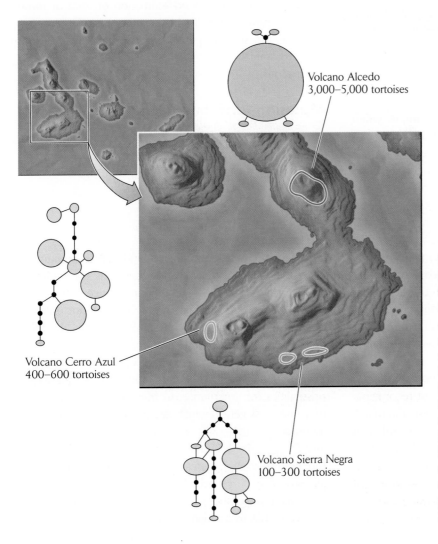

Volcano Alcedo
3,000–5,000 tortoises

Volcano Cerro Azul
400–600 tortoises

Volcano Sierra Negra
100–300 tortoises

FIGURE 13.11 Reduced genetic variation indicates a founder event in a Galápagos tortoise population. Mitochondrial haplotype frequencies for three populations of Galápagos tortoises on Isabela Island show that Volcano Alcedo was colonized relatively recently by one or a few individuals bearing a single haplotype, which is also found in populations to the south. Each step in each network diagram represents a single nucleotide substitution. The size of each oval is proportional to the number of individuals harboring the haplotype it represents. Small open circles represent inferred mutational steps producing haplotypes not found in the population. After L. B. Beheregaray et al., *Science* 302:75 (2003).

the individuals had the same mitochondrial haplotype, and the few variants were only a single mutational step away. This is the characteristic signature of a recently expanded population. The common mitochondrial haplotype in the Alcedo population occurred at lower frequencies in the two populations to the south. Thus, it appears that the Alcedo population was founded by one or a small number of individuals from the south and has recently grown to its present large size. Using information about the rate of mutation in mitochondrial DNA, the investigators estimated that the population was founded between 72,000 and 119,000 years ago—which corresponds to the estimated date of a massive explosive eruption of Alcedo Volcano that covered the area with meters of hot ash and undoubtedly wiped out the local tortoise population.

Coalescence, mutation rate, and time

We saw earlier that all the copies of a gene in a population can trace their ancestry back to a single copy at a time in the past that is equivalent to about $4N$ generations ago for nuclear genes and N generations ago for mitochondrial or chloroplast genes. We know that, given sufficient time, each lineage leading from the ancestral gene copy to the present is liable to accumulate mutations. Therefore, knowing the mutation rate and the length of a generation, one can calculate the coalescence time in years from the accumulated genetic variation in a population without having to know the population size. For example, when analyzed in this manner, all the copies of the mitochondrial genome in the present-day human population can be traced back to a single copy that existed roughly 140,000 years ago. Some people misinterpreted this result to mean that all of us descended from a single woman alive at that time, understandably dubbed "mitochondrial Eve." Of course, this finding applies only to the mitochondrial genome. We obtained the rest of our genome from many of mitochondrial Eve's male and female contemporaries. In addition, the coalescence time for each of our nuclear genes would be 4 times that of the mitochondrial genome, on average, and so the present human population retains much of the nuclear genetic variation present at the time of mitochondrial Eve.

Like the mitochondrial genome, the human Y chromosome is inherited as a single unit without recombination and through only one parent (in this case, the male). Thus, its coalescence time should be identical to that of the mitochondrial genome, assuming an equal number of males and females in the population. However, the coalescence time calculated for the Y chromosome suggests that the Y-chromosome "Adam" from whom all our Y chromosomes descended lived only 60,000 years ago—roughly

half of the mitochondrial coalescence time. Although human populations have roughly equal numbers of males and females, the discrepancy in coalescence time probably reflects the greater variation in reproductive success among males and their correspondingly lower effective population size (N_e).

Loss of variation by genetic drift is balanced by mutation and migration

As we have seen, genetic drift and coalescence ultimately cause a population to become genetically uniform. However, this inexorable process can be reversed by mutation or immigration, both of which introduce new genetic variation into a population. These processes balance each other, and a population eventually achieves an equilibrium level of genetic variation.

Showing the mathematical relationship between population size, mutation or immigration rate, and genetic variation was an important step in the study of population genetics because it allowed estimation of each of these variables from knowledge of the other two. Thus, as we shall see, measures of genetic variation can provide estimates of mutation rates or rates of migration between populations.

Mutation–drift balance

As we have seen, genetic drift is opposed by mutation. The two processes eventually come into equilibrium, at which point their relationship can be described by the equation

$$\hat{F} = \frac{1}{4N\mu + 1}$$

where $\hat{F}$ is the equilibrium fixation index, N is the number of individuals in the population, and μ is the mutation rate per allele copy per generation. Notice that N and μ influence $\hat{F}$ as a product, and so one cannot estimate one without knowing the other. What is the "4" doing there? Think of it this way: each individual has two copies of each nuclear gene, and each copy can be inherited through either the mother or the father, so there are four possibilities for inheritance. In the case of mitochondrial or chloroplast genes, which are present in a single copy inherited through only one parent, the "4" disappears, and $\hat{F}_{mitochondrial} = 1/(N\mu + 1)$. In either case, if there is no mutation ($\mu = 0$), $\hat{F}$ increases to 1—complete fixation of all genes. Of course, populations do experience mutation, and larger populations result in a lower fixation index (Figure 13.12), as do higher mutation rates.

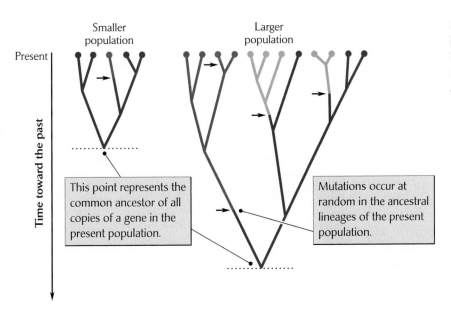

FIGURE 13.12 The drift–mutation balance preserves more genetic variation in larger populations. Larger populations have longer coalescence times, which allows more time for mutations to introduce new variation.

Let's do a simple calculation of $\hat{F}$ for a nuclear gene. Suppose that a population has a million individuals ($N = 10^6$) and that the mutation rate is one per million alleles per generation, or 10^{-6}. In this case, $\hat{F} = 1/(4 \times 10^6 \times 10^{-6} + 1) = \frac{1}{5}$, or 0.20. For a mitochondrial or chloroplast gene, the effective population size is smaller and the equilibrium fixation index is higher ($\hat{F} = 0.50$). This difference also reflects the 4 times more rapid genetic drift for mitochondrial or chloroplast versus nuclear genomes.

If genetic drift were a strong force in natural populations, we would expect to see a relationship between the size of isolated populations and the genetic variation they contain. Consider the populations of Galápagos hawks (*Buteo galapagoensis*) on islands of different sizes in the Galápagos archipelago. The hawks rarely fly between islands, so each island can be considered to harbor a relatively isolated population. Genetic variation in these populations was estimated by the sharing of minisatellite alleles (Figure 13.13). If a hawk population lacked genetic variation and all the individuals therefore had the same allele, the level of sharing would be 100%. Population size is related to the area of an island, so the smaller the island, the less genetic variation in its hawk population.

Migration–drift balance

When populations become divided into isolated subpopulations, those subpopulations undergo independent genetic change through mutation, genetic drift, and selection. As a result, their gene pools begin to diverge from one other. This process is opposed by **gene flow,** the process by which migration of individuals between isolated subpopulations brings new genetic variation to those subpopulations. We can define *migration (m)* quantitatively as the proportion of gene copies in a subpopulation brought

(a)

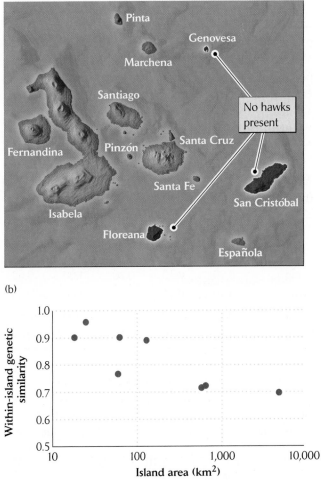

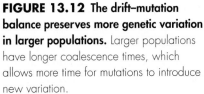

(b)

FIGURE 13.13 Smaller populations of Galápagos hawks harbor less minisatellite variation. (a) The Galápagos archipelago. Hawk population size varies in proportion to island area. (b) Individuals on smaller islands have higher average genetic similarity owing to the loss of minisatellite alleles by genetic drift. After J. L. Bollmer et al., *Auk* 122:1210–1224 (2005).

from outside in each generation. This figure is equivalent to the proportion of individuals in a subpopulation that were born elsewhere.

Just as drift and mutation come into balance, drift and migration also come into balance, at which point

$$\hat{F}_{ST} = \frac{1}{4Nm + 1}$$

This equation is directly comparable to the one relating $\hat{F}$ to population size and the rate of mutation. However, F must be calculated differently because migration occurs between subpopulations, whereas mutation occurs within subpopulations. When considering migration, the expected number of heterozygotes in an undivided population serves as the baseline for comparison. Thus,

$$F_{ST} = \frac{H_{total} - \overline{H}_{subpopulation}}{H_{total}}$$

where H_{total} is the heterozygosity expected if all the subpopulations were completely mixed and $\overline{H}_{subpopulation}$ is the average proportion of heterozygotes observed in the subpopulations. If the population were undivided, H_{total} and $\overline{H}_{subpopulation}$ would be the same, and F_{ST} would equal 0. If a different allele were fixed in each subpopulation, then all values of $\overline{H}_{subpopulation}$ would be 0, and F_{ST} would equal $H_{total}/H_{total} = 1$.

To illustrate the calculation of F_{ST}, consider three island subpopulations that have the following frequencies of three alleles:

| | Allele | | | | |
	A_1	A_2	A_3	Total	H
Island 1	0.50	0.30	0.20	1.00	0.62
Island 2	0.30	0.40	0.30	1.00	0.66
Island 3	0.10	0.50	0.40	1.00	0.58
Combined	0.30	0.40	0.30	1.00	0.66

The expected proportion of heterozygotes in each population is 1 minus the sum of the homozygote frequencies, which are the squared proportions of each of the alleles; that is, $H = 1 - p_{A_1}^2 + p_{A_2}^2 + p_{A_3}^2$. These values are

$$H_{Island\ 1} = 1 - 0.50^2 + 0.30^2 + 0.20^2 = 0.25 + 0.09 + 0.04 = 0.62$$

$$H_{Island\ 2} = 0.66;$$

$$H_{Island\ 3} = 0.58;$$

$$H_{total} = 0.66.$$

The average of the subpopulations is $\overline{H}_{subpopulation} = (0.62 + 0.66 + 0.58)/3 = 0.62$. Thus, $F_{ST} = (0.66 - 0.62)/0.66 = 0.061$. Now, we can rearrange the equation $\hat{F}_{ST} = 1/(4Nm + 1)$ to obtain

$$Nm = \frac{1}{4}(1 - F_{ST})/F_{ST}$$

For this example, $Nm = 3.8$. Because Nm is the product of the number of individuals in the subpopulation and the proportion of those individuals born elsewhere, it estimates the absolute number of immigrants per generation in each of the subpopulations. As a general rule of thumb, as few as one immigrant per generation ($F_{ST} = 0.20$, $Nm = 1$) is considered sufficient to prevent genetic differentiation of subpopulations with respect to genes that have no effect on fitness.

Selection in spatially variable environments can differentiate populations genetically

Natural selection can lead to genetic differentiation of populations even in the face of considerable gene flow. When the difference in selection pressures between two localities is strong relative to the rate of gene flow between them, differences in allele frequencies can be maintained by differential natural selection. This situation often results in a gradual change in allele frequencies, or in phenotypic characters under genetic influence, over some distance.

Botanists have long recognized that individuals of a species growing in different habitats may exhibit varying forms corresponding to local conditions. In many cases, these differences result from developmental responses (see Chapter 6). However, experiments on some species have revealed genetic adaptations to local conditions. Early in the twentieth century, the Swedish botanist Göte Turesson collected seeds from several species of plants, each of which lived in a variety of habitats—for example, alkaline versus acidic soils—and grew them together in his garden. This method is referred to as a **common garden experiment.** He found that even when grown under identical conditions, many of the plants exhibited differing forms that depended on their habitat of origin. Turesson suggested that these forms, which he called **ecotypes,** represent genetically differentiated lineages of a population, each restricted to a specific habitat. Because Turesson grew his plants under identical conditions, he realized that the differences between ecotypes must have had a genetic basis, and that they must have resulted from

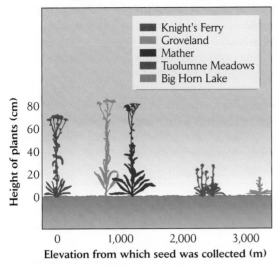

← Ecotypic differentiation among populations →

FIGURE 13.14 Individuals of a species may show genetically based geographic variation. Ecotypic differentiation among populations of yarrow (*Achillea millefolium*) was demonstrated by collecting seeds from different elevations and growing them under identical conditions in a common garden. The plants' phenotypes differed even when they were grown under identical conditions. After J. Clausen, D. D. Keck, and W. M. Hiesey, *Carnegie Inst. Wash. Publ.* 58:1–129 (1948).

evolutionary differentiation within the species according to habitat.

Experiments on yarrow (*Achillea millefolium*) have also revealed ecotypic variation. This plant grows in many habitats, ranging from sea level to more than 3,000 meters in elevation. Plants raised from seed collected at various elevations but grown together at sea level at Stanford, California, retained the distinctive sizes and levels of seed production typical of the populations from which they came (Figure 13.14).

In heterogeneous environments, whether natural or artificial, adaptation to particular habitat patches enhances fitness. Thus, mating with distant individuals that are adapted to different habitat conditions may reduce the fitness of progeny. Several studies have reported an **optimal outcrossing distance** in populations of plants. Nearby individuals are likely to be close relatives. Distant individuals are likely to be adapted to different conditions. The optimal outcrossing distance should be somewhere in between. In a study conducted in central Colorado, Mary Price and Nicolas Wasser fertilized flowers of the larkspur *Delphinium nelsoni* with pollen from the same plant and with pollen from individuals located at distances of 1, 10, 100, and 1,000 meters. The number of seeds

set per flower was greatest when the pollen came from individuals 10 meters distant and smallest for selfed pollen and for pollen from individuals 1,000 meters distant. Furthermore, when these seeds were planted, survival to 1 and 2 years was greatest among the offspring of matings across the optimal outcrossing distance of 10 meters (Figure 13.15).

Ecotypic differentiation even over distances of only a few meters has been observed where contrasting selection pressures are strong enough to overcome the migration of individuals, seeds, or pollen between habitat patches. Selection pressures may differ strongly across short distances where soil properties reflect sharp boundaries in underlying geology. Such boundaries occur naturally at the edges of serpentine outcrops, for example, but human activities also can create sharp environmental transitions. Soils that develop on mine tailings can exert strong selection pressure for tolerance to toxic metals, such as copper, lead, zinc, and arsenic, which contrasts sharply with the selection pressure on adjacent natural soils (Figure 13.16). Where suitable genetic variation exists, high metal concentrations in the soil select for plant populations that toler-

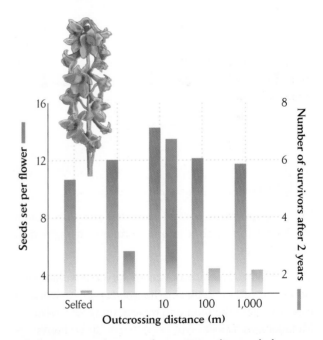

FIGURE 13.15 The optimal outcrossing distance balances the risks of inbreeding and of mating with individuals adapted to different conditions. The number of seeds set per flower and the number of seeds (out of 98) that produced plants surviving after 2 years varied depending on the distance from which pollen was obtained. The data show an optimal outcrossing distance of about 10 meters. After M. V. Price and N. M. Wasser, *Nature* 277:294–297 (1979).

(a)

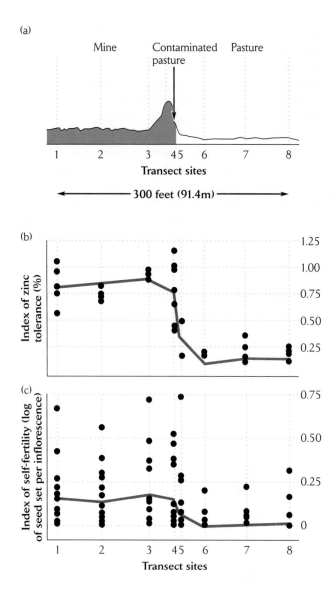

(b)

(c)

FIGURE 13.16 Evolution of zinc tolerance in plants has occurred on small spatial scales. Sweet vernal grass (*Anthoxanthum odoratum*) growing on soils derived from mine tailings at Trelogan, North Wales, United Kingdom, evolved tolerance for high concentrations of zinc in a few tens of generations. (a) A sharp gradient in zinc concentrations in the soil exists at the boundary between the mine and an adjacent pasture. (b) Zinc tolerance in plants varies sharply across the boundary. (c) Self-fertility has increased in plants on the mine tailings as a mechanism to forestall gene flow from intolerant plants in the pasture. After J. Antonovics and A. D. Bradshaw, *Heredity* 25:349–362 (1970).

ate those concentrations in a matter of a few generations. In such circumstances, gene flow across environmental boundaries would introduce intolerant genotypes into tolerant populations. Investigators have found evidence for increased self-compatibility in tolerant populations, which prevents outcrossing and maintains the integrity of the locally adapted gene pool.

Clearly, ecotypic differentiation and gene flow can influence the diversity of genotypes, and the genetic traces that these processes, as well as genetic drift, mutation, and migration, leave in patterns of genetic variation can tell us much about the structure of populations, particular those at high risk of extinction. However, for most traits important for the ecological relationships of organisms to their environments, natural selection remains the dominant force.

SUMMARY

1. All populations contain abundant genetic variation, which provides valuable information for studying population processes.

2. The ultimate source of genetic variation within populations is mutation. Mutations are changes in the sequence of the nucleotide subunits that make up DNA molecules. Mutations with phenotypic effects tend to be harmful to the organism, but most mutations have no functional significance for the organism and are thus "silent" or "neutral" with respect to individual fitness.

3. Useful genetic markers in population studies include variation in nucleotide sequences in nongenetic parts of the genome, such as microsatellites. Genetic markers can be used to study rates of dispersal of individuals within

and between populations, patterns of mating, and the history of change in population size and distribution, as well as the results of evolution by natural selection.

4. Genetic variation is maintained in populations by mutation, migration, and frequency-dependent selection in variable environments.

5. The frequencies of homozygous and heterozygous genotypes in a large, randomly mating population in the absence of selection, mutation, and migration can be estimated by the Hardy–Weinberg law. This law states that alleles with frequencies p and q will form homozygous genotypes with frequencies p^2 and q^2 and heterozygotes with frequency $2pq$.

6. Deviations from Hardy–Weinberg equilibrium may be caused by mutation, migration, nonrandom mating, small population size, or natural selection.

7. Nonrandom mating occurs when individuals mate preferentially with other individuals with similar genotypes (positive assortative mating), or with close relatives (inbreeding), or avoid such matings (negative assortative mating). Nonrandom mating changes the frequencies of genotypes within a population, but not the frequencies of alleles.

8. Inbreeding tends to reduce the frequency of heterozygotes in a population. A high frequency of homozygotes can lead to inbreeding depression.

9. The inbreeding coefficient, also known as the fixation index (F), measures the departure of the observed frequency of heterozygotes in a population from Hardy–Weinberg equilibrium. F increases with each generation of selfing or inbreeding.

10. In small populations, just by chance, all copies of a gene are descended from a single common ancestor existing at some time in the past, known as the coalescence time. The loss of genetic variation by the stochastic processes that cause coalescence is known as genetic drift.

11. Because genetic drift is faster and coalescence time is shorter in smaller populations, genetic variation decreases in proportion to population size. Populations can lose genetic variation through founder events, in which new populations are formed by a small number of individuals, and population bottlenecks, in which populations exist at small sizes for long periods.

12. Effective population size (N_e), which is the size of an ideal population that undergoes genetic drift at the same rate as an observed population, takes into account temporal variation in numbers of individuals and deviation from an even sex ratio.

13. Newly founded populations and populations recovering from bottlenecks tend to have relatively little genetic variation for their size, whereas declining populations retain more genetic variation. Thus, genetic variation can be used to reconstruct the history of population size.

14. The tendency of a population to lose genetic variation by genetic drift is balanced by mutation. These two processes eventually reach equilibrium, at which time the fixation index for a gene is inversely related to the product of the population size and the mutation rate.

15. Population subdivision results in genetic differentiation of subpopulations, which is opposed by migration of individuals between them. The rate of migration can be estimated by comparing heterozygosity in subpopulations with the expected heterozygosity of a completely mixed population (F_{ST}).

16. Differential selection pressures on subpopulations may result in variation in allele frequencies within a geographic range, even when opposed by gene flow.

REVIEW QUESTIONS

1. Compare and contrast the process that serves as the source of genetic variation with the processes that maintain genetic variation.

2. How is it that natural selection can reduce genetic variation within a given environment but maintain genetic variation across environments?

3. Under the assumptions of the Hardy–Weinberg law, why does the process of sexual reproduction produce no evolutionary change?

4. If two alleles for a trait (A_1 and A_2) are present at frequencies of 0.6 and 0.4, respectively, what will be the frequencies of the three possible genotypes at Hardy–Weinberg equilibrium?

5. Explain how assortative mating alters the frequencies of genotypes without altering the frequencies of alleles.

6. How do genetic drift, founder events, and population bottlenecks affect genetic diversity differently?

7. If the expected frequency of heterozygotes in a population (based on the Hardy–Weinberg law) is 0.4, but the observed frequency is 0.2, what is the inbreeding coefficient?

8. Why does inbreeding produce offspring that survive less well?

9. If a population of elephants is small but retains high genetic diversity, what can you conclude about the history of that population's size, and why?

10. What processes can make spatially separated populations possess genetic differences?

SUGGESTED READINGS

Avise, J. C. 2000. *Phylogeography: The History and Formation of Species.* Harvard University Press, Cambridge, Mass.

Avise, J. C. 2004. *Molecular Markers, Natural History, and Evolution.* Sinauer Associates, Sunderland, Mass.

Bollmer, J. L., et al. 2005. Population genetics of the Galápagos hawk (*Buteo galapagoensis*): Genetic monomorphism within isolated populations. *Auk* 122:1210–1224.

Charlesworth, D., and B. Charlesworth. 1987. Inbreeding depression and its evolutionary consequences. *Annual Review of Ecology and Systematics* 18:237–268.

Gilbert, D. A., et al. 1990. Genetic fingerprinting reflects population differentiation in the California Channel Island fox. *Nature* 344:764–767.

Hedrick, P. W., and S. T. Kalinowski. 2000. Inbreeding depression in conservation biology. *Annual Review of Ecology and Systematics* 31:139–162.

Heschel, M. S., and K. N. Page. 1995. Inbreeding depression, environmental stress, and population size variation in scarlet gilia (*Ipomopsis aggregata*). *Conservation Biology* 9:126–133.

Ingman, M., et al. 2000. Mitochondrial genome variation and the origin of modern humans. *Nature* 408:708–713.

Keller, L. F. 1998. Inbreeding and its fitness effects in an insular population of song sparrows (*Melospiza melodia*). *Evolution* 52:240–250.

Keller, L. F., and D. M. Waller. 2002. Inbreeding effects in wild populations. *Trends in Ecology and Evolution* 17:230–241.

Merola, M. 1994. A reassessment of homozygosity and the case for inbreeding depression in the cheetah, *Acinonyx jubatus:* Implications for conservation. *Conservation Biology* 8:961–971.

O'Brien, S. J., et al. 1985. Genetic basis for species vulnerability in the cheetah. *Science* 227:1428–1434.

Ouborg, N. J., Y. Piquot, and J. M. Van Groenendael. 1999. Population genetics, molecular markers and the study of dispersal in plants. *Journal of Ecology* 87:551–568.

Price, M. V., and N. M. Wasser. 1979. Pollen dispersal and optimal outcrossing in *Delphinium nelsoni. Nature* 277:294–297.

Ralls, K., J. D. Ballou, and A. Templeton. 1988. Estimates of the cost of inbreeding in mammals. *Conservation Biology* 2:185–193.

Rohde, D. L. T., S. Olson, and J. T. Chang. 2004. Modelling the recent common ancestry of all living humans. *Nature* 431:562–566.

Vilà, C., et al. 2002. Rescue of a severely bottlenecked wolf (*Canis lupus*) population by a single immigrant. *Proceedings of the Royal Society of London* B 270:91–97.

CHAPTER 14

Species Interactions

When prickly pear cactus (*Opuntia*) was introduced into Australia as an ornamental plant and to establish "living fences" for pastures, it spread rapidly over the island continent, covering thousands of acres of valuable pasture and rangeland. After several unsuccessful attempts to eradicate the plant, the cactus moth (*Cactoblastis cactorum*) was introduced from South America in the 1920s. The caterpillar of the cactus moth feeds on growing shoots of the prickly pear and quickly destroys the plant—literally nipping it in the bud and inoculating it with various pathogens and rot-causing organisms.

Once they became established in Australia, cactus moths exerted such effective control that within a few years, prickly pear had become a pest of the past (Figure 14.1). Cactus moths have since been introduced into South Africa and Hawaii to control introduced *Opuntia* species. Since their introduction into the West Indies for control purposes, however, the moths have invaded Mexico and Florida, where they are threatening native species of *Opuntia*.

The cactus moth has not eradicated the prickly pear in Australia because the cactus still manages to disperse to moth-free areas, thereby keeping one jump ahead of the moth. Thus, the cactus population maintains a low-level equilibrium in a continually shifting mosaic of isolated patches, as in a metapopulation. Indeed, a casual observer would probably never guess that the cactus moth keeps the prickly pear at its present low population levels because the moths are scarce in the remaining stands of cactus in Australia today. (The same moth probably controls prickly pear populations in some areas of its native range in South America, but its decisive role might have gone unnoticed if the appropriate experiment had not been performed in Australia.)

The cactus–cactus moth example shows the potentially strong influence of consumers on resource populations. Consumer–resource interactions are just one of the many types

(a)

(b)

FIGURE 14.1 The prickly pear cactus population is controlled by its predator, the cactus moth. Photographs of a pasture in Queensland, Australia, (a) 2 months before and (b) 3 years after the introduction of the cactus moth to control the prickly pear cactus. Main photos from A. P. Dodd, in A. Keast, R. L. Crocker, and C. S. Christian (eds.), *Biogeography and Ecology in Australia,* W. Junk, The Hague (1959), courtesy of W. H. Haseler, Department of Lands, Queensland, Australia. Inset photos by (a) D. Habeck and F. Bennet, University of Florida and (b) Peggy Greb/ Agricultural Services/U.S. Department of Agriculture.

of interactions between species that influence species' populations and guide their evolution through natural selection. This chapter provides a brief overview of the many kinds of species interactions in nature, including both antagonistic and mutually beneficial relationships. These interactions define the structure of biological communities and ecosystems and influence the functioning of those systems. They also emphasize the overriding influence of the biological environment on the behavior of individuals, the demography of populations, and the evolution of species.

CHAPTER CONCEPTS

- All organisms are involved in consumer–resource interactions

- The dynamics of consumer–resource interactions reflect mutual evolutionary responses

- Parasites maintain a delicate consumer–resource relationship with their hosts

- Herbivory varies with the quality of plants as resources

- Competition may be an indirect result of other types of interactions

- Individuals of different species can collaborate in mutualistic interactions

Predator–prey, herbivore–plant, and parasite–host relationships are all examples of **consumer–resource interactions,** which organize biological communities into food chains, along which food energy is passed through the ecosystem. It is typical of consumer–resource interactions that consumers benefit individually and their numbers may increase, while resource populations are decreased. Thus, while energy and nutrients move up a food chain, populations are controlled both from below by resources and from above by consumers.

Although consumer–resource interactions constitute the most fundamental ecological relationship between

species, these interactions are the basis for two additional types of interactions: competition and mutualism. When two consumers share the same resource, each reduces the availability of the resource to the other, and they are said to be engaged in *competition.* As we shall see in Chapter 16, competition influences population processes and can determine whether a population can persist in a particular environment.

Mutualisms are interactions between two species that benefit both. They take on many forms, but partners in mutualisms generally supply complementary resources or services. For example, many insects pollinate plants in

TABLE 14.1 A classification of the types of interactions between species based on their mutual effects

Effect on species 1	Effect on species 2	Type of interaction
+	−	Consumer–resource interactions, including predator–prey, herbivore–plant, and parasite–host interactions
−	−	Competition
+	+	Mutualism
+	0	Commensalism
−	0	Amensalism, perhaps mostly incidental

return for nectar or pollen rewards; bacteria in the roots of plants provide nitrogen to their hosts in return for carbon sources; ruminant mammals, such as sheep and cattle, maintain bacteria in specialized compartments in their stomachs, and in return the bacteria digest compounds in plants that the ruminant cannot.

Species interactions can be classified conveniently by the effect of each species on the other. When we consider that a species can benefit from an interaction (+), suffer (−), or be unaffected (0), the possible combinations of effects are +/−, −/−, +/+, +/0, or −/0 (0/0 represents the absence of any consequential interaction), as summarized in Table 14.1.

The interactions +/0 (commensalism) and −/0 (amensalism) are common in ecology and are important for many populations, but are not often considered in experimental and theoretical studies because of the absence of a mutual dynamic between the two participants. For example, when a bird places its nest in a tree, or a hermit crab uses the shell of a long-dead snail, the bird and the crab gain protection from predators (+), but the populations of trees and snails are unaffected (0). When an elephant crushes a grasshopper underfoot, the population of grasshoppers suffers (−), but the elephant is unaffected (0). We shall have little more to say about such interactions in nature.

Another term commonly used to describe some kinds of species interactions is **symbiosis** (literally, "living together"), which refers to individuals of different species that live in close association. Many cases of symbiosis involve partners in mutualisms whose lives are intimately intertwined, such as the algae and fungi that constitute lichens. However, the term *symbiosis* also extends to parasites that live within their hosts; both parties to such relationships are often specifically adapted to maintain the delicate balance between life and death. As we encounter the various relationships between species in more detail, you should keep in mind the endless complexity of these interactions and the indistinct boundaries between

predation, parasitism, and mutualism. To begin our discussion, we'll turn to the elemental consumer–resource relationship.

All organisms are involved in consumer–resource interactions

Consumer–resource interactions are the most fundamental interactions in nature because all non-photosynthetic organisms must eat, and all organisms are at risk of being eaten. Consumers go by many names. The most familiar are predator, parasite, parasitoid, herbivore, and detritivore. From the standpoint of species interactions, some of these distinctions are useful, but others can be confusing. Let's start with **predator.** Images of an owl eating a mouse or of a spider eating a fly capture the essentials of predation (Figure 14.2). Predators capture individuals and consume them, thereby removing them from the prey population and gaining nutrition to support their own reproduction.

In contrast, a **parasite** consumes parts of a living prey organism, or **host.** Parasites attach themselves to, or invade the bodies of, their hosts and feed on their tissues, their blood, or partially digested food in their intestines. Parasites that cause disease symptoms are called *pathogens*. Although parasitism may increase the probability of a host's dying from other causes or reduce its fecundity, a parasite generally does not, by itself, remove an individual from the host population. Indeed, it would be contrary to a parasite's best interests to kill the host upon which it feeds and depends for its survival.

Parasitoid is the term applied to species of wasps and flies whose larvae consume the tissues of living hosts—usually the eggs, larvae, or pupae of other insects. This strategy inevitably leads to the host's death, but not until the parasitoid larvae have completed their development and pupated (Figure 14.3). Parasitoids resemble parasites, because they reside within and eat the tissues of a

FIGURE 14.2 African lions are specialized for pursuing large prey. With their powerful legs and jaws, lions can subdue prey somewhat larger than themselves. But because they cannot maintain speed over long distances, successful hunting relies on stealth and surprise. Photo by Peter Blackwell/naturqpl.com.

living host, and predators, because they inevitably kill their host. Not surprisingly, parasitoids have their own parasites, which are called *hyperparasitoids*.

Herbivores eat whole plants or parts of plants. From the standpoint of consumer–resource relationships, herbivores function as predators when they consume whole plants, and as parasites when they consume living plant tissues but do not kill their victims. Thus, a deer browsing on a few leaves and stems functions as a parasite, while a sheep that consumes an entire plant, pulling it up by the roots and macerating it into lifeless shreds, functions as a predator. Consumption of a portion of a plant's tissues is referred to as **grazing** (when applied to grasses and other herbaceous vegetation, and to algae) or **browsing** (when applied to woody vegetation).

Detritivores consume dead organic material—such as leaf litter, feces, and carcasses—and therefore have no direct effect on the populations that produce those resources. In other words, detritivory is a commensal (+/0) interaction. Because they live off the wastes of other species, detritivores do not directly affect the abundance of their food supplies, and their activities do not usually influence the evolution of the living sources of their food. Detritivores are important in the recycling of nutrients within ecosystems, as we will see in Chapter 22. However, because detritivore populations generally are not dynamically coupled to their resource populations, detritivores will not be considered further in this chapter.

The various kinds of consumer–resource interactions we have just described can be organized usefully according to the duration and "intimacy" of the relationship between the interacting species and the probability that the interaction will lead to the death of resource individuals (Table 14.2).

The dynamics of consumer–resource interactions reflect mutual evolutionary responses

Resource organisms have as many tactics to avoid being eaten as their consumers have to hunt them. Because the fitness of both is at stake, evolutionary responses constantly adjust relationships between consumers and their resources, as we shall describe in more detail in Chapter 17. The commonplace images of cat and mouse or spider and fly might lead one to think that consumers have the upper hand. However, hiding, escape, and many other kinds of defensive tactics can be effective, depending on the particular circumstances of a consumer–resource relationship. For example, grasslands offer few hiding places for deer, antelope, and other grazers, so their escape depends on early detection of predators and swift movement. Plants cannot flee like animals, but many produce thorns and defensive chemicals that dissuade herbivores.

Where animals are able to hide or seek refuge in safer microhabitats, they are often sensitive to the presence of predators and adjust their behavior accordingly. Small fish

FIGURE 14.3 Parasitoid wasps develop inside the larvae or pupae of other insects. Photo by Scott Bauer.

TABLE 14.2   A classification of consumer–resource interactions

| Duration and/or intimacy of the association | Probability of death of resource organism | |
	Low	High
Short and casual	Grazers and browsers	Predators; seed predators
Long and close	Parasites and many arthropod herbivores	Parasitoids

Source: A. J. Pollard, in R. S. Fritz and E. L. Simms (eds.), *Plant Resistance to Herbivores and Pathogens: Ecology, Evolution, and Genetics,* University of Chicago Press, Chicago (1992), pp. 216–239.

living in ponds with larger predatory fish avoid the best feeding areas in open water and spend most of their time lurking in safer weed beds close to the water's edge. The following case study shows, however, that staying out of harm's way can have costs.

ECOLOGISTS IN THE FIELD   Predator avoidance and growth performance in frog larvae.

Staying safe from predators requires trade-offs. When prey individuals must remain in poor feeding areas to avoid predators, for example, their growth rates can be depressed. Slowly growing prey take longer to mature, extending the period during which an individual is most vulnerable to predation. They are also smaller as adults and therefore produce fewer offspring.

The effect of predation risk on the growth of frog larvae was demonstrated in laboratory and field experiments on bullfrog (*Rana catesbiana*) tadpoles by Rick Relyea and Earl Werner at the University of Michigan. They conducted laboratory experiments in which newly hatched tadpoles were placed in aquaria with caged dragonfly larvae or fish, which prey on small bullfrog tadpoles in natural ponds. The tadpoles reduced their activity in the presence of the predators, especially the dragonfly larvae, and also avoided the side of the aquarium where the caged predator was located (Figure 14.4). Similar experiments within enclosures set in a natural pond further demonstrated that the presence of caged dragonfly larvae reduced growth rates significantly in some species of frogs.

Other studies (see Figure 7.20, for example) emphasize that the ability to perceive predation risk is widespread in the animal world and has a strong effect on the behavior and habitat selection, as well as on the demography, of prey organisms. Of course, prey organisms respond to the presence of predators because such responses have been strongly selected over the evolutionary history of the consumer–resource interaction.

When prey cannot hide or escape, they often adopt protective defenses. These defenses rarely involve physical combat because few types of prey can match their

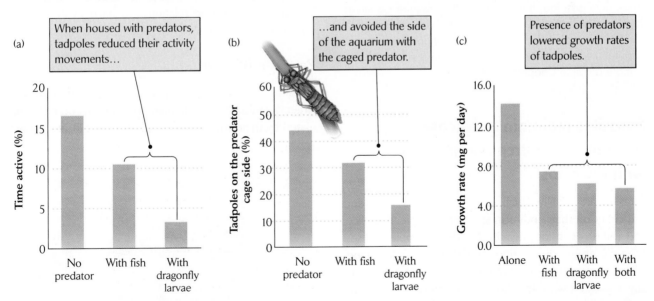

(a) When housed with predators, tadpoles reduced their activity movements…

(b) …and avoided the side of the aquarium with the caged predator.

(c) Presence of predators lowered growth rates of tadpoles.

FIGURE 14.4 Avoiding predators may result in reduced growth rates. Relyea and Werner housed bullfrog tadpoles with and without caged predators (fish and dragonfly larvae) and recorded (a) the activity levels of the tadpoles, (b) the number of tadpoles found on the side of the aquarium where the predator cage was located, and (c) the growth rates of the tadpoles. The dragonfly larvae, in particular, led to reduced activity, avoidance of areas near predators, and reduced growth rates on the part of the tadpoles. After R. A. Relyea and E. E. Werner, *Ecology* 80:2117–2124 (1999).

FIGURE 14.5 Many organisms have evolved chemical defenses to ward off predators. A bombardier beetle sprays a noxious liquid at the temperature of boiling water toward a predator. *Courtesy of Thomas Eisner, Cornell University.*

predators, and predators carefully avoid those that can. Instead, many seemingly defenseless organisms produce foul-smelling or stinging chemical secretions to dissuade predators. For example, whip scorpions and bombardier beetles direct sprays of noxious liquids at threatening animals (Figure 14.5). Many plants and animals contain chemical substances that make them unpalatable or poisonous. Slow-moving animals, such as porcupines and armadillos, protect themselves with spines or armored body coverings. These defenses, too, have costs, as they require resources that could otherwise be allocated to growth and reproduction. At the same time, predator populations are evolving adaptations to circumvent prey defenses, as we shall see in more detail in Chapter 17. ▌

Parasites maintain a delicate consumer–resource relationship with their hosts

Parasites are usually much smaller than their hosts and live either on their body surfaces (for example, ticks, lice, and mites) or inside their bodies (for example, viruses, bacteria, protozoans, various roundworms, flukes, tapeworms, and arthropods). Many parasites are only casually associated with their hosts, as in the case of mosquitoes that "graze" blood meals. Others remain within their hosts throughout their entire life cycles, and may even be transmitted between host mother and offspring through the host's eggs. One such parasite is the symbiotic bacterium *Wolbachia*, which infects the cells of many insects

and other invertebrates (Figure 14.6). Although *Wolbachia* infect many types of host cells, it is their presence in cells of the ovaries and testes that reduces host fitness most profoundly, principally by modifying sexual function. Infected males may be killed, develop as females, or be rendered incapable of mating with any female not already infected by the same strain of *Wolbachia*. In some host species, *Wolbachia* infections cause females to reproduce parthenogenetically, without having to mate with males.

The effects of parasites on host fitness vary dramatically, ranging in humans, for example, from the passing inconvenience of a cold virus to the deadly effects of HIV and the avian H5N1 influenza virus. Whereas *Wolbachia* is a serious threat to fitness, other symbionts that might have been parasitic in the past may evolve to become beneficial to their hosts. One example is *Buchnera*, a beneficial bacterial symbiont of insects, particularly aphids. *Buchnera* and its hosts are mutualists. The *Buchnera* symbionts are maintained in specialized cells, called bacteriocytes, and although they obtain carbohydrates and other nutrients from their hosts, they provide essential amino acids in return. Aphids feed on the phloem of plants, which contain virtually no amino acids, and so without the amino acids from the *Buchnera*, the aphids could not grow and reproduce. The genome of *Buchnera*, like those of many symbionts, is greatly reduced, and the symbiont relies on its host for many of its essential functions. The contrast between *Wolbachia* and *Buchnera* emphasizes the range of interactions that exist between symbionts and their hosts, from strict parasitism to mutualism. Indeed, it is thought that many mutualistic relationships have evolved from host–parasite interactions, and perhaps vice versa.

Parasite life cycles

Parasites that live inside, or in close association with, a larger organism enjoy a benign physical environment regulated by their host. Tapeworms, for example, are bathed

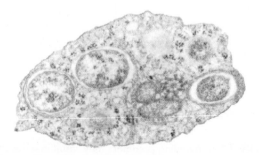

FIGURE 14.6 *Wolbachia* **is a common bacterial parasite of insects.** In this electron micrograph, *Wolbachia* are visible in an insect cell (arrow). *Courtesy of Scott O'Neill, from PLoS Biology 2(3): e76 (2004), doi:10.1371/journal.pbio.0020076.*

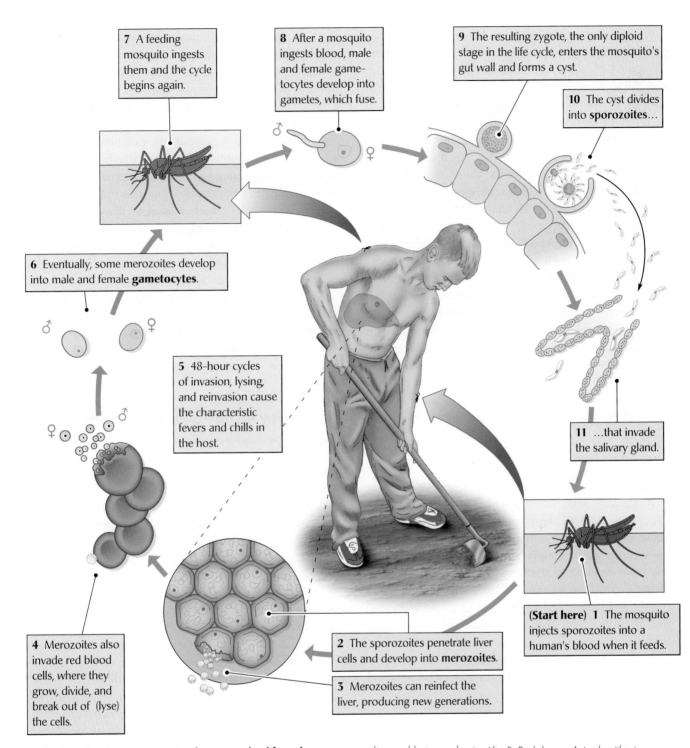

7 A feeding mosquito ingests them and the cycle begins again.

8 After a mosquito ingests blood, male and female gametocytes develop into gametes, which fuse.

9 The resulting zygote, the only diploid stage in the life cycle, enters the mosquito's gut wall and forms a cyst.

10 The cyst divides into **sporozoites**…

6 Eventually, some merozoites develop into male and female **gametocytes**.

5 48-hour cycles of invasion, lysing, and reinvasion cause the characteristic fevers and chills in the host.

11 …that invade the salivary gland.

4 Merozoites also invade red blood cells, where they grow, divide, and break out of (lyse) the cells.

(Start here) 1 The mosquito injects sporozoites into a human's blood when it feeds.

2 The sporozoites penetrate liver cells and develop into **merozoites**.

3 Merozoites can reinfect the liver, producing new generations.

FIGURE 14.7 Many parasites have complex life cycles. Different stages in the life cycle of the malaria parasite *Plasmodium* are adapted to life in two different hosts and to dispersal between hosts. After R. Buchsbaum, *Animals without Backbones,* 2nd ed., University of Chicago Press, Chicago (1948); M. Sleigh, *The Biology of Protozoa,* American Elsevier, New York (1973).

in a predigested food supply and retain for themselves little more than a highly developed capacity to produce eggs. Nonetheless, the life of a parasite is not easy. Host organisms have a variety of mechanisms to recognize invaders and destroy them. Moreover, parasites must disperse through a hostile environment to get from one host to another. Many accomplish this via complicated life cycles, often involving two or more hosts and at least one stage that can cope with the external environment.

The life cycle of the protozoan parasite *Plasmodium,* which causes malaria in humans, is a common textbook example (Figure 14.7). This parasite has two hosts, one a mosquito and the other a human or some other mammal, bird, or reptile. When an infected mosquito bites a human, cells called *sporozoites* are injected into the bloodstream with the mosquito's saliva. The sporozoites at first proliferate by mitosis in liver cells, then enter red blood cells as *merozoites,* where they feed on hemoglobin and grow. When a merozoite becomes large enough, it undergoes a series of divisions (asexual reproduction), and the daughter merozoites break out of the red blood cell. Each merozoite can enter a new red blood cell, grow, and repeat the cycle, which takes about 48 hours. (When the infection has built up to a high level, the emergence of daughter cells corresponds to periods of high fever resulting from the inflammation reaction of the host's immune system.) After several of these cycles, some of the merozoites that enter red blood cells change into sexual forms called *gametocytes.* If gametocytes are swallowed by a mosquito along with a meal of blood, they are transformed into eggs and sperm, and fertilization (sexual reproduction) takes place. The resulting zygotes penetrate the mosquito's gut wall and then undergo a series of divisions to produce sporozoites. These work their way into the salivary glands of the mosquito, from which they may enter a new host.

Parasite virulence and host resistance

The complex life histories of parasites involve a variety of interactions with hosts, and different sets of factors affect each stage of the parasite life cycle. The balance between parasite and host populations is influenced by the virulence of the parasite and by the immune response and other defenses of the host. **Virulence** is a measure of the capacity of a parasite to invade host tissues and proliferate in them. The virulence of an invading parasite may be reduced by actions of the host's immune system, including inflammation responses and the production of antibodies. Antibodies recognize and bind to foreign proteins, such as those on the outer surfaces of bacteria and protozoans, targeting them for attack by macrophage cells, which bind to the parasites and engulf them. The disabled parasites are then transported to the spleen and cleared from the body.

An immune response takes time to develop, however, and the delay gives the parasite a chance to multiply within a host. Parasites also have ways of circumventing the host's immune system. Some parasites produce chemical factors that suppress the immune system; this is the most troublesome feature of HIV. Others have surface proteins that mimic the host's own proteins and thus escape notice by the host's immune system. Trypanosomes, which are flagellate protists that cause sleeping sickness in humans, escape

the immune system by continually coating their surfaces with novel proteins produced by gene rearrangements.

Some schistosomes (trematode worms of the genus *Schistosoma*) excite an immune response when they enter a host, but do not succumb to antibody attack because they coat themselves with the host's proteins before its antibodies become numerous. As a consequence, other schistosomes that subsequently infect that host face a barrage of antibodies stimulated by the earlier entrance of the now entrenched individuals. When this response targets schistosome species closely related to the original parasite, it is known as **cross-resistance.** For example, many people in tropical regions are infected by schistosomes. One extremely virulent schistosome species, found only in humans, causes a debilitating disease called schistosomiasis or bilharziasis. But when a person has been infected previously by other schistosome species from wild game or domestic livestock, some of which have little effect on humans, the effect of infection by the bilharziasis parasite is moderated considerably.

Herbivory varies with the quality of plants as resources

Plants cannot hide or flee, so they rely on other tactics to escape their consumers. Plant defenses against herbivores include the inherently low nutritional value of most plant tissues as well as toxic compounds that plants produce and sequester for their defense. Plants also employ structural defenses, such as spines, hairs, tough seed coats, and sticky gums and resins (Figure 14.8).

The nutritional quality and digestibility of plants is critical to herbivores. Herbivores usually select plant food according to its nutrient content, preferring young leaves because of their low proportion of indigestible cellulose. Fruits and seeds are particularly nutritious compared with leaves, stems, and buds because of their higher nitrogen, fat, and sugar contents. Many plants use chemicals to reduce the availability of their proteins to herbivores, and thereby reduce their nutritional quality. Oaks and other plants sequester compounds called **tannins** in vacuoles in their leaves, which bind to plant proteins and inhibit their digestion. As a consequence, tannins can slow the growth of caterpillars and other herbivores that feed on the plant. However, insects that feed on tannin-rich plants can reduce the effects of tannins by producing detergent-like surfactants in their gut fluids, which tend to disperse tannin–protein complexes.

Tannins are an example of a **secondary compound:** a compound used by plants not for metabolism, but for

(a)
(b)

FIGURE 14.8 Structural and chemical defenses protect the stems and leaves of many plants from herbivores. (a) This cholla cactus (*Opuntia*) from Arizona is protected by sharp spines. (b) The white latex sap oozing from the stem of this milkweed plant (*Asclepias syriaca*) is toxic to most herbivores. Photos by (a) R. E. Ricklefs and (b) Bill Beatty/ Animals Animals, Earth Scenes.

other purposes—chiefly defense. Whereas tannins react with proteins of all types, many secondary compounds interfere with specific metabolic pathways or physiological processes of herbivores. Secondary compounds fall into three major classes based on their chemical structure: nitrogen compounds (ultimately derived from amino acids), terpenoids, and phenolics. The nitrogen compounds include indigestible structural compounds, such as lignin; alkaloids, including morphine (derived from poppies), atropine, and nicotine (from various members of the tomato family); nonprotein amino acids such as L-canavanine; and cyanogenic glycosides, which produce hydrogen cyanide (HCN). **Terpenoids** include essential oils, latex, and resins. Among the **phenolics,** many simple phenols have antimicrobial properties.

Some types of defensive chemicals are maintained at high levels in plant tissues at all times; these are called **constitutive defenses.** Others, known as **induced defenses,** are activated by herbivore damage in a manner analogous to the way foreign proteins induce an immune response in vertebrate animals (Figure 14.9). These chemicals increase dramatically in many plants following defoliation by herbivores (or the clipping of leaves by investigators). Wounding may cause the production of toxic, noxious, or nutrition-reducing compounds—locally in the area of a wound or systemically throughout the plant—that reduce subsequent herbivory. In some cases, these responses take only minutes or hours; in others, they require a new season of growth. When shoots of aspen, poplar, birch, and alder are heavily browsed by snowshoe hares, shoots produced during the following

Induction of chemical defenses in cotton plants following exposure to one mite species resulted in reduced populations of adult mites of another species...

...and their eggs.

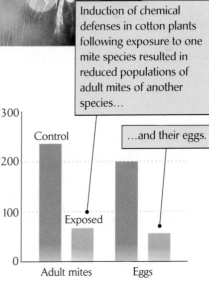

FIGURE 14.9 Plant defenses can be induced by herbivory. Mean numbers of the mite *Tetranychus urticae* were lower on cotton plants that had been previously exposed to a closely related mite species, *T. turkestani,* than on control plants with no previous mite exposure. This finding suggests that exposure to *T. turkestani* induced chemical defenses in the plants. From R. Karban and J. R. Carey, *Science* 225:53–54 (1984); photo by J. K. Clark.

year have exceptionally high concentrations of terpenes and phenolic resins, which are extremely unpalatable to hares. Some defenses can be induced even in untouched plants by volatile compounds released by neighboring plants that have been damaged—chemical communication among plants!

Inducibility suggests that some chemical defenses are too costly to maintain when herbivory is light or absent. Several studies have shown trade-offs between the production of defensive chemicals and plant growth. In addition, where soils are low in the nutrients required for the production of defensive chemicals, the costs of defense are relatively high. Undoubtedly, the strategies herbivores use to counter plant defenses are also costly, emphasizing once more the close connection between consumer–resource interactions and life history optimization.

Competition may be an indirect result of other types of interactions

Up to this point, we have considered direct interactions in which individuals of one species directly influence the well-being of individuals of another species. Interactions between resource and consumer individuals are the primary case in point. However, such interactions have indirect consequences for other species in an ecological system.

Consider a simple food chain, in which a predator species feeds on an herbivore species, which in turn feeds on a plant species. The interaction between the predator and its prey is a straightforward consumer–resource interaction; that is,

$$\text{consumer } (+) \rightarrow \text{resource } (-)$$

However, the herbivore also is a consumer, and when its population is reduced by its predator, the plant species enjoys the benefit of reduced herbivory. Thus, the predator and the plant are engaged in an indirect interaction:

$$\text{predator } (+) \rightarrow \text{herbivore } (-) \rightarrow \text{plant } (+)$$

Because such indirect interactions are felt across multiple trophic levels, they are often called *trophic cascades*. We shall consider interactions of this kind in more detail in Part 5 of this book.

Another kind of indirect interaction results from the use of a single resource by two or more consumers. This interaction can be diagrammed simply as

$$\text{consumer 1 } (+) \rightarrow \text{resource } (-) \leftarrow \text{consumer 2 } (+)$$

However, because each of the consumers reduces the availability of the resource to the other, the indirect interaction between the two consumer populations can be seen as

$$\text{consumer 1 } (-) \leftrightarrow \text{consumer 2 } (-).$$

Thus, even though individuals of consumer 1 and consumer 2 may never come into contact with each other, their populations are nonetheless affected by their use of a common resource. This interaction is referred to as *exploitation competition,* or *indirect competition,* and we shall consider it in detail in Chapter 16. However, individuals of different species that share a common resource may also interact directly through the kinds of antagonistic behaviors that we have already discussed in Chapter 9 in the context of interactions among individuals within species.

A competition–facilitation continuum

Interactions between species are not rigid; indeed, it is common for these interactions to change over the life cycle, and even for a direct relationship to become an indirect one. An increasingly important theme in the ecology of species interactions is the often-blurred distinction between competition and various forms of facilitation, such as commensalism and mutualism. A prominent case in point is the phenomenon of so-called nurse plants, in which individuals of one species facilitate the germination and growth of a second species. In desert environments of western North America, for example, several species of small trees, including paloverde and ironwood, provide protected sites for the establishment of various species of columnar cacti (Figure 14.10). The trees and the cacti may compete for water and soil nutrients, but small seedlings of the cacti have little effect on the resources available to established nurse plants, which fortuitously provide the seedlings with shade from the sun and protection from herbivores among their dense branches. Thus, early in its life cycle, the cactus is a commensal of the tree. As the cactus grows, it may have an increasingly negative effect on the resources available to its nurse plant, but the growth forms of the two species often differ so much that their growth is limited by different factors, and they can coexist.

Facilitation is a common theme in the development of biological communities, as we shall see in Chapter 19. The initial colonizers of newly exposed soil, for example, are often plants that can tolerate the heat and water stress of a completely open environment. The shade and organic soil materials provided by these pioneers allow other

each party to a mutualism is specialized behaviorally or physiologically to perform a function lacking in the other. Many mutualisms are symbioses, such as the relationship between the algae and fungi that make up lichens (see Figure 1.8). In other instances, such as seed dispersal mutualisms, the partners are only loosely associated and may interact mutualistically with a variety of species.

Mutualisms are widespread in the living world, and humans are no exception. Humans have beneficial gut bacteria that aid us in digestion and assimilation of nutrients. An unusual mutualism involves the honeyguides of southern Africa, which lead various mammals, including humans, to beehives. The birds eat the wax left behind after the humans have extracted the honeycombs. The honeyguides depend on a further mutualism with bacteria in their guts that secrete enzymes needed to digest the wax. We humans also have created our own mutualistic associations through the domestication of various pet, livestock, and crop species.

In fact, it is fair to say that all organisms are engaged in mutualistic interactions. The cells of all eukaryotic organisms are the product of ancient mutualisms between symbiotic bacteria. One partner in these relationships became a cellular organelle: the mitochondrion and, in plants, the chloroplast. These organelles even retain some of the original genetic material of their free-living ancestors.

Mutualisms occur in every kind of environment, but may be particularly important under stressful conditions. For example, the giant tubeworms living close to hydrothermal vents thousands of meters below the ocean surface depend entirely on symbiotic bacteria that can use energy from hydrogen sulfide in the hot vent water to synthesize organic molecules. Indeed, the worms lack mouths and digestive tracts as adults. Instead, they form special structures called trophosomes, which the bacteria invade through the skin from the surrounding water. Neither the bacteria nor the worms can live and reproduce without the other species.

In very general terms, mutualisms fall into three categories: trophic, defensive, and dispersive. The partners in **trophic mutualisms** are usually specialized in complementary ways to obtain energy and nutrients; hence the term *trophic*, which pertains to feeding relationships. We have just seen such a mutualisms between bacteria and giant tubeworms. Similarly, bacteria in the rumens of cows and other ungulates can digest the cellulose in plant fibers, which a cow's own digestive enzymes cannot break down. The cows benefit because they assimilate some of the by-products of bacterial digestion and metabolism for their own use (they also digest some of the bacteria). The

FIGURE 14.10 Nurse plants provide shade and protection for other species. This saguaro cactus established itself under a nurse plant, a paloverde tree. Without the nurse plant, it might not have survived the seedling stage in its harsh desert environment. Initially a commensal, the cactus may now be competing with its nurse plant for water and other resources. Photo by Bill Pogue/bpogue@billpogue.com.

species to invade the area, much as nurse plants facilitate the establishment of some desert cacti. However, as the new colonists grow, they capture progressively more sunlight and soil nutrients and shift from being commensals of the initial colonizers to being competitors.

Individuals of different species can collaborate in mutualistic interactions

Many interactions between species benefit both participants. For example, flowers provide honeybees with nectar, and the bees carry pollen between plants. These kinds of interactions are known as mutualisms. In most cases, bacteria benefit by having a steady supply of food in a warm, chemically regulated environment that is optimal for their own growth. In these cases, each of the partners supplies a limiting nutrient or energy source that the other cannot obtain by itself.

Species in **defensive mutualisms** receive food or shelter from their partners in return for defending those partners against their consumers. For example, in some marine ecosystems, specialized fishes and shrimps clean parasites from the skin and gills of other fish species (Figure 14.11). These cleaners benefit from the food value of parasites they remove, and the groomed fish are unburdened of some of their parasites. Cleaning mutualisms are most highly developed in clear, warm tropical waters. Many cleaners display their striking colors at locations, called cleaning stations, to which other fish come to be groomed. As might be expected, a few predatory fish species mimic the cleaners: when other fish come to them and expose their gills to be groomed, they get a bite taken out of them instead.

FIGURE 14.11 Cleaning mutualisms benefit both participants. The prawn *Lysmata amboinensis* is removing parasites from a moray eel. From this interaction, the prawn gets food, and the eel gets rid of some of its external parasites. Photo by Doug Perrine/DRK Photo.

ECOLOGISTS IN THE FIELD   Acacias house and feed the ants that protect them from herbivores.

Each population constantly adapts to evolutionary changes in other populations with which it interacts. These interactions shape the adaptations of all species, occasionally producing exceedingly intricate and complex living arrangements. One example, a mutualistic relationship between ants and acacias in Central America, drives home the lesson of this complexity.

Acacias and ants engage in mutually beneficial relationships in which the plants provide food and nesting sites for the ants, while the ants provide the plants with protection from herbivores and competing plants. The bull's-horn acacia (*Acacia cornigera*) has large, swollen, hornlike thorns with a tough woody covering and a soft pithy interior. To start a colony in an acacia, a queen ant of the species *Pseudomyrmex ferruginea* bores a hole in the base of one of these thorns and clears out some of the soft material inside to make room for her brood. In addition to housing the ants, the acacia provides carbohydrate-rich food for them in nectaries at the bases of the leaves as well as fats and proteins in the form of nodules, called Beltian bodies, at the tips of some leaves. The ants, in turn, protect their host plant from other insects. As the colony grows, ants occupy more and more of the acacia's thorns. A colony may increase to more than a thousand workers within a year, and may eventually include tens of thousands of workers. At any one time, about a quarter of the ants are outside their nests, actively gathering food and defending the acacia.

The relationship between *Pseudomyrmex* and *Acacia* is so tight that neither the ant nor the acacia can survive without the other. The ants require the nutrients provided by the acacia's Beltian bodies and cannot be reared on other foods. Acacias lacking ants are quickly devoured by insect herbivores. This last point was made dramatically by ecologist Daniel Janzen in a set of experiments conducted in Oaxaca, in southern Mexico. Janzen cut acacias to stimulate growth of new shoots and then kept ants from colonizing one set of plants. After 10 months of one such experiment, the shoots lacking ants weighed less than one-tenth as much as those with intact ant colonies, and they produced fewer than half the leaves and one-third the number of swollen thorns (Figure 14.12).

The mutualism between ants and acacias has been accompanied by adaptations in both species to increase the effectiveness of their association. For example, *Pseudomyrmex* is active both night and day—an unusual trait for ants—and thereby can provide protection for the acacia at all times. In addition, these ants have a true sting, like their wasp relatives, and will swarm vertebrate herbivores that attempt to feed on their host plant. The ants also clear away potential plant competitors by attacking

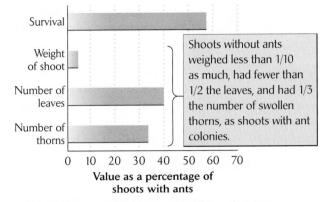

FIGURE 14.12 Some mutualists need their partners to survive and grow. Removal of ant (*Pseudomyrmex*) colonies from shoots of bull's-horn acacia (*Acacia cornigera*) in Oaxaca, Mexico, decreased their growth and survival rates. Data from D. H. Janzen, *Evolution* 20:249–275 (1966).

seedlings near their host plant's base as well as any vines or overhanging branches of other plants. In a similar adaptive gesture, the acacia retains its leaves throughout the year, although rainfall in its environment is extremely seasonal. By doing so, the bull's-horn acacia provides a year-round source of food for the ants. Most related plant species lose their leaves during the dry season.

Relegating defense against herbivores to a mercenary apparently makes good evolutionary sense under some circumstances. The fidelity of the defender is guaranteed by its absolute dependence on the host for food. One can imagine, however, that such an intimate relationship must be the result of a long history of increasingly close association. Acacia species that lack ants provide their own defenses against herbivores by sequestering toxic chemicals in their leaves. This strategy can be costly compared with providing food and nesting places for ants, but it also

allows the plant to shed its leaves during the dry season and therefore to live in more arid environments than ant-dependent acacias can tolerate. ▌

Animals participating in **dispersive mutualisms** include those that transport pollen between flowers in return for rewards such as nectar, or that eat nutritious fruits and disperse the seeds they contain to suitable habitats. Dispersive mutualisms rarely involve close living arrangements. Seed dispersal mutualisms are not usually highly specialized; for example, a single bird species may eat many kinds of fruit, and each kind of fruit may be eaten by many kinds of birds. Plant–pollinator relationships tend to be more restrictive because it is in a plant's interest that a flower visitor carry its pollen to another plant of the same species.

MORE ON THE WEB *Seed Dispersal.* The seeds of many plant species are widely distributed by animals, often to habitats favorable for germination and growth.

MORE ON THE WEB *Pollination.* Plants have many ways of manipulating their pollinators so as to increase the efficiency of pollen transfer between individuals.

In the next chapter, we shall consider the dynamics of interactions between consumers and their resources, building on the development of models for individual populations. A key insight is that a resource and its consumer have a unique interaction that can result in cyclic changes in populations of both species. Such insights reinforce the idea that ecological systems create their own dynamic properties that cannot be inferred from the individual components of the system, but rather depend on the ways in which species interact.

SUMMARY

1. Species interactions influence population dynamics, define the structure of ecological systems, and provide a rich context for evolution.

2. The primary types of species interactions are consumer–resource interactions, competition, and mutualism, although many variations on these themes blur the boundaries between them.

3. Species can benefit from (+) or suffer from (−) an interaction, or be unaffected by it (0). We can represent

consumer–resource interactions as +/−, competition as −/−, and mutualism as +/+.

4. Individuals of different species that live in close association are said to form a symbiosis. Symbioses may be mutualistic relationships, as in the case of the algae and fungi that form lichens, or consumer–resource relationships, as in the case of many parasites and their hosts.

5. Predators are consumers that remove individuals from prey populations as they consume them. Parasites consume portions of living host organisms, but usually do not kill them. Parasitoids (mostly small flies and wasps) kill the hosts they consume, but not immediately.

6. Herbivores can function either as predators, in the case of consumers that remove whole plants, or parasites, in the case of grazers or browsers that remove only a portion of a plant's tissues.

7. Consumer–resource interactions can also be classified by the duration and intimacy of the association between individuals, as well as by the probability that resource individuals will be killed.

8. Organisms avoid being consumed through tactics such as avoiding detection, remaining in safe refuges, or fleeing predators, as well as by means of chemical and structural defenses. All of these tactics impose costs on their users.

9. Many parasites are characterized by complex life cycles that may include multiple hosts and stages specialized to make the difficult journey from one host to another.

10. Parasite–host interactions often evolve a delicate balance between the immune response and other defenses of the host and parasite virulence.

11. Plants have numerous structural and chemical defenses to deter herbivores. These defenses include factors that influence the nutritional quality and digestibility of plant parts as well as specialized chemicals—secondary compounds—that have negative effects on herbivores.

12. Plant defenses may be constitutive, meaning that they are always present, or they may be induced by herbivory. The existence of induced defenses suggests that defenses are costly and impose trade-offs.

13. When two or more consumers utilize the same resource, each reduces resource availability for the other. This type of indirect interaction is called exploitation competition.

14. Mutualisms between species may involve trophic, defensive, and dispersive interactions. Each partner in a mutualism usually provides some product (such as nutrients) or service (such as defense against consumers) that the other cannot provide for itself.

REVIEW QUESTIONS

1. Compare and contrast the feeding habits of predators, parasites, and parasitoids.

2. If animal prey can reduce their risk of predation by hiding, why not hide all the time?

3. In parasite–host relationships, what abilities might natural selection favor in the parasite and in the host?

4. Some animals defer reproduction so that they can put more of their energy into rapid growth to become too large for predators to eat. What trade-off are these animals making?

5. Given the ubiquity of the defenses that have evolved in plants, what would you predict about the evolved responses of their herbivores?

6. When a plant population possesses inducible defenses, what might this suggest about the frequency of herbivory on this population?

7. Why is competition between two species for a shared resource considered an indirect interaction rather than a direct interaction?

8. What benefits do trophic mutualisms, defensive mutualisms, and dispersive mutualisms have in common?

SUGGESTED READINGS

Abrahamson, W. G. 1989. *Plant–Animal Interactions*. McGraw-Hill, New York.

Armbruster, W. S. 1992. Phylogeny and the evolution of plant–animal interactions. *BioScience* 42:12–20.

Barbosa, P., P. Gross, and J. Kemper. 1991. Influence of plant allelochemicals on the tobacco hornworm and its parasitoid, *Cotesia congregata*. *Ecology* 72:1567–1575.

Bryant, J. P. 1981. Phytochemical deterrence of snowshoe hare browsing by adventitious shoots of four Alaskan trees. *Science* 213:889–890.

Douglas, A. E. 1998. Nutritional interactions in insect–microbial symbioses: Aphids and their symbiotic bacteria *Buchnera*. *Annual Review of Entomology* 43:17–38.

Fritz, R. S., and E. L. Simms (eds.). 1992. *Plant Resistance to Herbivores and Pathogens: Ecology, Evolution, and Genetics*. University of Chicago Press, Chicago.

Godfray, H. C. J. 1994. *Parasitoids: Behavioral and Evolutionary Ecology*. Princeton University Press, Princeton, N.J.

Hay, M. E. 1991. Marine–terrestrial contrasts in the ecology of plant chemical defenses against herbivores. *Trends in Ecology and Evolution* 6:362–365.

Horton T. R., T. D. Bruns, and T. Parker. 1999. Ectomycorrhizal fungi associated with *Arctostaphylos* contribute to *Pseudotsuga menziesii* establishment. *Canadian Journal of Botany* 77:93–102.

Janzen, D. H. 1966. Coevolution of mutualism between ants and acacias in Central America. *Evolution* 20:249–275.

Johnson N. C., J. H. Graham, and F. A. Smith. 1997. Functioning of mycorrhizal associations along the mutualism–parasitism continuum. *New Phytologist* 135:575–585.

Karban, R., et al. 2000. Communication between plants: Induced resistance in wild tobacco plants following clipping of neighboring sagebrush. *Oecologia* 125:66–71.

Karban, R., et al. 2006. Damage-induced resistance in sagebrush: Volatiles are key to intra- and interplant communication. *Ecology* 87:922–930.

Martinsen, G. D., E. M. Driebe, and T. G. Whitham. 1998. Indirect interactions mediated by changing plant chemistry: Beaver browsing benefits beetles. *Ecology* 79:192–200.

Pérez-Brocal, V. et al. 2006. A small microbial genome: The end of a long symbiotic relationship? *Science* 314(5797):312–313.

Raghu, S., and C. Walton. 2007. Understanding the ghost of *Cactoblastis* past: Historical clarifications on a poster child of classical biological control. *BioScience* 57(8):699–705.

Relyea, R. A., and E. E. Werner. 1999. Quantifying the relation between predator-induced behavior and growth performance in larval anurans. *Ecology* 80:2117–2124.

Riegler, M., and S. L. O'Neill. 2007. Evolutionary dynamics of insect symbiont associations. *Trends in Ecology and Evolution* 22(12):625–627.

Sagers, C. L., and P. D. Coley. 1995. Benefits and costs of defense in a Neotropical shrub. *Ecology* 76:1835–1843.

Stein, B. A. 1992. Sicklebill hummingbirds, ants, and flowers. *BioScience* 42:27–33.

van Dam, N. M., K. Hadwich, and I. T. Baldwin. 2000. Induced responses in *Nicotiana attenuata* affect behavior and growth of the specialist herbivore *Manduca sexta*. *Oecologia* 122:371–379.

Werner, E. E., and M. A. McPeek. 1994. Direct and indirect effects of predators on two anuran species along an environmental gradient. *Ecology* 75:1368–1382.

Werner, E. E., et al. 1983. An experimental test of the effects of predation risk on habitat use in fish. *Ecology* 64:1540–1548.

Werren, J. H. 1997. Biology of *Wolbachia*. *Annual Review of Entomology* 42:587–609.

CHAPTER 15

Dynamics of Consumer–Resource Interactions

Population cycles were part of the lore of population ecology even before 1924, when Charles Elton's paper "Periodic Fluctuations in the Numbers of Animals: Their Causes and Effects" was published in the *British Journal of Experimental Biology*. Most of Elton's data concerned fur-bearing mammals in the Canadian boreal forest and tundra, where the Hudson's Bay Company had kept detailed records of the numbers of furs it purchased from trappers each year (Figure 15.1). Data for the lynx and its principal prey, the snowshoe hare, revealed dramatic and regular population fluctuations (Figure 15.2). Each cycle lasted approximately 10 years, and the cycles of the two species were highly synchronized, with peaks in lynx abundance tending to trail those in hare abundance by a year or two. The cycles were so predictable that Elton once remarked that an Eskimo hunter "might have reflected that his good luck and his bad luck chased each other with sufficient regularity to amount to a natural law." Furthermore, the lynx–hare population cycles were strongly correlated over large areas across Canada, with synchrony recognizable at distances of more than 1,000 km.

What causes these cycles? Early in the twentieth century, many hypotheses were proposed to explain population cycles, including the regular waxing and waning of sunspots—large magnetic storms on the sun that influence the earth's upper atmosphere. As population ecology grew as a discipline, however, it became evident that the cyclic behavior of many populations could be explained by dynamic interactions between predators and their prey.

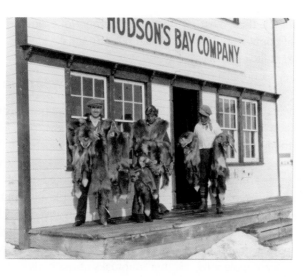

FIGURE 15.1 Fur trapping records revealed population cycles. Canadian fur trappers sold pelts to the Hudson's Bay Company in Manitoba, whose records of purchases provided the data that made it possible for Charles Elton to document pronounced population cycles of fur-bearing mammals. Courtesy of Hudson's Bay Company Archives, Provincial Archives of Manitoba.

CHAPTER CONCEPTS

- Consumers can limit resource populations

- Many predator and prey populations increase and decrease in regular cycles

- Simple mathematical models can reproduce cyclic predator–prey interactions

- Pathogen–host dynamics can be described by the S-I-R model

- The Lotka–Volterra model can be stabilized by predator satiation

- A number of factors can reduce oscillations in predator–prey models

- Consumer–resource systems can have more than one stable state

---

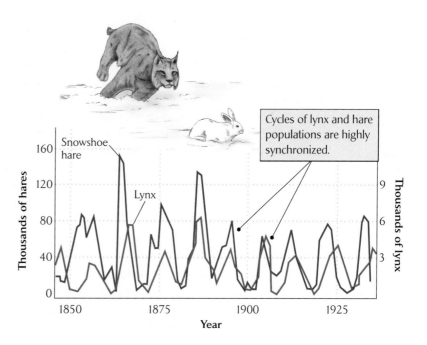

FIGURE 15.2 Population cycles of predators and their prey may be highly synchronized. According to the records of the Hudson's Bay Company, population cycles of the lynx and the snowshoe hare track each other closely. After D. A. MacLulich, *University of Toronto Studies*, Biol. Ser. No. 43 (1937).

The basic question of population biology is this: What factors influence the size and stability of populations? In Chapter 12, we saw how density-dependent factors modify the responses of birth and death rates to population density, and how delays in those responses can destabilize populations and cause fluctuations in their numbers. We shall see in the next chapter that competition for resources from other species can depress the growth rate of a population, leading to extinction in extreme cases. Most species are both consumers and resources for other consumers, however, so it is also important to ask whether populations are limited primarily by what they eat or by what eats them. In this regard, studies of predator–prey interactions attempt to answer at least two fundamental questions: First, do predators reduce the size of prey populations substantially below the carrying capacities set by resources for the prey? Second, do consumer–resource interactions cause populations to fluctuate independently of variation in the environment? The first question is of great practical concern to those interested in the management of crop pests, game populations, and endangered species. It also has far-reaching implications for understanding the interactions among species that share resources and, therefore, for understanding the structure of biological communities. The second question is motivated by observations of predator–prey cycles in nature and directly addresses the issue of stability in natural systems. Ecologists have tried to answer these questions with a combination of observation, theory, and experimentation.

Consumers can limit resource populations

According to the theory of density dependence developed in Chapter 11, populations are self-limiting because resources become scarcer as consumer populations grow. When resources are limited, mortality increases because individuals are more likely to die from starvation, predation, and disease, and a population may not be able to produce enough offspring to compensate for those deaths. Thus, populations of consumers are self-regulated because of their effect on their own resources. However, because consumers reduce their resources, it is likely that consumers contribute to the regulation of resource populations as well. Thus, populations are regulated from above and below. An important question for ecologists is the degree to which consumers influence the sizes of the populations that are their resources. As we shall see

through a number of examples, the influence of consumers can be substantial.

Predation on cyclamen mites

The cyclamen mite is a pest of strawberry crops in California. Populations of these mites are usually kept under control by a species of predatory mite of the genus *Typhlodromus*. Cyclamen mites typically invade a strawberry crop shortly after it is planted, but their populations usually do not reach damaging levels until the second year. *Typhlodromus* mites usually invade fields during the second year. Because they are such efficient predators, these mites rapidly reduce cyclamen mite populations and prevent further outbreaks of the pests.

Greenhouse experiments conducted during the early years of research on biological control of pests demonstrated the role of predation in keeping cyclamen mites in check. One group of strawberry plants was stocked with both predator and prey mites; a second group was kept predator-free by regular applications of parathion, an insecticide that kills the predator but does not affect the cyclamen mite. Throughout the study, populations of cyclamen mites remained low in plots they shared with *Typhlodromus*, but grew to damaging numbers in predator-free plots (Figure 15.3). In field plantings of strawberries, cyclamen mites also reached damaging levels where predators were eliminated by parathion (a good example of an insecticide having an undesired effect), but they were effectively controlled in untreated plots. When a cyclamen mite population began to increase in an untreated planting, the predator population quickly mushroomed and reduced the outbreak. On average, cyclamen mites were about 25 times more abundant in the absence of predators than in their presence.

Typhlodromus owes its effectiveness as a predator to several factors besides its voracious appetite. Most importantly, its populations can increase as rapidly as those of its prey. Cyclamen mites lay three eggs per day over the 4 or 5 days of their reproductive life span; *Typhlodromus* lay two or three eggs per day for 8–10 days. Another critical attribute of *Typhlodromus* is its ability to survive in the absence of accessible prey. During winter, when cyclamen mite populations dwindle to a few individuals hidden in crevices and folds of leaves in the crowns of strawberry plants, the predatory mites subsist on honeydew produced by aphids and whiteflies. When predators appear to control prey populations—and *Typhlodromus* is no exception—the predators usually exhibit a high reproductive capacity compared with that of their prey,

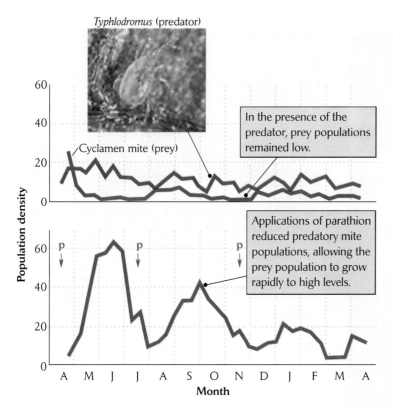

Typhlodromus (predator)

Cyclamen mite (prey)

In the presence of the predator, prey populations remained low.

Population density

Applications of parathion reduced predatory mite populations, allowing the prey population to grow rapidly to high levels.

A M J J A S O N D J F M A

Month

FIGURE 15.3 Predators can control prey populations. Infestations of strawberry plots by cyclamen mites (*Tarsonemus pallidus*) were tracked in the presence (*above*) and in the absence (*below*) of the predatory mite *Typhlodromus*. Prey populations are expressed as numbers of mites per leaf; predator levels are the numbers of leaflets (out of 36) on which one or more *Typhlodromus* were found. Parathion treatments are indicated by "p." After C. B. Huffaker and C. E. Kennett, *Hilgardia* 26:191–222 (1956). Photo courtesy of IPM Program, Cornell University.

combined with strong dispersal powers and an ability to switch to alternative food resources when their primary prey are unavailable.

Strong consumer effects—even as great as the level of control exerted by *Typhlodromus* on the cyclamen mite—are widespread, and they are not unique to terrestrial ecosystems—quite the contrary. Experiments have demonstrated that sea urchins in rocky shore communities control the populations of the algae on which they feed. The simplest experiments consisted of removing sea urchins and following the subsequent growth of their algal prey. When urchins were kept out of tide pools and off subtidal rock surfaces, the biomass of algae quickly increased, indicating that herbivory by urchins reduced algal populations below the level that the environment could support in the absence of consumers.

Different kinds of algae also appeared after herbivore removal. Large brown algae flourished and began to overgrow and replace other kinds of algae more able to withstand intense grazing by sea urchins. Among those that disappeared were coralline algae, whose hard, shell-like structure deters grazers (see Figure 23.8), and small green algae, whose short life cycles and high reproductive rates enable them to keep ahead of consumers in population growth. Most small species of algae could not grow under the deep shade of the thick stands of large brown algae. Thus, consumers influence not only the size of resource populations, but also the species composition of the biological community.

Herbivores and plant populations

We saw in Chapter 14 that cactus moths effectively control populations of the prickly pear cactus in Australia. Herbivorous insects have been used in other situations to control introduced weeds. Consider the example of Klamath weed, a European species that is toxic to livestock, which accidentally became established in northern California in the early 1900s (Figure 15.4). By 1944, the weed had spread over nearly a million hectares of rangeland in 30 counties. Biological control specialists borrowed an herbivorous beetle of the genus *Chrysolina* from an Australian control program. Within 10 years after the first beetles were released, Klamath weed had been all but obliterated as a range pest. Range biologists estimated its abundance to have been reduced by more than 99%.

In grasslands, native herbivores (mostly insects and grazing mammals) typically consume 30%–60% of the aboveground vegetation (Figure 15.5). Their influence on

(a)

(b)

FIGURE 15.4 Herbivores can control plant populations.
(a) Klamath weed, or St. John's wort (*Hypericum perforatum*),
became a widespread pest following its introduction into the
western United States. (b) The infestation of Klamath weed
was finally brought under control by introduced beetles of the
genus *Chrysolina*. Klamath weed contains high concentrations
of the alkaloid hypericin, which has therapeutic effects in
small quantities, but is dangerous to cattle and sheep. Photo
(a) by David Sieren/Visuals Unlimited; photo (b) courtesy of Verein für
Naturwissenschaftliche Heimatforschung zu Hamburg.

**FIGURE 15.5 Herbivory has dramatic effects on plant
production.** The area at the left, on the slope of Mauna Loa in
Hawaii, is protected from cattle grazing by a barbed-wire fence.
Photo by R. E. Ricklefs.

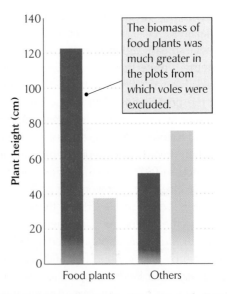

> The biomass of
> food plants was
> much greater in
> the plots from
> which voles were
> excluded.

**FIGURE 15.6 The effects of herbivores on plant production
can be measured by exclosure experiments.** Researchers
surrounded some grassland plots with fences that excluded voles.
The biomasses (summed plant height per 100 cm^2) of food
plants used by voles and of other plants were measured in the
fenced plots and in unfenced control plots after 2 years. After
G. O. Batzli and F. A. Pitelka, *Ecology* 51:1027–1039 (1970).

plant production is revealed by exclosure experiments. In one study in California, wire fences were constructed to keep voles out of small areas of grassland. At the end of the 2-year study, the food plants of the voles (mostly annual grasses) had grown more and produced more seeds within the fenced plots than outside the exclosures, where voles continued to graze. Perennial grasses and herbs not included in the voles' diet were not directly affected by the exclosures (Figure 15.6).

Many predator and prey populations increase and decrease in regular cycles

The population cycles of hares and their lynx consumers, described at the opening of this chapter, have periods of 9 to 10 years, on average. The periods of population cycles (usually measured from peak to peak) vary from species to species, and may even vary within a species. In Canadian boreal forest and tundra, cycles in resource populations of large herbivores, such as snowshoe hares, muskrat, ruffed grouse, and ptarmigan, are typically 9–10 years long. Four-year cycles are typical of resource populations of small herbivores, such as voles, mice, and lemmings. Predators that feed on short-cycle herbivores (Arctic foxes, rough-legged hawks, snowy owls) themselves have short population cycles; predators of larger herbivores (red foxes, lynx, marten, mink, goshawks, horned owls) have longer cycles.

The closely synchronized population cycles of some predators and their prey suggest that these oscillations could result from the way in which predator and prey populations interact with each other. In simple terms, predators eat prey and reduce their numbers. Consequently, predators go hungry, and their numbers drop as well. With fewer predators around, the remaining prey survive better, and their populations begin to increase. With increasing numbers of prey, the predators also begin to increase again, thereby completing the cycle.

Time delays and population cycles

We saw in Chapter 12 that population cycles can result from time delays in the responses of birth and death rates to changes in the environment. Most predator–prey interactions also have built-in time delays because of the time required by both populations to produce offspring. Population models predict that the period of a population cycle should be about four to five times the time delay. Thus, the 4-year and 9–10-year population cycles

of mammals inhabiting boreal forest and tundra environments are consistent with time delays of 1 and 2 years, respectively. Such time delays probably result from the intervals between birth and sexual maturity in both populations. In other words, the influence of prey availability in a particular year may not be felt strongly in a predator population until young born in that year are themselves old enough to reproduce.

In natural systems, one rarely finds a single species of consumer feeding on a single resource species. Even in the case of the lynx–hare interaction, additional predators, including coyotes and various birds of prey, feed on hares, and those predators also feed on a variety of small mammals, including ground squirrels and grouse. One field study in southern Yukon followed populations of hares on an island in a large lake and in adjacent mainland areas surrounding the lake. The island population was exposed to relatively few predator species and low predator numbers. The hare population on the island cycled in synchrony with mainland hare populations, but with lower amplitude (peak-to-trough variation) (Figure 15.7). While the peak population densities in the two areas were similar, the lowest densities of the hare population on the island were well above those on the mainland, suggesting that greater predator pressure increases the intensity of population cycles, but not their period. Predators moving back and forth from island to mainland, especially birds of prey, might have maintained the synchrony of the island and mainland populations.

Long-term observations show that synchronized population fluctuations may continue more or less unchanged over many cycles, so this dynamic behavior appears to represent a stable interaction between predators and prey. However, the period and intensity of a cycle also depend on the physical environment. For example, owl (predator) and vole (prey) populations cycle dramatically over 4-year periods in northern Scandinavia, but fluctuate annually in the milder climate of southern Sweden. One explanation for this difference is that the prolonged, heavy snow cover in the north protects the voles from the owls during the long winter, thus creating a delay in the effects of owl populations on vole survival. In the south, owls can hunt voles throughout much of the winter, so the effects of their numbers on vole survival are felt more quickly. With climates warming in northern Europe, however, winter snow cover in the north is briefer, and the cyclic behavior of the system appears to be breaking down (Figure 15.8). Another important difference is that northern vole populations support several highly specialized predators, such as weasels, whereas predators in the south tend to feed on a variety of prey organisms. As we shall see below, the

FIGURE 15.7 Hare populations fluctuated less on an island with few predators than on the surrounding mainland. Hare populations were tracked on Jacquot Island (5 km²), in a lake in southern Yukon, Canada, and on the nearby mainland during spring in 1977–2001. After C. J. Krebs et al., *Canadian Journal of Zoology* 80:1442–1450 (2002).

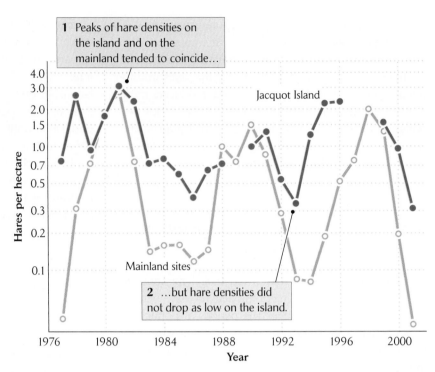

1 Peaks of hare densities on the island and on the mainland tended to coincide...

Jacquot Island

Mainland sites

2 ...but hare densities did not drop as low on the island.

use of alternative prey tends to stabilize predator–prey systems and may reduce cycling by eliminating the dependence of predators on a single type of prey.

Periodicity in pathogen–host relationships

Although pathogens are not strictly predators, their interactions with host species are also influenced by time delays. Most importantly, the development of immune responses removes host individuals from the susceptible population (because they are now immune) and slows the further spread of a pathogen. Measles, a highly contagious viral disease that nonetheless stimulates lifelong immunity, typically produces epidemics at 2-year intervals in unvac-cinated human populations (Figure 15.9). Two years are required for a population to accumulate a high enough density of newly susceptible infants to sustain a measles outbreak.

Host population density also influences the rate of pathogen transmission between hosts. At high host densities, pathogens can generate population cycles when they increase host mortality or impair host reproduction. Pathogens infect individuals more readily in dense than in sparse populations because they come into contact with new hosts more frequently. Under these conditions, high disease prevalence can drive host populations to low levels. As host population density decreases, however, the chain of contagion breaks, fewer host individuals become sick, and the host population can grow once more.

FIGURE 15.8 Population cycles of voles in northern Scandinavia have damped out with warmer winter temperatures. To estimate vole population densities, investigators trapped voles and recorded the total number of individuals of three vole species per 100 trap nights (the vole density index) during spring between 1972 and 2003. After B. Hörnfeldt, T. Hipkiss, and U. Uklund, *Proc. R. Soc. London* B 272:2045–2049 (2005).

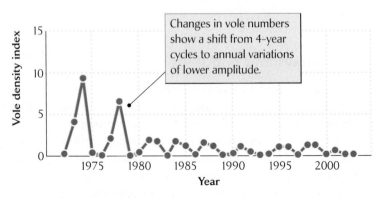

Changes in vole numbers show a shift from 4-year cycles to annual variations of lower amplitude.

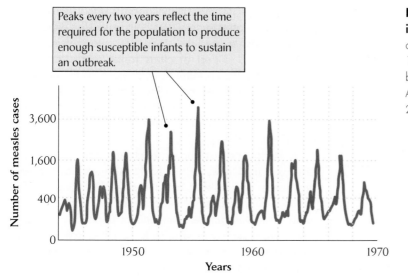

Peaks every two years reflect the time required for the population to produce enough susceptible infants to sustain an outbreak.

FIGURE 15.9 Development of host immunity influences pathogen population cycles. Cases of measles reported in London, England, between 1944 and 1968 (before a measles vaccine had been developed) peaked about every 2 years. After P. Rohani, D. J. D. Earn, and B. T. Grenfell, *Science* 286:968–971 (1999).

This pattern is evident in populations of the forest tent caterpillar, which periodically increase to such high densities that they can defoliate stands of trees over thousands of square kilometers. These infestations are usually brought under control by a virulent pathogen, the nuclear polyhedrosis virus, which causes high mortality of tent caterpillars at high population densities. In many regions, tent caterpillar infestations last about 2 years before the virus brings its host population under control. In other regions, however, infestations may last up to 9 years. Jens Roland, at the University of Alberta, discovered that forest fragmentation, which creates abundant forest edge habitat, tends to prolong outbreaks of the tent caterpillar (Figure 15.10). Apparently, caterpillars living at forest edges are exposed to more intense sunlight, which inactivates the nuclear polyhedrosis virus. In this case, habitat structure clearly has important secondary effects on population cycles.

Creating predator–prey cycles in the laboratory

During the 1920s, biologists became interested in using predators and pathogens to control populations of crop and forest pests. These initiatives created the need to develop laboratory models that biologists could use to explore predator–prey cycles under controlled conditions and identify their potential causes. Developing realistic models turned out to be difficult, however, because of the simplicity of laboratory cultures.

Some of the earliest experimental studies were carried out by the Russian biologist G. F. Gause on protists. His

initial results were disappointing because, in the simple environment of a test tube, predators were extremely efficient: they typically ate their prey populations to extinction and then died out themselves. Nonetheless, Gause found that this hopeless situation could be stabilized by providing refuges where some of the prey could escape predation. In one experiment, Gause introduced *Paramecium* as the prey and another ciliated protist, *Didinium,* as the predator into a nutritive medium in a plain test tube. By creating such a simple environment, Gause had stacked the deck against the prey; the predators readily found all of them, and when the last *Paramecium* had been consumed, the predators starved. In a second experiment, Gause added some structure to the environment by placing glass wool, in which *Paramecium* could find refuge from the predators, at the bottom of the test tube. The tables having been turned, the *Didinium* population starved after consuming all the readily available prey, but the *Paramecium* population was restored by individuals concealed in the glass wool.

Gause finally achieved recurring oscillations in predator and prey populations by periodically adding small numbers of predators—restocking the pond, so to speak. Repeatedly adding individuals to an experimental culture corresponds to natural recolonization of a habitat patch by individuals from other areas. This pattern of extinction and recolonization is reminiscent of the interaction between the cactus moth and the prickly pear: recall that the cactus escapes annihilation by dispersing to predator-free areas. In the end, however, Gause was unable to achieve predator–prey population cycles in a closed system. His failure underscores the significance of C. B. Huffaker's experiments, which we explore next.

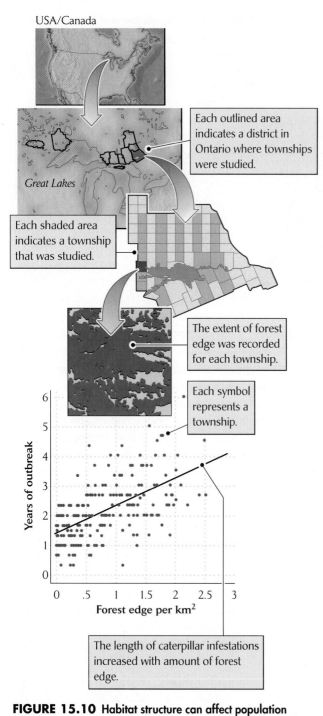

USA/Canada

Each outlined area indicates a district in Ontario where townships were studied.

Great Lakes

Each shaded area indicates a township that was studied.

The extent of forest edge was recorded for each township.

Each symbol represents a township.

The length of caterpillar infestations increased with amount of forest edge.

FIGURE 15.10 Habitat structure can affect population cycles. Researchers recorded the lengths of infestations of forest tent caterpillars in a number of townships in Ontario, Canada, with different amounts of forest edge. After J. Roland, *Oecologia* 93:25–30 (1993).

ECOLOGISTS IN THE FIELD

Huffaker's experiments on mite populations. C. B. Huffaker, a biologist at the University of California at Berkeley who pioneered the biological control of crop pests, attempted to produce an environment in the laboratory that would allow predator and prey to persist without restocking either population. Huffaker used the six-spotted mite (*Eotetranychus sexmaculatus*), a pest of citrus fruits, as the prey and another mite, *Typhlodromus occidentalis*, as the predator. He established experimental populations on trays within which he could vary the number, exposed surface area, and dispersion of the oranges he provided as food for the prey (Figure 15.11).

In Huffaker's first studies, each tray had 40 positions arranged in 4 rows of 10 each. At each position was an orange or a rubber ball the size of an orange. The exposed surface area of the oranges was varied by covering them with different amounts of paper, the edges of which were sealed in wax to keep mites from crawling underneath. In most experiments, Huffaker first established a prey population of 20 females per tray, then introduced two female predators 11 days later. (Both species reproduce parthenogenetically, so males were not required.)

When six-spotted mites were introduced to the trays alone, their populations leveled off at between 5,500 and 8,000 mites per tray. When predators were added, their numbers increased rapidly, and they soon wiped out the prey population, then became extinct themselves. Although the predators always eliminated the prey, the distribution of the exposed areas of the oranges influenced the course of extinction. When the exposed areas were in adjacent positions, the predators could travel quickly between them. In those cases, prey populations reached maxima of only 113–650 individuals and were driven to extinction within 23–32 days. When the same amount of exposed orange area was dispersed randomly throughout the 40-position tray, however, prey populations reached maxima of 2,000–4,000 individuals and persisted for 36 days on average. Thus, Huffaker could prolong the survival of the prey population by providing it with remote areas of suitable habitat to which predators dispersed slowly.

Huffaker reasoned that if predator dispersal could be further retarded, the two species might coexist. To accomplish this, he made the environment more spatially complex and introduced barriers to predator dispersal. The number of possible orange positions was increased to 120, and an exposed feeding area equivalent to six oranges was dispersed over all 120 positions. A mazelike pattern of Vaseline barriers was placed among the oranges to slow dispersal of the predators. *Typhlodromus* predators get around by walking, but six-spotted mites spin a silk line that they use like a parachute to float on wind currents. To take advantage of this behavior, Huffaker placed vertical wooden pegs throughout the trays, which the six-spotted mites used as jumping-off points in their wanderings. This arrangement finally produced a series of three population cycles over the 8 months of the experiment (Figure 15.12). The distribution of predators and prey throughout the trays continually shifted as the prey, on the way to extermination

(a)

(b)

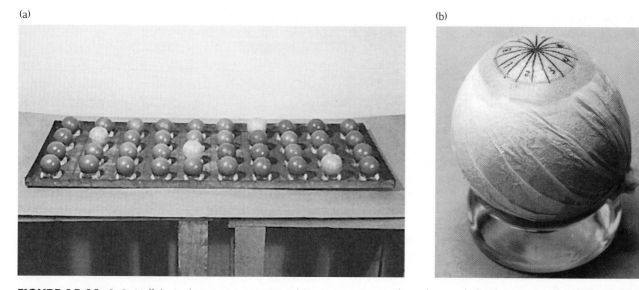

FIGURE 15.11 C. B. Huffaker's classic experiment tested the parameters of predator–prey coexistence. (a) In each experimental tray, four oranges, half exposed, are distributed at random among the 40 positions in the tray. Other positions are occupied by rubber balls. (b) Each orange is wrapped with paper and its edges sealed with wax. The exposed area has been divided into numbered sections to facilitate counting the mites. Courtesy of C. B. Huffaker, from C. B. Huffaker, *Hilgardia* 27:343–383 (1958).

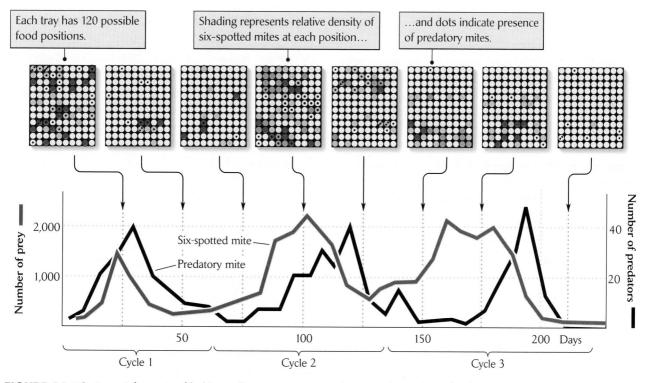

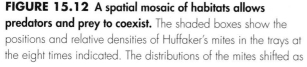

FIGURE 15.12 A spatial mosaic of habitats allows predators and prey to coexist. The shaded boxes show the positions and relative densities of Huffaker's mites in the trays at the eight times indicated. The distributions of the mites shifted as the prey colonized new feeding areas, staying a jump ahead of their predators. From C. B. Huffaker, *Hilgardia* 27:343–383 (1958).

in one feeding area, recolonized the next a jump ahead of their predators. Thus, Huffaker had effectively created a metapopulation in the laboratory.

Despite the tenuousness of the predator–prey cycle that was achieved, Huffaker's experiment demonstrated that a spatial mosaic of suitable habitats could enable predator and prey populations to coexist through time. Two kinds of time delays caused the populations to cycle: one resulting from the slow dispersal of predators between food patches, and the other resulting from the time needed for predator numbers to increase. As in Gause's experiments with protozoans, predator and prey could not coexist in the absence of suitable refuges for the prey. Huffaker created those refuges by dispersing food patches and creating barriers to predator movements. When the environment is complex enough that predators cannot easily find scarce prey, stable populations or stable population cycles can be achieved. ▌

Simple mathematical models can reproduce cyclic predator–prey interactions

Even before Huffaker's experimental creation of predator–prey cycles in the laboratory, theoretical biologists had developed mathematical models in an attempt to reproduce this population phenomenon on paper. Alfred J. Lotka and Vito Volterra independently developed the first mathematical descriptions of predator–prey interactions during the 1920s. The **Lotka–Volterra model** predicts oscillations in the abundances of predator and prey populations, with predator numbers lagging behind those of their prey.

The Lotka–Volterra model calculates the rate of change in the prey population and the rate of change in the predator population as each is influenced by the abundance of the other. Following a common convention, we designate the number of predator individuals by P and the number of prey individuals by V (think of V for "victim"). The rate of change in the prey population can be written in words as

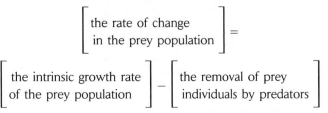

The first term is the unrestricted exponential growth of the prey population in the absence of predators, which we find by multiplying the exponential growth rate (r) by

the number of prey individuals (V). The second term is the removal of prey by predators, over and above other causes of death, cVP. The Lotka–Volterra model assumes that predation varies in direct proportion to the probability of a random encounter between a predator and a prey individual, which is the product of the prey and predator populations, VP. Accordingly, the rate of change of the prey population is given by

$$\frac{dV}{dt} = rV - cVP$$

where c is a coefficient expressing the efficiency of predation (think of c for "capture efficiency").

The growth rate of the predator population also has two components: (1) the birth rate, which depends on number of prey captured; and (2) a death rate imposed from outside the system:

$$\frac{dP}{dt} = acVP - dP$$

The birth rate is the number of prey captured (cVP) multiplied by a coefficient (a) for the efficiency with which food is converted into population growth. The death rate is a constant (d) multiplied by the number of predator individuals. Thus, predators have the same probability of dying during a time interval regardless of the predator population density. The expressions for the growth rates of the prey and predator populations are referred to as differential equations because they describe changes in numbers (dV or dP) with respect to a change in time (dt). The Lotka–Volterra model is therefore a continuous-time model.

For the prey in this model, when the term for population increase (rV) exceeds the removal of individuals by predators (cVP)—that is, when $rV > cVP$—the prey population increases. We can rearrange this inequality to give $P < r/c$. Thus, when the predator population is less than the ratio r/c, the prey increase in number. The inequality represents the number of predators that the population of prey can support and still increase. As you can see, this number is higher when the growth potential (r) of the prey population is higher (as when the prey themselves have more food) and when the predators are less efficient (c) at capturing them. When the terms for prey population increase and removal by predators are exactly balanced, the prey population neither increases nor decreases, and is said to be at equilibrium. At this point, $dV/dt = 0$ and $P = r/c$.

The predator population can increase when its own growth potential exceeds its death rate: $acVP > dP$, which can be rearranged to $V > d/ac$. This inequality represents

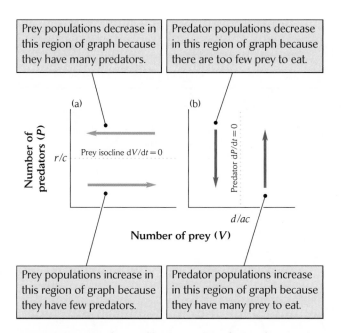

FIGURE 15.13 The equilibrium isoclines for predator and prey populations delineate regions of population increase and decrease. (a) The prey isocline (dV/dt = 0 when $P = r/c$) separates regions of prey population increase (low predator numbers) and decrease (high predator numbers). (b) The predator isocline (dP/dt = 0 when $V = d/ac$) separates regions of predator population increase (high prey numbers) and decrease (low prey numbers). The two graphs can be superimposed, as in Figure 15.14, to show the pattern of simultaneous change in both populations.

the number of prey required to support the growth of the predator population. This number is higher when the death rate of predators (*d*) is higher, and it is lower when predators are more efficient at capturing prey (*c*) and converting them into offspring (*a*). The predator population achieves an equilibrium size, dP/dt = 0, when $V = d/ac$.

Trajectories of predator and prey populations and the joint equilibrium point

The relationship between predators and prey can be portrayed as a graph with axes representing the sizes of the two populations, as shown in Figure 15.13. By convention, predator numbers increase along the vertical axis and prey numbers along the horizontal axis. The horizontal dotted line at $P = r/c$ in Figure 15.13a represents the condition dV/dt = 0 and is called the **equilibrium isocline** (or *zero growth isocline*) for the prey. At any combination of predator and prey numbers that lies in the region below this line, the prey population increases because there are

relatively few predators. In the region above the equilibrium isocline, the prey population decreases because predators remove them faster than they can reproduce.

The predator population can increase only when the abundance of prey lies to the right of the vertical dotted line at $V = d/ac$, the equilibrium isocline for the predator (Figure 15.13b). To the right of this line, prey are abundant enough to sustain the growth of the predator population. To the left of the isocline, the predator population decreases because prey are scarce. Thus, the criteria for both predators (*P*) and prey (*V*) to remain at equilibrium partition the graph into four regions.

The change in predator and prey populations together follows a closed cycle that combines the individual changes in the predator and prey populations (Figure 15.14). This cycle, called a **joint population trajectory,** can be traced through the four regions of the graph. In the lower right-hand region, for example, both predators and prey increase, and their joint population trajectory moves up and to the right. In the upper right-hand region, prey are still abundant enough that predators can increase, but the increasing number of predators depresses the prey population. Accordingly, the joint population trajectory moves up (more predators) and to the left (fewer prey).

The trajectories in the four regions together define a counterclockwise cycling of predator and prey populations one-fourth cycle out of phase, with the prey population increasing and decreasing just ahead of the predator population (Figure 15.15). Referring back to Figures 15.2

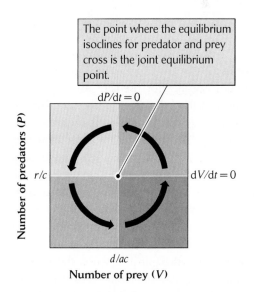

FIGURE 15.14 A joint population trajectory combines the individual changes in predator and prey populations. This trajectory shows the cyclic nature of the predator–prey interaction.

FIGURE 15.15 The Lotka–Volterra model predicts a regular cycling of predator and prey populations. The curves show how predator and prey populations continually cycle out of phase with each other.

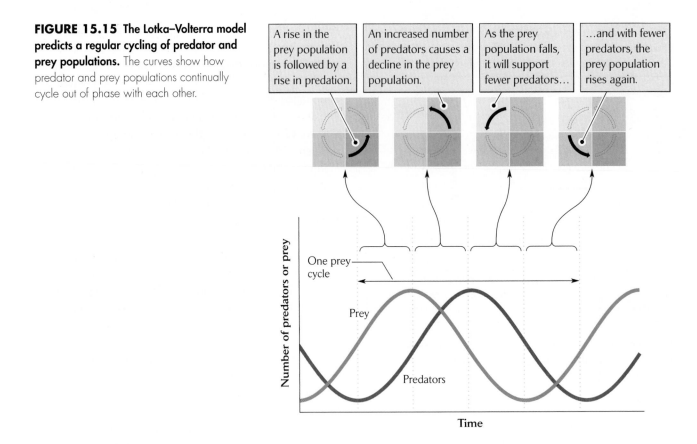

and 15.12, for example, you can see that in each cycle, prey populations tend to peak just ahead of predator populations.

LIVING GRAPHS To access an interactive tutorial on the Lotka–Volterra model, go to http://www.whfreeman.com/ricklefs6e.

The point in the center of Figure 15.14, at which the equilibrium isoclines for predator and prey populations cross, is called the **joint equilibrium point.** A combination of predator and prey populations that falls exactly at this point will not change over time. However, in the Lotka–Volterra model, when either of the populations strays ever so little from the joint equilibrium point, they oscillate around it in a continuous cycle rather than returning to it. For this reason, the Lotka–Volterra model is said to exhibit **neutral stability.** The system stays where it is, either at the joint equilibrium point or cycling around it, until it is perturbed. In this sense, the Lotka–Volterra model has no intrinsic stabilizing force. The period of the oscillation (T) is approximately $2\pi/\sqrt{(rd)}$, where π (pi) is a constant, approximately 3.14. For example, if the prey population growth rate were $r = 2$ (200%) per year and the predator death rate were $d = 0.5$ per year, the period

of the cycle would be 6.3 years. With a higher prey population growth rate or a higher predator death rate—that is, with a higher rate of population turnover—T would be shorter, and the system would oscillate more rapidly. The amplitude of the cycle depends only on how far the predator and prey populations are displaced from the joint equilibrium point.

It is important to point out that the Lotka–Volterra model is a set of differential (continuous-time) equations, meaning that the populations' responses to change are immediate. Thus, the cycling dynamic of the predator–prey interaction is not caused by time delays in responses, but rather reflects the time required for predator and prey populations to change in size; population responses are immediate, but they are unable to return the system exactly to the joint equilibrium point. The Lotka–Volterra model can also be written in a difference (discrete-time) form that introduces response time delays, but this form of the model produces unstable population cycles and eventual demise of the system. Other models based on difference equations, particularly the Nicholson–Bailey model of parasitoid–host interactions, produce stable cycles, but we will not consider them here.

Returning to the Lotka–Volterra model, the equilibrium isocline for the predator is the minimum number of prey ($V = d/ac$) that can sustain the growth of the predator population. The equilibrium isocline for the prey is the largest number of predators ($P = r/c$) that the prey population can sustain. If the reproductive rate of the prey (r) were to increase, or the capture efficiency of the predators (c) were to decrease, or both, the equilibrium isocline for the prey ($P = r/c$) would move upward. That is, the prey population would be able to bear the burden of a larger predator population, and it would increase. If the death rate of the predators (d) increased and either the prey capture efficiency (c) or the reproductive efficiency (a) of the predators decreased, the equilibrium isocline for the predator ($V = d/ac$) would move to the right, and more prey would be required to support the predator population. Increased predation efficiency (c) alone would simultaneously reduce both isoclines: fewer prey would be needed to sustain a given capture rate (the predator isocline would decrease), but the prey population would be less able to support the more efficient predators.

DATA ANALYSIS MODULE *Simulation Models of Predator–Prey Interactions.* Try changing variables in the Lotka–Volterra model to see the effects on the period and the amplitude of predator–prey cycles. You will find this module at http://www.whfreeman.com/ricklefs6e.

ECOLOGISTS IN THE FIELD **Testing a prediction of the Lotka–Volterra model.** One of the more surprising predictions of the Lotka–Volterra model is that an increase in the birth rate of the prey (r) should lead to an increase in the population of predators (P), but not in the prey population (V). It is as if the benefit to the prey of some improvement in their environment—a better supply of their own food, for example—is passed directly to their predators.

This prediction was tested by Brendan Bohannan and Richard Lenski of Michigan State University in a simple microcosm experiment. The prey in their system was the bacterium *Escherichia coli,* and the predator was the bacteriophage T4 (a virus that infects bacteria). Populations of bacteria and phage were maintained in a chemostat, a device in which the culture medium is continually replaced by a fresh supply as old medium is removed. In these experiments, the reproductive rate of *E. coli* was limited by the availability of glucose, which was supplied in concentrations of either 0.1 or 0.5 mg per liter of medium. Because a constant influx of new medium was balanced by removal of old medium, the bacteria and phage populations soon reached equilibrium levels. Consistent with the predictions of the Lotka–Volterra model, the higher rate of food provisioning to the bacteria led to an increase in the population of the phage, but not of the bacteria themselves (Figure 15.16). More rapid food provisioning

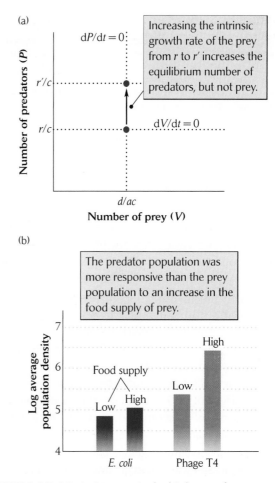

FIGURE 15.16 An increase in the birth rate of prey increases the predator population, but not the prey population. (a) According to the Lotka–Volterra model, an increase in the intrinsic growth rate of the prey population (r) raises the equilibrium isocline for the predator population (r/c), but does not change the equilibrium number of prey. (b) This prediction of the Lotka–Volterra model was tested by increasing the rate of resource (glucose) provisioning to cultures of *E. coli* bacteria in chemostats containing the bacteria and their predators, T4 bacteriophage. After B. J. M. Bohannan and R. E. Lenski, *Ecology* 78:2303–2315 (1997).

also increased the amplitude of the population cycles by supporting a more rapid rate of increase of the bacterial population, which carried it to higher densities before the phage could catch up. ∎

Pathogen–host dynamics can be described by the S-I-R model

Relationships between pathogens and their hosts can be built into models that are similar to the Lotka–Volterra predator–prey model. Such models add to our

FIGURE 15.17 The S-I-R model simulates pathogen–host interactions. Individuals in a host population are initially susceptible to a new pathogen (S). They become infected (I), during which time they can infect other individuals, then recover (R) and become resistant to further infection.

understanding of infectious diseases. Pathogens, unlike predators, do not always remove host individuals from a population. However, because hosts may develop immune responses that make some individuals resistant to the pathogen, the pathogen–host interaction can develop time delays that lead to population cycling.

The simplest model of infectious disease transmission that incorporates immunity is the **S-I-R model.** The S in S-I-R stands for susceptible individuals, I for infected individuals, and R for recovered individuals with acquired immunity (Figure 15.17). We can use this model to examine the course of a short-lived epidemic as it moves through a population.

A host individual infected by a pathogen (the primary case of the disease) will spread the disease to others, creating secondary cases. The course of the epidemic depends on two opposing factors: the rate of transmission (b) and the rate of recovery (g). The variable b includes the rate of contact of susceptible individuals with an infectious individual as well as the probability of infection given contact. The variable g determines the period over which an individual is infectious. The reproductive ratio, R_0, is defined as the number of secondary cases produced by a primary case during its period of infectiousness, where $R_0 = (b/g) S$. Thus, R_0 is the ratio of the rate of transmission (b) to the rate of recovery (g) times the number of susceptible individuals in the population (S).

Using the S-I-R model, we can ask whether the introduction of a small number of infectious individuals into a susceptible population at time 0 will cause an epidemic of the disease. If $R_0 > 1$, then a chain reaction will occur, and an epidemic will ensue, because each infected individual infects more than one other host individual before it recovers from the disease. When $R_0 < 1$, the infection fails to take hold in the host population because infected individuals fail to generate a single new infection, on average, before they recover. Even when an epidemic begins, as more individuals are infected and subsequently recover

and become resistant (R), the number of susceptible individuals (S) decreases, and the value of R_0 decreases in parallel. When it reaches $R_0 < 1$, the epidemic can no longer sustain itself.

Typical values for R_0 in childhood diseases of humans (measles, chicken pox, mumps, etc.) range from 5 to as high as 18. HIV, which is limited primarily by its mode of transmission by direct sexual contact or blood transfusion, has an R_0 value of 2–5. At the other extreme, malaria, which is transmitted by a mosquito vector, has an R_0 value greater than 100 in crowded human populations; infected people remain infectious for long periods, and mosquitoes are efficient vectors. The course that a typical disease epidemic might take is illustrated in Figure 15.18.

The basic S-I-R model includes no births of new susceptible individuals, nor loss of resistance among previously infected individuals, so the epidemic simply runs its course until all the individuals in the population are

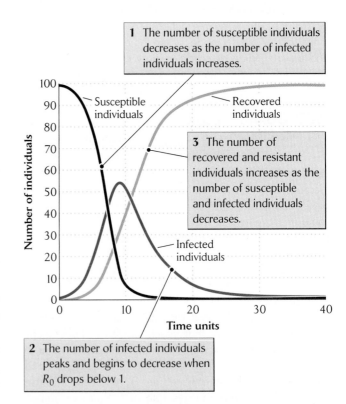

FIGURE 15.18 The S-I-R model can predict the spread of an epidemic through a host population. For this simulation, the size of the host population is arbitrarily set at 100 individuals, so that S, I, and R are expressed as numbers of individuals and also as percentages of the host population. R_0 is expressed ×100. At the beginning of the epidemic, when S is close to 1, $R_0 = 5$. The infection rate (b) is 1, the recovery rate (g) is 0.2, and the duration of infectiousness ($1/g$) is 5 time units.

recovered and resistant or too few susceptible individuals remain to sustain the spread of the disease. Influenza viruses spread through the human population in this manner. The effect of vaccination in this model is to remove individuals from the susceptible population, thus reducing the value of R_0 and reducing the probability that an epidemic can sustain itself.

Other factors can be added to the model, including births of susceptible infants, latency between infection and infectiousness, disease-dependent host mortality, host population dynamics, and vertical transmission of disease from parent to offspring. When recovered individuals lose their immunity and become susceptible again, the pathogen can produce periodic epidemics within the host population. For example, if an infected individual is infectious for 1 week and retains immunity for 5 years thereafter, the period between outbreaks of a disease with $R_0 = 5$ is almost exactly 1 year. Births of newly susceptible individuals have the same effect of increasing the number of susceptible individuals and creating periodic epidemics, as shown in Figure 15.9 for measles.

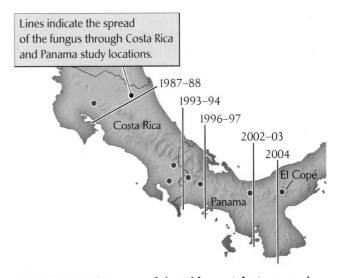

Lines indicate the spread of the fungus through Costa Rica and Panama study locations.

1987–88
1993–94
1996–97
2002–03
2004

Costa Rica

El Copé

Panama

FIGURE 15.19 A wave of chytrid fungus infection spread from the northwest to the southeast through Costa Rica and Panama from 1987 to 2004. The red dots indicate locations sampled for infected amphibians. From K. Lips, et al., *Proc. Natl. Acad. Sci. USA* 103:3165–3170 (2006).

ECOLOGISTS IN THE FIELD

The chytrid fungus and the global decline of amphibians. Most species of amphibians are declining worldwide, and many have already gone extinct. Amphibians are particularly sensitive to pollutants and changes in climate, but the most important cause of this population decline appears to be a fungal pathogen. As predicted by the S-I-R model, we would expect a typical pathogen to cause a host species to decline until there are too few susceptible hosts to support the continued spread of the pathogen. Accordingly, we would not expect a pathogen to drive a host species to extinction. However, the recently discovered pathogen in amphibians doesn't seem to follow this expectation.

In 2006, Karen Lips, of Southern Illinois University, and her colleagues documented the spread of the pathogenic fungus *Batrachochytrium dendrobatidis*, commonly called the chytrid fungus, throughout Central America. The origin of this fungus is not yet known, but it appears to be a recent arrival in Central America. Unlike many other pathogens, *B. dendrobatidis* can infect a wide variety of amphibian species. Thus, if the fungus kills off one host species, it can persist by infecting alternative host species. Such a pathogen poses a major threat to the persistence of entire groups of species.

Lips's team of researchers decided to document the spread of *B. dendrobatidis* among amphibians in Central America (Figure 15.19). They sampled more than 1,500 amphibians at a site in El Copé, Panama, where the fungus had not yet arrived. Between 2000 and July 2004, not one individual tested positive for *B. dendrobatidis*. By October 2004, however, 21 of 27 species sampled had a greater than 10% prevalence of the fungus in their populations. By December 2004, 40 species tested positive for the fungus.

During their years of testing species for the presence of the fungus, the researchers had also been counting amphibians in El Copé as they walked transects through amphibian habitat to estimate the population sizes of each species. At the end of 2004, however, coincident with the arrival of the fungus, the numbers of live amphibians that were counted along the transects declined sharply (Figure 15.20), while the number of dead amphibians increased. The dead amphibians included 38 different frog species, and 99% of the 318 dead individuals collected had moderate to severe chytrid infections.

While it remains unknown exactly how *B. dendrobatidis* kills its hosts, the fungus clearly has been responsible for the massive die-off of frogs in Central America. Extinction is always difficult to prove because a few individuals may still exist in an area undetected. However, many of the species involved in the die-off have not been seen for several years, and they are almost certainly extinct. The fungus is also appearing in other parts of the world, with similar effects. Hence, *B. dendrobatidis* poses a major threat to amphibian conservation around the world. An important message that emerges from this research is that when a pathogen is not restricted to a single host species, it has the ability to persist and spread even after it drives one of its hosts extinct. ∎

FIGURE 15.20 The chytrid fungus can cause rapid declines in amphibian populations. Numbers of live amphibians (natural-log transformed) were observed by biologists walking along transects in El Copé, Panama, from 1998 to 2005. Shortly after the chytrid fungus was first found there, the number of observations declined sharply. From K. Lips, et al., *Proc. Natl. Acad. Sci. USA* 103:3165–3170 (2006).

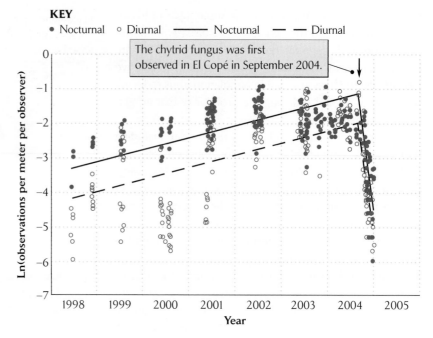

KEY
• Nocturnal   ○ Diurnal   —— Nocturnal   — — Diurnal

The chytrid fungus was first observed in El Copé in September 2004.

The Lotka–Volterra model can be stabilized by predator satiation

The Lotka–Volterra model provided an explanation for population cycles, but the model is so simple that it fails to represent nature in some important ways. We have already mentioned the absence of time delays from the continuous-time form of the model. As a result, when either the predator or the prey population is displaced from the joint equilibrium point, the system will undergo persistent oscillations in a closed cycle. Any further perturbation of the system will give these population oscillations a new amplitude and period until some other outside influence acts on them. This type of dynamic behavior is neutrally stable because no internal forces act to restore the populations to the joint equilibrium point. Therefore, random perturbations will eventually increase the oscillations to the point at which the trajectory reaches one of the axes of the predator–prey graph (Figure 15.14, where V or $P = 0$), and one or both populations will die out. This property in itself suggests that the Lotka–Volterra model greatly oversimplifies nature.

Other concerns about the adequacy of the model focus on the predation term (cVP). At a given density of predators (P), the rate at which prey are captured (cVP) increases in direct proportion to prey density (V). Accordingly, predators cannot be satiated; they just keep on eating, no matter how many prey they capture. Clearly, this aspect of the model is unrealistic. How would adding a bit more realism here affect the behavior of the model?

The functional response

The relationship of an individual predator's rate of food consumption to the density of its prey has been labeled the **functional response** by entomologist C. S. Holling. There are three potential types of functional responses, and the Lotka-Volterra model is based on the least realistic of these. According to the model, predators consume prey at a rate cVP, so the rate of consumption per individual predator is cV. This relationship, called a **type I functional response,** is illustrated in Figure 15.21. This means that the fecundity of individual predators, which in the model is proportional to the number of prey consumed (acV), increases without limit in direct proportion to the number of potential prey. In other words, when prey are numerous, the fecundity of individual predators is high regardless of their own numbers. Thus, the predator population grows rapidly, and prey numbers can be brought under control. In other words, predation has no limit.

Two factors dictate that the functional response should, instead, reach a plateau. First, predators may become satiated—constantly full—at which point their rate of feeding is limited by the rate at which they can digest and assimilate food. Second, as a predator captures more prey, the time it spends handling and eating each one

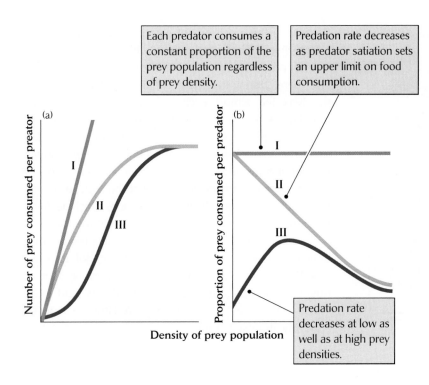

Each predator consumes a constant proportion of the prey population regardless of prey density.

Predation rate decreases as predator satiation sets an upper limit on food consumption.

Predation rate decreases at low as well as at high prey densities.

FIGURE 15.21 Predators can exhibit three types of functional responses to increasing prey density. These functional responses are shown in terms of (a) the number of prey consumed per predator and (b) the proportion of prey consumed per predator.

cuts into its searching time. Eventually, these two factors should reach a balance, and the prey capture rate should level off.

The type I functional response can be modified to take these limitations into account. An obvious modification is the **type II functional response,** in which the number of prey consumed per predator initially rises quickly as the density of prey increases, but then levels off with further increases in prey density. A **type III functional response** resembles the type II response in placing an upper limit on the rate of prey consumption, but differs in that predators consume relatively fewer prey at low prey densities.

At high prey densities, the type II and type III functional responses differ little: they are both inversely density-dependent. In other words, as the density of prey increases, the proportion of those prey consumed by a given number of predators decreases. Type III responses differ from type II responses in that the proportion of the prey consumed also decreases at lower prey densities.

Several circumstances might cause a type III functional response. First, a heterogeneous habitat affords a limited number of safe hiding places for prey, and those refuges protect a larger proportion of the prey when there are fewer of them. Second, when predators encounter prey frequently, they form a **search image** that helps them to identify and locate suitable prey—a mental image that focuses their attention, so to speak. At low prey densities, predators encounter prey less often and so do not learn to hunt them as efficiently. Third, predators may switch to alternative sources of food when particular prey are scarce, reducing pressure on the prey population. Such **switching** produces a type III response because consumption at low prey densities is reduced as predators switch to more abundant alternative food sources.

Many field and laboratory studies have demonstrated type III functional responses. For example, when the predatory water bug *Notonecta glauca* was presented with two types of prey, isopods and mayfly larvae, it consumed the more abundant type of prey, whichever it was, in a proportion greater than its percentage of occurrence (Figure 15.22). Predators switched to the more abundant prey because the success of their attacks was higher on prey present at greater densities. When the water bugs encountered mayfly larvae infrequently, fewer than 10% of their attacks were successful. At higher densities, and therefore higher encounter rates, attack success rose to almost 30%, showing that practice improves predator performance. *Notonecta* exhibited no innate preference for either type of prey, only a preference for the more abundant of the two.

The numerical response

Individual predators can increase their consumption of food items only to the point of satiation. Once all predators are sated, the only way consumption can keep up with a prey population of increasing density is for the

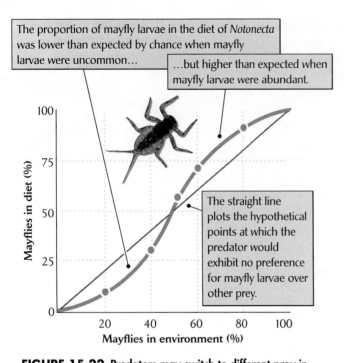

The proportion of mayfly larvae in the diet of *Notonecta* was lower than expected by chance when mayfly larvae were uncommon...

...but higher than expected when mayfly larvae were abundant.

The straight line plots the hypothetical points at which the predator would exhibit no preference for mayfly larvae over other prey.

FIGURE 15.22 Predators may switch to different prey in response to fluctuations in prey density. Researchers presented the predatory water bug *Notonecta glauca* with two types of prey, isopods and mayfly larvae, in different proportions. From M. Begon and M. Mortimer, *Population Ecology*, 2nd ed., Blackwell Scientific Publications, Oxford (1981); after J. H. Lawton, J. R. Beddington, and R. Bonser, in M. B. Usher and M. H. Williamson (eds.), *Ecological Stability*, pp. 141–158, Chapman & Hall, London (1974).

Because of the synchrony of hare population cycles over a large geographic area, most of this increase was due to local population growth rather than immigration from elsewhere. During the phase of hare population increase, the lynx fed almost exclusively on hares. After hare populations began to decline, the predators switched to alternative prey types, particularly red squirrels and other small mammals (Figure 15.23b). However, even though the populations of these smaller mammals were stable or increasing during the decline phase of the cycle, evidently they could not sustain the lynx population, which declined in parallel with the snowshoe hare, its preferred food.

The numerical response of the predator tends to lag behind changes in the population density of its prey, whether prey density is increasing or decreasing. Consequently, when prey are increasing, predators tend to be scarce; when they are decreasing, predators tend to be relatively abundant (Figure 15.24).

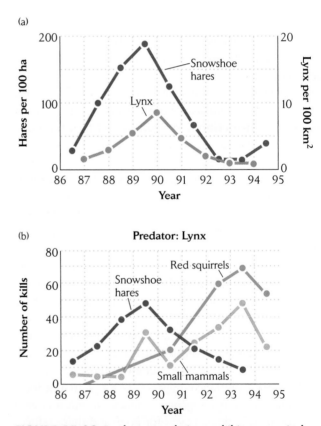

FIGURE 15.23 Predator populations exhibit a numerical response to changes in prey density. (a) In southern Yukon, the population densities of lynx closely tracked those of their preferred prey, snowshoe hares, through a hare population cycle. (b) Red squirrels and other small mammals were eaten by lynx in large numbers only after the densities of hares fell to a low level. After M. S. O'Donoghue et al., *Oikos* 82:169–183 (1998).

number of predators to increase, either by immigration or by population growth. Together, predator immigration and population growth constitute the **numerical response.** Populations of most predators grow slowly relative to populations of their prey, especially when the reproductive potential of a predator is lower than that of its prey and the predator's life span is longer.

Mobile predators migrating from surrounding areas may opportunistically congregate where resources become abundant. For example, local populations of bay-breasted warblers, small insectivorous birds of eastern North America, increase dramatically during periodic outbreaks of the spruce budworm. During outbreak years, their populations can reach 300 breeding pairs per km^2, compared with about 25 pairs per km^2 during non-outbreak years. This behavior shows how a predator can take advantage of a shifting mosaic of prey abundance.

In the study area in southern Yukon mentioned earlier, numbers of lynx increased 7.5-fold in response to increasing snowshoe hare populations (Figure 15.23a).

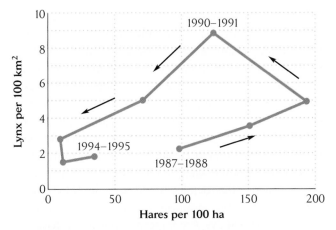

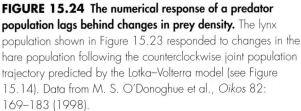

FIGURE 15.24 The numerical response of a predator population lags behind changes in prey density. The lynx population shown in Figure 15.23 responded to changes in the hare population following the counterclockwise joint population trajectory predicted by the Lotka–Volterra model (see Figure 15.14). Data from M. S. O'Donoghue et al., *Oikos* 82: 169–183 (1998).

A number of factors can reduce oscillations in predator–prey models

In population biology, the term **stability** is usually applied to the achievement of an unvarying equilibrium size, often referred to as the carrying capacity of the environment for a particular population. We now know that this term is too restrictive because predator and prey populations can fluctuate in stable cycles over long periods. Destabilizing factors must be present for cycling to occur, particularly time delays in the responses of populations to changes in their food supplies. Stable cycles can nonetheless be achieved because other factors balance these destabilizing forces and constrain the amplitude of predator–prey cycles. Among these stabilizing factors are the following:

1. Predator inefficiency (or enhanced prey escape or defense strategies)

2. Density-dependent limitation of either the predator or the prey population by factors external to their relationship

3. Alternative food sources for the predator

4. Refuges for the prey at low prey densities

5. Reduced time delays in predator responses to changes in prey abundance

Several of these factors deserve special comment. Predator inefficiency (low *c* in the Lotka–Volterra model)

results in higher equilibrium levels for both prey and predator populations (more predators can be supported by the larger prey populations) and in lower birth and death rates for both at equilibrium. Both of these consequences would seem to enhance the stability of a predator–prey system. Alternative food sources stabilize predator populations because individuals can switch between food types in response to changing prey abundances. Similarly, refuges from predation allow prey populations to maintain themselves at higher levels than they otherwise could in the face of intense predation, thereby facilitating the recovery phase of the population cycle. Indeed, so many factors tend to stabilize predator–prey interactions that the cyclic behavior of some systems is possible only because of the overriding influence of destabilizing time delays.

Time delays are ubiquitous in nature: they arise from the developmental periods of animals and plants, the time required for predators to immigrate from other areas, and the time course of immune responses by animals and of induced defenses in plants. In some circumstances—perhaps in less complex ecological systems such as tundra and boreal forest ecosystems—these factors outweigh stabilizing influences and result in population cycles.

Consumer–resource systems can have more than one stable state

The size of any population is influenced by the abundances of its resources and of its consumers. At one extreme, a resource population might be limited primarily by its own food supply while consumers remove an inconsequential number of resource individuals. At the other extreme, efficient consumers might depress a resource population to levels below its carrying capacity. As we have seen, the equilibrium size of a population often reflects a balance between the limiting influences of food supplies and consumers. Under some circumstances, however, a population may have two or more stable equilibrium points, only one of which may be occupied at a given time. These multiple equilibrium points are called **alternative stable states.**

Alternative stable states can arise when different factors limit populations at low and at high densities. At low densities, individuals might have access to refuges that make it difficult for their consumers to locate or capture them. In general, low population densities might make individuals so difficult to locate and capture that consumers switch to other resources that give higher returns on the time invested in foraging. In addition, at low population densities, individuals tend to increase in number faster than their consumers remove them because they are not so limited by their own resources. As population density increases, however, consumers might be drawn to the increasingly abundant food supply and eventually bring the population under control at a low stable equilibrium point well below its carrying capacity. The result is a **consumer-imposed equilibrium.**

Now let's consider a population considerably above its consumer-imposed equilibrium. At first, consumer efficiency goes up as the density of the population increases; consumers continue to keep the population in check, driving it back down to the consumer-imposed equilibrium. Eventually, however, the consumers themselves become limited—either consumer individuals become satiated by their resources and can no longer consume them at an increasing rate (type II and III functional responses), or the consumer population becomes limited by factors other than the resource population, such as suitable nest or den sites or their own predators. At this point, the resource population can escape consumer control and continue to increase to the carrying capacity set by its own resources—a **resource-imposed equilibrium.**

Under this scenario, a population would have two alternative equilibrium states. Because these are both stable equilibria, the population cannot easily escape one and move to the other, at least under constant environmental conditions. However, environmental perturbations that reduce a consumer population might release its resource population from consumer control and allow it to increase to its carrying capacity. Conversely, environmental changes that depress the resource population might bring it back into the range of consumer control, which would then drive it back down to the consumer-imposed equilibrium.

Alternative stable states have practical implications for the control of many populations, including those of crop and forest pests. For example, a heavy frost or an introduced disease might reduce a predator population long enough to allow a crop pest population to slip out of consumer control. The pest population would then continue to increase until it neared the higher, resource-imposed equilibrium. To farmers, this means that the population of a crop pest that is normally kept at harmless levels by predators or parasites suddenly explodes in a menacing outbreak and competes for their crops. After such a change, consumers exert little influence on the pest population until some quirk of the environment brings pest numbers back within the realm of consumer control.

Outbreaks of the winter moth, a forest pest in eastern North America that defoliates and sometimes kills trees, can be managed by introducing parasitoids that attack the caterpillars. When the winter moth population is reduced to a low level, it can be kept low for some time by small mammals that prey on the pupae in the leaf litter on the forest floor. However, the winter moth may escape predator control when climate or disease limits its predators. The population may then increase back toward its resource-imposed equilibrium until parasitoid populations increase enough to regain control of their host population.

MORE ON THE WEB   *Predator–Prey Dynamics in a Metapopulation of the Cinnabar Moth.* The stability of this herbivore–plant interaction depends on isolated refuges for the plant population.

MORE ON THE WEB   *Three-Level Consumer Systems.* When predators themselves have predators, their prey may benefit. Birds and wasps reduce the numbers of herbivorous insects on trees, and the trees benefit from the reduced damage by maintaining faster growth and achieving larger size.

DATA ANALYSIS   DATA ANALYSIS MODULE 4 *Maximum sustainable yield.* What is the maximum sustainable level of predation on a prey population? Do predators limit their own populations to achieve the maximum sustainable yield from their prey? You will find this module on page 324.

SUMMARY

1. Early observations of fur-bearing mammal populations revealed cyclic changes that stimulated theoretical and experimental investigations of the dynamics of consumer–resource interactions.

2. Experimental studies of pest species and their natural predators have demonstrated that, in many cases, consumers, including herbivores, can limit resource populations.

3. Predator and prey populations in natural systems often increase and decrease in regular synchronized cycles. These oscillations result from the interaction of predator and prey populations.

4. The underlying causes of synchronized predator–prey population cycles are time delays in the response of each population to changes in the size of the other.

Pathogen–host interactions also incorporate time delays resulting from immune responses and, as a result, pathogens can exhibit periodic outbreaks.

5. Predator and prey populations can be made to oscillate in the laboratory. Maintenance of population cycles usually requires a complex environment in which prey populations can find refuges from predation.

6. Alfred J. Lotka and Vito Volterra devised a simple mathematical model of predator–prey interactions that predicted population cycles. The Lotka–Volterra model uses differential equations in which the rate of prey removal is directly proportional to the product of the predator and prey population sizes.

7. The period of the Lotka–Volterra cycle depends on the birth and death rates of the predators and their prey—essentially the turnover rates of the populations. The model exhibits neutral stability, meaning that when the cycles are influenced by external perturbations, the system does not return to its original state.

8. A surprising prediction of the Lotka–Volterra model is that increased productivity of the prey should increase the size of the predator population, but not the prey population. This prediction has been verified in experimental studies.

9. The S-I-R model of pathogen–host interactions describes changes in the numbers of susceptible, infected, and recovered individuals in a population. The S-I-R model describes the course of a single disease epidemic, but if births of susceptible host individuals or loss of resistance to the pathogen are added, such models predict periodic, repeated epidemics.

10. The functional response describes the relationship between the rate at which an individual predator consumes prey and prey density. Whereas the Lotka–Volterra model, which assumes a type I functional response, produces a neutrally stable cycle, type II and III functional responses can lead to stable regulation of prey populations at low densities.

11. Type III functional responses can result from a higher proportion of prey finding refuges, lack of a search image, or switching by predators from their preferred prey to a more abundant alternative food source at low prey densities.

12. The numerical response describes the response of a predator population to increasing prey density by population growth and immigration.

13. Stability in predator–prey interactions is promoted by low predator efficiency, by density-dependent limitation of either predator or prey, by the availability of alternative resources for the predator, and by refuges for the prey. Stable population cycles in nature apparently represent a balance between these stabilizing factors and the destabilizing influence of time delays in population responses.

14. Consumer–resource systems can have two equilibrium points (alternative stable states), between which resource populations may shift depending on environmental conditions. The lower equilibrium point is determined by consumer pressure; the upper equilibrium point lies at the carrying capacity of the resource population. Environmental perturbations can shift a resource population from one to the other of these equilibrium points, resulting in successive outbreaks followed by periods during which the population is controlled by its consumers.

REVIEW QUESTIONS

1. What characteristics enable predator species to control populations of their prey, and why are these characteristics important?

2. If you wished to determine whether herds of African antelope affect the plant community on which they graze, what type of experiment could you conduct?

3. Which factors determine the duration of population cycles and which determine the magnitude of change in population sizes?

4. Compare and contrast the underlying causes of time delays in predator–prey interactions and in pathogen–host interactions.

5. In the classic experiments of C. F. Huffaker with mites and oranges, what mechanisms allowed the predator and prey populations to persist?

6. According to the Lotka–Volterra model of predator–prey interactions, why do predator and prey populations cycle?

7. According to the S-I-R model of pathogen–host interactions, what effect could one have on the spread of a disease by immunizing many (but not all) individuals in a population?

8. How do search image formation and prey switching behavior lead to a type III functional response in predators?

9. How could increased prey defenses and alternative food sources for predators reduce oscillations in predator–prey cycles?

10. How might a predator's type II functional response prevent it from controlling a large prey population, allowing that prey population to reach a resource-imposed equilibrium?

SUGGESTED READINGS

Anderson, R. M., and R. M. May. 1979. Population biology of infectious diseases: Part I. *Nature* 280:361–367.

Bohannan, B. J. M., and R. E. Lenski. 1997. Effect of resource enrichment on a chemostat community of bacteria and bacteriophage. *Ecology* 78:2303–2315.

Crawley, M. J. 1997. Plant–herbivore dynamics. In M. J. Crawley (ed.), *Plant Ecology,* 2nd ed., pp. 401–474. Blackwell Scientific, Oxford.

DeBach, P., and D. Rosen. 1991. *Biological Control by Natural Enemies,* 2nd ed. Cambridge University Press, New York.

Dobson, A. 1995. The ecology and epidemiology of rinderpest virus in Serengeti and Ngorongoro conservation areas. In A. R. E. Sinclair and P. Arcese (eds.), *Serengeti II: Dynamics, Management, and Conservation of an Ecosystem,* pp. 485–505. University of Chicago Press, Chicago.

Errington, P. L. 1963. The phenomenon of predation. *American Scientist* 51:180–192.

Hanski, I., et al. 2001. Small rodent dynamics and predation. *Ecology* 82:1505–1520.

Heesterbeek, J. A. P., and M. G. Roberts. 1995. Mathematical models for microparasites of wildlife. In B. T. Grenfell and A. P. Dobson (eds.), *Ecology of Infectious Diseases in Natural Populations,* pp. 90–122. Cambridge University Press, Cambridge.

Hörnfeldt, B. 1994. Delayed density dependence as a determinant of vole cycles. *Ecology* 75:791–806.

Hörnfeldt, B., T. Hipkiss, and U. Eklund. 2005. Fading out of vole and predator cycles? *Proceedings of the Royal Society of London* B 272:2045–2049.

Jansen, V. A. A., et al. 2003. Measles outbreaks in a population with declining vaccine uptake. *Science* 301:804.

Jeschke, J. M., M. Kopp, and R. Tollrian. 2002. Predator functional responses: Discriminating between handling and digesting prey. *Ecological Monographs* 72:95–112.

Jeschke, J. M., M. Kopp, and R. Tollrian. 2004. Consumer–food systems: Why type I functional responses are exclusive to filter feeders. *Biological Reviews* 79:337–349.

Korpimäki, E., et al. 2004. The puzzles of population cycles and outbreaks of small mammals solved? *BioScience* 54(12):1071–1079.

Krebs, C. J., et al. 1995. Impact of food and predation on the snowshoe hare cycle. *Science* 269:1112–1115.

Krebs, C. J., et al. 2001. What drives the 10-year cycle of snowshoe hares? *BioScience* 51(1):25–35.

Krebs, C. J., et al. 2002. Cyclic dynamics of snowshoe hares on a small island in the Yukon. *Canadian Journal of Zoology* 80:1442–1450.

Lips, K. R., et al. 2006. Infectious disease and global biodiversity loss: Pathogens and enigmatic amphibian extinctions. *Proceedings of the National Academy of Sciences USA* 103:3165–3170.

May, R. M. 1983. Parasite infections as regulators of animal populations. *American Scientist* 71:36–45.

Myers, J. H. 1993. Population outbreaks in forest lepidoptera. *American Scientist* 81:240–281.

O'Donoghue, M., et al. 1998. Behavioral responses of coyotes and lynx to the snowshoe hare cycle. *Oikos* 82:169–183.

Pech, R. P., et al. 1992. Limits to predator regulation of rabbits in Australia: Evidence from predator removal experiments. *Oecologia* 89:102–112.

Roland, J. 1993. Large-scale forest fragmentation increases the duration of tent caterpillar outbreak. *Oecologia* 93:25–30.

DATA ANALYSIS MODULE 4

Maximum Sustainable Yield: Applying Basic Ecological Concepts to Fisheries Management

Biologists are deeply concerned about the depletion of fish populations, particularly those of the highly productive coastal marine fisheries that are important food sources for much of the world. The growing human population, more efficient fishing methods, and government subsidies for world fisheries have steadily increased the pressure on these fish stocks, making it increasingly difficult to maintain a balance between harvest and production. The UN Food and Agriculture Organization (2001, 2002) recently estimated that about 70% of commercially important marine fish stocks were fully exploited (harvested at or near the maximum sustainable level with no room for expansion), overexploited (harvested at or above a long-term sustainable level), or depleted (catches are well below historic levels at the same fishing effort). The percentage of global fish stocks harvested at or below maximum sustainable levels has steadily decreased since the 1970s, while the proportion of overexploited fish populations increased from about 10% in the 1970s to nearly 30% in the 1990s.

When predators, including humans, consume or otherwise remove prey individuals from a population faster than they can be replaced by recruitment of offspring, the prey population decreases. Ideally, consumers should remove no more prey than can be replaced through reproduction; where possible, they should "manage" the prey population to provide the highest possible replacement rate. The replacement rate is highest when prey populations are at intermediate densities; that is, when they are large enough to produce offspring at a high rate, but not so large as to become self-limiting by reducing their own resources to a low level.

The largest number of individuals that can be removed or harvested from a prey population without depressing its population growth rate is known as the **maximum sustainable yield (MSY).** When a prey population grows according to the logistic model, the MSY is achieved when the prey have reached half the carrying capacity and the population growth rate is at its peak (see Figures 11.14 and 11.15). In the case of fish and game populations exploited by humans, the MSY represents the highest yield or harvest in a given period that can be replenished by prey production. Harvesting of the MSY is sometimes referred to as full exploitation because all "excess" production is removed. Management plans for many fish and game populations, which strive to maximize harvest while maintaining healthy populations, are based on this concept.

By using the logistic growth equation (see page 237), we can estimate the point at which the growth of a fish population is greatest by plotting the absolute growth rate,

$$\frac{dN}{dt} = r_0 N\left(1 - \frac{N}{K}\right)$$

against the size of the population (N). The resulting plot (DA Figure 4.1) shows population growth rate as a function of population size. Peak population growth, or production, occurs at the top of the curve at an intermediate population size.

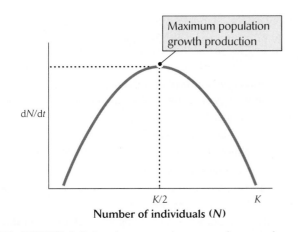

DA FIGURE 4.1 Population growth rate as a function of population size.

We can incorporate harvesting, or *yield,* into the logistic growth equation as follows:

$$\frac{dN}{dt} = r_0 N\left(1 - \frac{N}{K}\right) - Y$$

where Y, the yield, is subtracted from population growth. Fisheries managers often substitute biomass for numbers of individuals, in which case population growth and yield are measured in tons of fish, which matches the units in which catches are reported.

The equilibrium point of the yield model based on the logistic growth equation (at which the population size or biomass does not change over time) occurs when surplus production (over and above the replacement level) equals harvest such that

$$Y = r_0 N\left(1 - \frac{N}{K}\right)$$

The model assumes that this equilibrium can persist through time—a questionable assumption for many exploited populations. Nonetheless, the fundamental ecological principle embodied in this function is the foundation for the most common and well-studied production models used in fisheries management. It is a valuable tool for investigating sustainability in a variety of situations.

Yield is also related to fishing effort (for example, number of hours and area of nets or trawls), the catchability of the target fish (how efficiently they are captured with a given method), and population size, such that

$$Y = fqN$$

where f is fishing effort and q is catchability. If we substitute this relationship into the yield equation, we can solve for the equilibrium population size (N^*):

$$N^* = K\left(1 - \frac{fq}{r_0}\right)$$

At this point the yield will be

$$Y = fqN^* = fqK\left(1 - \frac{fq}{r_0}\right)$$

This expression shows that yield and fishing effort have the same parabolic relationship shown in DA Figure 4.1 for numbers and population growth, with MSY at the top of the curve where production is highest. It also shows that yield increases in direct relation to the carrying capacity of the fish population and to its intrinsic exponential growth rate (r_0), which might be thought of as the efficiency with which it converts its own resources into biomass.

Cod are the basis for many important commercial fisheries and are an important food source for many marine mammals. Cod stocks around the world are heavily fished, and some overexploited populations have dwindled to such low levels that the fisheries have been closed to protect the remaining stock. Pacific cod (*Gadus macrocephalus*) in the Gulf of Alaska are currently considered fully exploited, and this important fishery is tightly regulated. Estimates from 2005 indicated that the Pacific cod population in the Gulf of Alaska was roughly 140,000,000 individuals (Thompson and Dorn, 2005). We shall use that estimate as a basis for calculating the maximum sustainable yield of this fishery.

Step 1: Complete DA Table 4.1 using the yield equation and the fishing effort values provided in column 1.

DA TABLE 4.1	Yield and effort relationships for Pacific cod	
Effort	Yield ($r_0 = 0.2$)	Yield ($r_0 = 0.3$)
0.0	0	
0.1	5,400,000	
0.2	9,600,000	
0.3		
0.4		
0.5		
0.6		
0.7		
0.8		
0.9		
1.0		

Use an estimate of 0.2 for the r_0 of Pacific cod (FAO, 2001), a hypothetical catchability of 0.2 (indicating 20% vulnerability for the type of gear used), and a hypothetical K of 300,000,000 individuals (assuming that the current population is well below carrying capacity).

As an example, at an effort of 0.1 (row 2 in the table), yield is

$Y = fqK(1 - fq/r_0)$
$= (0.1 \times 0.2 \times 300,000,000) \times [1 - (0.1 \times 0.2)/0.2]$
$= 5,400,000$

• What happens to yield as effort is initially increased? At what point does yield begin to drop off as effort increases?

Step 2: Graph yield (*y*-axis) versus effort (*x*-axis) using the values in the completed first two rows of the table to visualize the relationship between these two variables.

• Your graph should reflect a parabolic relationship between the two variables. Based on your calculations and graph, what is the maximum sustainable fishing effort (f_{MSY}) and the MSY for this cod population?

The f_{MSY} can also be estimated using the derivative of the yield equation set to zero (when the change in yield over the change in fishing effort = 0), which is

$$f_{MSY} = \frac{r_0}{2q}$$

You should be able to substitute this formula into the equation for yield as a function of fishing effort to calculate the yield at the maximum sustainable fishing effort.

Step 3: Using the estimates of r_0 and q for Pacific cod provided in the example, calculate f_{MSY} using the derivative of the yield equation and compare it with the value you derived from the table and graph.

If your calculations are correct, the values will be the same.

Step 4: Using the same formulas in DA Table 4.1, calculate yield per unit effort using $r_0 = 0.3$ instead of 0.2 and fill in row 3 of the table.

• Compare the yield per unit effort values obtained using $r_0 = 0.2$ and $r_0 = 0.3$. How does a higher intrinsic exponential growth rate of the cod population influence the relationship between yield and effort, and ultimately MSY?

• How might changing other parameters, such as catchability (*q*) or the carrying capacity (*K*), affect the relationship between effort and yield?

Although the concept of maximum sustainable yield is based on well-established ecological principles, in practice it requires a level of precision in quantifying population abundance or stock biomass that is often not realistic. For example, although we can use field data and production models to estimate it, how do we know for sure when the "top" of the hump-shaped production curve has been reached? The only way to know for sure is to find the inflection point by increasing harvest until production begins to drop. At this point, the resource has been overexploited.

Literature Cited

Schaefer, M. B. 1954. Some aspects of the dynamics of populations important to the management of the commercial marine fisheries. *Bulletin of the Inter-American Tropical Tuna Commission* 1:27–56.

Thompson, G. G., and M. W. Dorn. 2005. Assessment of the Pacific cod stock in the Gulf of Alaska. Chapter 2 in *Report to the U.S. Department of Commerce, National Oceanic and Atmospheric Administration, National Marine Fisheries Service*. Alaska Fisheries Science Center, Seattle, Wash.

UN Food and Agriculture Organization (FAO). 2001 (online). *Report of the Second Technical Consultation on the Suitability of the CITES Criteria for Listing Commercially-Exploited Aquatic Species: A Background Analysis and Framework for Evaluating the Status of Commercially-Exploited Aquatic Species in a CITES Context*. Windhoek, Namibia, 22–25 October. FAO Fisheries Report 667. http://www.fao.org/fishery/nems/11030/en.

UN Food and Agriculture Organization (FAO). 2002. *The state of world fisheries and aquaculture*. UNFAO, Rome. http://www.fao.org/docrep/005/y7300e/y7300e00.htm.

Competition

The British botanist A. G. Tansley (1917) provided the first demonstration of competition between species in an experimental study. Tansley prefaced his report with the observation that closely related plant species living in the same region often grow in different habitats or on different types of soil. Tansley's observation was not new, nor was his suggestion that such ecological segregation might have resulted from competition for resources, leading to the exclusion of one species or the other. Nonetheless, no one had experimentally tested that hypothesis, or its alternative—namely, that the two species had such different ecological requirements that each could not grow where the other flourished.

Tansley selected two species of bedstraws (genus *Galium*), which are small, perennial, herbaceous plants. One species, *G. saxatile*, normally lives on acidic, peaty soils; the other, *G. sylvestre*, inhabits the alkaline soils of limestone hills and pastures. Tansley planted seeds of each species, both singly and together, in plots with soils taken from areas where each species grew naturally. Because the seeds were planted together in a common garden, the only differences in the plots were soil type and the presence or absence of the other species (Figure 16.1).

Like many ecological studies, Tansley's experiments were plagued by such technical problems as poor germination and lapses in watering. His results were nonetheless quite clear. When planted alone, each of the species grew and maintained itself on both types of soil, although germination and growth were most vigorous on the soil type on which the species grew naturally. When the two species were grown together on calcareous (limestone) soils, *G. sylvestre* plants overgrew and shaded *G. saxatile*. The reverse occurred on the more acidic, peaty soils that are typical of *G. saxatile* habitat.

Tansley concluded that *G. saxatile* is at a disadvantage on calcareous soils, and is thus unable to compete effectively with *G. sylvestre* on that soil type. Similarly, *G. sylvestre* grows less well on peat, and consequently is an inferior competitor to *G. saxatile* on that soil type. Both species, however, were able to establish themselves on either soil. These results suggested to Tansley that species are generally restricted to the most

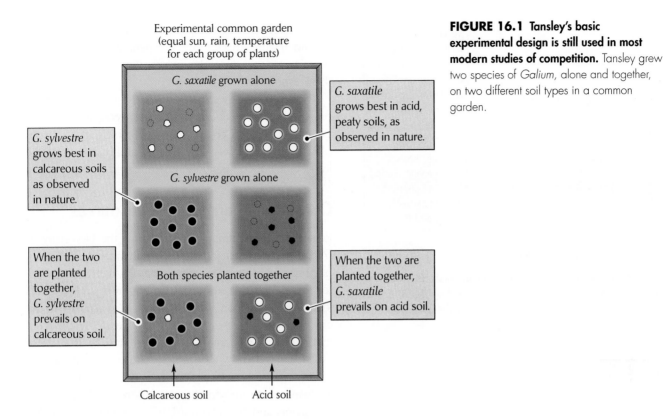

Experimental common garden
(equal sun, rain, temperature
for each group of plants)

G. saxatile grown alone

G. sylvestre grown alone

Both species planted together

Calcareous soil   Acid soil

G. sylvestre grows best in calcareous soils as observed in nature.

When the two are planted together, *G. sylvestre* prevails on calcareous soil.

G. saxatile grows best in acid, peaty soils, as observed in nature.

When the two are planted together, *G. saxatile* prevails on acid soil.

FIGURE 16.1 Tansley's basic experimental design is still used in most modern studies of competition. Tansley grew two species of *Galium*, alone and together, on two different soil types in a common garden.

favorable soil types when competing species are present, but may be broadly distributed over other soil types in the absence of competition. A species that grew poorly on a particular soil—as in the case of *G. saxatile* on calcareous soils—would probably not survive competition and thus would be absent from that soil type throughout its range.

In his brief paper, Tansley put on record (1) that the presence or absence of a species could be determined by competition with other species; (2) that the conditions of the environment affected the outcome of competition; and (3) that the present ecological segregation of species might have resulted from competition in the past. Although ecologists did not take up studies of competition again for more than 15 years, Tansley's approach, or some modification of it, is used in most modern studies of competition between species.

CHAPTER CONCEPTS

- Consumers compete for resources
- Failure of species to coexist in laboratory cultures led to the competitive exclusion principle
- The theory of competition and coexistence is an extension of logistic growth models
- Asymmetric competition can occur when different factors limit the populations of competitors

- Habitat productivity can influence competition between plant species
- Competition may occur through direct interference
- Consumers can influence the outcome of competition

Competition is any use or defense of a resource by one individual that reduces the availability of that resource to other individuals. Competition is one of the most important ways in which the activities of

individuals affect the well-being of others, whether they belong to the same species or different species. Competition between individuals of the same species is called **intraspecific competition,** and competition between

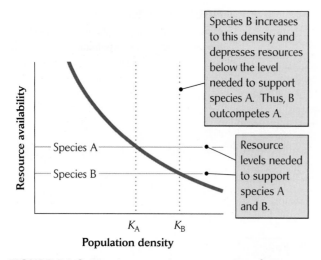

Species B increases to this density and depresses resources below the level needed to support species A. Thus, B outcompetes A.

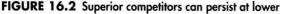

Resource levels needed to support species A and B.

FIGURE 16.2 Superior competitors can persist at lower resource levels. As resources are consumed, they decline to levels that no longer support the further growth of the consumer population, and the population may reach an equilibrium size (K). If species A can continue to grow at a resource level that curtails the growth of species B, species A will outcompete, and will eventually replace, species B.

individuals of different species is called **interspecific competition.**

As we saw in Chapter 14, the more crowded a population, the stronger the effects of competition between individuals. Intraspecific competition, like consumer–resource interactions, regulates population growth in a density-dependent manner. Furthermore, when genetic factors influence the efficiency of resource use, evolution tends to increase competitive ability within a population.

Competition between individuals of different species can depress the populations of both; in this case, each species contributes to the regulation of the other population as well as its own. Under some conditions, particularly when interspecific competition is intense, the population of one species may decline and finally die out. Because of this potential, competition is an important factor in determining which species can coexist within a habitat.

The outcome of competition between species depends on how efficiently individuals within each species exploit shared resources. Individuals in every population consume resources. When resources are scarce relative to demand for them, each act of consumption by one individual makes a resource less available to others as well as to itself. As consumption continues, resources decline to levels that no longer support the further growth of the consumer population, and the population may reach a stable equilibrium. When one population can continue to grow at a resource level that curtails the growth of a

second population, the first will eventually replace the second (Figure 16.2). Thus, competition and its various outcomes depend on the relationship of consumers to their resources.

In this chapter, we shall consider some of the general principles of competition between species, illustrate the potential effects of competition by examining the results of laboratory and field experiments, and demonstrate the importance of competition in natural systems.

Consumers compete for resources

Ecologist David Tilman, a professor at the University of Minnesota and a leading researcher on interactions among plant species, defined a **resource** as any substance or factor that is both consumed by an organism and supports increased population growth rates as its availability in the environment increases. Three things are key to this definition. First, a resource is consumed, and its amount or availability is thereby reduced. Second, a consumer uses a resource for its own maintenance and growth. Thus, food is always a resource, and water is a resource for terrestrial plants and animals. Third, when resource availability is reduced, biological processes are affected in such a way as to reduce consumer population growth.

Consumption includes more than just eating. Sessile animals—those permanently attached to a substrate—need open, available sites, so for those animals, space is a resource. Barnacles attached to rocks within the intertidal zone, for example, need space to grow, and barnacle larvae require space to settle and develop into adults (Figure 16.3). Crowding increases adult mortality and reduces fecundity by limiting the growth of adults and the recruitment of larvae. Open space fosters reproduction and recruitment. Thus, individual barnacles "consume" open sites as they colonize and grow on them.

Refuges and other safe sites are another kind of resource. Each habitat has a limited number of holes, crevices, or patches of dense cover in which an individual may escape predation or seek refuge from inclement weather. As some individuals occupy the best sites, others must settle for less favorable places; consequently, those individuals have a higher risk of impairment or death.

What kinds of factors are not resources? Temperature is not a resource. Higher temperatures may raise reproductive rates, but individuals do not consume temperature, and one individual does not change the temperature of the environment for another. Temperature and other nonconsumable physical and biological factors can limit the growth of populations, of course, but they must be considered separately from resources.

(a)

(b)

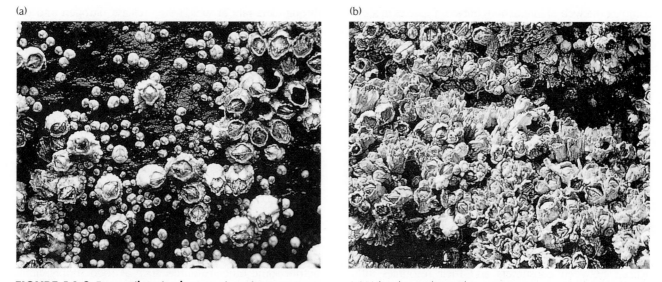

FIGURE 16.3 For sessile animals, space is an important resource. (a) When barnacles on the rocky coast of Maine are living above their optimal range in the intertidal zone, their density is low, and larvae can settle in the bare patches. (b) Lower in the intertidal zone, larvae can settle only on an older individual, and dense crowding precludes further population growth. *Courtesy of the American Museum of Natural History.*

Competition between closely and distantly related species

Charles Darwin emphasized that competition should be most intense between closely related species. In *On the Origin of Species,* he remarked, "As species of the same genus have usually, though by no means invariably, some similarity in habits and constitution, and always in structure, the struggle will generally be more severe between species of the same genus, when they come into competition with each other, than between species of distinct genera." Darwin reasoned that similar structure indicates similar function, especially with respect to resources consumed. This comment inspired Tansley to examine competition between two closely related species in the same genus of plants.

Although Darwin's insight is generally correct, distantly related organisms may also use many of the same resources. Barnacles and mussels, as well as algae, sponges, bryozoans, tunicates, and others, occupy space in the intertidal zone and actively compete with one another by preemption of space and overgrowth. Both fish and aquatic birds prey on aquatic invertebrates. Krill (*Euphausia superba*), shrimplike crustaceans that abound in subantarctic waters (Figure 16.4), are fed on by virtually every type of large marine animal, including fish, squid, diving birds, seals, and whales. Seal and penguin populations in the Southern Ocean have recently increased, apparently because commercial exploitation has decimated populations of a major competitor group, the whales. In terres-

trial habitats, spiders, ground beetles, salamanders, and birds consume invertebrates living in forest litter. In desert ecosystems, birds and lizards eat many of the same insect species, and ants, rodents, and birds consume the seeds of many of the same plants. These examples illustrate the strong potential for competition between distantly related organisms and remind us once again how extensive and

FIGURE 16.4 Common food sources bring distantly related organisms into competition. In the Southern Ocean, krill (*Euphausia superba*) supply food for a wide range of marine animal species. *Courtesy of Dr. Uwe Kils.*

complex the web of interactions between species in a biological community can be.

Renewable and nonrenewable resources

Resources can be classified according to whether they can be regenerated or not. **Nonrenewable resources,** such as space, are not regenerated. Once occupied, space becomes unavailable; it is "replenished" only when the consumer leaves. In contrast, **renewable resources** are constantly regenerated, or renewed. Births in a prey population continually supply food items for predators, just as the continuous decomposition of organic detritus in the soil provides a fresh supply of nitrate for plant roots.

Because competitors consume shared resources, interactions between consumers and their resources influence competitive relationships. This is particularly evident when we consider consumer–resource relationships in the broader context of the entire ecological system, including evolutionary responses among its component species. For example, some renewable resources originate outside the system, beyond the influence of consumers. Sunlight strikes the surface of the earth regardless of whether plants "consume" it, and local precipitation is largely independent of the uptake of water by plants. Thus, consumers reduce the immediate availability of such resources to others, but not their supply. Moreover, such resources do not respond in any way to consumption.

A second type of renewable resource is generated within the ecological system. Consumers directly depress the abundance of such resources. Most predator–prey, herbivore–plant, and parasite–host interactions involve renewable resources because the supply of prey, plants, and hosts is constantly regenerated. Yet, by reducing populations of their resources, consumers potentially reduce the rate of renewal of their food supply.

Some renewable resources regenerated within the ecosystem are linked only indirectly to their consumers, either through other links in a food chain or through abiotic processes. In the nitrogen cycle of a forest, for example, plants assimilate nitrate from the soil. Herbivores and detritivores consume plants or plant remains, returning large quantities of organic nitrogen compounds to the soil. These nitrogen compounds are further broken down by microorganisms, which release the nitrogen as nitrate, a form that plants can use. The uptake of nitrate by plants, however, has little direct effect on its renewal by detritivores and microorganisms. Similarly, consumption of detritus by detritivores does not immediately influence plant production. Clearly, however, detritivores and microorganisms do influence plant production indirectly through the rate at which they release nitrogen into the soil.

Limiting resources

By diminishing their resources, consumers limit their own population growth. As a population grows, its overall resource requirements grow as well. When its requirements increase so much that the decreasing supplies of a resource can no longer fulfill its needs, the population's size levels off, or even begins to decrease. However, whereas consumers reduce both renewable and nonrenewable resources, not all resources limit consumer populations. All terrestrial animals require oxygen, for example, but they do not noticeably depress its level in the atmosphere before some other resource, such as food supply, limits their population growth.

At one time, ecologists believed that populations were limited by the single resource that was most scarce relative to demand. This principle has been called **Liebig's law of the minimum,** after Justus von Liebig, a German chemist who articulated the idea in 1840. According to this law, each population increases until the supply of some resource—the **limiting resource**—no longer satisfies the population's need for it.

The supply of a resource needed to sustain a population's growth under a given set of conditions is unique to each resource. For example, David Tilman discovered that when the diatom *Cyclotella meneghiniana* is grown with limited supplies of silicate and phosphate in a laboratory culture, population growth ceases when phosphate levels are reduced to 0.2 millimolar (mM) or silicate levels are reduced to 0.6 mM. According to Liebig's law of the minimum, whichever of these resources is reduced to this limiting value first regulates the growth of the *Cyclotella* population.

Liebig's law applies strictly only to resources having an independent influence on the consumer population. In many cases, however, two or more resources interact to determine the growth rate of a consumer population; that is, the growth rate of a population at a particular level of one resource depends on the level of one or more other resources. When two resources together enhance the growth of a consumer population more than the sum of both individually, the resources are said to be *synergistic* (from the Greek roots *syn,* "together," and *ergon,* "work").

The principle of synergism can be illustrated with a study by the British ecologists W. J. H. Peace and P. J. Grubb on the small herbaceous plant *Impatiens parviflora,* which is common in woodlands of England. In one experiment, Peace and Grubb sowed *Impatiens* seeds in

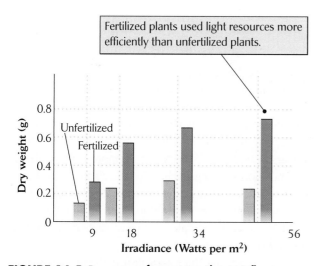

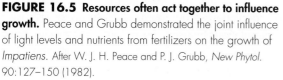

FIGURE 16.5 **Resources often act together to influence growth.** Peace and Grubb demonstrated the joint influence of light levels and nutrients from fertilizers on the growth of *Impatiens*. After W. J. H. Peace and P. J. Grubb, *New Phytol.* 90:127–150 (1982).

fertilized (with nitrate and phosphate) and unfertilized (control) soils, then exposed the resulting seedlings to different levels of light for 5 weeks after seed germination. Added light enhanced the growth of the fertilized plants more than that of the controls (Figure 16.5). Thus, the

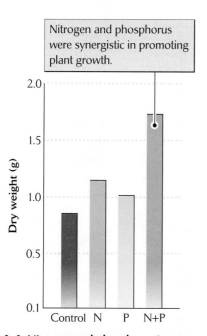

FIGURE 16.6 **Nitrogen and phosphorus interact synergistically to promoting plant growth.** Peace and Grubb demonstrated the joint influence of nitrogen (N) and phosphorus (P) fertilization on the growth of *Impatiens*. After W. J. H. Peace and P. J. Grubb, *New Phytol.* 90:127–150 (1982).

ability of *Impatiens* to use light depends on the availability of other resources. Plant growth requires both the carbon assimilated by photosynthesis, as a source of energy and for structural carbohydrates, and nitrogen and phosphorus, which are needed for the synthesis of proteins, phospholipids, and nucleic acids. At the highest light intensities used in the experiment, nitrogen and phosphorus were also shown to interact synergistically to enhance plant growth (Figure 16.6).

Failure of species to coexist in laboratory cultures led to the competitive exclusion principle

Many of the early studies of population dynamics were designed to determine the effects of one species on the population growth of another. In these experiments, two species were first grown separately, under controlled conditions and resource levels, to determine their carrying capacities in the absence of interspecific competition. The two species were then grown together under the same conditions to determine the effect of each on the other. The difference between the population growth of one species in the presence and in the absence of the other was taken as a measure of the intensity of competition between them.

Experiments of this kind on protists by the Russian biologist G. F. Gause, who also conducted early studies on predation (see Chapter 15), were among the first and most influential on subsequent work in population biology. When Gause grew *Paramecium aurelia* and *P. caudatum* separately on the same type of nutritive medium, both populations grew rapidly to limits imposed by their food supply (a particular type of bacteria growing in the culture medium). When the two species were grown together, however, only *P. aurelia* persisted (Figure 16.7). Similar experiments with fruit flies, mice, flour beetles, and annual plants typically produced the same result: one species persisted and the other died out, usually after 30 to 70 generations.

The accumulating results of laboratory experiments on competition were eventually summarized by the **competitive exclusion principle,** which states that two species cannot coexist indefinitely when the same resource limits both species. Similar species do, of course, coexist in nature. But as we shall see in later chapters, observations often reveal subtle differences between them in habitat or diet preference. These observations prompt us to ask how much difference in resource requirements is sufficient to allow coexistence. Although this question has been very

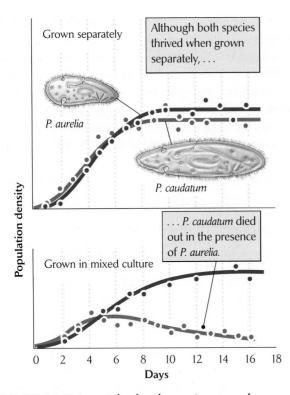

Grown separately

Although both species thrived when grown separately, . . .

P. aurelia

P. caudatum

Population density

. . . *P. caudatum* died out in the presence of *P. aurelia.*

Grown in mixed culture

0   2   4   6   8   10   12   14   16   18
Days

FIGURE 16.7 Gause's landmark experiments on the coexistence of species in laboratory cultures led to the competitive exclusion principle. In one such experiment, two species of *Paramecium* were grown in separate cultures and grown together under the same conditions. The vertical scales of the two graphs are the same. After G. F. Gause, *The Struggle for Existence*, Williams & Wilkins, Baltimore (1934).

difficult to answer, theoretical models of competition suggest some of the general conditions under which species may coexist.

The theory of competition and coexistence is an extension of logistic growth models

Most competition theory springs from mathematical models developed by A. J. Lotka, V. Volterra, and G. F. Gause, who used the logistic equation for population growth as their starting point. Much of modern population biology has been erected on the foundation provided by these models. Although they are now considered overly simplistic, these models still provide useful insights into the outcome of competitive interactions and the conditions under which species can coexist.

According to the logistic equation (see Chapter 11), the exponential rate of increase of a population of size N (r, or equivalently, $1/N \, dN/dt$) is expressed by

$$\frac{1}{N} \frac{dN}{dt} = \frac{r_0}{K}(K - N)$$

where r_0 is the exponential rate of increase when the population size is close to 0—that is, in the absence of intraspecific competition—and K is the number of individuals that the environment can support (the carrying capacity). In this equation, intraspecific competition appears as the term $(K - N)$; as N approaches K (that is, as population size approaches the carrying capacity), $K - N$ approaches 0, and so the growth rate also approaches 0. As we have seen before, a stable equilibrium is reached when $N = K$ and population size has reached the carrying capacity (Figure 16.8).

When modeling the interaction between two species whose populations grow logistically, we distinguish each of the populations by using a subscript 1 for species 1 and a subscript 2 for species 2. Now we can include the effect of competition from species 2 on the population growth of species 1 by subtracting the term $a_{1,2}N_2$ from the quantity within the parentheses. Hence,

$$\frac{1}{N_1} \frac{dN_1}{dt} = \frac{r_1}{K_1}(K_1 - N_1 - a_{1,2}N_2)$$

where r_1 is the exponential rate of increase of species 1 when its population size is close to 0, N_2 is the number of individuals of species 2, and $a_{1,2}$ is the competition coefficient—that is, the effect of an individual of species 2 on the exponential rate of increase of species 1. You can think of the competition coefficient $a_{1,2}$ as the degree to

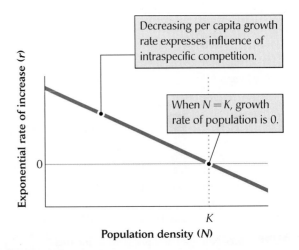

Decreasing per capita growth rate expresses influence of intraspecific competition.

When $N = K$, growth rate of population is 0.

Exponential rate of increase (r)

0

K

Population density (N)

FIGURE 16.8 The exponential rate of increase of a logistically growing population is a function of its density. Even as intraspecific competition depresses r, the population continues to increase until its size (N) is equal to the carrying capacity (K).

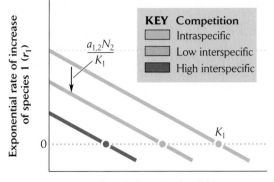

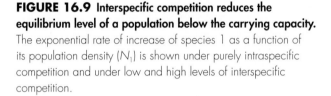

FIGURE 16.9 Interspecific competition reduces the equilibrium level of a population below the carrying capacity. The exponential rate of increase of species 1 as a function of its population density (N_1) is shown under purely intraspecific competition and under low and high levels of interspecific competition.

which each individual of species 2 uses the resources of species 1 relative to an individual of species 1. Thus, $a_{1,2}N_2$ represents the effect of population 2 on the growth rate of population 1 (Figure 16.9).

Because each competing species exerts an effect on the other, the mutual relationship between them requires two equations. One, presented above, incorporates the effect of species 2 on species 1. The second equation incorporates the effect of species 1 on species 2 and is similar to the first, but with the subscripts 1 and 2 reversed.

If two species are to coexist, the populations of both must reach a stable equilibrium size greater than zero. That is, both dN_1/N_1dt and dN_2/N_2dt must equal zero at some combination of positive values of N_1 and N_2. From the previous equation, we see that $dN_1/N_1dt = 0$ when

$$\hat{N}_1 = K_1 - a_{1,2}N_2$$

and $dN_2/N_2dt = 0$ when

$$\hat{N}_2 = K_2 - a_{2,1}N_1$$

The little hat (ˆ) over the Ns indicates that they are equilibrium values. In the absence of interspecific competition ($a_{1,2} = 0$), the equilibrium population size $\hat{N}_1$ is equal to K_1, a measure of the resources available to species 1. Interspecific competition reduces the effective carrying capacity of the environment for species 1 by the amount $a_{1,2}\hat{N}_2$. Thus, the carrying capacity for species 1 is reduced in proportion to the population size and competition coefficient of species 2 (see Figure 16.9).

In general, these equations tell us that coexistence is most likely (both $\hat{N}_1$ and $\hat{N}_2 > 0$) when interspecific

competition is relatively weak compared with intraspecific competition—that is, when the competition coefficients $a_{1,2}$ and $a_{2,1}$ are less than 1. In other words, to coexist, species must limit themselves more than they limit each other. Any number of species may coexist as long as this criterion is met for all pairs of them. Competition coefficients are most likely to be less than 1 when competitors are specialized to use different resources most efficiently. Notice that the outcome of competition does not depend on the exponential growth rates of the populations when they are small.

> **HELP ON THE WEB** *Go to Living Graphs.* Use the interactive tutorial on competition better to understand the dynamics between species.

Coexistence on multiple resources

In Gause's experiments on competition in *Paramecium,* two species were forced to utilize a single limiting resource (a particular kind of food). When two species compete for *two* limiting resources needed in different amounts by each species, certain combinations of resource abundance allow the two species to coexist. David Tilman first demonstrated this principle in a series of elegant experiments on competition between two diatom species in the genera *Cyclotella* and *Asterionella*. Both species require silicon to produce their glassy outer shells, and both require phosphorus to synthesize membrane phospholipids, DNA, and other molecules, but their relative requirements for these resources differ. Initial experiments demonstrated that the population growth of *Cyclotella* was equally limited by silicon (Si) and by phosphorus (P) when the two nutrients occurred at a ratio of about Si/P = 6. *Asterionella,* which has a much more extensive silicate shell relative to its cell size, was equally limited by both nutrients at a ratio of Si/P = 90. At nutrient ratios below these levels, each of the species is silica-limited; above these ratios, they are phosphorus-limited (Figure 16.10).

Although both species could grow in single-species cultures on the entire range of Si/P ratios Tilman used, *Cyclotella* uses silicon more efficiently and is able to increase in numbers on lower supplies of silicate. In contrast, *Asterionella* utilizes phosphorus more efficiently and can increase on lower supplies of phosphate. Between Si/P ratios of 6 and 90, the population growth of *Cyclotella* is limited by phosphorus and the population growth of *Asterionella* is limited by silicon. Tilman predicted that the two species should be able to coexist in culture within this intermediate range of Si/P ratios because each of the species would

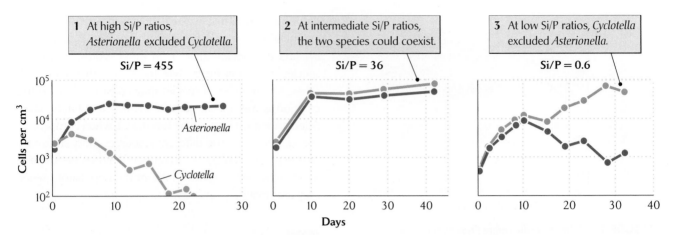

FIGURE 16.10 Tilman's experiments with diatoms showed that two species could coexist if they were limited by different resources. In cultures containing intermediate ratios of silicon to phosphorus (Si/P = 36), *Cyclotella* is limited by phosphate concentration and *Asterionella* by silicate concentration. After D. Tilman, *Ecology* 58:338–348 (1977).

be limited by a different resource. When he grew the species together in chemostat cultures, with continual input of fresh medium, he found that the two species persisted between Si/P ratios of 6 and 90 regardless of the rate of nutrient supply. At Si/P ratios above 90, *Asterionella* excluded *Cyclotella,* and at Si/P ratios below 6, *Cyclotella* excluded *Asterionella.*

Asymmetric competition can occur when different factors limit the populations of competitors

In many cases of competition, the relationship between competitors is asymmetrical in the sense that each has an advantage with respect to different factors in the environment. For example, one species might exploit resources more efficiently, while the other is better at tolerating stressful conditions or avoiding consumers. In many such cases of **asymmetric competition,** the competitors coexist locally in different microhabitats. The following study shows how two species of barnacles can persist in the rocky intertidal zone because each is better at doing a different thing.

Barnacles are sessile invertebrates that may form dense, continuous populations on rocky shores. Barnacles feed on plankton that they filter from the water that washes over them. Food is not a limiting resource for them because their feeding cannot substantially reduce the vast numbers of plankton in coastal waters. Instead, populations of barnacles in many areas are limited by space for settling and growth (see Figure 16.3).

Joseph Connell, a biologist at the University of California at Santa Barbara, performed a series of classic experiments on two species of barnacles within the intertidal zone of the rocky coast of Scotland. Adults of *Chthamalus stellatus* normally occur higher in the intertidal zone than those of *Balanus balanoides,* the more northerly of the two species. Although the vertical distributions of newly settled larvae of the two species overlap broadly within the intertidal zone, the line between the vertical distributions of adults is sharply drawn (Figure 16.11).

Connell demonstrated that adult *Chthamalus* are restricted to upper regions of the intertidal zone above *Balanus* not because of physiological tolerance limits, but because of interspecific competition. When Connell removed *Balanus* from rock surfaces, *Chthamalus* thrived in the lower regions of the intertidal zone where they normally do not occur. The two barnacle species compete directly for space. *Balanus* have heavier shells and grow more rapidly than *Chthamalus;* as individuals expand, the shells of *Balanus* edge underneath those of *Chthamalus* and literally pry them off the rock! *Chthamalus* can live in the upper intertidal zone because they are more resistant to desiccation than *Balanus;* even when surfaces in the upper intertidal zone are kept free of *Chthamalus, Balanus* do not invade.

Within each region of the intertidal zone, one of the barnacle species studied by Connell is the superior competitor. Such asymmetry in competition leading to local competitive exclusion reflects an imbalance in the ecological relationships between two species. The superior competitor for resources is almost always more strongly limited by some other factor, such as environmental stress

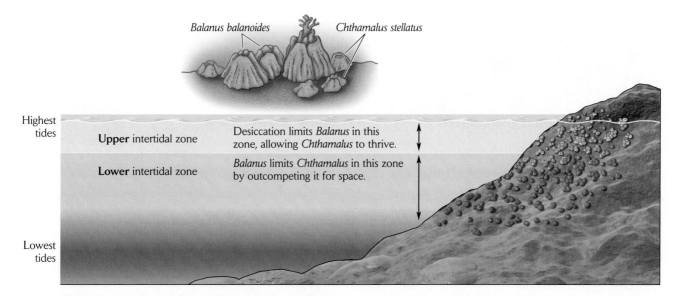

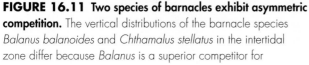

FIGURE 16.11 Two species of barnacles exhibit asymmetric competition. The vertical distributions of the barnacle species *Balanus balanoides* and *Chthamalus stellatus* in the intertidal zone differ because *Balanus* is a superior competitor for space, but cannot withstand the stress of desiccation as well as *Chthamalus. After* J. H. Connell, *Ecology* 42:710–723 (1961) and *Ecol. Monogr.* 31:61–104 (1961).

or predators. In this case, *Balanus,* the superior competitor for space, is limited by its lack of resistance to desiccation. Each of the two barnacle species has made a different trade-off between stress tolerance and competitive ability.

The results of field experiments on species that coexist in nature suggest that asymmetric competition is common. These experiments are generally conducted by growing individuals of each species in the presence and in the absence of the other. Interspecific competition is assessed by differences in the growth of individuals, reproductive success, population density, or other measures related to population growth. According to one survey of interactions between 98 pairs of species, neither species was negatively affected by the other in 44 cases (no competition), both species were affected by similar amounts in 21 cases (reciprocal competition), and only one species was affected in 33 of the interactions (asymmetric competition). The large proportion of cases in the last category indicates that asymmetric competition is more common than reciprocal competition, and that species often coexist in nature because they are limited by different factors.

DATA ANALYSIS MODULE *Asymmetry in Competition.* Use data for barnacles and algae to explore the effects of environmental conditions on the outcome of competition between two species. You will find this module at http://www.whfreeman.com/ricklefs6e.

Habitat productivity can influence competition between plant species

Plants can compete for both water and nutrients in the soil and sunlight above ground, all of which increase the productivity of vegetation. Several ecologists have suggested that the outcome of plant competition depends on nutrient levels in soils because nutrient levels control the growth of plant shoots and their ability to shade competitors. According to one hypothesis, promoted by plant ecologists P. J. Grubb and David Tilman, plants compete more intensely when mineral nutrients are less abundant in the soil. High nutrient levels are less likely to limit plant populations, and therefore interspecific competition should be weaker. The opposite point of view, espoused by J. P. Grime and Paul Keddy, suggests that competition is *less* intense when water and mineral nutrients are less abundant. The reason, according to Grime and Keddy, is that competition for light is more important than competition for soil nutrients. Lack of water and nutrients should limit populations to the point that individual plants are widely spaced and do not compete for light. The difference between the Grubb–Tilman and Grime–Keddy hypotheses lies in the relative importance placed on belowground and aboveground competition for resources—that is, for nutrients and light, respectively.

These two hypotheses can be tested by competition experiments in high-productivity and low-productivity environments, but the results often depend on the particular

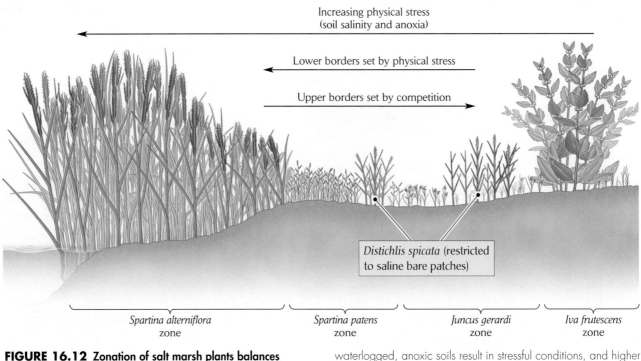

Increasing physical stress
(soil salinity and anoxia)

Lower borders set by physical stress

Upper borders set by competition

Distichlis spicata (restricted to saline bare patches)

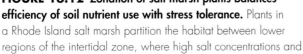

| *Spartina alterniflora* zone | *Spartina patens* zone | *Juncus gerardi* zone | *Iva frutescens* zone |

FIGURE 16.12 Zonation of salt marsh plants balances efficiency of soil nutrient use with stress tolerance. Plants in a Rhode Island salt marsh partition the habitat between lower regions of the intertidal zone, where high salt concentrations and waterlogged, anoxic soils result in stressful conditions, and higher regions with well-drained, salt-free soils. After N. C. Emery et al., *Ecology* 82:2471–2485 (2001).

species used and the experimental design. For example, one experimental study, conducted by S. D. Wilson and David Tilman in a mesic prairie habitat in Minnesota with three species of prairie grasses, found little relationship between nutrient levels and competition. The researchers established low-, medium-, and high-productivity plots by adding ammonium nitrate fertilizer to the soil. Aboveground biomass varied twofold to threefold among the plots, but competition intensity did not vary significantly among the plots. Wilson and Tilman suggested that plants competed primarily below ground for nutrients on low-nutrient plots and primarily above ground for light on high-nutrient plots, resulting in intense competition at all nutrient levels.

In another study, Nancy Emery and her collaborators Patrick Ewanchuk and Mark Bertness manipulated nutrient levels across a gradient of soil salinity and soil saturation (both stressful conditions for terrestrial plants) in a salt marsh in Rhode Island. The plant species in the marsh were strongly zoned according to their stress tolerance (Figure 16.12). When the investigators added fertilizer to otherwise unmanipulated plots at the junctions of these zones, the outcome of competition was reversed: the more stress-tolerant species was released from nutrient limitation and outcompeted the other species by overgrowth (Figure 16.13). Additional experiments involving a combination of nutrient addition and aboveground clipping of plants indicated that competition occurred primar-

ily below the soil surface under low nutrient conditions and above ground in fertilized plots, as Tilman and his colleagues in Minnesota argued for the prairie system. In the marsh setting, Emery and her collaborators suggested that plants traded off belowground competitive ability for the ability to tolerate stressful saline and anoxic conditions in the lower levels of the salt marsh environment.

This salt marsh plant community provides another example of asymmetric competition. As in the earlier example of barnacles in the intertidal zone, the superior competitor is limited by its lesser tolerance of stress. *Juncus gerardi*, for example, excludes *Spartina patens* from the upper, less stressful part of the intertidal zone, but cannot withstand the salinity and anoxia of soils lower down.

What are we to make of such studies? Competition appears to be pervasive, but how it is manifested depends very much on the characteristics of the interacting species and their habitats.

Competition may occur through direct interference

In many of the examples that we have seen so far, individuals compete indirectly through their mutual effects on shared resources. This kind of competition is called **exploitative competition.** Less frequently, competitors interact directly by aggressively defending resources.

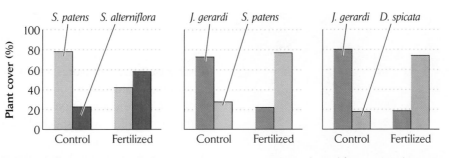

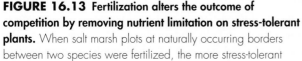

FIGURE 16.13 Fertilization alters the outcome of competition by removing nutrient limitation on stress-tolerant plants. When salt marsh plots at naturally occurring borders between two species were fertilized, the more stress-tolerant species were released from nutrient limitation, at which point their superior stress tolerance increased their competitive ability. After N. C. Emery et al., *Ecology* 82:2471–2485 (2001).

This behavior is referred to as **interference competition,** although distinctions between the two are sometimes blurred. Hummingbirds chase other hummingbirds, not to mention bees and moths, from flowering bushes. Encrusting sponges use poisonous chemicals to overcome other sponge species as they expand to fill open space on rock surfaces. Many shrubs release toxic chemicals into the soil that depress the growth of competitors. Even bacteria wage chemical warfare with one another to tip the balance of their competitive interactions.

Interference competition is often evident in experimental manipulations of competing animal species. For example, two species of voles (small mouselike rodents of the genus *Microtus*) are both present in some areas of the Rocky Mountains in North America. In western Montana, the meadow vole (*M. pennsylvanicus*) normally lives in wet habitats surrounding ponds and watercourses, whereas the mountain vole (*M. montanus*) is restricted to dry habitats. Ecologists believed that this spatial partitioning resulted from competition between the two species, but the nature of the interaction was not known. However, when meadow voles were experimentally trapped and removed from an area of wet habitat, mountain voles immediately began to move in from surrounding dry habitats. This result suggested that meadow voles excluded mountain voles from the wetter areas by direct, aggressive encounters—a case of interspecific territoriality. Interestingly, when mountain voles were trapped and removed from a dry habitat that they normally occupied exclusively, meadow voles began to move in. Thus, each species is behaviorally dominant in its preferred habitat, illustrating the principle of home field advantage in rodents.

Allelopathy

Chemical competition, or **allelopathy,** has been reported most frequently in terrestrial plants, in which such interactions may take on a variety of forms. In most cases, a toxic substance causes injury (-*pathy*) to other (*allelo-*) individuals directly. For example, black walnut (*Juglans nigra*) trees produce juglone, an aromatic organic compound that inhibits certain enzymes in other plants. As a result, few plant species are capable of germinating and becoming established under black walnut trees. A somewhat different mechanism of action has been suggested to explain the abundant oils in the leaves and bark of eucalyptus trees of Australia—namely, that they promote frequent fires in the leaf litter, which kill the seedlings of competitors (Figure 16.14).

In shrub habitats in southern California, several species of sage (genus *Salvia*) produce volatile terpenes, a class of organic compounds that includes camphor and gives foods spiced with sage part of their distinctive taste. Terpenes inhibit the growth of other vegetation in the laboratory, so investigators proposed an allelopathic function for these compounds in nature. Clumps of shrubby *Salvia* plants are usually surrounded by bare zones separating the sage from neighboring grassy areas (Figure 16.15). When observed over long periods, *Salvia* can be seen to expand into the grassy areas.

Regardless of whether *Salvia* plants release inhibiting compounds, their interaction with potential competitors is more complicated. In subsequent experiments, investigators fenced parts of the bare zones surrounding *Salvia* patches to exclude such herbivores as rabbits and ground squirrels. With no herbivores present, other plants became established right up to the edge of the sage patches. Counts of scats in the bare zones showed that rabbits rarely ventured more than a meter from the safe haven under the cover of the *Salvia* shrubs. These observations implied that the rabbits must be feeding on the more nutritious competitors of sage, but not on the sage itself. The observation that bare zones also surround patches of other shrub species, such as rabbitbrush (*Baccharis*), that do not produce

FIGURE 16.14 Some plants compete by chemical means.
The leaves and bark of the eucalyptus trees of Australia have a high oil content, which promotes fires that kill the seedlings of potentially competing species, but leave the eucalyptus trees unharmed. Photo by R. E. Ricklefs.

toxic volatile chemicals suggests that herbivores mediate competition between plant species. One might consider the bare zones as the result of a kind of allelopathy in which plants use herbivores, rather than toxic chemicals, to fight their battles with competitors.

Interference competition can be an important factor in the success of invasive species. For example, Australian ironwood trees (*Casuarina equisetifolia*) have been introduced into many tropical and subtropical regions of the world, including Florida and the Hawaiian Islands, where they rapidly invade and exclude other vegetation. The nearly complete absence of germination or establishment of native vegetation in soils covered by the needle-like *Casuarina* leaves suggests an allelopathic effect (Figure 16.16).

Consumers can influence the outcome of competition

Charles Darwin was the quintessential naturalist. Among his many seminal observations, he noted that grazing can maintain a high diversity of plants in grasslands. In the absence of grazers, dominant competitors grow rapidly and exclude other species. Similar results have been obtained from experiments on marine algal communities grazed by limpets, snails, and urchins. These studies indicate that predation has a strong hand in shaping the struc-

(a)

(b)

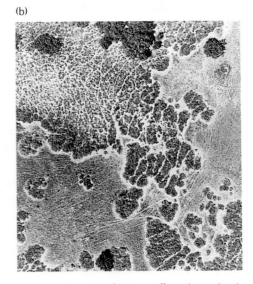

FIGURE 16.15 Some plants compete at a distance by producing airborne toxins or by harboring consumers.
(a) The bare zone at the edge of a clump of sage includes a 2 m wide strip with no plants (A to B) and a wider area of inhibited grassland (B to C) lacking wild oat and bromegrass, which are

found with other species to the right of C in unaffected grassland. (b) An aerial view shows sage and California sagebrush invading annual grassland in the Santa Ynez Valley of California. Courtesy of C. H. Muller, from C. H. Muller, *Bull. Torrey Bot. Club* 93:332–351 (1966).

FIGURE 16.16 The leaf litter under introduced ironwood (*Casuarina equisetifolia*) trees suppresses the germination and growth of native species. This woodland was photographed on Kauai, in the Hawaiian Islands, where the species is invasive. Courtesy of Eric Guinther.

ture of biological communities by influencing the outcome of competitive interactions between prey species.

University of Washington ecologist Robert Paine was one of the first investigators to demonstrate this point experimentally. On the exposed rocky coast of Washington, the intertidal zone harbors several species of barnacles, gooseneck barnacles, mussels, limpets, and chitons (a kind of grazing mollusk). All of these animals are preyed on by the sea star *Pisaster* (Figure 16.17). Paine removed sea stars from an area of rock 8 m long and 2 m high;

an adjacent area of similar size was left undisturbed as a control.

Following the removal of the sea stars, the number of prey species in the experimental plot decreased rapidly, from 15 at the beginning of the study to 8 at the end. Those remaining were principally the mussel *Mytilus* and other invertebrates living on, and in the cracks between, individual mussels. Diversity declined in the experimental plot because populations of mussels and barnacles increased and crowded out many of the other species.

(a)　　　　　　　　　　(b)　　　　　　　　　　(c)

FIGURE 16.17 Predators can maintain prey species diversity by reducing populations of superior competitors. (a) A congregation of sea stars (*Pisaster*) at low tide on the coast of the Olympic Peninsula, Washington. The sea star (b) is an important predator on mussels. (c) In the absence of sea stars, diversity decreases rapidly until only mussels, and barnacles living in the cracks between them, remain. Photo (a) by Ken Lucas/Visuals Unlimited; photo (b) by Daniel W. Gotshall/Visuals Unlimited; photo (c) by Francis & Donna Caldwell/Visuals Unlimited.

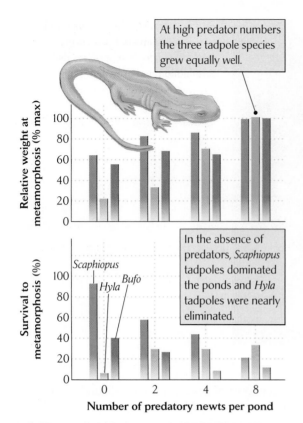

At high predator numbers the three tadpole species grew equally well.

In the absence of predators, *Scaphiopus* tadpoles dominated the ponds and *Hyla* tadpoles were nearly eliminated.

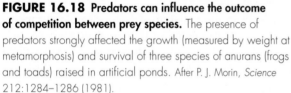

Number of predatory newts per pond

FIGURE 16.18 Predators can influence the outcome of competition between prey species. The presence of predators strongly affected the growth (measured by weight at metamorphosis) and survival of three species of anurans (frogs and toads) raised in artificial ponds. After P. J. Morin, *Science* 212:1284–1286 (1981).

Paine concluded that sea stars maintain the diversity of the intertidal zone community by limiting populations of mussels and barnacles, which are superior competitors for space in the absence of predators. This is another case of asymmetric competition in which the superior competitors are less well protected against predation.

Studies conducted in artificial ponds have shown that predators can reverse the outcome of competition among anuran (frog and toad) tadpoles. In one experiment conducted by Peter Morin of Rutgers University, ponds were supplied with 200 larvae of the spadefoot toad (*Scaphiopus holbrooki*), 300 of the spring peeper (*Hyla crucifer*; a frog species), and 300 of the southern toad (*Bufo terrestris*). Each of the ponds, which were identical in all other respects, also received 0, 2, 4, or 8 individuals of the predatory broken-striped newt (*Notophthalmus viridescens*).

In the absence of newt predation, *Scaphiopus* tadpoles grew rapidly, survived well, and dominated the ponds along with smaller numbers of *Bufo*; *Hyla* tadpoles were

all but eliminated by competition (Figure 16.18). However, the newts apparently preferred toad tadpoles, and at higher numbers of predators, survival of both *Scaphiopus* and *Bufo* decreased markedly. With fewer toad tadpoles per pond, supplies of food increased, and survival and growth of *Hyla* tadpoles improved immensely, as did the growth of the surviving *Scaphiopus* and *Bufo* tadpoles.

Apparent competition

Interactions between competing species that are mediated by consumers are often referred to as **apparent competition**—apparent because the depressing effect of one competitor species on the other resembles exploitative or interference competition, but represents the action of a different mechanism. We have seen apparent competition in the interactions between sage and grasses and between mussels and other rocky intertidal organisms: in both cases, consumers determined the competitive balance. The outcome of competition depended less on the ability of competitors to utilize food or other resources efficiently than on their ability to avoid or tolerate their own consumers. In the case of the sage–grass interaction, the shrubs provide protective cover for herbivores that prefer to eat the more nutritious grasses.

Pathogens may play a similar role in tipping the balance between competitors. Certainly, as Europeans colonized other continents, particularly North and South America, they brought with them diseases, such as smallpox, that their own populations could more or less tolerate, but which were devastating to Native Americans. Many invasive plants and animals may owe their success to similar mechanisms. For example, the Eurasian garlic mustard (*Alliaria petiolata*) has become widespread in North America, where it resists herbivory by native American insects and deer. Because it resists consumers that feed on native vegetation, garlic mustard probably gains its edge through apparent competition, rather than any ability it might have to utilize resources efficiently.

ECOLOGISTS IN THE FIELD   **Apparent competition between corals and algae mediated by microbes.** Apparent competition mediated by consumers illustrates the complexity of interactions in nature, but it is only one kind of indirect competitive interaction. Consider the interaction between algae and corals on tropical reefs. We have seen that some algae live symbiotically within the tissues of corals, but other kinds of algae have a less benign influence on corals. Marine biologists had observed that when those algae become established near corals, the corals decline in health and eventually die. Several mechanisms of this apparently competitive interaction had been proposed, including shading and allelopathy. Jennifer Smith from the University of California at Santa Barbara and her colleagues suspected, however, that there might be an alternative explanation. Previous research had shown that corals were more likely to die in the presence of dissolved organic carbon, and that those deaths were correlated with an increase in the growth of microbes that normally live on the surfaces of corals. Perhaps the algae were secreting dissolved organic carbon (in the form of excess polysaccharides produced by photosynthesis) and this carbon was supporting microbial growth on the corals. One consequence might be that the microbes—mostly bacteria—would block the diffusion of oxygen to the underlying coral tissues and suffocate the corals.

Smith and her colleagues grew corals (Figure 16.19) in containers, either alone or paired with algae that were separated from them by a fine mesh, which allowed dissolved organic carbon to pass through but excluded the movement of microbes. They found that the corals survived well when the algae were absent, but suffered high mortality when the algae were present. In another set of containers, they conducted the same experiment, but added an antibiotic to the water. In this case, the corals survived well with or without algae present, confirming that the coral mortality was due to microbes that could be controlled by an antibiotic. The investigators repeated this experiment with a larger number of coral and algal species. In containers without antibiotic treatment, 95% of the coral species experienced declining health. With antibiotic treatment,

FIGURE 16.19 Corals can be indirectly harmed by the presence of algae. The health of coral species such as this *Pocillopora verrucosa* suffers when algae become established nearby. Nature Picture Library/Alamy.

however, only 20% of the coral species were negatively affected by the presence of algae (Figure 16.20).

In all these experiments, it remained unclear why microbes that naturally live on corals killed corals when the algae were nearby. The investigators conducted further experiments in which they measured microbial growth on the surfaces of corals. They found that when algae secreted excess polysaccharides, the microbes grew faster and used up the oxygen in the water immediately surrounding the corals, causing the corals to suffocate. The fact that the health of some coral species, such as *Pocillopora*, was

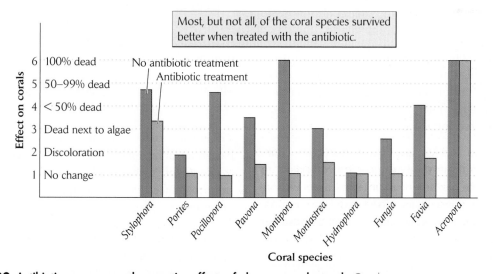

FIGURE 16.20 Antibiotics can reverse the negative effects of algae on coral growth. Corals were grown in the presence of algae with and without antibiotic treatment. The results suggested that the algae were exerting their negative effects on the health of the corals by supporting microbial overgrowth on their surfaces. After J. E. Smith et al., *Ecology Letters* 9:835–845 (2006).

restored by antibiotic treatment, whereas that of others, such as *Acropora*, was not, hints at even more complexity in this system.

This example shows that a negative association between two species can occur without direct competition for a shared resource. In the case of algae and corals, the negative interaction is caused by a chain of events that is initiated by the algae emitting polysaccharides. Those carbon compounds favor the overgrowth of coral-associated microbes, which reduce oxygen concentrations in the water to levels that harm the corals. It is important to recognize that pollution caused by human activities is also a source of organic carbon compounds that promote microbial growth on coral reefs and is thus an important factor in the deterioration of coral reefs now occurring in many parts of the world. ▌

As we have seen in Chapters 14–16, the populations making up biological communities are engaged in a variety of interactions that determine their relative abundances and ecological distributions, and even their ability to persist, within the ecological system. These interactions are also a component of the environment that selects the genetic traits that give each population an advantage in its various interactions. Evolutionary responses to natural selection imposed by species interactions are the subject of the next chapter.

SUMMARY

1. Competition is any use or defense of a resource by one individual that reduces the availability of that resource to other individuals. When the individuals belong to the same species, their interaction is called intraspecific competition; when they belong to different species, it is called interspecific competition.

2. A resource may be defined as anything that is consumed and used by the consumer for its own maintenance and growth and whose availability promotes consumer population growth. Thus, food, water, light, mineral nutrients, and space are resources, but temperature and other such conditions are not.

3. Competition tends to be most intense among closely related species, but distantly related species also compete for resources.

4. Resources may be classified as nonrenewable (space) or renewable (light, water, and food). Renewable resources may be further distinguished according to the influence of the consumer on their rate of supply: no influence, direct influence, or indirect influence.

5. According to Liebig's law of the minimum, limiting resources are those that are most scarce relative to demand. Of all the resources consumed by a population, only one or a few normally limit its growth. Two or more resources can interact synergistically to limit a consumer population.

6. The difference between the population growth of one species in the presence and in the absence of another is a measure of the intensity of competition between them.

7. Theoretical investigations and laboratory studies have led to the conclusion that no two competing species can coexist when the same resource limits both species. This generalization has come to be known as the competitive exclusion principle.

8. Some mathematical treatments of competition are based on the logistic equation for population growth, to which a term is added for the effect of interspecific competition. The strength of this effect is specified by the competition coefficient.

9. The equilibrium population sizes of two competing species can be described by an equation including the carrying capacities and competition coefficients for each of the species. In the most general terms, coexistence requires that the product of the competition coefficients of the first and second species be less than 1. In other words, to coexist, species must limit themselves more than they limit each other.

10. When different resources limit the populations of two species, those species can coexist.

11. In some cases, two species can coexist because each has an advantage with respect to different factors in the environment, such as efficiency of resource use versus stress tolerance. Such asymmetric competition is common in nature.

12. Transplant experiments with plants under varying conditions of intraspecific and interspecific competition illustrate differences in the mechanisms of competition between habitats of high and low productivity. Some experiments have shown that plants compete belowground for nutrients when nutrient levels are low, and aboveground for sunlight when nutrient levels are high.

13. When individuals compete indirectly through their effects on shared resources, competition is said to be exploitative; when they compete directly by defending resources against one another, their behavior is known as interference competition.

14. Rapid invasion of a habitat by one mobile animal species following the removal of another demonstrates interference competition through aggressive behavior.

15. Many plants compete directly by producing chemical substances that impair the growth and survival of indi-

viduals of other species. This mechanism is known as allelopathy. Plants can also reduce populations of their competitors by supporting consumers in various ways, such as by providing safe refuges.

16. Consumers can alter the outcome of competitive interactions and promote coexistence if they selectively prey on superior competitors. Interactions between consumer species that are mediated by their own predators or pathogens are referred to as apparent competition.

REVIEW QUESTIONS

1. What is the difference between renewable and nonrenewable resources?

2. How does Liebig's law of the minimum explain limits on a population's growth?

3. If two species require the same limiting resource, what would you predict about their ability to coexist?

4. When two species are at equilibrium, why do we set each species' population growth rate equal to zero?

5. Compare and contrast the Grubb–Tilman and the Grime–Keddy hypotheses regarding how soil nutrient levels affect the intensity of interspecific competition.

6. Why is allelopathy considered a form of interference competition?

7. What might be an argument for the hypothesis that closely related species should experience more intense competition than distantly related species?

8. How is it that the presence of consumers can cause an increase in the diversity of competing species?

SUGGESTED READINGS

Bartholomew, B. 1970. Bare zone between California shrub and grassland communities: The role of animals. *Science* 170:1210–1212.

Bruno, J. F., J. J. Stachowicz, and M. D. Bertness. 2003. Inclusion of facilitation into ecological theory. *Trends in Ecology and Evolution* 18:119–125.

Connell, J. H. 1961. The influence of interspecific competition and other factors on the distribution of the barnacle *Chthamalus stellatus. Ecology* 42:710–723.

Emery, N. C., P. J. Ewanchuk, and M. D. Bertness. 2001. Competition and salt-marsh plant zonation: Stress tolerators may be dominant competitors. *Ecology* 82:2471–2485.

Goldberg, D. E., and A. M. Barton. 1992. Patterns and consequences of interspecific competition in natural communities: A review of field experiments with plants. *American Naturalist* 139:771–801.

Grace, J. B., and D. Tilman (eds.). 1990. *Perspectives on Plant Competition.* Academic Press, San Diego.

Halligan, J. P. 1973. Bare areas associated with shrub stands in grasslands: The case of *Artemisia californica. BioScience* 23:429–432.

Halsey, R. W. 2004. In search of allelopathy: An eco-historical view of the investigation of chemical inhibition in California coastal sage scrub and chamise chaparral. *Journal of the Torrey Botanical Society* 131:343–367.

Hardin, G. 1960. The competitive exclusion principle. *Science* 131:1292–1297.

Kadmon, R. 1995. Plant competition along soil moisture gradients: A field experiment with the desert annual *Stipa capensis. Journal of Ecology* 83:253–262.

Keddy, P. 1989. *Competition.* Chapman & Hall, London.

Maksimowich, D. S., and A. Mathis. 2000. Parasitized salamanders are inferior competitors for territories and food resources. *Ethology* 106:319–329.

Muller, C. H., W. H. Muller, and B. L. Haines. 1964. Volatile growth inhibitors produced by aromatic shrubs. *Science* 143:471–473.

Paine, R. T. 1974. Intertidal community structure: Experimental studies on the relationship between a dominant competitor and its principal predator. *Oecologia* 15:93–120.

Schenk, J. J. 2006. Root competition: Beyond resource depletion. *Journal of Ecology* 94:725–739.

Schenk, H. J., R. M. Callaway, and B. E. Mahall. 1999. Spatial root segregation: Are plants territorial? *Advances in Ecological Research* 28:145–180.

Schoener, T. W. 1983. Field experiments on interspecific competition. *American Naturalist* 122:240–285.

Smith, J. E., et al. 2006. Indirect effect of algae on coral: Algae-mediated microbe-induced coral mortality. *Ecology Letters* 9:835–845.

Stachowicz, J. J. 2001. Mutualism, facilitation, and the structure of ecological communities. *BioScience* 51:235–246.

Tilman, D. 1977. Resource competition between planktonic algae: An experimental and theoretical approach. *Ecology* 58:338–348.

Tilman, D. 1982. *Resource Competition and Community Structure.* Princeton University Press, Princeton, N.J.

Wilson, S. D., and D. Tilman. 1993. Plant competition and resource availability in response to disturbance and fertilization. *Ecology* 74:599–611.

CHAPTER 17

Evolution of Species Interactions

S hortly after a few pairs of European rabbits were released on a ranch in Victoria in 1859, rabbits became a major pest in Australia. Rabbit populations increased so quickly that within a few years, local ranchers were erecting rabbit fences and organizing rabbit brigades—shooting parties—in vain attempts to keep their numbers under control. Eventually, hundreds of millions of rabbits ranged throughout most of the continent, destroying sheep pasturelands and threatening wool production. The Australian government tried poisons, predators, and other control measures, all without success.

After much investigation, the answer to the rabbit problem seemed to be a myxoma virus (a relative of smallpox) discovered in populations of a related South American rabbit. The myxoma virus produced a small, localized fibroma (a fibrous cancer of the skin) without severe effects in South American rabbits, but European rabbits infected by the virus lacked resistance and died quickly.

In 1950, the myxoma virus was introduced into Victoria. An epidemic of myxomatosis broke out among the introduced European rabbits and spread rapidly. The virus was transmitted primarily by mosquitoes, which bite infected areas of the skin and carry the virus on their mouthparts. The first myxomatosis epidemic killed 99.8% of the infected rabbits, reducing their populations to very low levels. But during the next outbreak of the disease, only 90% of the remaining rabbits were killed. During the third outbreak of the disease, only 40%–60% of infected rabbits succumbed, and their populations began to grow again.

The decline in the lethality of the myxoma virus resulted from evolutionary responses in both the rabbit and the virus populations. Before the introduction of the virus, a few rabbits had genes that conferred resistance to the disease. Although nothing had previously promoted an increase in the frequency of those genes, they were strongly selected by the myxomatosis epidemic, until most of the surviving rabbit population consisted of resistant animals (Figure 17.1). At the same time, less virulent virus strains became more

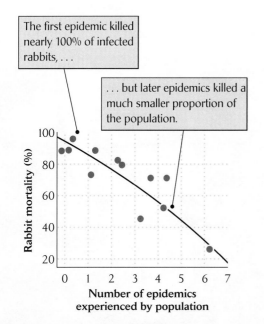

The first epidemic killed nearly 100% of infected rabbits, . . .

. . . but later epidemics killed a much smaller proportion of the population.

FIGURE 17.1 Interacting populations evolve in response to each other. The susceptibility of European rabbits in Australia to the introduced myxoma virus declined after the first epidemic. After F. Fenner and F. N. Ratcliffe, *Myxomatosis,* Cambridge University Press, London (1981).

prevalent because they did not kill their hosts as quickly and were therefore more readily dispersed to new hosts (mosquitoes bite only living rabbits).

Left on its own, the Australian rabbit–virus system would probably evolve to an equilibrium state of benign, endemic disease, as it had in the population of South American rabbits from which the myxoma virus was isolated. However, pest management specialists keep the system out of equilibrium by finding new strains to which the rabbits have yet to evolve resistance. In this way, they maintain the effectiveness of the myxoma virus as a pest control agent.

The less virulent strains of myxoma virus have a higher rate of growth in rabbit populations as a whole, if not within individual rabbits. This pattern is unlike that of some highly contagious human diseases, such as influenza and cholera, that spread directly through the atmosphere or water. Such pathogens do not depend on long-term survival of hosts for their dispersal, and they often exhibit high levels of virulence, with debilitating or even fatal consequences for their hosts. Similarly, most predators do not rely on a third party to find prey, and rather than evolving toward a benign equilibrium of restraint and tolerance, predator and prey tend to become locked in an evolutionary struggle of persistent intensity. The outcome of that struggle depends on which population gets the evolutionary upper hand.

CHAPTER CONCEPTS

- Adaptations in response to predation demonstrate selection by biological agents

- Antagonists evolve in response to each other

- Coevolution in plant–pathogen systems reveals genotype–genotype interactions

- Consumer and resource populations can achieve an evolutionary steady state

- Competitive ability responds to selection

- Coevolution involves mutual evolutionary responses by interacting populations

Plants and animals use a variety of structures and behaviors to obtain food and to avoid being eaten or parasitized. Much of this diversity is the result of natural selection acting on the ways in which plants and animals procure resources and escape predation. Wing markings that blend artfully into the background help moths escape the notice of predators. Flowers, by their insistent colors and fragrances, attract the notice of insects and birds that carry pollen from one flower to the next.

The agents whose influence has shaped such adaptations are biological: they are other living organisms. Their effects differ from those of physical factors in the environment in two ways. First, biological factors stimulate mutual evolutionary responses in the traits of interacting populations. For example, through natural selection and evolution, predators shape their prey's adaptations for escape,

but their own adaptations for pursuit and capture also respond to attributes of their prey. In contrast, adaptations of organisms in response to changes in the physical environment have no reciprocal effect on the environment.

Second, biological agents foster diversity of adaptations, rather than promoting similarity. In response to biological factors, organisms tend to specialize, pursuing unique assortments of prey, striving to avoid unique combinations of predators and pathogens, and engaging in mutually beneficial arrangements with other species. In response to similar physical stresses in the environment, however, many kinds of organisms evolve similar adaptations. We have seen this phenomenon, which is called *convergence,* in the reduced or finely divided leaves that minimize heat stress and water loss in many desert plants (see Chapter 5).

(a)

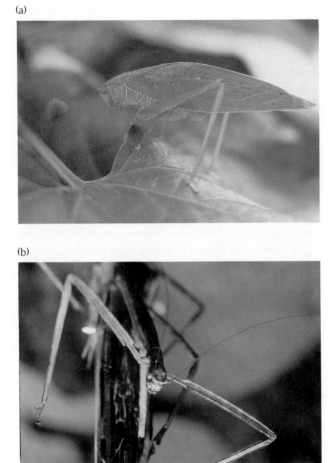

(b)

(c)

FIGURE 17.2 Many palatable organisms have evolved cryptic appearances to avoid detection by predators. (a) A katydid matches leaves; (b) a stick insect matches twigs; and (c) a lantern fly blends in with the bark of a tree. *Photos by R. E. Ricklefs.*

When populations of two or more species interact, each may evolve in response to those characteristics of the other that affect individual fitness. This process is referred to as **coevolution** when the evolved responses are reciprocal—that is, when adaptations in one population promote the evolution of adaptations in the other. This would be the case when an herbivore evolves a way to detoxify a noxious chemical that has evolved in a plant to protect it against that same herbivore. These adaptations represent a sequence of evolutionary responses resulting directly from the interaction between the two populations.

In its broadest sense, the term *coevolution* applies to the evolutionary responses of each species to all others with which it interacts (sometimes referred to as *diffuse coevolution*). However, many biologists restrict the application of the term more narrowly to the reciprocal evolution of related structures and functions in two interacting populations. Identifying unambiguous cases of such strict coevolution can be difficult. For example, hyenas have jaws and associated muscles that are strong enough to crack the bones of their prey. These modifications clearly are adaptations for eating their prey. However, the hyena's powerful jaws cannot be considered an example of

coevolution because the properties of the bones of their prey did not evolve to resist being eaten by hyenas, or any other predator. By the time a hyena has reached that part of its meal, bone structure has no consequence for the prey's survival. In contrast, when an herbivore evolves the ability to detoxify chemicals produced by a plant specifically to deter it, the requirements of the narrow definition of coevolution are more likely to be met.

In this chapter, we shall explore some of the consequences of evolutionary responses to interactions between predators and their prey, between competitors, and within mutualistic associations. When the evolutionary relationship between two species is antagonistic, as it is between predator and prey or between parasite and host, the species can become mired in an evolutionary struggle to increase their own fitness, each at the other's expense. Such a struggle may lead to an evolutionary stalemate in which both antagonists continually evolve in response to each other, but the net outcome of their interaction is a steady state. Alternatively, when one of the antagonists cannot evolve fast enough, it may be driven to extinction. In contrast, evolutionary relationships between species in mutually beneficial associations

(a)

(b)

(c)

FIGURE 17.3 Many unpalatable organisms have evolved warning coloration. (a) Predators learn to avoid brightly colored food items such as this caterpillar. (b, c) Some unpalatable insects aggregate to enhance the warning signal. Photo (a) by J. Burgett, photos (b) and (c) by Carl C. Hansen, courtesy of the Smithsonian Tropical Research Institute.

can lead to stable arrangements of complementary adaptations that promote their interaction.

Adaptations in response to predation demonstrate selection by biological agents

Coloration is an example of a trait that can evolve in prey under selection from predators. The changes in coloration favored by predation can then feed back on the adaptations of predators, enabling them either to find increasingly well-camouflaged prey or to avoid prey with coloration that signals noxious qualities. The evolution of form and coloration in many animals to avoid predation shows us one-half of the coevolution equation and emphasizes the strength of selection by biological agents.

Crypsis versus warning coloration

To avoid detection by predators, some prey organisms adopt a camouflaged appearance and resting position.

Predators selectively favor prey individuals better able to avoid them because those prey individuals that hide less effectively are discovered and eaten.

Many organisms achieve **crypsis,** or blending in with their backgrounds, by matching the color and pattern of bark, twigs, or leaves (Figure 17.2). Various animals resemble sticks, leaves, flower parts, or even bird droppings. These organisms are not so much concealed as they are mistaken for inedible objects and passed over. Of course, if an insect is to mimic a stick or a leaf convincingly, it has to behave like one. A leaf-mimicking insect resting on bark, or a stick insect moving rapidly along a branch, would not fool many predators.

Crypsis is a strategy of palatable, or edible, animals. Other animals take a bolder approach to antipredator defense: they produce noxious chemicals or accumulate them from food plants, and they advertise the fact with conspicuous color patterns. This strategy is known as **warning coloration,** or **aposematism** (Figure 17.3). Predators learn quickly to avoid markings such as the black and orange stripes of monarch butterflies, which taste so bitter that a single experience is well remembered. It is

(a)   (b)   (c)

FIGURE 17.4 Batesian mimics are palatable prey organisms that resemble noxious ones. Here, a harmless, palatable mantid (b) and moth (c) have both evolved to resemble a wasp (a). Photos by Larry Jon Friesen/Saturdaze.

FIGURE 17.5 Müllerian mimics are unpalatable organisms that share a pattern of warning coloration. The top two rows illustrate five aposematic color morphs in local populations of *Heliconius numata* (second row) and comimetic forms of the distantly related ithomiine butterflies *Melinaea menophilus* (one form), *M. ludovica* (one form), and *M. marsaeus* (three forms) in northern Peru. The bottom two rows portray geographic variation in the related Müllerian mimics *H. melpomene* and *H. erato* throughout tropical South America. Photos © 2006 Mathieu Joron.

no coincidence that many noxious animals adopt similar patterns. Black and either red or yellow stripes adorn such diverse animals as yellow-jacket wasps and coral snakes. These color combinations so consistently advertise noxiousness that some predators have evolved innate aversions to them and need not learn to avoid such prey by experience.

Why aren't all potential prey species noxious or unpalatable? Part of the answer is that chemical defenses may use a large portion of an individual's energy or nutrients that might otherwise be allocated to growth or reproduction. Furthermore, many noxious organisms rely on their food plants to supply toxic organic compounds that they cannot manufacture themselves, and not all food plants contain such compounds. When they do, the consumers must themselves avoid the toxic effects of the chemicals in order to use them effectively against their potential predators.

Mimicry

Unpalatable animals and plants that display warning coloration often serve as models for palatable ones, which evolve to resemble the noxious organisms. In this case, consumers are the agents of selection when they confuse well-matched edible mimics with unpalatable models. Such relationships are collectively referred to as **Batesian mimicry,** named after its discoverer, the nineteenth-century English naturalist Henry Bates. In his journeys to the Amazon region of South America, Bates found numerous cases of palatable insects that had forsaken the cryptic patterns of their close relatives and had come to resemble brightly colored, unpalatable species (Figure 17.4).

Experimental studies have demonstrated that mimicry does confer an advantage on mimics. For example, toads that were fed live bees, and were stung on the tongue, thereafter avoided palatable drone flies, which mimic bees. But when naive toads were fed only dead bees from which the stings had been removed, they relished the drone fly mimics (as well as the now harmless bees). Thus, toads learned to associate the conspicuous and distinctive color patterns of live bees with an unpleasant experience.

Another type of mimicry, called **Müllerian mimicry** after its discoverer, the nineteenth-century German zoologist Fritz Müller, occurs when several unpalatable species adopt a single pattern of warning coloration. Predators learn to avoid these mimics more efficiently because a predator's bad experience with one species confers protection on all the other members of the mimicry complex.

For example, most of the bumblebees and wasps that co-occur in mountain meadows share a pattern of black and yellow stripes. In the tropics, dozens of species of unpalatable butterflies, many of them distantly related, share patterns of black and orange "tiger stripes" or black, red, and yellow coloration patterns (Figure 17.5).

Antagonists evolve in response to each other

The term *coevolution* was coined by Charles Mode in a paper published in the journal *Evolution* in 1958. Mode was concerned with the relationship between agricultural crops and their fungal pathogens, especially rusts, which cause millions of dollars worth of crop losses each year. He developed a model of continual evolution of a pathogen and its host in response to evolutionary changes in each other. Mode's model assumed that pathogen virulence and host resistance were each controlled by a single dominant gene (*V* and *R*, respectively), and that both virulence and resistance were, by themselves, costly to the organism. Thus, the fitness of the host and the fitness of the pathogen were each contingent on the genotype of the other. In these circumstances, frequencies of virulence and resistance genes should tend to oscillate over time in a pattern similar to a predator–prey population cycle (see Figure 15.2).

Mode's model worked as follows: When the host is susceptible (genotype *rr*), selection favors virulent pathogens (genotype *VV* or *Vv*). Virulent pathogens cause selection for host resistance (genotype *RR* or *Rr*), which then increases in the host population. When the host is resistant, selection favors avirulent pathogens (genotype *vv*) because virulence is costly. When the pathogen is avirulent, selection favors susceptible hosts (genotype *rr*) because resistance is costly. These mutual responses cause a pattern of continual cycling: *r* (host) → *V* (pathogen) → *R* (host) → *v* (pathogen) → *r* (host), and so on.

In 1964, Paul Ehrlich and Peter Raven, who at the time were assistant professors at Stanford University, published an article, also in *Evolution,* that placed coevolution in a more ecological context and greatly popularized the term. Ehrlich and Raven noted that closely related groups of butterflies tended to feed on closely related species of host plants. For example, species of butterflies in the tropical genus *Heliconius* feed exclusively on passionflower vines of the genus *Passiflora* (Figure 17.6). Such tight consumer–resource relationships suggested a long evolutionary history linking butterflies and their host

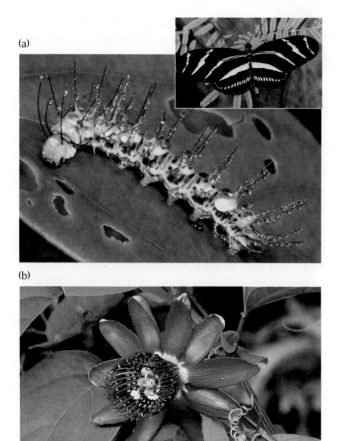

(a)

(b)

FIGURE 17.6 The taxonomic specificity of some predator–prey relationships suggests a long evolutionary history. Larvae of *Heliconius* butterflies (a) feed only on passionflower (*Passiflora*) vines (b). Photo (a) © Michael and Patricia Fogden/Corbis; inset photo courtesy of Andy McGregor; photo (b) by Ray Coleman/Photo Researchers.

plants, undoubtedly involving the evolution of the butterflies to tolerate the particular defenses of their host plants, and possibly the evolution of the passionflower vines to minimize herbivory by butterfly larvae.

Thus, the study of coevolution, and of evolutionary relationships between interacting species more generally, initially went in two directions. On one hand, Mode used modeling to address the genetic and evolutionary mechanisms underlying the relationships between consumer and resource populations. On the other hand, Ehrlich and Raven observed patterns of relationships in nature and interpreted them as outcomes of evolutionary interactions. Most recently, these two approaches have found common ground in analyses of the evolutionary history of traits directly involved in the relationships between species, as we shall see in the analysis of the coevolutionary

relationship between yucca plants and yucca moths presented later in this chapter. However, early experimental studies on evolution in laboratory populations had already demonstrated the powerful role of selection by one species on the evolved adaptations of another.

ECOLOGISTS IN THE FIELD **Evolution in houseflies and their parasitoids.** In a series of experiments conducted during the 1960s, David Pimentel and his colleagues at Cornell University explored the evolution of parasitoid–host relationships. They used the pupal stage of the housefly (*Musca domestica*) as the host and a wasp, *Nasonia vitripennis* (Figure 17.7), as the parasitoid. In one population cage (the control cage; Figure 17.8a), the wasps were allowed to parasitize a fly population that was kept at a constant number. Individual flies were added from a stock that had not been exposed to the wasp. Any flies that escaped attack by the wasps and emerged from their pupae were removed from the cage, so that the wasps were provided only with evolutionarily "naive" hosts. In a second population cage (the experimental cage; Figure 17.8b), the fly population was kept at the same constant number, but emerging flies were allowed to remain in the cage, so that the fly population could evolve resistance to the wasps. The population cages were maintained for about 3 years, long enough for evolutionary change to occur.

Over the course of the experiment, the reproductive rate of wasps in the experimental cage that permitted evolution dropped from 135 to 39 progeny per female, and longevity decreased from 7 to 4 days. In the control cage, where parasitoids were provided with naive fly pupae each generation, the wasps remained fecund and long-lived. The average parasitoid population in the experimental cage also decreased relative to the population in the control

FIGURE 17.7 Pimentel's study of coevolution used a parasitoid–host system. The wasp *Nasonia vitripennis*, a parasitoid of the housefly, is shown here laying eggs in a fly pupa. Courtesy of D. Pimentel, from D. Pimentel, *Science* 159:1432–1437 (1968).

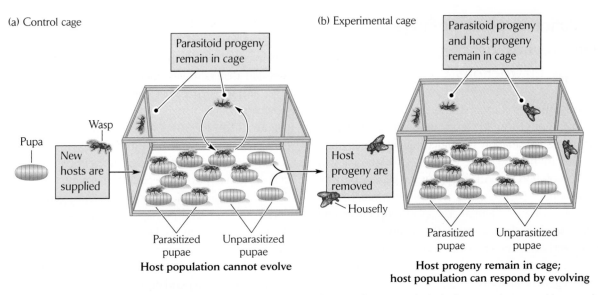

(a) Control cage

Parasitoid progeny remain in cage

Pupa

Wasp

New hosts are supplied

Parasitized pupae   Unparasitized pupae

Host population cannot evolve

(b) Experimental cage

Parasitoid progeny and host progeny remain in cage

Host progeny are removed

Housefly

Parasitized pupae   Unparasitized pupae

Host progeny remain in cage; host population can respond by evolving

FIGURE 17.8 Pimentel's classic experiment tested for a host evolutionary response to a parasitoid. The difference in parasitoid population size at the end of the experiment between the control cage in which the host population could not evolve (a) and the experimental cage in which it could evolve (b) indicated the effectiveness of the host's evolutionary response.

cage, and population size was more nearly constant than in the nonevolving control cage. These results suggest that the flies evolved resistance to the parasitoids when subjected to intense parasitism.

Experiments were then established in new population cages in which the numbers of flies were allowed to vary freely. The control cage started with flies and wasps that had had no previous contact with each other, and the experimental cage was established with individuals from the evolving population described above. In the control cage, the wasps were efficient parasitoids, and the system underwent dramatic oscillations. In the experimental cage, however, the wasp population remained low, and the flies attained a high and relatively constant population level (Figure 17.9). This result strongly reinforced the conclusion, drawn from the earlier experiments, that the fly hosts had evolved resistance to the wasp parasitoids. Unfortunately, no information was gathered in this experiment on the response of the experimental wasps to increased resistance in their fly hosts. ∎

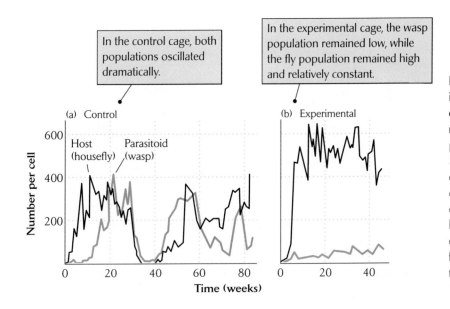

In the control cage, both populations oscillated dramatically.

In the experimental cage, the wasp population remained low, while the fly population remained high and relatively constant.

(a) Control

Host (housefly)   Parasitoid (wasp)

Number per cell

600

400

200

0   20   40   60   80

(b) Experimental

0   20   40

Time (weeks)

FIGURE 17.9 Population changes in Pimentel's parasitoid–host system demonstrated that populations evolve in response to each other. Houseflies and parasitoid wasps were placed together in 30-cell population cages. Numbers of flies and wasps per cell, as well as the pattern of population cycling, differed between the control cage (a), in which the fly population had no previous experience with the wasp, and the experimental cage (b), in which the fly population had been previously exposed to wasp parasitism. After D. Pimentel, *Science* 159:1432–1437 (1968).

Coevolution in plant–pathogen systems reveals genotype–genotype interactions

The suggestion that consumer and resource populations evolve in response to each other presupposes that each contains genetic variation for traits that influence their interactions. In the case of the wasp–fly interaction, it was clear that evolution had occurred, but the genetic basis of the evolutionary change could not be determined. This has been less of a problem in studies of crop plants and their pathogens. In these systems, the difference between virulence and avirulence may depend on a single gene, as Mode's model assumed, and thus is amenable to simple Mendelian genetic analysis.

Plant geneticists have developed strains of crops, such as wheat, that are resistant to particular genetic strains of pathogens, such as rusts (teliomycetid fungi). These crop strains differ from one another by being either susceptible or resistant to infection by particular strains of rust. Over the course of crop improvement programs, when new strains of rust appear in an area, crop geneticists select new resistant strains of the crop by exposing experimental populations to the pathogen. However, new strains of the pathogen continue to appear, either by migration or by mutation, creating continual evolutionary change in the system.

Genetic races of wheat rust are distinguished both by their physiological characteristics and by their virulence when tested on strains of wheat containing different resistance alleles. Most of the virulence strains within a single physiological race of rust differ by only one gene. A survey of wheat rust (*Puccinia graminis*) in Canada revealed that new virulence genes appear from time to time, and that when that happens, the altered rust strain sweeps through plantings of the crop within a few years (Figure 17.10). The rust–wheat system contains the essential element of coevolution envisioned by Mode: an interaction between the fitnesses of the genotypes of the host and the genotypes of the pathogen. The system is kept in flux by the introduction of new virulence genes by mutation in the rust, and perhaps by new resistance genes in the wheat, although the latter are pretty much controlled by plant geneticists nowadays.

Differences in the expression (and fitness) of genotypes in one species depending on the genotypes of another species are called **genotype–genotype interactions.** Such interactions have been found in many natural systems, and they may turn out to be the rule in populations of plants and herbivores and of hosts and pathogens. The genetics of most plant defenses are difficult to work out in as much detail as has been done for wheat resistance genes, but genetic effects can nonetheless be detected. D. N. Alstad and G. F. Edmunds, Jr., at the University of Minnesota, showed that variation in defenses against herbivores among individual ponderosa pine trees is paralleled by variation in the genotypes of the scale insects that infest them (Figure 17.11). Scale insects are extremely sedentary; they exhibit so little movement that local populations evolve independently on individual trees. Alstad and Edmunds drew this conclusion from the differing success

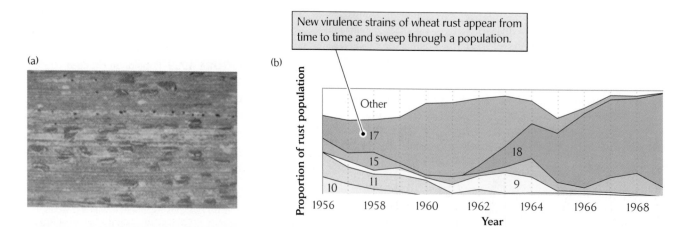

FIGURE 17.10 Coevolution involves an interaction between the genetically influenced fitness of a host and that of its parasite or pathogen. (a) Wheat rust (*Puccinia graminis*) growing on wheat. (b) The relative proportions of different virulence strains in this rust (indicated by different numbers within the graph) infecting Canadian wheat have changed over time. After G. J. Green, *Can. J. Bot.* 53:1377–1386 (1975); photo courtesy of Gary Munkvold.

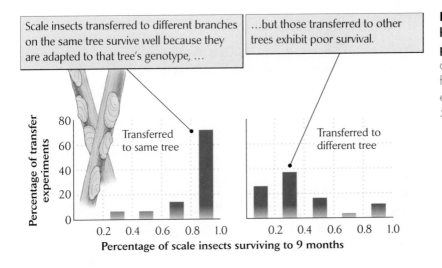

Scale insects transferred to different branches on the same tree survive well because they are adapted to that tree's genotype, …

…but those transferred to other trees exhibit poor survival.

Transferred to same tree

Transferred to different tree

Percentage of scale insects surviving to 9 months

FIGURE 17.11 Genetic variation in a host may parallel genetic variation in a pathogen. The survival rate of scale insects decreases markedly when they are moved from the tree on which their population has evolved. After G. F. Edmunds and D. N. Alstad, *Science* 199:941–945 (1978).

of scale insects experimentally transferred between trees and between branches on the same tree. The survival rate of scale insects transferred between trees was much lower than that of controls transferred between branches. Differences between individual trees, and between local populations of scale insects, are probably genetic, so this finding probably represents a case of genotype–genotype interaction. It could also represent a case of strict coevolution if the trees respond genetically to infestations of the scale insects. It is important to emphasize that not all evolutionary responses represent strict coevolution between two populations. Nonetheless, we can conclude that the evolution of most species is driven in part by their interactions with their consumers, resources, competitors, and mutualists.

Consumer and resource populations can achieve an evolutionary steady state

Because consumer and resource populations evolve continually in response to selection by their antagonists, we might wonder about the end result of these interactions: does evolution ever stop? In the case of the wheat–rust interaction, which produces strict coevolution between virulence and resistance, the system seems destined to cycle endlessly through different genotypes of each species. In contrast, when a species interacts with many others simultaneously, no single virulence or resistance factor is likely to convey a unique advantage over all the others. In such a case, the ability of virulent pathogens to switch to a more abundant host species, giving a reduced host population a chance to recover, might lead to an equilib-

rium state of maintained genetic diversity. Strict coevolution can produce time delays because each population responds to only one other population, and cyclic changes in gene frequencies can result, just as predator–prey cycles are more prevalent in simpler ecological systems. When multiple consumer and resource populations affect one another simultaneously, time delays are less important than those typical of one-to-one interactions.

Regardless of whether coevolutionary relationships are strict or diffuse, neither consumer nor resource population is likely to get the upper hand in the long run. Most ecological systems evolve toward a steady state in which evolution continues, but rates of resource exploitation by consumers remain more or less constant. As the rate at which resource populations are exploited increases, so does the potential strength of selection on those populations for new adaptations for escaping or avoiding consumers, at least up to limits set by the availability of genetic variation. The rate at which such new adaptations evolve in a resource population should vary in direct proportion to the rate at which it is exploited (Figure 17.12). Hence, any advantage that a consumer evolves over its resource populations should be only temporary.

The strength of selection for new adaptations in consumers for exploiting their resources should vary in the opposite fashion. When a particular resource population is not heavily exploited, adaptations of consumers that enable them to use that resource are selectively favored, and their exploitation of that resource population increases. As exploitation increases, however, that resource population is reduced, and further increases in consumer efficiency have less selective value. Very high rates of consumption could conceivably favor consumer individuals that shifted their diets toward other, more abundant resource populations. Hence, evolution might favor *less* efficient use of a

FIGURE 17.12 The rate of exploitation influences the rate of evolution in consumer and resource populations. Consumer–resource coevolution achieves a steady state when the rate of change in the rate of exploitation equals zero; that is, when the population consequences of consumer and resource adaptations balance.

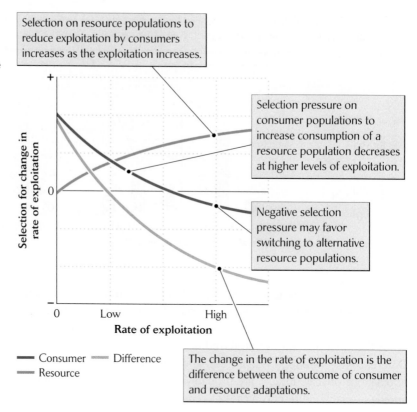

Selection on resource populations to reduce exploitation by consumers increases as the exploitation increases.

Selection pressure on consumer populations to increase consumption of a resource population decreases at higher levels of exploitation.

Negative selection pressure may favor switching to alternative resource populations.

The change in the rate of exploitation is the difference between the outcome of consumer and resource adaptations.

particular resource population by a consumer population as a consequence of adaptations to exploit another, more abundant resource population.

In the simple model shown in Figure 17.12, the rates of consumer and resource adaptation can achieve an evolutionary steady state in which selection on the resource population for adaptations to reduce consumption balances selection on the consumer population for adaptations to increase consumption. When consumer adaptations are relatively effective and the resource population is exploited at a high rate, selection on the resource population tends to improve its avoidance mechanisms faster than selection on the consumer population improves its ability to exploit the resource population. Conversely, when the exploitation rate is low, the resource population evolves more slowly than the consumer population. A balance between these influences should result in a relatively constant rate of exploitation regardless of the specific consumer and resource adaptations. As in any steady state, both antagonists continually evolve to maintain this balance, just as nations continually develop new weapons and defenses to maintain a stalemate in an arms race. This model is an example of the Red Queen hypothesis, which we discussed in the context of the evolutionary maintenance of sexual reproduction in populations (see Chapter 8).

Pimentel's experiments on parasitoid–host interactions, discussed above, illustrate the dynamics of this consumer–resource steady state. The housefly (host) and the wasp *Nasonia* (parasitoid) undoubtedly had achieved an evolutionary steady state in their natural habitat. When brought into the laboratory, the wasps were able to exploit housefly populations at a greatly increased rate because they required little time to search out hosts in the simplified environment of the population cages. Setting up these experimental conditions was equivalent to shifting the exploitation rate of houseflies by the wasps far above the steady-state level in Figure 17.12. This shift increased selection on the housefly to escape parasitism much more than it increased selection on the wasp to further increase its host exploitation rate. Consequently, the ability of houseflies to escape parasitoids increased, and the level of exploitation by wasps decreased, toward a new steady state.

Competitive ability responds to selection

Competitors, like predators and prey, exert selection pressure on each other. Under one scenario, competitors are selected to diverge from one another in terms of the resources they consume. An individual that uses resources not sought after by another species may enjoy greater resource availability and thus greater fitness. This scenario, to which we shall return in the following section, differs from coevolution in antagonistic and mutualistic relationships because the evolution of one species is not a response to particular adaptive changes in its competitor. Instead, it is indirect effects applied through resources (or through consumers in the case of apparent competition) that drive the evolution of competitors.

Ultimately, however, the most important force driving improvements in competitive ability is selection for increased efficiency of resource use. Evolution of resource use efficiency is unlike coevolution in the sense that it would happen in the absence of a competing species just because of competition within the population. However, competing species influence the exact ways in which efficiency can be increased because of their effects on resource availability.

Demonstrating genetic variation in competitive ability

Sometimes genetic changes that influence competitive ability express themselves in the phenotype so subtly that we cannot detect them by directly examining the traits of individuals. Instead, they must be inferred from changes in the outcome of competition in response to changes in the competitive environment. Several experiments have used this approach to demonstrate genetic variation in competitive ability, and hence the potential for competitive ability to evolve.

In one early competition experiment, population geneticist Francisco Ayala established two fruit fly species, *Drosophila serrata* and *D. nebulosa*, in population cages in the laboratory. The populations quickly achieved a pattern of stable coexistence, with 20%–30% *D. serrata* and 70%–80% *D. nebulosa* in each cage. In one cage, however, the frequency of *D. serrata* began to increase after the 20th week and attained about 80% by the 30th week, reversing the initial predominance of *D. nebulosa*.

In a second experiment, Ayala removed individuals of both species from the competing populations after the 30th week and tested them against stocks of flies that had been maintained in single-species cultures. The competitive ability of each species was found to have increased after exposure to the other in the first competition experiment. When the competitive ability of *D. serrata* individuals from the one cage in which that species predominated was tested against that of unselected stocks of *D. nebulosa*, *D. serrata* again showed superior competitive ability.

The particular adaptations responsible for the changes in competitive ability were not determined. They could conceivably include an increase in the efficiency of use of a food resource, the number of offspring produced per unit of food consumed, resistance to a common pathogen (apparent competition), or survival at any stage of the life cycle.

One generalization to come from this and similar experiments was that sparse populations can evolve the ability to compete against other species more rapidly than dense populations. Why? One possibility is that the adaptations needed to compete well against individuals of the *same* species conflict with those needed to compete well against *other* species. Sparse populations are less in need of adaptations for intraspecific competition, so the rarer of two competitors experiences stronger selection for increased interspecific competitive ability. We return to the work of David Pimentel for evidence that a rare competitor can evolve a competitive advantage (judged by relative population density) over a formerly superior adversary.

ECOLOGISTS IN THE FIELD **Back from the brink of extermination.** David Pimentel and his colleagues conducted laboratory experiments with flies to determine whether species show frequency-dependent evolutionary changes in their competitive ability. In other words, can one species, as it is being excluded by another species and becoming rare, evolve increased interspecific competitive ability rapidly enough to regain the upper hand? For their experiments, the investigators chose the housefly (*Musca domestica*) and the blowfly (*Phaenicia sericata*) (Figure 17.13), which have similar ecological

FIGURE 17.13 Two fly species were used in Pimentel's studies of competition. The blowfly (shown here) and housefly are often found on the same food resources in nature. Courtesy of L. Higley, University of Nebraska, Lincoln.

requirements and comparable life cycles (about 2 weeks). Both species feed on dung and carrion in nature, and they are often found together on the same food resources. The flies were raised in small population cages, with a mixture of agar and liver provided as food for the larvae and sugar for the adults.

The outcomes of four initial competition experiments using individuals from wild populations of houseflies and blowflies were split, with each species winning twice. The mean extinction time for the blowfly, when the housefly won, was 92 days; it was 86 days for the housefly when the blowfly won. The investigators concluded that the two species had similar competitive ability, but that the small cages did not allow enough time for evolutionary change before one of the species was excluded.

To prolong the housefly–blowfly interaction, Pimentel and his colleagues started a mixed population in a 16-cell population cage, which consisted of single cages in four rows of four with connections between them (Figure 17.14). Under these conditions, populations of houseflies and blowflies coexisted for almost 70 weeks. The houseflies were more numerous initially, but the two species showed a striking reversal of numbers at about 50 weeks, and the blowflies had excluded the houseflies by the end of the experiment (Figure 17.15).

After 38 weeks, when the blowfly population was still low, and just a few weeks prior to its sudden increase, individuals of both species were removed from the population cage and tested in competition with each other and with wild strains of the housefly and blowfly. Captured wild blowflies turned out to be inferior competitors against both wild houseflies and experimental houseflies from the population cage. But blowflies that had been removed from the population cage at 38 weeks consistently excluded both wild and experimental populations of the housefly in com-

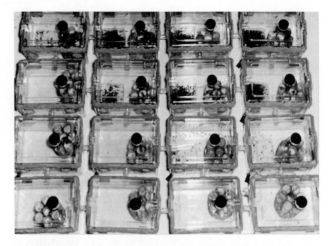

FIGURE 17.14 Pimentel used a 16-cell population cage to study competition between fly species. Note the vials with larval food in each cage and the passageways connecting the cells. The dark objects concentrated in the upper right-hand cells are fly pupae. Courtesy of D. Pimentel, from D. Pimentel et al., *Am. Nat.* 99:97–109 (1965).

petition experiments. Apparently, the experimental blowfly population had evolved superior competitive ability while it was rare and on the brink of extermination. ▌

Subsequent laboratory studies of model organisms, such as fruit flies (*Drosophila*), flour beetles (*Tribolium*), and mouse-ear cress (*Arabidopsis*), have consistently demonstrated evolutionary responses, and thus genetic variation, in competitive ability. However, the particular traits involved in most of these cases are not well understood. Because the outcome of competition depends on how

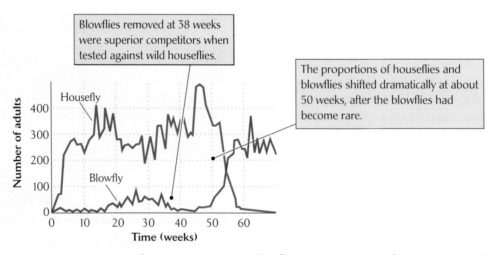

FIGURE 17.15 A rare competitor can evolve superior competitive ability. When Pimentel grew populations of houseflies and blowflies in a 16-cell population cage, the blowflies were on the brink of extermination at 38 weeks, but outnumbered the houseflies by the end of the experiment. After D. Pimentel et al., *Am. Nat.* 99:97–109 (1965).

efficiently each species exploits shared resources, many traits have the potential to influence competitive ability.

Studies of the evolution of competitive ability have practical applications to agricultural sciences, in which breeding good competitors is an important goal. Growing crop strains with superior competitive abilities could reduce crop losses to competing weeds while reducing the need for herbicides and other expensive and environmentally unfriendly interventions. For example, rice plants compete with other plant species by exuding allelopathic secondary compounds from their roots into the surrounding soils. Strains of rice vary in their ability to inhibit the growth of other plants in this way, and this variation has a genetic basis. Thus, it might be possible to select strains of rice that suppress weeds through direct competition yet retain a high level of seed production.

Character displacement

As we have seen, theory suggests that if resources are sufficiently varied, competitors should diverge to specialize on different resources. Specialization should reduce the degree to which each species uses the same resources its competitors use. Thus, specialization reduces competition and promotes coexistence. If competition exerts selection pressure in nature, then we should find evidence that competitors have influenced each other's adaptations toward divergence.

Although related species that live together tend to differ in the way they use the environment (using different food resources, for example), we cannot assume that these differences evolved as a result of their prior history of interaction. An alternative explanation is that each of the species became adapted to different resources in different places in the absence of competition between them, and when their populations subsequently overlapped as a result of range extensions, those ecological differences remained.

We might avoid this objection by comparing the ecology of a species where it co-occurs with a competitor with its ecology where that competitor is absent. Where two species coexist within the same geographic area, they are said to be **sympatric;** where their geographic ranges do not overlap, they are said to be **allopatric.** Suppose that species 1 occurs in areas A and B, and species 2 occurs in areas B and C (Figure 17.16). The populations of the two species in area B are sympatric; the population of species 1 in area A is allopatric with respect to the population of species 2 in area C. If areas A, B, and C all had similar environmental conditions and habitats, and if competition caused divergence, we would expect the sympatric

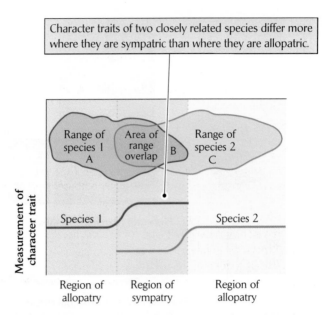

FIGURE 17.16 Character displacement is the evolutionary divergence of competing populations.

populations of species 1 and 2 in area B to differ more from each other than the allopatric populations of those species in areas A and C. This pattern is called **character displacement.**

Ecologists disagree on the prevalence of character displacement in nature. Some examples do seem to fit the pattern, however. One of these involves the ground finches (*Geospiza*) of the Galápagos archipelago (see Chapter 6). On islands with more than one finch species, the beaks of the species usually differ in size, indicating different ranges of preferred food size. For example, on Marchena Island and Pinta Island, the beak size ranges of the three resident species of ground finches do not overlap (Figure 17.17). On Floreana and San Cristobal, the two resident species, *G. fuliginosa* and *G. fortis,* have beaks of different sizes. On Daphne Island, however, where *G. fortis* occurs alone, its beak is intermediate in size between those of the two species on Floreana and San Cristobal. On Los Hermanos Island, *G. fuliginosa* occurs alone, and its beak is also intermediate in size.

The Galápagos ground finches clearly illustrate the diversifying influence of competition because the different species are distributed differently on small islands within the archipelago: some islands have two or three species and some only one. In many other cases, however, it is difficult to know whether differences between two species arose because of competition between them or evolved in response to selection by other environmental factors in different places, then were retained when the populations re-established contact. In most cases, genetic differences

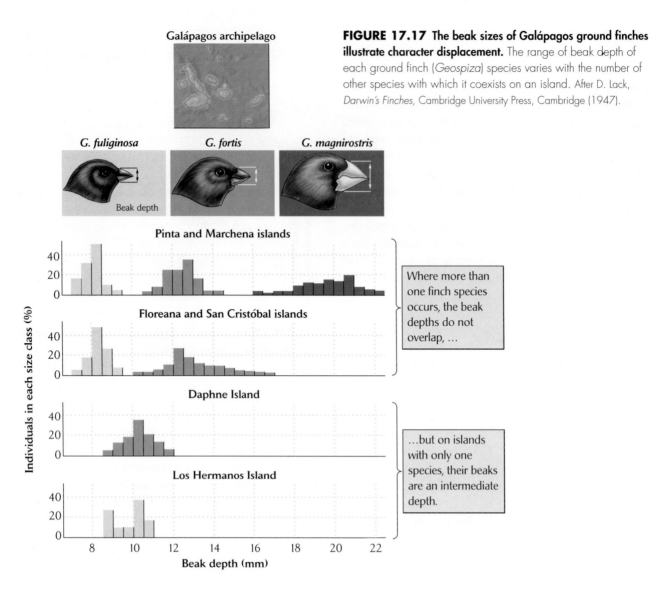

FIGURE 17.17 The beak sizes of Galápagos ground finches illustrate character displacement. The range of beak depth of each ground finch (*Geospiza*) species varies with the number of other species with which it coexists on an island. After D. Lack, *Darwin's Finches*, Cambridge University Press, Cambridge (1947).

Where more than one finch species occurs, the beak depths do not overlap, …

…but on islands with only one species, their beaks are an intermediate depth.

associated with the formation of new species evolve in allopatry, so why not differences that allow two species to avoid strong competition? In either case, coexistence depends on some degree of ecological difference between competing species, whether it is achieved in allopatry or as an evolutionary consequence of competition in sympatry.

Coevolution involves mutual evolutionary responses by interacting populations

Coevolution implies reciprocal evolutionary responses between pairs of populations, as we have seen, for example, in the genotype–genotype interaction between wheat and its rust pathogens. Such cases provide the most straightforward examples of coevolution because the traits are simple and we understand the genetic changes involved. Coevolution can also link changes in entire suites of traits

in interacting species, such as the adaptations required of flowers and their pollinators to form a tight pollination mutualism. In such cases, it becomes difficult to discern the order in which adaptations occurred or the specific changes in one of the interacting species that selected for a responding change in the other. Sometimes, in fact, complementary adaptations among pairs or small groups of species have been attributed to coevolution without any evidence for the evolutionary history of the relationship. As in the case of divergence between competing species, a close association between the adaptations of different species does not necessarily mean that they evolved as a consequence of reciprocal interactions.

Consider the mutualism in which ants protect aphids and leafhoppers from predators and, in return, harvest the nutritious honeydew that those insects excrete. This ant–homopteran mutualism has all the elements of coevolution, but how can we be sure that the adaptations of all the participants evolved in response to each other? Most insects that suck plant juices produce large volumes of excreta from which they either do not or cannot extract all the nutrients. Therefore, their honeydew production may simply reflect their diet, rather than having evolved to encourage protection by ants. For their part, many ants are voracious generalists that are likely to attack any insect they encounter; they may need no special adaptations to deter the predators of aphids and leafhoppers. Why, then, don't the ants eat the aphids and leafhoppers they tend? Perhaps this restraint is an evolved trait of ants that facilitates the ant–homopteran mutualism. Alternatively, it might have arisen as an extension of the common ant behavior of defending plant structures that produce nectar, such as flowers or specialized nectaries.

Plant defenses and herbivore responses

The best evidence for coevolution comes from reconstructing the evolutionary histories of traits in coevolving groups of organisms. Consider the chemical give-and-take between the larvae of bruchid beetles and the seeds of legumes (members of the pea family) that they consume. Adult bruchids lay their eggs on developing seedpods. The larvae then hatch and burrow into the seeds, which they consume as they grow. Most legume seeds contain secondary compounds that inhibit the digestive enzymes of insect herbivores. Although these toxins provide an effective biochemical defense against most insects, many bruchid beetles have metabolic pathways that either bypass the toxins or are insensitive to them. Among legume species, however, soybeans stand out as being resistant to attack

even by most bruchid species. When bruchids lay their eggs on soybeans, the larvae die soon after burrowing beneath the seed coat. Chemicals isolated from soybeans have been shown to inhibit the development of bruchid larvae in experimental situations.

Seeds of the tree-sized tropical legume *Dioclea megacarpa* contain a nonprotein amino acid called L-canavanine that is toxic to most insects. It is incorporated into an insect's proteins in place of the amino acid arginine, which it closely resembles. However, one species of bruchid that feeds on this plant, *Caryedes brasiliensis,* possesses enzymes that discriminate between L-canavanine and arginine during protein formation, as well as enzymes that degrade L-canavanine to forms that can be used as a source of nitrogen. Thus, it seems that for every defense, a new counterattack can be devised. Because *Dioclea megacarpa* evolved in a group of legumes that lacked L-canavanine, and because *Caryedes brasiliensis* evolved in a group of beetles that cannot discriminate between normal and toxic amino acids, their adaptations appear to represent reciprocal evolution.

ECOLOGISTS IN THE FIELD **A counterattack for every defense.** To evaluate whether relationships between insects and their host plants are examples of coevolution, University of Illinois biologist May Berenbaum studied a plant–herbivore system in New York State with some similarities to the legume–bruchid beetle interaction. Umbellifers (members of the parsley family; Figure 17.18a)

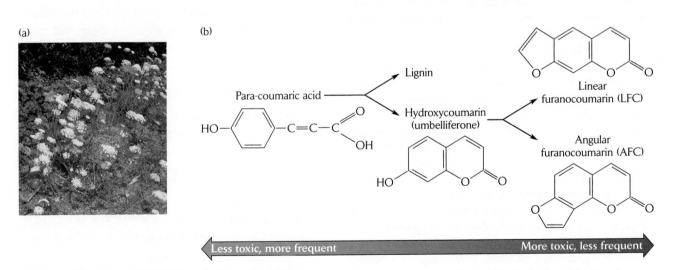

FIGURE 17.18 Plant secondary compounds and herbivore resistance may have coevolved. The taxonomic relationships among certain umbellifers, which produce defensive chemicals called furanocoumarins, and among insects that can feed on these plants suggest that these plants and herbivores have coevolved. (a) Queen Anne's lace (*Daucus carota*) is a familiar umbellifer. (b) As one proceeds down the biosynthetic pathway leading to the furanocoumarins, the toxicity of the chemicals increases, and the number of plant species synthesizing them decreases. Photo (a) by Alfred Brousseau, courtesy of Saint Mary's College of California.

produce many defensive chemicals, among the most prominent of which are the furanocoumarins. The biosynthetic pathway of these compounds leads from para-coumaric acid (which, being a precursor of lignin, is found in virtually all plants) to hydroxycoumarins and finally to the furanocoumarins, which occur in two chemical forms, linear furanocoumarins (LFCs) and angular furanocoumarins (AFCs) (Figure 17.18b). As one proceeds down this biosynthetic pathway, toxicity increases. Hydroxycoumarins have some properties that are toxic to herbivores; LFCs interfere with DNA replication in the presence of ultraviolet light; and AFCs interfere with herbivore growth and reproduction quite generally.

The most toxic of these chemicals occur among the fewest plant families. Para-coumaric acid is widespread among plants, occurring in at least a hundred families, while only thirty-one families possess hydroxycoumarins. LFCs are restricted to eight plant families and are widely distributed in only two: Umbelliferae and Rutaceae (the citrus family). AFCs are known only from two genera of legumes and ten genera of Umbelliferae.

Among the herbaceous umbellifer species in New York State, some (especially those growing in woodland sites with low levels of ultraviolet light) lack furanocoumarins, others contain LFCs only, and some contain both LFCs and AFCs. Berenbaum's surveys of herbivorous insects collected from these plant species revealed several interesting patterns: (1) host plants containing both AFCs and LFCs were, somewhat surprisingly, attacked by more species of insect herbivores than were plants with only LFCs or with no furanocoumarins; (2) insect herbivores found on AFC plants tended to be extreme diet specialists, most being found on no more than three genera of plants; and (3) these specialists tended to be abundant compared with the few generalists found on AFC plants and compared with all herbivores found either on LFC plants or on umbellifers lacking furanocoumarins.

Although LFCs and (especially) AFCs effectively deter most herbivorous insects, some insect genera that have evolved to tolerate these chemicals have obviously become successful specialists. One can make a strong case for coevolution here. The taxonomic distribution of hydroxycoumarins, LFCs, and AFCs across the Umbelliferae suggests that plants containing LFCs are a subset of those containing hydroxycoumarins, and that those containing AFCs are an even smaller subset of those containing LFCs. This pattern is consistent with an evolutionary sequence of increasingly toxic umbellifer defenses progressing from hydroxycoumarins to LFCs and AFCs. Furthermore, the insects that specialize on plants containing LFCs belong to groups that characteristically feed on plants containing hydroxycoumarins, and those that specialize on plants containing AFCs have close relatives that can feed on plants containing LFCs. These taxonomic patterns are consistent with coevolution within the system.

This story of the evolution of chemical defenses by plants and resistance to those defenses by certain groups of insects is somewhat conjectural, based on the logic of the evolutionary relationships of the taxa involved. We have no means of directly watching such evolutionary interactions unfold; evolution occurs too slowly in natural systems. Berenbaum's inferences about evolution build on the idea that evolutionarily older or younger traits (such as absence and presence of AFCs) should be found among close relatives if those traits are linked by evolution. This logic has been elaborated into a branch of evolutionary biology known as *phylogenetic reconstruction,* which uses similarities and differences among species to determine their evolutionary relationships.

MORE ON THE WEB *Inferring Phylogenetic History.* How can one reconstruct evolutionary relationships among species from their traits?

In some cases, long associations between groups of interacting organisms set the stage for coevolution. In other cases, the relationships are recent and changeable. For example, larvae of butterflies of the family Pieridae feed on a variety of host plants. When M. F. Braby and J. W. H. Trueman overlaid the distribution of host plants on a diagram of the evolutionary relationships among pierid butterfly species, they found many cases of evolutionary lineages that had switched from typical pierid host plants in the Brassicaceae (members of the mustard and cabbage family) to plants from distantly related families, such as mistletoes, heaths, and even pines. Clearly, the adaptations that enabled pierids to switch to these host plants could not have been coevolved with the Brassicaceae.

The yucca moth and the yucca

The application of phylogenetic reconstruction to the problem of coevolution is probably best illustrated by the curious pollination mutualism between yucca plants (Yucca, a member of the agave family) and moths of the genus *Tegeticula* (Figure 17.19). This relationship was first described more than a century ago, but its details have been worked out only during the past few years, largely through studies by Olle Pellmyr, of the University of Idaho, and his colleagues.

Adult female yucca moths carry balls of pollen between yucca flowers by means of specialized mouthparts. During the act of pollination, a female moth enters a yucca flower, makes cuts in the ovary with her ovipositor, and deposits one to fifteen eggs. After each egg is laid, the moth crawls

(a)

(b)

FIGURE 17.19 The relationship between yucca and yucca moth is an obligate mutualism. The Mojave yucca (a, *Yucca shidigera*) is pollinated only by a yucca moth of the genus *Tegeticula* (b). The moth larvae develop only on yucca plants. Photo (a) by Alfred Brousseau, courtesy of Saint Mary's College of California; photo (b) by Larry Jon Friesen/Saturdaze.

to the top of the pistil of the flower and deposits a bit of pollen on the stigma. This behavior ensures that the flower is fertilized and that the moth's offspring will have developing seeds to feed on. After the moth has laid her eggs, she may scrape some pollen off the anthers and add it to the ball she carries in her mouthparts before flying to another flower. Male moths also come to the flowers to mate with the females, but only the females carry pollen.

The relationship between the moth and the yucca is an *obligate mutualism*. *Tegeticula* larvae can grow nowhere else; *Yucca* has no other pollinator. In return for the pollination of its flowers, the yucca seemingly tolerates the moth larvae feeding on its seeds, but the extent of this loss of potential reproduction is small, rarely exceeding 30% of the plant's seed crop.

The moth's apparent restraint concerning the number of eggs laid per flower is a puzzling aspect of the yucca–moth relationship. Over the short term, it would seem that moths laying larger numbers of eggs per flower might have higher individual reproductive success and evolutionary fitness, even though such behavior over the long term might lead to extinction of the yucca. In fact, it is the yucca that regulates the number of eggs laid per flower. When too many eggs are laid in the ovary of a particular flower—too many being enough to eat a majority

of the developing seeds—the flower is aborted and the moth larvae die. While this strategy might also seem to reduce the seed production of the yucca, resources that would have supported the production of seeds in the now aborted flower are diverted to other flowers. Selective abortion of insect-damaged fruit occurs widely among plants, and yuccas use this mechanism to keep their moth pollinators in line.

The moth and the yucca have many adaptations that support their mutualistic interaction. On the yucca's part, its pollen is sticky and can be easily formed into a ball that the moth can carry, and the stigma is specially modified as a receptacle to receive pollen. On the moth's part, individuals visit flowers of only one yucca species, mate within the flowers, lay their eggs in the ovary within the flower, exhibit restraint in the number of eggs laid per flower, and have specially modified mouthparts and behaviors to obtain and carry pollen. Because the mutualism of *Tegeticula* and *Yucca* is so tight, one might expect all these traits to be a result of coevolution between the two.

In fact, however, many of these traits are present in the larger lineage of nonmutualistic moths (the family Prodoxidae) within which *Tegeticula* evolved. A diagram of the evolutionary relationships among species, known as a *phylogenetic tree,* can reveal such patterns. Examination

GLOBAL CHANGE

Invasive plant species and the role of herbivores

The spread of invasive species is one of several important ways in which our world is changing. It is difficult to know just how many organisms are carried by humans to distant parts of the world, either intentionally or inadvertently, but most introduced species fail to become established in their new location, and among those that do gain a foothold, few spread widely. Those few, however, that become sufficiently widespread and abundant to have major effects on local ecosystems still count in the hundreds of species. The success of these species raises the question, what conditions favor the establishment of invasive species? If we knew the answer, we would be in a better position to control invasive species.

Over the years, many ecologists have investigated invasive plant species and the conditions that favor their establishment. Some of their results suggest that invasive plants can spread through a new region because they leave behind their natural enemies, including herbivores, parasites, and pathogens. Other studies that have tested this "enemy-escape" hypothesis have been less conclusive. When the results of hypothesis testing are mixed, ecologists find it helpful to determine whether a hypothesis is supported, *on average,* across all studies. One approach to finding a consensus among studies is to conduct a *meta-analysis,* which considers all the relevant data and quantifies the average strength of effects—in this case, those of antagonists on invasive species.

John Parker and his colleagues at Georgia Tech University searched the literature for studies that assessed how native and introduced herbivores affected the abundances of native and introduced plants. They found 63 studies, examining more than 100 species of introduced plants, in which researchers manipulated the presence and absence of herbivores. Most of these herbivores were vertebrate generalists such as bison, deer, and rabbits. Parker and his colleagues quantified the effect of herbivores in each study as the ratio of plant abundance with herbivores present ($+H$) to plant abundance with herbivores absent ($-H$). They then averaged the effects of each category of herbivores across all the plant species in each category. Their results point to the importance of coevolutionary relationships in determining how well a species fares in its interactions with antagonists.

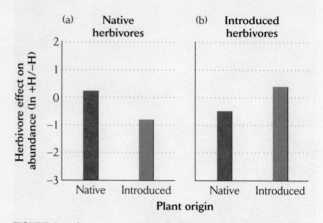

FIGURE 1 When we compare results across a large number of studies, native herbivores tend to increase the abundances of native plant species and decrease the abundances of introduced plant species (a). In contrast, introduced herbivores have a positive effect on the abundances of introduced plant species and a negative effect on those of native species. The zero line represents no herbivore effect. The y-axis is the natural logarithm of the ratio of plant abundance with herbivores present ($+H$) to plant abundance with herbivores absent ($-H$). After J. D. Parker, D. E. Burkepile, and M. E. Hay, Science 311:1459–1461 (2006).

The meta-analysis revealed an unexpected outcome (Figure 1): When the herbivores were also introduced—often from the introduced plant's native range—the introduced plant species were more abundant than the native plant species. However, in the presence of native herbivores, the introduced species were less abundant than the native species.

One of the studies included in the meta-analysis, presented here as an example of this type of research, involved pampas grass, a popular landscape plant in California that was introduced to California from South America and has become invasive. To evaluate the effect of herbivory on the abundance of pampas grass, John Lambrinos of the University of California placed cages around some plots to exclude jackrabbits—a native herbivore in California—and left other plots uncaged to allow herbivory. When jackrabbits were excluded, pampas grass survival was approximately 60%. When the jackrabbits were given free rein, survival plummeted to about 5% (Figure 2).

(a)

(b)

FIGURE 2 Native jackrabbits in California (a) can cause a substantial decline in the survival of the invasive pampas grass (b). Photo (a) by John Cancalosi/Peter Arnold; photo (b) by Patricia Head/Animals Animals Enterprises.

What do the results of the meta-analysis tell us about the enemy-escape hypothesis? Recall that introduced plants were predicted to do better in their new locations because they were leaving their historical antagonists behind. Parker and his colleagues argued that what really matters is the origin of the herbivores a plant encounters when it arrives in a new area. While the introduced plants may leave their antagonists behind at home, they may face a new suite of generalist herbivores against which they have not evolved defenses. As a result, these plants do not fare well against native herbivores. However, when an introduced plant's original herbivores are also present in the new area, the plants find themselves among antagonists against which they have evolved defenses, and as a result, they fare better. This result suggests that escaping their native herbivores is probably not the prime reason for the success of invasive plants in many cases.

This same logic applies to native plants. Native plants suffer high herbivory by introduced herbivores because the two groups have no shared evolutionary history during which the native plants might have evolved effective defenses. The native plants do much better against the native herbivores against which they have evolved a variety of defenses. These studies demonstrate that the ability to invade a new region of the world is more complex than it might first seem.

The meta-analysis by Parker and colleagues holds two important lessons for ecologists, one about methods and the other about plant–herbivore relationships. First, the distinction between the effects of native and of introduced herbivores could not have been discerned in a single study, such as that on pampas grass, unless both native and introduced herbivores had been used in the same study—which is rarely the case. The meta-analysis permitted the researchers to make the comparison between the different types of herbivores by averaging effects over many studies. Second, in the plant–herbivore systems included in the meta-analysis, we can conclude that the introduced plants have specific adaptations to defend themselves against herbivores with which they have had a long association. Whereas many ecologists consider rabbits and ungulates to be generalist feeders wherever they occur, it is clear that native and introduced herbivores feed in different ways or express preferences for plant species that have not evolved specific defenses against them. In this way, herbivores can tip the balance of competitive interactions to favor species of plants from their own native areas.

FIGURE 17.20 Phylogenetic trees can reveal preadaptations. The phylogenetic tree of the moth family Prodoxidae shows when traits critical to the yucca moth–yucca mutualism in moths of the genus *Tegeticula* evolved. After O. Pellmyr and J. N. Thompson, *Proc. Natl. Acad. Sci. USA* 89:2927–2929 (1992).

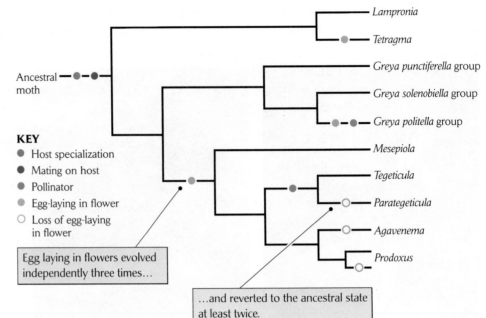

KEY
- ● Host specialization
- ● Mating on host
- ● Pollinator
- ● Egg-laying in flower
- ○ Loss of egg-laying in flower

Egg laying in flowers evolved independently three times…

…and reverted to the ancestral state at least twice.

of a phylogenetic tree of the Prodoxidae (Figure 17.20) shows that several of the highly specialized traits of *Tegeticula* are found in other members of the family. Indeed, host specialization and mating on the host plant are evolutionarily old features of the family—features found in all its members. The trait of egg laying in flowers has evolved independently at least three times in the family and has reversed (reverted to the ancestral state) at least twice, in *Parategeticula* and *Agavenema*. Of the species that lay eggs in flowers, only *Tegeticula* and one species of *Greya* actually function as pollinators; the others are strictly parasites of the plants in which their larvae grow. Thus, the *Tegeticula–Yucca* mutualism probably evolved from a parasite–host relationship. It should also be mentioned that *Greya politella* pollinates *Lithophragma parviflorum*, a plant in the saxifrage family, which is not closely related to the yuccas. We can see from this phylogenetic tree that many of the adaptations that occur in the yucca–yucca moth mutualism appear to have been present in the moth lineage before the establishment of the mutualism itself, just as flower abortion occurs widely among plants and is not

unique to this mutualism. Such traits that become useful for a purpose other than that for which they evolved are often referred to as **preadaptations.**

Where does this leave us with regard to coevolution? The consensus among ecologists is that species interactions strongly affect evolution and shape the adaptations of consumer and resource populations alike. Diffuse coevolution is common in that populations simultaneously respond to an array of complex interactions with many other species. Coevolution in the narrow sense, in which changes in one evolving lineage stimulate evolutionary responses in the other, and vice versa, may be seen most readily in symbioses, including both antagonistic and mutualistic relationships, in which strong interactions are limited to a pair of species. Even in such cases as that of the yucca and its moth pollinator, some traits that appear to be coevolved may have been preadaptations that were critical to the establishment of the obligate mutualism in the first place. No subtlety of definition can, however, detract from the reality that interactions among species are major sources of selection and evolutionary response.

SUMMARY

1. Interactions between species select for traits that provide an advantage in those interactions. Such selection by biological agents stimulates reciprocal evolutionary responses in the traits of the interacting populations and fosters diversity of adaptations.

2. Coevolution is the reciprocal evolution of related structures or functions in species that interact ecologically.

The interactions may be antagonistic (consumer–resource, competition) or cooperative (mutualism).

3. We see evidence of evolutionary responses arising from species interactions throughout nature, but nowhere more conspicuously than in antipredator adaptations involving coloration. Some species are cryptic and blend in with their backgrounds to avoid detection. In contrast, naturally unpalatable or noxious prey species advertise this property with warning coloration.

4. Palatable species often evolve to resemble noxious species to fool predators, a phenomenon known as Batesian mimicry. Sometimes, many noxious species evolve to resemble one another, thereby reinforcing learning and avoidance by predators. This strategy is called Müllerian mimicry.

5. Early studies of coevolution emphasized, on one hand, theoretical interactions between hosts and pathogens controlled simply by virulence and resistance genes, and on the other hand, observations on the specialization of insect herbivore species on a narrow range of host plants in nature.

6. Experimental evidence of evolutionary changes that affect the outcome of consumer–resource interactions has been obtained in laboratory studies of parasitoid–host interactions. After periods of coevolution, parasitoid populations decreased and host populations increased, apparently following selection for improved host resistance to parasitoids.

7. Studies on pathogens of crop plants have revealed a simple genetic basis for virulence and resistance. Evolutionary responses of herbivores to variation in their host plants have been demonstrated by local adaptation of herbivorous insects to particular host plant individuals.

8. Because selection on a resource population for adaptations to avoid consumption increases in proportion to the rate at which that population is exploited by consumers, and because selection for consumer efficiency decreases as the exploitation rate increases, consumer and resource populations can achieve an evolutionary steady state at which those two selection pressures are balanced.

9. The most important force driving improvements in competitive ability is selection for increased efficiency of resource use. Experiments on competition between species of flies have revealed reversals of competitive ability after one species became rare. By testing selected populations against unselected controls, investigators have demonstrated genetic changes in the competing populations.

10. Specialization on different resources should reduce competition and promote coexistence. Such evolutionary divergence between competitors is referred to as character displacement. One may test whether specialization results from evolutionary divergence by comparing the traits of a population in the presence and in the absence of a competitor.

11. Analysis of the biosynthetic pathways of plant secondary compounds has shown that plants may evolve increasingly toxic chemical defenses in response to insect herbivory. When variations in these pathways and in insect resistance to the chemicals are overlaid on taxonomic relationships within each group, ecologists can infer the evolutionary history of a plant–insect interaction.

12. The interaction between yucca moths and yuccas is an obligate mutualism in which the moth pollinates the plant and its larvae feed on the plant's developing seeds. Both the moth and the yucca have adaptations that promote this relationship, but phylogenetic analysis shows that some of the moth's adaptations are also present in close relatives that are not mutualists of yuccas. Such traits are called preadaptations.

REVIEW QUESTIONS

1. What makes coevolution a unique type of evolution?

2. Why might many palatable prey species evolve crypsis as part of their defense against predators, whereas many unpalatable prey species evolve conspicuousness?

3. Compare and contrast Batesian and Müllerian mimicry.

4. Why is coevolution considered a genotype–genotype interaction?

5. In simple models of coevolution between consumer and resource species, how is it possible to achieve an evolutionary steady state while both species continue to evolve?

6. If you observe that two related species differ in the food they consume, why can you not necessarily conclude that this difference is the product of a history of competition between the two species?

7. What conditions favor the evolution of an obligate mutualism?

8. How might a preadaptation make it difficult to demonstrate coevolution between two species?

SUGGESTED READINGS

Armbruster, W. S. 1992. Phylogeny and the evolution of plant–animal interactions. *BioScience* 42:12–20.

Bais, H. P., et al. 2003. Allelopathy and exotic plant invasion: From molecules and genes to species interactions. *Science* 301:1377–1380.

Berenbaum, M. R. 1983. Coumarins and caterpillars: A case for coevolution. *Evolution* 37:163–179.

Bogler, D. J., J. L. Neff, and B. B. Simpson. 1995. Multiple origins of the yucca–yucca moth association. *Proceedings of the National Academy of Sciences USA* 92:6864–6867.

Braby, M. F., and J. W. H. Trueman. 2006. Evolution of larval host plant associations and adaptive radiation in pierid butterflies. *Journal of Evolutionary Biology* 19:1677–1690.

Brodie, E. D. 1999. Predator–prey arms races. *BioScience* 49:557–568.

Bshary, R. 2002. Biting cleaner fish use altruism to deceive image-scoring client reef fish. *Proceedings of the Royal Society of London B* 269:2087–2093.

Davies, N. B., and M. Brooke. 1991. Coevolution of the cuckoo and its hosts. *Scientific American* 264:92–98.

Day, T., and K. A. Young. 2004. Competitive and facilitative evolutionary diversification. *BioScience* 54:101–109.

Ehrlich, P. R., and P. H. Raven. 1964. Butterflies and plants: A study in coevolution. *Evolution* 18:586–608.

Ewald, P. W. 1994. *Evolution of Infectious Disease.* Oxford University Press, Oxford.

Feldman, R., D. F. Tomback, and J. Koehler. 1999. Cost of mutualism: Competition, tree morphology, and pollen production in limber pine clusters. *Ecology* 80:324–329.

Futuyma, D. J., and M. Slatkin (eds.). 1983. *Coevolution.* Sinauer Associates, Sunderland, Mass.

Janz, N., and S. Nylin. 1998 Butterflies and plants—a phylogenetic study. *Evolution* 52:486–502.

Janzen, D. H. 1985. The natural history of mutualisms. In D. H. Boucher (ed.), *The Biology of Mutualism,* pp. 40–99. Croom Helm, London.

Kawano, K. 2002. Character displacement in giant rhinoceros beetles. *American Naturalist* 159:255–271.

Lambrinos, J. G. 2002. The variable invasive success of *Cortaderia* species in a complex landscape. *Ecology* 83:518–529.

Losos, J. B. 2007. Detective work in the West Indies: Integrating historical and experimental approaches to study island lizard evolution. *BioScience* 57(7):561–572.

Mode, C. J. 1958. A mathematical model for the co-evolution of obligate parasites and their hosts. Evolution 12:158–165.

Mueller, U. G., et al. 2005. The evolution of agriculture in insects. *Annual Review of Ecology, Evolution, and Systematics* 36:563–595.

Parker, J. D., D. E. Burkepile, and M. E. Hay. 2006. Opposing effects of native and exotic herbivores on plant invasions. *Science* 311:1459–1461.

Pellmyr, O., and C. J. Huth. 1994. Evolutionary stability of mutualism between yuccas and yucca moths. *Nature* 372:257–260.

Pellmyr, O., J. Leebens-Mack, and C. J. Huth. 1996. Non-mutualistic yucca moths and their evolutionary consequences. *Nature* 380:155–156.

Pellmyr, O., et al. 1996. Evolution of pollination and mutualism in the yucca moth lineage. *American Naturalist* 148:827–847.

Pimentel, D. 1968. Population regulation and genetic feedback: Evolution provides foundation for control of herbivore, parasite, and predator numbers in nature. *Science* 159:1432–1437.

Scriber, J. M. 2002. Evolution of insect–plant relationships: Chemical constraints, coadaptation, and concordance of insect/plant traits. *Entomologia Experimentalis et Applicata* 104:217–235.

Thompson, J. N. 1994. *The Coevolutionary Process.* University of Chicago Press, Chicago.

Tomback, D. F., and Y. B. Linhart. 1990. The evolution of bird-dispersed pines. *Evolutionary Ecology* 4:185–219.

Weiner, J. 1994. *The Beak of the Finch.* Knopf, New York.

CHAPTER 18

Community Structure

To most ecologists, the term *community* means an assemblage of species that occur together in the same place. Ecologists also agree that the species living together within a community can interact strongly as consumers and resources or as competitors. Indeed, much of this chapter and the next three chapters concerns the consequences of these interactions for the diversity and distributions of species, and for the functioning and stability of ecological systems. Yet ecologists have not always agreed on the meaning of "community," and much of the early history of the discipline consisted of sharp debates between the supporters of very different views. In fact, in some form, these debates are still with us. Some ecologists have asserted that the community is a unit of ecological organization having recognizable boundaries and whose structure and functioning are regulated by interactions among species. Others have regarded the community as a loose assemblage of those species that can tolerate the conditions of a particular place or habitat, but which has no distinct boundary where one type of community meets another.

The idea that communities are organized ecological units reached its extreme in the concept of communities as *superorganisms*. From this perspective, the functions of various species are connected like those of parts of the body and have evolved so as to enhance their interdependence. This viewpoint requires that communities be discrete entities that can be distinguished from one another, in the sense that we distinguish individuals within populations or different species within a community. The most influential advocate of the organismal viewpoint was the American plant ecologist Frederic E. Clements (1874–1945). Clements's idea of the community was closely tied to vegetation

FIGURE 18.1 The boundaries of some communities are clearly defined. Hillsides in southern California have chaparral vegetation at higher elevations, grassland on the lower, hotter slopes, and live oaks in the moister valleys between ridges. Photo by Christi Carter/Grant Heilman Photography.

types. He pointed out that some community boundaries—for example, between deciduous forest and prairie in the midwestern United States, or between broad-leaved forest and needle-leaved forest in southern Canada—are clearly defined and respected by most species of plants and animals (Figure 18.1).

Clements's **holistic concept** of the community feels right in some ways. We cannot ponder the significance of a kidney's functioning apart from the organism to which it belongs. Many ecologists have argued that soil bacteria make no sense without reference to the detritus they feed on, their consumers, and the plants nourished by their wastes. Accordingly, one can understand each species only in terms of its contribution to the dynamics of the whole system. Most importantly, according to the holistic concept, ecological and evolutionary relationships among species enhance community properties, such as the stability of energy flow and nutrient cycling patterns, making a community much more than the sum of its individual parts.

In response to Clements, the botanist Henry A. Gleason (1882–1975) championed an **individualistic concept** of community organization. Gleason believed that a community, far from being a distinct unit like an organism, is merely a fortuitous association of species whose adaptations and requirements enable them to live together under the physical and biological conditions of a particular place. A plant association, he said, is "not an organism, scarcely even a vegetational unit, but merely a coincidence." Accordingly, community structure and functioning simply express the interactions of individual species that make up local associations, and do not reflect any organization, purposeful or otherwise, above the species level. Remember that natural selection acts on the fitness of individuals, and so each population in a community evolves so as to maximize the reproductive success of its individual members, and not to benefit the community as a whole.

As we shall see, modern ecology integrates the individualistic premise that most species assemblages lack distinct boundaries and the holistic premise that attributes of community structure and function arise from interactions among species.

CHAPTER CONCEPTS

- A biological community is an association of interacting populations

- Measures of community structure include numbers of species and trophic levels

- Feeding relationships organize communities in food webs

- Food web structure influences the stability of communities

- Communities can switch between alternative stable states

- Trophic levels are influenced from above by predation and from below by production

Every place on earth—each meadow, each pond, each rock at the edge of the sea—is shared by many coexisting organisms. These organisms are linked to one another by their feeding relationships and other interactions, forming a complex whole often referred to as a **community.** Interrelationships within communities govern the flow of energy and the cycling of elements within the ecosystem. They also influence population processes, and in doing so determine the relative abundances of species.

The members of a community must be compatible in the sense that the outcomes of all their interactions allow their survival and reproduction. Although the theory of species interactions, as we have seen in Part 4 of this book, tells us when predator and prey populations, or two competitor populations, can coexist, this theory cannot be applied easily to large numbers of interacting species. Thus, ecologists still debate the factors that determine numbers of coexisting species and still argue about why those numbers vary from place to place. Moreover, it is also important to understand how species interactions influence the structure and functioning of communities.

Species assume different functional roles in communities, and their relative abundances reflect how they fit into the entire web of interactions within the community. Assemblages of species also change over time, whether in response to a disturbance or following some intrinsic dynamic process.

A biological community is an association of interacting populations

Throughout the development of ecology as a science, *community* has often meant an assemblage of plants and animals occurring in a particular locality and dominated by one or more prominent species or by some physical characteristic. When we speak of an oak community, a sagebrush community, or a pond community, we refer to all the plants and animals found in a particular place dominated by the community's namesake (Figure 18.2). Used in this way, the term is unambiguous: a community

(a)

(b)

(c)

FIGURE 18.2 Communities are often named after their most conspicuous members or physical characteristics. (a) A ponderosa pine community in the Santa Catalina Mountains of Arizona. (b) A riparian forest community bordering a stream coursing through dry mountains in southern Arizona. (c) A young deciduous forest community in the Great Smoky Mountains of Tennessee. Photos by R. E. Ricklefs.

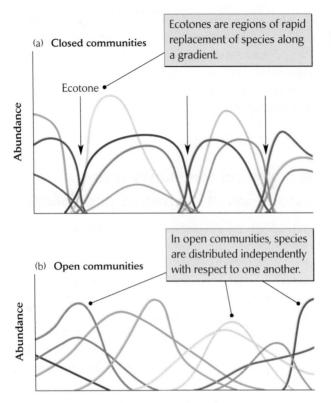

(a) **Closed communities**

Ecotones are regions of rapid replacement of species along a gradient.

Ecotone •

Abundance

(b) **Open communities**

In open communities, species are distributed independently with respect to one another.

Abundance

Environmental gradient

FIGURE 18.3 Closed community structure is distinguished from open community structure by the presence of ecotones. Hypothetical distributions of species along an environmental gradient (a) when the species are organized into distinct assemblages (closed communities) and (b) when they are distributed independently along the gradient (open communities). Arrows indicate ecotones between closed communities. Each curve represents the abundance of a different species along the environmental gradient.

is spatially defined and includes all the populations within its boundaries. Each community can be named. Indeed, many European ecologists use a complex taxonomy of communities—the Braun–Blanquet system—based on a rigidly defined method of sampling plant species composition, which places each community in a hierarchy of types organized by their similarity.

When populations extend beyond arbitrary spatial boundaries and one assemblage of species blends gradually into the next, both the concept and the reality of the community become more difficult to pin down. Migrations of birds between temperate and tropical regions link assemblages of species in each area. Salamanders, which complete their larval development in streams and ponds but pursue their adult existence in the surrounding woods, tie together aquatic and terrestrial assemblages. So do trees when they shed their leaves into streams and contribute to an aquatic food chain based on detritivores.

A complex array of interactions directly or indirectly ties together all members of a community into an intricate web. The influence of each population extends to ecologically distant parts of the community. Insectivorous birds, for example, do not eat trees, but they do influence trees by preying on many of the insects that feed on foliage or pollinate flowers. The ecological and evolutionary effects of a population extend in all directions throughout the structure of a community by way of its influence on predators, competitors, and prey.

One way to visualize the geographic organization of biological communities is to plot the abundances of species along a spatial transect or a gradient of environmental conditions—for example, from dry to wet soils. We can

FIGURE 18.4 Ecotones are often associated with abrupt changes in the physical environment. In this section of the coast of the Bay of Fundy, New Brunswick, seaweeds extend only to the high-tide mark. Between the high tide mark and the spruce forest, waves wash soil from the rocks and salt spray kills pioneering land plants, leaving the area devoid of vegetation. Photo by R. E. Ricklefs.

imagine two extreme types of patterns, shown schematically in Figure 18.3, in which the distribution of each species is plotted on a gradient of environmental conditions. In one case (Figure 18.3a), the distributions of several species coincide closely, but are largely separated from those of other sets of species. Ecologists have called this case the **closed community** concept. Each set of species with overlapping distributions is a closed community, a discrete ecological unit with distinct boundaries. This pattern is consistent with a holistic view in that the species belonging to a community are closely associated with one another and share ecological tolerance limits. The boundaries of such communities, called **ecotones,** are regions of rapid replacement of species along a spatial transect or ecological gradient.

Alternatively, the distribution of each species may not coincide closely with the distributions of others, so that the species appear to be independently distributed along a spatial transect or a gradient of ecological conditions. This pattern is referred to as the **open community** concept (Figure 18.3b). Such communities have no natural boundaries, so their extent is arbitrary. The distribution of

each member of a local assemblage may extend independently into other associations of species. As we shall see, the concepts of open and closed communities both have validity in nature.

Ecotones

Ecotones are places where many species reach the edges of their distributions. Ecotones are especially prominent where sharp physical differences separate distinct communities. Such differences occur at the transition between most terrestrial and aquatic (especially marine) environments (Figure 18.4), between north-facing and south-facing slopes of mountains (see Figure 4.19), and where underlying geologic formations cause the mineral content of soils to change abruptly. Sharp community boundaries may also arise where one species or growth form so dominates the environment that the edge of its range determines the distribution limits of many other species.

An ecotone between plant associations on serpentine and non-serpentine soils in southwestern Oregon is represented in Figure 18.5. Levels of nickel, chromium, and

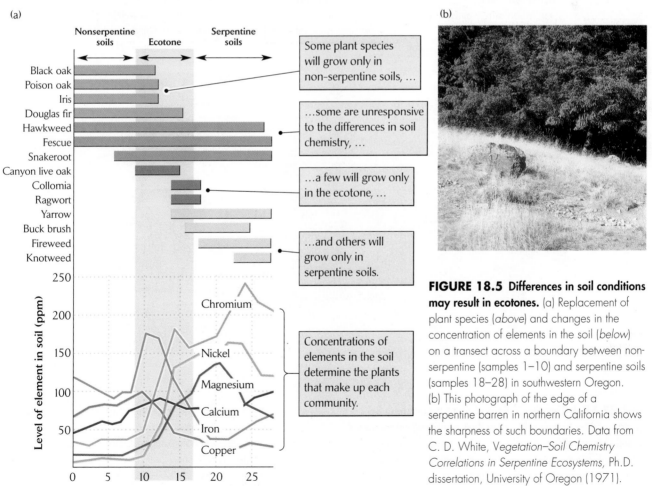

FIGURE 18.5 Differences in soil conditions may result in ecotones. (a) Replacement of plant species (*above*) and changes in the concentration of elements in the soil (*below*) on a transect across a boundary between non-serpentine (samples 1–10) and serpentine soils (samples 18–28) in southwestern Oregon. (b) This photograph of the edge of a serpentine barren in northern California shows the sharpness of such boundaries. Data from C. D. White, *Vegetation–Soil Chemistry Correlations in Serpentine Ecosystems*, Ph.D. dissertation, University of Oregon (1971). Photo by R. E. Ricklefs.

magnesium increase as we move across the boundary into the serpentine soil; copper and iron concentrations in the soil drop off. The edge of the serpentine soil marks the boundaries of many species that either cannot invade the communities on serpentine soils, such as black oak, or are restricted to them, such as buckbrush and fireweed. A few species, such as collomia and ragwort, exist only within the narrow zone of transition; others, such as hawkweed and fescue, which are seemingly unresponsive to the variations in soil chemistry, extend across the ecotone. Thus, the transition between serpentine and non-serpentine soils only partly conforms to the concept of closed communities; the ecotone is recognized by many, but not all, species.

The fact that plant species are restricted to particular soils in nature does not mean that their distributions are determined solely by their physiological tolerance of soil characteristics. In fact, many plants that are restricted to serpentine or other poor-quality soils grow better in

normal fertile soils. For example, in the coastal ranges of northern California, several species of pines and cypresses are restricted to serpentine soils, while others are present only on extremely acidic soils. When planted on different types of soils in a common garden, seedlings of many of these species grew best on soil from their native habitat (Figure 18.6). However, the Sargent cypress, which is restricted to serpentine soils in nature, grew almost as well on normal soils, and pygmy cypress, which is restricted to acidic soils, grew much better on "normal" and serpentine soils. In fact, the ecological limits of these species are determined by the *interaction* between their ability to grow on different soil types and their ability to compete with other species that cannot tolerate serpentine or acidic soils.

Dominant plants can reinforce or even create ecotones by mechanisms other than competition when they alter their environment. Consider the sharp boundaries between broad-leaved and coniferous needle-leaved forests in some regions, which develop even though spatial

Lodgepole pine grows on acidic (A) soils in nature and did not survive on other soil types.

Pygmy cypress is restricted to acidic (A) soils in nature but grew much better on normal (N) and serpentine (S) soils.

Sargent cypress is restricted to serpentine (S) soils in nature but grew almost as well in normal (N) soils.

Lodgepole pine

Pygmy cypress

Sargent cypress

FIGURE 18.6 The distributions of plant species may be determined by factors other than their physiological tolerance of soil characteristics. Seedlings of lodgepole pine (*Pinus bolanderi*), pygmy cypress (*Cupressus pygmaea*), and Sargent cypress (*Cupressus sargentii*) were grown in a common garden on acid (A), normal (N), and serpentine (S) soils. The observation that some of the species grew well on soil types on which they do not normally grow in nature suggests that some factor other than soil chemistry is restricting their distributions. After C. McMillan, *Ecol. Monogr.* 26:177–212 (1956).

changes in temperature and precipitation are gradual. The decomposition of conifer needles produces abundant organic acids, thus increasing soil acidity. Furthermore, because needles tend to decompose slowly, a thick layer of partly decayed organic material accumulates at the soil surface. This dramatic shift in conditions between broad-leaved and needle-leaved forests marks the edges of distributions of many shrub and herb species that grow best on the soils within one forest type or the other. Similarly, at boundaries between grassland and shrubland or between grassland and forest, sharp changes in soil surface temperature, soil moisture, light intensity, and fire frequency result in many species replacements. Boundaries between grasslands and shrublands are often sharp because when one or the other growth form holds a slight competitive edge, it dominates the community. For example, grasses can prevent the growth of shrub seedlings by reducing the moisture content of surface layers of soil; shrubs can depress the growth of grass seedlings by shading them. Fire maintains a sharp ecotone between prairies and forests in the midwestern United States. The perennial grasses in prairies resist fire damage that kills tree seedlings outright, but fires cannot penetrate deeply into the moister forest habitats.

The continuum concept and gradient analysis

Although distinct ecotones often form where environmental conditions change abruptly, they are less likely to occur along gradients of gradual environmental change. The broad-leaved deciduous forests of eastern North America are bounded at conspicuous ecotones to the north, where they are replaced by cold-tolerant coniferous needle-leaved forests, to the west by drought- and fire-resistant grasslands, and to the southeast by fire-resistant pine forests that can grow on highly weathered, nutrient-poor soils. The deciduous forests themselves are not homogeneous, however. Different species of trees and other plants occur in different areas within the deciduous forest biome. The species of trees found in any one region—for example, those native to eastern Kentucky—have geographic ranges that only partly coincide, suggesting that they have partially independent evolutionary backgrounds and ecological relationships (Figure 18.7). Some of the species reach their northern limits in Kentucky, some their southern limits. Because few species have broadly overlapping geographic ranges, the assemblage of plant species that is found in any given spot does not form a closed community.

A more detailed view of the forests of eastern Kentucky would reveal that many tree species segregate along

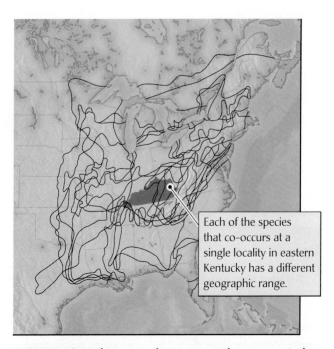

Each of the species that co-occurs at a single locality in eastern Kentucky has a different geographic range.

FIGURE 18.7 The species that occur together at a particular place may have different geographic distributions. None of the twelve tree species that occur together in plant associations in eastern Kentucky have the same geographic range. After H. A. Fowells, *Silvics of Forest Trees of the United States*, U.S. Department of Agriculture, Washington, D.C. (1965).

local gradients of environmental conditions. Some grow along ridgetops, others along moist river bottoms; some on poorly developed, rocky soils, others on rich, organic soils. At a larger scale, within broadly defined habitat types, such as forest, grassland, or estuary, species replace one another continuously along gradients of physical conditions. The environments of the eastern United States form such a continuum, with a north–south temperature gradient and an east–west rainfall gradient. The distribution of species along an environmental gradient is often referred to as the **continuum** concept.

The continuum concept can be visualized by a **gradient analysis,** in which the abundance of each species is plotted on a continuous gradient of one or more physical conditions, such as moisture, temperature, salinity, exposure, or light level. Cornell University ecologist Robert Whittaker pioneered gradient analysis in North America, and his work was influential in putting to rest Clements's extreme view of closed communities. Whittaker conducted most of his work in mountainous areas, where moisture and temperature vary over short distances according to elevation, slope, and exposure. These variables in turn determine light, temperature, and moisture levels at a particular site.

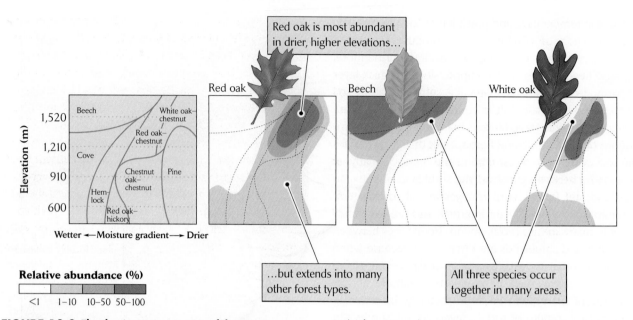

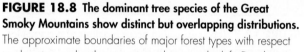

FIGURE 18.8 The dominant tree species of the Great Smoky Mountains show distinct but overlapping distributions. The approximate boundaries of major forest types with respect to elevation and soil moisture are shown at the left. Distributions of red oak, beech, and white oak, however, are not limited to the forest types bearing their names. Relative abundances were measured as the percentage of stems more than 1 cm in diameter of the focal species in samples of approximately 1,000 stems. After R. H. Whittaker, *Ecol. Monogr.* 26:1–80 (1956).

In the Great Smoky Mountains of Tennessee, Whittaker found that the dominant tree species have distinct but partially overlapping ecological distributions on a gradient of moisture and elevation, and also occur widely outside the plant associations that bear their names (Figure 18.8). For example, red oak grows most abundantly in relatively dry sites at high elevations, but its distribution extends into forests dominated by beech, white oak, chestnut, and even hemlock (an evergreen conifer).

In fact, red oak is found throughout the entire range of elevation in the Great Smoky Mountains. Beech prefers moister conditions than red oak, and white oak reaches its greatest abundance in drier conditions, but all three species occur together in many areas. In the Santa Catalina Mountains of southern Arizona, Whittaker found that plant species occupied unique ecological distributions, with their peaks of abundance scattered along a moisture gradient (Figure 18.9). These distributions are shaped by the adaptations of the species to environmental conditions and the interactions among species that compete with one another along the gradient.

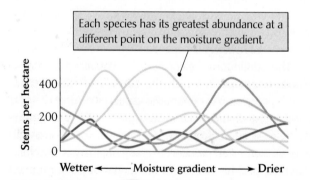

FIGURE 18.9 Many gradient analyses have revealed open community structures. Whittaker found such a pattern when he recorded the distributions of plant species along a soil moisture gradient in the Santa Catalina Mountains of southeastern Arizona at 1,830–2,140 m elevation. After R. H. Whittaker and W. A. Niering, *Ecology* 46:429–452 (1965).

Measures of community structure include numbers of species and trophic levels

Understanding how communities vary from place to place is the first step to understanding the processes that influence the structure and functioning of ecological systems. One of the simplest and most revealing measures of a community's structure is the number of species it includes. This measure is often referred to as **species richness.**

Biologists have hardly cataloged all the species of plants and animals on earth, let alone microbes. About 1.5 million species have been described and named worldwide; estimates of the total run upward into the tens of millions, not including bacteria and viruses, whose enormous variety has only recently been revealed by DNA sequencing of soils and waters. Because many species of plants and animals—many of them unknown to science—are becoming rare or extinct, ecologists are urgently trying to learn why some communities are more biologically diverse than others and to find ways to preserve as much of this natural heritage as possible.

Naturalists have known for centuries that more species live in tropical regions than in temperate and boreal zones. For example, when ecologists counted all the trees, shrubs, and saplings on a 50-hectare plot on Barro Colorado Island, a 16 km² island in Gatun Lake, Panama (Figure 18.10), they found more than 300 species among the 240,000 individuals with a diameter of 1 centimeter or more. This number exceeds the number of tree species found in all of Canada. Plots of only 1 hectare in some parts of Amazonian Peru and Ecuador contain more than 300 species; every other individual tree in such a plot belongs to a different species! Like forest trees, most types of organisms exhibit their highest species richness in the tropics.

Even the simplest biological communities contain overwhelming numbers of species. To manage this complexity and to characterize the structure of communities more fully, ecologists often partition diversity into numbers of species fulfilling different roles in the functioning of communities and ecosystems. One simple way to partition species is with respect to their feeding relationships. Each species can be placed on one of several **trophic levels** within the community, so named because those levels correspond to different points in the chain of consumer–resource feeding relationships (*trophos* is Greek for "nourishment"). Plants and other **autotrophic** ("self-nourishing") organisms, which are the **primary producers** in the ecosystem, occupy the bottom level. At the next level are consumers of primary producers—herbivores from ants to zebras—which are called **primary consumers.** Several levels of carnivores—**secondary consumers,** tertiary consumers, and so on—reside on the levels above primary consumers. Detritivores are difficult to place on a particular trophic level, but they nonetheless fill a distinct role in ecosystems and can be considered an ecological group.

Within trophic levels, ecologists have used the types of resources consumed and the methods or locations of foraging to place species in **guilds,** which are groups of species that feed on similar resources, and often have similar ways of life. The members of guilds need not be closely related. For example, herbivore species might be placed in guilds of leaf eaters, stem borers, root chewers, nectar sippers, or bud nippers. In deserts of the southwestern United States, many species of rodents, ants, and birds consume seeds and thus are placed together in a seed-eating guild.

FIGURE 18.10 Tropical forests harbor the greatest species richness of any communities. The number of different tree species on Barro Colorado Island, Panama, is obvious even in this aerial photograph. Photo by Carl C. Hansen, courtesy of the Smithsonian Tropical Research Institute.

Feeding relationships organize communities in food webs

Ecologists use feeding relationships to describe the structure of a community. Consider a single food chain consisting of several trophic levels: At the bottom is a plant, which is fed on by a particular type of caterpillar. The caterpillar is in turn eaten by a robin, which may be preyed on by a cat. Each of these consumers obtains the energy and nutrients it needs from its food resources. Thus, energy and nutrients can be said to travel up the food chain, from the plant to the cat. Feeding relationships are more complex than this simple picture suggests, however. Most consumers eat only a few of the many kinds of resources available to them, yet rare is the resource eaten by only a single type of consumer. Energy and nutrients follow many different, interconnected paths through the ecosystem, collectively referred to as a **food web.** The feeding interactions represented by the food web have profound effects on the species richness of the community and on its productivity and stability.

Effects of species richness on food web structure

As we have seen, two important attributes of communities are species richness and the feeding relationships captured in food webs. Is there any relationship between these two attributes? The two food webs in Figure 18.11 portray similar numbers of species organized in strikingly different relationships. The intertidal mudflat community (Figure 18.11a) is relatively simple, having seven links among the seven species portrayed in the diagram, with only one species consuming resources on more than one trophic level. By contrast, the plant–insect–parasitoid community (Figure 18.11b) is complex; it exhibits twelve links among eight species and several cases of **omnivory** (feeding on more than one trophic level).

We can rank the complexity of a food web by its number of feeding links and trophic levels. Comparisons of food webs suggest that the number of feeding links *per species* is independent of the species richness of the community, as shown in Figure 18.12 for assemblages of invertebrates that form in the water trapped by pitcher plants. Thus, the number of interactions that each species has with other species is independent of the overall diversity of the community. However, another generalization that has emerged from these comparisons is that the number of trophic levels, and the number of guilds within trophic levels, increases with species richness. This trend is also

apparent in the food webs of pitcher-plant communities. Thus, increasing species richness is usually associated with increasing food web complexity. How these attributes of communities are determined and their consequences for community functioning remain active areas of ecological research.

Effect of food web structure on species diversity

We have just seen that species diversity can increase food web complexity, but, conversely, feeding relationships can affect species diversity within a community. For example, when a predator controls the population of an otherwise dominant competitor, it may allow inferior competitors to persist. Thus, the number of species on a particular trophic level within a food web may depend on consumption by species at higher trophic levels.

Robert Paine of the University of Washington was one of the first ecologists to address the relationship between food web organization and community diversity. Recall from Chapter 16 the experiment in which he removed predatory sea stars from areas of rocky shoreline along the coast of Washington. In response, the primary prey of the sea stars, the mussel *Mytilus,* spread rapidly, crowding other organisms out of the experimental plots and reducing the diversity and complexity of local food webs, particularly the diversity of herbivores (see Figure 16.17). In another experiment, Paine showed that the same principle applies to the diversity of primary producers. Removal of the sea urchin *Strongylocentrotus,* an herbivore, allowed a small number of competitively superior algae to domi-

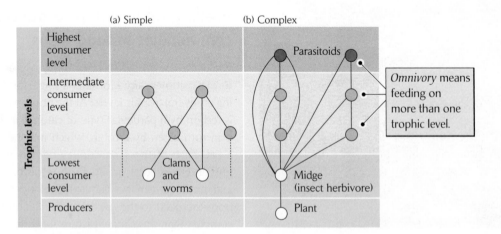

FIGURE 18.11 Different communities of similar species richness may have different food web structures. (a) An intertidal mudflat community containing gastropods, bivalves, and their prey has a relatively simple food web, involving only one omnivorous species. (b) A food web based on the plant

Baccharis, its insect herbivores, and their parasitoids is more complex, involving several omnivorous species. In food web diagrams such as these, lines connect the resources below to the consumer above. Not all prey species are depicted. After S. L. Pimm, *Food Webs,* Chapman & Hall, London and New York (1982).

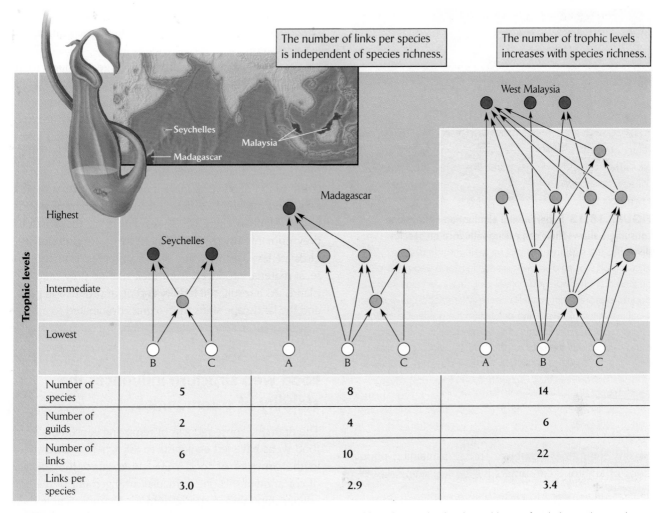

FIGURE 18.12 Increasing species richness is associated with increasing food web complexity. These food web diagrams for communities of invertebrates living in *Nepenthes* pitcher plants in different regions bordering on the Indian Ocean show increasing ecological diversity (more trophic levels and guilds within trophic levels) and longer food chains, but similar numbers of feeding links per species, with increasing species richness. Sources of food are live insects (A), recently drowned insects (B), and older organic debris (C). After R. A. Beaver, *Ecol. Entomol.* 10:241–248 (1985).

nate an area, crowding out many ephemeral or grazing-resistant species.

John Terborgh, of Duke University, and his colleagues described a spectacular example of this consumer effect in an unintended predator removal experiment within the rain forests of Venezuela. Waters rising behind a dam built for hydroelectric power isolated a number of very small patches (0.25–0.9 ha) of forest on hills, now islands surrounded by water. These patches were too small to support predators of the larger herbivores, including howler monkeys and green iguanas, so populations of those herbivore species skyrocketed. Howler monkeys reached densities equivalent to 1,000 individuals per km^2 on predator-free islands, compared with 20–40 per km^2 on adjacent mainland sites. Armadillos disappeared from the islands,

so leaf-cutter ants were also freed of predators, and their densities increased from less than 1 colony per 4 hectares to between 1 and 7 colonies per hectare. This ecosystem "meltdown" strikingly affected the regeneration of trees. Undisturbed mainland forests contain 200–400 small saplings per 500 m^2; on the isolated islands, saplings were reduced to as few as 39 stems per 500 m^2 and averaged only 136 stems. As a result, forest regeneration was severely disrupted on the islands, and both productivity and plant diversity plummeted.

Many such predator removal experiments show that some consumers can maintain diversity among resource species and thereby influence the structure of a community (Figure 18.13). Such species are called **keystone consumers** because when they are removed, the edifice

FIGURE 18.13 Experimental elimination of keystone consumers shows their controlling influence on species diversity. The experimental plot on the right side of the photograph was sprayed with insecticide for 8 years; the plot on the left is an unsprayed control plot. The insecticide kept populations of the chrysomelid beetle *Microrhopapla vittata* from reaching outbreak levels and defoliating the goldenrod *Solidago altissima,* its preferred food plant. Consequently, goldenrod came to dominate the sprayed plot and shaded out the many other species growing in the more diverse control plot. Courtesy of Walter Carson, from W. P. Carson and R. B. Root, *Ecol. Monogr.* 70:73–99 (2000).

of the community tumbles. Thus, maintaining populations of keystone consumers is vital to the stability of a community.

A variety of food web types

The consumer–resource relationships represented in food webs are key to understanding community organization. Robert Paine distinguished different types of food webs that describe different ways in which species influence one another within communities. **Connectedness webs** emphasize feeding relationships among species, portrayed as links in a food web. **Energy flow webs** represent an ecosystem viewpoint, in which the connections between species are quantified by the flux of energy between a resource and its consumer. In **functional webs,** the importance of each species in maintaining the integrity of a community is reflected in its influence on the growth rates of other species' populations. This controlling role, which can be revealed only by experiments, need not correspond to the amount of energy flowing through a particular link in the food web, as shown for the intertidal food web in Figure 18.14. Note that some consumers, such as the limpets *Acmaea pelta* and *A. mitra* and the chiton *Tonicella,* ingest considerable food energy, but removal of these consumers has no detectable effect on the abundance of their resources. The most effective control was

exerted by the sea urchin *Strongylocentrotus* and the chiton *Katharina,* which can be considered keystone consumers in this system.

Some influences travel unusual routes through food webs. For example, in one study of periwinkle snails (*Littorina*) grazing on algae on rocky surfaces, investigators noticed that snails infected with trematode worms became less efficient grazers, probably because the parasites interfered with digestion. The presence of parasites resulted in much higher algal densities and a change in the species composition of algae. A more indirect sequence of events explained the effect of foxes on the vegetation of islands in the Aleutian chain of Alaska. Foxes on these islands prey primarily on seabirds. Where foxes are present, seabirds are less common, and they transfer less nutrient-rich fecal material produced from their marine prey onto the islands. As a result, soil fertility and plant production drop, and the landscape shifts from a grass-dominated to a forb- and shrub-dominated plant community (Figure 18.15).

Food web structure influences the stability of communities

The dramatic consequences of removing consumers from food webs have led ecologists to ask whether differences in the structure of food webs could affect the stability of communities. Is one particular arrangement of feeding relationships among species intrinsically more stable than a different arrangement among the same number of species? How important is food web stability to the structure of communities?

Stability, of course, has many meanings. Stability over ecological time has two essential components, constancy and resilience. **Constancy** is a measure of the ability of a system to resist change in the face of outside influences; indeed, constancy is sometimes referred to as *resistance.* **Resilience** is the ability of a system to return to some reference state after a disturbance. Resilience, like density-dependent population regulation, implies that the system has internal processes that can compensate for disturbance-induced change. In the case of populations, increasing birth rates help to restore a population to its carrying capacity after a drop in numbers. Resilience in communities also depends on changes in birth and death rates, but with the additional influence of interactions among species.

An experiment conducted on northern California grasslands illustrates the complex responses of communities to environmental perturbations. This experiment tested the resilience of the grassland community in response to climate change. Kenwyn Suttle and Meredith Thomsen, in

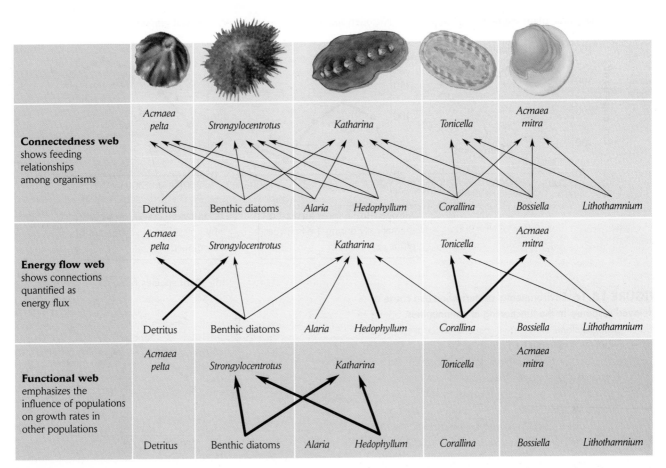

	Acmaea pelta	Strongylocentrotus	Katharina	Tonicella	Acmaea mitra
Connectedness web shows feeding relationships among organisms	Detritus	Benthic diatoms	Alaria Hedophyllum	Corallina	Bossiella Lithothamnium
Energy flow web shows connections quantified as energy flux	Detritus	Benthic diatoms	Alaria Hedophyllum	Corallina	Bossiella Lithothamnium
Functional web emphasizes the influence of populations on growth rates in other populations	Detritus	Benthic diatoms	Alaria Hedophyllum	Corallina	Bossiella Lithothamnium

FIGURE 18.14 Ecologists use three approaches to depicting trophic relationships. Three types of food web diagrams, here applied to the species of a rocky intertidal zone on the coast of Washington, depict different ways in which species influence one another within communities. The thickness of an arrow reflects the strength of that relationship. After R. T. Paine, *J. Anim. Ecol.* 49:667–685 (1980).

(a)

(b)

FIGURE 18.15 Fox predation on seabirds transforms plant communities on a subarctic island. (a) On Buldir Island, where foxes are absent, the plant community is dominated by grasses. (b) On Ogangan Island, where foxes are present, the community is dominated by shrubs and forbs. From D. A. Croll et al., *Science* 307:1959–1961 (2005).

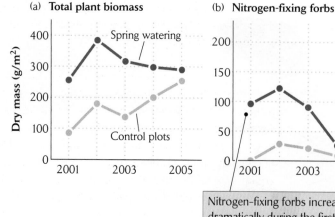

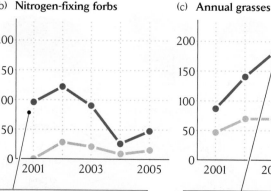

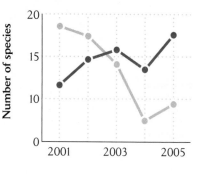

FIGURE 18.16 Environmental perturbation can cause a delayed response in the functioning of communities. Spring watering of grassland plots growing in a Mediterranean climate zone of northern California had a number of effects. Watering increased production (measured as plant biomass; a) and initially shifted dominance to nitrogen-fixing forbs (b) from annual grasses (c). After several years, production returned to control levels, and the dominance of grasses was restored, but the number of plant species (d) remained greatly reduced by the perturbation. After Suttle, Thomsen, and Power, *Science* 315:640–642 (2007).

the laboratory of Mary Power at the University of California at Berkeley, watered grassland plots in the late spring over a 5-year period, in effect extending the normal rainy season in this Mediterranean climate zone. Production, measured as plant biomass at the end of the growing season, increased greatly in the watered plots over that in unwatered control plots. The initial surge in production was particularly evident in nitrogen-fixing forbs, which became more productive than annual grasses in the system. After 4 or 5 years, however, production returned to the level in the control plots, and annual grasses regained their dominance (Figure 18.16). Evidently, adjustments in the species composition of the communities caused a return to the original balance of production by nitrogen-fixing forbs and by grasses. However, plant species richness on the experimental plots dropped markedly. In this experiment, the system required several years to settle into a new pattern of ecosystem function. Regardless of the apparent resilience of some attributes of the ecosystem, the results of the experiment were alarming in suggesting that any extension of the rainy season resulting from climate change is likely to lead to a reduction of species richness in the grassland habitat.

ECOLOGISTS IN THE FIELD **Does species diversity help communities bounce back from disturbance?** As human activities increasingly alter the species diversity and food web structure of biological communities, it is important to know the effects of these changes on the stability of ecological systems. When a community is perturbed—for example, by a fire or an epidemic disease sweeping through it—how rapidly can the system return to its original, unperturbed state? Ecologists are especially interested in knowing whether the removal of species from a community, particularly keystone species, reduces its resilience in the wake of perturbation.

Chris Steiner and his colleagues at Rutgers University tackled this question by assembling communities of organisms in small water bottles in the laboratory. The communities were simple, consisting of producers (algae), detritivores (bacteria), consumers of algae and bacteria (protozoans), and omnivorous predators (rotifers). To vary the species diversity and food web structure of the communities, the researchers selected one, two, or four species from each trophic level, and then raised these low-, medium-, and high-diversity communities with low and high nutrient supplies. After 3 weeks, the number of species in each of the diversity treatments had settled to about

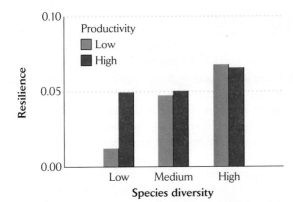

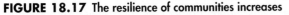

FIGURE 18.17 The resilience of communities increases with diversity. The resilience of communities of algae, bacteria, protozoans, and rotifers in laboratory microcosms under conditions of low nutrient availability is greater in communities with more species. Resilience is measured as the daily rate of return to control (unperturbed) biomass levels. After C. F. Steiner et al., *Ecology* 87(4):996–1007 (2006).

four, six, and eight species, respectively, in both the high-productivity (high-nutrient) and low-productivity (low-nutrient) microcosms. Next, to quantify resilience, the investigators simulated a major episode of mortality by removing 90% of the organisms from some of the microcosms in each treatment group, then measured how rapidly the total biomass of those communities returned to the levels of the undisturbed controls.

In the high-productivity treatments, resilience differed little between the low-, medium-, and high-diversity communities (Figure 18.17). In contrast, in the low-productivity treatments, resilience was positively related to the number of species in the community. Resilience appeared to depend primarily on the rapid reproductive rates of some producer species of algae, which were able to increase quickly, providing an abundant food resource to rebuild populations at higher trophic levels. Most producers attained high growth rates under the high-productivity treatments, but the nutrient concentrations in the low-productivity treatments limited the growth of some algal species. The greater resilience of the more diverse communities might not reflect species diversity per se, in the sense that different species complement one another in contributing to the functioning of the system. Instead, more diverse communities might just by chance include one or more species that can maintain rapid growth under low-nutrient conditions.

Such controlled experiments with laboratory microcosms offer useful insights into the role that species diversity plays in determining community resilience. However, because laboratory experiments greatly simplify nature, ecologists need to examine these relationships further under more natural conditions. Species losses in many

parts of the world heighten the importance of understanding how diversity influences the ability of communities to respond to perturbations. ▍

There are a number of ways in which species diversity could influence community stability. On one hand, a more complex food web structure might increase the stability of a community if predators have alternative prey, in which case their population sizes should depend less on fluctuations in the numbers of a particular prey species. In addition, where energy can take many routes through a system, disruption of one pathway should merely shunt more energy through another. On the other hand, as communities become more diverse, species exert greater influences on one another through their various interactions; these biological links, in turn, might create destabilizing time delays in population processes (see Chapters 12 and 15).

The structures of natural food webs, such as those shown in Figures 18.11 and 18.12, vary tremendously. Yet we presume that each of these food webs has persisted over long periods of ecological and even evolutionary time, meaning that all are essentially stable, perhaps with different balances of constancy and resilience. Does variation in food web structure mean that the rules of food web stability depend on particular organisms and ecological circumstances? Or is stability not an important consideration, in which case food web structure, rather than being selected for stability, merely reflects the feeding relationships of the individual species that form a community?

Communities can switch between alternative stable states

A resilient system is able to return to a "reference" state following a perturbation. Sometimes, however, a system can have more than one stable reference state. We discussed alternative stable states in Chapter 15 in reference to population regulation. There, we saw that a population could have an upper equilibrium state determined primarily by its resources and a lower equilibrium state determined by its predators and parasites. At either point, the population is stable, meaning that small perturbations are followed by return to the reference point.

Biological communities, which, after all, consist of multiple populations, can also exhibit alternative stable states. Switching an entire community between alternative stable states requires a more dramatic external perturbation, such as the removal of a keystone consumer. A community might have alternative stable states when members differ in their responses to an important environmental factor. For example, if one keystone species thrives over a lower temperature range than another keystone species, the character of the community might change in a stepwise fashion with climate warming as one keystone species replaces the other.

Such transitions between alternative stable states occur at the prairie–forest boundary in the midwestern United States. During years of abundant rain, fire is suppressed, and trees extend out onto the prairies, eventually shading out the prairie species and replacing the grassland with forest if moist conditions persist. During stretches of dry years, forests become drier and prairie fires more frequent. These fires can enter the forests and open them up for colonization by prairie plants. Thus, an environmental driving mechanism can shift the community between alternative states. These states tend to be stable once formed because forests suppress fire by retaining moisture, and prairie plants, whose underground rhizomes and root crowns resist the effects of burning, encourage fire through the accumulation of abundant flammable organic material above ground.

ECOLOGISTS IN THE FIELD

Mimicking the effects of ice scouring on the rocky coast of Maine. Several kinds of perturbations have been shown to shift communities between alternative stable states. For example, working in the intertidal zone on the rocky coast of Maine, University of Pennsylvania ecologists Peter Petraitis and Steve Dudgeon cleared plots dominated by the rockweed alga *Ascophyllum nodosum* to mimic the effect of winter ice scouring, which occurs frequently in the area. The cleared plots were 1, 2, 4, and 8 m² in area. The clearings—especially the largest ones—were quickly colonized by the brown alga *Fucus vesiculosus* and the barnacle *Semibalanus balanoides*, and their presence prevented the re-establishment of rockweed over the 5 years of the experiment.

The new community that formed following perturbation depended on the exposure of the rocky shore (Figure 18.18). Northward-facing sites became dominated by barnacles and southward-facing sites by *Fucus*. In this case, the switch to an alternative state depended on the poor colonization ability of *Ascophyllum* compared with *Fucus* and barnacles. It remained to be seen whether the cleared patches would return to stands dominated by

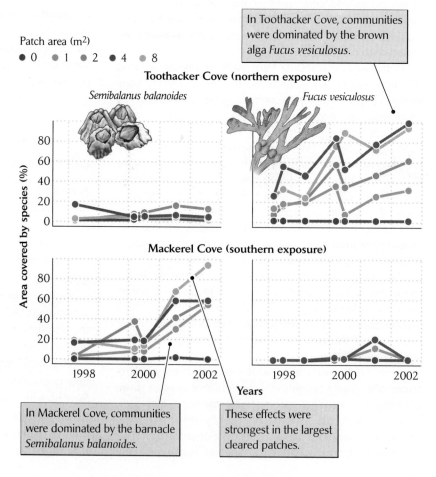

FIGURE 18.18 Removal of organisms in a New England rocky intertidal zone community results in replacement by one of several possible new communities. In this experiment, researchers cleared rockweed (*Ascophyllum nodosum*) from patches 1 m² to 8 m² in area and observed the communities that developed in those patches over 5 years. After P. S. Petraitis and S. R. Dudgeon, *J. Exp. Mar. Biol. Ecol.* 326:14–26 (2005).

Ascophyllum over longer periods as the rockweed gradually competitively excluded the *Fucus* and barnacles. However, periodic ice scouring should clearly maintain the region as a mosaic of different types of communities existing under otherwise identical environmental conditions. ▌

Trophic levels are influenced from above by predation and from below by production

We have seen in Chapters 14 and 15 that predators can depress populations of their prey species dramatically. This principle can apply equally well to entire trophic levels. In a classic paper published in 1960, three University of Michigan ecologists, Nelson Hairston, Frederick Smith, and Larry Slobodkin, suggested that the earth is green because carnivores depress the populations of herbivores that would otherwise consume most of the vegetation. When the indirect effects of consumer–resource interactions extend through additional trophic levels of a community, this phenomenon is called a **trophic cascade** (Figure 18.19). When higher trophic levels determine the sizes of the trophic levels below them, the situation is referred to as **top-down control.** When the size of a trophic level is determined by the rate of production of its food resource, the situation is referred to as **bottom-up control.**

Ecologists have debated the relative strengths of top-down and bottom-up control mechanisms for years. When experimenters remove predators from a community, herbivore populations often increase so rapidly that they decimate the plant resources they feed on. We saw an example earlier in this chapter on islands isolated by rising waters in Venezuela that were too small to sustain top predators. Alternatively, plants may control herbivore populations from the bottom up, so to speak, by resisting consumption by means of various secondary compounds (see Chapter 14). Accordingly, we might expect to find top-down control in many aquatic ecosystems because aquatic plants and algae, especially phytoplankton, are much more edible than most terrestrial vegetation (see Chapter 22).

A survey of zooplankton and phytoplankton densities in natural lakes by Mathew Leibold and his colleagues at the University of Chicago showed that the density of zooplankton (the primary consumers) varied in parallel with the density of phytoplankton (the producers), a pattern that is consistent with bottom-up control (Figure 18.20a). When the researchers added predatory fish to the experimental lakes, however, they decreased the density of zooplankton, and phytoplankton abundance increased in most cases, sometimes by a factor of more than 10, indicating top-down control (Figure 18.20b). These results suggest that production generally determines the density of populations feeding at higher trophic levels in aquatic

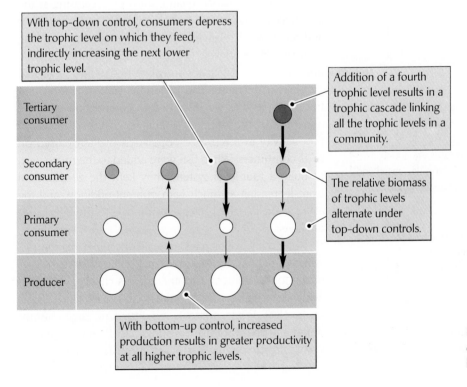

With top-down control, consumers depress the trophic level on which they feed, indirectly increasing the next lower trophic level.

Addition of a fourth trophic level results in a trophic cascade linking all the trophic levels in a community.

The relative biomass of trophic levels alternate under top-down controls.

With bottom-up control, increased production results in greater productivity at all higher trophic levels.

FIGURE 18.19 The trophic structure of a community may be determined by bottom-up or top-down control.

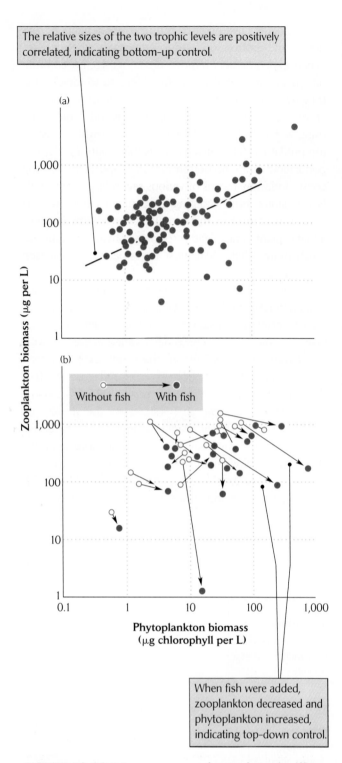

The relative sizes of the two trophic levels are positively correlated, indicating bottom-up control.

When fish were added, zooplankton decreased and phytoplankton increased, indicating top-down control.

FIGURE 18.20 Primary consumer density shows the effects of both bottom-up and top-down influences. (a) Relationship between zooplankton biomass and phytoplankton biomass in natural lakes sampled over a range of productivity. (b) Introducing predatory fish to lakes reduces zooplankton populations and results in an increase in phytoplankton biomass. Arrows connect measures for the same lakes before and after the addition of fish. *After M. A. Leibold et al., Annu. Rev. Ecol. Syst. 28:467–494 (1997).*

ecosystems, but that top-down interactions can nonetheless adjust the sizes of trophic levels within a narrower range.

Lars-Anders Hansson and his colleagues at the University of Lund in Sweden investigated bottom-up effects on community structure in aquatic ecosystems by adding inorganic nutrients (phosphorus and nitrogen) to experimental microcosm communities to boost their productivity. The experimental systems were established in hundreds of large cylindrical tanks in a greenhouse and stocked with either three trophic levels (bacterial detritivores, photosynthetic flagellates and algae, and zooplankton) or four trophic levels (adding fish as predators on zooplankton) (Figure 18.21a). The results of the experiment revealed both bottom-up and top-down control. In both the three-level and four-level systems, adding inorganic nutrients increased the densities of most of the trophic levels in the system. However, when fish were added (the fourth trophic level), zooplankton levels decreased in both the low- and high-productivity treatments, and densities of producers increased (Figure 18.21b).

Thus, like Leibold's comparative survey and experiments on natural lakes, Hansson and colleagues' microcosm experiments provided evidence that increased productivity tends to increase the density of all higher trophic levels. However, the experiments also showed that consumers could depress the size of the trophic level immediately below them and bump up populations two levels below. In the three-level system, zooplankton grazing shifted the dominance relationships of organisms at the producer trophic level. At low nutrient levels, flagellates and algae were relatively more abundant than bacteria; at high nutrient levels, increasing zooplankton populations depressed flagellates and algae and allowed bacterial densities to increase. This experiment also demonstrates the principle that we discussed in conjunction with the dynamics of predation (see Chapter 15), that increased productivity in a resource population is often passed on to its consumers. When fish were added to the microcosm to build a four-level system, they kept the zooplankton from increasing as much with nutrient addition, and the algae as well as the bacteria responded to the high nutrient levels.

Top-down trophic cascades also occur in marine systems. Copepods, for example, may switch between alternative food resources as nutrient levels change, setting off a massive change in the entire food web (Figure 18.22, on p. 388). When nutrients are abundant, large diatoms dominate the producer trophic level, and copepods switch from feeding on ciliate protists to feeding on

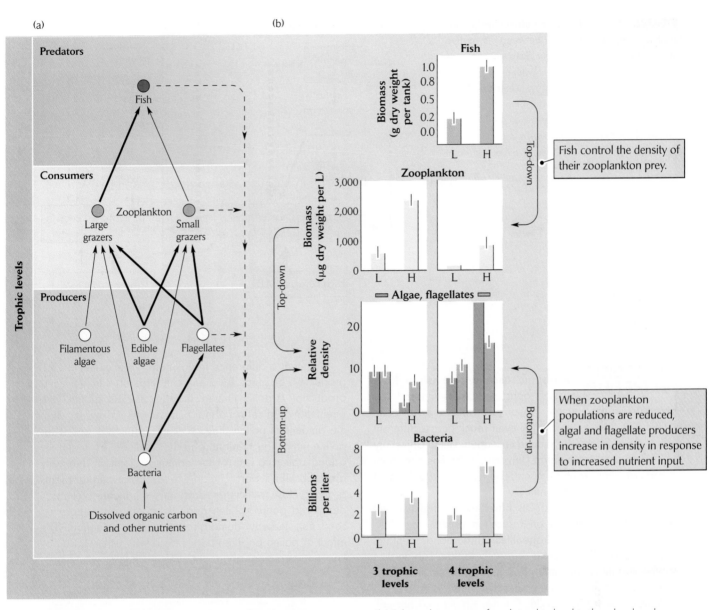

FIGURE 18.21 Community structure and its response to changes in productivity depend on the number of trophic levels. (a) Interactions between the trophic levels in Hansson et al.'s microcosms. Thick arrows represent strong interactions; thin arrows, weak interactions. Dashed arrows show excretion.

(b) Relative biomasses of each trophic level in three-level and four-level experimental microcosms with low (L) and high (H) nutrient inputs. After L.-A. Hansson et al., *Proc. R. Soc. Lond.* B 265:901–906 (1998).

diatoms. As a result, ciliate populations increase, and their principal food resources, small types of algae, are greatly reduced. Low nutrient levels favor the growth of small algae over diatoms, so copepods switch to feeding on ciliates, and the drop in the ciliate population releases the small algae from consumer pressure. Hence, changes in nutrient levels can switch the system between alternative stable states. Copepods themselves suffer predation by jellyfish, whose top-down effects also depend on the

nutrient status of the system. At high nutrient levels, jellyfish predation on copepods reduces their consumption of large diatoms, which then dominate the system and maintain high ecosystem productivity. At low nutrient levels, jellyfish predation on copepods reduces their consumption of ciliates, which results in a reduction of small algal biomass. In this case, the top predator—the jellyfish—has contrasting effects on the productivity and the producer biomass of the system.

FIGURE 18.22 Change in nutrient levels can switch a marine community between alternative stable states. (a) High nutrient levels favor large diatoms as primary producers, which are fed on directly by copepods. (b) At low nutrient levels, copepods feed on ciliates, which themselves feed on small algae. Dashed arrows indicate indirect effects of predators on algae. Arrow thickness is proportional to effect size. After H. Stibor et al., *Ecol. Lett.* 7:321–328 (2004).

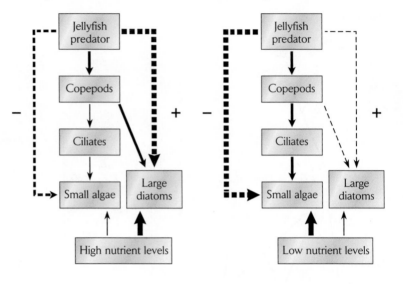

(a) **Large diatoms dominate**

(b) **Small algae dominate**

  ECOLOGISTS IN THE FIELD **A trophic cascade from fish to flowers.** Trophic cascades caused by predators in food webs appear to be pervasive in communities, but their effects may even reach across different ecosystems. Tiffany Knight and colleagues at the University of Florida and Washington University in St. Louis recently asked whether trophic cascades in ponds could affect nearby terrestrial communities.

For some time, ecologists have known that fish act as important predators on insects in ponds, including the aquatic larvae of dragonflies. Thus, ponds containing fish tend to have fewer dragonfly larvae. Larvae that survive eventually metamorphose into flying adults, which are important predators on other flying insects. In a study conducted near Gainesville, Florida, Knight and her colleagues compared both larval and adult dragonfly abundances in and around four ponds that had been stocked with fish and four ponds that lacked fish. The ponds were separated by an average distance of 1,000 meters. Larvae were sampled by sweeping the ponds with nets, and adult population densities were assessed by visual observations. As expected, ponds with fish produced fewer larval and adult dragonflies than ponds without fish (Figure 18.23).

But does this difference in dragonfly abundance have any effect on the terrestrial community? Many of the prey caught by adult dragonflies, including bees, flies, and butterflies, are pollinators of plants. The investigators reasoned that if dragonflies depressed the populations of these pollinators, flowers in the vicinity of ponds without fish would receive fewer pollinator visits than flowers close to ponds stocked with fish. That is exactly what they found (Figure 18.24).

Whether a difference in pollinator visits influences the production of seeds by plants depends on whether seed production is pollen-limited. In many studies, plants have been shown to set a normal number of seeds even when the amount of pollen they receive is reduced. To test for pollen limitation, Knight and her colleagues manually added pollen to one of the common plants in the study area, St. John's wort (*Hypericum fasciculatum*). For plants growing close to fish-free, dragonfly-enhanced, pollinator-depressed ponds, pollen addition significantly improved seed set, thereby demonstrating pollen limitation. The effect of pollen addition near ponds with fish, where pollinators were more abundant, was much less.

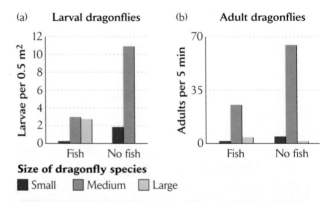

FIGURE 18.23 The presence or absence of fish in ponds influences dragonfly densities. Ponds with fish produced fewer larval dragonflies (a) and fewer adult dragonflies (b) than ponds without fish. After T. M. Knight et al., *Nature* 437:880–883 (2005).

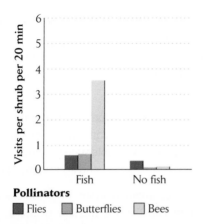

FIGURE 18.24 The presence or absence of fish in ponds can influence nearby terrestrial communities. Pollinators paid more visits to individuals of a common plant species (St. John's wort) that lived on the edges of ponds with fish. After T. M. Knight et al., *Nature* 437:880–883 (2005).

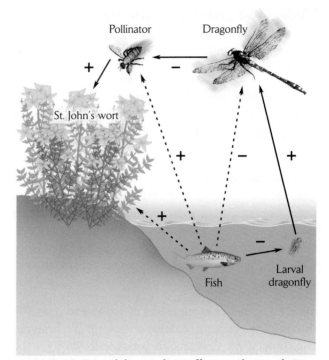

FIGURE 18.25 Fish have indirect effects on the populations of several species in and around ponds. The solid arrows represent direct effects, and the dashed arrows indirect effects; the nature of the effect is indicated by a + or −. Fish have indirect effects, through a trophic cascade, on several terrestrial species: dragonfly adults (−), pollinators (+), and plants (+). After T. M. Knight et al., *Nature* 437:880–883 (2005).

From this series of experiments, the researchers were able to demonstrate that the presence of fish in a pond reduced the abundance of dragonfly larvae, which reduced the abundance of dragonfly adults, and thereby increased the abundance of pollinators and the number of seeds produced by nearby plants. In short, via a complex trophic cascade, adding fish to a pond improved the reproductive success of a plant on land (Figure 18.25). ▌

SUMMARY

1. A biological community is an association of interacting populations. Questions about communities address the evolutionary origins of community properties, relationships between community organization and stability, and the regulation of species diversity.

2. Ecologists characterize communities in terms of the number of species, their organization into guilds of species using similar resources, and food webs portraying feeding relationships among species.

3. Communities may form discrete units separated by abrupt transitions in species composition along spatial transects or gradients of ecological conditions. This pattern is known as a closed community structure. More commonly, however, species are distributed over ecological gradients independently of the distributions of other species. Ecologists refer to this pattern as an open community structure.

4. Regions of rapid replacement of species, called ecotones, sometimes occur at sharp physical boundaries or

accompany changes in the growth forms that dominate a habitat. The aquatic–terrestrial transition provides an example of the first kind of ecotone, the prairie–forest transition an example of the second.

5. The distribution of species along an environmental gradient, referred to as the continuum concept, can be visualized by a gradient analysis, in which the abundance of each species is plotted on a gradient of one or more environmental conditions. The results of gradient analyses emphasize the open structure of most communities.

6. A simple measure of a community is its number of species, or species richness. Within communities, species can be organized into trophic levels, which correspond to different points in the chain of consumer–resource feeding relationships. Within trophic levels, species can be further organized into guilds based on similar food resources and ways of life.

7. Community structure can be depicted by means of food web diagrams showing the feeding relationships among species within a community. The complexity of a food web can be characterized by the number of feeding links per species and the average number of trophic levels on which a species feeds.

8. Consumer removal experiments have shown that keystone consumers can maintain diversity among resource species and thereby influence the structure of a community.

9. The influences of species on one another in communities can be described by three different types of food webs. Connectedness webs portray feeding relationships among species, energy flow webs show the energy flux between a resource and its consumer, and functional webs show the influence of a species on the growth rates of other species' populations.

10. The stability of communities in response to change in the environment or the addition or removal of species has two components. Constancy depends on the ability of the community to resist change, whereas resilience is its ability to return to some reference state following perturbation. Losses of species and global climate change due to human activities are likely to alter community composition and structure.

11. Loss of a keystone consumer, a major disturbance, or a change in environmental conditions can shift a community from one alternative stable state to another, often with dramatic change in structure and species composition.

12. Experimental manipulations of trophic levels show that consumers can depress the size of the trophic level immediately below them, which indirectly increases populations two trophic levels below. This effect is referred to as top-down control. When the productivity of one trophic level affects the productivity of higher trophic levels, the effect is known as bottom-up control. When the indirect effects of consumer–resource interactions extend through four or more trophic levels, the result is a trophic cascade.

REVIEW QUESTIONS

1. What are the differences between the community perspectives of Frederic E. Clements and Henry Gleason?

2. What are the conceptual parallels between the holistic community concept and closed communities? Between the individualistic community concept and open communities?

3. Given that the distributions of many animals are determined by the species composition of the plant community, what might you predict about the diversity of animals in the area surrounding an ecotone?

4. How did Robert Whittaker's data on tree species composition along gradients of temperature and moisture in mountains provide evidence against the idea of closed communities?

5. Compare and contrast connectedness food webs, energy flow food webs, and functional food webs.

6. Why might you expect that communities with fewer species would also contain fewer trophic levels?

7. If a community can exhibit alternative stable states, how might this affect our perspective on community resilience?

8. What does it mean when a community is said to exhibit top-down control as opposed to bottom-up control?

SUGGESTED READINGS

Brown, J. H., and E. J. Heske. 1990. Control of a desert–grassland transition by a keystone rodent guild. *Science* 250:1705–1707.

Carson, W. P., and R. B. Root. 2000. Herbivory and plant species coexistence: Community regulation by an outbreaking phytophagous insect. *Ecological Monographs* 70:73–99.

Croll, D. A., et al. 2005. Introduced predators transform subarctic islands from grassland to tundra. *Science* 307:1959–1961.

Grimm, V., and C. Wissel. 1997. Babel, or the ecological stability discussions: An inventory and analysis of terminology and a guide for avoiding confusion. *Oecologia* 109:323–334.

Halpern, B. S., et al. 2005. Predator effects on herbivore and plant stability. *Ecology Letters* 8:189–194.

Hansson, L.-A., et al. 1998. Consumption patterns, complexity and enrichment in aquatic food chains. *Proceedings of the Royal Society of London B* 265:901–906.

Knight, T. M., et al. 2005. Trophic cascades across ecosystems. *Nature* 437:880–883.

Leibold, M. A., et al. 1997. Species turnover and the regulation of trophic structure. *Annual Review of Ecology and Systematics* 28:467–494.

Litzow, M. A., and L. Ciannelli. 2007. Oscillating trophic control induces community reorganization in a marine system. *Ecology Letters* 10:1124–1134.

Pace, M. L., et al. 1999. Trophic cascades revealed in diverse systems. *Trends in Ecology and Evolution* 15:483–488.

Paine, R. T. 1980. Food webs: Linkage, interaction strength and community infrastructure. *Journal of Animal Ecology* 49:667–685.

Pimm, S. L. 1991. *The Balance of Nature?* University of Chicago Press, Chicago.

Polis, G. A., and D. R. Strong. 1996. Food web complexity and community dynamics. *American Naturalist* 147:813–846.

Power, M. E., et al. 1996. Challenges in the quest for keystones. *BioScience* 46:609–620.

Price, J. E., and P. J. Morin. 2004. Colonization history determines alternate community states in a food web of intraguild predators. *Ecology* 85:1017–1028.

Risser, P. G. 1995. The status of the science of examining ecotones. *BioScience* 45:318–325.

Schröder, A., L. Persson, and A. M. De Roos. 2005. Direct experimental evidence for alternative stable states: A review. *Oikos* 110:3–19.

Shurin, J. B., et al. 2002. A cross-ecosystem comparison of the strength of trophic cascades. *Ecology Letters* 5:785–791.

Shurin, J. B., et al. 2007. Diversity–stability relationship varies with latitude in zooplankton. *Ecology Letters* 10:127–134.

Steiner, C. F., et al. 2006. Population and community resilience in multitrophic communities. *Ecology* 87(4):996–1007.

Stibor, H., et al. 2004. Copepods act as a switch between alternative trophic cascades in marine pelagic food webs. *Ecology Letters* 7:321–328.

Suttle, K. B., M. A. Thomsen, and M. E. Power. 2007. Species interactions reverse grassland responses to changing climate. *Science* 315:640–642.

Terborgh, J., et al. 2001. Ecological meltdown in predator-free forest fragments. *Science* 294:1923–1926.

Whittaker, R. H. 1953. A consideration of climax theory: The climax as a population and pattern. *Ecological Monographs* 23:41–78.

Whittaker, R. H. 1967. Gradient analysis of vegetation. *Biological Reviews* 42:207–264.

Wilson, W. G., et al. 2003. Biodiversity and species interactions: Extending Lotka–Volterra community theory. *Ecology Letters* 6:944–952.

Wood, C. L., et al. 2007. Parasites alter community structure. *Proceedings of the National Academy of Sciences USA* 104:9335–9339.

CHAPTER 19

Ecological Succession and Community Development

On August 27, 1883, the island of Krakatau, in the Sunda Strait of present-day Indonesia, exploded after months of volcanic activity. Most of the island was blown away, and all life was obliterated. Huge tsunamis swept the coasts of nearby Sumatra and Java, killing tens of thousands of people. Immense quantities of ash filled the atmosphere, dimming the sun and creating spectacular red sunsets all over the globe, and sending temperatures to the coldest levels in years.

Once the sobering effects of the enormous catastrophe had waned, scientists recognized the immense value of Krakatau as a natural laboratory for studying the development of biological communities on a newly formed, raw terrain of volcanic ash. Expeditions to the remaining island fragments were mounted, and reports were filed on the appearance and establishment of plants and animals over the ensuing century. The nearest sources of colonists for Krakatau were on Sumatra and Java, about 40 kilometers distant.

As one might have expected, several sea-dispersed plants common on tropical shores throughout the region were the first plant species to show up on the island, making up

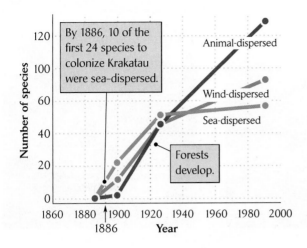

By 1886, 10 of the first 24 species to colonize Krakatau were sea-dispersed.

Animal-dispersed

Wind-dispersed

Sea-dispersed

Forests develop.

FIGURE 19.1 Plants dispersed by physical forces are the first to arrive in primary succession. This graph shows the number of plant species dispersed by sea, wind, and animals present on the island of Krakatau from 1883 through 1990. Data from R. J. Whittaker, http://www.geog.ox.ac.uk/research/biogeography/krakatau-2.htm.

10 of the 24 species that had colonized Krakatau by 1886 (Figure 19.1). Among the other pioneers were wind-dispersed grasses and ferns whose seeds and spores could be blown across the ocean. These grasses and ferns dominated the first plant communities to develop away from the beach on Krakatau.

Eventually, wind-dispersed tree species arrived. By the 1920s, closed forest had developed over most of Krakatau, and some of the pioneering species were pushed to marginal habitats or disappeared from the island. As forests developed, birds and bats were attracted to the island. Some of these were fruit-eating species that brought the seeds of animal-dispersed trees and shrubs with them. Indeed, most new arrivals in the flora of Krakatau after the 1920s were animal-dispersed plants, which today outnumber sea- and wind-dispersed species.

The vegetation of Krakatau will continue to change for many years as more plants invade the island and distinct plant communities develop in different habitats. Moreover, the island fragments that now make up Krakatau are constantly changing as a result of continuing volcanic eruptions, erosion of soft ash deposits, and storms that pass through the region. Krakatau will continue to be an important laboratory for studying the dynamics of community change.

CHAPTER CONCEPTS

- The concept of the sere includes all the stages of successional change

- Succession ensues as colonists alter environmental conditions

- Succession becomes self-limiting as it approaches the climax

Communities exist in a state of continuous flux. Organisms die and others are born to take their places; energy and nutrients pass through the community. Yet the appearance and composition of most communities do not change appreciably over time. Oaks replace oaks and squirrels replace squirrels in continual self-perpetuation. But when a habitat is **disturbed**—a forest cleared, a prairie burned, a coral reef obliterated by a hurricane, an island covered by volcanic ash—the community slowly rebuilds. Pioneering species adapted to disturbed habitats are successively replaced by other species as the com-

munity attains its former structure and composition (Figure 19.2). This is the classic view of community dynamics that dominated ecology through much of the twentieth century.

The sequence of changes initiated by disturbance is called **succession,** and the ultimate association of species achieved is called a **climax community.** These terms describe natural processes that caught the attention of early ecologists, including Frederic E. Clements. By 1916, Clements had outlined the basic features of succession, supporting his conclusions with detailed studies of change

(a)   (b)   (c)

(d)   (e)   (f)

FIGURE 19.2 Species successively replace one another in the process of succession. Stages of succession in an oak–hornbeam forest in southern Poland are shown from (a) immediately after clearing to (b) 7, (c) 15, (d) 30, (e) 95, and (f) 150 years thereafter. Photos by Z. Glowacinski, courtesy of O. Jarvinen. From Z. Glowacinski and O. Jarvinen, *Ornis Scand.* 6:33–40 (1975).

in plant communities in a variety of environments. Since then, the study of community development has grown to include detailed studies of the processes that underlie succession, the adaptations of organisms to the different conditions of early and late succession, interactions between colonists and the species that replace them, alternative pathways of succession depending on initial conditions, and continual change in the climax community itself.

The concept of the sere includes all the stages of successional change

The creation of any new habitat—a plowed field, a sand dune at the edge of a lake, an elephant's dung, a temporary pond left by a heavy rain—attracts a host of species particularly adapted to be good pioneers. These colonizing species change the environment of the new habitat. Plants, for example, shade the earth's surface, contribute detritus to the soil, alter soil moisture content, and some-

times exude toxic chemicals into the soil. These changes may inhibit the colonizing species that caused them, but may make the environment more suitable for the species that follow. In this way, the character of the community changes with time.

The opportunity to observe succession presents itself conveniently in abandoned agricultural fields of various ages. On the Piedmont of North Carolina, bare fields are quickly covered by a variety of annual plants (Figure 19.3). Within a few years, herbaceous perennials and shrubs replace most of the annuals. Shrubs are followed by pines, which eventually crowd out earlier successional species; pine forests are invaded in turn and then replaced by a variety of deciduous species that constitute the last stage of the successional sequence. Change comes rapidly at first. Crabgrass quickly enters an abandoned field, hardly allowing time for the plow's furrows to smooth over. Horseweed and ragweed dominate the field in the first summer after abandonment, aster in the second summer, and broomsedge in the third. The pace of suc-

FIGURE 19.3 Abandoned agricultural fields undergo a series of successional changes. This old field on the Piedmont of North Carolina is an example of the communities that develop after abandonment of agricultural land. Here, shrubs are beginning to replace annual plants and perennial herbs. Photo by Stephen Collins/Photo Researchers.

cession falls off as slower-growing plants appear: the transition to pine forest requires 25 years, and another century must pass before the developing deciduous forest begins to resemble the natural climax vegetation of the area.

The transition from abandoned field to mature forest is only one of several successional sequences that may lead to similar climax communities within a given biome. Each of these successional sequences is called a **sere.** The course of community development in each sere depends on its beginning. For example, the sequence of species that develops on newly formed sand dunes at the southern end of Lake Michigan in Indiana differs from the sequence of species that develops on abandoned fields a few miles away. Sand dunes are first invaded by marram and bluestem grasses. Individuals of these species growing at the edge of a dune send out rhizomes (runners) under the surface of the sand, from which new shoots sprout (Figure 19.4). These perennial grasses stabilize the dune's surface and add organic detritus to the sand. Annual herbs

(a)

(b)

(c)

(d)

FIGURE 19.4 Primary succession on sand dunes begins with the invasion of perennial grasses. These scenes are from Indiana Dunes State Park, on the south shore of Lake Michigan. Marram grass is the first invader of dunes, spreading from the

edge by means of underground rhizomes (a). Once the dunes have been settled by marram grass (b) and organic nutrients begin to accumulate, shrubs can become established (c). The shrubs are eventually replaced by trees (d). Photos by R. E. Ricklefs.

follow these grasses onto the dunes, further enriching and stabilizing the sandy soil and gradually creating conditions suitable for the establishment of shrubs: sand cherry, dune willow, bearberry, and juniper. These shrubs are followed by pines, but the pines do not reseed themselves well after their initial establishment. After one or two generations, the pines give way to the forests of beech, oak, maple, and hemlock that are characteristic of other soils in the region. In the same area, succession beginning in a marsh also ends in beech–maple forest as the wetland fills in with sediment and plant detritus and progressively dries out. Thus, although the initial stages of the sere depend on the habitat where it begins, the influence of the starting conditions—whether sand dune, marsh, or abandoned field—wanes with time as the seres converge on similar climax communities.

Primary succession

Beginning with Clements's classic work on succession, published in 1916, ecologists have classified seres into two types according to their origin. **Primary succession** is the establishment and development of communities in newly formed or disturbed habitats previously devoid of life—sand dunes, lava flows, rock bared by erosion or landslides or exposed by receding glaciers. Primary succession is the process that took place on Krakatau after all life was obliterated there. The regeneration of a community following a disturbance is called **secondary succession.** The distinction between the two types of seres blurs, however, because disturbances vary in the degree to which they destroy the fabric of a community and its physical support systems. A tornado that levels a large area of forest usually leaves untouched soil nutrients, seeds, and living roots, so succession follows quickly. In contrast, a severe fire may burn through organic layers of the forest soil, destroying the results of hundreds or thousands of years of community development.

A striking example of primary succession is the natural conversion of ponds in north-temperate and boreal climates to dry land. Retreating glaciers left deep ponds called *kettle holes* where large chunks of ice formed depressions and then melted. Even today, new ponds are formed behind beaver dams. These ponds undergo a char-

(a)

(b)

(c)

FIGURE 19.5 Some ponds undergo bog succession. In this bog forming behind a beaver dam in Algonquin Provincial Park, Ontario, Canada (a), the open water in the center (b) is stagnant, poor in minerals, and low in oxygen. These conditions result in accumulation of detritus and lead to a gradual filling in of the bog, which passes through stages dominated by shrubs and, later, black spruce (c). Photos by R. E. Ricklefs.

acteristic pattern of change known as *bog succession,* which begins when rooted aquatic plants become established at the edge of a pond (Figure 19.5). Some species of sedges form mats on the water surface extending out from the shoreline. Occasionally these mats grow completely over a pond, producing a more or less firm layer of vegetation over the water surface—a so-called "quaking bog."

Detritus produced by the sedge mat accumulates as layers of organic sediment on the bottom of the pond, where the stagnant water contains little or no oxygen to sustain microbial decomposition. Eventually these sediments become peat, which is used by humans as a soil conditioner and sometimes as a fuel for heating (Figure 19.6). As a bog accumulates sediments and detritus, sphagnum moss and shrubs, such as Labrador tea and

cranberry, become established along the edges, themselves adding to the development of a soil with progressively more terrestrial qualities. At the edges of the bog, shrubs may be followed by black spruce and larch, which eventually give way to birch, maple, and fir, depending on the locality. In this way, what started as an aquatic habitat is transformed over thousands of years through the accumulation of organic detritus until the soil rises above the water table and a terrestrial habitat emerges.

Secondary succession

Secondary succession rapidly follows a disturbance that leaves some organisms in place. Breaks in the canopy of a forest tend to close over as surrounding individuals grow toward the light in the opening. A small gap, such as that left by a falling limb, is quickly filled by the growth of branches from surrounding trees. A big gap left by a fallen tree may provide saplings in the understory a chance to reach the canopy and claim a permanent place in the sun. A large area cleared by fire may have to be colonized anew by seeds blown or carried in from surrounding intact forest.

Even when reseeding initiates a secondary sere, the type of disturbance and the size of the gap it creates influence which species become established first. Some plants require abundant sunlight for germination and establishment, and their seedlings are intolerant of competition from other species. These species usually have strong powers of dispersal; they often have small seeds that are easily blown about and can reach the centers of large gaps inaccessible to members of the climax community.

FIGURE 19.6 Bog succession fills aquatic habitats with organic detritus. A 1-meter vertical section through a peat bed in a filled-in bog in Quebec, Canada, reveals accumulations of organic detritus from the plants that successively colonized the bog as it was filled. The peat beds are several meters thick. Photo by R. E. Ricklefs.

ECOLOGISTS IN THE FIELD   **Gap size influences succession on marine hard substrata.** The influence of gap size and gap isolation on succession has been investigated in marine habitats, where disturbance and recovery occur quickly. We saw in Chapter 18 how experimental creation of gaps to mimic ice scouring shifted community dominance from rockweed to either barnacles or *Fucus* seaweeds. In a similar study of marine community development off the coast of South Australia, University of Melbourne ecologist Michael Keough investigated the colonization of artificially created bare patches by various sessile subtidal invertebrates that grow on hard surfaces. These invertebrates have different colonizing and competitive abilities, which are generally inversely related (Table 19.1). Keough created bare patches ranging in size from 25 to 2,500 cm^2 (5–50 cm on a side). Some were cleared areas within larger areas of rock occupied

TABLE 19.1	Life history attributes of the major sessile marine invertebrates at Edithburgh, South Australia			
Taxon	Growth form	Colonizing ability	Competitive ability	Capacity for vegetative growth
Tunicates	Colonial	Poor	Very good	Very extensive, up to 1 m²
Sponges	Colonial	Very poor	Good	Very extensive, up to 1 m²
Bryozoans	Colonial	Good	Poor	Poor, up to 50 cm²
Serpulid polychaetes	Solitary	Very good	Very poor	Very poor, up to 0.1 cm²

Source: M. J. Keough, *Ecology* 65:423–437 (1984).

by sessile invertebrates; others were artificial hard surfaces, such as ceramic tiles, placed in sand some distance from any source of colonists.

The gaps surrounded by intact communities were quickly filled in by such highly successful competitors as tunicates and sponges. In this case, gap size had little influence on succession because the distances from the edges to the centers of the gaps (less than 25 cm) were easily spanned by growth. The many bryozoan and polychaete larvae that attempted to colonize these patches were quickly overgrown. Among the isolated patches, gap size had a much greater effect on the pattern of colonization. Tunicates and sponges, which do not disperse well, tended not to colonize small isolated gaps, thereby giving bryozoans and polychaetes a chance to obtain a foothold. Because larger gaps make bigger targets, many of these were settled by small numbers of tunicates and sponges, which then spread rapidly and eliminated other species that had colonized along with them. As a result, tunicates and sponges predominated on the larger isolated gaps, but bryozoans and polychaetes—which, once established, can prevent the colonization of tunicate and sponge larvae—dominated many of the smaller gaps.

In this system, bryozoans and polychaetes are disturbance-adapted species—what botanists call **weeds,** or *ruderals.* They colonize open patches quickly, mature and produce offspring at an early age, and then are often eliminated by more slowly colonizing but superior competitors. Such weedy species require frequent disturbances to stay in the system.

The size of a gap also influences whether predators and herbivores will be active there. Consumers can affect the course of succession as well as the trophic structure of the community. Some consumers feed in large gaps because large gaps are easy to find and they contain abundant resources, reducing the need to find new areas often. Other consumers that are themselves vulnerable to predation may require the cover of intact habitat, from whose edges they venture to feed. (Recall from Chapter 16 the rabbits that fed on grasses only within range of the cover of shrubs.) These consumers are likely to graze small

gaps and gap edges more intensively than the centers of large gaps (Figure 19.7). ▌

The climax community

Ecologists have traditionally viewed succession as leading to the ultimate expression of community development, the climax community. Early studies of succession demonstrated that many seres found within a region, each developing under a particular set of local environmental conditions, progress toward similar climax states. Such

FIGURE 19.7 Consumers can affect the course of succession. A natural cleared path in a bed of mussels (*Mytilus californianus*) on the central coast of California, has been colonized by a heavy growth of the green alga *Ulva*. Note the distinct grazed zone around the perimeter of the patch. It is created by limpets, which feed only short distances away from refuge in the mussel bed. Courtesy of W. P. Sousa, from W. P. Sousa, *Ecology* 65:1918–1935 (1984).

observations led Clements to his concept of mature communities as natural units—even as closed systems (see Chapter 18). He stated this view clearly in 1916:

> The developmental study of vegetation necessarily rests upon the assumption that the unit or climax formation is an organic entity. As an organism the formation arises, grows, matures, and dies. Its response to the habitat is shown in processes or functions and in structures which are the record as well as the result of these functions. Furthermore, each climax formation is able to reproduce itself, repeating with essential fidelity the stages of its development. The life history of a formation is a complex but definite process, comparable in its chief features with the life history of an individual plant. (*Carnegie Inst. Wash. Publ.* 242:1–512)

Clements recognized fourteen climaxes in the terrestrial vegetation of North America: two types of grassland (prairie and tundra), three types of scrub (sagebrush, desert scrub, and chaparral), and nine types of forest, ranging from pine–juniper woodland to beech–oak forest. He believed that climate alone determined the nature of the local climax, and that the different climax states were discrete, recognizable, and separate from one another. Aberrations in community composition caused by soils, topography, fire, or animals (especially grazers) represented interrupted stages in the transition toward the local climax—immature communities.

In recent years, the concept of the climax as a closed system has, of course, been greatly modified—to the point of outright rejection by most ecologists. Communities are more commonly viewed as open systems whose composition varies continuously over environmental gradients. In addition, various factors may result in alternative "climax" communities. These factors include the intensity of a disturbance and the size of the gap it produces, as well as physical conditions during early succession. Whereas in 1930 plant ecologists described the climax vegetation of much of Wisconsin, for example, as a sugar maple–basswood forest, by 1950 ecologists placed this forest type on an open continuum of climax communities. To the south, beech increased in prominence; to the north, birch, spruce, and hemlock were added to the climax community; in drier regions bordering prairies to the west, oaks became prominent. Locally, quaking aspen, black oak, and shagbark hickory, long recognized as successional species on moist, well-drained soils, came to be accepted as climax species on drier upland sites.

Mature forest communities in southwestern Wisconsin, representing the end points of local seres, range from forests on dry sites dominated by oak and aspen to forests on moist sites dominated by sugar maple, ironwood, and basswood. These communities were ordered by J. T. Curtis and R. P. McIntosh along a **continuum index,** which is the scale of an environmental gradient based on changes in physical characteristics or community composition along that gradient. They calculated the continuum index for Wisconsin forests from the relative abundances in each forest type of several species whose presence indicated different ranges of environmental conditions. Values for the index ranged between arbitrary extremes of 300 for a pure stand of bur oak to 3,000 for a pure stand of sugar maple. Although stages in the seres leading to the sugar maple climax community have intermediate values, low and intermediate values may also represent local climax communities determined by topographic or soil conditions. Thus, the so-called climax vegetation of southwestern Wisconsin actually represents a continuum of forest (and, in some areas, prairie) types (Figure 19.8).

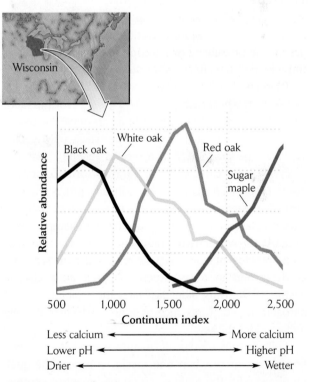

FIGURE 19.8 Climax communities represent a continuum of vegetation types. This continuum index of forest communities of southwestern Wisconsin was constructed from the relative abundances of several species of trees. Soil moisture, exchangeable calcium, and pH increase toward the right. After J. T. Curtis and R. P. McIntosh, *Ecology* 32:476–496 (1951).

Succession ensues as colonists alter environmental conditions

Two factors determine a species' presence in a sere: how readily it invades a newly formed or disturbed habitat, and its response to changes that occur in the environment over the course of succession. Organisms that disperse and grow rapidly have an initial advantage over species that disperse slowly, and they dominate the early stages of a sere. Other species disperse slowly, or grow slowly once established, and therefore become established late in the sere. Early successional species sometimes modify environments in ways that allow later-stage species to become established. The growth of herbs on a cleared field, for example, shades the soil surface and helps the soil to retain moisture, providing conditions more congenial to the establishment of less drought-tolerant plants. Conversely, some colonizing species may inhibit the entrance of others into a sere, either by competing more effectively for limiting resources or by direct interference.

Ecologists Joseph Connell and R. O. Slatyer classified this diverse array of processes governing the course of succession into three categories: facilitation, inhibition, and tolerance. These categories describe the effect of one species on the probability of a second becoming established and whether that effect is positive, negative, or neutral.

Clements viewed succession as a developmental sequence in which each stage paves the way for the next, just as structure follows structure as an organism develops or a house is built. Processes by which one species increases the probability of a second species becoming established are categorized as **facilitation.** Colonizing plants enable later-stage species to invade, just as wooden forms are essential to the pouring of a concrete wall, but have no place in the finished building. For example, alder trees (*Alnus*), which harbor nitrogen-fixing bacteria in their roots, provide nitrogen to soils developing on sandbars in rivers and in areas exposed by retreating glaciers (Figure 19.9). Thus, alder facilitates the establishment of nitrogen-limited plants such as spruce, which eventually replace alder thickets.

Soils do not develop in marine systems, but facilitation occurs there when one species enhances the quality of a site for the settling and establishment of another. Working with experimental panels placed below the tide level in Delaware Bay, T. A. Dean and L. E. Hurd found that although some sessile invertebrate species inhibited the establishment of others, hydroids facilitated settlement of tunicates, and both hydroids and tunicates facilitated settlement of mussels. In southern California, early-

FIGURE 19.9 Alder facilitates succession by adding nitrogen to soils. An alder tree and (*inset*) alder cones. Main photo by K. Ward/Bruce Coleman; inset photo by Gilbert S. Grant/Photo Researchers.

arriving, fast-growing algae provide dense protective cover for re-establishment of kelp following disturbance by winter storms. In areas experimentally kept clear of early successional algae, grazing fish quickly removed settling kelp plants.

Inhibition of one species by the presence of another is a common phenomenon, as we have seen in the context of competition (Chapter 16) and predation (Chapter 15). Individuals of one species can inhibit those of other species by eating them, reducing their resources below the subsistence level, or attacking them with noxious chemicals or antagonistic behavior. In the context of succession, inhibition in the early stages of a sere can prevent movement toward a climax. Of course, climax species, by definition, inhibit the pioneering and transitional species of a sere. Because inhibition is so intimately related to species replacement, it forms an integral part of the orderly succession from the early stages of a sere through the climax.

Inhibition can create an interesting situation—what is called a **priority effect**—when the outcome of an interaction between two species depends on which becomes established first. Colonists are often seeds or larvae, which are vulnerable stages in the life history. Thus, it sometimes happens that neither of a pair of species can become established in the presence of competitively superior adults of the other. In this case, the course of succession depends on precedence. Precedence, in turn, may be strictly random, depending on which species reaches a new habitat or disturbed site first, or it may depend on certain properties of a disturbed site—its size, its location, the season, and so on. We have seen such a case in the subtidal habitats of South Australia, where bryozoans, when they become established first, prevent the establishment of tunicates and sponges. Because of their stronger powers of dispersal, bryozoans are more likely to become established in small, isolated gaps than elsewhere.

The establishment in a sere of a species showing **tolerance** is not influenced by its interactions with other species, but depends only on its dispersal ability and its tolerance of the physical conditions of the environment. Once established, such species are then subject to interactions with other species, and from then on, competitive ability and life span determine their position and dominance within the sere. Poor competitors that have short life cycles but tolerate stressful conditions often become established quickly and dominate early stages of succession, only to be replaced by better competitors.

ECOLOGISTS IN THE FIELD **Plant life histories influence old-field succession.** Tolerance and inhibition interact with life history characteristics to shape details of the species sequence during succession. During the 1940s

and 1950s, Duke University plant ecologists Henry J. Oosting and Catherine Keever observed how these factors combine to influence the early stages of plant succession on old agricultural fields in the Piedmont region of North Carolina (see Figure 19.3). The first 3 to 4 years of old-field succession are dominated by a small number of species that replace one another in rapid sequence: crabgrass, horseweed, ragweed, aster, and broomsedge. All of these species can tolerate the stressful conditions of cleared agricultural land; however, the life cycle of each species partly determines its place in the successional sequence (Figure 19.10).

Crabgrass, a rapidly growing annual, is usually the most conspicuous plant in a cleared field during the year in which the field is abandoned. Horseweed is a winter annual whose seeds germinate in autumn. Through winter, the plant exists as a small rosette of leaves; it blooms by the following midsummer. Because horseweed disperses well and develops rapidly, it usually dominates 1-year-old fields. But because its seedlings require full sunlight, horseweed is quickly replaced by shade-tolerant species. Thus, early succession is dominated by tolerance—the colonizing species disperse readily and can cope with the harsh conditions of newly exposed ground—but rapidly shifts to inhibition.

Decaying horseweed roots stunt the growth of horseweed seedlings, so the species is self-limiting in the sere. Such growth inhibitors presumably are by-products of other adaptations that increase the fitness of horseweed during the first year of succession. Regardless of how it arises, however, self-inhibition is common in early stages of succession.

Ragweed is a summer annual; its seeds germinate early in spring, and the plants flower by late summer. In fields that are plowed under in late autumn, ragweed, not horseweed, dominates the first summer of succession. Aster and broomsedge are biennials that germinate in spring and early summer, exist through winter as small plants, and bloom for the first time in their second autumn. Broomsedge persists and flowers during the following autumn

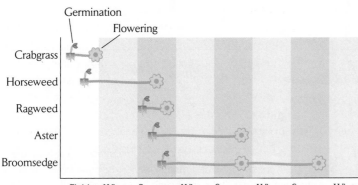

FIGURE 19.10 The life histories of plants influence their place in successional sequences. A schematic summary of the life histories of five early successional plants that colonize abandoned agricultural fields in North Carolina shows their place in the successional sequence.

as well, when it overgrows aster and other early colonizers and dominates the sere until the arrival of shrubs and trees. ▌

Facilitation, inhibition, and invasive species

The establishment and spread of non-native invasive species is governed by many of the same mechanisms that operate during succession. Indeed, invasive species often dominate plant communities during early succession. Although the factors that can contribute to the success of invasive plants are many, their success often hinges on their interactions with fungi and other soil organisms that may either facilitate or inhibit their establishment.

Some fungi form mutualistic symbioses with the roots of plants, known as *mycorrhizae,* that help them to extract mineral nutrients from the soil; in return, the plants provide the fungi with carbohydrate energy. (Mycorrhizal associations will be described in much greater detail in Chapter 24.) The same mycorrhizal fungi that benefit some plants are, however, parasitic or pathogenic when associated with other plants. John Klironomos established a large experiment at the University of Guelph in Ontario

to determine how ten different species of mycorrhizal fungi influenced the growth of ten species of plants in old fields.

Klironomos obtained samples of each species from two different areas, one in Ontario ("home") and the other in Quebec ("foreign"), so that he could test the role of local adaptation in the species' interactions. Klironomos found that whether the fungi inhibited or facilitated plant growth, and by how much, depended on the particular combination of plant species and fungal species tested, and that results were not consistent between "home–home" and "home–foreign" combinations of the same two species (Figure 19.11). Thus, the interactions between plants and fungi appear to depend strongly on local evolutionary adaptation. In addition, both the facilitating and the inhibiting effects were strongest in "home–home" combinations, emphasizing the importance of such evolutionary responses in fine-tuning local interactions between species, whether mutualistic or antagonistic.

Klironomos's experiments with plants and fungi suggest that introduced species might escape some of the strongest inhibiting effects of soil organisms, but might also fail to benefit from other soil organisms that would have facilitated them in their native ranges. Many ecolo-

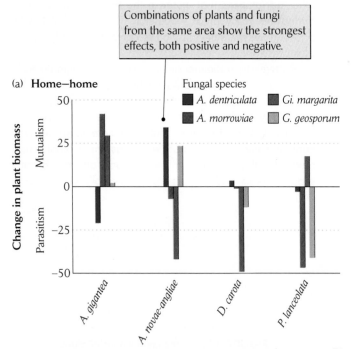

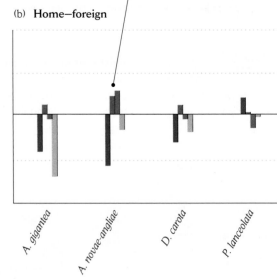

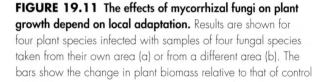

FIGURE 19.11 The effects of mycorrhizal fungi on plant growth depend on local adaptation. Results are shown for four plant species infected with samples of four fungal species taken from their own area (a) or from a different area (b). The bars show the change in plant biomass relative to that of control plants grown with no fungi present. The effects of the fungi are highly idiosyncratic and range from parasitic (negative influence) to mutualistic (positive influence). After J. Klironomos, *Ecology* 84:2292–2301 (2003).

gists studying invasive species believe that some plants and animals have become invasive because they have escaped their native predators and pathogens. Such cases include the prickly pear cactus and European rabbit in Australia, which have been controlled successfully through the introduction of antagonists from their native range (see Chapters 14 and 17).

Planting of seeds of invasive plants in soils obtained from their native and non-native ranges has demonstrated the role of soil organisms in the success of some of these plants. For example, black cherry trees (*Prunus serotina*) are self-inhibiting within their native range in North America, but not in areas of Europe where they have become invasive. Within the native range, cherry seedlings close to parent trees rarely survive, whereas in Europe, seedlings grow readily close to the parent tree. Sterilizing the soils has shown that soil pathogens are responsible for this self-inhibition. Similar experiments with two maple species, one native to North America and the other native to Europe, showed that both species were similarly inhibited when seedlings were grown in soil obtained from beneath the same species in its native range. However, soil from the non-native region obtained from under other tree species *increased* the growth of seedlings of both maple species. These results suggest that soil organisms associated with dominant native species facilitate the establishment of non-native maples. In native soils, facilitating interactions of this kind are apparently overshadowed by the inhibitory influence of natural enemies in the soil.

The differing adaptations of early and late successional species

Succession in terrestrial habitats often exhibits a regular progression of plant forms. Early colonizers and later inhabitants tend to have different strategies of growth and reproduction. Early-stage species capitalize on their dispersal ability to colonize newly created or disturbed habitats quickly. Climax species disperse and grow more slowly, but their shade tolerance as seedlings and their large size as mature plants give them a competitive edge over species that arrive early in the sere. Early-arriving species are typically adapted to colonize unexploited habitats and to tolerate the often stressful conditions in those habitats. Plants of climax communities are typically adapted to grow and prosper in the environments created by the early arrivals. The progression of species is therefore accompanied by a shift in the balance between adaptations promoting dispersal, rapid growth, and early reproduction and adaptations enhancing competitive ability (Table 19.2).

Most early-arriving species produce many small seeds that are usually wind-dispersed (dandelion and milkweed

TABLE 19.2	General characteristics of early and late successional plants	
Characteristic	Early	Late
Number of seeds	Many	Few
Seed size	Small	Large
Dispersal	Wind, stuck to animals	Gravity, eaten by animals
Seed viability	Long, latent in soil	Short
Root:shoot ratio	Low	High
Growth rate	Rapid	Slow
Mature size	Small	Large
Shade tolerance	Low	High

are examples). Their seeds can remain dormant in the soils of forest and shrub habitats for years, in what are called **seed banks,** until fires or treefalls create the bare-soil conditions required for their germination and growth. In contrast, the seeds of most climax species are relatively large, providing their seedlings with ample nutrients to get started in the light-limited environment of the forest floor (Figure 19.12). Seedlings of species that survive well in the shade tend to grow poorly in direct sunlight. Climax species allocate a large proportion of their production to root and stem tissue to support growth to large size and ensure their competitive ability; thus, they grow slowly

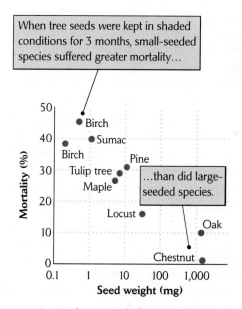

FIGURE 19.12 The survival of tree seedlings in shade is directly related to seed weight. This graph shows the relationship between seed weight and seedling mortality after 3 months under shaded conditions. After J. P. Grime and D. W. Jeffrey, *J. Ecol.* 53:621–642 (1965).

compared with pioneer species that must produce seeds quickly and abundantly. Plants balance shade tolerance and growth rate against each other; each species adopts a compromise between those adaptations that best suits individuals for survival and reproduction at a particular point in a sere.

Succession becomes self-limiting as it approaches the climax

Succession continues until the addition of new species to the sere and the exclusion of established species no longer change the environment of the developing community. The progression from small to large growth forms modifies conditions of light, temperature, moisture, and soil nutrients. Conditions change more slowly, however, after the vegetation achieves the largest growth form that the environment can support. The final biomass dimensions of a climax community are limited by climate independently of events during succession.

Once forest vegetation establishes itself, new tree species change patterns of light intensity and soil moisture less dramatically. To be sure, the species composition of a community may change even after a vegetation structure similar to that of the climax is reached. For example, beech and maple replace oak and hickory in northern deciduous forests because their seedlings are better competitors in the shade of the forest floor (Figure 19.13). Beech and maple seedlings develop as well under their parents as they do under the oak and hickory trees they replace, possibly because these species lack self-inhibition by soil pathogens or better tolerate browsing by dense populations of deer.

The time required for succession to proceed from a new or disturbed habitat to a climax community varies with the nature of the climax and the initial quality of the habitat. A mature oak–hickory climax forest will develop within 150 years on an old field in North Carolina. Climax stages of grasslands in western North America are reached in 20 to 40 years of secondary succession. In the humid tropics, forest communities regain most of their climax elements within 100 years after clear-cutting, provided that the soil is not abused by farming or prolonged exposure to sun and rain. However, several more centuries might pass before a tropical forest achieves a fully mature structure and species composition. Primary succession usually proceeds more slowly. For example, radiocarbon dating methods suggest that beech–maple climax forest requires up to 1,000 years to develop on Lake Michigan sand dunes.

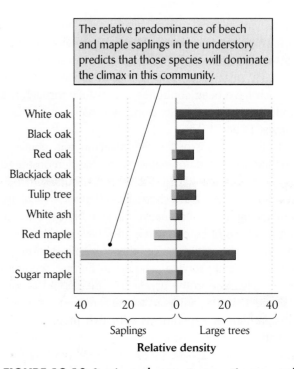

FIGURE 19.13 Species replacement can continue even after a vegetation structure similar to that of the climax community is attained. The species composition of a forest undisturbed for 67 years near Washington, D.C., predicts a gradual successional change beyond the present oak–beech stage. After R. L. Dix, *Ecology* 38:663–665 (1957).

One must not forget, however, that the climax is an elusive concept. Biological communities also change in response to long-term climate change, making the climax a moving target at best. The practical reality for much of the earth is that human activities keep most communities from reaching any possible steady state. In North America, land use has changed continuously since the continent's initial occupation by humans. Hunting, fire, and logging have had lasting effects on communities. Communities are continuously transformed by the disappearance of keystone consumers, such as the wolf and the passenger pigeon (a seed predator), and of entire species of forest trees, including the American chestnut and now the eastern hemlock, as well as by climate change and the introduction of invasive species. Although communities tend toward equilibrium, their most common state is one of dynamic response to changing conditions.

Climax communities under extreme environmental conditions

Many factors determine the composition of a climax community, among them soil nutrients, moisture, slope, and exposure. As we have seen, fire is an important feature

(a)

(b)   (c)

FIGURE 19.14 Many plant species are adapted to frequent fires. (a) A stand of longleaf pine in North Carolina shortly after a fire. Although the seedlings may be badly burned (b), the growing shoot is protected by the long, dense needles (c, shown here on an unburned individual) and often survives. In addition, the slow-growing seedlings have extensive roots that store nutrients to support the plant's growth following fire damage. Photo (a) by R. E. Ricklefs; photo (b) by Jeffrey LePore/Photo Researchers; photo (c) by David Sieren/Visuals Unlimited.

of many climax communities, favoring fire-resistant species and excluding species that would otherwise dominate. The vast southern pine forests along the Gulf Coast and southern Atlantic coast of the United States are maintained by periodic fires. Pines have become adapted to withstand scorching that destroys oaks and other broadleaved species (Figure 19.14). Some species of pines do not even shed their seeds unless triggered by the heat of a fire passing through the understory below. After a fire, pine seedlings grow rapidly in the absence of competition from other understory species.

Any habitat that is occasionally dry enough to create a fire hazard but is normally wet enough to produce and accumulate a thick layer of plant detritus is likely to be influenced by fire. Chaparral vegetation in seasonally dry habitats in California is a fire-maintained climax that gives way to oak woodland in many areas when fire is prevented. The forest–prairie edge in the midwestern United States separates "climatic climax" and "fire cli-

max" communities—these terms refer to the dominant physical factors that determine their species composition. Recall that frequent burning kills seedlings of deciduous trees, but perennial prairie grasses sprout from their roots after a fire (see Figure 7.14). As we saw in Chapter 18, the forest–prairie boundary occasionally shifts back and forth across the countryside, depending on the intensity of recent drought and the extent of recent fires.

Grazing pressure can also modify a climax community. Grassland can be turned into shrubland by intense grazing. Herbivores may kill or severely damage perennial grasses and allow shrubs and cacti that are unsuitable for forage to invade. Most herbivores graze selectively, suppressing favored species of plants and bolstering competitors that are less desirable as food. On African savannas, a regular succession of grazing ungulate species make their way through an area, each feeding on different types of forage. When wildebeests, the first in the succession, were experimentally excluded from some areas, the subsequent

(a)

(b)

FIGURE 19.15 Some grazers prefer to feed in areas previously grazed by others. Zebras (a) and Thomson's gazelles (b) both feed in the Serengeti ecosystem of East Africa, but eat different plants. The gazelles prefer to feed in areas previously grazed by wildebeests and other large herbivores. Photos by R. E. Ricklefs.

wave of Thomson's gazelles preferred to feed in other areas, which *had* been previously grazed by wildebeests or other large herbivores (Figure 19.15). Apparently, heavy grazing by wildebeests stimulates the growth of food plants that gazelles prefer and reduces cover within which predators of the smaller ungulates could conceal themselves. In western North America, cattle grazing allows invasion by the alien cheatgrass (*Bromus tectorum*), which promotes fires and may lead succession to an alternative stable state—that is, a different climax community in an anthropogenically altered landscape.

Transient and cyclic climaxes

We usually view succession as a series of changes leading to a stable climax, whose character is determined by the local environment. Once established, a beech–maple forest perpetuates itself, and its general appearance changes little despite constant replacement of individuals within the community. Yet not all climaxes are persistent. Simple cases of **transient climaxes** include the communities in seasonal ponds—small bodies of water that either dry up in summer or freeze solid in winter. These extreme seasonal changes regularly destroy the communities that become established in the ponds each year. Each spring the ponds are restocked from larger, permanent bodies of water or from resting stages left by plants, animals, and microorganisms before the habitat disappeared in the previous year, starting succession over again.

Succession recurs whenever a new environmental opportunity appears. Excreta and dead organisms, for example, provide resources for a variety of scavengers and detritus feeders. On African savannas, carcasses of large mammals are devoured by a succession of vultures (Figure 19.16). The first are large, aggressive species that gorge themselves on the largest masses of flesh. These are followed by smaller species that glean smaller bits of meat from the bones, and finally by a kind of vulture that cracks open bones to feed on marrow. Scavenging mammals, maggots, and microorganisms enter the sequence at different points and ensure that nothing edible remains. This succession has no local climax because all the scavengers disperse when the feast concludes. However, the scavengers form part of a larger climax: the entire savanna community.

In simple communities, the particular life history characteristics of a few dominant species can create a **cyclic climax.** Suppose, for example, that plant species A can germinate only under species B, B only under C, and C only under A. These relationships create a regular cycle of species dominance in the order A, C, B, A, C, B, A, . . . , in which the length of each stage is determined by the life span of the dominant species. Cyclic climaxes usually follow such a scheme, often with one stage being bare substratum. Many such cycles are driven by harsh environmental conditions, such as wind or frost heaving (Figure 19.17).

When high winds damage heaths and other types of vegetation in northern Scotland, shredded foliage and broken twigs create openings for further damage, and the process becomes self-accelerating. Soon a wide swath is opened in the vegetation. Regeneration occurs on the

FIGURE 19.16 Scavengers may form a transient successional sequence. These vulture species are feeding in turn on a wildebeest carcass in Masai Mara National Reserve, Kenya. Photo by R. E. Ricklefs.

protected side of the damaged area while wind damage further encroaches on exposed vegetation. Consequently, waves of damage and regeneration move through the community in the direction of the wind. If we watched the sequence of events at any one location, we would witness a healthy heath being reduced to bare earth by wind damage and then regenerating in repeated cycles (Figure 19.18). If we looked at the whole heath, we would see a mosaic of plants and bare soil.

Such mosaic patterns of vegetation types typify any climax community where deaths of individuals alter the environment. Treefalls open a forest canopy and create patches of habitat that are drier, hotter, and sunnier than the forest floor under the unbroken canopy. These openings are often invaded by early colonizing specialists, which persist until the canopy closes again. Thus, treefalls create a shifting mosaic of successional stages within an otherwise uniform forest community. Indeed, adaptation by different species to growing in particular conditions created by different-sized gaps could enhance the overall diversity of a climax community.

Our concept of the climax community must include cyclic patterns of change, mosaic patterns of distribution, and alternative stable states. The climax is a dynamic state, self-perpetuating in its composition, even if by regular cycles of change. Persistence is the key to the climax, and a persistent cycle defines a climax as well as an unchanging steady state does.

FIGURE 19.17 Cyclic succession is usually driven by stressful environmental conditions. Waves of wind damage and regeneration in fir forests move across the slopes of Mount Shimigare, Japan. Photo by M. E. Dodd/Amanita Photo Library.

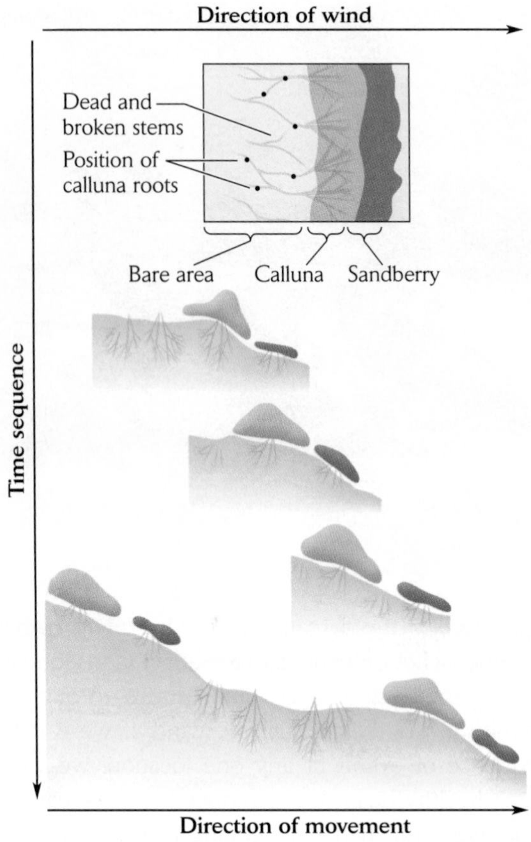

FIGURE 19.18 Cyclic succession may involve a sequence of damage and regeneration. This pattern is seen in dwarf heath communities of northern Scotland. At the top is a diagram of a heath from above, showing a band of sandberry (*Arctostaphylos uva-ursi*) growing in the protected lee of the calluna (*Calluna vulgaris*) heath as wind damages to the heath on its upwind side (to the left). Below is a side view of the band of sandberry and heath as it appears to "migrate" downwind over time. After A. S. Watt, *J. Ecol.* 35:1–22 (1947).

Direction of wind

Dead and broken stems
Position of calluna roots

Bare area   Calluna   Sandberry

Time sequence

Direction of movement

Succession emphasizes the dynamic nature of biological communities. By upsetting their natural balance, disturbance reveals to us the forces that determine the presence or absence of species within a community and the processes responsible for regulating community structure. Succession also emphasizes the idea that communities often comprise patchwork mosaics of successional stages and reminds us that community studies must consider disturbance and environmental change on many scales of time and space.

SUMMARY

1. Succession is community change following either habitat disturbance or the formation of new habitat. The particular sequence of communities at a given location is referred to as a sere, and the ultimate association of plants and animals is called a climax community.

2. Succession on newly formed or disturbed habitat that is devoid of life is referred to as primary succession. Early colonists in primary succession modify the environment for the species that follow them. More moderate disturbances, which leave much of the physical structure of the ecosystem intact, are followed by secondary succession.

3. The initial stages of the sere depend on the intensity and extent of the disturbance, but its end point reflects climate—that is, within a region, seres tend to converge on a single climax. However, variations in the area of gaps created by disturbance and in conditions during early stages of succession may lead to alternative climax states.

4. According to Curtis and McIntosh's continuum concept of the climax, the nature of the climax varies continuously over gradients of climate and other environmental conditions.

5. Joseph Connell and R. O. Slatyer categorized the processes that govern succession as facilitation, inhibition, and tolerance. These processes are distinguished by the effect of one established species on the probability of colonization by a second potential colonist.

6. Facilitation refers to processes by which earlier successional species make conditions in a sere more favorable for the establishment of later species. Inhibition refers to processes by which species in the sere make conditions less favorable for the establishment or persistence of others. Tolerance characterizes species whose establishment is not influenced by the presence of other species in the sere, but rather by environmental conditions there.

7. Characteristics of species vary according to their place in a sere. Pioneering species tend to have many small seeds that are easily dispersed, produce shade-intolerant seedlings that grow rapidly, and reach maturity quickly; late-stage species have the opposite features. The features of early-stage species tend to make them good colonizers, while those of late-stage species tend to make them strong competitors.

8. Succession continues until the addition of new species to the sere and the exclusion of established species no longer change the environment of the developing community.

9. The character of the climax may be influenced by extreme conditions, such as fire and intense grazing, that alter interactions among species in a sere.

10. Transient climaxes develop on ephemeral resources and habitats, such as temporary ponds and the carcasses

of individual animals. In such cases, we may think of a regional climax as including transient successional sequences.

11. Cyclic local climaxes may develop in simple communities where each species can become established only in association with some other species. Cyclic climaxes are often driven by harsh physical conditions, such as frost and strong winds.

REVIEW QUESTIONS

1. How do ecologists distinguish between primary and secondary succession?

2. How would a trade-off between dispersal ability and competitive ability affect which types of species could colonize small as opposed to large gaps in a community?

3. Why do most ecologists no longer consider a climax community to correspond to Clements's concept of a closed community?

4. Compare and contrast the concepts of facilitation, inhibition, and tolerance in the context of ecological succession.

5. If two plant species have similar dispersal and competitive abilities, what factor might help determine which species occupies an early stage in a sere?

6. How do mycorrhizal fungi play roles in facilitation and inhibition?

7. Why do early and late successional species tend to possess different adaptations?

8. What factors can prevent a climax community from remaining in a steady state?

9. Why are transient climaxes not stable?

SUGGESTED READINGS

Berkowitz, A. R., C. D. Canham, and V. R. Kelly. 1995. Competition vs. facilitation of tree seedling growth and survival in early successional communities. *Ecology* 76:1156–1168.

Bever, J. D., et al. 2001. Arbuscular mycorrhizal fungi: More diverse than meets the eye, and the ecological tale of why. *BioScience* 51(11):923–932.

Briggs, J. M., et al. 2005. An ecosystem in transition: Causes and consequences of the conversion of mesic grassland to shrubland. *BioScience* 55(3):243–254.

Bruno, J. F., J. J. Stachowwicz, and M. D. Bertness. 2003. Inclusion of facilitation into ecological theory. *Trends in Ecology and Evolution* 18:119–125.

Callaway, R. M., and F. W. Davis. 1993. Vegetation dynamics, fire, and the physical environment in coastal central California. *Ecology* 74:1567–1578.

Callaway, R. M., and L. R. Walker. 1997. Competition and facilitation: A synthetic approach to interactions in plant communities. *Ecology* 78:1958–1965.

Christensen, N. L., and R. K. Peet. 1984. Convergence during secondary forest succession. *Journal of Ecology* 72:25–36.

Connell, J. H., and R. O. Slatyer. 1977. Mechanisms of succession in natural communities and their role in community stability and organization. *American Naturalist* 111:1119–1144.

Foster, B. L., and K. L. Gross. 1999. Temporal and spatial patterns of woody plant establishment in Michigan old fields. *American Midland Naturalist* 142:229–243.

Foster, B. L., and D. Tilman. 2000. Dynamic and static views of succession: Testing the descriptive power of the chronosequence approach. *Plant Ecology* 146:1–10.

Grubb, P. J. 1977. The maintenance of species diversity in plant communities: The importance of the regeneration niche. *Biological Reviews* 52:107–145.

Halpern, C. B., et al. 1997. Species replacement during early secondary succession: The abrupt decline of a winter annual. *Ecology* 78:621–631.

Howe, H. F., and M. N. Miriti. 2004. When seed dispersal matters. *BioScience* 54(7):651–660.

Keever, C. 1950. Causes of succession on old fields of the Piedmont, North Carolina. *Ecological Monographs* 20:230–250.

Keough, M. J. 1984. Effects of patch size on the abundance of sessile marine invertebrates. *Ecology* 65:423–437.

Klironomos, J. N. 2003. Variation in plant response to native and exotic arbuscular mycorrhizal fungi. *Ecology* 84:2292–2301.

Knowlton, N. 1992. Thresholds and multiple stable states in coral reef community dynamics. *American Zoologist* 32:674–682.

Nijjer, S., W. E. Rogers, and E. Siemann. 2007. Negative plant–soil feedbacks may limit persistence of an invasive tree due to rapid accumulation of soil pathogens. *Proceedings of the Royal Society of London* B 274:2621–2627.

Prach, K., P. Pysek, and P. Smilauer. 1997. Changes in species traits during succession: A search for pattern. *Oikos* 79:201–205.

Reinhart, K. O., and R. M. Callaway. 2004. Soil biota facilitate exotic *Acer* invasions in Europe and North America. *Ecological Applications* 14:1737–1745.

Reinhart, K. O., and R. M. Callaway. 2006. Soil biota and invasive plants. *New Phytologist* 170:445–457.

Reinhart, K. O., et al. 2003. Plant–soil biota interactions and spatial distribution of black cherry in its native and invasive ranges. *Ecology Letters* 6:1046–1050.

Riggan, P. J., et al. 1988. Interaction of fire and community development in chaparral of southern California. *Ecological Monographs* 58:155–176.

Shafroth, P. B., et al. 2002. Potential responses of riparian vegetation to dam removal. *BioScience* 52(8):703–712.

Sousa, W. P. 1984. Intertidal mosaics: Patch size, propagule availability, and spatially variable patterns of succession. *Ecology* 65:1918–1935.

Turner, M. G., V. H. Dale, and E. H. Everham. 1997. Fires, hurricanes, and volcanoes: Comparing large disturbances. *BioScience* 47:758–768.

Turner, M. G., et al. 2003. Disturbance dynamics and ecological response: The contribution of long-term ecological research. *BioScience* 53(1):46–56.

Whittaker, R. J., M. B. Bush, and K. Richards. 1989. Plant recolonization and vegetation succession on the Krakatau Islands, Indonesia. *Ecological Monographs* 59:59–123.

Whittaker, R. J., S. H. Jones, and T. Partomihardjo. 1997. The rebuilding of an isolated rain forest assemblage: How disharmonic is the flora of Krakatau? *Biodiversity and Conservation* 6:1671–1696.

Zobel, D. B., and J. A. Antos. 1997. A decade of recovery of understory vegetation buried by volcanic tephra from Mount St. Helens. *Ecological Monographs* 67:317–344.

CHAPTER 20

Biodiversity

The great naturalist–explorers of the nineteenth century—Charles Darwin, Henry W. Bates, Alfred Russel Wallace, and others—traveled to the tropics and discovered there a great store of species unknown to European scientists. As we saw in Chapter 18, the numbers of species in most groups of organisms—plant, animal, and perhaps microbial—increase markedly toward the equator (Figure 20.1). For example, a hectare of forest typically has fewer than 5 kinds of trees in boreal regions, 10–30 in temperate regions, and 100–300 in tropical regions. These latitudinal trends in diversity are pervasive, and extend even to the greatest depths of the oceans, where conditions were once thought to be unvarying over the globe.

Why do so many different kinds of organisms live in the tropics (and so few toward the poles)? The factors that regulate the diversity of biological communities are the subject of this chapter. Historically, biologists have held two views on such questions. One maintains that diversity increases without limit over time, barring catastrophes such as meteorite impacts that cause mass extinctions of species. Tropical environments, being much older than temperate and polar environments, have had time to accumulate more species. Accordingly, diversity is simply a matter of history. The second view holds that diversity reaches an equilibrium, at which the appearance of new species balances the loss of already existing species. In either case, factors that add species would seemingly weigh more heavily in the balance—or factors that remove species, less heavily—closer to the tropics.

Throughout the first half of the twentieth century, the first, historical viewpoint enjoyed the broader favor. Ecologists understood that tropical environments have dominated the earth's surface over much of its history, whereas changes in climate (particularly during the Ice Age) have occasionally ravaged most temperate and polar biotas, resetting the diversity clock, so to speak. More recently, however, with the integration of population

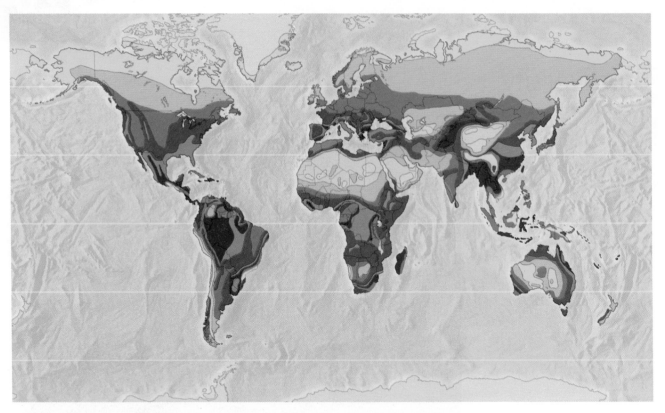

Number of species per 10,000 km²

☐ <100	☐ 200–500	■ 1,000–1,500	■ 2,000–3,000	☐ 4,000–5,000
☐ 100–200	■ 500–1,000	■ 1,500–2,000	■ 3,000–4,000	■ ≥5,000

FIGURE 20.1 Species richness varies over the surface of the earth. Estimated numbers of plant species in 100 × 100 km grid squares show the general trend of increasing species richness toward the tropics, as well as the augmenting effect of mountains and the depressing effect of arid areas. After W. Barthlott, W. Lauer, and A. Placke, *Erdkunde* 50:317–326 (1996).

ecology into community theory, ecologists accept that diversity might be regulated at a steady state. Accordingly, opposing diversity-dependent processes balance each other, just as density-dependent birth and death rates match at a population's carrying capacity. This viewpoint challenges ecologists to identify the processes responsible for adding and removing species at local and regional scales, and to discover why the balance between these processes differs from place to place.

CHAPTER CONCEPTS

- Variation in the relative abundance of species influences concepts of biodiversity
- The number of species increases with the area sampled
- Large-scale patterns of diversity reflect latitude, environmental heterogeneity, and productivity
- Diversity has both regional and local components

- Diversity can be understood in terms of niche relationships
- Equilibrium theories of diversity balance factors that add and remove species
- Explanations for high tree species richness in the tropics focus on forest dynamics

The general trend of increasing species richness toward the tropics obscures the fact that variation also occurs within latitudinal belts. In particular, species richness varies between different continents and between areas of distinct climate and topography. For example, temperate forests of eastern Asia support twice the number of tree species found in North America and three times the number found in Europe. Mangrove wetlands, in contrast to other tropical habitats, typically have few species, rarely more than 10 species of trees and shrubs in any particular place and only about 50 worldwide.

These variations suggest that a variety of processes influence the number of species found in a particular place. The predominant latitudinal gradient in species richness suggests processes acting uniformly over the entire globe. Patterns in diversity that reflect variations in temperature and precipitation within a latitudinal belt imply the influence of local processes. The exceptionally high diversity associated with the complex topography of the Andes of South America and the Himalaya of southern Asia indicates that geographic factors also affect species richness. The processes that influence diversity clearly act over a variety of spatial and temporal scales, and ecologists have used many approaches to characterize these processes. An important first step has been to find ways to describe species diversity that allow ecologists to compare diversity across regions and quantify patterns of species richness.

Variation in the relative abundance of species influences concepts of biodiversity

The general term **biodiversity** refers to the variation among organisms and ecological systems at all levels, including genetic variation within populations, morphological and functional differences between species, and variation in biome structure and ecosystem processes in both terrestrial and aquatic systems. Because biodiversity is so all-encompassing, ecologists often study one of the simplest and most general indices to biodiversity: the number of species within an area, often called *species richness*. Yet, by any criterion one chooses, all the species in an area are not equal. Some are abundant, others rare. Some have important effects on population dynamics in the community; others scarcely make themselves noticed. How does one measure the "presence" of a species—its *gravitas* in the community? That depends on one's purpose, and also on the practical considerations involved in sampling species in nature.

One important way of characterizing species diversity is by the relative abundances of species. Abundance can be quantified by the number, density, or biomass of individuals within a sample area, by the frequency of sample plots in which a particular species is recorded, or by *cover* (the proportion of the area of the habitat covered by a species), a measure often employed by botanists. Numbers of individuals alone could certainly misrepresent the ecological influence of species if a sample included, say,

both ants and elephants. In that case, it would be tempting to quantify the total mass of a species or its total rate of energy metabolism. Adopting a measure of abundance is a necessary first step in the effort to determine why particular patterns of relative abundance characterize regions or communities in nature.

The **relative abundance** of a species is its proportional representation in a sample or community. A common portrayal of relative abundances is the **rank-abundance plot**, which displays the abundances of species, usually on a logarithmic scale, ranked from the most common to the rarest (Figure 20.2). Such plots drive home the nearly universal observation that ecological systems contain a few abundant species—often called **dominant** species—and many more rare ones.

Because many species are rare, almost any sample, no matter how large, is bound to miss infrequently encountered species. Thus, in comparing diversity between samples, it is important either to use identical sampling techniques or to take sample size into account by using an index of species richness. The latter can be done only

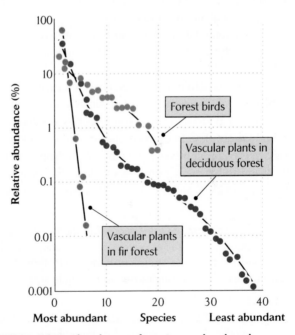

FIGURE 20.2 Abundances of species can be plotted on a logarithmic scale. Rank-abundance curves are shown for three natural communities: birds in a deciduous forest in West Virginia; vascular plants in a subalpine fir forest in the Great Smoky Mountains, Tennessee; and vascular plants in a deciduous cove forest in the Great Smoky Mountains. Each dot represents one species. Abundance is represented by number of individuals for birds and by net primary production for plants. After R. H. Whittaker, *Communities and Ecosystems* (2nd ed.), Macmillan, New York (1975).

when species richness (S) and sample size (N, number of individuals or some other measure of abundance) have a known relationship. Two indices of species richness are noteworthy here. Margalef's index, $D = (S - 1)/\ln(N)$, assumes that species richness increases as the logarithm of sample size. (Margalef's index is described in more detail in Data Analysis Module 5.) Menhinick's index, $D = S/\sqrt{N}$, uses the square root of sample size as a reference point. Both indices normalize species richness in relation to the size of the sample.

Several measures of species diversity, referred to as *heterogeneity measures*, combine species richness with variation in the abundances of species, also known as *species evenness*. One such measure is **Simpson's index**, γ (gamma) $= \Sigma p_i^2$, where p_i is the proportion of individuals in a sample that belong to species *i*. Thus, γ is the sum of the squared proportion of each species in the sample, and it is the probability that any two individuals drawn randomly from the sample will be the same species. Simpson's index is more conveniently expressed as its inverse, $S_i = 1/\gamma$, where S_i is always less than the number of species in the sample because the each species is weighted by its relative abundance in the index. In fact, the value of S_i is identical to the value that one would obtain for a sample of S_i equally common species.

This section is hardly an exhaustive treatment of diversity indices, the discussion of which fills volumes. However, it should emphasize to you the variation in abundances among species within a community. This variation poses problems for the description of diversity and challenges ecologists to understand why some species are common and others are rare. In practice, species richness and various diversity and heterogeneity indices are strongly correlated, and so for many purposes simple counts of species richness provide a sufficient basis for studying diversity.

MORE ON THE WEB *The Lognormal Distribution.* Variation in the abundances of species within a community can be described by a simple statistical relationship that shows how number of species increases with sample size.

DATA ANALYSIS DATA ANALYSIS MODULE 5 *Quantifying Biodiversity.* Learn methods for quantifying species diversity on page 435.

The number of species increases with the area sampled

As a rule, more species are found within large areas than within small areas. The Swedish agricultural chemist Olaf

Arrhenius first formalized this **species–area relationship** in 1921. Since then, it has been common practice to compare species richness (S) with area (A) using a power function of the form

$$S = cA^z$$

where *c* and *z* are constants fitted to the data. Ecologists portray species–area relationships graphically by plotting the logarithm of species richness against the logarithm of area, as shown in Figure 20.3. After log transformation, the species–area relationship becomes

$$\log S = \log c + z \log A$$

which is the equation for a straight line with slope *z*.

Species–area relationships in many groups of organisms are linear on logarithmic axes and often have slopes within the range $z = 0.20$–0.35—that is, the number of species increases in proportion to the one-fifth to one-third power (fifth root to cube root) of area. This relationship reflects the outcome of many processes, and over local to global scales, the changing influences of these processes alter the slope of the relationship in predictable ways.

Over the smallest areas, beginning with samples that include only a few individuals, the number of species

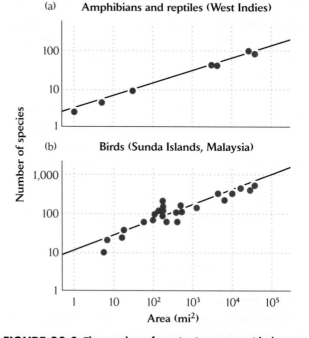

FIGURE 20.3 The number of species increases with the area sampled. Species–area relationships for (a) amphibians and reptiles in the West Indies and (b) birds in the Sunda Islands of the Malay Archipelago show this pattern. After R. H. MacArthur and E. O. Wilson, *Evolution* 17:373–387 (1963), and *The Theory of Island Biogeography*, Princeton University Press, Princeton, NJ. (1967).

increases with area simply because more individuals are sampled. The slope of this relationship is usually relatively high. A sample of one individual can have only one species. A second individual is likely to be another species. Thus, species richness initially doubles, or nearly so, as the sample size doubles ($z = 1$).

When areas are large enough so that samples pick up most of the local species, the species–area relationship depends primarily on new habitat types being included in progressively larger areas. For example, a single hectare might include a single habitat type, and a county-level sample several habitat types. A state-level sample would introduce still more habitat types as a variety of large-scale geographic features, such as mountains, are added, and a continent-level sample would include all the habitats formed in a wide range of climate zones as well as in association with diverse geographic features. On a global scale, the inclusion of different continents would introduce still another factor increasing species richness: the evolution of distinct evolutionary lineages on isolated continents.

A study of species–area relationships in temperate grasslands in North Carolina, Sweden, and the Netherlands illustrates these changes in the slope of the relationship as a function of scale (Figure 20.4). The smallest sample areas in each location were 10 cm² (about an inch and a half on a side), and generally included two or three plant species at most. As the sample area increased above 1 m², the species–area relationship assumed a slope that remained constant up to an area the size of a state or a European country, more than a billionfold larger. Above that point, the slope began to rise rapidly again as samples on the global scale combined species from different continents. Thus, three processes, acting on three scales, determine the overall shape of the species–area relationship: local sampling, the formation of a variety of habitat types within an environmentally heterogeneous region, and the evolution of distinct lineages on isolated continents.

Species–area relationships on islands

Species–area relationships on islands are influenced by a fourth process, species extinction. Some ecologists have found the slope of the species–area relationship to be higher when islands of different sizes are compared than it is across continental areas over a comparable size range. The range of an endemic island species (a species found on that island and nowhere else) can be no larger than the island it inhabits, whereas the ranges of continental species generally exceed the borders of a particular sampling area. As a result, dispersal of individuals prevents populations within small continental areas from going extinct locally,

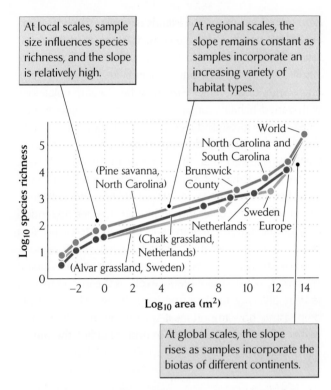

FIGURE 20.4 The slope of the species–area relationship is influenced by different processes on different scales. Species–area relationships are shown for temperate grasslands in (a) North Carolina, (b) The Netherlands, and (c) Sweden. At all three locations, the number of grassland plant species increases with sample size over small areas, with inclusion of new habitat types over intermediate areas, and with incorporation of different continents at the largest (global) scale. Note that the species–area relationships for the three locations are nearly superimposable. After J. Fridley et al., *Am. Nat.* 168:133–143 (2006).

whereas islands of similar size are more likely to lose species. Thus, smaller and larger areas within continents have more similar complements of species than do smaller and larger islands, and the slope of the species–area relationship is consequently lower.

This greater difference in species richness between small islands and large islands must reflect differences in their intrinsic qualities. Factors that are likely to increase species richness include habitat diversity, which tends to increase with the size (and resulting topographic heterogeneity) of an island, and size per se, as larger islands make better targets for potential immigrants from the mainland. In addition, larger islands support larger populations, which may persist longer owing to their greater genetic diversity, broader distributions over habitats, and numbers large enough to prevent stochastic extinction.

Whether island size itself or habitat diversity is more important to the species–area relationship can be determined by comparing islands of similar size but different habitat variety, and vice versa, to tease apart the separate effects of these two factors statistically. The Lesser Antilles, for example, contain islands in a range of sizes, but islands of any given size may be oceanic volcanoes with varied habitats ranging from coastal mangrove wetlands to rain forest and cloud forest, or they may be low areas of raised seabed dominated by dry forest and scrub. Analyses showed that the number of species of bats, which are not habitat specialists, was sensitive to island area, but not to habitat diversity. At the other extreme, the number of reptile and amphibian species depended only on habitat diversity and was unrelated to island area per se. Reptiles and amphibians have very large populations on these islands, but they tend to be narrow habitat specialists, which undoubtedly explains the importance of habitat diversity to their species richness. This example emphasizes how important it is to study underlying processes when one is interested in understanding the meaning of patterns.

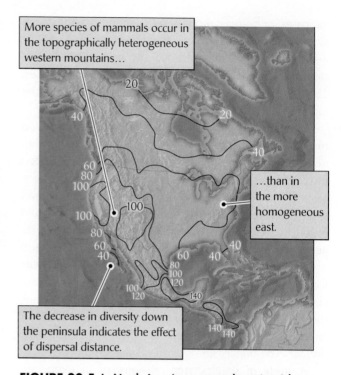

FIGURE 20.5 In North America, mammal species richness increases toward the equator and in regions of high habitat diversity. The contour lines on the map indicate the numbers of mammal species found in sample blocks 150 miles on a side. After G. G. Simpson, *Syst. Zool.* 13:57–73 (1964).

Large-scale patterns of diversity reflect latitude, environmental heterogeneity, and productivity

In the Northern Hemisphere, the number of species in most groups of animals and plants increases from north to south. Ecologists have visualized these patterns of diversity by mapping the numbers of species found within sample areas delimited by latitude and longitude. For example, within the area of North America extending to the Isthmus of Panama, the number of mammal species occurring in square sample blocks 150 miles on a side increases from fewer than 20 in northern Canada to more than 140 in Central America (Figure 20.5). These increases in species richness from north to south parallel increasing temperatures, as discussed later in this section.

Numbers of mammal species also increase from east to west in North America. In this case, the increases in species richness reflect the influence of geographic heterogeneity. Within a narrow range of latitude across the middle of the United States, more mammal species live in the topographically heterogeneous western mountains (90–120 species per block) than in the environmentally more uniform eastern states (50–75 species per block). Presumably, the greater heterogeneity of environments in the West provides suitable conditions for a greater number of species. In addition, notice that diversity decreases toward the south along the peninsula of Baja California. In

other words, fewer species occur at a progressively greater distance from the southwestern United States, suggesting that dispersing individuals are unlikely to extend the limits of a species range down a narrow peninsula.

The pattern of species richness for breeding land birds in North America resembles that for mammals, but trees, reptiles, and amphibians present strikingly different patterns (Figure 20.6). Reptile species richness decreases fairly uniformly as temperature decreases toward the north, while trees and, especially, amphibians are more diverse in the moister eastern half of North America than in the drier, more mountainous western regions.

Ecological heterogeneity and habitat productivity

Within a region, heterogeneity of soils, vegetation structure, and other habitat measures exerts a strong influence on diversity. Censuses of breeding birds in 5–20 ha blocks of various habitat types in North America reveal an average of about 6 species in grasslands, 14 in shrublands, and 24 in floodplain deciduous forests. More productive habitats tend to harbor more species, but habitats with simple vegetation structure, such as grasslands and marshes, have

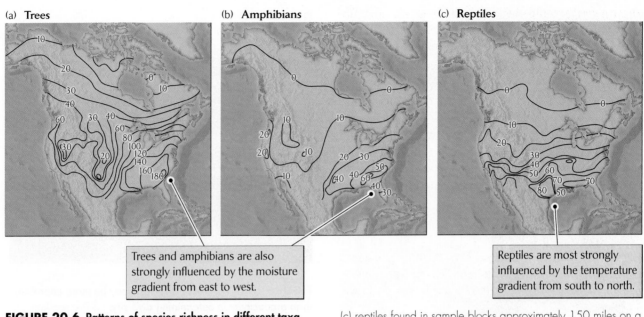

(a) Trees

(b) Amphibians

(c) Reptiles

Trees and amphibians are also strongly influenced by the moisture gradient from east to west.

Reptiles are most strongly influenced by the temperature gradient from south to north.

FIGURE 20.6 Patterns of species richness in different taxa show different environmental influences. The contour lines indicate the numbers of species of (a) trees, (b) amphibians, and (c) reptiles found in sample blocks approximately 150 miles on a side. After D. J. Currie, *Am. Nat.* 137:27–49 (1991).

fewer species than more complex habitats with similar productivity (Figure 20.7). This principle also applies to plants. Marshes, for example, are highly productive, but are ecologically uniform, and thus have relatively few species of plants. Desert vegetation is less productive than marsh vegetation, but its greater structural heterogeneity makes room for more kinds of inhabitants (Figure 20.8).

Structural complexity and diversity have always gone together in the minds of bird-watchers and other naturalists, but Robert and John MacArthur were the first to place this relationship in a quantitative framework that made it accessible to analysis. They did this simply by plotting the species richness of birds observed in different habitats in relation to diversity in foliage height, a measure of the

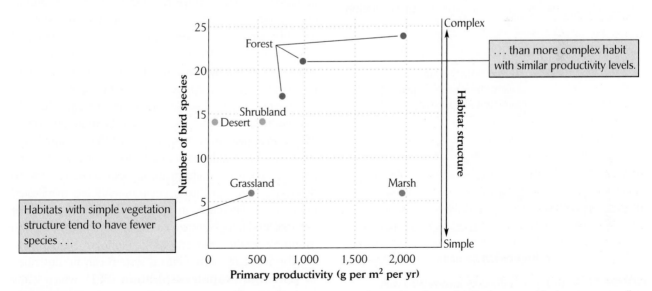

... than more complex habit with similar productivity levels.

Habitats with simple vegetation structure tend to have fewer species ...

FIGURE 20.7 Diversity is higher in structurally complex habitats. Among seven habitat types in temperate regions, the average number of bird species tends to increase with habitat productivity. Species richness is lowest, however, in structurally simple marsh and grassland habitats, and highest in structurally complex forest habitats. After E. J. Tramer, *Ecology* 50:927–929 (1960); productivity data from R. H. Whittaker, *Communities and Ecosystems*, 2nd ed., Macmillan, New York (1975).

(a)

(b)

FIGURE 20.8 **Vegetation structure may be more important than productivity in determining diversity.** (a) The Sonoran Desert of Baja California and (b) a salt marsh in Eastham, Massachusetts, illustrate extremes of habitat productivity. Nevertheless, the highly productive marsh, with its simple vegetation structure, has fewer plant species than the desert. Photo (a) by R. E. Ricklefs; photo (b) by David Weintraub/Photo Researchers.

structural complexity of vegetation (Figure 20.9). Other ecologists soon demonstrated similar relationships. Among web-building spiders, for example, species richness varies in direct relation to heterogeneity in the heights of the tips of vegetation to which spiders attach their webs. In desert habitats of the southwestern United States, lizard species richness closely parallels the total volume of vegetation per unit of area, which parallels plant species richness and structural heterogeneity.

Solar energy input and precipitation

Energy input from the sun and water input from precipitation predict species richness reasonably well in most groups of organisms. University of California at Irvine ecologist Brad Hawkins and his colleagues surveyed many published studies and concluded that solar energy input (or, more generally, environmental temperature) provides the best prediction of animal species richness in regions north of 15°N, while water availability has more influence throughout the tropics and into the Southern Hemisphere (Figure 20.10). Precipitation is more influential than energy in the tropics and in south-temperate regions because temperatures are relatively uniform throughout those regions, so differences between those environments are manifested primarily as differences in precipitation. In contrast, solar radiation and temperature increase dramatically from north to south over the continents of the Northern Hemisphere.

The input of energy into a system can be described by **potential evapotranspiration** (PET), which is the amount of water that could be evaporated from the soil and transpired by plants, given the average temperature and humidity. This measure integrates temperature and solar radiation and is thus an index to the overall energy

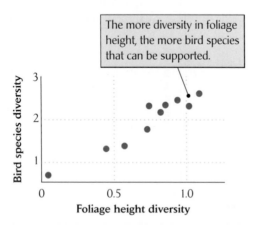

The more diversity in foliage height, the more bird species that can be supported.

FIGURE 20.9 **Bird species diversity is correlated with foliage height diversity.** Robert and John MacArthur found this relationship in areas of deciduous forest, old fields, and regenerating forest habitats in eastern North America. After R. H. MacArthur and J. MacArthur, *Ecology* 42:594–598 (1961).

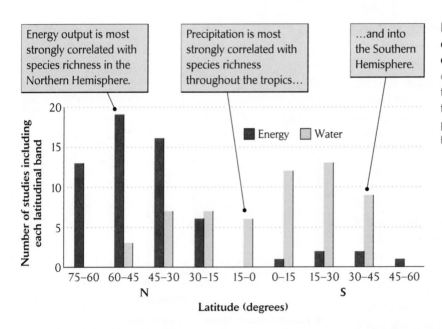

FIGURE 20.10 The climatic variables associated with species richness vary across the globe. Bars indicate the number of studies in which variables related either to solar energy input (such as environmental temperature) or to water input were significant predictors of species richness. After B. A. Hawkins et al., *Ecology* 84:3105–3117 (2003).

input into the environment. PET predicts species richness in North American vertebrates relatively well, although the relationships are complex (Figure 20.11). In each vertebrate class, species richness increases as PET rises, at least over the lower part of the PET range, reflecting increasing diversity from north to south within the continent (see Figures 20.5 and 20.6). However, above a certain PET threshold—about 500 mm per year for birds and mammals, 1,000 mm per year for amphibians, and 1,500 mm per year for reptiles—species richness levels off. Above these thresholds, increasing temperature fails to improve

the capacity of the environment to support additional vertebrate species, particularly in the arid western parts of the continent, where increasing temperature becomes a stress.

Nonetheless, the general correlation between PET and species richness has given rise to the idea that there is a causal relationship between the two, known as the *energy–diversity hypothesis*. One possibility is that a larger amount of energy in an ecosystem can be shared by a larger number of species. Greater energy input and the resulting higher biological productivity might also support larger population sizes, thereby reducing rates of extinction, so that species are able to persist that would not be able to maintain populations at a lower energy level. High energy input might also accelerate evolutionary change and increase the rate of species formation. Although these ideas are attractive, none of these mechanisms has been verified experimentally, and some doubt that the energy–diversity correlation either represents a general pattern or explains species richness.

As we shall see, the factors that influence species richness are complex. So many factors play roles at different scales that we should not expect to find a simple comprehensive explanation for all observed patterns.

Diversity has both regional and local components

Diversity can be measured at any spatial scale. **Local diversity** (or *alpha diversity*) is the number of species in a small area of homogeneous habitat. Clearly, local diversity is sensitive to how one defines *local* and *habitat* and to how intensively one samples the community.

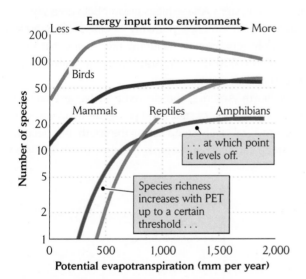

FIGURE 20.11 Species richness is correlated with energy input into the environment. This graph shows the relationship of potential evapotranspiration (PET) to species richness for birds, mammals, amphibians, and reptiles in North America. After D. J. Currie, *Am. Nat.* 137:27–49 (1991).

Regional diversity (or *gamma diversity*) is the total number of species observed in all habitats within a geographic area that includes no significant barriers to the dispersal of organisms. Thus, how we define *region* depends on the type of organism we are interested in. The important point is that within a region, distributions of species should reflect the distribution of suitable habitats, rather than the species' ability to disperse to a particular locality.

If each species occurred in all habitats within a region, then local and regional diversities would be the same. However, if each habitat had a unique biota, regional diversity would equal the sum of the local diversities of all the habitats in the region. Ecologists refer to the difference, or *turnover*, in species from one habitat to another as **beta diversity.** The greater the difference in species between habitats, the greater is beta diversity.

Ecologists have devised many indices to quantify beta diversity, each of them useful in particular applications. Sørensen's similarity index, which compares the species in two communities (1 and 2), is the number of species held in common (C) divided by the average number of species in each community (S_1 and S_2):

$$\text{Sørensen similarity} = \frac{C}{(S_1 + S_2)/2}$$

Sørensen similarity ranges from 0 (completely dissimilar) to 1 (all species shared). A related measure, Jaccard similarity (which also ranges from 0 to 1), is $J = C/(C + U_1 + U_2)$, where U_1 and U_2 are the number of unique (unshared) species in communities 1 and 2.

Another measure of beta diversity is the rate at which similarity decreases with the distance between two samples. An example illustrating this pattern for state- and province-level floras in temperate regions of eastern Asia and eastern North America (Figure 20.12) shows that beta diversity is higher in Asia with respect to both latitude and longitude. As one might expect from the higher turnover of species between areas in eastern Asia, regional (gamma)

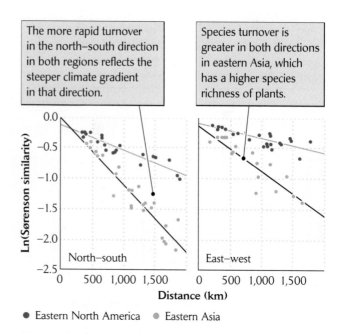

The more rapid turnover in the north–south direction in both regions reflects the steeper climate gradient in that direction.

Species turnover is greater in both directions in eastern Asia, which has a higher species richness of plants.

● Eastern North America   ● Eastern Asia

FIGURE 20.12 Turnover of species with distance is shown by a decrease in the Sørensen similarity index. Here, the logarithm of Sørensen similarity is plotted as a function of the distance between state- or province-level floras in eastern Asia and eastern North America. After H. Qian et al., *Ecol. Lett.* 8:15–22 (2005).

diversity also is higher there. Of course, distance is a stand-in for environmental change—primarily temperature in the north–south (latitudinal) direction and precipitation in the east–west (longitudinal) direction—and Sørensen similarity also decreases with climate differences in both directions. Eastern Asia exhibits greater beta diversity than eastern North America primarily because of its greater topographic complexity, including mountains that rise to more than 6,000 m in the south and the high-elevation Plateau of Tibet in the west.

In North America, the decrease in Jaccard similarity between the floras of states and provinces in an east–west direction is relatively steep across the south, but becomes less steep farther to the north (Figure 20.13). The steep

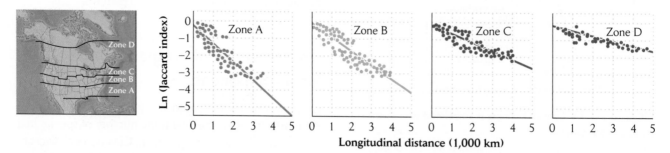

FIGURE 20.13 The beta diversity of plants decreases from south to north in North America.
The decrease in Jaccard similarity between large-scale floras as a function of longitudinal distance is greater in the south than in the north. After H. Qian and R. E. Ricklefs, *Ecol. Lett.* 10:737–744 (2007).

decrease in similarity with distance in the south is strongly related to the contrast in climate between the wet east and the dry west. In the north, distance itself, rather than climate, plays the dominant role. Because climates are more uniform across more northern latitudinal belts, the more important factor may be that not all species have moved back into northern regions that were covered by ice during the Pleistocene glacial cycles.

Other indices of beta diversity take into consideration data from many local communities. For example, Whittaker's measure is the total regional species richness (S_{total}) divided by the average local species richness ($\bar{\alpha}$); that is, $\beta = S_{total}/\bar{\alpha}$. Because $\bar{\alpha}$ can vary only between 1 and S_{total}, $\beta = 1$ when all species occur in all communities ($\bar{\alpha} = S_{total}$) and $\beta = S$, when all communities have a single species. Notice that with Whittaker's measure, regional diversity = local diversity × beta diversity ($\gamma = \bar{\alpha}\,\beta$).

Local communities and the regional species pool

The species that occur within a region are referred to as its **species pool.** All the members of the regional species pool are potential members of each local community. Yet not all of those species are found everywhere in the region. A central concept of ecology is that membership in local communities is restricted to those species that can coexist in the same habitat. Thus, each local community is a subset of the regional species pool.

Any species present within a local community must be able to tolerate the conditions of its environment and find suitable resources for survival and reproduction. As we saw in Chapter 10, the *fundamental niche* of a species is the range of conditions and resources within which individuals of that species can persist. Interactions with other species, however, may restrict the distribution of that species to those parts of its fundamental niche where it is most successful. This more restricted range of conditions and resources is referred to as the species' *realized niche.*

Whether a species occurs in a local community depends both on its adaptations to local conditions and resources and on its interactions with competitors, consumers, and pathogens. The species present within the regional species pool are thus sorted into different communities based on their adaptations and interactions. This process is referred to as **species sorting.**

Brown University ecologist Mark Bertness and his colleagues have described a striking case of species sorting, observed in the distributions of mussels and coralline algae on the rocky coast of Patagonia, in southern Argentina. On wave-exposed headlands, rock surfaces at all levels in the intertidal zone are covered with the mussel *Perumytilus purpuratus.* On nearby rocky shores in protected bays, the upper parts of the intertidal zone are bare, the middle level is dominated by mussels, and the lower levels are covered with the erect coralline alga *Corallina officinalis* (Figure 20.14). Both species can settle at any of these levels. Species are sorted between exposed and protected

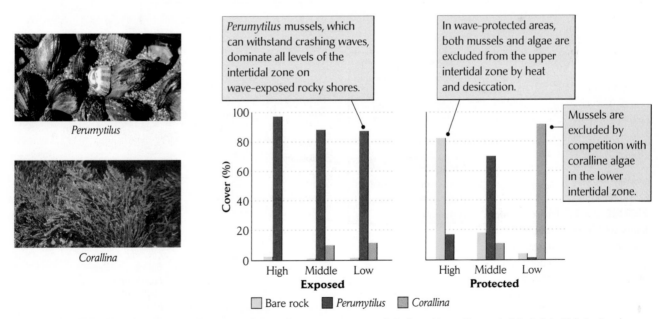

FIGURE 20.14 Species sorting on the Patagonian rocky coast depends primarily on physical factors. After M. D. Bertness et al., *Ecol. Monogr.* 76:439–460 (2006). Photos by Dr. Dirk Schories/ Instituto de Biologia Marina Universidad Austral de Chile (top) and Daniel W. Gotshall/Visual Unlimited (bottom).

sites by physical factors. Coralline algae cannot withstand wave battering, but mussels secrete strong byssal threads that provide a firm attachment to the rocks. In protected areas, rock surfaces heat up so much in the upper intertidal zone that neither species can survive without the constant splashing of waves; in the lower part of the intertidal zone, coralline algae overgrow the mussels and exclude them through direct competition.

Species sorting can also be demonstrated experimentally by bringing together many species from a regional species pool in a variety of habitats. Over time, species interactions will cause the elimination of some species from these experimental communities, although which species disappear will vary from habitat to habitat depending on the particular adaptations of the species to local environmental conditions and resources.

ECOLOGISTS IN THE FIELD **Species sorting in wetland plant communities.** Evan Weiher and Paul Keddy of the University of Ottawa established a large

experiment to investigate the principles of species sorting. Working with wetland plants, they sowed seeds of 20 species in 120 wetland microcosms that differed with respect to soil fertility, water depth, fluctuation in water depth, soil texture, and organic leaf litter. These artificial communities were followed for 5 years. Although the total biomass of vegetation increased over the course of the experiment, the number of species in each of the communities decreased as dominant competitors excluded other species. Diversity in each community fell from 15 species at the beginning of the experiment to an average of 3 or 5 species, depending on the fertility of the soil, at the end. Moreover, several distinct communities developed under different environmental conditions, so that the species composition of the communities differed.

Weiher and Keddy thought of the environmental variables as filters. If a species was unable to tolerate or compete effectively under a particular set of environmental conditions, then it did not join communities having those conditions (Figure 20.15). One of the 20 species in the original species pool failed to germinate under any of the combinations of conditions in the experimental micro-

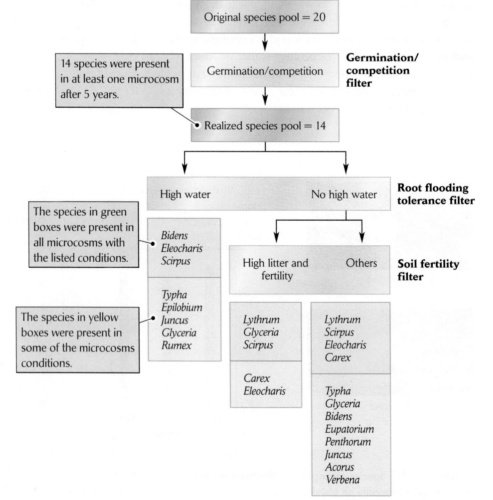

FIGURE 20.15 Environmental filters contribute to species sorting. Certain environmental conditions in a local community eliminate those species that cannot tolerate or compete under those conditions. After E. Weiher and P. A. Keddy, *Oikos* 73:323–335 (1995).

cosms, and 5 others were unable to persist under any of those combinations. The most important environmental filters sorting out the remaining 14 species were high versus low water levels and high versus low soil fertility. The experiment clearly demonstrated the sorting of species from a regional pool in relation to the particular habitat conditions of local communities.

Similar long-term experiments on prairie grasses and forbs have been performed by David Tilman and his colleagues at the University of Minnesota. As in Weiher and Keddy's experiment, plant species were sorted with respect to soil nutrient levels, but the investigators showed that the number of species that coexisted in local experimental communities also depended on the number of species seeded in the plots. This result demonstrates that local diversity also reflects the size of the regional species pool. ▌

Species interactions and ecological release

As we have just seen, competitive interactions between species play a major role in species sorting. Thus, for a given range of habitats, species sorting should be greatest where the regional pool contains the most species. In such a situation, each species should be able to maintain itself over only a narrow range of habitats—those to which it is best adapted—and beta diversity should be high.

This relationship has been documented in studies comparing islands and neighboring continental regions. In such studies, one can compare species richness among regions having a similar climate and range of habitats, but different degrees of geographic isolation. Islands usually have fewer species than comparable mainland areas, but island species often attain greater densities than their mainland counterparts. In addition, they expand into habitats that would be filled by other species on the mainland. Collectively, these phenomena are referred to as **ecological release.**

Ecological release can be seen in surveys of bird communities over the same range of habitat conditions in two continental regions and five islands of various sizes within the Caribbean basin. These surveys show that where fewer species occur, each is likely to be more abundant and to live in more habitats (Figure 20.16). Thus, as the size of the regional species pool decreases, the realized niche of each species becomes broader. Because the range of habitats is the same in each of the regions, the differences among the regions in average species abundance and number of habitats occupied by each species can be attributed to local interactions among species within the pool of each region.

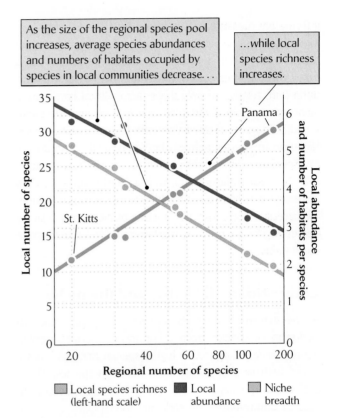

FIGURE 20.16 Populations in regions with few species show ecological release. The species richness of bird communities was surveyed in seven regions of the Caribbean basin with species pools of different sizes, ranging from Panama, a large region with a relatively large species pool, to St. Kitts, a small, isolated island with a small species pool. After G. W. Cox and R. E. Ricklefs, *Oikos* 29:60–66 (1977); J. M. Wunderle, *Wilson Bull.* 97:356–365 (1985).

Diversity can be understood in terms of niche relationships

The fundamental niche represents the range of environmental conditions, habitat structures, and resource qualities within which an individual or species can survive and reproduce. Thus, for example, the boundaries of a particular species' niche might extend between temperatures of 10°C and 30°C, prey sizes of 4 and 12 mm, perches on branches with diameters between 5 and 20 mm, or daytime light levels between 10 and 50 W per m². Of course, the niche of any species would include many more variables than these, and ecologists often cite the multidimensional nature of the niche to acknowledge the complexity of species–environment relationships.

The degree to which the niches of two species overlap determines how strongly those species might compete. Thus, the niche relationships of species provide an informative measure of community structure. Every biological community can be thought of as having a total niche space within which its member species must fit. Thus, communities with different numbers of species can differ with respect to only three factors: total community niche space, niche overlap among species, and the niche breadth (degree of specialization) of individual species. Variation in species richness between local communities can represent differences in their total niche space. However, in a particular locality, with a fixed volume of total niche space, species can be added to the community only by increasing niche overlap (sharing of niche space) or by decreasing niche breadth (partitioning total niche space more finely).

Competition, diversity, and the niche

As we saw in Chapter 16, intense competition leads to exclusion of species from a community. Consequently, many ecologists argue that higher species richness is associated with weaker competition between species. There are several mechanisms that could increase species richness in communities by reducing interspecific competition. Essentially, competitors would have to either avoid competition through ecological specialization or be limited by predators rather than by resources.

Most ecologists agree that the high species richness in the tropics results at least in part from the presence of a greater variety of ecological resources. For example, the number of bird species increases at lower latitudes in part because there are more fruit-feeding species, nectar-feeding species, and insectivorous species that hunt by searching for prey while quietly sitting on perches. These feeding behaviors are uncommon among birds in temperate regions because suitable resources are lacking there. Among mammals, the tropics are species-rich primarily because of the many bat species in tropical communities, which can feed on fruits, nectar, and night-flying insects. Nonflying mammals are no more diverse at the equator than they are in temperate regions. Similarly, herbivorous insects have diversified to take advantage of the immense variety of plant species in the tropics. Epiphytes and lianas (woody vines), which are generally absent from forests at higher latitudes because of freezing winter temperatures, augment tropical plant diversity.

Species diversity and niche diversity

Species richness is generally paralleled by the functional diversity, or niche diversity, of the species in a community.

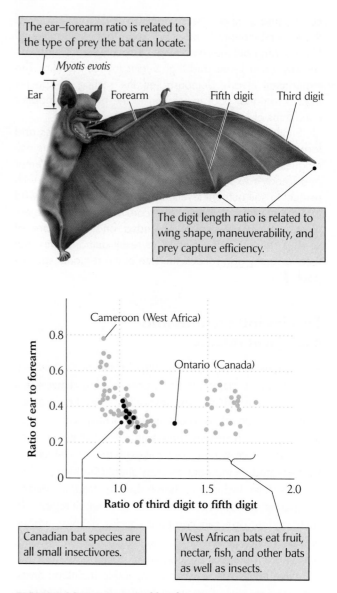

FIGURE 20.17 A tropical bat fauna occupies more morphologically defined niche space than a temperate bat fauna. The niche relationships among the species in each of two bat faunas, one from southeastern Ontario, Canada, and one from Cameroon, West Africa, are visualized by plotting their distribution in morphological space. After M. B. Fenton, *Can. J. Zool.* 50:287–296 (1972).

One way to assess niche diversity is to use morphology as an indicator of ecological role—that is, to assume that differences in morphology among related species reveal different ways of life. For example, the sizes of prey captured vary in relation to the body sizes of consumers, and different shapes of appendages can be related to different methods of locomotion for hunting and escaping predators. Morphological analyses of communities have consistently revealed that the number of species packed into a certain amount of morphologically defined niche space is relatively constant—in other words, the average niche breadth remains the same in communities with different numbers of species. Therefore, as species diversity increases, so does niche diversity. This finding suggests that more diverse communities have a greater variety of ecological resources (more niche space).

To illustrate this principle, let's compare bat communities in temperate and tropical localities. A morphological space defined by two axes represents many of the important ecological properties of bats. The first axis—the ratio of ear length to forearm length—is a measure of ear size relative to body size. This ratio is related to the bat's sonar system and thus to the type of prey it can locate. The second axis—the ratio of the lengths of the third and fifth digits of the hand bones in the wing—describes whether the wing is long and thin or short and broad. Therefore, this axis determines how a bat flies, and hence the types of prey it can pursue and the habitats within which it can capture prey efficiently.

To visualize the niche relationships among species, we plot each bat species in each community on a graph whose axes are these two morphological ratios (Figure 20.17). In the less diverse community in Ontario, in southern Canada, all the bat species are small insectivores having similar morphology. The bats of the morphologically more diverse community in Cameroon, in tropi-

cal West Africa, exploit a far greater variety of ecological resources. In addition to small insectivorous species, there are fruit eaters, nectar eaters, fish eaters, and large, predatory bat eaters, to name a few.

Another example of the relationship between species diversity and niche diversity comes from freshwater fish communities. In streams and rivers, the numbers of species in most taxonomic groups increase from the headwaters to the mouth. One presumes that as a stream increases in size, it offers a greater variety of ecological opportunities, more abundant resources, and more stable and therefore more reliable physical conditions. Local communities reflect these changes. For example, a headwater spring in the Rio Tamesi drainage of east central Mexico was found to support only one fish species, a detritus-feeding platyfish (*Xiphophorus*) (Figure 20.18). Farther downstream, two more species were added to the community: a detritus-feeding molly (*Poecilia*) that prefers slightly deeper water, and a mosquito fish (*Gambusia*) that eats mostly insect larvae and small crustaceans. Fish communities even farther downstream included additional carnivores—among them, fish eaters—and other fish that feed primarily on filamentous algae and vascular plants. Downstream communities had all the species found in upstream communities plus additional ones restricted to downstream localities. Thus, species diversity increases as a stream becomes larger and presents more kinds of habitats and a greater variety and abundance of food items.

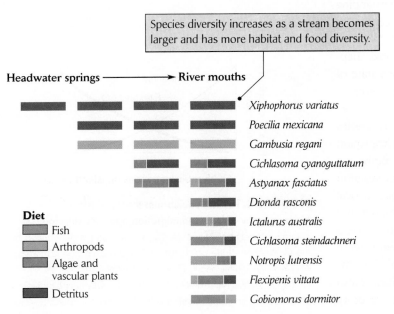

FIGURE 20.18 **Fish exhibit more ecological roles in more diverse communities.** Fish were sampled at four locations in the Rio Tamesi drainage of east central Mexico, from a headwater spring to a downstream community at the river mouth. After R. M. Darnell, *Am. Zool.* 10:9–15 (1970).

Equilibrium theories of diversity balance factors that add and remove species

Our survey of diversity patterns suggests several general conclusions. At a global scale, species richness increases dramatically from high latitudes toward the equator. Within latitudinal belts, diversity appears to be correlated with temperature, ecosystem productivity, topographic heterogeneity within a region, and the structural complexity of local habitats. Isolated islands exhibit species impoverishment. Everywhere, higher diversity is associated with greater ecological variety.

How do we explain these patterns of diversity? The ultimate source of diversity is speciation: the production of new species by the splitting of evolving lineages. We shall discuss the mechanisms by which new species appear in Chapter 21. Species also disappear; indeed, most of the species that have ever existed are extinct. Changes in diversity within regions over time reflect the balance of speciation and extinction. If new species are produced faster than old ones disappear, diversity should tend to increase until a mass extinction event occurs. Alternatively, diversity might achieve a stable equilibrium at a point at which speciation and extinction exactly balance.

Most mechanisms of speciation require large areas, so any speciation–extinction balance would have to be explained, at least in part, by regional processes; local communities would reflect only sorting from the regional species pool. Differences in diversity between regions could arise as a consequence of differences in rates of speciation, extinction, or both. Therefore, understanding these large-scale processes is essential to understanding patterns of species diversity on a global scale. We shall explore these processes in more detail in the next chapter. Here, we turn to the general idea of a steady state of species richness, illustrated, in particular, by patterns of diversity on oceanic islands.

Steady-state, or equilibrium, models of local species richness resemble models of density-dependent regulation of population size, discussed in Chapter 11. Births are analogous to the formation of new species or colonization by species from elsewhere; deaths are analogous to local extinctions of species. Accordingly, each community has a steady-state number of species, just as a habitat has a carrying capacity for a particular species. Ecologists were attracted to this view because it helped explain what was known about species diversity within local habitats. It also placed at least part of the problem of species diversity

within the domain of ecology: present-day processes taking place within small areas.

Species richness on islands

During the 1960s, Robert MacArthur, then at the University of Pennsylvania, and E. O. Wilson, at Harvard University, developed their famous **equilibrium theory of island biogeography,** which states that the number of species on an island balances regional processes governing immigration against local processes governing extinction. This theory can be illustrated with simple model.

Consider a small offshore island. It is too small for new species to form locally, so the number of species on the island can increase only by immigration from other islands or from a continental landmass. The biota of the closest continental area makes up the species pool of potential colonists. As the number of species on the island increases through colonization, the rate of immigration of new species decreases, because as more species from the mainland pool of potential colonists become established on the island, fewer colonizing individuals belong to new species. When all mainland species occur on the island, the immigration rate of new species must be zero. Species also go extinct on the island, and the rate of extinction increases with the number of species at risk. Thus, the colonization rate decreases, and the extinction rate increases, as a function of the number of species on an island. The number of species at which the immigration and extinction curves cross represents a steady state (Figure 20.19).

Immigration and extinction rates are unlikely to be strictly proportional to the number of potential colonists

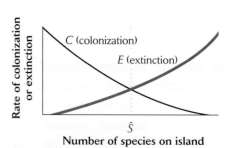

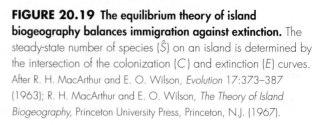

FIGURE 20.19 The equilibrium theory of island biogeography balances immigration against extinction. The steady-state number of species ($\hat{S}$) on an island is determined by the intersection of the colonization (C) and extinction (E) curves. After R. H. MacArthur and E. O. Wilson, *Evolution* 17:373–387 (1963); R. H. MacArthur and E. O. Wilson, *The Theory of Island Biogeography,* Princeton University Press, Princeton, N.J. (1967).

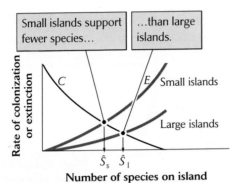

FIGURE 20.20 Smaller islands support fewer species because of higher extinction rates. $\hat{S}_s$ = equilibrium number of species on small islands; $\hat{S}_l$ = equilibrium number of species on large islands.

and the number of species established on an island. Some species undoubtedly colonize more easily than others, and those species reach the island first. The remaining species in the pool of potential colonists have a lower probability of reaching the island and becoming established, so the rate of immigration initially decreases more rapidly with increasing island diversity than it would if all mainland species had equal colonization potential. Accordingly, the immigration rate should follow a curve. Furthermore, as the number of species on the island increases, competition between species increases the probability of extinction by reducing the abundances of individual species (see Figure 20.16); accordingly, the extinction curve should rise ever more rapidly as species diversity increases.

If smaller populations have a higher probability of extinction, then extinction curves should be higher for small islands than for large islands, and small islands should support fewer species at the steady state than large islands (Figure 20.20). Furthermore, if the rate of immigration to

islands decreases with increasing distance from mainland sources of colonists, then far islands should receive fewer immigrants than near islands and should therefore support fewer species (Figure 20.21).

LIVING
GRAPHS   To access an interactive tutorial on island biogeography and the dynamics of species diversity, go to http://www.whfreeman.com/ricklefs6e.

The equilibrium theory of island biogeography describes a dynamic and resilient steady state. Therefore, if some disaster exterminated a part of an island's biota—or all of it, as in the case of Krakatau (see Chapter 19)—new colonists would, over time, restore diversity to its pre-disturbance equilibrium. This prediction was tested by Daniel Simberloff, at that time a graduate student at Harvard University and presently at the University of Tennessee, along with his advisor, E. O. Wilson. After first counting the arthropod species present on each of four small mangrove islands in Florida Bay, Simberloff removed the entire arthropod fauna by fumigating the islands with methyl bromide. (This was a major accomplishment achieved by erecting metal scaffolds covered with plastic sheeting over entire red mangrove trees.) The islands were then resampled at regular intervals for a year (Figure 20.22).

As predicted, arthropod species richness increased more rapidly on islands nearer to sources of colonists than farther away. Numbers of species on both near and far islands began to level off before the end of the experiment, indicating that a steady state had been reached in each case. In addition, as predicted, the new equilibrium numbers of species were similar to the numbers of species present on the islands prior to defaunation. These results supported the equilibrium theory and suggested that ecological processes, as opposed to long-term evolutionary processes, could explain patterns of variation in local species richness.

Equilibrium theory in continental communities

The equilibrium view of diversity also applies to large islands and continents. The main difference between small islands and larger "continental" islands is that new species can form on large islands, in addition to arriving from outside by immigration. Because new species can form there, larger islands have higher species richness than would be expected from colonization rates alone. We can see this effect of species formation in the diversities of

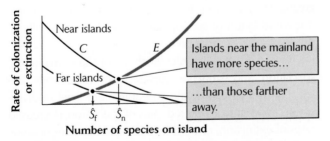

FIGURE 20.21 Islands close to the mainland support more species because of higher immigration rates. $\hat{S}_f$ = equilibrium number of species on far islands; $\hat{S}_n$ = equilibrium number of species on near islands.

FIGURE 20.22 Patterns of recolonization on four small islands supported the equilibrium theory of island biogeography. The entire arthropod faunas of four small mangrove islands in the Florida Keys were exterminated by methyl bromide fumigation. Estimated numbers of species present before defaunation are indicated at left. After D. S. Simberloff and E. O. Wilson, *Ecology* 50:278–296 (1970).

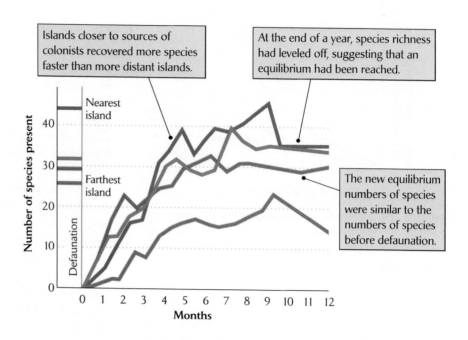

Islands closer to sources of colonists recovered more species faster than more distant islands.

At the end of a year, species richness had leveled off, suggesting that an equilibrium had been reached.

The new equilibrium numbers of species were similar to the numbers of species before defaunation.

Anolis lizards on islands in the West Indies. Among small islands, the increase in species richness with area reflects the shifting colonization–extinction equilibrium. However, islands exceeding about 1,000 km^2 in area (Puerto Rico, Jamaica, Hispaniola, and Cuba) are large enough for new species to form, and the slope of the species–area relationship abruptly increases from $z = 0.06$ to $z = 0.76$ at that island size. Larger islands—essentially continents—such as Madagascar, New Guinea, and Aus-

tralia, have supported evolutionary radiations in many endemic animal and plant lineages.

Does higher species richness lead to higher rates of extinction or species formation? There is no single answer to this question. Curves relating rates of speciation and extinction to diversity within large regions might look like those in Figure 20.23. Whether the probability of extinction per species increases or decreases at higher species numbers could depend on several factors. Increasing species numbers might encourage the extinction rate to rise if competitive exclusion increased with species richness, whereas the extinction rate might decrease if mutualistic relationships and alternative pathways of energy flow buffered processes that lead to extinction. The rate of speciation might level off once most opportunities for further diversification had been taken advantage of by some species, whereas it might increase if diversification led to greater specialization and a higher probability of reproductive isolation of subpopulations, leading eventually to the formation of separate new species.

Regardless of the particular shape of the immigration, speciation, and extinction curves, many biologically reasonable models can define an equilibrium level of diversity. In every case, increased rates of speciation, decreased rates of extinction, or both lead to higher species richness at the equilibrium point. Once this principle is understood, the challenge is to identify the factors responsible for variation in the rates of species appearance and disappearance in different places.

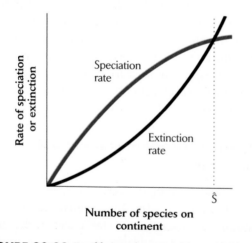

FIGURE 20.23 Equilibrium theory can be applied to continental regions. On a continent, new species are added to the regional pool by the evolutionary process of speciation as well as by immigration from elsewhere. After R. H. MacArthur, *Biol. J. Linn. Soc.* 1:19–30 (1969).

Explanations for high tree species richness in the tropics focus on forest dynamics

Because the number of animal species can be related to the diversity of their plant resources, explaining global patterns of plant diversity poses the greater challenge to ecologists. This challenge is particularly evident when we consider that all plants compete for the same essential resources—light, soil nutrients, and water—and so their opportunities for partitioning niche space may be quite limited. One way of looking at the issue of plant species diversity is to distill it into a single question: "Why are there so many different kinds of trees in the tropics?"

There are several plausible mechanisms that could lead to high tree species richness in a region:

1. Environmental heterogeneity allows species to coexist because they can specialize on different parts of the niche space.

2. Disturbances, such as tree falls, are important sources of environmental heterogeneity in forests.

3. Herbivores and pathogens afflict common species more than they do rare ones, and the resulting rare species advantage allows many species to coexist.

4. Because tree species are closely matched ecologically, competitive exclusion takes a long time, so species added to a community are likely to remain there.

Each of these possible mechanisms is an example of local interactions. Regional processes of species formation and extinction also influence global patterns of species richness by influencing the regional species pools from which local communities are drawn. However, we shall defer consideration of those processes to the next chapter, which deals with history and biogeography.

Environmental heterogeneity

Many ecologists have argued that the number of tree species varies in proportion to the heterogeneity of the environment. All tree species, whether tropical or extra-tropical in distribution, are specialized with respect to soil and climate. Some species flourish on well-drained soils on slopes, while others reach their highest abundances on low-lying wet soils. Tree species also replace each other across environmental gradients. Could greater variation in the physical environment in the tropics account for the tenfold (or more) greater diversity of trees in tropical

than in temperate forests? It seems unlikely, unless trees recognize much finer habitat differences in the tropics than they do in temperate regions, especially considering that temperate regions contain greater heterogeneity in some climatic factors. For example, seasonal variation in temperature and differences between north-facing and south-facing slopes are more marked in temperate regions. Thus, although many tropical trees are specialized for their habitat or microhabitat, this factor probably does not explain the latitudinal gradient in species diversity.

Disturbance and gap dynamics

Several ecologists, particularly Joseph Connell of the University of California at Santa Barbara, have related the high species richness of tropical rain forests to habitat diversity created by disturbance. We have already touched on the role of disturbance in Chapter 19: disturbance of communities by physical conditions, predators, or other factors opens up space for colonization and initiates a cycle of succession by species adapted for colonizing disturbed sites. With a moderate level of disturbance, a community becomes a mosaic of habitat patches at different stages of succession; together, these patches contain the full diversity of species characteristic of a sere. This idea is known as the **intermediate disturbance hypothesis.**

For the intermediate disturbance hypothesis to account satisfactorily for differences in species richness between regions, especially of the magnitude of the latitudinal differences in tree species diversity, there would have to be comparable latitudinal differences in levels of disturbance. The most frequent disturbance of consequence in forests is the death of an individual tree, which opens space for the establishment of new individuals. Yet death rates of forest trees do not differ appreciably between temperate and tropical areas; about 0.5%–2% of the individuals that make up the canopy die each year in both regions. Nor is it likely that major disturbances such as storms and fires are more frequent in the tropics.

Although disturbances may be equally frequent in tropical and temperate regions, their effects may vary with latitude. The tropics experience more precipitation, soils have less organic matter, and the sun beats down from directly overhead for much of the day. These factors create greater differences between forest gaps and the rest of the environment, and seemingly should provide more opportunities for habitat specialization in the tropics.

Many species in both tropical and temperate forests are gap specialists whose persistence depends on the occasional formation of new gaps in the forest canopy.

In the tropics, many of these gap specialists are lianas, which contribute a special character to the structure of tropical forests. However, only about half the tree species in tropical forests depend on gaps for regeneration. There-fore, while gap formation contributes to the diversity of any forest, it is doubtful that disturbances of this kind can explain the tenfold difference in tree species diversity between the tropics and temperate regions.

Steve Hubbell, of the University of California at Los Angeles, and his colleagues at the Smithsonian Tropical Research Institute have studied recruitment of tree seed-lings in gaps in a tropical rain forest on Barro Colorado Island, Panama, for more than 25 years. Within their 50 ha study plot, differences in gap size and frequency of gap formation did not explain variation in tree species richness. The numbers of species represented as seedlings were the same in gaps and in control sites outside gaps. Furthermore, gaps were colonized by shade-tolerant as well as gap-dependent species.

These findings led Hubbell to suggest that even though tree species might be specialized for different kinds of ger-mination sites, which species actually invade a gap depends more on the vagaries of recruitment than on environmental conditions in the gap. Accordingly, competition for germi-nation sites is reduced simply because not all species reach germination sites for which they can compete effectively. As a result, more species can coexist within the region because gap sites do not all contain the same suite of spe-cies. However, such recruitment limitation is partly a *conse-quence,* rather than a *cause,* of the high species diversity and low average abundance of species in tropical forests. Thus, recruitment limitation may not explain why tropical forests became so diverse, but it does suggest that competitive exclusion decreases as diversity increases. In other words, biodiversity may be self-accelerating in this regard.

One of the ways in which tree species differ is in their growth rates. Pioneering species grow rapidly, but many require the high-light environments of gaps to become established. Other species grow slowly, but can germinate under the shade of the forest canopy. This range of life his-tories extends from species that disperse well to colonize isolated forest gaps, grow up quickly, and die young to species that become established less frequently but persist for long periods in the forest canopy. If species richness reflected this niche diversity, then one would expect a correlation between species richness and variation among species in these life history traits. When Richard Condit and his colleagues associated with the Center for Tropi-cal Forest Science applied this test to data from ten plots within the tropics, they found the opposite trend: spe-cies in the more diverse forests exhibited less variation in

growth and survival rates. In this respect, higher species richness was associated with more similar life histories.

Herbivore and pathogen pressure

When consumers reduce resource populations below their carrying capacities, they can reduce competition and promote the coexistence of many resource species. More-over, when consumers feed selectively on superior com-petitors, which are often more abundant, competitively inferior species may persist in a community. We saw an example of this effect in Chapter 16 in the case of preda-tory sea stars that reduced populations of mussels on the rocky coast of Washington, thereby allowing the persis-tence of less competitive marine invertebrates.

From Darwin's time to the present, naturalists have believed that both selective and nonselective herbivory can influence the diversity of plant species. Daniel Jan-zen, of the University of Pennsylvania, suggested many years ago that herbivory could be responsible for the high species richness in tropical forests. He argued that herbi-vores feed on buds, seeds, and seedlings of abundant spe-cies efficiently enough to reduce their densities, allowing other, less common species to grow in their place. The key to this idea is that abundance per se, rather than any particular quality of individuals as resources, makes a spe-cies vulnerable to consumers. Consumers locate abundant resource species easily, and therefore their own popula-tions grow to high levels. Abundant resource populations also exert strong selection pressure on potential consum-ers to specialize on them.

Several lines of evidence support this "pest pressure" hypothesis. For example, attempts to establish plants in monoculture frequently fail because of infestations of her-bivores. Dense plantations of rubber trees in their native habitats in the Amazon basin, where many species of her-bivores have evolved to exploit them, have met with a singular lack of success. Nevertheless, rubber tree planta-tions thrive in Malaysia, where specialist herbivores are not (yet) present. Attempts to grow many other commer-cially valuable crops in single-species stands in the tropics have met the same disastrous end. Escape from predators, herbivores, and pathogens may be one reason why some introduced species become invasive outside their native distributions.

The pest pressure hypothesis predicts that seedlings should be less likely to become established close to adults of the same species than at a distance from them. Adult individuals may harbor populations of specialized her-bivores and pathogens that could readily infest nearby seedlings. Furthermore, because most seeds fall close to

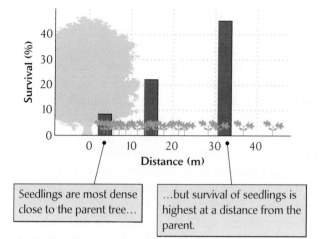

FIGURE 20.24 Seedling survival varies with distance from the parent tree. Investigators tracked the survival rates of seedlings of the Neotropical tree *Dipteryx panamensis* to 18 months of age. After D. H. Janzen, *Am. Nat.* 104:501–528 (1970); D. A. Clark and D. B. Clark, *Am. Nat.* 124:769–788 (1984).

their parent plant, herbivores may be attracted to the abundance of seedlings there while overlooking the few that disperse to a more distant location. The prediction that success in germination and establishment should decrease with seedling density and increase with distance from the parent has been tested in a number of studies, which have yielded varied but generally supportive results (Figure 20.24).

In an experiment on seedling survival in *Sebastiana longicuspis*, a common tree species in Belize, Thomas Bell and his colleagues at the University of Oxford established 0.25 m² study plots under mature trees that contained naturally established seedlings. They eliminated pathogenic fungi in some of the plots by treating them with a selective fungicide, and left others untreated as control plots. In half of the plots in each treatment, the investigators thinned seedlings to a density of 100 individuals per square meter; unmanipulated plots were left with natural densities of 400–1,100 individuals per square meter. Thinning improved seedling survival in the untreated plots, and fungicide application substantially increased survival in both low-density and high-density plots over the month-long experiment (Figure 20.25). Similarly, detailed observations on a 50 ha forest plot on Barro Colorado Island have shown that seedling survival exhibits similarly strong density dependence in most species, favoring seedlings at greater distance from adult trees of the same species, where seedling density is lower.

In the temperate zone, few seeds escape predation by squirrels and weevils, and herbivores and pathogens attack seedlings just as they do in the tropics. If pest pressure does promote greater diversity in the tropics, it must operate differently in different latitudinal belts. In particular, tropical herbivores and plant pathogens must be either more specialized with respect to host plant species or more sensitive to the density and dispersion of host populations. Few studies have examined pest pressure in both tropical and temperate latitudes. However, recent experiments by Carol Augspurger and Henry Wilkinson of the University of Illinois tested the basic premise of host specificity in tropical soil pathogens.

The researchers took samples of the soil pathogen *Pythium*, an oomycete fungus that causes damping-off disease, from natural environments in Panama and maintained these isolates in the laboratory. In addition, temperate isolates of *Pythium* known to be pathogenic to crop plants were obtained from stock cultures. They exposed seedlings of eight tropical tree species to 75 tropical isolates (of unknown pathogenicity) and 7 temperate isolates. Three of the temperate isolates and three of the tropical isolates turned out to be pathogenic to the tropical tree seedlings, and each of these isolates infected seedlings of

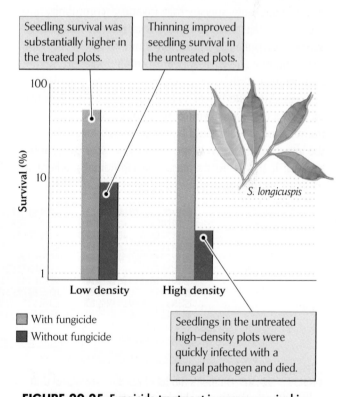

FIGURE 20.25 Fungicide treatment improves survival in seedlings of a tropical tree. Researchers treated some plots containing seedlings of *Sebastiana longicuspis* with fungicide and left others untreated. In addition, they thinned half of the plots in each treatment to create low-density plots. After T. Bell et al., *Ecol. Lett.* 9:569–574 (2006).

up to five of the species tested. Thus, *Pythium* does not appear to be specialized with respect to its host species, and several of the temperate isolates were able to infect tropical tree species even though these pathogens and hosts had no previous evolutionary history together. If soil pathogens were responsible for maintaining tropical tree diversity, we would have expected to find a higher degree of host specialization. Of course, ecologists need to conduct much more research on this issue. At the present time, however, although self-limitation appears to be very powerful in populations of tropical forest trees, the underlying mechanisms are not well understood, and their role in maintaining latitudinal diversity gradients has not been confirmed.

Random ecological drift

Pest pressure and recruitment limitation can reduce the consequences of interspecific competition for community membership. Steve Hubbell, in his book *The Unified Neutral Theory of Biodiversity and Biogeography,* has argued that these factors, along with limited ecological specialization, make most tropical tree species competitively equivalent. Accordingly, new species that invade a community are likely to remain there for long periods, if not indefinitely. Species disappear by extinction more or less at random, just as neutral alleles disappear from a population by random genetic drift (see Chapter 13). Large populations of trees distributed over extensive areas of tropical forest would be relatively immune to such extinction, so the numbers of species could build up to high levels through the production of new species within large geographic regions. If this were the case, the latitudinal gradient in species diversity would reflect region size and regional rates of species production more than the ability of species to coexist in local communities.

Imagine a tropical forest with a fixed number of individual trees belonging to S number of species, all of which are competitively equivalent (that is, "neutral"). When an individual dies, it is replaced by the progeny of one of the other trees in the forest at random (Figure 20.26). Thus, the fitnesses of all the trees in the forest are identical— their probabilities of death and reproduction are the same. Because death and reproduction are stochastic processes under these circumstances, a forest with N individuals will eventually consist of the progeny of a single ancestral individual, and thus be reduced to a single species, in an average of N generations. This tendency toward random ecological drift would be countered within small areas by immigration of individuals from other areas (recall the role of colonization in maintaining species richness on islands) and within a region-sized area (the *metacommunity*) by speciation. Under one model of speciation, the equilibrium species richness in the metacommunity (regional diversity) would be $S = J_M v$, where J_M is the number of individuals in the metacommunity and v (the Greek letter "nu") is the rate of speciation expressed per individual.

Thus, in Hubbell's model, species richness is a function of region size (hence number of individuals, which determines the extinction rate) and the rate of species formation. When Hubbell incorporated a rate of migration between local populations, his theory successfully predicted species richness, species abundance distributions, species–area relationships, and beta diversity. It is a wonderfully comprehensive model that has attracted much attention in ecology. There is a problem with this model, however:

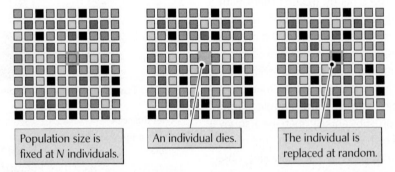

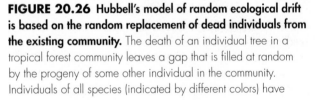

| Population size is fixed at N individuals. | An individual dies. | The individual is replaced at random. |

FIGURE 20.26 Hubbell's model of random ecological drift is based on the random replacement of dead individuals from the existing community. The death of an individual tree in a tropical forest community leaves a gap that is filled at random by the progeny of some other individual in the community. Individuals of all species (indicated by different colors) have equal probabilities of dying or filling openings in the forest during each time interval. Without the formation of new species or immigration from outside the community, a forest with N individuals would be filled with the descendants of a single individual in an average of N generations.

random ecological drift is a slow process. The time in generations required for a neutral system to achieve equilibrium is about the same as the number of individuals in the metacommunity. The average time (in generations) to extinction of a species is about the same as the size of its population. Consider some back-of-the-envelope calculations for forest trees of the Amazon basin, which contains about $J_M = 10^{10}$ (10 billion) individual trees of about 10^4 (10,000) species. Thus, the average population per species is about 10^6 (1 million) individuals; average generation time is about 10^2 (100) years. According to Hubbell's theory, equilibrium would be attained in about 10^{12} (a trillion) years, and the average time to species extinction would be 10^8 (100 million) years. Random ecological drift is clearly too slow to account for diversity on this scale.

How well Hubbell's model explains patterns of latitudinal diversity for trees, or for any other group, remains to be seen. The basic premise of neutrality and stochastic change might apply better to the local scale—perhaps 50 ha of forest—than to large regions because on that smaller scale, random events would become more important. Nonetheless, Hubbell's theory emphasizes the potential importance of large-scale regional processes for understanding ecological patterns. Certainly, from the standpoint of regional diversity, region size and rates of species production and extinction must be important considerations.

But the descriptions of local and regional processes in this chapter do not tell us the whole story of biodiversity. The species present in any community at a particular time represent millions of years of adaptation and diversification on a global scale. In the next chapter, we shall consider some of the basic issues that history and geography raise for the study of ecological systems.

SUMMARY

1. A conspicuous pattern revealed by studies of biological communities is the tendency of species diversity in tropical regions to greatly exceed that at higher latitudes. Explaining patterns of species diversity has been a major challenge for ecologists.

2. The simplest measure of biodiversity is species richness. But in any community, some species are common and others are rare. Relative abundances in a sample or community may be characterized by a rank-abundance plot.

3. Because the number of species increases with the number of individuals sampled, several indices of species richness, such as Margalef's and Menhinick's indices, take into account the size of the sample. Other indices of diversity, most notably Simpson's index, account for variation in the abundances of species when comparing diversity between samples.

4. The number of species in a sample increases in proportion to the area sampled. Three processes determine the slope of this species–area relationship: sample size at local scales, habitat diversity at regional scales, and the evolution of distinct lineages at global scales.

5. When islands of different sizes are compared, the slope of the species–area relationship tends to be higher than on the mainland because larger islands tend to have more habitat diversity than smaller ones, because they are better targets for colonization, and because larger populations better resist extinction.

6. Species richness reflects environmental heterogeneity within a region as well as local environmental conditions. Species richness tends to increase with habitat productivity, but the complexity of vegetation structure has an even greater influence on species richness.

7. Variation in solar energy input is strongly associated with species richness in north-temperate areas, whereas variation in precipitation is strongly associated with species richness throughout the tropics and the Southern Hemisphere.

8. Local, or alpha, diversity is the number of species in a small area of homogeneous habitat. The total diversity of species within a region containing many habitats is regional, or gamma, diversity. The difference, or turnover, in species from one habitat to another is beta diversity.

9. Local communities contain a subset of the regional species pool. Membership in a local community is determined by the adaptations of species to local conditions and resources and by their interactions with other species. The processes that determine local community composition are collectively referred to as species sorting.

10. Ecological release, which is an increase in the population densities and habitat distributions of species in less diverse communities, provides strong evidence that competition for resources structures biological communities and limits diversity.

11. Attempts to measure niche space consistently show that more diverse communities have more niche space and that average niche breadth is independent of local species richness. Accordingly, species diversity increases with niche diversity.

12. Recent thinking about diversity has been dominated by equilibrium theories, which state that diversity reflects a balance between processes that add species to a community and those that remove species. Thus, differences among communities in species richness reflect differences in the relative rates of these processes.

13. Differences in the numbers of species on islands emphasize the importance of regional processes—immigration from a continent or from other islands—to the maintenance of local species diversity. On continents, the addition of species to local communities reflects, in part, the rate of production of new species, which is also a regional process.

14. Several explanations for high tree species richness in tropical forests focus on environmental heterogeneity and the role of disturbance in creating heterogeneous conditions within gaps in the forest canopy. However, there is no evidence that these processes are more important in tropical than in temperate regions.

15. Consumers may enhance diversity by reducing resource populations (and hence competition for resources), thereby making competitive exclusion less likely. Density-dependent predation should favor the persistence of rare species and enhance diversity. Evidence that predators and pathogens may act in a density-dependent manner supports this pest pressure hypothesis.

16. Steve Hubbell's theory of random ecological drift emphasizes the stochastic nature of death and reproduction in populations of competitively equivalent individuals. Accordingly, the number of species in a region is a function of the total number of individuals, which determines the extinction rate, and the rate of formation of new species. One problem with this theory is that such random processes are too slow to account for observed patterns of diversity.

REVIEW QUESTIONS

1. When quantifying diversity, why might you want to incorporate both the number of species and the relative abundance of each species?

2. How do local sampling, regional habitat diversity, and the isolation of continents affect the slope of the species–area relationship?

3. Why might regions with greater habitat diversity contain higher species richness?

4. How do alpha, beta, and gamma diversity differ?

5. Discuss potential differences between tropical and temperate regions with respect to habitat heterogeneity and the distribution of species across gradients of environmental conditions.

6. How does the concept of the realized niche help explain the phenomenon of ecological release?

7. Why might we expect that regions with high species diversity will also exhibit high niche diversity?

8. Why do small islands that are distant from a mainland typically have fewer species than large islands that are close to a mainland?

9. How might the pest pressure hypothesis explain the high diversity of trees in the tropics?

SUGGESTED READINGS

Augspurger, C. K. 1984. Seedling survival of tropical tree species: Interactions of dispersal distance, light-gaps, and pathogens. *Ecology* 65:1705–1712.

Augspurger, C. K., and H. T. Wilkinson. 2007. Host specificity of pathogenic *Pythium* species: Implications for tree species diversity. *Biotropica* 39:702–708.

Bertness, M. D., et al. 2006. The community structure of western Atlantic Patagonian rocky shores. *Ecological Monographs* 76:439–460.

Brokaw, M., and R. T. Busing. 2000. Niche versus chance and tree diversity in forest gaps. *Trends in Ecology and Evolution* 15:183–188.

Case, T. J., and M. L. Cody. 1987. Testing island biogeographic theories. *American Scientist* 75:402–411.

Connell, J. H. 1978. Diversity in tropical rain forests and coral reefs. *Science* 199:1302–1310.

Cornell, H. V., and J. H. Lawton. 1992. Species interactions, local and regional processes, and limits to the richness of ecological communities: A theoretical perspective. *Journal of Animal Ecology* 61:1–12.

Currie, D. J. 1991. Energy and large-scale patterns of animal- and plant-species richness. *American Naturalist* 137:27–49.

Fridley, J. D., et al. 2005. Connecting fine- and broad-scale species–area relationships of Southeastern U.S. flora. *Ecology* 86:1172–1177.

Fridley, J. D., et al. 2006. Integration of local and regional species–area relationships from space–time species accumulation. *American Naturalist* 168:133–143.

Gaston, K. J. 2000. Global patterns in biodiversity. *Nature* 405:220–227.

Givnish, T. J. 1999. On the causes of gradients in tropical tree diversity. *Journal of Ecology* 87:193–210.

Heywood, V. H. (ed.). 1996. *Global Biodiversity Assessment.* Cambridge University Press, Cambridge.

Hillebrand, H. 2004. On the generality of the latitudinal diversity gradient. *American Naturalist* 163:192–211.

Hubbell, S. P. 2001. *The Unified Neutral Theory of Biodiversity and Biogeography.* Princeton University Press, Princeton, N.J.

Hubbell, S. P., et al. 1999. Light-gap disturbances, recruitment limitation, and tree diversity in a Neotropical forest. *Science* 283:554–557.

Janzen, D. H. 1970. Herbivores and the number of tree species in tropical forests. *American Naturalist* 104:501–528.

Kalmar, A., and D. J. Currie. 2007. A unified model of avian species richness on islands and continents. *Ecology* 88:1309–1321.

Lewinsohn, T. M., and T. Roslin. 2008. Four ways towards tropical herbivore megadiversity. *Ecology Letters* 11:398–416.

Lomolino, M. V. 2000. Ecology's most general, yet protean pattern: The species–area relationship. *Journal of Biogeography* 27:17–26.

MacArthur, R. H. 1965. Patterns of species diversity. *Biological Reviews* 40:510–533.

MacArthur, R. H. 1972. *Geographical Ecology: Patterns in the Distribution of Species.* Harper & Row, New York.

Magurran, A. E. 2004. *Measuring Biological Diversity.* Blackwell Publishing, Oxford.

Mills, K. E., and J. D. Bever. 1998. Maintenance of diversity within plant communities: Soil pathogens as agents of negative feedback. *Ecology* 79:1595–1601.

Nekola, J. C., and P. S. White. 1999. The distance decay of similarity in biogeography and ecology. *Journal of Biogeography* 26:867–878.

Packer, A., and K. Clay. 2000. Soil pathogens and spatial patterns of seedling mortality in a temperate tree. *Nature* 404:278–281.

Pärtel, M., et al. 1996. The species pool and its relation to species richness: Evidence from Estonian plant communities. *Oikos* 75:111–117.

Purvis, A., and A. Hector. 2000. Getting the measure of biodiversity. *Nature* 405:212–219.

Ricklefs, R. E. 1987. Community diversity: Relative roles of local and regional processes. *Science* 235:167–171.

Ricklefs, R. E. 2006. The unified neutral theory of biodiversity: Do the numbers add up? *Ecology* 87:1424–1431.

Ricklefs, R. E., and D. Schluter (eds.). 1993. *Species Diversity in Ecological Communities: Historical and Geographical Perspectives.* University of Chicago Press, Chicago.

Rosenzweig, M. 1995. *Species Diversity in Space and Time.* Cambridge University Press, Cambridge.

Suding, K. N., et al. 2005. Functional- and abundance-based mechanisms explain diversity loss due to N fertilization. *Proceedings of the National Academy of Sciences USA* 102:4387–4392.

Terborgh, J., R. B. Foster, and P. Nuñez. 1996. Tropical tree communities: A test of the non-equilibrium hypothesis. *Ecology* 77:561–567.

Weiher, E., and P. Keddy. 1995. The assembly of experimental wetland plant communities. *Oikos* 73:323–335.

Wills, C., et al. 1997. Strong density- and diversity-related effects help to maintain species diversity in a Neotropical forest. *Proceedings of the National Academy of Sciences USA* 94:1252–1257.

Wright, S. J. 2002. Plant diversity in tropical forests: A review of mechanisms of species coexistence. *Oecologia* 130:1–14.

Zobel, M. 1997. The relative roles of species pools in determining plant species richness: An alternative explanation of species coexistence? *Trends in Ecology and Evolution* 12(7):266–269.

DATA ANALYSIS MODULE 5

Quantifying Biodiversity

Measures of species diversity are central to describing the structure of biological communities and to formulating policies for management and conservation of natural resources. Maintaining and promoting species diversity are common goals of natural resource management, in part because species diversity is often linked to ecosystem health and function.

Although species diversity has many meanings, it has two essential components: *richness,* or the number of species present, and *evenness,* the variation in relative species abundances. There are numerous ways to measure richness and evenness; Magurran (2004) provides a compre-

hensive review of the concepts and methods involved. Some commonly used indices are discussed on pages 413–414 of this textbook.

The first component of diversity, richness, provides a straightforward common currency for comparing ecological systems. Although richness is a simple measure, several considerations make it more complex than it initially seems. In particular, because we can rarely account for every individual in a given area, we must employ various methods for estimating richness. A related problem is that the number of species in a sample generally increases with the number of individuals sampled, particularly for species-rich groups, such as arthropods, microbes, and tropical plants. Hence, both the actual abundance of individuals and the sampling effort made to estimate abundance can greatly influence richness estimates (Gotelli and Colwell, 2001).

Richness indices are estimates of species richness that take into account the size of the sample. Fair comparisons of species richness should be based on the number of species observed adjusted by the size of the sample, which removes any bias in observed richness due to the number of individuals sampled. For example, **Margalef's index** is based on the idea that the number of species (S) increases in direct relationship to the logarithm of the number of individuals sampled (N). Simply put, this index assumes that the chance of finding new species in a group of individuals decreases as more of the individuals are examined. This assumption is based on the common pattern of a few abundant species and many rare ones in communities (see Figure 20.2). DA Figure 5.1 shows a typical logarithmic relationship between number of species and number of individuals examined; this relationship becomes linear when species are plotted against the natural logarithm of individuals (inset). The y-intercept for this relationship is always 1 because the first individual examined represents a species (1 on the y-axis) and $\ln(1) = 0$ (0 on the x-axis). Thus, only the slope of the line (D) varies in an ideal Margalef distribution, such that

$$S = 1 + D \times \ln(N)$$

which can then be rearranged into Margalef's index:

$$D = \frac{(S - 1)}{\ln(N)}$$

Indices such as Margalef's are generally calculated for data from different habitats or experimental plots and then compared to assess richness. An advantage of this approach is that richness can be estimated from a single observation drawn from the relationship between species and numbers of individuals. The index assumes a single slope in the relationship between the number of species and the natural logarithm of the number of individuals in each habitat or plot, which is the basis for comparing species richness. Although simple to calculate and useful for some comparisons, Margalef's and other richness indices do not completely overcome problems arising from variation in sampling effort and the number of individuals recorded because the assumption of a linear relationship between ($S - 1$) and $\ln(N)$ is not always met.

The effect of sampling effort or sample size can be visualized with species sampling curves. Species sampling curves allow for meaningful comparisons of species richness among habitats or experimental treatments because they account for relationships between sampling effort, actual abundance or number of individuals, and richness. Sampling curves are based on empirical relationships and make fewer assumptions than indices such as Margalef's. **Accumulation curves** portray the increase in the number of species observed as individuals are added to a sample. Raw accumulation curves show an irregular pattern of species accumulation because sampling cannot be instantaneous in time and space, so seasonal and habitat variation tend to be sampled in succession. In contrast, **rarefaction curves** are based on repeated random subsampling of the total samples, which produces a smooth curve.

DA Figure 5.1 portrays both raw accumulation and rarefaction curves for the number of species in the soil seed bank of a tropical forest. Both curves indicate that the number of species in the sample levels off (reaches

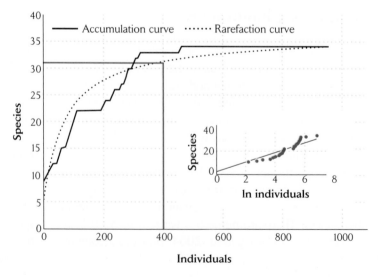

DA FIGURE 5.1 Raw accumulation and rarefaction curves for the soil seed bank of a tropical rain forest (Butler and Chazdon, 1998). Curves were constructed using EstimateS 7.5 software (Colwell, 2005). The red line indicates the rarefaction estimate for richness, in this case 31 species, when 400 individuals are counted. The inset graph shows the number of species plotted against the natural logarithm of the number of individuals for the same data set.

an asymptote) at about 34 when more than 500 seeds are counted. This observation suggests that all the species present are included in the sample of 500 seeds. In other cases, involving highly diverse or relatively small samples, raw accumulation curves may not reach an asymptote. Raw accumulation curves can assess richness accurately only when they do reach an asymptote because the leveling of the curve shows that the vast majority of species have been accounted for. The raw accumulation curve also has an irregular steplike pattern because it is the actual accumulation of species recorded as individuals are successively examined. This feature of accumulation curves can be problematic, as the location of the asymptote can sometimes be deceiving. Consider, for example, portions of the raw accumulation curve in DA Figure 5.1 where species accumulation levels off for a while, but then increases again.

- What would the richness estimate have been if it were assumed that the asymptote was reached with 180 individuals?

In contrast to the raw accumulation curve, the rarefaction curve is smooth because it is produced by repeatedly subsampling the pool of individuals in a sample and calculating the average number of species present for a given number of individuals. Rarefaction curves begin with the entire sample and tell us the richness of a random subsample of any designated size. For example, in DA Figure 5.1, a richness estimate can be derived from any point on the rarefaction curve, as is shown for a sample of 400 individuals (= 31 species). Hence, rarefaction curves factor out sampling effort, allowing for more meaningful comparisons among habitats or study plots where sampling effort may vary.

To compare richness in samples from different plots or habitats, rarefaction of the larger samples down to the size of the smallest sample is a useful approach that accounts for differences in sampling effort. Before fast computers made rarefaction by subsampling practical, one could approximate richness in rarified samples using equations based on probability theory developed by Hurlbert (1971) and Simberloff (1972):

$$\text{expected number of species } (s) = \sum_{i=1}^{s} \left\{ 1 - \frac{\binom{N - N_i}{n}}{\binom{N}{n}} \right\}$$

In this equation, N is the total number of individuals in the sample to be rarefied, N_i is the number of individuals in the ith species in the sample to be rarefied, and n is the total number of individuals in the smallest sample (the

standard). This equation uses the probability of a given species being present, based on numbers in the large sample (N and N_i), to estimate the number of species that would be present in a sample of a smaller size (n).

$$\binom{N - N_i}{n} \text{ is calculated as } \frac{(N - N_i)!}{n!(N - N_i - n)!}$$

where ! indicates a factorial (for example, $4! = 4 \times 3 \times 2 \times 1 = 24$).

Let's apply these methods to a study of aquatic insect species richness in intermittent wetlands along the central Platte River in Nebraska. This investigation, which was conducted to assess the influence of hydroperiod—the duration of inundation by floodwaters—on aquatic insect communities, was part of a larger study examining how water levels in the central Platte River valley influence the health and function of floodplain wetlands. DA Table 5.1 shows data from aquatic insect emergence traps placed in two intermittent wetlands. Intermittent wetland A holds water for a longer period each year (about 330 days) than intermittent wetland B (about 295 days). Each trap collected a different number of individual insects, so rarefaction can be used to standardize the data to the trap with the lower number of individuals. In this case, data from wetland A (total number of individuals caught, $N = 22$) can be standardized to the number of individuals caught in wetland B (total number of individuals caught, $n = 13$) using the equations above.

Step 1: Use Hurlbert's and Simberloff's equation to calculate the probability that species 1 would have been collected in wetland A if only 13 individuals were collected in the trap there.

DA Table 5.2 provides values for each component of the equation for species 1. These parameters are then plugged into the original equation to yield

$$\text{probability of species 1 occurring} = \left\{ \frac{1 - \dfrac{20!}{13!(20 - 13)!}}{\dfrac{22!}{13!(22 - 13)!}} \right\} =$$

$$\left\{ 1 - \frac{77520}{497240} \right\} = 0.844$$

Step 2: Calculate rarefaction estimates for each species in wetland A and then sum them to estimate richness for wetland A. In making your calculations, zero values are ignored.

Complete DA Table 5.3 by calculating the expected value for each species and then summing the expected values

DA TABLE 5.1	Data from aquatic insect emergence traps	
Species number	Wetland A	Wetland B
1	2	0
2	1	0
3	2	0
4	1	0
5	1	0
6	1	0
7	7	4
8	4	0
9	1	0
10	1	0
11	0	1
12	0	2
13	0	2
14	1	0
15	0	0
16	0	2
17	0	1
18	0	1
Total taxa	11	7
Total individuals	22	13

Source: M. R. Whiles and B. S. Goldowitz, *Ecological Applications* 11:1829–1842 (2001).

to estimate the total number of species that would be expected in wetland A if only 13 individuals had been collected there. Note that only N_i will change for each subsequent species, so only the parameters in rows 3, 4, and 5 of the calculations for species 1 in DA Table 5.2

DA TABLE 5.2	Values to use when estimating the probability for species 1
Parameter	Values for rarefaction of species 1 in wetland A
N	22
n	13
N_i	2
$N - N_i$	20
$\binom{N - N_i}{n}$	$\dfrac{20!}{13!(20 - 13)!}$
$\binom{N}{n}$	$\dfrac{22!}{13!(22 - 13)!}$

DA TABLE 5.3	Rarefaction estimates for each species in wetland A	
Species	N_i	Expected value
1	2	0.844
2	1	
3	2	0.844
4	1	
5	1	
6	1	
7	7	
8	4	
9	1	
10	1	
14	1	
Total		

will change when calculating the estimates for the other species.

In this case, the richness estimate you obtain for wetland A by rarefaction should be very close to the richness estimate for wetland B (7 total species).

• Based on just these two samples, does species richness vary appreciably between these two intermittent wetlands?

• Would your answer have differed if you had considered only the original richness values for the two wetlands (11 versus 7)?

Margalef's index can also be calculated for the two wetlands using the data sets in DA Table 5.1. For wetland A, Margalef's index is

$$D = \frac{10}{\ln(22)} = 3.2$$

Step 3: Calculate Margalef's index for wetland B and compare the values for each wetland.

• Are these values more similar or more different than the estimates based on rarefaction?

• Based on this exercise, which technique is likely to be more accurate for estimating richness: calculation of a richness index or rarefaction? Why?

Richness is one of two metrics that ecologists use to assess species diversity. Evenness, or variation in the abundances of species, is the other. Heterogeneity measures such as Simpson's index (page 414) account for both species richness and species evenness.

Step 4: Calculate Simpson's index for the trap catches from the two wetlands.

For this index, you will need to calculate the proportion of each species (p_i) in the sample from each wetland, which is simply the number of individuals of the ith species divided by the total number of individuals collected. For example, for species 1 in wetland A, p_i is $2/22 = 0.091$.

• Do the values for Simpson's index follow the same trends as those for richness alone? Describe a situation in which richness and heterogeneity values might show very different trends.

Literature Cited

Butler, B. J., and R. L. Chazdon. 1998. Species richness, spatial variation, and abundance of the soil seed bank of a secondary tropical rainforest. *Biotropica* 30:214–222.

Colwell, R. K. 2005. EstimateS: Statistical estimation of species richness and shared species from samples. Version 7.5. User's Guide and application published at http://purl.oclc.org/estimates.

Gotelli, N. J., and R. K. Colwell. 2001. Quantifying biodiversity: Procedures and pitfalls in the measurement and comparison of species richness. *Ecology Letters* 4:379–391.

Hurlbert, S. H. 1971. The nonconcept of species diversity: A critique and alternative parameters. *Ecology* 52:577–586.

Magurran, A. E. 2004. *Measuring Biological Diversity*. Blackwell, Oxford.

Simberloff, D. 1972. Properties of rarefaction diversity measures. *American Naturalist* 106:414–415.

Whiles, M. R., and B. S. Goldowitz. 2001. Hydrologic influences on insect emergence production from central Platte River wetlands. *Ecological Applications* 11:1829–1842.

CHAPTER 21

History, Biogeography, and Biodiversity

I n Chapter 20, we saw that the number of species on a small island depends on the regional pool of potential colonists as well as on local processes, such as species sorting and species interactions. Thus, the structure and composition of local communities depend on the broader geographic context. We also know that adaptation to environmental change requires evolution over hundreds or thousands of generations. Thus, past environments provide the historical setting for the evolution of populations and diversification of species within regions.

The origin and maintenance of the earth's biodiversity is one of ecology's central issues. Understanding the history and geography of biodiversity helps us to understand the role of ecological systems in the large-scale processes responsible for generating species within regions. Are we living at a time of maximum species richness over the globe? Ignoring the effect of human activities, would the number of species continue to increase in the future? If the past holds the key to the future, not to mention our understanding of the present, then ecologists should pay attention to the lessons of history.

Many groups of organisms have left fossil records, albeit fragmentary ones, in sedimentary rocks near the earth's surface (Figure 21.1). These fossils suggest that during the past 600 million years, a period encompassing most of the evolution of multicellular organisms, the sizes of regional species pools have varied considerably. At times they have increased through biological diversification; at other times they have declined, sometimes precipitately, because of catastrophic events or competition from newly evolving life forms. To the degree that local communities reflect the regional species pool, we must question whether ecological systems ever truly achieve equilibrium.

Interpretation of the fossil record has also suggested historical causes for the high species richness in the tropics. According to one hypothesis, tropical conditions appeared

(a)

(b)

FIGURE 21.1 Fossils of animals and plants reveal the history of biodiversity. Photo (a) by Newman & Flowers/Photo Researchers; photo (b) by James L. Amos/Photo Researchers.

on the earth's surface earlier than colder environments, allowing time for the evolution of a greater variety of tropical plants and animals. A version of this "time hypothesis" was fully stated as early as 1878 by the English naturalist Alfred Russel Wallace, co-discoverer with Darwin of the theory of evolution by natural selection:

> The equatorial zone, in short, exhibits to us the result of a comparatively continuous and unchecked development of organic forms; while in the temperate regions there have been a series of periodical checks and extinctions of a more or less disastrous nature, necessitating the commencement of the work of development in certain lines over and over again. In the one, evolution has had a fair chance; in the other, it has had countless difficulties thrown in its way. The equatorial regions are then, as regards their past and present life history, a more ancient world than that represented by the temperate zones, a world in which the laws which have governed the progressive development of life have operated with comparatively little check for countless ages, and have resulted in those wonderful eccentricities of structure, of function, and of instinct—that rich variety of colour, and that nicely balanced harmony of relations which delight and astonish us in the animal productions of all tropical countries. (A. R. Wallace, *Tropical Nature and Other Essays*, Macmillan, New York and London)

Because the tropical zone girdles the earth about its equator—the earth's widest point—tropical latitudes include more area, both land and sea, than temperate and polar latitudes. For this reason alone, it is not surprising that regions in the tropics should harbor more species than regions of similar size in temperate or boreal climate zones. During the early part of the Cenozoic era, 65 million–35 million years ago, the earth's climate was much warmer than it is now, and tropical and subtropical environments extended to Canada and Russia, squeezing temperate and boreal climate zones into smaller areas closer to the poles. During the last 35 million years, the climate of the earth has become cooler and drier, and tropical environments have contracted.

Both high and low latitudes have experienced drastic fluctuations in climate, particularly during the Ice Age of the last 2 million years. Temperate and polar regions witnessed the expansion and retreat of glaciers, which caused major habitat zones to be displaced geographically and, possibly, to disappear. Periods of glacial expansion were coupled with low rainfall and reduced temperatures in the tropics. The Amazonian rain forest, which today covers most of the Amazon River's vast drainage basin, is thought to have been repeatedly restricted to small, isolated refuges during dry periods correlated with glacial expansion in the north. Restriction and fragmentation of rain forest habitat could have driven many species to extinction; conversely, isolation of populations in patches of rain forest could have facilitated the formation of new species.

CHAPTER CONCEPTS

- Life has unfolded over millions of years of geologic time
- Continental drift influences the geography of evolution
- Biogeographic regions reflect long-term evolutionary isolation
- Climate change influences the distributions of organisms
- Organisms in similar environments tend to converge in form and function

- Closely related species show both convergence and divergence in ecological distributions
- Species richness in similar environments often fails to converge between different regions
- Processes on large geographic and temporal scales influence biodiversity

---

The earth provides an ever-changing backdrop for the development of biological communities. The millions of years of earth history have encompassed changes in climate and other physical conditions, rearrangements of continents and ocean basins, growth and wearing down of mountain ranges, catastrophic impacts with extraterrestrial bodies, and continual evolution of new forms of life. The history of life reveals itself to us in the geochemical record of past environments, in fossil traces left by long-extinct taxa, and in the geographic distributions and evolutionary relationships of living species.

The most obvious consequence of earth history is the heterogeneous distribution of animal and plant forms over the earth's surface. Australia, for example, has many unique forms—koalas, kangaroos, and eucalyptus trees (Figure 21.2)—because of its long isolation as an island continent surrounded by ocean barriers to the dispersal of terrestrial organisms. Every part of the earth has its own distinctive fauna and flora. Even the major ocean basins, interconnected as they are by continuous corridors of water, have partly differentiated biotas, isolated over millions of years by ecological barriers of temperature and salinity.

The structure and functioning of organisms reflects their ancestry as much as it does the local environment. For example, the marsupial mode of reproduction (involving, among other characteristics, early birth and subsequent development of young in a pouch) is uniquely a property of the marsupial line of mammalian evolution. It is not the result of unique ecological properties of the continent of Australia, where marsupials now are most

diverse. Biologists refer to such characteristics shared by a lineage irrespective of environmental factors as **phylogenetic effects.** These effects reflect the inertia of evolution—the lack of change of some attributes in the face of change in the environment.

Ecologists recognize that phylogenetic effects can influence ecological systems, although this is difficult to demonstrate experimentally. Imagine that the plants and animals of Australia were replaced by a similar number of taxa from other regions with similar climates, perhaps the southwestern United States and the Middle East. Would the new ecosystems function in the same manner as the ones they replaced, with similar levels of biological productivity and responses to environmental perturbation?

History and geography also affect the diversification of species. Each region of the earth has a different history, and those histories have effects that extend down to local communities, with the result that each region has a unique level of species diversity. In this chapter, we shall first briefly examine some historical processes that have shaped the distribution and development of ecological systems. Then we shall examine the principle of convergence, which states that inhabitants of similar environments with disparate historical origins often resemble one another because they adapt to similar ecological factors. This principle can also be applied to the diversity of biological communities. We shall see that history and biogeography have indeed influenced the character of local communities and have played an important role in the development of patterns of diversity.

(a)

(b)

(c)

(d)

FIGURE 21.2 Australia has many unique terrestrial animals and plants. Pictured here are (a) leaves and (b) flowers of a *Eucalyptus* species, (c) the inflorescence of a species of *Banksia*, and (d) a red kangaroo (*Macropus rufus*). These distinctive life forms evolved in isolation in Australia and are found nowhere else on earth. Photos by R. E. Ricklefs.

Life has unfolded over millions of years of geologic time

The earth formed about 4.5 billion years ago, and life arose within its first billion years. Over most of the history of the earth, life forms remained primitive. Physical conditions at the earth's surface, and the ecological systems that developed, were strikingly different from those of the present. The atmosphere had little oxygen, and early microbes used strictly anaerobic metabolism. Indeed, it was not until photosynthetic microbes evolved and began producing oxygen as a by-product of their metabolism that oxygen levels in the atmosphere began to increase. To a large degree, life has created its own environment. At some point, the atmospheric concentration of oxygen became high enough to sustain oxidative metabolism and made it possible for more complex life forms to evolve.

The eukaryotic cell, which is the basic building block of all modern complex organisms, is a product of the last billion years of evolution. We have scant records of the early development of multicellular animals because most ancient life forms lacked the hard skeletons or shells that

fossilize most readily. Much evidence of early complex life forms consists of tracks and burrows in the mud in which they lived.

All of this changed about 540 million years ago (Mya), when most of the modern phyla of invertebrate organisms suddenly appear in the fossil record. Echinoderms, arthropods, mollusks, and brachiopods rose to prominence in the oceans of that time, as did other life forms—evolutionary experiments, so to speak—that are no longer with us (Figure 21.3). No one knows for sure why animals began to protect themselves with hard shells or outer skeletons at that moment in history, but paleontologists regard their appearance as the beginning of life in its modern form. The interval between that critical moment and the present, occupying about one-eighth of the total history of the earth, has been divided into three major eras, with numerous periods and epochs nested within them.

The divisions of geologic time coincide with changes in the fauna and flora of the earth that are easily perceived in the fossil record. The first major division, the Paleozoic era (the name means "old animals"), extends from

FIGURE 21.3 Many inhabitants of early Paleozoic seas had hardened outer shells. This reconstruction shows a number of invertebrates representing sponges, segmented worms, and arthropods as well as many forms that left no descendants in later faunas. Painting by D. W. Miller; from D. Erwin, J. Valentine, and D. Jablonski, *Am. Sci.* 85(2):126–137 (1997).

the earliest appearance of animals with hard skeletons 542 Mya to a mass extinction event at the end of the Permian period 251 Mya. The Mesozoic era ("middle animals," also known as the Age of Reptiles after the animals that were dominant on land at that time), extended from 251 to 65 Mya. Its end also coincided with mass extinctions of animal taxa, including the dinosaurs. We are living in the Cenozoic era ("recent animals," also known as the Age of Mammals), which extends from 65 Mya to the present. The time from the beginning of the Cenozoic era to the beginning of the Ice Age 2 Mya is often called the Tertiary period.

Most types of organisms present during the Paleozoic era are gone. The Mesozoic era was the time of the early evolution of many prominent contemporary groups, such as the flowering plants, mammals, and modern insects, but most contemporary biodiversity is the product of the recent past. The evolutionary diversification of life has continued uninterrupted, for the most part, during the past 65 million years of the Cenozoic era, the time most relevant to contemporary ecological systems.

Continental drift influences the geography of evolution

The earth's surface has been restless over its history. The continents are islands of low-density rock floating on the denser material of the earth's interior. Giant convection currents in the semi-molten material of the underlying mantle carry continents along like gigantic logs on the sur-

face of the ocean. At some times in the past continents have coalesced, and at other times they have drifted apart. This movement of landmasses across the surface of the earth, called **continental drift,** has two important consequences for ecological systems. First, the positions of continents and major ocean basins profoundly influence climatic patterns. Second, continental drift creates and breaks down barriers to dispersal, alternately connecting and separating evolving biotas in different regions of the earth.

Toward the end of the Paleozoic era, about 250 Mya, the continents came together in a giant landmass known as **Pangaea** (Figure 21.4). By 150 Mya, Pangaea had separated into a northern landmass, known as **Laurasia,** and a southern landmass, known as **Gondwana,** with the Tethys Ocean in the space between them. By 100 Mya, Gondwana itself had begun to break up into three parts: West Gondwana, including present-day Africa and South America, which were themselves beginning to pull apart; East Gondwana, including Antarctica and Australia; and India, which had separated from present-day Africa and was drifting toward a collision with Asia, which finally occurred at about 45 Mya.

By the end of the Mesozoic era (65 Mya), South America and Africa were widely separated. The connection between Australia and South America through a temperate Antarctica had finally dissolved by 50 Mya. At about the same time, in the Northern Hemisphere, a widening Atlantic Ocean finally separated Europe and North America, but a land bridge had already formed by 70 Mya on the other side of the world between North America and Asia. More recent events of significance were the closing of the Tethys Ocean by the joining of Europe and Africa about 17 Mya and completion of a land bridge between North and South America 3–6 Mya.

Continental movements have profoundly affected climate as well as biogeography. Australia, for example, has drifted northward through different climate zones, moving from a wetter, more temperate climate in the late Mesozoic into its present-day, primarily subtropical, arid climate zone. More broadly, the changing positions of the continents have altered circulation patterns in the oceans and the distribution of heat over the earth's surface. In the early Cenozoic era, Antarctica drifted over the South Pole, and the Arctic Ocean became mostly enclosed between North America and Eurasia, leading to colder temperatures at high latitudes. Separation of Antarctica from Australia and South America established a circumpolar ocean current around Antarctica that further cooled the region. The formation of the Isthmus of Panama about 3 Mya finally shut off water movement between the tropical Pacific and Atlantic oceans.

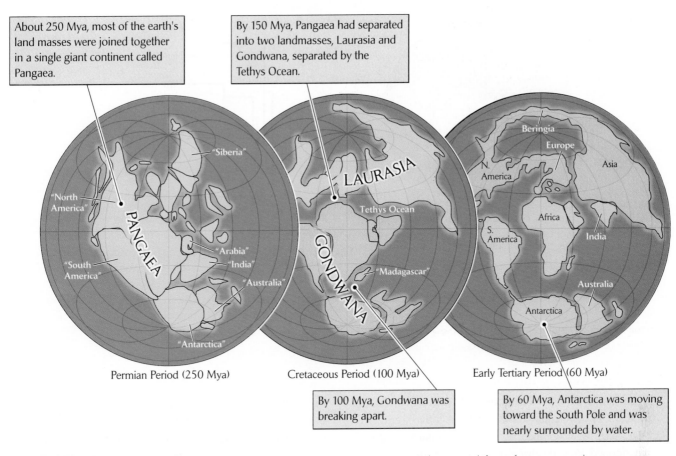

About 250 Mya, most of the earth's land masses were joined together in a single giant continent called Pangaea.

By 150 Mya, Pangaea had separated into two landmasses, Laurasia and Gondwana, separated by the Tethys Ocean.

Permian Period (250 Mya)

Cretaceous Period (100 Mya)

Early Tertiary Period (60 Mya)

By 100 Mya, Gondwana was breaking apart.

By 60 Mya, Antarctica was moving toward the South Pole and was nearly surrounded by water.

FIGURE 21.4 The positions of the continents have changed over geologic time. At the end of the Paleozoic era, about 250 Mya, the continents formed a single landmass, known as Pangaea. Subsequent drifting of continents to their present positions has isolated biotas in distinct biogeographic regions. After E. C. Pielou, *Biogeography*, Wiley, New York (1979).

Biogeographic regions reflect long-term evolutionary isolation

Many details of continental drift have yet to be resolved, particularly in such complicated areas as the Caribbean Sea, Australasia, and the Mediterranean Sea–Persian Gulf region. Nevertheless, the history of connections between the continents endures in the distributions of animals and plants. We have only to look at the distribution of the flightless ratite birds to see the connection between the southern continents that made up Gondwana. Emus and cassowaries in Australia and New Guinea, rheas in South America, ostriches in Africa, and the extinct moas of New Zealand all descended from a common ancestor that inhabited Gondwana before its breakup (Figure 21.5). The splitting of a widely distributed ancestral population by continental drift, or some other barrier to dispersal, is referred to as **vicariance.**

The distributions of animals led Alfred Russel Wallace to delineate six major zoogeographic regions that are still recognized today (Figure 21.6). We now know that these regions correspond to landmasses isolated many millions of years ago by continental drift. Over the course of that isolation, animals and plants in each region developed distinctive characteristics independently of evolutionary changes in other regions. Botanists recognize six major biogeographic regions with boundaries that closely coincide with those of the zoogeographic regions. In addition, botanists distinguish the unusual flora of the Cape Region of South Africa as the Cape Floristic Province.

The **Nearctic region** and the **Palearctic region,** corresponding roughly to North America and Eurasia, respectively, maintained connections across either what is now Greenland or the Bering Strait between Alaska and Siberia through most of the past 100 million years. Consequently, these two areas share many groups of animals and plants. European forests seem familiar to travelers from North America, and vice versa; few species are the same, but both regions have representatives of many of the same genera and families.

The continents of the Southern Hemisphere, including Africa (the **Afrotropical region** or *Ethiopian region*),

(a)

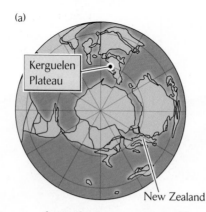

(b)

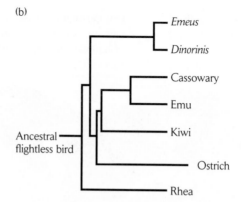

FIGURE 21.5 Lineages of ratite birds were separated by the fragmentation of Gondwana. (a) The Southern Hemisphere, 80 Mya, showing the positions of landmasses as Gondwana was breaking apart. (b) The ancestors of ratite birds once ranged over the continuous landmass of Gondwana. The evolutionary relationships of these flightless birds have been reconstructed from DNA sequences, including "ancient" DNA obtained from fossils of the extinct moas of New Zealand. Rheas became isolated about 89 Mya with the separation of South America from the Gondwanan landmass. The formation of three lineages leading to (1) moas (*Emeus* and *Dinorinis*), (2) emu, cassowaries, and kiwis, and (3) ostriches occurred about 82 Mya with the separation of New Zealand from Gondwana. Kiwis apparently arrived there later by overwater dispersal. Ostriches must have island-hopped across the Kerguelen Plateau (now mostly under water), India, and Madagascar to Africa. After A. Cooper et al., *Nature* 409:704–707 (2001).

South America (the **Neotropical region,** including tropical Central America and the West Indies), and Australia and New Guinea (the **Australasian region**), experienced long histories of isolation from the rest of the terrestrial world, during which many distinctive forms of life evolved. For example, a distinctive lineage of mammals, referred to as the Afrotheria, evolved and diversified on the continent of Africa while it was isolated from other landmasses from about 100 to 20 Mya. This group includes, among living mammals, the elephants and elephant shrews, tenrecs, aardvarks, hyraxes, and manatees. After Africa drifted northward to connect with the Eurasian landmass, elephants dispersed to other regions of the world, including North America (mammoths and mastodons). Of course, other animal species moved in the other direction to populate Africa with primates, carnivores, ungulates, and rodents.

The **Indomalayan region** or *Oriental region* comprises the biota of Southeast Asia and India, which was isolated from tropical areas of Africa and South America. As one might expect, temperate (Palearctic) and tropical (Indomalayan) Asia have closer affinities than temperate

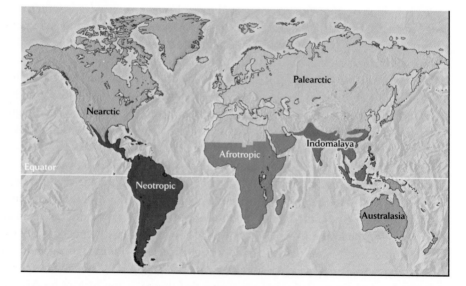

FIGURE 21.6 The major zoogeographic regions of the earth are based on the distributions of animals. This scheme, which is widely accepted today, originated with Alfred Russel Wallace in 1876. Biogeographic regions based on plant distributions are similar.

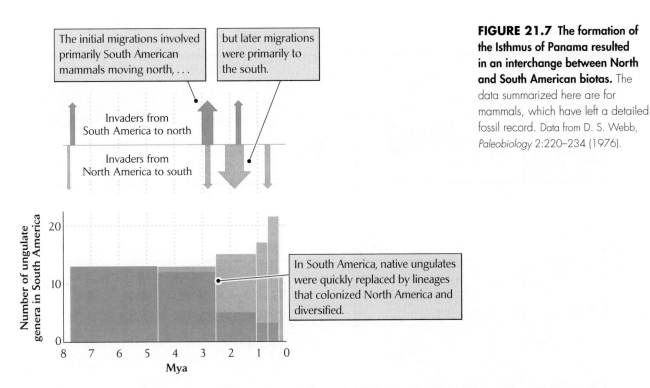

The initial migrations involved primarily South American mammals moving north, . . .

but later migrations were primarily to the south.

Invaders from South America to north

Invaders from North America to south

In South America, native ungulates were quickly replaced by lineages that colonized North America and diversified.

FIGURE 21.7 The formation of the Isthmus of Panama resulted in an interchange between North and South American biotas. The data summarized here are for mammals, which have left a detailed fossil record. Data from D. S. Webb, *Paleobiology* 2:220–234 (1976).

North America (Nearctic) and tropical South America (Neotropical) because of the continuous land connection between them. Indeed, the temperate forests of Asia contain a high percentage of tree species descended from tropical lineages, whereas those of temperate North America have few such species (among those few, however, are the catalpa and paw-paw).

A land connection between North and South America through the Isthmus of Panama was formed during the Pliocene epoch, about 3 Mya. Although some taxa had island-hopped between the continents before that time, the land bridge allowed the exchange of many taxa, as shown for mammals in Figure 21.7. The exchange was uneven, however. More North American lineages entered South America than the reverse, and some of the North American groups diversified and may have caused the extinction of many South American endemics, including a rich fauna of marsupial mammals.

Climate change influences the distributions of organisms

The earth's climatic patterns ultimately depend on energy from the sun, which warms land and seas and evaporates water. Ocean currents, which are constrained by the positions of the continents, distribute that heat over the surface of the earth (see Chapter 4). When polar regions are occupied by landmasses or landlocked oceans, as they are at present, they can become very cold. Ice has a high

albedo and reflects most of the light and heat it receives, further intensifying the cold. But the polar regions have not always been as cold as they are today. Polar regions always receive relatively little solar energy. In the past, however, when they were covered by oceans that extended to tropical areas, ocean currents distributed heat more evenly, and temperate climates extended to near the poles.

Between 50 and 35 Mya, large portions of North America and Europe were tropical. We know from fossil remains of plants that tropical forests reached into Canada, and that warm temperate forests covered the Bering land bridge. The Antarctic land connection between South America and Australia supported luxuriant temperate vegetation and animal life. However, as Antarctica drifted over the South Pole during the last half of the Cenozoic era, and as North America and Eurasia gradually encircled the northern polar ocean, the earth's climates became more strongly differentiated. Tropical environments contracted into a narrow zone near the equator, and temperate and boreal climate zones expanded.

These climate changes had profound effects on the geographic distributions of plants and animals. One consequence of the cooling trend at high latitudes was the retreat of plants and animals that could not tolerate freezing to lower latitudes. This shift in distributions resulted in a greater distinction between temperate and tropical biotas. Thirty-five million years ago, what is now temperate North America supported a mixture of tropical and temperate forms growing side by side. Today these plants and animals occupy different climate zones. Thus, greater

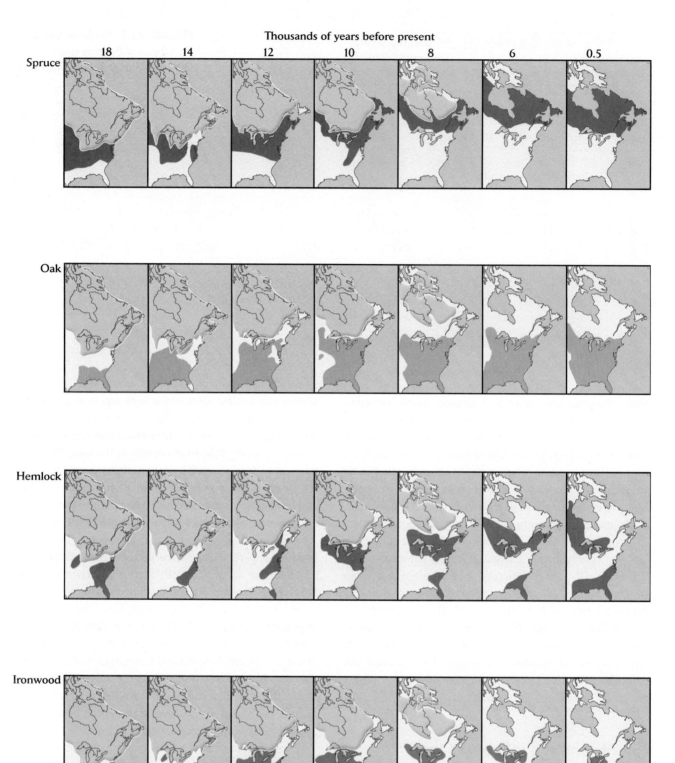

FIGURE 21.8 Climate change following the most recent glaciation caused shifts in tree distributions. The maps show migration routes for five types of trees in eastern North America from the glacial refuges they occupied 18,000 years ago to their present distributions. Note that the migration routes for the five species differ. As a result, groups of species that differ from present-day communities would have been found growing together at different times. After G. L. Jacobson, T. Webb, III, and E. C. Grimm, in W. F. Ruddiman and H. E. Wright, Jr. (eds.), *North America during Deglaciation*, Geological Society of America, Boulder, CO (1987), pp. 277–288.

stratification of climate was matched by greater stratification of the biota.

About 2 Mya, the gradual cooling of the earth gave way to a series of dramatic oscillations in climate that had dramatic effects on habitats and organisms in most parts of the world. This was the Ice Age, or Pleistocene epoch. Alternating periods of cooling and warming led to the advance and retreat of ice sheets at high latitudes over much of the Northern Hemisphere and caused cycles of cool, dry climates and warm, wet climates in the tropics. Ice sheets came as far south as Ohio and Pennsylvania in North America and covered much of northern Europe, driving vegetation zones southward, possibly restricting tropical forests to isolated refuges where conditions remained moist, and generally disrupting biological communities all over the world.

A striking example of this disruption is the migration of forest trees in eastern North America and Europe. At the peak of the most recent glacial period, many tree species were restricted to southern refuges, but after the ice began to retreat, about 18,000 years ago, forests began to spread north again. Pollen grains deposited in the lakes and bogs left by retreating glaciers record the coming and going of plant species. These records show that the composition of plant associations changed as species migrated over different routes across the landscape.

The migrations of some representative tree species from their southern refuges are mapped in Figure 21.8. The distribution of spruces shifted northward just behind the retreating glaciers. Oaks expanded out of their southern refuges to cover most of the eastern part of temperate North America, from southern Canada to the Gulf Coast.

Some pines had glacial refuges in the Carolinas, and their distributions shifted to the north and west, where they are currently centered near the Great Lakes. The pines now distributed across the Atlantic and Gulf Coast lowlands of the southeastern United States endured the glacial period farther to the south, in Florida or the Bahamas. Hemlocks had had a more restricted refuge in the valleys of the Appalachian Mountains, and extended northward through the mountains into Pennsylvania, New York, and New England. Ironwood expanded out of small refuges in the Gulf states to cover most of eastern North America 12,000 to 10,000 years ago, then contracted with further climatic warming to its present range centered in Michigan and southern Ontario.

As a result of this movement in response to climate change, the composition of forests over the past 18,000 years has included combinations of species that do not occur anywhere in eastern North America today; conversely, some of the combinations of species that do occur at present did not occur in the past. For some species, the environment changed too rapidly through the Pleistocene cycles of glacial expansion and retreat, and they disappeared altogether.

The forests of Europe suffered from the spread of glaciers even more than those in North America because populations were blocked from shifting southward by the Alps and Mediterranean Sea. A number of northern European tree species went extinct. Many species that survived were restricted to refuges in southern Europe, from which they expanded after the glaciers receded beginning about 18,000 years ago (Figure 21.9). Danish ecologists Jens-Christian Svenning and Fleming Skov

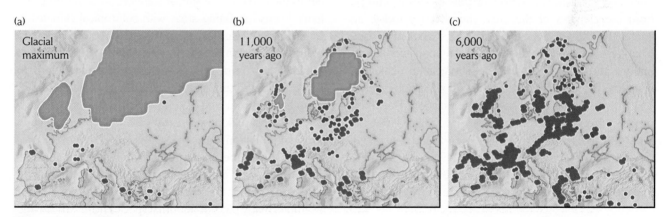

■ Oaks present   ■ Oaks absent

FIGURE 21.9 Deciduous oaks in Europe shifted their distributions following the end of the most recent glacial period. (a) Oaks were present only in a few refuges in the Mediterranean region during the peak of the most recent glaciation 18,000 years ago, and were absent from other parts of Europe sampled. (b) Oaks expanded from those refuges after the glaciers receded. (c) The expansion of the oaks reached its maximum extent about 6,000 years ago. The records are based on pollen deposits in shallow lakes and ponds. After P. Taberlet and R. Cheddadi, *Science* 297:2009–2010 (2002).

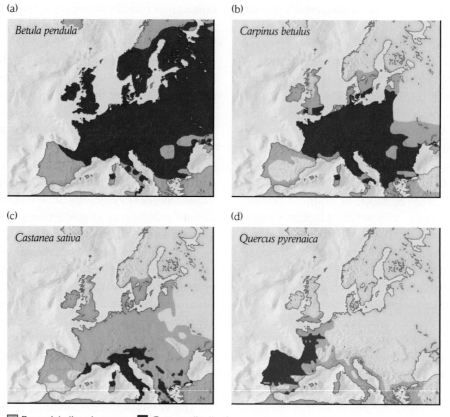

(a) Betula pendula

(b) Carpinus betulus

(c) Castanea sativa

(d) Quercus pyrenaica

▨ Potential climatic range   ■ Current distribution

FIGURE 21.10 Tree species in Europe have expanded into ecologically suitable areas to a varying extent following the most recent glaciation. The silver birch (*Betula pendula*, a) has been able to expand as rapidly as suitable environments developed to the north, and the common hornbeam (*Carpinus betulus*, b) nearly so. However, the sweet chestnut (*Castanea sativa*, c) and especially the Pyrenean oak (*Quercus pyrenaica*, d) have failed to disperse to all areas of suitable climate. Both produce large seeds that cannot be transported long distances. After J.-C. Svenning and F. Skov, *Ecol. Lett. 7*:565–573 (2004).

estimated the potential ranges of tree species in Europe from the climates of the areas they occupy today, an approach known as ecological niche modeling (see Chapter 10). Using a combination of population distributions and climatic data, they estimated an ecological envelope—that is, combinations of conditions under which each species could sustain a population. They then mapped the present distributions of conditions suitable for those species (Figure 21.10). They found that many tree species have not yet expanded fully into their potential ranges. These findings suggest that the European flora has not returned to an equilibrium state.

Organisms in similar environments tend to converge in form and function

Just as long periods of isolation have led to the evolution of unique life forms in many regions of the earth, similar environmental conditions in each of these regions have led to the evolution of similar solutions to common problems. Plants inhabiting areas with subtropical climates in Mexico and in eastern Africa have different evolutionary origins reflecting more than 100 million years of isolation, but they share similar growth forms and adaptations to arid conditions (see Figure 5.1). Thus, the different evolutionary histories and taxonomic affinities of the biota of the earth's regions are partly veiled by convergence in form and function.

Convergence is the process whereby unrelated species living under similar ecological conditions evolve to resemble one another more than their ancestors did. For example, many pairs of mammal species from African and from South American rainforests show close resemblances despite their different evolutionary histories (Figure 21.11). Plants and animals of North and South American deserts resemble each other morphologically more than one would expect based on their different phylogenetic origins. Similarities have also been noted in the behavior and ecology of Australian and North American lizards, despite

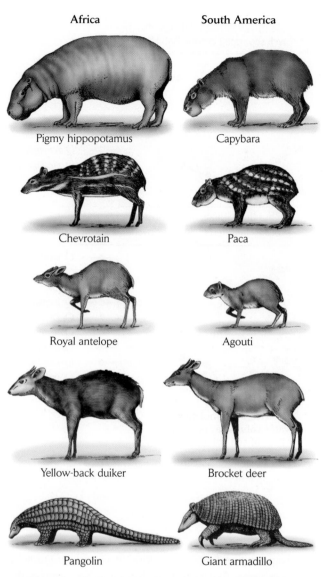

Africa

Pigmy hippopotamus

Chevrotain

Royal antelope

Yellow-back duiker

Pangolin

South America

Capybara

Paca

Agouti

Brocket deer

Giant armadillo

FIGURE 21.11 Pairs of unrelated African and South American rain forest mammals with similar lifestyles and adaptations show striking convergence. Each pair is drawn to the same scale. After F. Bourliere, in B. J. Meggars, E. S. Ayensu, and W. D. Duckworth (eds.), *Tropical Forest Ecosystems in Africa and South America: A Comparative Review*, Smithsonian Institution Press, Washington, D.C. (1973), pp. 279–292.

communities, for example, the ancient Monte Desert of South America lacks bipedal, seed-eating, water-independent rodents like the kangaroo rats of North America and the gerbils of Asia. Among frogs and toads, several South American forms have carried adaptation to desert environments a step further than their North American counterparts: they construct nests of foam to protect their eggs from drying out.

The relative balance between evolutionary history and environment in determining the outcome of evolution and the assembly of biological communities has not been fully resolved. Generally speaking, however, convergence of form and function under similar environmental conditions is a broadly applicable principle in ecology and evolutionary biology. This principle can be tested locally by examining the adaptations of species with different evolutionary histories in the same environment. For example, David Ackerly, at the University of California at Berkeley, compared leaf size and standard leaf area (SLA, leaf area per gram of leaf tissue) in twelve evolutionary lineages of chaparral shrubs and their closest non-chaparral relatives. Chaparral shrubs typically have small, thick leaves. As the lineages made the transition to the Mediterranean climate of the chaparral biome, standard leaf area, but not leaf size, was significantly reduced in several of them. Most of these adaptive shifts occurred in lineages originally from cool, moist habitats; lineages of plants from warmer subtropical environments characteristically have small leaves, which were therefore preadapted for moving into Mediterranean climates as they formed in California.

Closely related species show both convergence and divergence in ecological distributions

Organisms with similar adaptations tend to thrive in the same type of habitat. But, as we have seen, the coexistence of species in the same community depends in part on their not competing so intensely that one species is excluded. Closely related species share a large proportion of their adaptations because of their common ancestry. Therefore, we might expect such species to compete intensely when they are members of the same community. Less closely related species might have slightly more divergent adaptations that would allow them to partition the resources of a community and thus compete less intensely. Accordingly, we can predict that less closely related species should be able to coexist in the same community. More closely related species might exhibit character displacement as divergent adaptations that would allow them to persist as different environments evolved.

the fact that they belong to different families and have evolved independently for perhaps 100 million years. Dolphins and penguins both evolved from terrestrial ancestors, but both have body shapes more closely resembling those of tuna, whose swimming lifestyle they share.

Convergence reinforces our belief that adaptations conform to certain general rules governing structure and function in relation to the environment. However, detailed studies often turn up remarkable differences between the plants and animals in superficially similar environments. Despite the striking convergence among desert-dwelling

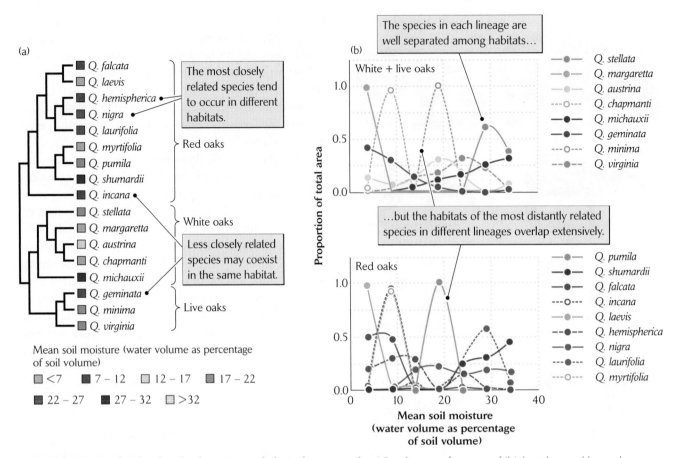

FIGURE 21.12 Closely related oak species exclude each other from local communities. (a) Diagram of the evolutionary relationships of oak (*Quercus*) species found in habitats with varying soil moisture conditions in north central Florida.

(b, c) Distributions of species of (b) the white and live oak lineage and (c) the red oak lineage along the moisture gradient. After J. Cavender-Bares et al., *Am. Nat.* 163:823–843 (2004).

University of Minnesota ecologist Jeannine Cavender-Bares and her colleagues tested these predictions with species of oaks across a moisture gradient in Florida. The species belonged to two evolutionary lineages: red oaks, and white oaks plus live oaks. Species from each of these lineages were distributed widely across the moisture gradient (Figure 21.12). The investigators found that different species of red oaks were not present in the same habitats, but different species of red and white oaks were. Thus, red oaks are too similar to one another to coexist easily, whereas red and white oaks differ enough that they can live together in the same habitats. The evolutionarily conserved traits associated with the different positions of species within each lineage along the environmental gradient are primarily structural characteristics of leaves and wood that influence vulnerability to freezing. Adaptable attributes that converge readily and allow some species of red oaks and white oaks to live together include traits such as rhizome resprouting and growth rate. Both of these traits influence how quickly trees can recover from fire. Because

the frequency of fires varies along the moisture gradient, it makes sense that species in the same position on the moisture gradient would have similar adaptations to fire.

The influence of evolutionary relationships on community assembly also surfaced in an experiment by Hafiz Maherali and John Klironomos at the University of Guelph. These investigators grew plantains (*Plantago lanceolata*) in soils together with eight species of mycorrhizal fungi. The fungal species either belonged to a single taxonomic family (Gigasporaceae or Glomeraceae) or were mixed four and four from each family. The communities of mycorrhizal fungi retained more species after one year when the original soil had contained a mixture of species from both families. Adding species from a third family (Acaulosporaceae) resulted in even higher diversity in the fungal communities (Figure 21.13). Thus, among mycorrhizal fungi, only a limited number of species from the same family can coexist, but species from different families presumably exert weaker competitive effects on one another. Incidentally, the more fungal species were

(a)

Family	Experimental treatments

Family Gigasporaceae	1	2	3	4	5	6	7
Gigaspora albida		x		x	x		
G. gigantea		x		x	x		x
G. margarita		x		x		x	
G. rosea		x		x		x	x
Scutellospora calospora		x	x		x		x
S. dipurpurescens		x	x			x	x
S. heterogama		x	x	x			
S. pellucida		x	x			x	
Family Acaulosporaceae							
Acaulospora denticulata				x		x	
A. taevis				x			
A. morrowiae				x			
A. spinosa				x		x	
Family Glomeraceae							
Glomus aggregatum	x			x	x		
G. clarum	x			x			x
G. constrictum	x	x					
G. etunicatum	x	x		x			
G. hoi	x		x				
G. intraradices	x	x			x		
G. microaggregatum	x		x			x	
G. mosseae	x	x		x			

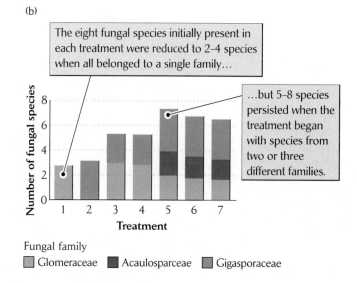

(b)

The eight fungal species initially present in each treatment were reduced to 2–4 species when all belonged to a single family...

...but 5–8 species persisted when the treatment began with species from two or three different families.

Fungal family
Glomeraceae Acaulosparceae Gigasporaceae

FIGURE 21.13 Less closely related species of mycorrhizal fungi have a higher probability of coexistence. Investigators grew different combinations of mycorrhizal fungal species with host plants (*Plantago lanceolata*). (a) Diagram of the evolutionary relationships of the mycorrhizal fungi from three taxonomic families used in the experimental treatments. (b) Numbers of fungal species present in each treatment after one year. After H. Maherali and J. N. Klironomos, *Science* 316:1746–1748 (2007).

present, the more they stimulated plant growth, demonstrating the complementary functions of fungal species in mycorrhizal–plant mutualisms.

Species richness in similar environments often fails to converge between different regions

Does the principle of convergence apply to communities as well as to species? If numbers of species and other aspects of community structure and function primarily reflect local environmental conditions, then we might expect independently derived communities in different regions that occupy similar habitats to have similar numbers of species, regardless of the number of species in the regional species pool. Furthermore, if local processes such as competition constrain the number of coexisting species, then local species richness in a particular type of habitat should reach an upper limit, or *saturation point*. Above that saturation point, further increases in the regional

species pool should not increase local species diversity (Figure 21.14). Conversely, if regional processes also influence local communities, then local diversity and regional diversity should vary in parallel.

We discussed the relationship between local (within-habitat) diversity and regional diversity in the West Indies in Chapter 20 (see Figure 20.16). In that setting, local diversity increased with increasing regional diversity, but turnover of species between habitats (beta diversity) also increased. Thus, species apparently can be added to local communities as the regional pool increases, but gaining membership in a local community also becomes more difficult as the number of species in that community increases.

Studies of the relationship between local and regional diversity have generally supported the idea that communities are open to invasion when additional species are produced within a region. For example, the number of fish species in short stretches of a stream (local communities) reflects the regional pool of species within entire river drainages in northern South America and West Africa (Figure 21.15). The fact that local communities have fewer

FIGURE 21.14 The saturation hypothesis can be tested by relating local to regional diversity.

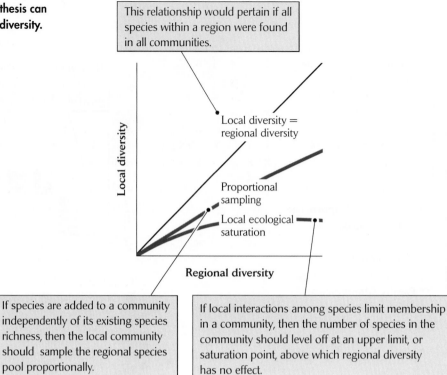

This relationship would pertain if all species within a region were found in all communities.

Local diversity = regional diversity

Proportional sampling

Local ecological saturation

Local diversity (vertical axis)

Regional diversity (horizontal axis)

If species are added to a community independently of its existing species richness, then the local community should sample the regional species pool proportionally.

If local interactions among species limit membership in a community, then the number of species in the community should level off at an upper limit, or saturation point, above which regional diversity has no effect.

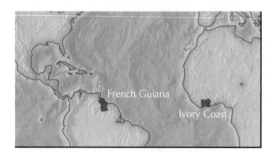

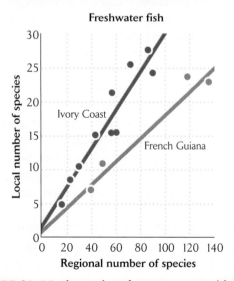

Freshwater fish

Ivory Coast

French Guiana

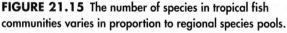

FIGURE 21.15 The number of species in tropical fish communities varies in proportion to regional species pools. Data from the Ivory Coast in West Africa and French Guiana in northern South America clearly indicate that these communities are not saturated. After B. Huegeny et al., *Oikos* 80:583–587 (1997).

species than the regional pool confirms the ideas of species sorting, competitive exclusion, and dispersal limitation.

ECOLOGISTS IN THE FIELD **Why are there so many more temperate tree species in Asia?** The regional species pool for temperate deciduous forests of eastern North America includes 253 species of trees, more than twice the number (124) found in similar habitats in Europe. Temperate eastern Asia, whose climate also resembles that of eastern North America, has 729 tree species (Figure 21.16a). These figures represent the total diversity of each region, but local diversity within small areas of uniform habitat exhibits parallel differences. While a graduate student at the University of Pennsylvania, Roger Latham sought to understand why species diversity varies by a factor of nearly 6 across these three regions even though environmental conditions and forest growth forms—primarily deciduous, broad-leaved trees—are similar.

Latham determined that these diversity patterns reflect the histories and unique geographic positions of the three regions. The greater diversity in Asia results partly from the proportion of its species (32%) that belong to predominantly tropical genera. Over evolutionary time, the continuous corridor of forest habitat from the tropics of Southeast Asia to the north has allowed tropical plants and animals to invade and adapt to temperate ecosystems. In the Americas, the humid tropics of Central America are separated

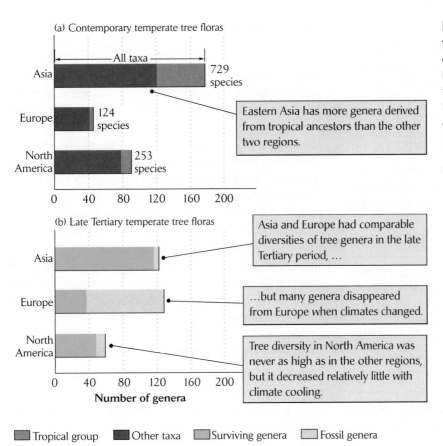

(a) Contemporary temperate tree floras

Eastern Asia has more genera derived from tropical ancestors than the other two regions.

Asia — 729 species
Europe — 124 species
North America — 253 species

(b) Late Tertiary temperate tree floras

Asia and Europe had comparable diversities of tree genera in the late Tertiary period, …

…but many genera disappeared from Europe when climates changed.

Tree diversity in North America was never as high as in the other regions, but it decreased relatively little with climate cooling.

☐ Tropical group   ☐ Other taxa   ☐ Surviving genera   ☐ Fossil genera

from the moist temperate areas of North America by a broad subtropical band of arid conditions. In Europe, the Mediterranean Sea and arid North Africa isolate temperate ecosystems from tropical Africa.

The fossil record suggests an ancient origin for the diversity difference among eastern North America, Europe, and eastern Asia. Almost twice as many genera of trees are found as fossils in eastern Asia as in North America (see Figure 21.16b), paralleling the difference in diversity observed today. It is likely that the more complex geography of eastern Asia compared with eastern North America, as well as the consistent connection of temperate eastern Asia with tropical Southeast Asia, resulted in a higher rate of species production over the past 60 million years. But notice that the fossil record of Europe includes many more genera of trees than that of North America, in contrast to the present. A large proportion of the European genera became extinct during the climatic cooling leading to the Ice Age, while few genera of North American trees disappeared. As Europe cooled, the Alps and the Mediterranean Sea posed barriers to southward movement (see Figure 26.10), and many cold-intolerant plant taxa died out (Figure 21.17). In North America, southward migration to areas bordering the Gulf of Mexico was always possible during cold periods (see Figure 21.8). ▌

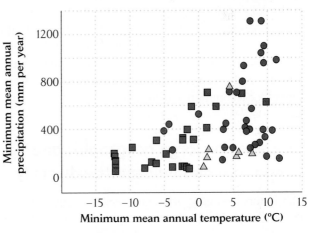

Current status in Europe
● Extinct   △ Relictual   ■ Widespread

FIGURE 21.17 Tree genera that disappeared from Europe during the late Tertiary could not tolerate low mean annual temperatures. Tree genera found in the Pliocene fossil record of Europe but which currently live outside that region cannot tolerate mean annual temperatures below about 0°C, although many of these genera are more drought-tolerant than extant species in Europe. After J.-C. Svenning, *Ecol. Lett.* 6:646–653 (2003).

Compared with these differences in temperate forests, the difference in species richness between mangrove wetlands of the Atlantic–Caribbean region and of the Indo–West Pacific region is even more striking. Mangroves are tropical trees that occur within tidal zones along coastlines and river deltas. Mangroves exhibit convergent adaptations to the high salt concentrations and anaerobic conditions in the water-saturated sediments in which they take root (see Figure 2.13). Fifteen lineages of terrestrial trees have independently colonized the mangrove habitat, and several of these have subsequently diversified there.

At present, the mangrove flora of the Atlantic and Caribbean includes 7 species in 4 genera, 3 of which are cosmopolitan (occur worldwide). In contrast, the mangrove flora of the Indo–West Pacific region includes at least 40 species in 17 genera, 14 of which are endemic to the region. Habitat extent cannot explain the greater diversity of Indo–West Pacific mangroves; both regions have roughly equal areas of mangrove wetlands. Rather, this large diversity difference appears to have resulted from plant taxa invading mangrove habitats more frequently in the Indo–West Pacific than in the Atlantic–Caribbean region. Much of the Malay Archipelago in the Indo–West Pacific region consists of islands of various sizes scattered on a shallow continental shelf, perhaps affording ideal conditions for populations to become isolated in mangrove habitats and new species of mangrove specialists to form. This type of geography scarcely exists in the Americas.

Ecologists agree that the regional species pools for various groups of organisms can differ among the major continental landmasses. However, the causes of these differences, and the degree to which they influence the structure of local communities, are active topics of discussion.

Processes on large geographic and temporal scales influence biodiversity

The history and geography of a region clearly influence the diversity of both the entire region and its local habitats. As we saw in Chapter 20, geography is important because large regions with varied climate and topography contain a wider variety of habitats, which support a greater variety of species. Barriers to dispersal such as mountains and oceans isolate populations long enough for distinct lineages to evolve. Ecological variation provides a template for specialization.

Let's revisit the question we asked in Chapter 20, "Why are there so many species in the tropics?" from the perspective of these historical processes.

Age and area

Over the Tertiary history of the earth, when contemporary geographic patterns in species richness were becoming established, tropical environments occupied far more area than temperate and boreal environments. As we have seen, tropical climates once extended outside of the present-day equatorial latitudes far to the north in Eurasia and North America, and the global area of the tropics was immense. Tropical environments have contracted considerably over the past 35 million years, but their former extent might still have left an imprint on diversity patterns. A recent analysis of the area and history of the major climate zones on earth has demonstrated that species richness is strongly connected to the age and area of these zones (Figure 21.18). Even though most species that lived 35 Mya are extinct, many of them became the ancestors of modern species.

Species production and extinction

The balance between speciation and extinction within a region also influences species richness. Several hypotheses have proposed reasons why speciation rates may be higher in tropical than in temperate regions. First, the equable climates of the tropics might favor more sedentary life histories. Populations within which individuals do not disperse far could more easily become isolated and evolve independently, eventually forming new species. Consistent with this hypothesis, large rivers and mountain valleys and passes appear to block dispersal for many species in tropical regions. Thus, one finds more locally distributed species and much greater turnover of species with distance (beta diversity) at low latitudes than at higher latitudes.

Second, equable tropical climates impose less environmental stress than temperate and boreal climates, so species interactions assume more prominent roles as selective factors. Because the partners in such interactions evolve in response to each other, evolution never stops, and this constant change is likely to speed divergence between populations and thus formation of new species. Accordingly, predator–prey, host–pathogen, and mutualistic interactions are the strongest drivers of evolution in the tropics, whereas adaptation to relatively fixed environmental stresses is more important at higher latitudes. The complexity of species interactions in the tropics also provides numerous ways in which populations can specialize, either as consumers or as resources defending themselves against consumers. We have seen how consumers can promote species richness by depressing populations of competitively superior species. The degree to which species interactions promote diversification remains to

Eocene

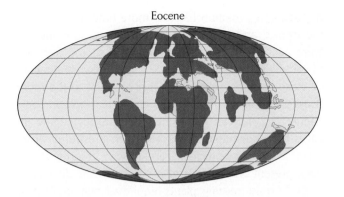

Oligocene

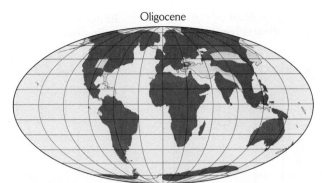

Miocene

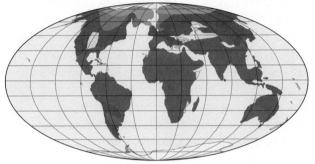

Last glacial maximum

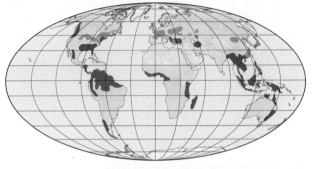

Present

☐ Boreal ■ Temperate ■ Tropical

FIGURE 21.18 The historical extent of climate zones helps explain global patterns of species richness. Tropical climates were much more extensive during the Eocene epoch (56–34 Mya) and have contracted since the Oligocene epoch (34–23 Mya). Boreal environments appeared only during the Miocene epoch (23–5 Mya). The present-day species richness of trees in eleven boreal, temperate, and tropical forest types is predicted better by the area occupied by those forest types during the mid- to late Tertiary than by their area at present. After P. V. A. Fine and R. H. Ree, *Am. Nat.* 168:796–804 (2006).

be determined, but ecologists are directing considerable effort to evaluating this hypothesis.

Is the rate of diversification (that is, speciation minus extinction) actually greater in the tropics? Phylogenetic trees could provide an answer because rapid diversification should appear as short branch lengths near the tips of a phylogenetic tree. More frequent speciation and less frequent extinction would reduce the divergence time between sister species and the lengths of terminal branches on the tree (Figure 21.19). University of British Columbia biologists Jason Weir and Dolph Schluter constructed phylogenetic trees for birds and mammals by computing genetic distances between sister species from the similarity of their DNA sequences. They found, contrary to expectation, that new species formed more frequently in temperate than in tropical regions, but because the difference in extinction was even greater than that in speciation, diversification was lower there. Their analysis focused on relatively recent periods dominated by climatic cooling at high latitudes, so these results may not represent long-term differences between rates of these processes at different latitudes. Nonetheless, they illustrate the principle that phylogenetic information can be used to test hypotheses based on long-term processes.

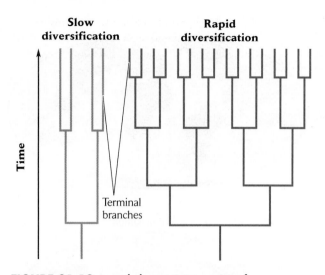

FIGURE 21.19 In a phylogenetic tree, rates of diversification can be recognized in genetic distances between sister species. When speciation proceeds rapidly compared with extinction, the number of species grows rapidly, and lineage splitting events are relatively recent.

Evolutionary conservatism and diversification

Evolution tends to be conservative; that is, small evolutionary changes are more likely than *adaptive shifts*— changes that allow an organism to occupy a different type of habitat. If evolution has had more time to operate in the tropics, and a large proportion of ancestral lineages were adapted to tropical conditions, then the decrease in species richness with latitude might reflect the failure of those primarily tropical lineages to adapt to the more stressful conditions in temperate and boreal environments (Figure 21.20). Even within the tropics, relatively few lineages have been able to adapt to such stressful environments as mangrove wetlands and deserts. Freezing conditions present another adaptive barrier that is difficult to cross. To resist freezing, plants require adaptations of their stem and trunk anatomy to prevent formation of gas bubbles in water-conducting elements, buds able to withstand freezing, and either deciduous or frost-protected leaves, among others. Although most temperate plants evolved within groups having primarily tropical distributions, only half the families of flowering plants have made this transition. In addition, many tropical groups of woody plants appear in temperate regions only as herbs, which overwinter as seeds or whose buds are protected from freezing at or below the surface of the soil.

The fossil record of diversity

With respect to species richness, the tropics have been favored by their old age and possibly by conditions that have sped the production of new species there. If these factors have been responsible for the difference in species richness between tropical and temperate regions, then species richness should increase continually over time. To determine the course of species richness through time in the past, we must turn to the fossil record.

Smithsonian scientist Carlos Jaramillo and his coworkers assembled a record of the diversity of flowering plants from about 65 Mya until 25 Mya using fossil deposits of pollen found in northwestern South America. Research on contemporary floras has shown that pollen morphotypes— that is, pollen grains with distinctive morphological features (see Figure 4.25)—indicate local diversity reasonably well. The pollen fossil record shows alternating periods of increase and decrease in plant diversity corresponding to periods of warming and cooling (Figure 21.21). Remarkably, this record contains about the same number of morphotypes at the end as at the beginning—a period of 40 million years, over which species replaced each other many times. None of the morphotypes present at the end of the sequence were present at the beginning. The overall picture, then, is one of relatively constant diversity over long periods in spite of continual evolutionary diversification and extinction.

Well-documented assemblages of fossil mammals furnish another opportunity to study the diversity of biologi-

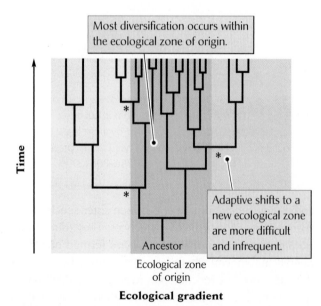

FIGURE 21.20 Diversification of a lineage within its ecological zone of origin reflects evolutionary conservatism. Occasional adaptive shifts (asterisks) to different ecological zones may be difficult and infrequent, giving rise to a gradient in contemporary diversity that favors the ecological zone of origin. After R. E. Ricklefs, *Ecology* 87:S3–S13 (2006).

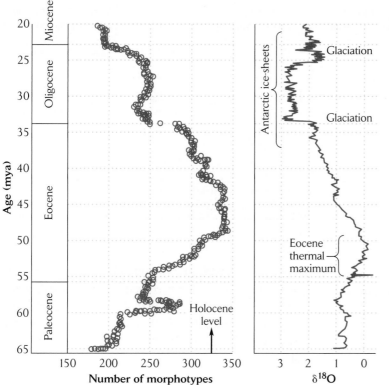

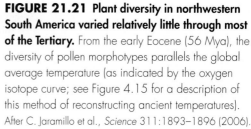

FIGURE 21.21 Plant diversity in northwestern South America varied relatively little through most of the Tertiary. From the early Eocene (56 Mya), the diversity of pollen morphotypes parallels the global average temperature (as indicated by the oxygen isotope curve; see Figure 4.15 for a description of this method of reconstructing ancient temperatures). After C. Jaramillo et al., *Science* 311:1893–1896 (2006).

cal communities over long periods, during which communities experience climate change, large-scale movements of species across continents, and evolutionary change among species. When one can sample enough fossils from many localities, it becomes possible to pose questions about changes in diversity in response to environmental change, as well as about the relationship between local and regional diversity.

Blaire Van Valkenburgh, of the University of California at Los Angeles, and Christine Janis, of Brown University, studied 115 fossil mammal assemblages from different locations in North America, mostly in the western United States. Their sampling covered the period from about 44 Mya to the present. The beginning of this period represents the zenith of warm, moist conditions in the Northern Hemisphere—a time when most of what is now the United States was covered by tropical forest. About 35 Mya, and again at about 25 Mya, the climate of North America became noticeably colder and drier, a trend that has continued until the present. The total numbers of herbivore and carnivore species in all the fossil assemblages show that herbivore diversity increased to a maximum 12–15 Mya, when grasslands had spread across much of the continent. Herbivore diversity then began to decline, steadily at first, then rapidly during the past 300,000 years as glaciers expanding and receding over much of North America dramatically altered the environment (Figure 21.22). This analysis of North American

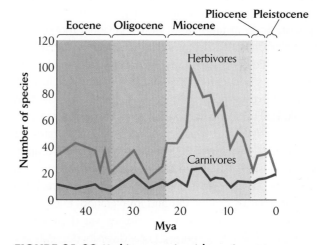

FIGURE 21.22 Herbivore species richness in western North America varied with changes in climate and vegetation during the Tertiary period. The number of herbivore species was relatively constant through the Eocene and Oligocene epochs, but increased to a maximum in association with drying climates and the spread of grasslands during the Miocene, finally declining with late Tertiary climatic cooling. Carnivore species richness varied relatively little over the same period. After B. Van Valkenburgh and C. M. Janis, in R. E. Ricklefs and D. Schluter (eds.), *Species Diversity in Ecological Communities. Historical and Geographical Perspectives,* University of Chicago Press, Chicago (1993), pp. 330–340.

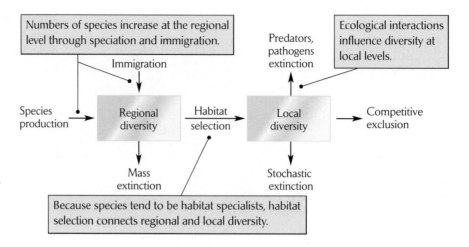

FIGURE 21.23 Many factors influence regional and local species diversity. Mass extinctions are caused by factors acting over large areas, whereas stochastic extinction independently affects small populations. After R. E. Ricklefs and D. Schluter, in R. E. Ricklefs and D. Schluter (eds.), *Species Diversity in Ecological Communities. Historical and Geographical Perspectives,* University of Chicago Press, Chicago (1993), pp. 350–363.

fossil mammals shows that diversity does not necessarily increase steadily with time and that the highest diversity often does not occur during the warmest periods.

Interaction of local and regional processes

As we have seen, many processes are important in the regulation of biodiversity, each at a different characteristic scale of time and space (Figure 21.23). Scale in space varies from the activity ranges of individuals, through the dispersal distances of individuals within populations, to the expansion and contraction of geographic ranges of species. Scale in time varies from individual movements (behavior), through death and replacement of individuals in populations (demography and population regulation), interactions between populations (competitive exclusion), and selective replacement of genotypes within populations (evolution), to formation of new species (speciation).

Local species diversity depends on local rates of extinction—resulting from predators, pathogens, competitive exclusion, changes in the physical environment, and stochastic changes in small populations—and regional rates of species production and immigration. Every location on earth has limited accessibility, via dispersal, to sources of colonizing species. Local diversity depends not only on the accessibility of a region to colonists, but also on the capacity of that region to support a variety of species, to generate new forms through speciation, and to sustain taxonomic diversity in the face of environmen-

tal variation (Figure 21.24). Although ecology has traditionally focused on local, contemporary systems, it has recently expanded its purview to embrace geographic and historical processes that traditionally belonged within the disciplines of systematics, evolution, biogeography, and paleontology.

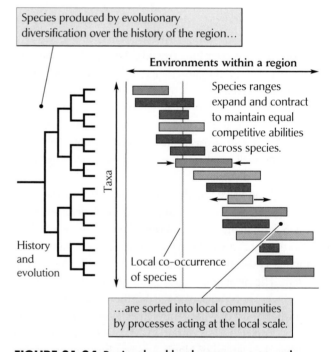

FIGURE 21.24 Regional and local processes act together to determine patterns of diversity and local species richness within regions. After R. E. Ricklefs, *Am. Nat.* 170:S56–S70 (2007).

SUMMARY

1. Ecological systems reflect a variety of regional processes as well as evolutionary and geologic history, in addition to the outcomes of local interactions of populations with their environment and with one another. Thus, the history and geography of life provide important contexts for understanding biological communities.

2. Life arose early in the earth's 4.5-billion-year history, but an abundant fossil record of modern life forms first appeared about 540 Mya, a point that marks the beginning of the Paleozoic era. The Mesozoic era, during which reptiles were dominant on land, began about 251 Mya; the age of mammals, the Cenozoic era, began 65 Mya.

3. Continental drift has changed the positions of the continents continually throughout the evolution of life, opening and closing pathways of dispersal between continental landmasses and ocean basins and greatly altering climates on earth.

4. Animals and plants have evolved independently on different continents during prolonged periods of geographic isolation. Consequently, we can distinguish six major biogeographic regions, with distinctive floras and faunas, that were once isolated from other landmasses.

5. The earth's climate cooled considerably during the last half of the Cenozoic era, causing tropical climate zones to contract to a narrower equatorial band and causing temperate and boreal climate zones to expand.

6. The Cenozoic cooling trend culminated in the Ice Age 2 million years ago, during which alternating periods of glacial advance and retreat caused shifts in distributions and extinctions of many species in the Northern Hemisphere.

7. Ecological niche modeling based on the present distributions of species reveals that many species do not occupy all environmentally suitable areas. This discrepancy suggests that many populations and communities exist out of equilibrium with changing climates.

8. The principle of convergence states that, despite their different histories of independent evolution, inhabitants of similar environments on different continents often resemble one another in form and function because they adapt to similar ecological conditions.

9. Close relatives may resemble one another too closely to coexist in local communities, and may be widely dispersed over ecological gradients. Conversely, less closely related species with greater ecological differences may partition resources in ways that allow them to coexist.

10. If community diversity were regulated only by local interactions among species, whose outcome is determined primarily by environmental conditions, then biodiversity would exhibit convergence between regions. The observation that it does not demonstrates that the unique histories and biogeographic settings of each continent influence local species diversity.

11. Biodiversity reflects a broad array of local, regional, and historical processes and events operating on a hierarchy of temporal and spatial scales. Diversity reflects the age and area of a region. Modern diversity has its roots primarily in tropical environments, and the latitudinal gradient in diversity partly reflects adaptive barriers to invading more stressful environments.

12. The fossil record of regional diversity sometimes reveals long-term stability in numbers of species, particularly in comparison to the turnover of species over time. This observation suggests that regional diversity is regulated within broad limits.

13. Understanding patterns of species diversity requires consideration of the history of a region and the integration of ecological study with the related disciplines of systematics, evolution, biogeography, and paleontology.

REVIEW QUESTIONS

1. Why do ecologists think that historical and evolutionary effects better explain the characteristics of mammal species in Australia than environmental effects?

2. When studying global patterns of diversity, why is it important to understand continental drift?

3. What is the relationship between the length of time the modern continents have been connected and the similarity among the species on each continent?

4. How does knowledge of historic climatic patterns affect our interpretation of present-day patterns of species diversity?

5. What does the observation of convergent traits among distantly related species tell us about evolution by natural selection?

6. Why does initial trait similarity between closely related species often favor the evolution of trait divergence?

7. In what way do the age and area of a region affect its species richness?

8. How can the fossil record inform us about current differences in diversity between temperate and tropical regions?

SUGGESTED READINGS

Ackerly, D. D. 2004. Adaptation, niche conservatism, and convergence: Comparative studies of leaf evolution in the California chaparral. *American Naturalist* 163:654–671.

Ackerly, D. D., D. W. Schwilk, and C. O. Webb. 2006. Niche evolution and adaptive radiation: Testing the order of trait divergence. *Ecology* 87:S50–S61.

Brown, J. H. 1995. *Macroecology*. University of Chicago Press, Chicago.

Cavender-Bares, J., et al. 2004. Phylogenetic overdispersion in Floridian oak communities. *American Naturalist* 163:823–843.

Farrell, B. D., C. Mitter, and D. J. Futuyma. 1992. Diversification at the insect–plant interface. *BioScience* 42:34–42.

Guisan, A., and N. E. Zimmermann. 2000. Predictive habitat distribution models in ecology. *Ecological Modelling* 135:147–186.

Jablonski, D., K. Roy, and J. W. Valentine. 2006. Out of the tropics: Evolutionary dynamics of the latitudinal diversity gradient. *Science* 314:102–106.

Latham, R. E., and R. E. Ricklefs. 1993. Continental comparisons of temperate-zone tree species diversity. In R. E. Ricklefs and D. Schluter (eds.), *Species Diversity: Historical and Geographical Perspectives*, pp. 294–314. University of Chicago Press, Chicago.

Latham, R. E., and R. E. Ricklefs. 1993. Global patterns of tree species richness in moist forests: Energy–diversity theory does not account for variation in species richness. *Oikos* 67:325–333.

Lomolino, M. V., B. R. Riddle, and J. H. Brown. 2006. *Biogeography*. 3rd ed. Sinauer Associates, Sunderland, Mass.

Maherali, H., and J. N. Klironomos. 2007. Influence of phylogeny on fungal community assembly and ecosystem functioning. *Science* 316:1746–1748.

Marshall, L. G. 1988. Land mammals and the Great American Interchange. *American Scientist* 76:380–388.

Mittelbach, G. G., et al. 2001. What is the observed relationship between species richness and productivity? *Ecology* 82:2381–2396.

Mittelbach, G. G., et al. 2007. Evolution and the latitudinal diversity gradient: Speciation, extinction and biogeography. *Ecology Letters* 10:315–331.

Orians, G. H., and R. T. Paine. 1983. Convergent evolution at the community level. In D. J. Futuyma and M. Slatkin (eds.), *Coevolution*, pp. 431–458. Sinauer Associates, Sunderland, Mass.

Pearson, R. G., and T. P. Dawson. 2003. Predicting the impacts of climate change on the distribution of species: Are bioclimatic envelope models useful? *Global Ecology and Biogeography* 12:361–371.

Pielou, E. C. 1991. *After the Ice Age*. University of Chicago Press, Chicago.

Pulliam, H. R. 2000. On the relationship between niche and distribution. *Ecology Letters* 3:349–361.

Qian, H., and R. E. Ricklefs. 2000. Large-scale processes and the Asian bias in species diversity of temperate plants. *Nature* 407:180–182.

Ricklefs, R. E. 2004. A comprehensive framework for global patterns in biodiversity. *Ecology Letters* 7:1–15.

Ricklefs, R. E. 2006. Evolutionary diversification and the origin of the diversity/environment relationship. *Ecology* 87:S3–S13.

Ricklefs, R. E. 2007. History and diversity: Explorations at the intersection of ecology and evolution. *American Naturalist* 170:S56–S70.

Ricklefs, R. E., and D. Schluter (eds.). 1993. *Species Diversity in Ecological Communities: Historical and Geographical Perspectives*. University of Chicago Press, Chicago.

Roy, K., and E. E. Goldberg. 2007. Origination, extinction, and dispersal: Integrative models for understanding present-day diversity gradients. *American Naturalist* 170:S71–S85.

Svenning, J. C., and F. Skov. 2004. Limited filling of the potential range in European tree species. *Ecology Letters* 7:565–573.

Van Valkenburgh, B., and C. M. Janis. 1993. Historical diversity patterns in North American large herbivores and carnivores. In R. E. Ricklefs and D. Schluter (eds.), *Species Diversity in Ecological Communities: Historical and Geographical Perspectives*, pp. 330–340. University of Chicago Press, Chicago.

Vermeij, G. J. 1991. When biotas meet: Understanding biotic interchange. *Science* 253:1099–1104.

Webb, C. O. 2000. Exploring the phylogenetic structure of ecological communities: An example for rain forest trees. *American Naturalist* 156:145–155.

Webb, C. O., et al. 2002. Phylogenies and community ecology. *Annual Review of Ecology and Systematics* 33:475–505.

Weir, J. T., and D. Schluter. 2007. The latitudinal gradient in recent speciation and extinction rates of birds and mammals. *Science* 315:1574–1576.

Wiens, J. J. 2007. Global patterns of diversification and species richness in amphibians. *American Naturalist* 170:S86–S106.

Wiens, J. J., and M. J. Donoghue. 2004. Historical biogeography, ecology and species richness. *Trends in Ecology and Evolution* 19:639–644.

Wiens, J. J., and C. H. Graham. 2005. Niche conservatism: Integrating evolution, ecology, and conservation biology. *Annual Review of Ecology Evolution and Systematics* 36:519–539.

CHAPTER 22

Energy in the Ecosystem

Ve humans consume a large proportion of the earth's biological production. Each year, the earth's plants, algae, and photosynthetic bacteria harness enough energy from sunlight to make 224 billion tons of dry biomass. Approximately 59% of this biomass is produced in terrestrial ecosystems. Of the terrestrial production, an astonishing 35%–40% is used by humans, either directly as food and fiber crops or indirectly as feed for animals.

The oceans, traditionally a source of food for people living near the coast, are now providing food for much of the world's human population. In 1950, the total catch of fish and other seafood was about 20 million tons. The total catch increased to 75 million tons by 1980 and has now leveled off at about 90 million tons annually. Annual production by fish farms has increased from about 5 million tons in 1980 to more than 40 million tons at present.

How much of the production of algae in the oceans is required to sustain the fisheries on which we humans depend? Can we hope to harvest yet more food from the oceans? In 1995, two marine ecologists, D. Pauly and V. Christensen, working at the International Center for Living Aquatic Resources Management in the Philippines, sought to apply their knowledge of energy flow in natural ecosystems to these questions.

Pauly and Christensen assumed that for each step in the food chain that leads from microscopic algae to the fish we eat, about 90% of consumed energy is used to maintain the consumer. This means that only 10% is converted through growth and reproduction into biomass, and thus potential food for other organisms. From studies of the diets of marine organisms, Pauly and Christensen estimated that the number of feeding steps leading from algae to fish varied from about 1.5, on average, for coastal and reef ecosystems to 3 for the open ocean. Knowing the number of feeding steps and assuming an energy transfer efficiency of 10% per step, they used a simple calculation to convert harvested fish into amounts of algae needed to sustain them. Using data for the early 1980s, Pauly and Christensen showed that for inshore fisheries, which produce most of the seafood consumed by humans, the algal growth required to sustain the harvest amounted to 24%–35% of the total production of the ecosystem. Because we don't eat everything that grows in the sea, Pauly and Christensen suggested that human harvesting might have been approaching its upper limit—a prediction borne out by the subsequent leveling off of the wild fish catch. Only in the open ocean, where we exploit the highly dispersed fishery inefficiently at the end of a longer food chain, do we usurp a small fraction (about 2%) of the total production.

CHAPTER CONCEPTS

- Ecosystem function obeys thermodynamic principles
- Primary production provides energy to the ecosystem
- Many factors influence primary production
- Primary production varies among ecosystems

- Only 5%–20% of assimilated energy passes between trophic levels
- Energy moves through ecosystems at different rates
- Ecosystem energetics summarizes the movement of energy

During the early part of the twentieth century, several new concepts emerged that led the study of ecology in novel directions. One of these was the realization that feeding relationships link organisms into a single functional entity, the biological community. Foremost among the proponents of this new ecological viewpoint during the 1920s was the English ecologist Charles Elton. Elton argued that organisms living in the same place not only had similar tolerances of physical factors in the environment, but also interacted with one another, most importantly in a system of feeding relationships that he called a *food web*. Every organism must feed in some manner to gain nourishment, and each may be fed on by some other organism.

A decade later, in 1935, the English plant ecologist A. G. Tansley took Elton's idea an important step further by considering organisms, together with the physical factors of their surroundings, as ecological systems. Tansley regarded this assembly, which he called the **ecosystem,** as the fundamental unit of ecological organization. Tansley envisioned the biological and physical parts of nature together, unified by the dependence of organisms on their physical surroundings and by their contributions to maintaining the conditions and composition of the physical world.

Ecosystem function obeys thermodynamic principles

Working independently of the ecologists of his day, Alfred J. Lotka, a chemist by training, was the first to consider populations and communities as energy-transforming systems. The most fundamental energy transformation in these systems is the conversion of light energy into chemical energy by photosynthesis. Further energy transformations take place as herbivores convert the energy in the carbon compounds in plants and other autotrophs into energy they can use for their own metabolism, activity, growth, and reproduction. Similarly, carnivores utilize energy from the carbon compounds in their prey.

Lotka believed that the size of a system and the rates of energy and material transformations within it obey certain **thermodynamic principles** that govern all energy transformations. Just as heavy machines and fast machines require more fuel to operate than lighter and slower ones, and inefficient machines require more fuel than efficient ones, the energy transformations of ecosystems grow in direct proportion to their size (roughly, the total mass of their constituent organisms), productivity (rates of transformations), and inefficiency. The earth itself is a giant thermodynamic machine in which the circulation of winds and ocean currents and the evaporation of water are driven by the energy in sunlight. Energy from the sun is also assimilated by plants, and that energy ultimately fuels most biological systems.

Lotka's ideas about ecosystems, published in 1925, were not widely understood or appreciated at the time. It remained for Raymond Lindeman, a young aquatic ecologist at the University of Minnesota, to bring the concept of the ecosystem as an energy-transforming system to the attention of many ecologists. On its publication in 1942, Lindeman's framework for understanding ecological

FIGURE 22.1 Lindeman visualized a pyramid of energy within the ecosystem. The breadth of each bar represents the amount of energy at that trophic level in the ecosystem. Energy is lost with each transfer to a higher trophic level.

systems on the basis of thermodynamic principles made a deep impression. He adopted Tansley's notion of the ecosystem as the fundamental unit in ecology and Elton's concept of the food web, including inorganic nutrients at the base, as the most useful expressions of ecosystem structure.

The food chain by which energy passes through the ecosystem has many links—plant, herbivore, and carnivore, for example—which Lindeman referred to as *trophic* levels. (Remember that the Greek root of the word *trophic* means "nourishment.") Furthermore, Lindeman visualized a **pyramid of energy** within the ecosystem, with less energy reaching each successively higher trophic level (Figure 22.1). Lindeman argued that energy is lost at each level because of the work performed by organisms at that level and because of the inefficiency of biological energy transformations. Thus, plants gather only a portion of the light energy available from the sun. Herbivores harvest even less of that light energy because plants use a portion of the energy they assimilate to maintain themselves, and that energy is not available to herbivores as plant biomass. The same may be said of the secondary consumers that feed on herbivores, and of each successively higher level of the food chain. Most of the energy in the food we humans consume is used to maintain ourselves, and little—Pauly and Christensen used a value of 10% for marine food chains—becomes biomass available to the next trophic level in the food chain.

By the 1950s, the ecosystem concept had fully pervaded ecological thinking and had given rise to a new branch of ecology, called **ecosystem ecology,** concerned with the cycling of matter and the associated passage of energy through an ecosystem. Energy and the masses of elements, such as carbon, provided a common "currency" that ecologists could use to compare the structure and functioning of different ecosystems. Measurements of energy and nutrient assimilation became the tools for exploring this new thermodynamic concept of the ecosystem.

With this new conceptual framework, ecologists began to measure energy flow and the cycling of nutrients. One of the strongest proponents of this approach was Eugene P. Odum of the University of Georgia, whose textbook *Fundamentals of Ecology,* first published in 1953, influenced a generation of ecologists. Odum depicted ecosystems as a series of simple energy flow diagrams (Figure 22.2) representing the use and transfer of energy by all the organisms at each trophic level. These diagrams simplified nature, but nonetheless conveyed the important principle that energy passes from one link in the food chain to the next, diminished by respiration and the shunting of unused organic materials to detritus-based food chains.

Unlike energy, most of which enters ecosystems as light and leaves as heat, nutrients are regenerated and retained largely within the system. Matter cycles through an ecosystem after it is taken up in inorganic forms and converted to biomass by plants. Some of that matter is passed up the food chain, but all of it eventually returns to inorganic forms by the process of decomposition. The major inorganic nutrients cycled through the ecosystem, besides hydrogen and oxygen, are the elements carbon, nitrogen, phosphorus, and sulfur. Nutrient cycling is dominated by carbon, which is assimilated in the form of carbon dioxide by plants through photosynthesis, converted to organic forms of carbon, and then returned to the environment as carbon dioxide through respiration. As concentrations of carbon dioxide in the atmosphere increase with human combustion of fossil fuels, the capacity of plants to remove the carbon in carbon dioxide and sequester it as organic carbon is of vital interest.

FIGURE 22.2 E. P. Odum developed a "universal" model of energy flow through ecosystems. The energy ingested by organisms at each trophic level is reduced by respiration and excretion, so that less energy is available for consumption by the next trophic level.

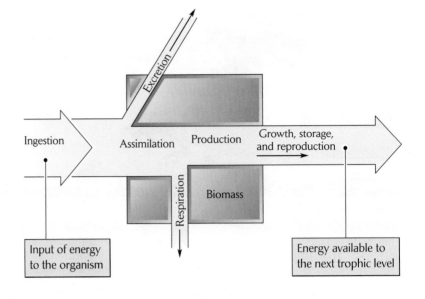

The flow of energy is difficult to measure directly, but a convenient index to energy flow is provided by measuring the movement of elements among ecosystem components. Because photosynthesis transforms light energy into the chemical energy of reduced carbon in organic molecules, one can follow the movement of energy through the ecosystem by tracking the movement of biological forms of carbon.

A second reason for the prominence of nutrient cycling in ecosystem ecology is the fact that, in many circumstances, the quantities of certain nutrients limit the production of biomass by photosynthetic organisms, which is the material and energetic base of the entire ecosystem. The open oceans, for example, are virtual deserts because of their scarce nutrients, particularly nitrogen. In contrast, availability of water, rather than sunlight or minerals in the soil, limits the productivity of desert plants. Understanding how nutrients cycle among components of the ecosystem is crucial to understanding the regulation of ecosystem structure and function.

Primary production provides energy to the ecosystem

Sunlight is the ultimate source of energy for most living things, including the power needs of human society. Fossil fuels such as oil and coal represent the remains of organisms accumulated over millions of years as deposits of organic matter in the earth's crust. Sunlight drives the air currents that we increasingly harvest as wind power and evaporates the water that eventually falls as precipitation and fills the rivers that we dam for hydroelectric power.

Without the energy of sunlight, biological systems would not exist as we know them, nor would we exist to know of their absence.

Plants, algae, and some bacteria capture light energy and transform it into the energy of chemical bonds in carbohydrates by photosynthesis. This process of energy assimilation, which underlies all ecosystem functions, is referred to as **primary production.** Without primary production, practically nothing that we could call life would exist.

As we saw in Chapter 3, photosynthesis chemically unites two common inorganic compounds, carbon dioxide (CO_2) and water (H_2O), to form the sugar glucose ($C_6H_{12}O_6$), with the release of oxygen (O_2). The overall chemical balance of the photosynthetic reaction is

$$6\ CO_2 + 6\ H_2O \rightarrow C_6H_{12}O_6 + 6\ O_2$$

Photosynthesis transforms carbon from an oxidized (low-energy) state in CO_2 to a reduced (high-energy) state in the chemical bonds of glucose. Because work is performed on the carbon atoms to increase their energy level, photosynthesis requires energy. That energy is provided by visible light. In quantitative terms, for each gram of carbon assimilated, a plant transforms 39 kilojoules (kJ) of light energy from the sun into chemical energy in carbohydrates.

The photosynthetic pigments that capture the energy of light actually absorb only a small fraction of the total incident solar radiation. In addition, because of inefficiencies in the many biochemical steps of photosynthesis, plants assimilate no more than a third (and usually much less) of the light energy absorbed by those photosynthetic pigments. The rest is lost as heat.

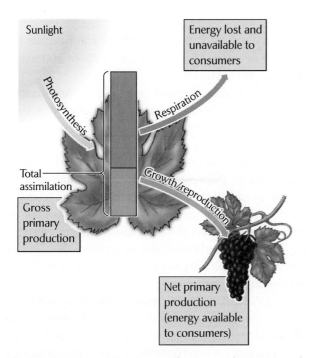

FIGURE 22.3 Gross primary production can be partitioned into respiration and net primary production.

Photosynthesis supplies the carbohydrates and energy that a plant needs to build tissues, grow, and reproduce. Rearranged and joined together, glucose molecules become fats, starches, oils, and cellulose. These carbohydrate compounds may be used to build the plant's tissues or stored as a source of energy for future needs. Combined with nutrient elements such as nitrogen, phosphorus, sulfur, and magnesium, simple carbohydrates derived from glucose produce an array of proteins, nucleic acids, and pigments. Plants cannot grow unless they have all these basic building materials, and they must be present in the right proportions. For example, the photosynthetic pigment chlorophyll contains an atom of magnesium, so even when all other necessary elements are present in abundance, a plant lacking sufficient magnesium cannot produce chlorophyll, and thus cannot engage in photosynthesis.

Plants and other photosynthetic autotrophs form the base of most food chains and are therefore referred to as the *primary producers* of the ecosystem. Ecologists are interested in the rate of primary production (referred to as *primary productivity*) because it determines the total energy available to the ecosystem. The total energy assimilated by photosynthesis represents **gross primary production.** Plants use some of this energy to maintain themselves and meet their metabolic needs through **respiration.** As a result, plant biomass contains substantially less energy than the total assimilated (Figure 22.3). The energy accu-mulated in plant biomass, and thus available to consumers, is referred to as **net primary production.**

Measuring primary production

Primary production involves fluxes of carbon dioxide, oxygen, minerals, and water and the accumulation of biomass (Figure 22.4). In principle, the rates of any of these flows could provide an index to primary productivity. It is worth discussing the measurement of primary production in some detail, as this will provide a better understanding of the processes involved in production and of the difference between gross and net production.

The unit of primary productivity is energy per unit of area per unit of time. When comparing productivity, ecologists often use kilojoules per square meter per year (kJ per m^2 per year) or watts per square meter (W per m^2). (The watt, which is the familiar unit used to rate the power consumption of light bulbs and appliances, is equal to one joule of energy per second.) Production need not be measured only in terms of energy, however. Net production can be quantified conveniently as grams of carbon assimilated, dry weight of plant tissues, or their energy equivalents. For example, ecologists often report annual net primary productivity in tons of carbon per hectare per year. Ecologists use such indices interchangeably because they are highly correlated. The energy equivalent of an organic compound can be calculated from its carbon content: organic compounds contain approximately 39 kJ of metabolizable energy per gram of carbon.

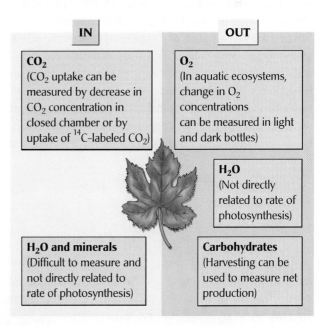

FIGURE 22.4 Many of the fluxes involved in primary production can be measured.

In terrestrial ecosystems, ecologists often estimate net production by the amount of plant biomass produced in a year. In areas of seasonal growth, they may measure net annual production by cutting, drying, and weighing plants at the end of the growing season (a method known as *harvesting*). The growth of roots, rhizomes, and tubers is often ignored because they are difficult to remove from most soils; thus, harvesting measures *annual aboveground net production* (AANP), the most common basis for comparing terrestrial communities. However, because underground production can be substantial and differs as a proportion of total production between ecosystems, comparisons of AANP should be viewed with caution.

Production of small plants or individual leaves can be quantified directly by carbon dioxide uptake. Because the atmosphere contains so little carbon dioxide (0.04%), plants can measurably reduce its concentration in an enclosed, sunlit chamber within a short period. The technology for measuring carbon dioxide concentration by absorption of infrared light (the basis for the "greenhouse" effect) is now so advanced that plant physiologists can measure the rate of carbon dioxide uptake on a few square centimeters of leaf under natural conditions in a matter of seconds. When a plant is exposed to light, carbon dioxide flux includes both assimilation (uptake) and respiration (output), and thus measures net production. Respiration can be measured separately by carbon dioxide production in the absence of light. Gross production can then be estimated by adding respiration to the net production (Figure 22.5).

The radioactive isotope carbon-14 (^{14}C) provides a useful variation on this method of measuring production. When a known amount of ^{14}C-labeled carbon dioxide is added to an airtight chamber, plants assimilate the radioactive carbon atoms roughly in proportion to their occurrence in the air inside the chamber. Thus, the rate of carbon assimilation can be calculated by dividing the amount of ^{14}C in the plant by the proportion of ^{14}C in the chamber at the beginning of the experiment. For example, if a plant takes up 10 mg of ^{14}C in an hour, and ^{14}C constitutes 5% of the carbon in the chamber, we can calculate that the plant assimilates carbon at a rate of about 200 mg per hour (10 divided by 0.05).

Scientists can measure assimilation and respiration of carbon dioxide over short periods on a larger spatial scale—for example, over a patch of forest or grassland—by using tower-mounted sensors that quantify the vertical movement of carbon dioxide in the atmosphere above the vegetation. This technique, which is known as *eddy flux covariance*, helps scientists to understand how temperature and precipitation control primary production and affect the respiration of plants and other organisms. The technique also helps them predict how ecosystem productivity might change in response to climate change in the future.

In aquatic systems, harvesting provides a convenient method for estimating the primary production of large photosynthetic organisms, such as kelps, but this technique is not practical for small organisms, such as phytoplankton. Because most waters contain such high concentrations of bicarbonate ions, measuring changes in carbon dioxide in aquatic systems is also impractical. However, because oxygen dissolves so poorly in water, one can measure small changes in oxygen concentrations in most aquatic systems. Remember that photosynthesis produces molecular oxygen (O_2) as a by-product. To estimate primary production, samples of water containing phytoplankton are suspended in pairs of sealed bottles beneath the surface of a body of water. One bottle (the "light bottle") is clear and allows sunlight to enter; the other (the "dark bottle") is opaque (Figure 22.6). In the light bottle, photosynthesis and respiration occur together, and part of the oxygen produced by the first process is consumed by the second. In the dark bottle, respiration consumes oxygen without its being replenished by photosynthesis. Thus, gross production can be estimated by adding the change in oxygen concentration in the dark bottle (respiration alone) to that in the light bottle (photosynthesis plus respiration).

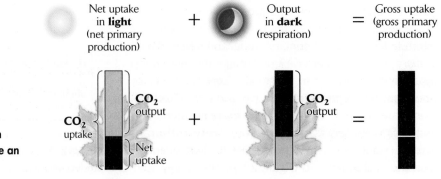

FIGURE 22.5 Measurements of carbon dioxide flux in dark and light can provide an estimate of gross primary production.

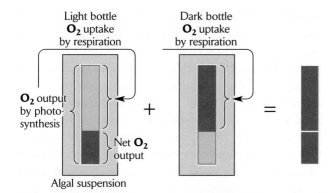

FIGURE 22.6 Paired light and dark bottles can be used to measure production by aquatic phytoplankton.

![DATA ANALYSIS] **DATA ANALYSIS MODULE** *Measuring Ecosystem Productivity Using Dissolved Oxygen to Estimate Stream Metabolism.* You will find this module at http://www.whfreeman.com/ricklefs6e.

Many factors influence primary production

Light and temperature

The rate of photosynthesis depends on the availability of light, at least under low-light conditions. In the shade of a forest, light intensity may be reduced to a hundredth of its level above the canopy, and light becomes the limiting factor in plant production. Low light intensity also limits production in the depths of lakes and the ocean. Above a certain light intensity, however, photosynthetic pigments become *saturated;* that is, they cannot absorb additional light energy or use it efficiently. At that point, production becomes limited by other factors, such as the availability of water, carbon dioxide, or nutrients.

Ecologists define the amount of light striking a surface as *irradiance,* which is commonly expressed in watts per square meter. In photosynthesis, the capture of light energy is proportional to the number of photons absorbed, and light is quantified in a corresponding manner. One speaks of a mole of photons (roughly 6×10^{23} photons) as one Einstein (E). About nine Einsteins of photons are required for each mole of oxygen (O_2) produced in photosynthesis. Accordingly, plant physiologists have adopted microEinsteins per square meter per second (μE per m^2 per second) as a measure of irradiance. However, because we shall follow energy as it moves through the ecosystem, we shall continue to use watts and joules to describe energy flux and energy content.

A flat surface above the earth's atmosphere directly facing the sun would receive an average of 1,366 W per m^2. This intensity of solar radiation—the energy reaching the outer limit of the atmosphere—is called the **solar constant.** In reality, the average irradiance at any area on the surface of the earth is far less. Nighttime periods without light, the low incidence of light early and late in the day and at high latitudes, absorption of light by the atmosphere, and reflection of light by clouds diminish light intensity at the earth's surface (see Chapter 3).

The rate of photosynthesis in plants varies in direct proportion to irradiance under low light intensities. With brighter light, however, the rate of photosynthesis increases more slowly or levels off as intensity increases. The response of photosynthesis to light intensity has two reference points (Figure 22.7). The first, called the **compensation point,** is the level of light intensity at which assimilation of energy by photosynthesis just balances loss of energy by respiration. Above the compensation point, the energy balance of the plant is positive; below the compensation point, the energy balance is negative. The second reference point, called the **saturation point,** is the level of light intensity above which the rate of photosynthesis no longer responds to increasing light intensity. Among terrestrial plants, the compensation points of species that normally grow in full sunlight (a maximum of about 500 W per m^2) occur between 1 and 2 W per m^2. The saturation points of such species are usually reached between 30 and 40 W per m^2—less than a tenth of the energy

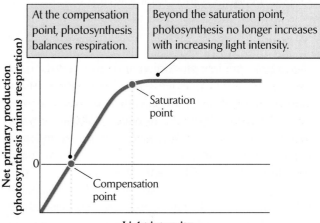

FIGURE 22.7 Photosynthesis increases with light intensity to the saturation point. The compensation point is the level of light intensity at which photosynthesis (measured here by carbon dioxide assimilation) balances respiration. The saturation point is the light intensity at which photosynthesis levels off. After M. G. Barbour, J. H. Burk, and W. D. Pitts, *Terrestrial Plant Ecology,* Benjamin Cummings, Menlo Park, Calif. (1980).

level of bright, direct sunlight. As one might expect, the compensation and saturation points of plants that typically grow in shade occur at lower light intensities.

Like the rates of most other physiological processes, the rate of photosynthesis generally increases with temperature, at least up to a point. The optimum temperature for photosynthesis varies with the prevailing temperature of the environment—from about 16°C in many temperate species to as high as 38°C in tropical species. Net production depends on the rate of respiration as well as the rate of photosynthesis: although net production rises with the rate of photosynthesis, it drops with an increasing rate of respiration. Respiration, in turn, generally rises with increasing leaf temperature. Thus, net production, and therefore net assimilation of CO_2, may actually decrease at higher temperatures.

Photosynthetic efficiency is the percentage of the energy in sunlight that is converted to net primary production during the growing season. This measure provides a useful index to rates of primary production under natural conditions. Where water and nutrients do not limit primary production severely, the photosynthetic efficiency of an ecosystem as a whole varies between 1% and 2%. What happens to the remaining 98%–99% of the light energy? Some of it is assimilated during photosynthesis and later respired. Leaves and other surfaces reflect anywhere from 25% to 75% of the light energy. Molecules other than photosynthetic pigments absorb most of the remainder, which is converted to heat and either radiated or conducted across the leaf surface or dissipated by the evaporation of water from the leaf (transpiration).

Water

Water limits primary production in many terrestrial habitats. As we saw in Chapter 2, the tiny openings (stomates) in leaves through which carbon dioxide and oxygen are exchanged with the atmosphere also allow water to escape the leaf by transpiration. When soil moisture approaches the wilting point, plants close their stomates to reduce water loss. Closing the stomates, however, prevents uptake of CO_2, and photosynthesis slows to a standstill. Consequently, the rate of photosynthesis depends on soil moisture availability, a plant's ability to tolerate water loss, and the influence of air temperature and solar radiation on the rate of transpiration.

Agronomists quantify the drought resistance of crop plants in terms of transpiration efficiency, also called **water use efficiency,** which is the number of grams of dry matter produced (net production) per kilogram of water transpired. In most plants, water use efficiencies are less than 2 g per kg, but they may be as high as 4–6 g per kg in drought-tolerant crops.

Because water use efficiency varies so little among a wide variety of plant species, ecologists can relate production directly to water availability in the environment. However, much of the precipitation received by an area is never taken up by plants. Groundwater, surface water (streams), and evaporation from the soil account for the remainder of the water budget of an ecosystem. For example, in perennial grasslands in southern Arizona, production varied in direct proportion to precipitation during the summer growing season, but achieved a rate of only about 200 kg per hectare for each 10 cm of precipitation. Ten centimeters of rainfall is equivalent to a million kilograms of water per hectare. Thus, the water use efficiency of the grassland ecosystem as a whole is only 0.2 g per kg, about one-tenth that expected based on the amount of water transpired. This finding suggests that only about 10% of the precipitation is taken up and transpired by plants in this habitat. Much of the annual precipitation comes during the summer months in extremely heavy thundershowers, after which most of the water quickly runs off the land.

The adaptations of photosynthetic mechanisms described in Chapter 3 can increase water use efficiency considerably and convey advantages to plants under hot, dry conditions. For example, C_4 photosynthesis, by maintaining high concentrations of carbon dioxide in the photosynthetic tissues of leaves, not only reduces transpiration by reducing the amount of time the stomates need to stay open, but also reduces photorespiration, which would otherwise undo some of the assimilatory work of photosynthesis and reduce water use efficiency. Measurements on 14 species of C_4 crop plants grown in Colorado revealed an average water use efficiency of about 3.1 g per kg, which was about twice the value observed in 51 species of C_3 crop plants (1.6 g per kg). Plants that use CAM photosynthesis partition carbon dioxide uptake and photosynthesis between nighttime and daytime and therefore take in carbon dioxide during the coolest period of the daily cycle, when the potential for water loss is lowest. This adaptation also increases water use efficiency.

Finally, as carbon dioxide concentrations in the atmosphere increase, CO_2 enters leaf tissues more readily, making water use more efficient. Thus, more rapid CO_2 assimilation will partially offset the negative effects of higher temperatures and reduced precipitation on ecosystem productivity expected in some regions due to global warming.

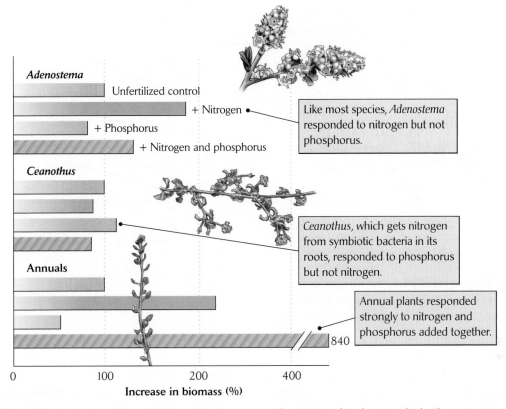

FIGURE 22.8 Fertilizers stimulate plant growth in natural habitats. Response of the chaparral shrubs *Adenostema* (a typical chaparral plant), *Ceanothus* (which harbors nitrogen-fixing bacteria), and annual grasses and forbs to fertilization with nitrogen, phosphorus, or both. The response to fertilization is given by the increase in biomass as a percentage of that of unfertilized controls. After G. S. McMaster, W. M. Jow, and J. Kummerow, *J. Ecol.* 70:745–756 (1982).

Nutrients

The observation that fertilizers stimulate plant growth in most environments suggests that nutrients limit primary production. Production in both terrestrial and aquatic environments can be enhanced by the addition of various nutrients, especially nitrogen and phosphorus. The addition of nutrients stimulates production the most in systems in which nutrient availabilities are lowest.

When nitrogen and phosphorus fertilizers were applied singly and in combination to chaparral habitat in southern California, most species responded with increased production to additions of nitrogen, but not phosphorus (Figure 22.8). This result suggests that availability of nitrogen limits production in most chaparral species. However, the growth of California lilac bushes (*Ceanothus greggii*), which harbor nitrogen-fixing bacteria in their root systems, responded to the addition of phosphorus, but not nitrogen. These plants obtain sufficient nitrogen from their symbiotic bacteria. The production of annual plants (forbs and grasses) increased when nitrogen was applied, but was depressed somewhat by the application of phosphorus alone. When equal amounts of nitrogen and phosphorus were applied together, however, production soared. Evidently, the annual plants could take advantage of increased phosphorus only in the presence of high levels of nitrogen. The relative availabilities of different nutrients have to match their requirements by plants to ensure their most efficient use.

Agronomists and ecologists calculate the **nutrient use efficiency (NUE)** of plants as the ratio of dry matter production to the assimilation of a particular nutrient element, usually expressed as grams per gram. The NUE for nitrogen, for example, is expressed as grams of dry matter produced per gram of nitrogen assimilated. Values of NUE for crop plants vary widely, but commonly fall within the range of 50–300 g per g for nitrogen and 300–1,500 g per g for phosphorus, reflecting the higher requirement of plants for nitrogen.

The nutrient use efficiencies of natural vegetation also vary widely depending on other limitations on plant production, including availability of water and other nutrients

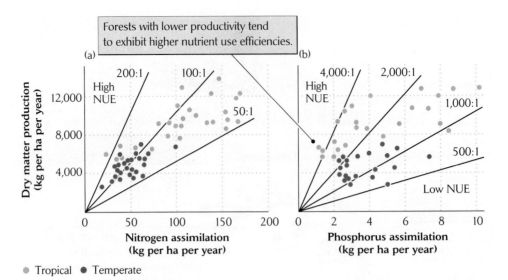

Forests with lower productivity tend to exhibit higher nutrient use efficiencies.

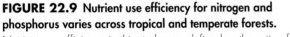

● Tropical ● Temperate

FIGURE 22.9 Nutrient use efficiency for nitrogen and phosphorus varies across tropical and temperate forests. Nutrient use efficiency in this study was defined as the ratio of the dry matter in leaf litter to (a) nitrogen or (b) phosphorus in the litter. After P. M. Vitousek, *Am. Nat.* 119:553–572 (1982).

that are required in small amounts. Peter Vitousek, a plant ecologist at Stanford University, attempted to pinpoint the influences of environmental factors on production in tropical and temperate forests by comparing the cycling of individual elements with the flow of energy through each system as a whole. He estimated the total dry mass of litter produced each year per unit of area by collecting falling leaves and branches. He then measured the amounts of several nutrient elements in the litter, including nitrogen and phosphorus. The ratio of the amount of dry matter to the amount of each element served as a measure of the nutrient use efficiency associated with producing leaves and other components of plant litter. He reasoned that the element most strongly correlated with litter production probably limited overall plant production.

Vitousek found that the nitrogen content of litter was more nearly constant than that of phosphorus (Figure 22.9) or calcium (not shown), indicating that production paralleled nitrogen assimilation. His analysis also revealed that production does not vary in direct proportion to nutrient assimilation. Rather, forest types with lower productivity tend to exhibit higher nutrient use efficiencies. Two factors could cause these higher NUEs: trees in the more efficient forests might assimilate more energy per unit of nutrient assimilated, or they might retain nutrients for reuse by drawing them back into their stems before dropping their leaves. Remember that Vitousek sampled leaves after the trees had shed them, rather than when they were first produced. Tropical trees evidently

retain phosphorus to a greater extent than do temperate trees, perhaps owing to its relative scarcity in highly weathered tropical soils.

Primary production varies among ecosystems

Primary production varies greatly with latitude (Figure 22.10). The favorable combination of intense sunlight, warm temperatures, abundant rainfall, and ample nutrients in most parts of the humid tropics results in the highest terrestrial productivity on earth. In temperate and polar ecosystems, low winter temperatures and long winter nights curtail production.

Within each latitudinal belt, where light does not vary appreciably from one locality to the next, net production is related directly to temperature and annual precipitation. Above a certain threshold of water availability, net production increases by 0.4 gram of dry matter per kilogram of water in hot deserts and by 1.1 g per kg in short-grass prairies and cold deserts. Thus, a given amount of water supports almost three times as much plant production in the cooler climates as in the hotter climates within a given latitudinal belt. However, when precipitation exceeds about 3 m per year—which occurs primarily in the humid tropics—net production actually decreases (Figure 22.11). In their studies across a precipitation gradient in Hawaii, University of Florida ecologist Edward

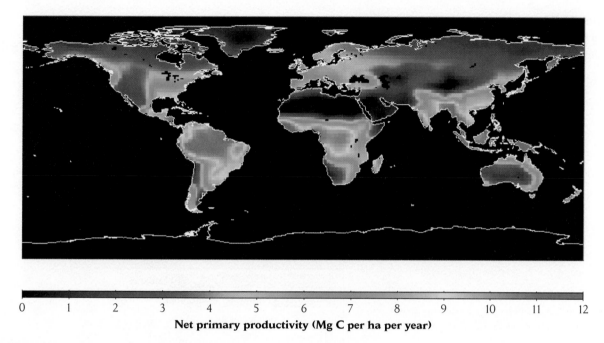

Net primary productivity (Mg C per ha per year)

FIGURE 22.10 Net primary productivity is highest in the humid tropics. This map was produced by extrapolation from global climatic data and from local studies describing the relationship of net primary production to annual mean temperature and precipitation. The scale is in metric tons (Mg; 10^6 g) of carbon per hectare per year. Courtesy of E. A. G. Schuur.

Production increases with precipitation up to about 3 m per year, but declines at higher precipitation rates.

(a)

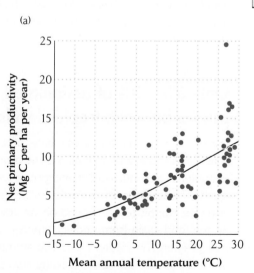

(b)

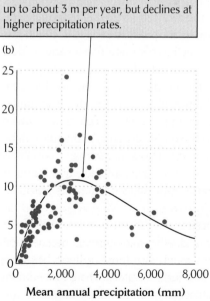

FIGURE 22.11 Net primary productivity in terrestrial ecosystems is influenced by temperature and precipitation. (a) Net production increases with temperature. (b) Net production also increases with precipitation up to a point, but declines at very high annual precipitation rates. The data points are from individual study sites throughout the world. After E. A. G. Schuur, *Ecology* 84:1165–1170 (2003).

FIGURE 22.12 Net primary productivity varies among habitat types. Data from R. H. Whittaker and G. E. Likens, *Human Ecol.* 1:357–369 (1973).

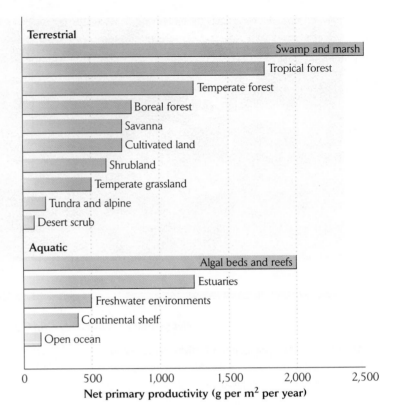

Schuur and his colleagues demonstrated that in areas of high precipitation, the decomposition of organic matter is depressed in the waterlogged soils. Thus, nitrogen and other nutrients are regenerated only slowly from organic detritus in the soil; the reduced rate of nutrient regeneration in turn depresses plant production.

Net primary production per unit of area varies widely among aquatic and terrestrial habitat types (Figure 22.12). As we have seen, the productivity of terrestrial vegetation is highest in the humid tropics; it is lowest in tundra and desert habitats. Swamp and marsh ecosystems, which occupy the interface between terrestrial and aquatic habitats, can produce more biomass annually than tropical forests because water is continuously available and nutrients are rapidly regenerated in the mucky sediments surrounding plant roots.

In the open ocean, the remains of dead organisms tend to sink to the depths, and nutrient regeneration occurs mostly in sediments at the ocean floor. Nutrients are scarce in the sunlit surface waters, where low nutrient levels limit productivity to a tenth that of temperate forests, or even less. Upwelling zones (where nutrients reach the surface from deeper waters) and continental shelf areas (where bottom sediments in shallow water rapidly exchange nutrients with surface waters) support greater

production. Estuaries, coral reefs, and coastal algal beds, where nutrients are generally abundant and tend to be recirculated locally, are the most productive marine ecosystems. Primary production in freshwater environments is considerably higher than that in the open ocean. Freshwater production is highest in rivers, shallow lakes, and ponds, where nutrients are most abundant, and lowest in clear streams and deep lakes.

Only 5%–20% of assimilated energy passes between trophic levels

Primary production forms the base of ecological food chains and is the source of all the chemical energy in the ecosystem. How much of this energy reaches higher trophic levels? The abundances and biological activity of organisms at higher trophic levels depend on the transfer of energy up the food chain from primary producers (see the pyramid of energy in Figure 22.1). The amount that reaches the top depends on how efficiently assimilated energy is converted into growth and reproduction—biomass that can be consumed by the next trophic level. Biochemical transformations in plants dissipate much of the energy of gross primary production before it can be consumed by herbivores feeding at the next trophic level. With each further step in the food chain, 80%–95% of energy is lost. All the grass in Africa piled together would dwarf a mound of all the grasshoppers, gazelles, zebras, wildebeests, and other animals that eat grass. That mound of herbivores, in turn, would overwhelm the pitiful heap of all the lions, hyenas, and other carnivores that feed on them.

As Raymond Lindeman first pointed out in 1942, the amount of energy reaching each trophic level depends on both net primary production at the base of the food chain and the efficiencies of energy transfers at each trophic level. Of the light energy assimilated by photosynthesis (gross primary production), plants use between 15% and 70% to maintain themselves, depending on their environment and growth form. Plants living in the tropics typically have higher rates of respiration relative to photosynthesis than plants in cooler environments. The respired energy used for maintenance is eventually lost as heat, thereby making that portion unavailable to consumers.

Herbivores and carnivores are more active than plants and expend correspondingly more of their assimilated energy on maintenance. As a result, production at each trophic level is typically only 5%–20% that of the level below it. Ecologists refer to the percentage of energy transferred from one trophic level to the next as **ecological efficiency** or *food chain efficiency.*

Many studies of ecological efficiencies have led to the generalization that 10% of energy is passed from one trophic level to another. This is not a fixed law, because a variety of influences can increase or decrease this percentage. Nonetheless, a simple and surprising consequence of this 10% rule of thumb is that only 1% of the total energy assimilated by primary producers ends up as production on the third trophic level. Very little energy is available to support consumers at even higher trophic levels. Thus, the pyramid of energy narrows very quickly as one climbs from one trophic level to the next. These observations suggest that humans, who already command such a large proportion of the earth's total primary production, can increase their food supplies primarily by eating lower on the food chain—that is, eating more plant products and fewer animal products.

To understand why ecological efficiencies are only 5%–20%, we must examine how consumers make use of the food energy they consume. Regardless of the source of its food, an organism uses the energy from that food to maintain itself, to fuel its activities, and to grow and reproduce. Once ingested, the energy in food follows a variety of paths through the organism. To begin with, many components of food are not easily digested, such as hair,

FIGURE 22.13 Not all components of food can be assimilated. The undigested fibrous plant material in this elephant dung represents egested energy. Photo by R. E. Ricklefs.

feathers, insect exoskeletons, cartilage, and bone in animal foods and structural materials such as cellulose and lignin in plant foods (Figure 22.13). These substances may be defecated or regurgitated, and the energy they contain is referred to as **egested energy.** What an organism digests and absorbs constitutes its **assimilated energy;** thus, ingested energy − egested energy = assimilated energy. The portion of this assimilated energy used to meet metabolic needs, most of which escapes the organism as heat, makes up **respired energy.** Animals excrete another, usually smaller, portion of assimilated energy in the form of nitrogen-containing organic wastes (primarily ammonia, urea, or uric acid) produced when the diet contains an excess of nitrogen; the portion of the assimilated energy that is excreted is called **excreted energy.** Assimilated energy retained by the organism becomes available for growth and reproduction. The new biomass produced by growth and reproduction becomes available to feed organisms at the next trophic level: assimilated energy − respiration − excretion = production.

Assimilation efficiency

The energy value of plants to their consumers depends on their food quality—that is, on how much cellulose, lignin, and other indigestible materials they contain. More formally, **assimilation efficiency** is the ratio of assimilated energy to ingested energy, usually expressed as a percentage. Herbivores assimilate as much as 80% of the energy in seeds and 60%–70% of that in young vegetation. Most grazers and browsers (elephants, cattle, grasshoppers) extract 30%–40% of the energy in their food. Millipedes,

(a)

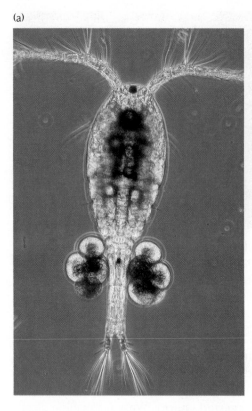

(b)

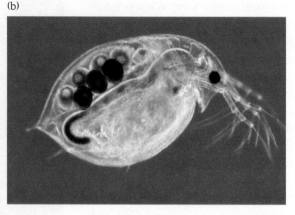

FIGURE 22.14 The nutrient requirements of consumers vary with their biology. (a) A marine calanoid copepod. (b) A freshwater *Daphnia*. Both of these consumers feed on algae, but *Daphnia* grows much more rapidly and thus requires proportionally more phosphorus. Photo (a) by Roland Birke/Photo Researchers; photo (b) by M. I. Walker/Science Source/Photo Researchers.

which eat decaying wood composed mostly of cellulose and lignin (and the microorganisms that occur in decaying wood), assimilate only 15% of the energy in their diet. Foods of animal origin are more easily digested than foods of plant origin. Assimilation efficiencies of most secondary consumers range from 60% to 90%.

All organisms require nutrients as well as energy. An imbalance develops when essential nutrients are present in different proportions in a consumer's food supplies than in its own tissues. In this case, assimilation efficiency drops because consumers must process larger amounts of food to obtain the most limiting nutrient. The study of the balance between what goes in and what comes out is referred to as **stoichiometry.** In ecology, principles of stoichiometry are often applied to understand the relationship between nutrients in consumed food and produced biomass. Imbalances represent unusable nutrients, which must be egested or excreted, and reduced ecological efficiency. Suppose, for example, that a consumer's tissues contained nitrogen and phosphorus in the ratio 10:1. If this consumer fed on prey with a N:P ratio of 20:1, then to take in enough phosphorus to build tissues, it would have to take in twice as much nitrogen as needed, and the excess nitrogen would have to be excreted. Stoichiometric considerations influence not only the efficiency of material and energy transfers up the food chain, but

also the production of waste products that may then be used by other consumers with different nutrient balance requirements.

The nutrient ratios required by different species vary depending on aspects of their biology. For example, diatoms have high requirements for silicon because they produce glass shells for protection (see Figure 2.6). Vertebrates require large amounts of calcium and phosphorus for bone growth. Birds and mammals that are fruit eaters often have to supplement their diets with snail shells or bits of limestone to achieve the right nutrient balance. Growth rates and other life history traits also can influence the nutrient balance of organisms. Slowly growing marine calanoid copepods (Figure 22.14a) have N:P ratios as high as 50:1. The more rapidly growing freshwater copepods of the genus *Daphnia* (Figure 22.14b) have N:P ratios below 15:1 because of the large amounts of phosphorus needed to synthesize the nucleic acids necessary for their rapid growth.

Each organism grows, and each produces offspring. The energy channeled into growth and reproduction as a percentage of the total assimilated energy is the **net production efficiency.** Active homeothermic animals exhibit low net production efficiencies: those of birds are less than 1% and those of small mammals with high reproductive rates range up to 6%. These organisms use most of their assimilated energy to maintain salt balance, circulate blood, produce heat for thermoregulation, and move. In contrast, sedentary poikilothermic animals, particularly aquatic species, channel as much as 75% of their assimilated energy into growth and reproduction.

Detritus food chains

Terrestrial plants, especially woody species, allocate much of their production to structures that are difficult to ingest, let alone digest. As a result, even though herbivores have specialized adaptations for extracting energy from plants, they still tend to have low assimilation efficiencies. Most of the production of terrestrial plants is consumed as **detritus**—dead remains of plants and indigestible excreta of herbivores—by organisms specialized to attack wood, leaf litter, and fibrous plant egesta. This partitioning between herbivory and detritus feeding establishes two parallel food chains in terrestrial communities. The first originates when relatively large animals feed on leafy vegetation, fruits, and seeds; the second originates when relatively small animals and microorganisms consume detritus in the litter and soil. These separate food chains sometimes mingle considerably at higher trophic levels, but the energy of detritus tends to move into the food chain more slowly than the energy assimilated by herbivores.

The relative importance of herbivore-based and detritivore-based food chains varies greatly among ecosystems. Herbivores predominate in the plankton communities of oceans and lakes, detritivores in terrestrial ecosystems. The proportion of net primary production that enters each of these food chains depends on the relative allocation of plant tissues between structural and supportive functions on one hand and growth and photosynthetic functions on the other. A variety of studies have shown that herbivores consume 1.5%–2.5% of the net primary production in temperate deciduous forests, 12% of that in old-field habitats, and 60%–99% of that in plankton communities.

Energy moves through ecosystems at different rates

Ecological efficiency determines the proportion of the energy assimilated by plants that eventually reaches each higher trophic level of an ecosystem. The *rate* of energy transfer between trophic levels or, inversely, the **residence time** of energy at each trophic level, provides a second index to the energy dynamics of an ecosystem. At a given rate of production, the residence time of energy

and the amount of energy stored in living biomass and detritus are directly related: the longer the residence time, the greater the accumulation of energy.

The average residence time of energy at a particular trophic level equals the energy stored in the tissues of organisms divided by the rate at which energy is converted into biomass, or net productivity:

$$\text{residence time (years)} = \frac{\text{energy stored in biomass (kJ per m}^2\text{)}}{\text{net productivity (kJ per m}^2 \text{ per year)}}$$

When biomass is substituted for energy, this equation expresses the **biomass accumulation ratio.** For example, plants in humid tropical forests produce dry matter at an average rate of 1.8 kg per m^2 per year and have an average living biomass of 42 kg per m^2. Inserting these values into the equation above gives a biomass accumulation ratio of 23 years (42/1.8), the average residence time of biomass in tropical forest plants. Average residence times for primary producers range from more than 20 years in forest ecosystems to less than 20 days in aquatic phytoplankton-based ecosystems (Figure 22.15). In all ecosystems, however, some energy persists, while some disappears quickly. For example, leaf eaters and root feeders consume much of the biomass accumulated by forest trees during the year of its production—some of it within days. In contrast, biomass accumulated in the cellulose and lignin in the trunks of trees may not be recycled for centuries.

Figure 22.15 underestimates the average residence time of energy in organic matter because it does not include the accumulation of dead organic matter in leaf litter. The residence time of energy in accumulated litter can be determined by an equation analogous to the one above:

$$\text{residence time (years)} = \frac{\text{litter accumulation (g per m}^2\text{)}}{\text{rate of litter fall (g per m}^2 \text{ per year)}}$$

In forested ecosystems, this value varies from 3 months in the humid tropics to 1–2 years in dry and montane tropical habitats, 4–16 years in the southeastern United States, and more than 100 years in temperate mountains and boreal regions. Warm temperatures and abundant moisture in lowland tropical regions create optimal conditions for rapid decomposition of litter. The accumulation of litter, humus, and other organic materials in soils at high latitudes, particularly in the extensive boreal forest and tundra ecosystems of the Northern Hemisphere, represents an important store of organic carbon and a sink for the atmospheric carbon produced by the combustion

FIGURE 22.15 Biomass accumulation ratios for primary producers vary among ecosystems. Data from R. H. Whittaker and G. E. Likens, *Human Ecol.* 1:357–369 (1973).

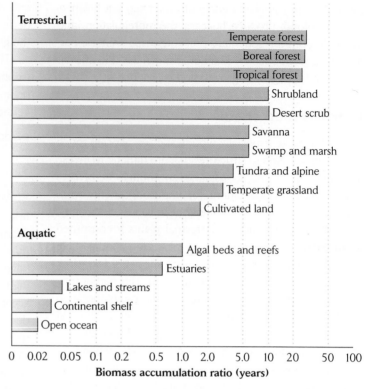

Ecosystem energetics summarizes the movement of energy

The flux of energy through an ecosystem and the efficiency of its transfer describe certain aspects of the structure of that ecosystem: number of trophic levels, relative importance of detritus feeding and herbivory, residence times for biomass and accumulated detritus, and turnover rates of organic matter. The importance of these measures to understanding ecosystem function was argued by Lindeman, who constructed the first energy budget for an entire biological community: that of Cedar Bog Lake in Minnesota (Figure 22.16). Subsequent energy flow studies have affirmed energy's usefulness as a universal currency in ecology, a common denominator to which all populations and their acts of consumption can be reduced.

The overall energy budget of an ecosystem reflects a balance between income and expenditure, just like a bank account. The ecosystem gains energy through the photosynthetic assimilation of energy from sunlight by autotrophs and through the transport of organic matter

of fossil fuels. Warming climates at high northern latitudes are beginning to accelerate the decomposition of these organic carbon sinks, releasing carbon dioxide into the atmosphere and potentially speeding climatic warming.

into the system from external sources. Organic materials produced outside the system are referred to as *allochthonous* inputs (from the Greek *chthonos*, "of the earth," and *allos*, "other"). Energy derived from photosynthesis within the system is referred to as *autochthonous* production. In a study at Root Spring, near Concord, Massachusetts, herbivores assimilated energy at a rate of 0.31 W per m^2, but the net productivity of aquatic plants and algae was only 0.09 W per m^2; the balance was derived from outside the system in the form of leaves dropped into the spring from nearby vegetation. In general, autochthonous production predominates in large rivers, lakes, and most marine ecosystems as well as in terrestrial ecosystems; allochthonous inputs make up the largest part of the energy budget in small streams and springs under the closed canopies of forests. Life in caves and in the abyssal depths of the oceans, where no light penetrates, subsists on energy transported in from outside.

Many ecosystems, including Root Spring, accumulate energy in the form of organic carbon in sediments or biomass. The balance of carbon gain and carbon loss in an ecosystem is referred to as **net ecosystem production,** a measure of net carbon accumulation. One suspects that the carbon budgets of ecosystems are either well balanced, in which case the ecosystem production is zero, or somewhat positive. A system cannot persist long with negative net ecosystem production. Systems with positive

FIGURE 22.16 Cedar Bog Lake in Minnesota has been the site of important studies on energy flow in aquatic ecosystems. Courtesy of G. David Tilman, University of Minnesota.

net ecosystem production serve as important carbon sinks. Carbon sequestered from the atmosphere in the form of organic compounds in living biomass and soil has no "greenhouse" warming effect on the atmosphere. Aquatic, especially marine, ecosystems are important accumulators of carbon in deep-water sediments, both in the form of organic carbon and as calcium carbonate.

Net ecosystem production is difficult to estimate on the global scale at which the numbers really matter. One recent set of calculations for terrestrial ecosystems suggests that about 1%–2% of the total gross primary production of the earth's land surface is retained as stored carbon. This amounts to about 2 billion tons of carbon per year—not nearly enough to offset the nearly 8 billion tons of carbon produced each year by human burning of fossil fuels, plus the additional billions of tons released by forest fires started by human activities. We shall take a closer look at the cycling of carbon in the next chapter.

SUMMARY

1. Charles Elton described biological communities as systems linked together by feeding relationships. A. G. Tansley took this idea a step further by describing organisms and their physical surroundings as a functional unit, which he called an ecosystem.

2. Alfred J. Lotka conceived the idea of ecosystems as energy-transforming systems that conform to thermodynamic principles. Raymond Lindeman, in 1942, brought together and popularized Lotka's, Tansley's and Elton's ideas.

3. Eugene P. Odum championed energy as a common currency for describing ecosystem structure and function.

4. Primary production is the process by which plants, algae, and some bacteria capture light energy and transform it into the energy of chemical bonds in carbohydrates by photosynthesis. Gross primary production is the total energy assimilated by photosynthesis. Net primary production is the energy accumulated in photosynthe-

sizer biomass; hence it is gross primary production minus respiration.

5. The unit of primary production is energy per unit of area per unit of time, but a number of other measures can be used as equivalents, including harvested biomass, carbon dioxide or oxygen flux, or assimilation of radioactive carbon (^{14}C).

6. The level of light intensity at which photosynthesis balances respiration is called the compensation point. The rate of photosynthesis increases with increasing light intensity up to a saturation point, above which it levels off.

7. Photosynthetic efficiency (the percentage of the energy in sunlight that is converted to primary production during the growing season) is 1%–2% in most habitats.

8. Plant production in many terrestrial environments is limited by, and varies with, the availability of water. Water use efficiency is the ratio of production (in grams of dry mass) to water transpired (in kilograms). Water use efficiency typically ranges between 1 and 2 g per kg; it occasionally reaches 4–6 g per kg in drought-adapted species.

9. Nutrient availability also limits production. Thus, production can be enhanced by the addition of various nutrients, especially nitrogen and phosphorus. The addition of nutrients stimulates production most greatly in systems where nutrient inputs are lowest.

10. Nutrient use efficiency (NUE) is the ratio of dry matter production to the assimilation of a particular nutrient element, such as nitrogen or phosphorus.

11. Primary production is greatest in the humid tropics. Net primary production generally increases with temperature and precipitation, but declines at the highest precipitation levels. Among aquatic ecosystems, estuaries, coral reefs, and coastal algal beds are the most productive.

12. The movement of energy through a food chain can be characterized by the efficiency of energy transfer from one trophic level to the next, known as ecological efficiency. Assimilation efficiency is the ratio of assimilation to ingestion, and net production efficiency is the ratio of production to assimilation.

13. Assimilation efficiency in consumers depends on the quality of the diet, particularly the amount of digestion-resistant structural material it contains. Assimilation efficiency varies from about 15% to 90%.

14. Assimilation efficiency also depends on the balance between the proportions of nutrients in the diet and the proportions of those nutrients required by the consumer.

15. Biomass that is not assimilated because it is not consumed or digested becomes part of the detritus food chain.

16. The average residence time of energy or biomass at a trophic level is the ratio of energy or biomass to net productivity. Average residence times for primary producers vary from 20 years in some forests to 20 days or less in aquatic plankton-based ecosystems.

17. Net ecosystem production is the balance of carbon gain and carbon loss in an ecosystem. Positive net ecosystem production represents a carbon sink that removes carbon dioxide from the atmosphere.

REVIEW QUESTIONS

1. Why is the efficiency of energy transfer between two trophic levels generally quite low?

2. Compare and contrast the movement of energy and of nutrients in ecosystems.

3. How do ecologists distinguish between gross primary production and net primary production?

4. Explain how light can limit growth at a plant's compensation point, but not above a plant's saturation point.

5. Why might plant growth be minimal when either nitrogen or phosphorus is applied alone, but much higher when the two nutrients are applied in combination?

6. At a particular latitude, what environmental factors determine the productivity of an ecosystem?

7. Why do different plant diets result in very different assimilation efficiencies for herbivores?

8. Why are residence times much longer in forest ecosystems than in aquatic phytoplankton-based ecosystems?

9. Why might the earth sustain a larger human population if humans ate plant products rather than animal products?

SUGGESTED READINGS

Cook, R. E. 1977. Raymond Lindeman and the trophic–dynamic concept in ecology. *Science* 198:22–26.

Dutta, K., et al. 2006. Potential carbon release from permafrost soils of Northeastern Siberia. *Global Change Biology* 12:2336–2351.

Elser, J. J., et al. 1996. Organism size, life history, and N-P stoichiometry. *BioScience* 46:674–684.

Elser, J. J., et al. 2007. Global analysis of nitrogen and phosphorus limitation of primary producers in freshwater, marine and terrestrial ecosystems. *Ecology Letters* 10:1135–1142.

Fenchel, T. 1988. Marine plankton food chains. *Annual Review of Ecology and Systematics* 19:19–38.

Field, C. B., et al. 1998. Primary production of the biosphere: Integrating terrestrial and oceanic components. *Science* 281:237–240.

Golley, F. B. 1994. *A History of the Ecosystem Concept in Ecology.* Yale University Press, New Haven, Conn.

Hessen, D. O., et al. 2004. Carbon, sequestration in ecosystems: The role of stoichiometry. *Ecology* 85:1179–1192.

Howarth, R. W. 1988. Nutrient limitation of net primary production in marine ecosystems. *Annual Review of Ecology and Systematics* 19:89–110.

Kay, A., et al. 2004. Stoichiometric relations in an ant–treehopper mutualism. *Ecology Letters* 7:1024–1028.

Laws, R. M. 1985. The ecology of the Southern Ocean. *American Scientist* 73:26–40.

Lawton, J. H. 1994. What do species do in ecosystems? *Oikos* 71:367–374.

Lindeman, R. 1942. The trophic–dynamic aspect of ecology. *Ecology* 23:399–418.

Lovett, G., J. Cole, and M. Pace. 2006. Is net ecosystem production equal to ecosystem carbon accumulation? *Ecosystems* 9:152–155.

Melillo, J. M., et al. 2002. Soil warming and carbon-cycle feedbacks to the climate system. *Science* 298:2173–2176.

Odum, E. P. 1968. Energy flow in ecosystems: A historical review. *American Zoologist* 8:11–18.

Pauly, D., and V. Christensen. 1995. Primary production required to sustain global fisheries. *Nature* 374:255–257.

Randerson, J. T., et al. 2002. Net ecosystem production: A comprehensive measure of net carbon accumulation by ecosystems. *Ecological Applications* 12:937–947.

Schade, J. D., et al. 2005. A conceptual framework for ecosystem stoichiometry: Balancing resource supply and demand. *Oikos* 109:40–51.

Schuur, E. A. G. 2003. Productivity and global climate revisited: The sensitivity of tropical forest growth to precipitation. *Ecology* 84:1165–1170.

Schuur, E. A. G., O. A. Chadwick, and P. A. Matson. 2001. Carbon cycling and soil carbon storage in mesic to wet Hawaiian montane forests. *Ecology* 82:3182–3196.

Stanhill, G. 1986. Water use efficiency. *Advances in Agronomy* 39:53–85.

Webb, W., et al. 1978. Primary productivity and water use in native forest, grassland, and desert ecosystems. *Ecology* 59:1239–1247.

Whittaker, R. H., and G. E. Likens. 1973. Primary production: The biosphere and man. *Human Ecology* 1:357–369.

Wiegert, R. G. 1988. The past, present, and future of ecological energetics. In L. R. Pomeroy and J. J. Albert (eds.), *Concepts of Ecosystem Ecology: A Comparative View*, pp. 29–55. Springer-Verlag, New York.

Zimov, S. A., E. A. G. Schuur, and F. S. Chapin. 2006. Permafrost and the global carbon budget. *Science* 312:1612–1613.

CHAPTER 23

Pathways of Elements in Ecosystems

Should you be worried about the change in carbon dioxide concentrations in the earth's atmosphere? Combustion of fossil fuels and the burning of forests has increased the atmospheric concentration of CO_2 from 280 to 385 parts per million since the beginning of the Industrial Revolution. Most of the change has been produced in recent decades, and projections show this trend increasing. Increases in CO_2 and other "greenhouse" gases in the atmosphere could bring about dramatic changes in climate through global warming, perhaps on the order of what we experience during extreme El Niño events. Such a scenario could decrease agricultural production and dislocate some of the human population. But major worries also include a rise in sea level due to melting ice caps and expansion of the surface waters of the oceans as they warm. These changes could flood coastal areas, causing economic hardship and shifts in the distributions of human populations.

Yet the earth has witnessed far greater changes in atmospheric carbon dioxide concentrations in the past. Before the Industrial Revolution, CO_2 concentrations probably were as low as they have ever been in geologic history. There is an important difference, however, between the present and the past: CO_2 levels are changing more rapidly than they ever have before.

Understanding why CO_2 levels are rising, and how ecosystem processes might counter this trend, depends on a basic understanding of the sources and sinks of carbon and other elements in the biosphere. These elements continually cycle through ecosystems. The routes they take are determined by the particular chemical transformations in which each element participates. Organisms—including humans—move elements through their cycles within ecosystems whenever they carry out biochemical energy transformations. This chapter shows how physical, chemical, and biological processes result in the cycling of elements within ecosystems. We shall see that many aspects of element cycling make sense only when one understands that chemical transformations and energy transformations go hand in hand.

Unlike energy, which is lost as heat, chemical elements remain within the biosphere, where they cycle continually between organisms and the physical environment. Organisms use inorganic compounds in the earth's crust or atmosphere to synthesize organic compounds, but once they are assimilated in biological forms, they are recycled over and over by organisms before being lost in sediments, streams, and groundwater or escaping to the atmosphere as gases. Although the solar energy assimilated by autotrophs is "new" energy received from outside the biosphere, most nutritive materials taken up by autotrophs have been used before. Plant roots may absorb ammonium from the soil that had leached out of decaying leaves on the forest floor that same day. Leaves may assimilate carbon dioxide produced recently by animal, plant, or microbial respiration. Regardless of the source, the individual atoms have all been around before.

Energy transformations and element cycling are intimately linked

Organisms help to move chemical elements through their cycles within ecosystems whenever they carry out the biochemical transformations needed to sustain their life processes. Transformations that incorporate inorganic forms of elements into the molecules of organisms are **assimilatory** processes. One example of an assimilatory transformation of an element is photosynthesis, in which plants use energy from the sun to change an inorganic form of carbon (carbon dioxide) into the organic form of carbon found in carbohydrates. In the overall cycling of carbon, photosynthesis is balanced by respiration, a complementary **dissimilatory** process that involves the transformation of organic carbon back to an inorganic form, accompanied by the release of energy.

Not all chemical transformations of elements in ecosystems take place within living organisms, nor do all involve the net assimilation or release of useful quantities of energy. Many chemical reactions occur in the air, soil, and water. Some of these reactions convert elements into forms that organisms can assimilate. The weathering of bedrock, for example, releases certain elements (potassium, phosphorus, and silicon, for example) from compounds in the rock and makes them available to organisms. Lightning storms produce small amounts of reduced nitrogen (ammonia, NH_3) from molecular nitrogen (N_2) and water vapor (H_2O) in the atmosphere, which plants and microbes can assimilate. Such reactions might have been involved in the origin of life itself. Other physical and chemical processes, such as precipitation of calcium carbonate in the oceans, remove elements from circulation and incorporate them into rocks in the earth's crust, where they may remain untouched for eons.

Most biochemical energy transformations are associated with the oxidation and reduction of carbon, oxygen, nitrogen, and sulfur. An atom is *oxidized* when it gives up electrons, and it is *reduced* when it accepts electrons. In a sense, the electrons carry with them a portion of the energy content of an atom. Upon being oxidized, an atom releases energy along with the electrons it gives up; upon being reduced, an atom gains energy along with the electrons it accepts.

In biochemical transformations, an energy-releasing oxidation is paired with an energy-requiring reduction, and energy shifts from the reactants in one reaction to the products in the other (Figure 23.1). Such coupled transformations are possible only when the oxidation side releases at least as much energy as the reduction side requires. The energy changes associated with various chemical transformations vary widely, however, depending on the compounds involved and the numbers of electrons exchanged. The nature of the physical world is such that the energies of two transformations rarely match. Energy supplied by an oxidation in excess of that required by a coupled reduction cannot be used and is lost in the form

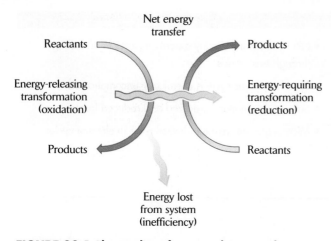

FIGURE 23.1 The coupling of energy-releasing and energy-requiring chemical transformations is the basis of energy flow in ecosystems. Energy is transferred from the reactants in an energy-releasing oxidation to the products of an energy-requiring reduction.

of heat. These imbalances account for the thermodynamic inefficiency of life processes and of the overall functioning of ecosystems.

A typical coupling of transformations is that of the oxidation of the carbon in a carbohydrate molecule (glucose, for example), which releases energy, with the reduction of the nitrogen in nitrate to the nitrogen in the amino acids of proteins, which requires energy. Like many biochemical transformations, this coupling links an energy-releasing reaction to the assimilation of an element—nitrogen, in this case—required for growth and reproduction. In animals, biochemical transformations provide the energy sources for maintaining the cellular environment and effecting movement. Some of these transformations involve many steps of the type shown in Figure 23.1, linked together into a biochemical pathway (Figure 23.2).

Photosynthesis accomplishes the initial input of energy into the ecosystem by an assimilatory reduction of carbon in which light, rather than a coupled dissimilatory process, serves as the source of energy. A portion of that energy is lost from the ecosystem with each subsequent transformation, creating a diminishing pyramid of energy with each link in the food chain (see Figure 22.1). The cycling of elements between the living and nonliving parts of the ecosystem is connected to energy flow by the coupling of the dissimilatory part of one set of coupled transformations to the assimilatory part of another.

Ecosystems can be modeled as a series of linked compartments

With each biochemical transformation, one or more chemical elements are changed from one form to another. Each form of an element within an ecosystem may be thought of as residing in a separate compartment, like a room of a house, into and out of which atoms move as physical and biological processes transform them. The entire ecosystem may be thought of as a set of compartments among which elements are cycled (Figure 23.3). For example, photosynthesis moves carbon from the inorganic compartment

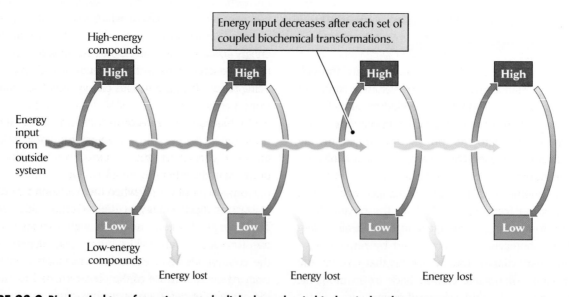

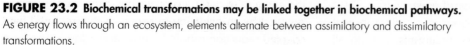

FIGURE 23.2 Biochemical transformations may be linked together in biochemical pathways. As energy flows through an ecosystem, elements alternate between assimilatory and dissimilatory transformations.

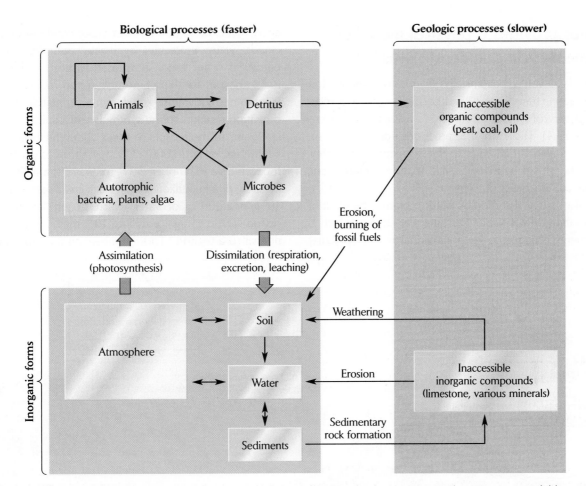

FIGURE 23.3 The cycling of elements through ecosystems may be modeled as a set of compartments. Within each compartment in this model, we can recognize subcompartments; for example, the compartment that represents available organic forms of elements is further subdivided into compartments represented by autotrophs, animals, detritus, and microbes.

to the compartment containing organic forms; respiration returns it to the inorganic compartment.

Such **compartment models** of ecosystems can be organized hierarchically, with subcompartments within compartments. The carbon in the inorganic compartment, for example, includes carbon dioxide in the atmosphere and dissolved in water, carbonate and bicarbonate ions dissolved in water, and calcium carbonate, mostly as a precipitate in the water column and in sediments. The organic compartment also has many subcompartments: autotrophs, animals, microorganisms, and detritus. As organisms feed on other organisms, they move carbon among these subcompartments.

The movement of elements within and between compartments may either require or release energy. Photosynthesis (an assimilatory process) adds energy to carbon, which we may think of as lifting the element to the second floor of a house. In descending the respiration "staircase," carbon releases this stored chemical energy, which an organism can then use for other purposes.

Elements cycle rapidly among some compartments of ecosystems and much more slowly among others. Movements of elements between living organisms and inorganic forms occur over periods ranging from a few minutes to the life spans of organisms or their subsequent existence as organic detritus. We saw in Chapter 22 that some of the organic matter in terrestrial environments has an average residence time on the order of centuries. Both organic and inorganic forms of elements may leave this rapid circulation for geologic compartments that are not readily accessible to transforming agents. For example, coal, oil, and peat contain vast quantities of organic carbon that has been removed from circulation, often for many millions of years. Inorganic carbon is removed from circulation in marine environments by the transformation of thick layers of calcium carbonate sediments into limestone. These forms of carbon are returned to the rapidly cycling biological compartments of ecosystems only by the slow geologic processes of volcanism, uplift, and erosion.

Water provides a physical model of element cycling in ecosystems

Water is involved chemically in photosynthesis, but evaporation, transpiration, and precipitation drive most movement of water through terrestrial ecosystems (Figure 23.4). These physical processes nonetheless couple the movement of water to transformations of energy. Thus, the global hydrologic cycle illustrates many basic features of the cycles of elements.

Solar energy absorbed by water performs the work of evaporation. Water vapor has potential energy, which is the energy required to keep individual water molecules apart. When atmospheric water vapor condenses to form clouds, water molecules aggregate, and the potential energy in water vapor is released as heat, which eventually escapes the biosphere as infrared radiation. Thus, from a thermodynamic standpoint, evaporation and condensation resemble photosynthesis and respiration.

Water in the biosphere totals about 1.4 billion cubic kilometers, or $1,400,000 \times 10^{18}$ g. It's hard to get a feeling for such a large number: 10^{18} g of water is a billion times a billion, or a quintillion, grams. Each cubic meter contains 10^6 g, or 1,000 kg, of water, which is equivalent to a metric ton (T). So 10^{18} g is a trillion (10^{12}) metric tons, a quantity called a teraton (TT). Numbers on the order of 10^{18}

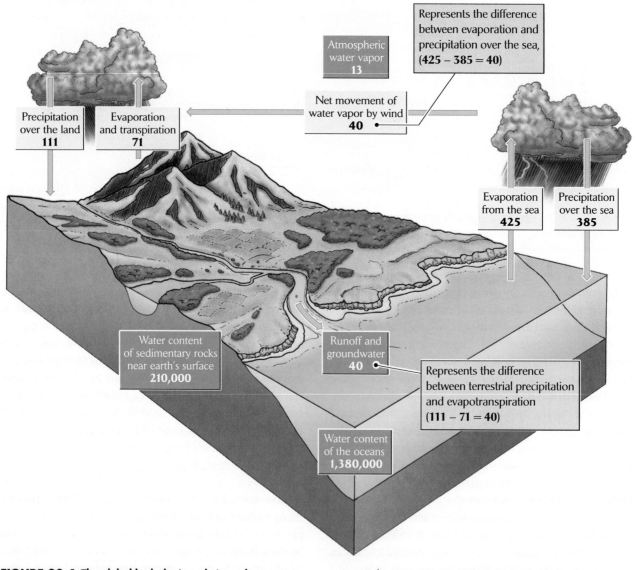

FIGURE 23.4 The global hydrologic cycle is analogous to the cycles of chemical elements. The estimated sizes of compartments (dark boxes) and transfers between compartments (light boxes) are expressed in teratons (TT) and TT per year, respectively. Compartment sizes from R. G. Barry and R. J. Chorley, *Atmosphere, Weather, and Climate*, Holt, New York (1970); fluxes from T. E. Graedel and P. J. Crutzen, *Atmosphere, Climate, and Change*, Scientific American Library, New York (1995).

are generally reserved for astronomy and the federal budget, but we'll use teratons here as a unit of measure to keep the number of zeros to a minimum.

More than 97% of the water in the biosphere resides in the oceans. Other reservoirs of water include ice caps and glaciers (29,000 TT), underground aquifers (8,000 TT), lakes and rivers (100 TT), soil moisture (100 TT), water vapor in the atmosphere (13 TT), and all the water in living organisms (1 TT). Each of these reservoirs may be regarded as a separate compartment in a compartment model.

Over land surfaces, precipitation (111 TT per year) exceeds evaporation and transpiration (71 TT per year). Over the oceans, evaporation exceeds precipitation by a similar amount. Much of the water that evaporates from the ocean surface is carried by winds over land, where it is captured as precipitation. This net flow of atmospheric water vapor from ocean to land (40 TT per year) is balanced by runoff from the land by way of rivers back into ocean basins.

Evaporation determines how fast water moves through the biosphere. The solar energy absorbed by liquid water to create water vapor is the energy source driving the hydrologic cycle. We can calculate the energy that drives the global hydrologic cycle by multiplying the mass of water evaporated (about 500 TT per year) by the energy required to evaporate 1 g of water (2.24 kJ). The product, just over 10^{21} kJ per year (about 32 billion megawatts), represents about one-fourth of the total solar energy striking the earth. Condensation of water vapor to form precipitation releases the same amount of energy as heat. Evaporation and precipitation are closely linked because the atmosphere has a limited capacity to hold water vapor; any increase in the evaporation of water creates an excess of water vapor in the atmosphere and causes an equal increase in precipitation. In other words, evaporation drives the movement of water through the atmospheric compartment of the hydrologic cycle. As the global climate warms, evaporation will increase, and the total precipitation over the surface of the earth will rise.

The average amount of water vapor in the atmosphere corresponds to about 2.5 cm of water spread evenly over the surface of the earth. An average of 65 cm of rain or snow falls each year, which is 26 times the average amount of water vapor in the atmosphere. Thus, the water in the atmospheric compartment replaces itself 26 times each year on average. (Conversely, water has an average residence time as vapor in the atmosphere of 1/26 of a year, or 2 weeks.) Together, soils, rivers, lakes, and oceans contain more than 100,000 times as much water as exists in the atmosphere. Fluxes through both compartments are the same, however, because evaporation balances precipitation. Thus, the average residence time of water in its liquid form at the earth's surface (about 2,800 years) is 100,000 times longer than its residence time in the atmosphere.

The carbon cycle is closely tied to the flux of energy through the biosphere

The carbon cycle resembles the hydrologic cycle in that energy from the sun is its driving force. The carbon cycle is much more complex, however, owing to the various chemical reactions that carbon undergoes. Three classes of processes cause carbon to cycle through ecosystems (Figure 23.5): (1) assimilatory and dissimilatory reactions, primarily in photosynthesis and respiration, (2) exchanges of carbon dioxide between the atmosphere and the oceans, and (3) precipitation of carbonate sediments in the oceans.

Photosynthesis and respiration

Photosynthesis and respiration are the major energy-transforming reactions of life. Approximately 85 billion metric tons (85×10^{15} g) of carbon enter into these reactions worldwide each year. (We will refer to a billion metric tons as a gigaton, using the abbreviation GT.) During photosynthesis, carbon gains electrons and is reduced (Figure 23.6). This gain of electrons is accompanied by a gain in chemical energy. An equivalent amount of energy is released by respiration, which results in a loss of electrons and a loss of chemical energy.

Although ecologists cannot estimate the total carbon in organic matter within the biosphere precisely, it probably adds up to something like 2,650 GT, including both that in living organisms and that in organic detritus and sediments. Considering that 85 GT of carbon is assimilated by photosynthesis each year, the average residence time of carbon in organic matter is approximately 2,650 GT divided by 85 GT per year, which equals 31 years.

Ocean–atmosphere exchange

The second class of carbon cycling processes involves physical exchanges of carbon dioxide between the atmosphere and oceans, lakes, and streams. Carbon dioxide dissolves readily in water; indeed, the oceans contain about 50 times as much CO_2 as the atmosphere does. Carbon dioxide is continuously exchanged across the boundary between the oceans and the atmosphere; as some molecules are dissolving in the oceans, others are

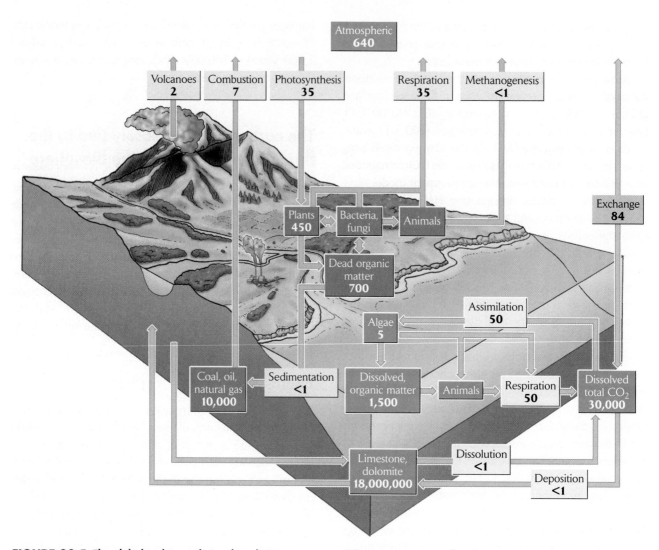

FIGURE 23.5 The global carbon cycle involves diverse biochemical and chemical transformations. The estimated sizes of compartments (dark boxes) and transfers between compartments (light boxes) are expressed in gigatons (GT) and GT per year, respectively. After T. Fenchel and T. H. Blackburn, *Bacteria and Mineral Cycling,* Academic Press, New York (1979); W. D. Grant and P. E. Long, *Environmental Microbiology,* Wiley, New York (1981).

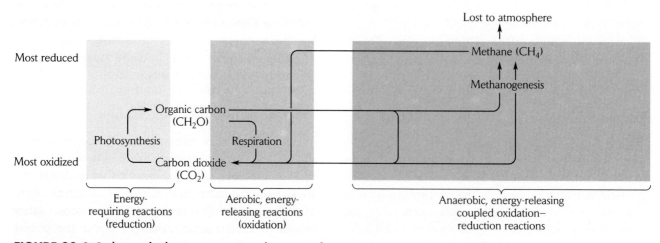

FIGURE 23.6 Carbon cycles between organic and inorganic forms. Methane is produced only by certain archaebacteria under anaerobic conditions.

escaping solution to enter the atmosphere. The total amount of carbon dioxide in the oceans remains constant, however, until new carbon dioxide enters from a source outside the atmosphere–ocean system——from the burning of fossil fuels, for example.

Exchange across the air–water boundary links the carbon cycles of terrestrial and aquatic ecosystems. In fact, the ocean is an important sink for the carbon dioxide produced by the burning of fossil fuels. As CO_2 concentrations in the atmosphere increase, the rate of dissolution of CO_2 in the oceans also increases, thereby reducing the rate of increase in atmospheric CO_2 below what it would be in the absence of air–water exchange.

Of the total carbon in the atmosphere in the form of carbon dioxide (640 GT), approximately 35 GT is assimilated by land plants and 84 GT is dissolved in the oceans and other surface waters each year. Respiration and the escape of dissolved carbon dioxide from surface waters to the atmosphere replace these amounts. Overall, the average residence time of carbon in the atmosphere is about 5 years. Because of this short residence time, the amount of carbon dioxide in the atmosphere is sensitive to the rate of CO_2 production, and it has increased very nearly in parallel with the burning of fossil fuels. By 2000, combustion of fossil fuels added nearly 7 GT of carbon to the atmosphere annually, an amount equivalent to almost 1% of the total atmospheric carbon dioxide and one-fifth of the amount assimilated by land plants.

Precipitation of carbonates

The third class of carbon cycling processes occurs only in aquatic systems. It involves the dissolution of carbon in water and the reverse process, the precipitation (deposition) of carbonate sediments, particularly limestone and dolomite. On a global scale, dissolution and precipitation approximately balance each other, although certain conditions favoring precipitation have led to the deposition of extensive layers of calcium carbonate sediments in the past. Carbon dissolution and deposition in ocean waters occur about 100 times more slowly than assimilation and dissimilation of carbon in biological systems. Thus, the exchange between sediments and the water column is relatively unimportant to the short-term cycling of carbon in ecosystems. Locally and over long periods, however, it can assume much greater importance; in fact, most of the earth's carbon is locked up in sedimentary rocks (Figure 23.7).

As we saw in Chapter 3, when carbon dioxide dissolves in water, it forms carbonic acid,

$$CO_2 + H_2O \rightarrow H_2CO_3,$$

FIGURE 23.7 Most of the earth's carbon is locked up in sedimentary rocks. These sedimentary deposits of limestone in the mountains of southern Texas represent calcium carbonate precipitated out of solution in the shallow seas that once covered the area. Photo by Gerald & Buff Corsi/Visuals Unlimited.

which readily dissociates into hydrogen, bicarbonate, and carbonate ions:

$$H_2CO_3 \rightarrow H^+ + HCO_3^- \rightarrow 2H^+ + CO_3^{2-}.$$

Calcium, when present, equilibrates with the carbonate ions to form calcium carbonate:

$$Ca^{2+} + CO_3^{2-} \rightleftharpoons CaCO_3.$$

Calcium carbonate has low solubility under most conditions, so it readily precipitates out of the water column to form sediments. This process effectively removes carbon from aquatic ecosystems, but the rate of removal is less than 1% of the annual cycling of carbon, and a similar amount is added back by input from rivers, which are naturally somewhat acidic and tend to dissolve carbonate sediments.

Dissolution and deposition may be affected locally by the activities of organisms. In the marine system, under approximately neutral pH conditions, carbonate and bicarbonate are in chemical equilibrium:

$$CaCO_3 \text{ (insoluble)} + H_2O + CO_2 \rightleftharpoons Ca^{2+} + 2 HCO_3^- \text{ (soluble)}.$$

Uptake of CO_2 for photosynthesis by aquatic algae and plants shifts the equilibrium to the left, resulting in the formation and precipitation of calcium carbonate. Many algae excrete this calcium carbonate to the surrounding water, but reef-building algae and coralline algae incorporate it into their hard body structures (Figure 23.8). When

FIGURE 23.8 The "skeletons" of coralline algae are made of calcium carbonate. These algae precipitate calcium carbonate in conjunction with their uptake of dissolved carbon dioxide during photosynthesis. Photo by L. Newman & A. Flowers/ Photo Researchers.

photosynthesis exceeds respiration in the system as a whole (as it does during algal blooms), calcium carbonate tends to precipitate out of the system.

Methanogenesis

In some kinds of habitats, such as waterlogged sediments in swamps or marshes, oxygen is not available to serve as a terminal electron acceptor for respiration. Certain archaebacteria that live in such anaerobic sediments have evolved the ability to use organic carbon itself to oxidize organic carbon when oxygen is not available. For example, they can use the carbon in methanol as an electron acceptor in a reaction that produces methane (CH_4) *and* as an electron donor in a reaction that produces carbon dioxide (see Figure 23.6). The overall reaction, which makes some energy available for other biochemical work, is

$$4\ CH_3OH\ (methanol) \rightarrow 3\ CH_4 + CO_2 + 2\ H_2O.$$

The resulting methane is released from the surface of the water and results in the phenomenon known as "swamp gas."

The factors that control methane production have received considerable attention recently because methane is an important greenhouse gas. In fact, one molecule of methane can absorb about 25 times as much infrared radiation as one carbon dioxide molecule. Moreover, methane production is increasing due to growing numbers of cattle and ever more land being converted to rice paddies (both the rumens of cattle and the sediments in rice paddies contain methanogenic archaebacteria).

Changes in atmospheric carbon dioxide concentrations

The physical environment of the earth has changed dramatically over time, in large part owing to the activities of organisms. Geologists can estimate the amounts of carbon removed from the atmosphere by burial of organic matter and by precipitation of carbonates in marine sediments, as well as when those sediments were formed. From this information, they can estimate ancient concentrations of carbon dioxide in the atmosphere and their changes over time (Figure 23.9).

These estimates indicate that during the early part of the Paleozoic era, roughly 550–400 million years ago (Mya), the atmosphere held 15 to 20 times more carbon dioxide than at present. This amount declined precipitately between 400 and 300 Mya to nearly its present levels. This decline was initiated by a sharp increase in the rate of weathering in terrestrial environments following the development of forest vegetation on land, and by the deposition of the vast accumulations of organic sediments that make up most of the earth's coal beds. Toward the end of the Paleozoic era, about 250 Mya, the CO_2 concentration in the atmosphere increased again to nearly 5 times its present level, remained high for approximately 100 million years through the early Mesozoic, and has been declining steadily ever since.

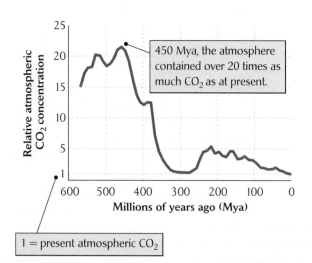

FIGURE 23.9 Concentrations of carbon dioxide in the atmosphere have decreased since the early Paleozoic era. Values are expressed in multiples of the concentration (approximately 300 parts per million) at the beginning of the Industrial Revolution. After R. A. Berner, *Science* 276:544–546 (1997).

The early Paleozoic and early Mesozoic eras were truly greenhouse times. Perhaps "hothouse" would be more apt. Average temperatures across the globe were high, and tropical life flourished even at high latitudes. The current increase in atmospheric CO_2, troubling as it is, will not return the earth to the hothouse conditions of former times, at least not any time soon.

Where did all the CO_2 in the Paleozoic atmosphere go? Most of the "geologic" carbon taken from the earth's primitive atmosphere is bound up in limestone sediments. This carbon is returned to the atmosphere very slowly as limestone is subducted below the edges of continental plates, carbonates are transformed to carbon dioxide under intense heat and pressure deep in the earth, and carbon dioxide is finally outgassed in volcanic eruptions at the relatively slow rate of 0.13–0.23 GT per year.

ECOLOGISTS IN THE FIELD **What caused the rapid decline in atmospheric carbon dioxide during the Devonian?** Why did concentrations of carbon dioxide in the atmosphere decline so rapidly over a period of 50 million years starting about 400 Mya? How can we infer events in the biosphere that occurred so long ago? Were ecological changes in ancient ecosystems involved? Geologist Gregory Retallack of the University of Oregon used several lines of evidence to provide a plausible explanation for this change. Retallack studied fossilized soils (paleosols) formed during the Devonian period (417–354 Mya) in what is now Antarctica. (At that time, the climate there was warm, and vegetation flourished.) By comparing paleosols with modern soils, it is possible to interpret many processes occurring in soils in the past and their consequences for the biosphere.

The beginning of the Devonian was marked by two striking changes (Figure 23.10). One was a change in soil chemistry, indicated by a marked increase in clay content. The other was a dramatic increase in the density and depth of plant roots. The middle of the Devonian witnessed a striking increase in the diameter of trunks, stems, roots, and rhizomes of plants, indicating the development of the first forests. These changes were followed at the end of the Devonian period and the beginning of the Carboniferous period by the appearance of thick peat deposits, which later turned into coal.

The atmospheric concentration of CO_2 itself can be estimated by the ratio of carbon-13 to carbon-12 isotopes in soil. Because the ^{13}C isotope is 8% heavier than the more common ^{12}C isotope, it enters into biochemical reactions less readily, and thus has a lower concentration in organic matter and respired CO_2 than in the atmosphere (where ^{13}C makes up about 1.1% of all carbon). Carbon diox-

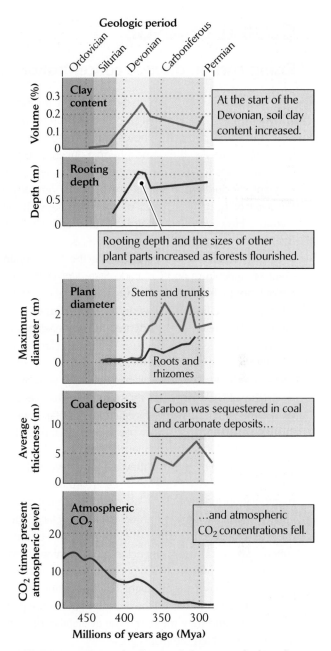

FIGURE 23.10 Fossil soils reveal changes in the biosphere. Changes in the features of paleosols during the middle of the Paleozoic era reveal increases in terrestrial vegetation and progressive weathering of soils. These changes would have led to the sequestration of atmospheric carbon dioxide in coal and carbonate sediments. After G. J. Retallack, *Science* 276:583–585 (1997).

ide in soil comes both from atmospheric sources directly and from respiration of soil organisms. The $^{13}C/^{12}C$ ratio indicates the relative amounts of atmospheric CO_2 and respired CO_2. With some assumptions about the production of CO_2 in soil, one can estimate the concentration of CO_2 in the atmosphere.

GLOBAL CHANGE

Rising carbon dioxide concentrations and the productivity of grasslands

As we have seen, carbon dioxide is an important component of the global carbon cycle: it is removed from the atmosphere by photosynthesis and returned by respiration. As the atmospheric concentration of CO_2 continues to increase over time, what will happen to plant production? If plants are currently limited by the availability of CO_2, rates of photosynthesis should increase. However, if they already have enough CO_2, then no change should occur.

Jack Morgan and his colleagues at the U.S. Department of Agriculture and Colorado State University were curious about how elevated CO_2 concentrations might affect the vegetation of the western Great Plains of Colorado. These semiarid grasslands are used as rangelands for grazing cattle (Figure 1). Because of the dry climate, the investigators suspected that the plants there were more likely to be limited by the availability of water than by the availability of CO_2. However, they also knew that elevated CO_2 concentrations would facilitate CO_2 uptake, which in turn would increase water use efficiency and lead to higher productivity at a given level of water availability.

To test their hypothesis, they assigned several 16 m^2 plots of grassland to one of three treatments: (1) no manipulation (control plots), (2) raising the atmospheric concentration of CO_2 within an enclosure placed over the plot to a level of 720 μmol of CO_2 per mol of atmospheric gas, or (3) using an enclosure containing normal (ambient) air (360 μmol CO_2 per mol) to control for any effect the enclosure itself might have on plant growth. The researchers monitored the growth of the entire plant community, and they measured the availability of water in the soil in which the plants grew.

FIGURE 1 The semiarid grasslands of Colorado. Photo by Jim Steinberg/Animals Animals—Earth Scenes.

The investigators found that the plant community accumulated 35% to 41% more biomass under elevated CO_2 concentrations than under ambient CO_2 concentrations (Figure 2). They also examined the growth of the three dominant grass species that composed 88% of the community: needle grass (*Stipa comata*), blue grama grass (*Bouteloua gracilis*), and western wheatgrass (*Pascopyrum smithii*). The increase in total community biomass was primarily due to the 84% increase in needle grass biomass (Figure 3).

Why did elevated CO_2 concentrations increase the growth of needle grass? When the researchers measured the availability of water in the soil, they noticed that the plots with elevated CO_2 contained more water. Because elevated CO_2 concentrations accelerated CO_2 assimilation, the plants closed their stomata at higher

Retallack surmised that the increase in terrestrial vegetation—particularly the penetration of soil by fine roots—would have dramatically increased the rate of weathering of soil, thus increasing its clay content. Roots and their associated microorganisms secrete organic acids to break down soil minerals into soluble forms; decomposition of organic detritus also forms organic acids. Roots hold clay particles in the soil and therefore enhance the soil's water-holding capacity, which further enhances chemical weathering. The resulting increase in weathering would have caused tremendous amounts of calcium and magnesium to be washed out of soil wherever terrestrial vegetation had developed. As dissolved calcium and magnesium ions entered the oceans, they would have formed insoluble compounds with the abundant bicarbonate ions there and precipitated out of the water as sediments. As bicarbonate was withdrawn from the oceans, it would have been replaced by carbon dioxide diffusing in from the atmosphere. Thus, as vegetation promoted weathering, new sedimentary rock was formed, partly from the

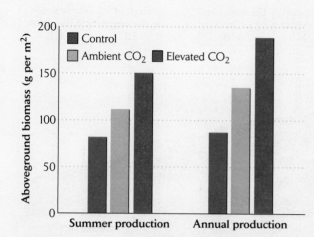

FIGURE 2 Community production in grassland plots without enclosures (controls), with enclosures containing ambient atmospheric CO_2 concentrations, and with enclosures containing elevated atmospheric CO_2 concentrations. Production was measured at the end of the summer. Production growth increased within the enclosures containing ambient CO_2, but increased even more when the enclosure contained elevated atmospheric CO_2 concentrations. After J. A. Morgan et al., *Ecol. Applic.* 14:208–219 (2004).

digestible than needle grass raised under ambient CO_2 concentrations. Thus, should the earth someday experience a doubling of ambient CO_2 concentrations, the potential increase in plant production in semiarid grasslands will not necessarily increase the forage available to cattle and other grazers. This finding illustrates the potentially wide-ranging effects of elevated CO_2 concentrations on natural ecosystems.

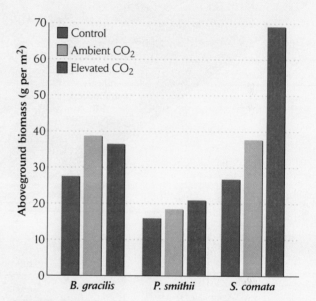

FIGURE 3 Production of the three dominant grass species in the grassland plots. These data represent summer production. Only needle grass (*S. comata*) actually responded to elevated CO_2 concentrations with an increase in production. After J. A. Morgan et al., *Ecol. Applic.* 14:208–219 (2004).

leaf water potentials, thereby reducing transpiration and withdrawal of water from the soil. Thus, plant production appeared to be limited partly by water availability, which was increased by elevated CO_2 concentrations.

Unexpectedly, elevated CO_2 concentrations also reduced the quality of needle grass as forage for cattle. By using bacterial slurries from the rumens of cattle to digest the grass, the investigators found that needle grass raised under elevated CO_2 concentrations was 16% less

mineral constituents of old continental crust and partly from the missing atmospheric carbon dioxide. ▮

Nitrogen assumes many oxidation states in its cycling through ecosystems

The ultimate source of nitrogen for life is molecular nitrogen (N_2) in the atmosphere, which constitutes the largest

pool of nitrogen on earth. This form of nitrogen dissolves to some extent in water, but is absent from rock. Lightning discharges convert some molecular nitrogen into forms that plants can assimilate, but most enters the biological pathways of the nitrogen cycle (Figure 23.11) through its assimilation by certain microorganisms in a process referred to as **nitrogen fixation.** Although nitrogen fixation and denitrification constitute only a small fraction of the earth's annual nitrogen flux, most biologically cycled

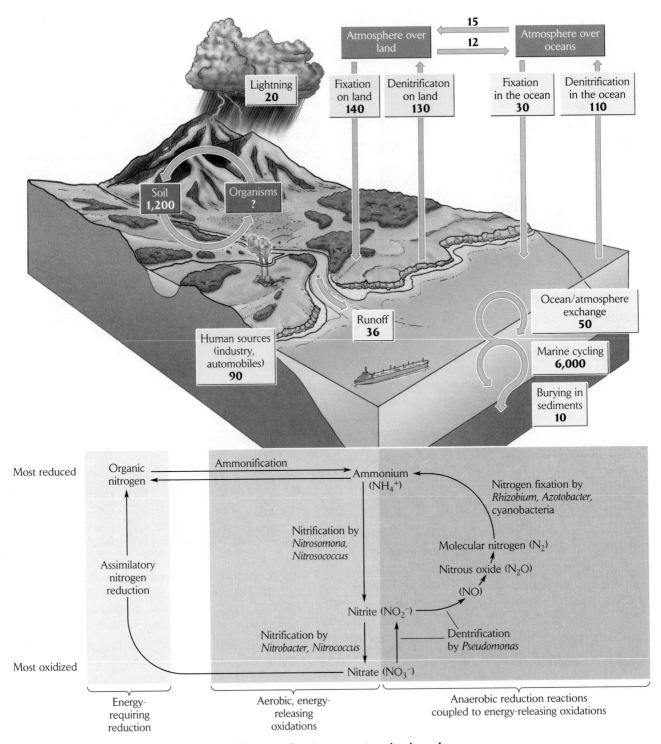

FIGURE 23.11 Nitrogen assumes several different oxidation states as it cycles through ecosystems. The estimated sizes of compartments (dark boxes) and transfers between compartments (light boxes) are expressed in gigatons (GT) and GT per year, respectively.

nitrogen can ultimately be traced back to nitrogen fixation. Once in the biological realm, nitrogen follows pathways more complicated than those of carbon because nitrogen atoms can take on a greater variety of oxidized and reduced forms.

Ammonification

Let's begin with the reduced (organic) nitrogen found in proteins. Plants obtain nitrogen from the soil, either as ammonium (NH_4^+) or as nitrate (NO_3^-), which they must then reduce to an organic form such as ammonium, with an input of energy. This organic nitrogen is the most reduced form of the nitrogen atom, with the highest potential chemical energy. It is used to construct proteins, both by the plants themselves and by consumers higher up the food chain.

Proteins are eventually metabolized, and excess nitrogen is excreted into the environment as waste. This step in the nitrogen cycle is **ammonification.**

Ammonification occurs through the breakdown of proteins into their component amino acids by hydrolysis and the oxidation of the carbon in those amino acids. This process results in the production of ammonia (NH_3), which is usually transformed to ammonium (NH_4^+). Ammonification is carried out by all organisms, including ourselves. Although carbon is oxidized and energy is released, the nitrogen atom itself is not oxidized, so its potential energy does not change during ammonification.

Nitrification and denitrification

The ammonia excreted as waste is further metabolized by microorganisms. **Nitrification** is the oxidation of nitrogen, first from ammonia to nitrite (NO_2^-), then from nitrite to nitrate (NO_3^-). During these oxidation reactions, nitrogen atoms are stripped of six, and then two more, of their electrons. These oxidation steps release much of the potential chemical energy of organic nitrogen. Each step is carried out only by specialized bacteria: $NH_3 \rightarrow NO_2^-$ by *Nitrosomonas* in the soil and by *Nitrosococcus* in marine systems; $NO_2^- \rightarrow NO_3^-$ by *Nitrobacter* in the soil and *Nitrococcus* in the oceans. The overall pathway for nitrification is thus

$$NH_3 \rightarrow NO_2^- \rightarrow NO_3^-.$$

Because both nitrification steps are oxidation reactions, they can occur only in the presence of a powerful oxidizing agent, such as molecular oxygen (O_2), that can act as an electron acceptor. However, in waterlogged, anaerobic soils and sediments and in oxygen-depleted bottom waters, nitrate and nitrite are more oxidized than the surrounding environment, and they themselves can act as electron acceptors (oxidizers). Under these conditions, reduction reactions are thermodynamically favored, and nitrogen may be reduced to nitric oxide (NO):

$$NO_3^- \rightarrow NO_2^- \rightarrow NO.$$

This reaction, called **denitrification,** is accomplished by bacteria such as *Pseudomonas denitrificans.* Denitrification is important for breaking down organic matter in oxygen-depleted soils and sediments, but it also results in the loss of nitrogen from soils as a gas. Additional chemi-

cal reactions under anaerobic, reducing conditions in soils and water can produce molecular nitrogen:

$$NO \rightarrow N_2O \rightarrow N_2,$$

which is also lost.

Denitrification may explain the low availability of nitrogen in marine systems and many freshwater habitats, including marshes and rice paddies. When organic remains of plants and animals sink to the depths of the oceans, their oxidation in deep waters and bottom sediments is often accomplished anaerobically by bacteria using nitrate as an oxidizer. Nitrate and nitrite are thereby converted to the dissolved gases NO and N_2, which cannot be assimilated by algae.

Nitrogen fixation

The loss of readily available nitrogen from ecosystems by denitrification is offset by nitrogen fixation. This reduction of nitrogen to biologically useful forms is accomplished by specialized bacteria such as *Azotobacter,* which is a free-living species; *Rhizobium,* which occurs in symbiotic association with the roots of some legumes (members of the pea family) and other plants (Figure 23.12); and some cyanobacteria. The enzyme responsible for nitrogen fixation by these microorganisms—nitrogenase—is inactivated by oxygen and works efficiently only under extremely low oxygen concentrations. This explains why *Azotobacter* bacteria, living freely in the soil, exhibit only a small fraction of the nitrogen-fixing capacity of *Rhizobium* bacteria, which live in the cores of root nodules. In these nodules, root cells infected by *Rhizobium* form membrane-bounded structures called symbiosomes, within which the bacteria live. Within a symbiosome, oxygen concentrations are kept very low so as not to interfere with the activity of nitrogenase. Although symbiosomes contain little free oxygen, they do have an abundant supply bound to a special kind of hemoglobin, called leghemoglobin, which has a high affinity for oxygen. Leghemoglobin keeps the concentration of free oxygen in the symbiosome very low while providing a continuous supply for *Rhizobium* respiration.

Nitrogen fixation requires energy, though no more than the conversion of an equivalent amount of nitrate to ammonium by plants. The reduction of one atom of molecular nitrogen to ammonium requires approximately the amount of energy released by the oxidation of an atom of organic carbon to carbon dioxide. Nitrogen-fixing bacteria obtain the energy they need to reduce N_2 to NH_4^+ by oxidizing sugars or other organic compounds. Free-living bacteria must obtain these resources by metabolizing organic detritus in the soil, sediments,

(a)

(b)

FIGURE 23.12 Some plants provide anaerobic environments for nitrogen-fixing bacteria. (a) Nodules on the roots of legumes, such as these soybeans, harbor symbiotic nitrogen-fixing *Rhizobium* bacteria. (b) The numerous brown "rods" in this scanning electron micrograph are *Rhizobium* bacteria within a symbiosome. Photo (a) courtesy of Thomas R. Sinclair; photo (b) by Simko/Visuals Unlimited.

or water column. More abundant supplies of energy are available to *Rhizobium,* whose plant partners supply them with carbohydrates.

On a global scale, nitrogen fixation approximately balances the production of N_2 by denitrification. On a local scale, nitrogen fixation can assume much greater importance, especially in nitrogen-poor habitats. When land is first exposed to colonization by plants—as, for example, are areas left bare by receding glaciers or newly formed lava flows—plant species with nitrogen-fixing symbionts dominate the colonizing vegetation.

 The fate of soil nitrate in a temperate forest. Ecologists are increasingly using rare isotopes of common chemical elements to follow the fates of those elements in natural ecosystems. Most elements have stable (nonradioactive) isotopes that investigators can add to ecosystems in excess of their natural occurrence and then measure as they move through various compartments. In addition to oxygen-18 and carbon-13, one of the most useful stable isotopes has been nitrogen-15 (^{15}N), which constitutes only 0.4% of the total nitrogen in the biosphere (most nitrogen has an atomic weight of 14).

Gregory Zogg and his colleagues conducted one such experiment to follow the fate of nitrate in a maple forest in northern Michigan. Nitrate is naturally produced in the soil by nitrification of ammonia, but it also enters the soil

in rainfall carrying dissolved products of fossil fuel combustion that have entered the atmosphere. The researchers simulated this precipitation input by adding dissolved sodium nitrate enriched with ^{15}N to natural forest soil in the amount of 29.5 mg ^{15}N per m^2, a level typical of local rainwater. They then measured ^{15}N in various components of the soil, including microbial biomass, organic matter, roots of various sizes, and inorganic soil components, at intervals over a period of 16 weeks (Figure 23.13).

A large proportion of the ^{15}N was immediately (within 2 hours) incorporated into microbial biomass, inorganic soil components, and soil organic matter. Most of the ^{15}N-labeled nitrate had disappeared within 2 weeks of the initial application. Small amounts were converted to ammonium (NH_4^+), which does not last long in soil before being taken up by bacteria or plant roots. The ^{15}N entered plant roots much more slowly than it was taken up by bacteria, not reaching its peak concentration in roots until a week after the initial application. The total ^{15}N added to the soil decreased to about one-third of the initial amount in 1 month and to one-fourth in 2 months. Some of the labeled nitrogen might have been washed deeper into the soil than the 10 cm sampling depth, or washed out in groundwater, and a small amount might have been lost by denitrification. However, the majority had, by this time, been taken up by plant roots and incorporated into stem and leaf tissue (which was not sampled in this study). The surprising result was that bacteria in the soil competed effectively with plant roots for nitrate. As a result, a large

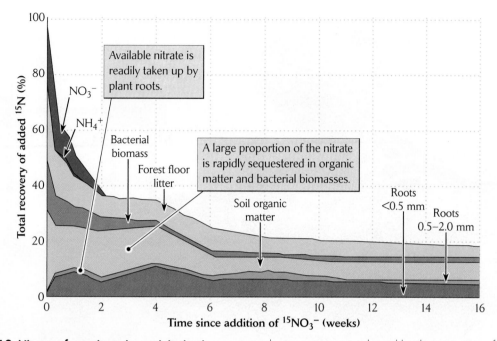

FIGURE 23.13 Nitrogen from nitrate in precipitation is quickly immobilized by bacteria and in organic matter in forest soil. The subsequent gradual disappearance of the stable isotope ^{15}N in this experiment reflects uptake from the soil by plants as nitrogen is released by decomposition of organic matter, including bacterial biomass. Most of the inorganic nitrate disappears within 2 days. After G. P. Zogg et al., *Ecology* 81:1858–1866 (2000).

part of the added nitrate was rapidly immobilized in organic compounds that were not immediately available to plants. ▌

The phosphorus cycle is chemically uncomplicated

Ecologists have studied the role of phosphorus in ecosystems intensively because organisms require this element at a relatively high level (though only about one-tenth that of nitrogen). Phosphorus is a major constituent of nucleic acids, cell membranes, energy transfer systems, bones, and teeth. It is thought to limit plant production in many aquatic habitats. Influxes of phosphorus into rivers and lakes in the form of sewage and runoff from fertilized agricultural lands can artificially stimulate production in aquatic habitats, which can upset natural ecosystem balances and alter the quality of these habitats. Pollution by phosphorus-containing detergents was a major contributor to this problem until phosphorus-free alternative products were developed.

The phosphorus cycle (Figure 23.14) has fewer steps than the nitrogen cycle because, except in a very few microbial transformations, phosphorus does not undergo oxidation–reduction reactions in its cycling through ecosystems. Plants assimilate phosphorus, in the form of phosphate ions (PO_4^{3-}), from soil or water and incorporate it directly into various organic compounds. Animals eliminate excess phosphorus in their diets by excreting phosphate ions in urine; phosphatizing bacteria also convert phosphorus in detritus into phosphate ions. Phosphorus does not enter the atmosphere in any form other than dust, so little phosphorus cycles between the atmosphere and other compartments of ecosystems.

Acidity greatly affects the availability of phosphorus to terrestrial plants. In acidic soils, phosphorus binds tightly to clay particles and forms relatively insoluble compounds with iron and aluminum. In basic soils, it forms other insoluble compounds—for example, with calcium. When both calcium and iron or aluminum are present in soils, the highest concentrations of soluble phosphate occur at a pH of between 6 and 7.

In well-oxygenated aquatic systems, phosphorus readily forms insoluble compounds with iron or calcium and precipitates out of the water column. Thus, marine and freshwater sediments act as a phosphorus sink, continually removing precipitated phosphorus from rapid circulation in ecosystems. Phosphorus compounds readily dissolve and enter the water column only in oxygen-depleted aquatic sediments and bottom waters. Under such conditions, iron tends to combine with sulfur rather than with phosphorus, forming soluble sulfides rather than insoluble phosphate compounds.

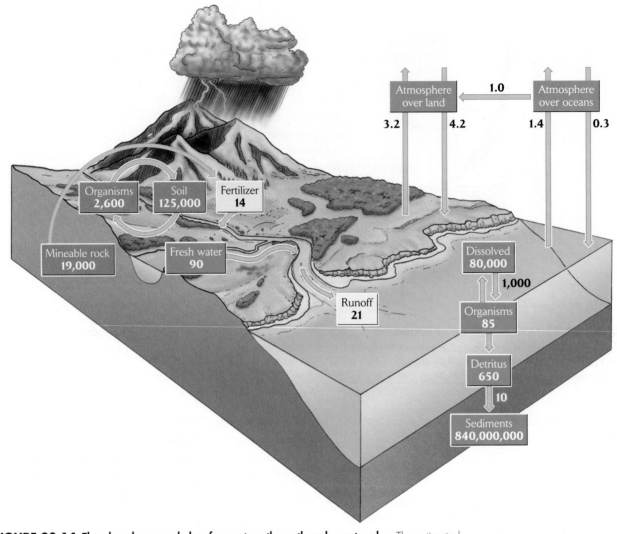

FIGURE 23.14 The phosphorus cycle has fewer steps than other element cycles. The estimated sizes of compartments (dark boxes) and transfers between compartments (light boxes) are expressed in gigatons (GT) and GT per year, respectively.

Because phosphorus tends to precipitate out of solution, productivity in the warm surface waters of temperate lakes often decreases during the summer months as phosphorus concentrations decrease. Phosphorus is replenished in these surface waters only when fall overturn occurs. Phytoplankton in the surface waters get plenty of sunlight and obtain their carbon from abundant dissolved carbon dioxide; thus, carbon and sunlight do not limit production as much as phosphorus does during the summer (Figure 23.15). Bacteria in the water column obtain their carbon and phosphorus in organic forms from detritus as dead organisms decompose in the surface layers of the lake. Phytoplankton have no other source of phosphorus, however, than that recycled through detritus, so phytoplankton and bacteria compete for this resource just as bacteria and plants compete for nitrate in soil.

After spring overturn, when phosphorus is abundant in the surface waters, the carbon:phosphorus (C:P) ratio of most organisms matches the needs of their consumers, and egested material has a similar or even lower C:P ratio than ingested food. By summer, however, phosphorus concentrations in the surface waters of lakes are so low that the food supplies consumers rely on become phosphorus-depleted. Furthermore, the potential growth rates of organisms increase under the higher summer temperatures, so their phosphorus demand also increases. Therefore, they must eat more food to obtain sufficient phosphorus for growth and reproduction. As a result, organisms extract as much phosphorus as possible from their diet and egest material that is further depleted in phosphorus. In other words, during summer, the ecological stoichiometry of the lake becomes imbalanced in that the composition of food does not match the requirements of the consumer (Figure 23.16).

(a) **Carbon**

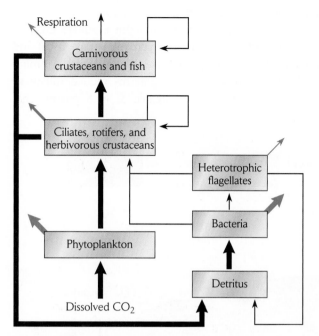

(b) **Phosphorus**

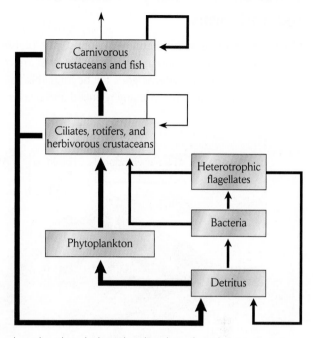

FIGURE 23.15 The cycling pathways of carbon and phosphorus differ in temperate lakes. These flow diagrams summarize element cycling in Lake Constance, Germany, during summer. (a) Phytoplankton obtain carbon in the form of dissolved carbon dioxide, which is abundant throughout the summer. (b) Their phosphorus, however, comes from organic detritus, which is severely depleted by late summer. After U. Gaedke, S. Hochstädter, and D. Straile, *Ecol. Monogr.* 72(2):251–270 (2002).

(b) **Summer**

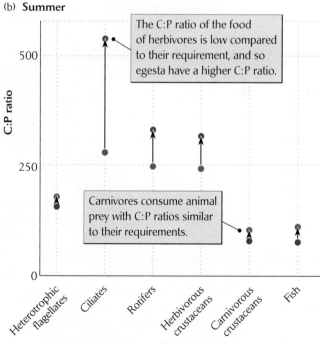

Consumers: ● Food   ● Egesta

(a) **Spring**

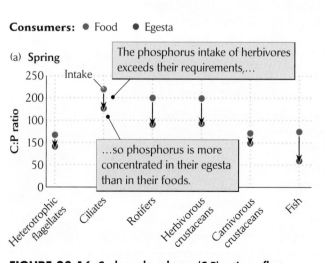

FIGURE 23.16 Carbon:phosphorus (C:P) ratios reflect the depletion of phosphorus in surface layers of temperate lakes in summer. C:P ratios in consumers, their food, and their egesta in (a) spring and (b) summer. Ciliates, rotifers, and herbaceous crustaceans feed on phytoplankton, which have lower concentrations of phosphorus in summer (C:P ratio about 250) than in spring (C:P ratio < 200). Heterotrophic flagellates, carnivorous crustaceans, and fish feed on bacteria and on other animals, which maintain fairly low and constant C:P ratios (100–125), so they experience little phosphorus limitation in summer. After U. Gaedke, S. Hochstädter, and D. Straile, *Ecol. Monogr.* 72(2):251–270 (2002).

Sulfur exists in many oxidized and reduced forms

Organisms require sulfur because it forms part of two amino acids, cysteine and methionine. But the importance of sulfur in ecosystems goes far beyond this role. Like nitrogen, sulfur exists in many reduced and oxidized forms, so it follows complex chemical pathways that affect the cycling of other elements (Figure 23.17).

The most oxidized form of sulfur is sulfate (SO_4^{2-}); the most reduced forms are hydrogen sulfide (H_2S) and organic forms of sulfur, such as those found in amino acids. To assimilate sulfur, organisms reduce sulfate to organic sulfur ($SO_4^{2-} \rightarrow$ organic S) in a process that

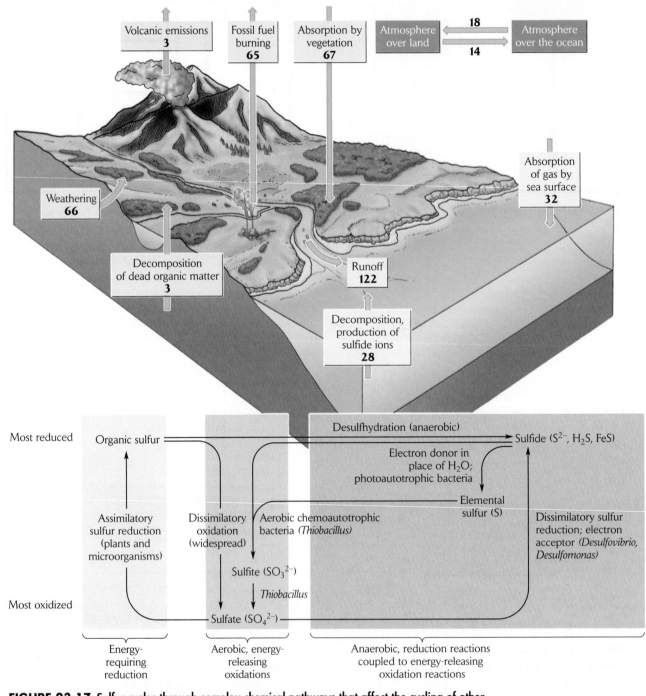

FIGURE 23.17 Sulfur cycles through complex chemical pathways that affect the cycling of other elements. The estimated sizes of compartments (dark boxes) and transfers between compartments (light boxes) are expressed in gigatons (GT) and GT per year, respectively.

FIGURE 23.18 Streams draining from the refuse of coal mines may be extremely acidic. When materials such as these mine spoils in Tioga County, Pennsylvania, are exposed to the air, the sulfides they contain are oxidized to sulfate, which then combines with water to produce sulfuric acid. Photo by Tim McCabe, courtesy of the U.S. Department of Agriculture, Soil Conservation Service.

requires energy input. In aerobic environments, reduction of sulfate to organic sulfur balances the oxidation of organic sulfur back to sulfate, which occurs either directly or with sulfite (SO_3^{2-}) as an intermediate step. This oxidation occurs when animals excrete excess dietary organic sulfur and when microorganisms decompose plant and animal detritus.

Under anaerobic conditions, as in waterlogged sediments, sulfate, like nitrate, may function as an oxidizer. In such reducing environments, the bacteria *Desulfovibrio* and *Desulfomonas* can use sulfate reduction to oxidize organic carbon. The coupling of these reactions makes some energy available to them. The reduced sulfur may then be used by photosynthetic bacteria to assimilate carbon by pathways analogous to photosynthesis in green plants. In these reactions, sulfur takes the place of the oxygen atom in water as an electron donor. As a result, elemental sulfur (S) accumulates unless the sediments are exposed to aeration or oxygenated water. Under those conditions, elemental sulfur may be further oxidized by aerobic chemoautotrophic bacteria, such as *Thiobacillus,* to sulfite and sulfate.

The fate of the reduced sulfur produced under anaerobic conditions depends on the availability of metals. Frequently, hydrogen sulfide (H_2S) forms; it escapes from shallow sediments and mucky soils as a gas having the characteristic smell of rotten eggs. Anaerobic conditions generally favor the reduction of ferric iron (Fe^{3+}) to ferrous iron (Fe^{2+}), which can combine with sulfide ions (S^{2-}) to form iron sulfide (FeS). Sulfides are commonly associated with coal and oil deposits, which lack free oxygen. When these materials are exposed to the atmosphere in mine wastes or burned for energy, the reduced sulfur oxidizes (with the help of *Thiobacillus* bacteria in mine wastes)

to sulfate. This oxidized sulfur combines with water to produce sulfuric acid (H_2SO_4), the major component of acid rain and acidic mine drainage (Figure 23.18).

Microorganisms assume diverse roles in element cycles

As you certainly have noticed, many of the transformations discussed in this chapter are accomplished mainly or entirely by bacteria. In fact, were it not for the activities of such specialized microorganisms, many element cycles would be altered drastically, and the productivity of the biosphere would be much reduced. For example, without the capacity of some microorganisms to use nitrogen, sulfur, and iron as electron acceptors, little decomposition would occur in anaerobic organic sediments, and the resulting accumulation of organic matter would reduce the amount of inorganic carbon available in ecosystems. Without nitrogen-fixing bacteria, denitrification under anaerobic conditions would slowly deplete ecosystems of available nitrogen.

To be sure, many of the chemical transformations carried out by microorganisms, such as metabolism of sugars and other organic molecules, are accomplished in similar ways by plants and animals. However, the bacteria and archaebacteria are distinguished physiologically by the ability of many species to metabolize substrates under anaerobic conditions and to use substrates other than organic carbon as energy sources.

As we have seen, every organism needs, above all, a source of carbon for building organic structures and a source of energy to fuel life processes. Organisms differ with respect to their sources of carbon. **Heterotrophs** obtain carbon in reduced (organic) form by consuming other organisms or organic detritus. All animals and fungi, and many bacteria, are heterotrophs. **Autotrophs** assimilate carbon as carbon dioxide and expend energy to reduce it to an organic form. **Photoautotrophs** use sunlight as their source of energy for this reaction (photosynthesis). All green plants and algae are photoautotrophs, as are cyanobacteria. All of these organisms use H_2O as an electron donor (reducing agent) and are aerobic. Purple and green bacteria are also autotrophs, but their light-absorbing pigments differ from those of green plants, they use H_2S or organic compounds as electron donors, and they are anaerobic.

Chemoautotrophs use CO_2 as a carbon source, but they obtain energy for its reduction by the aerobic oxidation of inorganic substrates: methane (for example, *Methanosomonas* and *Methylomonas*); hydrogen (*Hydrogenomonas* and *Micrococcus*); ammonia (the nitrifying bacteria *Nitrosomonas* and *Nitrosococcus*); nitrite (the nitrifying bacteria *Nitrobacter* and *Nitrococcus*); hydrogen sulfide, sulfur, and sulfite (*Thiobacillus*); or ferrous iron (*Ferrobacillus* and *Gallionella*). Chemoautotrophs are almost exclusively bacteria, which are apparently the only organisms that can become specialized biochemically to make efficient use of inorganic substrates in this way and dispose of the resulting waste products.

The highly productive communities of marine organisms that develop around deep-sea hydrothermal vents illustrate the special role of microorganisms in ecosystem function (Figure 23.19). Scientists from the Woods Hole Oceanographic Institution first discovered these miniature ecosystems in deep water off the Galápagos archipelago in 1977. Vent communities have since been found widely distributed in the ocean basins of the world. The most conspicuous members of these communities are giant white-shelled clams and tubeworms (pogonophorans) that grow up to 3 meters in length, but numerous crustaceans, annelids, mollusks, and fish also cluster at great densities around hydrothermal vents. The high productivity of vent communities contrasts strikingly with the desertlike appearance of the surrounding ocean floor.

How do these communities obtain energy? The vents are located well below the level of light penetration, making photosynthesis impossible. In fact, the productivity of the vent communities depends on the unique qualities of the water issuing from the vents themselves. This water is hot and loaded with hydrogen sulfide (H_2S), a reduced form of sulfur. Where vent water and seawater mix, conditions are ideal for chemoautotrophic sulfur bacteria. These bacteria use oxygen from seawater to oxidize the hydrogen sulfide in vent water ($H_2S \rightarrow SO_4^{2-}$ + energy), which

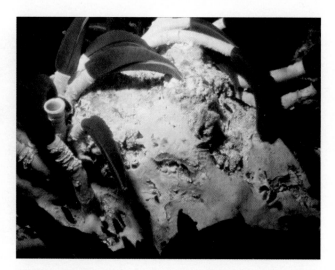

FIGURE 23.19 Chemoautotrophic sulfur bacteria form the base of the food chain in hydrothermal vent communities. Other vent organisms, such as these tubeworms (*Riftia pachyptila*) at a Pacific hydrothermal vent, rely on these bacteria to produce food. Photo by C. Van Dover, courtesy of OAR/National Undersea Research Program (NURP).

provides them with a source of energy for the assimilatory reduction of inorganic carbon and nitrogen from seawater (e.g., $NO_3^- \rightarrow NH_4^+$). All the other members of the vent community feed on these bacteria, which thus form the base of the local food chain. The pogonophorans have gone so far as to house symbiotic colonies of the bacteria within the tissues of a specialized organ, the trophosome, providing a protected place for the bacteria to live in return for a share of the carbohydrate and organic nitrogen they produce.

In this chapter we have examined the cycling of several important elements from the standpoint of their chemical and biochemical reactions. Elements are cycled through ecosystems primarily because the metabolic activities of organisms result in chemical transformations of these elements. The kinds of transformations that predominate depend on the physical and chemical conditions of the ecosystem. Each type of habitat presents a different chemical environment, particularly with respect to the presence or absence of oxygen and possible sources of energy. It stands to reason, therefore, that the patterns of element cycling should differ greatly among habitats and ecosystems. In the next chapter, we shall contrast element cycling in aquatic habitats and in terrestrial habitats by focusing on how some of the unique physical features of each of these environments affect the chemical and biochemical transformations involved in the production of organic compounds and the recycling of elements.

SUMMARY

1. Unlike energy, chemical elements are retained within the biosphere, where they are continually cycled between the physical and biological components of ecosystems.

2. The movement of energy through ecosystems parallels the paths of several elements, particularly carbon. Different biochemical transformations of carbon either require or release energy.

3. Most biochemical energy transformations are associated with coupled biochemical reactions. An energy-releasing oxidation reaction is paired with an energy-requiring reduction reaction, and energy shifts from the reactants in one reaction to the products in the other.

4. The cycling of chemical elements through ecosystems may be thought of as movement between compartments. The major compartments of ecosystems are living organisms and organic detritus, accessible inorganic forms, and inaccessible organic and inorganic forms, for the most part locked away in sediments.

5. The global hydrologic cycle provides an analogy for the cycling of elements in ecosystems. Energy is required to evaporate water because molecules of water vapor have a higher energy content than molecules of liquid water. This energy is released as heat when water vapor condenses in the atmosphere to produce precipitation. Overall, evaporation balances precipitation.

6. Three classes of processes cause carbon to cycle through ecosystems: photosynthesis and respiration, exchanges of carbon dioxide between the atmosphere and oceans, and precipitation of carbonate sediments in the oceans.

7. Carbon dioxide is continuously exchanged between the atmosphere and surface waters. Carbon dioxide dissolved in the oceans enters into a chemical equilibrium with bicarbonate and carbonate ions, which, in the presence of calcium, tend to precipitate and form sediments. Thick accumulations of these marine sediments can become limestone rock.

8. In anaerobic environments, such as waterlogged sediments in marshes and rice paddies, some archaebacteria can use organic carbon as an oxidizer. This process results in the production of methane (CH_4), a potent greenhouse gas. Archaebacteria in the rumens of cattle also produce significant quantities of methane.

9. During the early Paleozoic era, the atmosphere held 15–20 times as much carbon dioxide as at present, and global temperatures were much higher. The development of forest vegetation about 400 million years ago

led to increased weathering, dissolution of calcium and magnesium, and the formation of carbonate sediments in the oceans, greatly reducing CO_2 concentrations in the atmosphere.

10. Nitrogen has many reduced and oxidized forms and consequently follows complicated pathways through ecosystems. Most nitrogen that cycles through the biological compartments of ecosystems follows the cycle leading from nitrate through organic nitrogen (following assimilation by plants), ammonia, nitrite (following nitrification by bacteria), and then back to nitrate (following further nitrification). The last two steps are accomplished by specialized bacteria in the presence of oxygen.

11. Under anaerobic conditions in soils, sediments, and deep waters, certain bacteria can use nitrate in place of oxygen as an oxidizing agent. This process, called denitrification, transforms nitrate into nitrite and eventually into nitrous oxide and molecular nitrogen.

12. The loss of available nitrogen through denitrification is balanced by bacterial nitrogen fixation. Much nitrogen fixation in terrestrial systems is carried out by *Rhizobium* bacteria in nodules on the roots of legumes.

13. Ecologists can follow the movement of elements such as nitrogen using stable isotopes, such as nitrogen-15 (^{15}N), which serve as markers for the appearance of the element in different ecosystem compartments.

14. Plants assimilate phosphorus in the form of phosphate ions (PO_4^{3-}). The energy potential of the phosphorus atom does not change during its cycling through ecosystems, so it is not involved in energy-transforming reactions.

15. The availability of phosphorus, which often limits primary production by plants and phytoplankton, varies with the acidity and oxidation level of the soil or water. Ratios of carbon to phosphorus (C:P) are often used to determine the quality of food resources for consumers.

16. The reduction of sulfate to organic sulfur is balanced by the oxidation of organic sulfur to sulfate in aerobic environments. In anaerobic environments, it may act as an oxidizer in the form of sulfate or as a reducing agent (for photoautotrophic bacteria) in the forms of elemental sulfur and sulfide ions.

17. Many transformations of elements, particularly under anaerobic conditions, are accomplished by biochemically specialized microorganisms. These organisms therefore play important roles in the cycling of elements through ecosystems.

REVIEW QUESTIONS

1. How does energy from the sun drive the movement of water from the oceans to the continents and back to the oceans again?

2. How does the ocean ameliorate the effects of fossil fuel combustion on CO_2 concentrations in the atmosphere?

3. Why is methane gas commonly produced in swamps and rice paddies?

4. Why was a substantial drop in atmospheric CO_2 from 400 to 300 Mya associated with the development of forest vegetation on land?

5. Why is the availability of usable forms of nitrogen low at the bottom of the ocean?

6. How might nitrogen-fixing bacteria living in symbiosis with a plant affect the types of environments in which the plant could live?

7. How do low pH, high pH, and high oxygen availability in water affect the availability of phosphorus in water?

8. Compare and contrast photoautotrophs and chemo-autotrophs.

SUGGESTED READINGS

Arrigo, K. R. 2005. Marine microorganisms and global nutrient cycles. *Nature* 437:349–355.

Berner, R. A. 1997. The rise of plants and their effect on weathering and atmospheric CO_2. *Science* 276:544–546.

Berner, R. A., and A. C. Lasaga. 1989. Modeling the geochemical carbon cycle. *Scientific American* 260:74–81.

Coleman, D. C., C. P. P. Reid, and C. V. Cole. 1983. Biological strategies of nutrient cycling in soil systems. *Advances in Ecological Research* 13:1–55.

Fenchel, T., and B. J. Finlay. 1995. *Ecology and Evolution in Anoxic Worlds*. Oxford University Press, Oxford.

Gaedke, U., S. Hochstädter, and D. Straile. 2002. Interplay between energy limitation and nutritional deficiency: Empirical data and food web models. *Ecological Monographs* 72(2):251–270.

Grassle, J. F. 1985. Hydrothermal vent animals: Distribution and biology. *Science* 229:713–717.

Grünfeld, S., and H. Brix. 1999. Methanogenesis and methane emissions: Effects of water table, substrate type and presence of *Phragmites australis*. *Aquatic Botany* 64:63–75.

Hobbie, J. E., and E. A. Hobbie. 2006. [15]N in symbiotic fungi and plants estimates nitrogen and carbon flux rates in Arctic tundra. *Ecology* 87:816–822.

Howarth, R. W. 1993. Microbial processes in salt-marsh sediments. In T. E. Ford (ed.), *Aquatic Microbiology*, pp. 239–259. Blackwell Scientific Publications, Oxford.

Jannasch, H. W., and M. J. Mottl. 1985. Geomicrobiology of deep-sea hydrothermal vents. *Science* 229:717–725.

Morgan, J. A., et al. 2004. CO_2 enhances productivity, alters species composition, and reduces digestibility of shortgrass steppe vegetation. *Ecological Applications* 14:208–219.

Pearson, H. L., and P. M. Vitousek. 2002. Nitrogen and phosphorus dynamics and symbiotic nitrogen fixation across a substrate-age gradient in Hawaii. *Ecosystems* 5:587–596.

Post, W. M., et al. 1990. The global carbon cycle. *American Scientist* 78(4):310–326.

Retallack, G. J. 1997. Early forest soils and their role in Devonian global change. *Science* 276:583–585.

Schlesinger, W. H. 1991. *Biogeochemistry: An Analysis of Global Change*. Academic Press, San Diego.

Stacey, G., R. H. Burris, and H. J. Evans (eds.). 1992. *Biological Nitrogen Fixation*. Chapman & Hall, New York.

Vitousek, P. M., et al. 1997. Human alteration of the global nitrogen cycle: Causes and consequences. *Issues in Ecology* 1:1–17.

West, J. B., et al. 2006. Stable isotopes as one of nature's ecological recorders. *Trends in Ecology and Evolution* 21(7):408–414

Zogg, G. P., et al. 2000. Microbial immobilization and the retention of anthropogenic nitrate in a northern hardwood forest. *Ecology* 81:1858–1866.

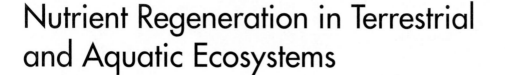

Nutrient Regeneration in Terrestrial and Aquatic Ecosystems

During the 1960s, scientists in the northeastern United States and central Europe became alarmed by the deaths of many trees in forests on poor soils. These deaths appeared to be correlated with acidified precipitation. So-called acid rain is caused by the products of fossil fuel combustion spewed into the atmosphere. Among these products are vast quantities of sulfur dioxide (SO_2) and nitrous oxide (N_2O), which are further oxidized in the atmosphere to sulfate (SO_4^{2-}) and nitrate (NO_3^-). These ions, in turn, form sulfuric acid and nitric acid when they come into contact with water droplets in the atmosphere. Consequently, the pH of rainwater can fall to as low as 4, and that of streams to almost 5.

It was clear that acid rain was harming forests. In 1970, the U.S. Congress passed the Clean Air Act to reduce emissions of SO_2 and particulate matter from factories and power plants. The effect of this legislation was striking. Sulfur oxides and, especially, particulate matter in the atmosphere began to decline almost immediately (Figure 24.1). Scientists were surprised to find, however, that the forests did not show signs of recovery.

Puzzled by this finding, ecologist Gene Likens and his colleagues began to examine records of soil and water chemistry and of tree growth that had been kept since 1963 at the Hubbard Brook Experimental Forest in New Hampshire. By examining inputs and outflows of various elements, they were able to piece together a three-part explanation for the failure of trees to respond to the change in rainwater inputs. First, sulfur emissions began to decline after the Clean Air Act, but remained relatively high, and nitrogen oxide emissions actually increased. Hence, the acidity of precipitation and streams

(a)

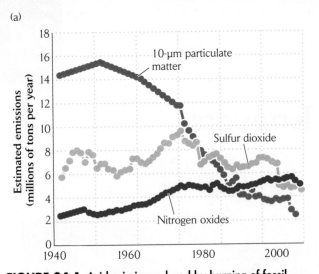

(b)

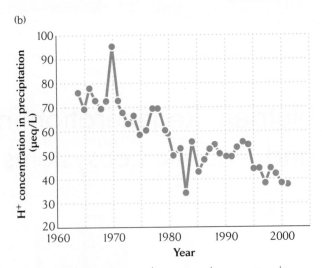

FIGURE 24.1 Acid rain is produced by burning of fossil fuels. (a) Particulate, sulfur dioxide, and nitrogen oxide emissions in the region upwind of the Hubbard Brook watershed are shown for the period before and after passage of the Clean Air Act in 1970. Note that nitrogen oxide emissions have continued to increase. (b) Hydrogen ion concentrations in precipitation (100 μeq per L = pH 4; 10 μeq per L = pH 5). After G. E. Likens, *Ecology* 85:2355–2362 (2004).

declined slowly. Second, particulate emissions dropped precipitately after 1970, but this also removed a major source of calcium for the Hubbard Brook ecosystem, accounting for perhaps half of the pre-1970 inputs. Calcium and other positively charged ions reduce the acidity of water and soil. Without this input of calcium, the system remained acidic.

The third factor preventing recovery was the long-term leaching of calcium and other positive ions from the soil by the hydrogen ions in the acidified rainwater. Trees require calcium and magnesium for proper growth, but concentrations of these elements in the soil had decreased dramatically. Acid rain was literally stealing nutrients from the soil. This problem was manifested in the concentrations of calcium in the trunks of red spruce trees over a 50-year period. At one locality, calcium concentrations increased during the 1950s as increasingly acidic rain solubilized calcium in the soil, which made it more readily available to the trees. After 1960, however, soil calcium concentrations decreased, and the amount of calcium assimilated began to decline and limit tree growth.

The Hubbard Brook study held several important lessons for forest ecologists. First, trees in regions of acid rain die not because of direct effects of high hydrogen ion concentrations, but because of long-term leaching of nutrients from the soil. Second, the natural recovery of forests growing on nutrient-poor soils will require restoration of soil nutrients through the slow process of weathering, which, as we shall see, could take a century or more. Thus, even when causes of environmental deterioration are addressed quickly, their overall effects on ecosystem function may remain for many years. More generally, the study demonstrated that an understanding of nutrient cycling and regeneration processes is crucial for understanding how ecosystems function.

CHAPTER CONCEPTS

- Weathering makes nutrients available in terrestrial ecosystems
- Nutrient regeneration in terrestrial ecosystems occurs in the soil
- Mycorrhizal associations of fungi and plant roots promote nutrient uptake
- Nutrient regeneration can follow many paths
- Climate affects pathways and rates of nutrient regeneration

- In aquatic ecosystems, nutrients are regenerated slowly in deep water and sediments
- Stratification hinders nutrient cycling in aquatic ecosystems
- Oxygen depletion facilitates regeneration of nutrients in deep waters
- Nutrient inputs control production in freshwater and shallow-water marine ecosystems
- Nutrients limit production in the oceans

Chemical elements can take on different forms as they cycle through ecosystems, but only some of those forms are useful to organisms. Nitrogen, for example, can exist in the form of nitrogen oxides, ammonium, or N_2, but only nitrogen in the first two forms is accessible to plant life. The chemical and biochemical reactions that create these compounds depend on an element's chemical properties, but are also uniquely modified by the physical and chemical conditions created in each type of terrestrial and aquatic ecosystem. Because all organisms rely on the presence of nutrients in forms they can use, the cycling and regeneration of nutrients is an important regulator of ecosystem function. In this chapter, we shall discuss how biochemical processes in soils, water, and sediments influence the productivity of ecosystems and the cycling of nutrient elements within them.

Once incorporated by primary producers, nutrients must be regenerated in usable forms by decomposition or replenished from outside the system for nutrient cycling to continue. The processes of nutrient regeneration are different in terrestrial and aquatic ecosystems. To be sure, similar chemical and biochemical transformations occur in both systems: oxidation of carbohydrates, nitrification, and oxidation of sulfur by chemoautotrophic bacteria, among many others. But regenerated nutrients are more easily recovered in terrestrial systems than in aquatic systems. In terrestrial ecosystems, most elements cycle through detritus at the soil surface, where plant roots and their associated mycorrhizal fungi have ready access to nutrients. In aquatic habitats, sediments are the ultimate source of regenerated nutrients, and sediments at the bottom of lakes and oceans are often far removed from sites of primary production in surface waters. Another difference is that most ecosystem metabolism in terrestrial ecosystems is aerobic, while in aquatic systems much of the cycling of nutrients results from anaerobic processes.

Weathering makes nutrients available in terrestrial ecosystems

In terrestrial ecosystems, most nutrient elements cycle rapidly through three compartments: the soil, plant biomass, and detritus. Nutrients are lost in streams and groundwater that leave the local ecosystem (Figure 24.2). Therefore, some source of new nutrients is needed if nutrient cycling is to continue at the same level and maintain the productivity of the ecosystem. New inorganic nutrients are added to the system by the weathering of the bedrock that underlies the soil as well as by input from the atmosphere as particulates, as ions dissolved in precipitation, or

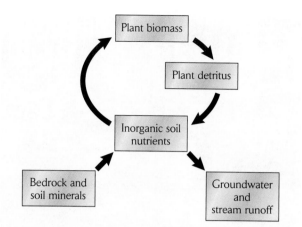

FIGURE 24.2 Nutrients in terrestrial ecosystems cycle through living organisms, detritus, and inorganic soil components. Nutrients are slowly added to soil by weathering and removed from ecosystems dissolved in runoff and groundwater.

as molecular elements (such as N_2) assimilated in organic forms by bacteria. Biological communities developing on newly exposed rock or sand are completely dependent on such external sources of nutrients.

The major source of new nutrients for terrestrial ecosystems is the formation of soil through the weathering of bedrock and other parent material (see Chapter 4). *Weathering* is the physical breakdown and chemical alteration of rocks and minerals near the earth's surface. Substances such as carbonic acid in rainwater and organic acids produced by the decomposition of plant litter react with minerals in the parent material and release various ions essential to plant growth.

How rapidly does weathering occur? Weathering normally takes place under deep layers of soil, where it cannot be measured directly. However, in many undisturbed areas, soil scientists can estimate the rate of weathering indirectly by quantifying the net loss of certain elements from a system. Positive ions, such as calcium (Ca^{2+}), potassium (K^+), sodium (Na^+), and magnesium (Mg^{2+}), are good candidates for such measurements because they dissolve readily in water and leave the soil in groundwater and eventually in streams, in which they can be measured easily. When a soil is at equilibrium, as it may be in undisturbed areas, the loss of an element from the soil equals the input of that element from weathering plus gains from other sources, such as precipitation and particulate matter. Thus, one can estimate weathering input from information about precipitation input and loss in streams.

Such measurements are conveniently made in some areas in watersheds. A **watershed** is the entire drainage area of a stream or river, which all surface water and groundwater leaves at a single point. Scientists have

FIGURE 24.3 Rain gauges are used to measure nutrient inputs. These rain gauges have been installed in a ponderosa pine stand in California to intercept precipitation falling through the canopy of the forest and running down the trunks of trees. Analyses of the nutrient content of water collected in gauges like these help researchers to determine the overall nutrient budget of the forest and the specific pathways of mineral cycles. Courtesy of the U.S. Forest Service.

obtained detailed nutrient budgets for several small watersheds by measuring nutrient inputs in rainwater collected at various locations in the watersheds (Figure 24.3) and nutrient outputs in the streams that drain them (Figure 24.4).

Perhaps the best-known watershed study comes from the Hubbard Brook Experimental Forest of New Hamp-

FIGURE 24.4 Stream gauges are used to measure nutrient outputs. This stream gauge has been placed at the lower end of a watershed at the Coweeta Hydrological Laboratory, North Carolina. The V-shaped notch is engineered so that the flow of water through the weir can be estimated from the water level in the basin behind the notch. Photo by Barry Near, USFS.

shire, described at the opening of this chapter. Scientists have monitored changes in the forest ecosystem there for more than 50 years. During the 1960s and 1970s, a period of highly acidified precipitation, the annual input of dissolved Ca^{2+} in precipitation at Hubbard Brook averaged 2 kilograms per hectare (kg per ha), while loss of dissolved Ca^{2+} in stream flow was 14 kg per ha. Therefore, the net loss to the system equaled 12 kg per ha per year. Living and dead plant biomass increased in the watershed during the study period because the forest was still recovering from clearing many years before. Incorporation of calcium into vegetation and detritus brought the overall removal of calcium from the mineral soil to 21 kg per ha per year.

Because calcium constitutes about 1.4% of the weight of the bedrock in the area, scientists at the time estimated that making up this annual loss of calcium would have required the weathering of about 1,500 kg (21/0.014) of bedrock per hectare, or approximately 1 mm of depth, per year. However, because this was a period of high acidity and rapid leaching of ions from soil, we now know that the soil system was not in equilibrium, and that inputs of calcium through weathering were undoubtedly much lower than that estimate. Nonetheless, this example illustrates how little the slow weathering of bedrock contributes to the annual uptake of nutrients by vegetation. The bulk of the nutrients made available to plants come from the breakdown of detritus and small organic molecules within the soil. Typically, weathering of bedrock provides only 10% of the soil nutrients taken up by vegetation each year. In other words, most nutrients in ecosystems are regenerated within those ecosystems.

Nutrient regeneration in terrestrial ecosystems occurs in the soil

Uptake of inorganic nutrients by plants and decomposition of detritus by microorganisms are biochemical processes influenced by temperature, moisture, pH, and other factors. Thus, the rate of nutrient cycling and the overall productivity of the ecosystem are sensitive to these physical influences. Because nutrients move between ecosystem compartments in turn—that is, from soil to plant to detritus and back to soil—the rate of cycling is limited by the slowest step. In most cases, this step is the decomposition of detritus.

Plants assimilate elements from soil far more rapidly than weathering generates them from parent material. Important nutrients, such as nitrogen, phosphorus, and sulfur, are typically scarce in parent material. Igneous

FIGURE 24.5 The decomposition of plant detritus regenerates nutrients. Plant detritus accumulating on a forest floor is broken down by soil organisms, and the nutrients it contains are released in forms that can be taken up and used by plants. Photo by R. E. Ricklefs.

rocks such as granite and basalt contain no nitrogen, only 0.3% phosphate, and only 0.1% sulfate by mass. Most sedimentary rocks contain little more. Hence, weathering adds little of these nutrients to soil; inputs from precipitation and nitrogen fixation are also small. The development of ecosystems on newly exposed substrata requires long periods of nutrient accumulation. In mature ecosystems, plant production depends on rapid regeneration of these nutrients from detritus and their retention within ecosystems.

Organic detritus accumulates everywhere, most conspicuously in the soil of terrestrial habitats. In these habitats, parts of plants not consumed by herbivores blanket the soil surface, along with animal excreta and other organic remains (Figure 24.5). Ninety percent or more of the plant biomass produced in forest ecosystems passes through this detritus compartment. The processes of decay break down the detritus, releasing the nutrients it contains in forms that can be reused by plants.

The breakdown of leaf litter in a forest occurs in four ways: (1) by water leaching out soluble minerals and small organic compounds; (2) by consumption by large detritivores (millipedes, earthworms, wood lice, and other invertebrates); (3) by breakdown of the woody components and other carbohydrates in leaves by fungi; and (4) by decomposition of almost everything by bacteria.

Between 10% and 30% of the substances in newly fallen leaves dissolve in cold water. Leaching by water rapidly removes most salts, sugars, and amino acids from the litter, making them available to soil microorganisms and the roots of plants; left behind are complex carbohydrates, such as cellulose, and other large organic compounds, including proteins and lignin. Large detritivores typically assimilate only 30% to 45% of the energy available in leaf litter, and even less from wood. They nonetheless speed decay because they macerate plant detritus in their digestive tracts, and the finer particles in their egested wastes expose new surfaces to feeding by fungi and bacteria.

The leaves of different tree species decompose at different rates depending on their composition. For example, in a forest in eastern Tennessee, weight loss from shed leaves during the first year after leaf fall ranged from 64% for mulberry to 39% for oak, 32% for sugar maple, and 21% for beech. Needles of pines and other conifers also decomposed slowly. These differences among species depend to a large extent on the lignin content of the leaves, which determines their toughness. Lignin fibers are long, complex chains of organic molecules. They lend wood many of its structural qualities and resist decomposition more than cellulose. In fact, only the so-called "white rot" fungi can break down lignin fibers; these fungi secrete a variety of oxidizing agents that attack the chemical structure of lignin. The decomposition rate of detritus also depends on its content of nitrogen, phosphorus, and other nutrients required by bacteria and fungi for their own growth. The higher the concentration of these nutrients, the faster microorganisms can grow, and the more rapidly they can decompose plant detritus.

The unique role of saprotrophic ("detritus eating") fungi in regenerating nutrients is to break down some types of litter that are resistant to decomposition by other organisms. Most fungi consist of a network of threadlike structures called hyphae, which can penetrate plant litter and wood where bacteria cannot reach. The familiar mushrooms and shelf fungi are merely fruiting structures produced by the mass of hyphae deep within litter or wood (Figure 24.6; see also Figure 1.7). Like bacteria, fungi secrete enzymes and other substances and absorb simple sugars and amino acids released by the action of these enzymes on organic matter. Fungi differ from most bacteria in being able to break down cellulose (only a few bacteria, protozoans in the guts of termites, and snails can accomplish this) and, especially, lignin.

Mycorrhizal associations of fungi and plant roots promote nutrient uptake

Some kinds of fungi grow on the surfaces of, or inside, the roots of plants. This symbiotic association of fungus and

FIGURE 24.6 Shelf fungi speed the decomposition of a fallen log. The visible fruiting structures are produced by fungal hyphae that are growing throughout the interior of the log, slowly destroying its structure. Photo by R. E. Ricklefs.

root is called a **mycorrhiza** (plural, **mycorrhizae;** literally, "fungus root"). Mycorrhizae enhance a plant's ability to extract less soluble nutrients, such as phosphorus, from soil and may greatly increase primary production, especially on poor soils.

Two principal forms of mycorrhizae are recognized: **arbuscular mycorrhizae (AM)** and **ectomychorrhizae (EcM).** AM fungi penetrate cell walls in root tissue and form vesicles or branched structures in intimate contact with root cell membranes. The name *arbuscular,* meaning "treelike," refers to the branched structures penetrating the root. AM are formed only by fungi in the taxonomic division Glomeromycota and are typically associated with the roots of herbaceous species, including many crop plants. EcM fungi are commonly associated with woody plants. They form a dense sheath around the outsides of small roots and penetrate the spaces between the cells of the root cortical layer (Figure 24.7). Most EcM fungi belong to the divisions Basidiomycota (typical mushrooms) and Ascomycota (morels and truffles, for example). Other specialized forms of non-AM fungi associated with orchids penetrate the walls of root cells, but do not form sheaths.

Mycorrhizae occur everywhere, but they promote plant growth most strongly in soils that are relatively depleted of nutrients (Figure 24.8). Mycorrhizae increase a plant's uptake of minerals by penetrating a greater volume of soil than the roots could accomplish alone and by increasing the total surface area available for nutrient assimilation. In addition, because the fungi secrete enzymes and acid (hydrogen ions) into the surrounding soil, mycorrhizae are more effective than plant roots alone at extracting certain inorganic nutrients from the soil. Mycorrhizae, especially EcM forms, may also protect plant roots from disease

by physically excluding pathogens or by producing antibiotics (antibacterial toxins). The main advantage of this association for the fungi appears to be that they gain a reliable source of organic carbon in the form of simple sugars transported from the leaves to the roots of their host plants.

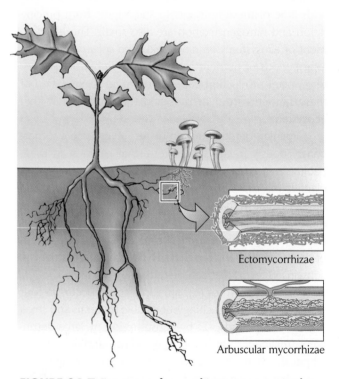

Ectomycorrhizae

Arbuscular mycorrhizae

FIGURE 24.7 Two types of mycorrhizae are recognized. In ectomycorrhizae, the fungus forms a sheath around a plant root and penetrates the spaces between the superficial cells of the cortex. In arbuscular mycorrhizae, the fungus penetrates cell walls and forms close associations with root cell membranes.

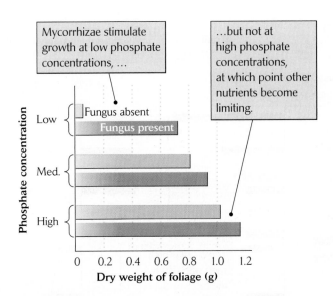

FIGURE 24.8 Mycorrhizae promote plant growth most strongly in poor soils. Adding different amounts of phosphate fertilizer to soil and inoculating soil with the AM fungus *Glomus macrocarpus* both influenced the growth of tomato plants (*Lycopersicon esculentum*). After J. L. Harley and S. E. Smith, *Mycorrhizal Symbiosis*, Academic Press, London (1983).

Nutrient regeneration can follow many paths

As we have seen, organic detritus is consumed by many types of organisms in soil, including invertebrates, bacteria, saprotrophic fungi, and mycorrhizal fungi. Various microorganisms and plants compete for the inorganic nutrients released by decomposition. As a result, the pathways of nutrient cycling can be extremely complex (Figure 24.9).

The first step in decomposition is the breakdown of large organic polymers such as structural carbohydrates, lignin, proteins, and DNA into their monomeric subunits, including amino acids and nucleic acids. As we have seen, microorganisms secrete enzymes and other reactive substances to accomplish this goal. Ecologists consider depolymerization to be the rate-limiting step in the decomposition of detritus and, consequently, the limit on the overall productivity of terrestrial ecosystems.

Microorganisms can degrade the organic monomers produced by depolymerization to inorganic forms when their intake of nutrients exceeds their requirements. When microorganisms incorporate inorganic nutrients into their own organic structures, those nutrients are removed from the available pool in the soil—a process known as **immobilization**—until the microorganisms die and are decomposed or are eaten by other consumers and digested.

Pathways of decomposition in soils shift depending on the relative availability of inorganic nutrients. For example, in nitrogen-poor soils, both microorganisms and plants tend to use amino acids as a nitrogen source. Because nitrogen is in short supply, microorganisms do not produce much ammonium or nitrate as a waste product of their metabolism. In soils with higher concentrations of nitrogen, microorganisms produce more ammonium and nitrate in microsites with high nitrogen availability, but plants and microorganisms in low-nitrogen microsites nearby compete intensely for this resource. At even higher levels of nitrogen availability, nitrogen ceases to limit growth, and microorganisms metabolize much of the organic nitrogen in amino acids to nitrate, which can be taken up directly by plants. Thus, with increasing levels of nitrogen in the soil, the primary source of nitrogen for plants shifts from amino acids and other small organic compounds to ammonium and, finally, to nitrate. In addition to shifting the system between nutrient cycling pathways, nutrient addition (fertilization) often increases the overall rate of decomposition of organic matter because

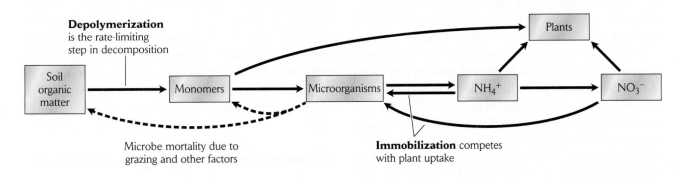

FIGURE 24.9 Depolymerization of large organic molecules in soil regulates nitrogen cycling. The breakdown of proteins, lignins, nucleic acids, and other macromolecules in soil produces small monomers, such as amino acids and nucleic acids, which microorganisms and plants take up and metabolize readily. After J. P. Schimel and J. Bennett, *Ecology* 85:591–602 (2004).

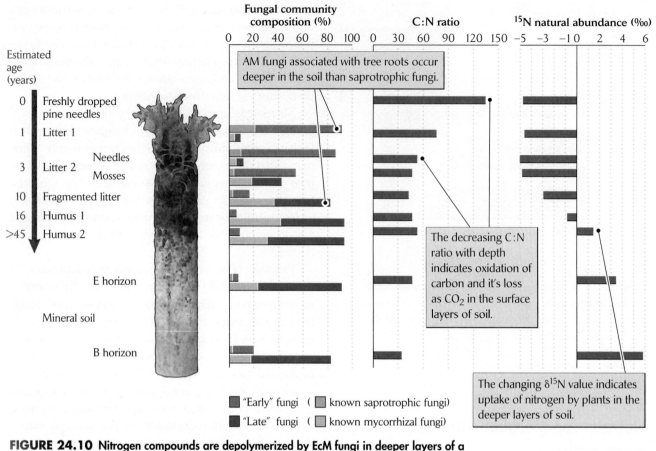

Fungal community composition (%)

C:N ratio

15N natural abundance (‰)

AM fungi associated with tree roots occur deeper in the soil than saprotrophic fungi.

The decreasing C:N ratio with depth indicates oxidation of carbon and it's loss as CO$_2$ in the surface layers of soil.

The changing δ^{15}N value indicates uptake of nitrogen by plants in the deeper layers of soil.

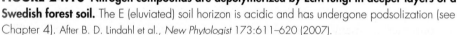

■ "Early" fungi (■ known saprotrophic fungi)
■ "Late" fungi (■ known mycorrhizal fungi)

FIGURE 24.10 Nitrogen compounds are depolymerized by EcM fungi in deeper layers of a Swedish forest soil. The E (eluviated) soil horizon is acidic and has undergone podsolization (see Chapter 4). After B. D. Lindahl et al., *New Phytologist* 173:611–620 (2007).

fertilization supports the growth of populations of bacteria and fungi.

The complex processes of decomposition are to some extent spatially separated within the soil: bacteria and saprotrophic fungi are more active in the surface layers of organic detritus, and mycorrhizae are concentrated in the deeper mineral layers of the soil. Björn Lindahl and his colleagues at the Swedish University of Agricultural Sciences observed this pattern when they studied the decomposition of litter at different soil depths in a Scots pine (*Pinus sylvestris*) forest in central Sweden (Figure 24.10). Their study showed that different processes of detritus decomposition are spatially separated in soil.

Oxidation of organic carbon is greatest in the surface layers, whereas nitrogen is released and taken up by plant roots deeper in the soil. The ratio of carbon to nitrogen in freshly fallen pine needles exceeded 150. The C:N ratio had dropped to about 50, however, after several years of decomposition by saprotrophic fungi that metabolized the organic carbon to obtain energy. The role of EcM fungi

in the decomposition of nitrogen-containing compounds was apparent in the natural concentrations of nitrogen-15 measured at different soil depths. Because this heavier isotope is not assimilated by plants as readily as the more common isotope ^{14}N, pine needles contain about 5 parts per thousand (0.5%) less ^{15}N than the atmosphere does. This difference is expressed as a δ^{15}N value of −5‰. In the deeper layers of soil dominated by AM fungi, the δ^{15}N value increased with depth to +5‰ as plant roots preferentially assimilated ^{14}N compounds. This study showed that differences in the concentrations of elements with depth in soil can provide a picture of the dynamics of nutrient mobilization and assimilation over time.

Climate affects pathways and rates of nutrient regeneration

Patterns of nutrient cycling differ across ecosystems in part because climates affect weathering, soil properties, and rates of decomposition of detritus by microorganisms. These differences are evident when we compare tropical and temperate ecosystems.

Tropical soils do not retain nutrients well because they tend to be so deeply weathered that they contain little clay (see Chapter 4). As a consequence, unless plants take up nutrients rapidly, they are washed out of the soil. Yet, in spite of their nutrient-poor soils, tropical forests often exhibit extremely high primary productivity. Their high productivity is supported by (1) rapid decomposition of detritus under warm, humid conditions, (2) rapid uptake of nutrients by plants and other organisms from the uppermost layers of soil, and (3) efficient retention of nutrients by plants and their mycorrhizal fungal associates.

Comparative studies of temperate and tropical forests show that in the tropics, detritus decomposes more rapidly and does not form a substantial nutrient pool. Litter on the forest floor constitutes an average of about 20% of the total biomass of vegetation (including trunks and branches) and detritus in temperate needle-leaved forests, 5% in temperate hardwood forests, and only 1%–2% in tropical rain forests. Of the total organic carbon in the system as a whole, more than 50% occurs in soil and litter in northern forests, but less than 25% in tropical rain forests, where the majority is in living biomass.

The same pattern applies to the relative proportions of other nutrients in soil and living vegetation. Distributions of phosphorus and nitrogen between the soil and living vegetation in a temperate and a tropical forest are compared in Figure 24.11. The soil:biomass ratio of both elements was much lower in the tropics.

Eutrophic and oligotrophic soils

While recognizing the general nutrient poverty of many tropical soils, we should also distinguish between nutrient-rich and nutrient-poor soils within the tropics. **Eutrophic,** or "well-nourished," soils can develop in geologically active areas where natural erosion is high and soils are relatively young. With bedrock closer to the surface, weathering adds nutrients more rapidly and soils retain nutrients more effectively. In tropical regions of the Western Hemisphere—the Neotropics—such eutrophic soils occur widely in the Andes, in Central America, and in the West Indies. By contrast, **oligotrophic,** or nutrient-poor, soils develop in old, geologically stable areas, particularly on sandy alluvial deposits (as in much of the Amazon basin), where intense weathering over long periods removes clay and reduces the capacity of soils to retain nutrients.

Especially in areas with oligotrophic soils, nutrient retention by vegetation is crucial to high productivity in

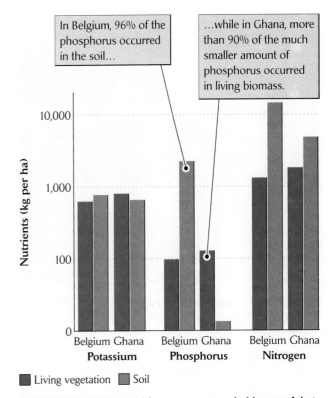

FIGURE 24.11 Tropical forest ecosystems hold most of their nutrients in living vegetation. In an ash–oak forest in Belgium, most of the phosphorus and nitrogen occur in the soil as detritus, decomposed organic molecules, or inorganic nutrients. In a tropical deciduous forest in Ghana, the quantities of nutrients in living vegetation are similar to those in Belgium, but soil:biomass ratios are much lower. From data in P. Duvigneaud and S. Denayer-de-Smet, in D. E. Reichle (ed.), *Analysis of Tropical Forest Ecosystems,* Springer-Verlag, New York (1970), pp. 199–225; D. J. Greenland and J. M. Kowal, *Plant Soil* 12:154–174 (1960); J. D. Ovington, *Biol. Rev.* 40:295–336 (1965).

tropical ecosystems. In these environments, plants retain nutrients by keeping their leaves for long periods and by withdrawing nutrients from them before they are dropped. They also grow dense mats of roots (and associated fungi) that remain close to the soil surface (where litter decomposes) and may even extend up the trunks of trees to intercept nutrients washing down from the forest canopy.

Habitat conversion and soil nutrients

The typical nutrient cycling pattern in tropical ecosystems, in which most nutrients are found in living biomass and nutrients are regenerated and assimilated rapidly, has important implications for tropical agriculture and conservation.

Over extensive regions of old, deeply weathered soils in the tropics, the planting of crops, such as corn, on

FIGURE 24.12 Large areas of tropical forest are cleared for agriculture each year. This lowland forest in Panama has been cut and burned to make room for planting crops. The fertility of the soil will decrease dramatically within 2 or 3 years. Photo by R. E. Ricklefs.

clear-cut land has predictable adverse consequences for soil fertility (Figure 24.12). The practice of cutting and burning trees releases many inorganic nutrients, which may support 2 or 3 years of crop growth, but these nutrients are quickly leached out of the soil when no natural vegetation remains to assimilate them. Consequently, concentrations of inorganic nutrients in the soil decline rapidly. Furthermore, as exposed tropical soils dry out, upward movement of water draws iron and aluminum oxides toward the surface, where they can form a bricklike substance called **laterite.** Surface runoff of water over the impenetrable laterite accelerates erosion, further depleting nutrients and choking streams with sediment. Traditional slash-and-burn agriculture on oligotrophic soils in the tropics usually alternates 2 or 3 years of crops with 50 to 100 years of forest regeneration to rebuild soil quality. Where human populations are too dense to allow this practice, soils cannot be replenished naturally, and they deteriorate rapidly without expensive and environmentally damaging inputs of fertilizers.

A comparison of forested ecosystems cleared for crops in Canada, Brazil, and Venezuela demonstrates that soils with abundant organic matter can maintain fertility longer under intensive agriculture. The carbon content of undisturbed soils was 8.8 kilograms per square meter (kg per m^2) at a prairie site in Canada, 3.4 kg per m^2 at a semi-arid thorn forest site in Brazil, and 5.1 kg per m^2 under a Venezuelan rain forest. After 65 years of cultivation, the carbon content of the Canadian soil had been reduced by 51%, which is equivalent to decline at an exponential rate of about 1% per year. In marked contrast, the carbon content of the Brazilian soil had decreased by 40% after 6 years of cultivation (9% per year), and that of the Venezuelan soil had decreased by 29% after 3 years of cultivation (11% per year). These results suggest that cultivated temperate soils retain organic matter ten times longer than tropical soils, and thereby provide a more persistent store of inorganic nutrients that can be released slowly by decomposition.

Obviously, vegetation is critical to the development and maintenance of soil fertility in many tropical ecosystems. Even in temperate zones, soils do not retain nutrients when vegetation is removed (Figure 24.13). In one study, a small watershed in the Hubbard Brook Experimental Forest was clear-cut and its nutrient flux compared with that in similar undisturbed forest systems. The clear-cutting increased stream flow severalfold because no trees were present to take up water. Losses of nutrients, particularly calcium, carried away by streams were 3–20 times greater than losses in comparable undisturbed systems.

The nitrogen budget of the clear-cut Hubbard Brook watershed sustained the most striking change. Plants assimilate available soil nitrogen so rapidly that the undisturbed forest gained nitrogen at the rate of 1–3 kg per hectare per year from precipitation and nitrogen fixation. By contrast, the clear-cut watershed had a net loss of nitrogen of 54 kg per ha per year—as much as vegetation in undisturbed forest can assimilate yearly and many times

FIGURE 24.13 Clear-cutting experiments have demonstrated the role of vegetation in nutrient retention. This clear-cut watershed at the Coweeta Hydrological Laboratory, North Carolina, was employed in studies of evapotranspiration and runoff in forest ecosystems. Courtesy of the USDA Forest Service.

the precipitation input (7 kg per ha per year). As in the undisturbed watershed, organic nitrogen was converted to nitrate by soil microorganisms. In the cleared watershed, however, trees were not present to take up nitrate. Thus, because nitrate ions do not bind well to particles of clay and humus, they were lost from the soil.

ECOLOGISTS IN THE FIELD **Will global warming speed the decomposition of organic matter in boreal forest soils?** Higher temperatures accelerate the regeneration of nutrients within soils. At one extreme, plant litter breaks down rapidly in tropical regions, and the released nutrients are either taken up by plants or leave the forest ecosystem in ground and surface waters. At the opposite extreme, decomposition is so slow in boreal forests and tundra that thick layers of organic matter accumulate in the soil. Decomposition is slow in part because soils are frozen much of the year; below a certain depth, soils may be permanently frozen. According to one estimate, the permanently and seasonally frozen soils of boreal forests worldwide hold 200–500 gigatons of carbon, which represents almost 80% as much as the amount of carbon in the atmosphere. If temperatures of boreal soils were to increase because of global warming, soil microorganisms and animals might metabolize a substantial fraction of this soil carbon, which would return to the atmosphere as respired carbon dioxide. Ecologists are concerned about this possibility, and several studies have sought to understand how decomposition of organic matter in boreal soils responds to temperature.

How does one measure the breathing of a forest? University of California at Irvine ecologist Michael L. Goulden led his team of researchers to a spruce forest near Thompson in Manitoba, Canada, in an attempt to answer this question. The spruce forest ecosystem has three main components: soil; aboveground vegetation, mostly spruce trees and moss (Figure 24.14); and the atmosphere above the forest with which it exchanges oxygen and carbon dioxide. The investigators determined carbon dioxide concentrations in air samples by measuring the absorption of a particular wavelength of infrared light by carbon dioxide gas. Carbon dioxide leaving the soil was trapped in containers at the soil surface. Carbon dioxide moving between the forest and the atmosphere was estimated by measuring the concentrations of CO_2 in air at a particular height (29 meters in this case) and the rate of vertical air movement at a large number of sites (the eddy flux covariance method). When these values were averaged over rising and falling currents of air for long periods, they provided a measure of net movement of carbon dioxide into or out of the forest as a whole.

Air temperatures in the forest were highly seasonal, with minimums averaging −10°C to −25°C in the win-

FIGURE 24.14 Organic matter is abundant and decomposes slowly in boreal forests. The floor of this boreal forest near Fairbanks, Alaska, has a thick layer of moss on the ground. Photo by R. E. Ricklefs.

ter months and 15°C to 25°C in summer. Seasonal variation in soil temperature was much less. It was greatest at the soil surface and decreased with depth, but followed the seasonal trend in air temperature with a lag of 1 to 2 months. Soils thawed in late spring and remained unfrozen late into the year. For the forest as a whole, net uptake of carbon dioxide began in early May and reached a peak of 10–15 kg C per hectare per day in June and early July, when air temperature reached its peak (Figure 24.15). Net uptake declined to near zero by August and September as increasing soil temperatures stimulated soil respiration. After the end of September, the forest lost carbon, at 6–8 kg per ha per day in October and 2–3 kg per ha per day from December through April. Overall, annual net carbon flux over the 4-year study varied from a loss of 0.7 metric tons (T) per ha to a gain of 0.1 T per ha.

These figures should be compared with the forest's annual primary production of 8.0 T per ha per year, and its total carbon content of 95 T in aboveground living and

FIGURE 24.15 Carbon flux in a boreal forest is sensitive to temperature. A spruce forest in Manitoba, Canada, assimilates carbon during the warm summer months when photosynthesis is occurring, but loses more carbon during the rest of the year as respiration continues in the soil, whose temperature remains above freezing into the late summer and early fall. After M. L. Goulden, S. C. Wofsy, J. W. Harden, et al., *Science* 279:214–217 (1998).

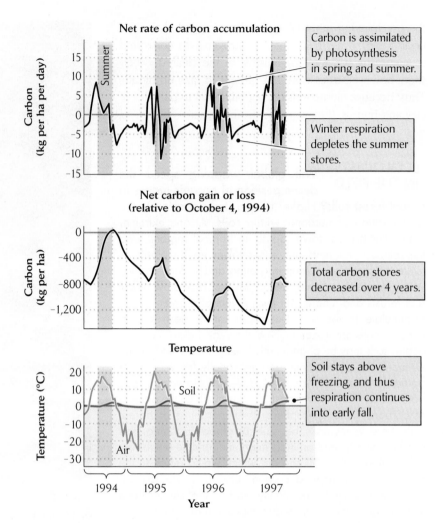

dead biomass and 200 T in the soil. Thus, observed losses were a small part of the total carbon in the system. However, soil respiration was clearly sensitive to soil thawing, and even a small increase in the depth of thawing would significantly increase the release of carbon dioxide into the atmosphere. How much thawing of frozen soils at high latitudes will accelerate global warming is uncertain at this point. What Goulden's study shows, however, is that many ecosystems are not in equilibrium, and that we can expect major readjustments in the sizes of the pools of elements in various compartments of the biosphere as global conditions change. ▌

In aquatic ecosystems, nutrients are regenerated slowly in deep water and sediments

Because most cycling of elements takes place in an aqueous medium, the chemical and biochemical processes involved do not differ markedly between terrestrial and aquatic ecosystems. What is distinctive about most rivers, lakes, and oceans is that organic matter sinks to the bottom and accumulates in deep layers of water and benthic sediment deposits, from which nutrients are regenerated and returned to zones of productivity relatively slowly.

Sediments in aquatic systems resemble terrestrial soils superficially, but the roles of soils and sediments in nutrient cycling differ in two important ways. First, regeneration of nutrients from terrestrial detritus takes place close to plant roots, where nutrients are assimilated. In contrast, aquatic plants and algae assimilate nutrients in the uppermost sunlit (photic) zone of the water column, often far removed from the sediments where nutrients are regenerated. Second, decomposition of terrestrial detritus occurs, for the most part, aerobically, and hence relatively rapidly. In contrast, aquatic sediments often become depleted of oxygen. The lack of oxygen greatly slows most biochemical transformations and changes the way in which some nutrients are regenerated.

Aquatic systems are able to maintain high productivity only when bottom sediments are not far below the photic

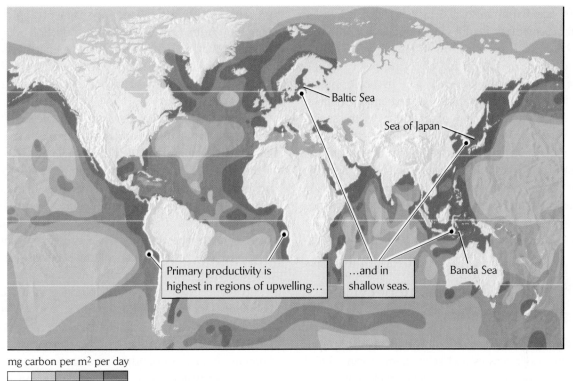

mg carbon per m² per day

100 150 250 500

FIGURE 24.16 Primary productivity in aquatic ecosystems is highest where nutrients regenerated in sediments can reach the photic zone. The map shows primary productivity in the world's oceans in milligrams of carbon fixed per square meter per day. After R. K. Barnes and K. H. Mann, *Fundamentals of Aquatic Ecosystems*, Blackwell, Oxford (1980).

zone at the water surface, or some means exists of bringing nutrients regenerated in those sediments back to the photic zone. A map of primary productivity in the oceans (Figure 24.16) reveals the locations of highly productive areas. Primary productivity is highest in shallow seas, both in the tropics (for example, the Coral Sea and the waters surrounding Indonesia) and at high latitudes (the Baltic Sea, the Sea of Japan). It is also high in areas having strong upwelling currents, such as the western coasts of Africa and the Americas.

Excretion and microbial decomposition regenerate some nutrients in the photic zone where nutrient assimilation and production take place, just as they do in terrestrial soils. Nutrients sometimes cycle rapidly within productive surface layers of the water column with little loss to sedimentation. This cycling depends on the establishment of steady-state conditions so that the growth of phytoplankton populations keeps up with grazing by herbivores; it also depends on the proper stoichiometry of required nutrients. A study conducted in deep water off the western coast of North America showed just such an equilibrium. Phytoplankton assimilated nitrogen about as rapidly as zooplankton excreted it. About half the nitrogen present was taken up directly as ammonium, and about half was first nitrified ($NH_4^+ \rightarrow NO_3^-$) by bacteria and then taken up by phytoplankton.

Sedimentation of nutrients is a prominent feature of most aquatic systems. Nitrogen budgets for the Bay of Quinte, Lake Ontario, illustrate this point. Canadian limnologists C. F.-H. Liao and D. R. S. Lean conducted studies within water columns enclosed by "limnocorrals," enclosures formed by sheets of plastic suspended by floats at the surface and entrenched in sediment at the bottom (in this case, 4 meters beneath the surface). Such enclosures make it possible to study the fluxes of nutrients by adding compounds labeled with stable isotopes. The limnocorrals had cross sections large enough that water at the surface and immediately over the sediments at the bottom could mix in parallel with the lake as a whole.

Measurements of nitrogen cycling in late spring (June 5) revealed uptake by phytoplankton at a rate of 18.5 micrograms per liter per day (μg per L per day), loss to grazing by herbivores at a rate of 9.7 μg per L per day, and sedimentation of particulate organic matter out of the surface waters at a rate of 2.6 μg per L per day. In late summer (September 5), the rate of uptake had increased to 129 μg per L per day, loss to grazing to 27 μg per L per day, and sedimentation to 63 μg per L per day. During both periods, nitrogen uptake by phytoplankton exceeded losses to zooplankton grazing and to sedimentation. Thus, total nitrogen concentrations in the surface water were increasing, which is generally the case through the summer growing season. Without vertical mixing and return of regenerated sediments to the surface, however, sedimentation would have quickly removed most of the nitrogen from the water. Nitrogen lost in sinking particulate matter was 14% as much as uptake in June and 28% as much in September.

Stratification hinders nutrient cycling in aquatic ecosystems

Vertical mixing of water requires an input of energy to accelerate water masses and keep them moving. Winds supply most of this energy, causing turbulent mixing of shallow water and upwelling currents along seacoasts, although variations in water density related to temperature and salinity establish vertical currents in other marine ecosystems.

Vertical mixing of water can be hindered when sunlight heats surface waters, establishing a *thermocline* (see Chapter 4), or when fresh water floats over denser salt water. The latter happens in estuaries, at the edges of melting ice, and in regions of extreme precipitation. Several processes can promote vertical mixing. In marine systems, when evaporation exceeds freshwater input, the surface layers of water become more saline, hence denser, and literally fall through the less dense water below them. Surface layers also become denser and sink when ice forms and salt is excluded from the crystallized water. Temperate lakes experience spring and fall overturn when seasonal temperature changes make their surface waters denser than the layers below (see Chapter 4).

Vertical mixing of water affects production in two opposing ways. On one hand, mixing can bring nutrient-rich water from the depths to the photic zone and thereby promote production. On the other hand, mixing can carry phytoplankton far below the photic zone, where they cannot maintain themselves, much less reproduce. Under such conditions, primary production may shut down altogether, resulting in the seeming contradiction of nutrient-rich water without primary production.

A more typical situation in temperate lakes and ponds is one in which thermal stratification during summer

prevents vertical mixing and, as sedimentation removes nutrients from surface layers, production decreases. Nutrients may be regenerated in the deeper layers of the lake, but they cannot reach the surface until stratification breaks down and vertical mixing ensues with cool fall temperatures.

Thermal stratification develops only weakly, if at all, in lakes at high and low latitudes (Figure 24.17). In polar and boreal regions, too little heat enters lakes to establish a thermocline. Thus, the water column tends to warm uniformly, to the extent that it warms at all. In the tropics, sun and constant high air temperatures warm water to the deepest parts of a lake. Nonetheless, even in tropical lakes, small temperature increases caused by sunlight in the upper layers of the water column can cause significant stratification.

In marine systems, currents produce more complex conditions. For example, the intermixing of two very different water masses, one stratified and the other not, may create excellent conditions for phytoplankton growth. Sometimes, at the boundary of a shallow-water system and a deep-water system, mixed (deep) and stratified (shallow) water masses are brought together. On the mixed side, nutrients may be abundant, but phytoplankton may have been carried below the photic zone. On the

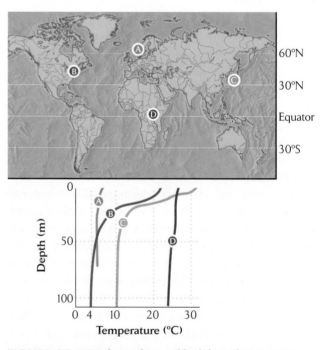

FIGURE 24.17 Lakes at low and high latitudes experience little thermal stratification. Temperature profiles are shown for four lakes at different latitudes at the height of summer. After G. E. Hutchinson, *A Treatise on Limnology,* Vol. 1, Wiley, New York (1957).

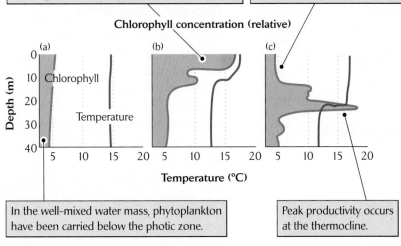

Where the two systems meet, some of the mixed water may enter the stratified water mass, carrying nutrients that stimulate production.

In the stratified water mass, production is low because nutrients in the surface waters have been depleted.

In the well-mixed water mass, phytoplankton have been carried below the photic zone.

Peak productivity occurs at the thermocline.

FIGURE 24.18 The meeting of water masses in marine ecosystems can affect productivity. Chlorophyll concentrations and temperatures are shown as a function of water depth for three locations in the western English Channel in July 1975. Chlorophyll concentrations provide an index to phytoplankton biomass and thus to the rate of primary production. (a) A well-mixed water mass from a deep-water system. (b) A "front" in the region of mixing between water masses a and c. (c) A stratified water mass from a shallow-water system. Productivity is greatest at the thermocline because of the presence of regenerated nutrients in the water below this level. After R. K. Barnes and K. H. Mann, *Fundamentals of Aquatic Ecosystems*, Blackwell, Oxford (1980).

stratified side, the surface waters may have been depleted of nutrients. Where the two systems meet, some of the nutrient-laden mixed water may enter the stratified water mass and stimulate production (Figure 24.18).

Oxygen depletion facilitates regeneration of nutrients in deep waters

During prolonged periods of stratification in freshwater lakes, bacterial respiration in the carbon-rich bottom sediments tends to deplete the oxygen supply in the hypolimnion (the cool, dimly lit layer of water below the thermocline). When bottom waters become anoxic, bacteria may continue to respire using sulfate as an oxidizer (see Chapter 23). This process results in increasing concentrations of reduced sulfur, primarily in the form of hydrogen sulfide.

In the oxygen-depleted environment of bottom sediments and the waters immediately over them, bacteria have insufficient oxygen to nitrify (oxidize) ammonium (see Chapter 23). Additionally, elements such as iron and manganese shift from oxidized to reduced forms, which increases their solubility. In particular, as ferric iron (Fe^{3+})

is reduced to ferrous iron (Fe^{2+}), insoluble iron–phosphate complexes become soluble, and both elements tend to move into the water column. Thus, nutrients accumulate under these reducing conditions.

The water chemistry of the hypolimnion of an English lake, Esthwaite Water, during the course of a single season shows the effects of anaerobic conditions as they develop over the summer (Figure 24.19). After vertical mixing shuts down in June, oxygen in the hypolimnion decreases gradually, while dissolved carbon dioxide increases. The water becomes depleted of oxygen by early July, and remains so until fall overturn in late September. During the period of oxygen depletion, levels of ferrous iron, phosphate, and ammonium (the reduced forms of iron, phosphorus, and nitrogen, respectively) increase dramatically in bottom sediments. At the sediment–water boundary, these materials become soluble and enter the water column. The return of oxidizing conditions in the fall reverses the chemistry of the bottom water, initially because it is replaced by surface water, but ultimately because several elements are oxidized in the presence of oxygen. The oxidized forms of these elements produce insoluble compounds, which precipitate out of the water column. Nitrogen is a conspicuous exception: in well-oxygenated waters, nitrifying bacteria convert ammonium to nitrate, which generally remains in solution.

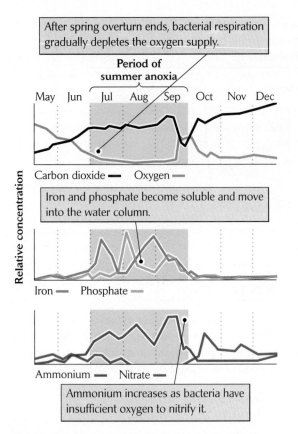

FIGURE 24.19 Oxygen depletion in the hypolimnion changes water chemistry. Concentrations of oxygen, carbon dioxide, and inorganic nutrients in the hypolimnion of Esthwaite Water, England, change with the seasons. After G. E. Hutchinson, *A Treatise on Limnology*, Vol. 1, Wiley, New York (1957).

Nutrient inputs control production in freshwater and shallow-water marine ecosystems

Natural lakes exhibit a wide range of productivity, depending on external inputs of nutrients from rainfall and streams and internal regeneration of nutrients in the lake. In shallow lakes lacking a hypolimnion, nutrients are supplied continuously through resuspension of bottom sediments. In somewhat deeper lakes in which stratification is weak, occasional strong winds or unusual periods of summer cold may initiate vertical mixing. Such mixing returns regenerated nutrients to the surface, which stimulates production. In very deep lakes, bottom waters rarely mix with surface waters, and production depends almost entirely on external nutrient sources. Aquatic ecologists classify lakes on a continuum ranging from oligotrophic (poorly nourished) to eutrophic (well-nourished), depending on their nutrient status and productivity.

Phosphorus is the most important element in determining the fertility of most lakes, and low levels of phosphorus limit production in these systems. Phosphorus is often scarce in well-oxygenated surface waters. In small lakes on the Canadian Shield, productivity increased dramatically in response to the experimental addition of phosphorus, but not nitrogen or carbon (Figure 24.20). Naturally eutrophic lakes have characteristic seasonal patterns of production and phosphorus cycling that maintain the system in a well-nourished, dynamic steady state.

FIGURE 24.20 Phosphorus is critical to the productivity of freshwater lakes. An experiment in a natural lake on the Canadian Shield demonstrated the crucial role of phosphorus in eutrophication. The near basin, fertilized with carbon (as sucrose) and nitrogen (as nitrates), exhibited no change in organic production. The far basin, separated from the first by a plastic curtain, was fertilized with phosphorus in addition to carbon and nitrogen, and was covered by a heavy bloom of photosynthetic cyanobacteria within 2 months. Courtesy of D. W. Schindler, from *Science*, 184:897–899 (1974).

FIGURE 24.21 Salt marshes are highly productive ecosystems. Salt marshes are a common feature of protected bays along most temperate coasts. Photo by R. E. Ricklefs.

it can be regenerated by photosynthesis. The oxidative breakdown of organic detritus, often augmented by raw sewage dumped into rivers and lakes, creates a condition of severe oxygen depletion. The problem is heightened in winter, when photosynthesis rates are low and little oxygen is generated within the water column. In its worst manifestations, this type of pollution can deplete oxygen all the way to the water surface, suffocating fish and other obligate aerobic organisms.

Estuaries and salt marshes

Shallow-water marine ecosystems are among the most productive on earth. Among these ecosystems are estuaries, which are semi-enclosed coastal regions at the mouths of rivers, and salt marshes, which are areas with emergent vegetation (plants rooted below the waterline) growing between the highest and lowest tide levels (Figure 24.21). The high productivity of these ecosystems results from their plentiful supplies of nutrients, which are brought in by rivers and tidal flow and are also rapidly regenerated within the system.

The effects of high production in estuaries and salt marshes extend to the surrounding marine ecosystems through their net export of organic matter and inorganic nutrients. A Georgia salt marsh was found to export nearly 10% of its gross primary production and almost half of its net primary production to marine systems in the form of organisms, particulate detritus, and dissolved organic molecules carried out with the tides (Figure 24.22). Because

Sewage and drainage from fertilized agricultural lands can greatly alter natural nutrient cycling in lakes by overloading them with inorganic nutrients and organic materials during times of the year when natural ecosystem processes cannot handle these inputs. Primary production may shoot upward in response to inputs of inorganic nutrients. Increased primary production is not bad in and of itself; indeed, many lakes and ponds are artificially fertilized to increase commercial fish production. But overproduction of organic matter within a lake or river (**eutrophication**) can lead to imbalance when decomposers of this excess organic matter consume oxygen faster than

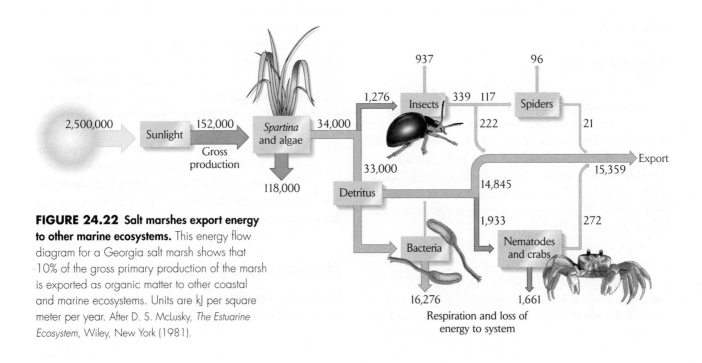

FIGURE 24.22 Salt marshes export energy to other marine ecosystems. This energy flow diagram for a Georgia salt marsh shows that 10% of the gross primary production of the marsh is exported as organic matter to other coastal and marine ecosystems. Units are kJ per square meter per year. After D. S. McLusky, *The Estuarine Ecosystem*, Wiley, New York (1981).

(a)

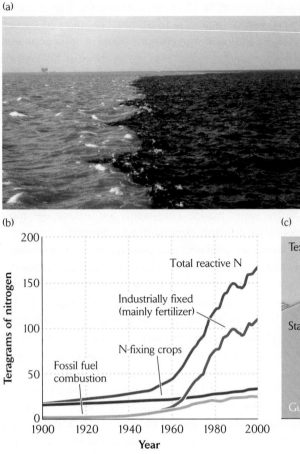

FIGURE 24.23 High concentrations of nutrients can create hypoxic "dead zones." (a) Nutrient-laden fresh water from the Mississippi River floats over the salt water of the Gulf of Mexico. (b) The amount of nitrogen from various sources entering the Gulf in river water has increased dramatically since 1960. Drainage from croplands carries nitrogen from artificial fertilizers and from nitrogen-fixing crops, particularly soybeans, into rivers. Drainage also carries nitrogen oxides from fossil fuel combustion washed out of the atmosphere by precipitation. (c) Large areas of the Gulf close to the Louisiana coast have hypoxic bottom waters (with oxygen concentrations < 2 mg per L). Photo (a) by Nancy Rabalais, Louisiana Universities Marine Consortium; part (b) after D. F. Boesch, *Estuaries* 25:886–900 (2002).

(b)

(c)

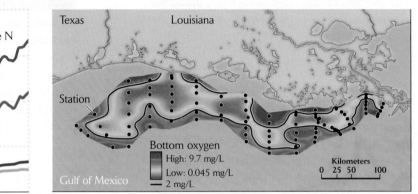

Hypoxic zones

Excess nutrient input can upset the balance of nutrient fluxes in estuaries and adjacent shallow marine ecosystems, just as it can in lakes and rivers. In estuaries, such imbalances may be exacerbated by lack of vertical mixing. Stratification is common where large rivers spread great volumes of fresh water over the surface of the ocean. Fresh water, being less dense than seawater, tends to float, so surface and bottom layers of water do not mix (Figure 24.23a).

Excess nutrients enter estuaries when rivers carry large amounts of inorganic nutrients and organic pollutants derived from sewage and fertilizer inputs (Figure 24.23b). The initial result is a burst of productivity in the surface layers of the water. Much of the newly produced organic material eventually sinks below the surface layers, and its decomposition by bacteria in the bottom waters depletes those waters of oxygen and makes them uninhabitable for many kinds of fish and invertebrates. The depletion of oxygen to the extent that aquatic organisms can no longer survive is called **hypoxia.** In the Gulf of Mexico, hypoxia is defined as less than 2 mg of oxygen per liter of water, the level below which bottom-dwelling fish and shrimp cannot survive. (Normal oxygen levels in these waters would exceed 10 mg per L.) Hypoxic "dead zones" have formed in most of the world's estuaries, including the Chesapeake Bay. A dead zone at the mouth of the Mississippi River extends over 20,000 km^2 in the Gulf of Mexico off the coast of Louisiana (Figure 24.23c). These hypoxic waters have damaged bottom fisheries over large areas, creating both environmental and economic disasters.

Nutrients limit production in the oceans

Primary productivity in marine ecosystems is closely related to the supply of nutrients in surface layers of water.

As a result, the highest levels of production occur in shallow seas, where vertical mixing reaches to the bottom, and in areas of strong upwelling. Primary productivity in the open ocean is typically low (see Figure 24.16).

Some areas of the open ocean have abundant nitrogen and phosphorus, but phytoplankton densities and primary production are low there as well. These conditions suggest that production in these areas is limited by shortages of other elements, including iron and silicon. Iron is an important component in many metabolic pathways, and silicon is the raw material for the silicate shells of diatoms (see Figure 2.6), which make up most of the phytoplankton in the oceans. Iron is lost from the photic zone when it combines with phosphorus and precipitates out of the water, and silicon is lost when diatoms die and their dense shells fall to the bottom.

Phytoplankton densities in the Southern Ocean are high in waters near continental sources of nutrients. High phytoplankton production is concentrated in waters downcurrent of Australia and New Zealand, South America and the Antarctic Peninsula, and southern Africa, where currents carry nutrients picked up from shallow-water sediments (Figure 24.24a). Not all of the Southern Ocean is equally productive, however. For example, the area west of southern South America between 40°S and 50°S appears to have too little silicon, probably because that element sinks below the photic zone more rapidly than nitrogen and phosphorus across the long stretch of

the southern Pacific Ocean (Figure 24.24b). Productivity also drops off close to the continent of Antarctica, partly owing to low water temperatures and extended winter darkness.

The Redfield ratio and nutrient limitation in the open ocean

We have seen the problems that can be caused by an unbalanced stoichiometric relationship between nutrient supplies and nutrient requirements. When nutrient concentrations in ocean waters do not match the nutrient requirements of photosynthetic organisms, primary production is reduced, and abundant nutrients go unused.

Many years ago, U.S. oceanographer Alfred Redfield noticed that the average ratio of nitrogen to phosphorus in phytoplankton (16:1) approximated the ratio of those elements in deep waters of the open ocean. Redfield proposed that this N:P ratio reflects the requirements of phytoplankton, which incorporate 16 times as much nitrogen as phosphorus into their biomass and release nitrogen and phosphorus to the environment at the same ratio as they decompose after death. The 16:1 N:P ratio became known as the Redfield ratio, to which carbon was subsequently added to make a C:N:P ratio of 106:16:1. Of course, the Redfield ratio is an average, and many types of phytoplankton deviate from it owing to different growth strategies (like the copepods described in Chapter 22).

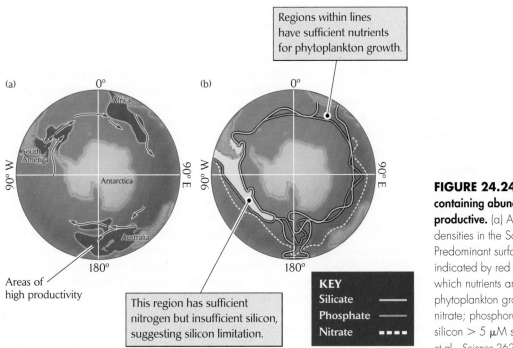

Regions within lines have sufficient nutrients for phytoplankton growth.

(a)

Areas of high productivity

This region has sufficient nitrogen but insufficient silicon, suggesting silicon limitation.

KEY
Silicate ———
Phosphate ———
Nitrate ▪▪▪▪

FIGURE 24.24 Not all ocean waters containing abundant nutrients are productive. (a) Areas of high phytoplankton densities in the Southern Ocean. Predominant surface current directions are indicated by red arrows. (b) Regions within which nutrients are sufficient for abundant phytoplankton growth (nitrogen > 10 μM nitrate; phosphorus > 1 μM phosphate; silicon > 5 μM silicate). After C. W. Sullivan et al., *Science* 262:1832–1837 (1993).

Several processes in the open ocean can influence N:P ratios. Where upwelling is weak, surface waters have few nutrients, and primary productivity is low. Under these conditions, nitrogen-fixing cyanobacteria are favored, and N:P ratios increase. In contrast, where upwelling is strong and nutrients are abundant in surface waters, primary productivity is high. Under these conditions, N:P ratios decrease as a result of a series of steps that eventually cause productivity to *decrease*. This sequence begins as deaths of the abundant photosynthetic green algae and diatoms remove substantial organic carbon from the waters. When these organisms die, their bodies sink toward the ocean depths, carrying their organic carbon compounds with them. As bacteria decompose this particulate matter by aerobic respiration, the oxygen concentration of the water decreases, and bacteria begin to use nitrate (NO_3^-) and nitrite (NO_2^-) as oxidizing agents. As a result of these denitrification reactions and of anaerobic ammonium oxidation ($NH_4^+ + NO_2^- \rightarrow N_2 + 2\ H_2O$) by bacteria, nitrogen is converted to molecular nitrogen gas (N_2). The result is a decrease in the N:P ratio. This decrease tends to upset the stoichiometric balance in the ocean and reduce primary production.

Of course, nitrogen and phosphorus are not the only nutrients required for phytoplankton production, and oceanographers have become increasingly interested in the roles of iron and other nutrients in marine productivity. Of special interest is the potential that adding nutrients holds for increasing the rate of carbon sequestration in the oceans.

ECOLOGISTS IN THE FIELD

Does iron limit marine productivity? Some 20% of the open ocean appears to have enough nitrogen and phosphorus to support high levels of primary production, but nonetheless supports low densities of phytoplankton. These regions are referred to as high-nutrient low-chlorophyll (HNLC) areas, and they have puzzled marine biologists for years. One hypothesis is that phytoplankton populations in these areas are kept low by zooplankton grazers, although it is unclear why this would happen in some regions of the ocean and not others. Another hypothesis is that production is limited by *micronutrients*—nutrients that are required by organisms in small amounts.

In the late 1980s, John H. Martin, of the Moss Landing Marine Laboratories in California, proposed that production in HNLC areas is limited by iron. In well-oxygenated surface waters, the oxidized ferric form of iron (Fe_3^+) forms complexes with other elements, including phosphorus, and precipitates out of the system. Whereas inshore areas receive iron from rivers, remote parts of the oceans receive much smaller iron inputs, almost exclusively from windblown dust. Martin suggested, in accordance with Liebig's law of the minimum (see Chapter 16), that if iron were the single nutrient that limited phytoplankton growth, additions of iron to HNLC areas would increase primary production dramatically.

In a large-scale experiment conducted in 1993 off the Pacific coast of South America, about 5° south of the equator, Martin and his colleagues fertilized a target area by distributing 450 kg of dissolved iron—roughly the amount in an automobile—over 64 km2 of ocean. This treatment increased the concentration of iron in that area almost a hundredfold. Within a few days, phytoplankton populations inside the fertilized patch, as measured by the concentration of chlorophyll in surface waters, tripled. This result clearly demonstrated that iron limits primary production in natural ocean waters. Additional laboratory experiments have shown that low concentrations of other micronutrients, such as manganese and copper, could potentially limit production in the open ocean. Manganese and copper are required as cofactors in important metabolic enzymes.

Several large-scale fertilization experiments have been conducted more recently in the Southern Ocean close to Antarctica. Investigators in one of these, the Southern Ocean Iron Experiment (SOFeX), also conducted by the Moss Landing Marine Laboratory, added iron to two areas of ocean. Both areas had high nitrate concentrations; one had abundant silicon (in the form of silicic acid, H_4SiO_4, about 60 μM), but the other was deficient in silicon (< 3 μM) (Figure 24.25). The addition of iron boosted primary production dramatically in both areas. Evidently, enough silicon was already present for diatom growth to increase in both regions. Nitrate and silicon levels in the water became depressed as these nutrients were taken up by phytoplankton.

More importantly, the results of this experiment suggested that the addition of iron could increase the capacity of the ocean to act as a carbon dioxide sink. Dissolved CO_2 concentrations decreased, and particulate organic carbon increased, after the addition of iron. Much of the particulate carbon eventually sank to the ocean depths, thereby removing CO_2 from surface waters and reducing CO_2 concentrations in the air above the water surface.

The original motivation for Martin's and subsequent experiments was to determine whether ocean waters fertilized with iron could quickly remove carbon dioxide from the atmosphere to counteract increasing anthropogenic inputs. Although the addition of iron increased primary production, its potential long-term effects on CO_2 sequestration are less clear. Martin's experiment was a failure in the sense that zooplankton populations increased along with the phytoplankton and regenerated much of the assimilated CO_2 by respiration. The results of SOFeX seem initially more promising: considerable particulate carbon

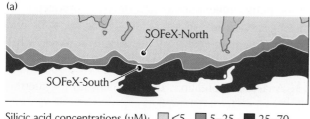

(a)

Silicic acid concentrations (μM): ☐ <5 ◻ 5–25 ■ 25–70

FIGURE 24.25 Fertilization of open ocean waters with iron greatly increases primary production. (a) Map showing the locations of iron additions in the Southern Ocean Iron Enrichment Experiment (SOFeX) in areas of high and low silicon concentrations. (b, c) False-color satellite images of each of the experimental areas showing greatly increased phytoplankton production (as indicated by chlorophyll *a* reflectance). The white areas represent cloud cover. After K. H. Coale et al., *Science* 304:408–414 (2004), parts (b) and (c) reprinted by permission from AAAS.

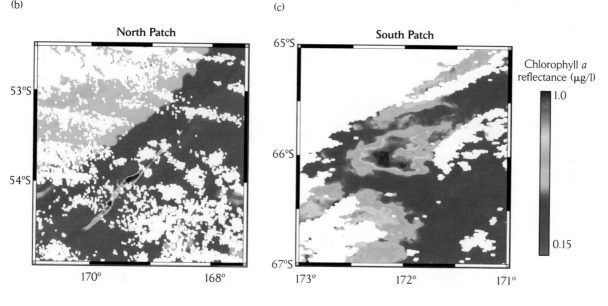

precipitated below the water mixing zone at the ocean surface (about 40 meters depth), and the concentration of CO_2 in the surface waters decreased by about 10% from that at the beginning of the experiment. Nonetheless, the long-term effects of iron enrichment on CO_2 sequestration remain unresolved, as does the question of whether large-scale applications of this approach could have potential adverse effects. ▌

DATA ANALYSIS MODULE *Consumer Populations and Energy Flow: Estimating Secondary Production.* You will find this module at http://www.whfreeman.com/ricklefs6e.

SUMMARY

1. Nutrient cycles in terrestrial and aquatic ecosystems result from similar chemical and biochemical reactions expressed in different physical and chemical environments.

2. Weathering of bedrock in terrestrial ecosystems and the associated release of new nutrients proceed slowly compared with the assimilation of nutrients from soil by plants. Therefore, the productivity of vegetation depends on nutrients regenerated from plant litter and other organic detritus.

3. Decomposition of organic detritus regenerates nutrients in terrestrial ecosystems. Nutrients are released from leaf litter by leaching of soluble substances; consumption by large detritivores; breakdown of cellulose, lignin, and other macromolecules by fungi; and the eventual release of phosphorus, nitrogen, and sulfur in mineral forms, primarily by bacteria.

4. Mycorrhizae are symbiotic associations of fungi with the roots of plants. Arbuscular mycorrhizae (AM) form close associations with the cell membranes in root tissues.

Ectomycorrhizae (EcM) form a dense sheath around a root and penetrate spaces between the cells of the cortex. Both types of mycorrhizae enhance a plant's uptake of soil nutrients.

5. The first step in the decomposition of detritus is depolymerization, whereby large organic molecules such as proteins and nucleic acids are broken down into subunits. The rate of depolymerization is thought to limit terrestrial ecosystem productivity.

6. Different processes of detritus decomposition in soil are spatially separated. In temperate forests, carbon oxidation is greatest in the surface layers, whereas decomposition of nitrogen-containing compounds and uptake of nitrogen by plant roots occurs deeper in the soil.

7. In many tropical ecosystems, soils are deeply weathered and retain nutrients poorly. In such environments, decomposition of organic matter and assimilation of nutrients by plants both proceed rapidly, so most nutrients, especially phosphorus, are in living vegetation.

8. When tropical forests are clear-cut for agriculture, the soils soon lose their fertility as nutrients that are not assimilated by vegetation are quickly washed out. Many tropical soils cleared for agriculture develop impenetrable surface layers of laterite.

9. Because low temperatures slow decomposition of organic matter, soils at high latitudes accumulate large quantities of organic carbon. Scientists are concerned that warming global temperatures will speed the decomposition of this material and contribute substantially to increasing atmospheric carbon dioxide concentrations.

10. Nutrient cycling in aquatic ecosystems differs from that in terrestrial ecosystems in that the aquatic sediments where nutrients accumulate are spatially removed from sites of nutrient assimilation in the photic zone.

11. The primary productivity of aquatic ecosystems is highest where nutrients from bottom sediments can be transported to the surface, as in shallow seas and zones of upwelling, and where nutrients are cycled rapidly within the photic zone.

12. Vertical mixing of water can increase production by bringing nutrients regenerated in aquatic sediments back to the photic zone, but can also decrease production by moving phytoplankton below the photic zone. Production in temperate lakes may decrease in summer, when thermal stratification prevents vertical mixing.

13. Anaerobic conditions develop beneath the thermocline because bacterial respiration depletes bottom waters of oxygen. These conditions promote regeneration of some nutrients.

14. Low levels of phosphorus limit production in most freshwater lakes. The addition of phosphorus and other nutrients to streams and lakes in sewage and agricultural runoff may alter patterns of nutrient cycling and production, upsetting natural balances in freshwater ecosystems.

15. Shallow-water marine communities, particularly estuaries and salt marshes, are extremely productive because of nutrient input from rivers and tidal flow as well as rapid nutrient regeneration within the system. Marshes and estuaries are major exporters of organic matter and inorganic nutrients to surrounding marine systems.

16. Excess nutrient inputs to estuaries and shallow coastal areas by rivers can upset the balance of nutrient fluxes and lead to hypoxic conditions. Pollution of the Mississippi River by sewage and fertilizer runoff has led to the formation of a "dead zone" at the mouth of the river in the Gulf of Mexico.

17. The primary productivity of marine systems is generally limited by nutrient availability. Silicon or iron may be limiting nutrients in the open ocean, where both elements tend to leave the water column as sediments.

18. Nutrient ratios in ocean waters must be balanced with nutrient requirements to promote high productivity. The nitrogen:phosphorus ratio in the deep waters of the open ocean is about 16:1 (the Redfield ratio) and reflects the average composition of marine phytoplankton. Nitrogen fixation, denitrification, and anaerobic ammonium oxidation can shift the ratio up or down.

19. Large-scale fertilization experiments have demonstrated that production is limited by iron in parts of the open ocean. This finding raises the possibility that adding iron to vast areas of the ocean might create a sink for carbon dioxide and help to reduce atmospheric CO_2 concentrations.

REVIEW QUESTIONS

1. Why is the weathering of bedrock responsible for such a small fraction of the nutrients available to plants?

2. In a mature ecosystem, what is the major source of soil nutrients for plants?

3. Compare and contrast the structure and function of arbuscular mycorrhizae and ectomycorrhizae.

4. Why do tropical and temperate soils differ in rates of nutrient regeneration?

5. Explain why agricultural soils in Canada retain their nutrients for many more years than agricultural soils in tropical South America.

6. How might global warming cause the release of CO_2 from boreal forest soils?

7. How does nutrient cycling differ between terrestrial and aquatic ecosystems?

8. How does the occurrence of stratification in freshwater lakes differ among latitudes, and how does such stratification affect the productivity of lake ecosystems?

9. By what chain of events does the dumping of raw sewage into the Mississippi River lead to fish kills in the "dead zone" in the Gulf of Mexico?

10. If the addition of iron to open ocean waters can triple the production of marine phytoplankton, what does this suggest about the relative importance of nitrogen and phosphorus in limiting productivity in these ecosystems?

SUGGESTED READINGS

Arrigo, K. R. 2005. Marine microorganisms and global nutrient cycles. *Nature* 437:349–355.

Baskin, Y. 1995. Can iron supplementation make the equatorial Pacific bloom? *BioScience* 45:314–316.

Bertness, M. D. 1992. The ecology of a New England salt marsh. *American Scientist* 80:260–268.

Binkley, D., and D. Richter. 1987. Nutrient cycles and H^+ budgets of forested ecosystems. *Advances in Ecological Research* 16:1–51.

Blain, S., et al. 2007. Effect of natural iron fertilization on carbon sequestration in the Southern Ocean. *Nature* 446:1070–1074.

Boesch, D. R. 2002. Challenges and opportunities for science in reducing nutrient over-enrichment of coastal ecosystems. *Estuaries* 25:886–900.

Buesseler, K. O., et al. 2004. The effects of iron fertilization on carbon sequestration in the Southern Ocean. *Science* 304:414–417.

Coale, K. H., et al. 2004. Southern Ocean Iron Enrichment Experiment: Carbon cycling in high- and low-Si waters. *Science* 304:408–414.

Goulden, M. L., et al. 1998. Sensitivity of boreal forest carbon balance to soil thaw. *Science* 279:214–217.

Hobbie, E. A., and T. R. Horton. 2007. Evidence that saprotrophic fungi mobilise carbon and mycorrhizal fungi mobilise nitrogen during litter decomposition. *New Phytologist* 173:447–449.

Hobbie, J. E., and E. A. Hobbie. 2006. ^{15}N in symbiotic fungi and plants estimates nitrogen and carbon flux rates in Arctic tundra. *Ecology* 87:816–822.

Hobbie, S. E., and P. M. Vitousek. 2000. Nutrient regulation of decomposition in Hawaiian forests. *Ecology* 81:1867–1877.

Kintisch, E. 2007. Should oceanographers pump iron? *Science* 318:1368–1370.

Klironomos, J. N., and M. M. Hart. 2001. Animal nitrogen swap for plant carbon. *Nature* 410:651–652.

Liao, C. F.-H., and D. R. S. Lean. 1978. Nitrogen transformations within the trophogenic zone of lakes. *Journal of the Fisheries Research Board of Canada* 35:1102–1108.

Likens, G. E. 2004. Some perspectives on long-term biogeochemical research from the Hubbard Brook Ecosystem Study. *Ecology* 85:2355–2362.

Likens, G. E., C. T. Driscoll, and D. C. Buso. 1996. Long-term effects of acid rain: Response and recovery of a forest ecosystem. *Science* 272:244–246.

Lindahl, B. D., et al. 2007. Spatial separation of litter decomposition and mycorrhizal nitrogen uptake in a boreal forest. *New Phytologist* 173:611–620.

Martin, J. H., et al. 1994. Testing the iron hypothesis in ecosystems of the equatorial Pacific Ocean. *Nature* 371:123–129.

McLusky, D. S. 1989. *The Estuarine Ecosystem,* 2nd ed. Chapman & Hall, New York.

Peers, G., and N. M. Price. 2004. A role for manganese in superoxide dismutases and growth of iron-deficient diatoms. *Limnology and Oceanography* 49:1774–1783.

Rabelais, N. N., et al. 1996. Nutrient changes in the Mississippi River and system responses on the adjacent continental shelf. *Estuaries* 19:386–407.

Rabelais, N. N., R. E. Turner, and W. J. Wiseman, Jr. 2002. Gulf of Mexico hypoxia, a.k.a. The dead zone. *Annual Review of Ecology and Systematics* 33:235–263.

Richards, B. N. 1987. *The Microbiology of Terrestrial Ecosystems.* Wiley, New York.

Schimel, J. P., and J. Bennett. 2004. Nitrogen mineralization: Challenges of a changing paradigm. *Ecology* 85:591–602.

Smith, S. E., and D. J. Read. 1997. *Mycorrhizal Symbiosis,* 2nd ed. Academic Press, London.

Stevenson, F. J. 1986. *Cycles of Soil: Carbon, Nitrogen, Phosphorus, Sulfur, Micronutrients.* Wiley, New York.

Sullivan, C. W., et al. 1993. Distributions of phytoplankton blooms in the Southern Ocean. *Science* 262:1832–1837.

Tunnicliffe, V. 1992. Hydrothermal-vent communities of the deep sea. *American Scientist* 80:336–349.

Van Cleve, K., et al. 1991. Element cycling in taiga forest: State-factor control. *BioScience* 41:78–83.

CHAPTER 25

Landscape Ecology

Throughout much of this book, we have considered ecology at a local scale. We have considered the roles of physical conditions and species interactions at a particular location, and we have seen how these factors can affect the ecology of individuals, populations, communities, and ecosystems. In doing so, we have often found it helpful to focus on a fairly homogeneous area of land or water. In this chapter, we shall consider ecology at much larger geographic scales that incorporate a realistic heterogeneity of habitats, and we shall see how our perspective can change at these broader regional scales.

Mazeika Sullivan and his colleagues at the University of Vermont demonstrated the importance of considering the entire landscape in their study of the birds that live on 27 streams in Vermont that feed into Lake Champlain. For each stream, they measured the physical characteristics of the stream itself (such as depth and width) as well as habitat characteristics in the floodplains and riparian areas out to 50 m beyond the stream banks. They then walked along the edges of the streams and recorded which bird species were using each of them. By combining their knowledge of the types of habitats that were present across the landscape of 27 streams and which birds were using each stream, the researchers could evaluate the importance of the habitats for bird species richness.

Sullivan and his colleagues observed 101 species of birds, including waterbirds (such as wood ducks), wading birds (such as great blue herons), fish-eating birds (such as ospreys), and insect-eating birds (such as northern rough-winged swallows). Across the landscape, different bird groups preferred different habitat characteristics. For example, the species richness and abundance of waterbirds was highest in shallow streams, whereas fish-eating birds were more abundant in larger streams that would provide more fish. In contrast, the richness and abundance of insect-eating birds was highest in areas containing a variety of habitat types, including shallow streams and abundant meadows and hardwood forests. Thus, the mixture of habitat types across the landscape is a key to maintaining insect-eating birds. Collectively, it is the heterogeneity of habitat types across the landscape of streams feeding into Lake Champlain that is important for supporting bird species richness. Thus, conserving this heterogeneity of habitats over a large area is likely to be critical to conserving bird species diversity.

CHAPTER CONCEPTS

- Landscape mosaics reflect both natural and human influences
- Landscape mosaics can be quantified using remote sensing, GPS, and GIS
- Habitat fragmentation can affect species abundance and species richness
- Habitat corridors and stepping stones can offset the effects of habitat fragmentation

- Landscape ecology explicitly considers the quality of the matrix between habitat fragments
- Different species perceive the landscape at different scales
- Organisms depend on different landscape scales for different activities and at different life history stages

We touched on the importance of considering geographic scales larger than the local community when we considered concepts such as metapopulation

FIGURE 25.1 Landscapes contain a variety of habitat types. This landscape in Holland contains a mosaic of aquatic and terrestrial habitats of different shapes and sizes. Photo by The Irish Image Collection/Corbis.

dynamics (in Chapter 12) and the theory of island biogeography (in Chapter 20). As you may recall, metapopulation models show that whether a species occurs in a given habitat patch is determined in part by how many other patches are occupied. In a similar way, the theory of island biogeography demonstrates that the number of species on a particular island can be understood only if we consider the size of the island and its distance from the mainland. In both cases, the broader spatial perspective helps us better understand the processes responsible for the dynamics of populations and the diversity of species in a community.

When we start considering larger areas of land, perhaps the most striking observation is that those larger areas contain a greater variety of habitats. A large area of many diverse habitat types is called a **landscape.** The diversity within a landscape includes not only the variety of terrestrial and aquatic habitat types within it, but also the ways in which those habitats are arranged. For example, if we were to fly over North America or Europe in an airplane, we would notice that many of the landscapes below us contain a mixture of forests, fields, rivers, lakes, and urban areas that take on a wide variety of sizes and shapes (Figure 25.1).

The study of the composition of landscapes and the spatial arrangement of habitats within them, and of how those patterns influence individuals, populations, communities, and ecosystems at different spatial scales, is known as **landscape ecology.** The mosaic of habitats in a landscape can reflect influences that are both historical and modern, both natural and anthropogenic. Of particular interest to landscape ecologists is how fragmentation of the landscape into isolated habitat patches of different sizes and shapes influences biodiversity, as well as how habitat corridors and the quality of the matrix between habitat fragments affect patterns of local species richness and species turnover. Taking a landscape approach also allows us to ask how different organisms experience their environment at different spatial scales. In short, a landscape approach strives to examine ecological patterns using a broader spatial picture.

Landscape mosaics reflect both natural and human influences

Influences from the past

Nature and humans have been shaping the diversity of habitats across landscapes for millennia. Some of the oldest influences include geologic events such as volcanic eruptions and the advance and retreat of glaciers across continents. These events have left their mark on the landscape by moving large amounts of rock and soil and by creating and changing the locations of water bodies. Such long-lasting influences of historical processes are known as **legacy effects.**

One of the more interesting legacy effects of glaciers that you can observe today is the presence of eskers, which are the remnants of long, winding streams of water that once flowed inside or under the glaciers. Over time, these glacial streams deposited soil and rock in their paths. Now that the glaciers have melted away, these old stream courses appear as long and winding hills (Figure 25.2). These hills may harbor unique microhabitats that favor particular communities.

Long-ago human activities may continue to influence landscapes as well. During the first century A.D., the Romans built small villages and farms in northern France. For reasons that are unclear, these farms were abandoned by the fourth century, and the land reverted to forest shortly thereafter. Archeological evidence suggests that the land farmed by the Romans was within 200 m of their farm buildings. With this knowledge, ecologists were able to investigate whether that ancient farming continues to affect soils and plants in the region. When Jan Plue and colleagues at Katholieke Universiteit [Catholic University] in Belgium compared farmed and unfarmed sites in 2002, they found that forested land on farmed sites had higher soil pH, more available phosphorous, and greater plant species richness, including many weedy species. These differences were probably caused by the slow breakdown of ancient building materials, which contributed calcium

FIGURE 25.2 A long, winding esker divides cropland in North Dakota. The esker was formed by a stream that ran through the base of a glacier that was once thousands of meters thick. Photo by Tom Bean/Corbis.

and phosphorus to the soil, and by the introduction of numerous plant species by the Romans. In short, human habitation from 1,600 years ago has continued to have strong legacy effects on the modern forest.

Influences in the present

Landscape mosaics continue to be shaped today. Catastrophes such as tornadoes, hurricanes, floods, mudslides, and fires can alter vegetation structure at both local and regional scales, and these changes to plant communities may, in turn, cause changes in other populations and communities that depend on them. Interestingly, the amount of destruction caused by a catastrophic event at different scales is influenced by a number of other factors, including weather, regional and local topography, and human land use practices throughout the landscape. For example, wildfires are more intense when fanned by wind and fueled by dead, dry plant litter; they also burn faster going uphill than downhill.

FIGURE 25.3 Fires burned more than a million acres around Yellowstone National Park in 1988. After fires went out, the landscape was composed of a mosaic of burned and unburned areas. Photo by Jonathan Blair/Corbis.

Although catastrophic events have occurred naturally for eons, fire is one event whose frequency and intensity has been influenced by humans. In the area in and around Yellowstone National Park, for example, natural fires were largely suppressed through much of the twentieth century. During the summer of 1988, a drought made the area susceptible to fire. Hundreds of fires were started that summer, ignited by both human activities and natural causes such as lightning strikes. Most fires burned relatively small areas of less than 100 acres, but a few of the fires burned much larger areas. In total, 1.2 million acres burned, and the pattern of burning created a mosaic of burned and unburned patches across the greater Yellowstone landscape (Figure 25.3). The patterns of burning across the landscape depended on where the fires started as well as how characteristics of the landscape (amounts of plant litter, slopes, and local wind intensities) helped to influence their spread.

A number of animals are well known for their ability to alter the landscape mosaic. The American beaver builds dams along streams to make the habitat more suitable for its way of life (see Figure 19.5). The ponds created by the dams not only help the beaver, but also provide aquatic habitat for a range of other species, including fish, amphibians, insects, and waterfowl. Beavers are not the only animals that have such large influences on their habitat. For example, many species rely on holes dug in wet areas by alligators to ensure permanent sources of water (Figure 25.4). Because of their disproportionately large effects on the landscape, such animals are often called **ecosystem engineers.**

Some ecosystem engineers can transform the landscape simply by eating large quantities of plants. Although most herbivores have relatively small effects on the plants they eat, every once in a while an herbivore population explodes and consumes most of the available vegetation. A number of herbivorous insect species, such as the spruce budworm, experience such irruptions. In sequential years of warm and dry spring weather, spruce budworm larvae have high survival rates, and the population increases dramatically each year. The larvae feed on the

FIGURE 25.4 Alligators are ecosystem engineers. The American alligator (*Alligator mississippiensis*) digs deep holes in wetlands to ensure a source of water into the dry season. Michael P. O'Neill/Photo Researchers.

FIGURE 25.5 Logging can produce a mosaic of habitat types. This forest in Olympic National Park, Washington, has been cut in distinct patches. Dan Lamont/Corbis.

needles of a variety of spruce and fir trees, causing part or all of the host tree to die. By creating patches of dead trees, spruce budworm irruptions can have a major effect on the landscape mosaic (not to mention a substantial economic impact on forest-related industries).

Without a doubt, humans are the most impressive ecosystem engineers. Wide-ranging human effects on the landscape include housing developments, the clearing of forests for agriculture, the construction of dams and irrigation channels, the channelization of waterways for improved navigation, and the logging of forests. Logging provides a particularly good example of a human activity that produces a mosaic of habitat types across the landscape. In the western United States, a common practice is to log medium-sized swaths of forest scattered throughout the landscape (Figure 25.5). This practice helps to minimize soil erosion and other damaging effects of large-scale clear-cutting. As you might imagine, the decision to log scattered forest patches quickly produces a mosaic of forest patches of different ages. Logging is just one of the many human activities with legacy effects that may persist for many years into the future.

Landscape mosaics can be quantified using remote sensing, GPS, and GIS

Approaching ecology from a landscape perspective may seem to be a daunting task due to the challenge of quantifying landscape mosaics across large expanses. Fortunately, modern mapping technologies can assist us in this effort. One very useful technology is **remote sensing.** As its name implies, remote sensing is the collection of geographic information from a distance. For our purposes, we can consider remote sensing to be the collection of landscape information based on photographs taken from airplanes or satellites.

Recall from Chapter 3 that solar radiation strikes the earth's surface and is reflected back into space. The reflected radiation spans a range of wavelengths, from ultraviolet to near-infrared. Because different objects on the earth's surface absorb and reflect different wavelengths, each element of the landscape—including forests, fields, and bodies of water—has a unique signature of reflected wavelengths (Figure 25.6). Indeed, even different types of vegetation can have unique wavelength signatures. Researchers use these signatures to identify various elements of the landscape in aerial or satellite images. In this way, they can collect vast amounts of landscape-level data from anywhere in the world, including places that are dangerous or difficult to visit. From these data, they can create digital maps of mountains, waterways, coastlines, and habitat patches across landscapes.

The **Global Positioning System (GPS)** is another technology that has proved useful to ecologists. Originally designed for military operations, it is now available to civilians, more and more of whom are using it in their automobiles and cell phones to help them navigate. Satellites orbiting around the earth send out signals that can be picked up by GPS receivers. When signals from at least four satellites can be detected, the receiver can calculate the latitude, longitude, and altitude of any location on earth. Ecologists can use GPS in many ways; for example, they can map the precise locations of trees across a landscape to within a few meters, or they can follow the long-distance movements of animals carrying radio transmitters.

(a)

(b)

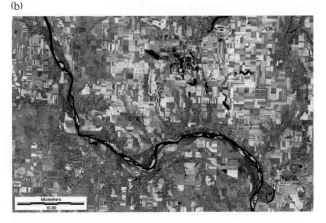

FIGURE 25.6 Landscape information can be collected from airplanes or satellites. (a) This satellite image of the Missouri River in North Dakota was obtained from a satellite that samples reflected wavelengths in the visible spectrum. The image looks very natural, with vegetation appearing green, bare soil appearing tan, and water appearing black. (b) This image of the same landscape was obtained from a satellite that samples reflected wavelengths in the visible and near-infrared spectrum. Here, growing vegetation stands out as bright red, making it much easier to detect. Landsat imagery courtesy of NASA Goddard Space Flight Center and U.S. Geological Survey.

Once ecologists have created a map of the landscape mosaic from remote sensing data and have identified the locations of organisms or habitat features using GPS, they can put all of that information together using a **geographic information system (GIS).** Simply put, GIS is a way to bring together diverse sets of geographic information, including maps of soils, elevations, land uses, water availability, plant distributions, and animal distributions. When all of these data are brought together, ecologists can use GIS computer programs to quantify characteristics of the landscape mosaic and to look for patterns in how organisms are affected by those characteristics.

ECOLOGISTS IN THE FIELD **Quantifying the habitat preferences of butterflies in Switzerland.** Gabriele Cozzi and his colleagues at the University of Zürich exploited the power of modern mapping technologies to assess the importance of local and regional habitat features for populations of endangered fritillary butterflies in the Swiss Alps. They began by entering maps of landscape features (created from remote sensing images) into a GIS program and identifying a number of those features, including the hundreds of wetlands in the region that are favorable butterfly habitat. From these maps, they randomly selected 36 wetlands to survey for the presence of any of three endangered butterfly species (Figure 25.7).

During their visits to each wetland, the researchers recorded its location and altitude using GPS receivers and then surveyed it for butterflies and the plants on which they feed. By combining their survey data with their landscape maps, they could see how altitude (a local habitat variable) and the proportion of wetlands in the landscape at different spatial scales (a regional habitat variable) affected the probability of each butterfly species occurring in the focal wetland that they surveyed (Figure 25.8).

The researchers found that the occurrence of each butterfly species depended on both local and regional habitat variables. At the local scale, the small pearl-bordered fritillary (*Boloria selene*) and Titania's fritillary (*Boloria titania*) were more common at higher altitudes, whereas the lesser marbled fritillary (*Brenthus eno*) was more common at low altitudes. These differences most likely reflect the tolerances of the three species for the colder temperatures at higher altitudes.

When the researchers considered the importance of regional habitat variables, they found that the occurrence of *Boloria selene*, which is capable of long-range dispersal to colonize new sites, was influenced by the proportion of wetland habitat at the largest spatial scale (within a 4,000 m radius of the focal wetland). The occurrence of *Brenthus eno*, which is a medium-range disperser, was influenced by the proportion of wetland habitat at the medium spatial scale (within a 2,000 m radius of the focal wetland). Finally, the occurrence of *Boloria titania*, which is a poor disperser, was influenced by the proportion of wetland habitat at a small spatial scale (within 1,000 m of the focal wetland).

In essence, the researchers found that the more wetlands in the area, the greater the probability that a species would be found in the focal wetland. Indeed, we would expect this pattern from our knowledge of the dynamics of butterfly metapopulations on the Åland Islands of Finland, described in Chapter 12. This study, however, showed that landscape scale made a difference. Only the wetlands within a short distance of the focal wetland

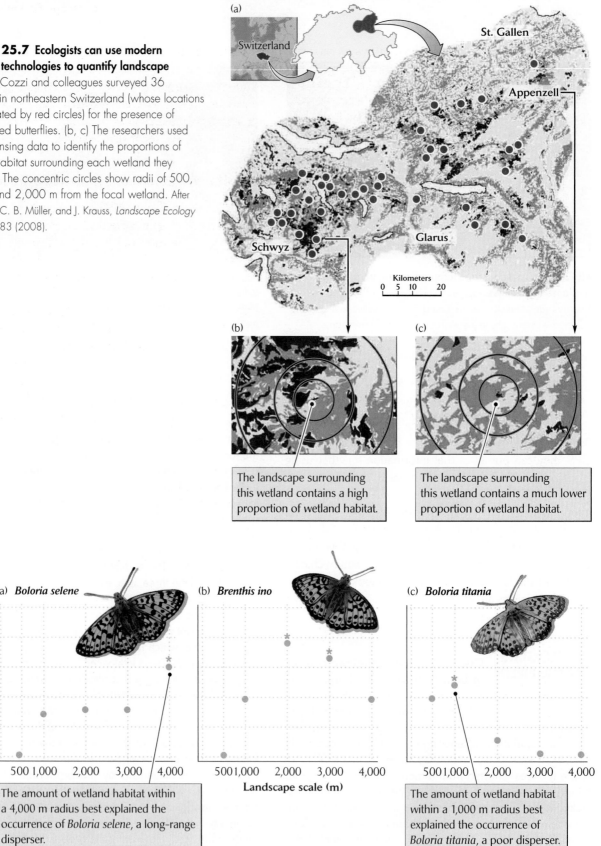

FIGURE 25.7 Ecologists can use modern mapping technologies to quantify landscape features. Cozzi and colleagues surveyed 36 wetlands in northeastern Switzerland (whose locations are indicated by red circles) for the presence of endangered butterflies. (b, c) The researchers used remote sensing data to identify the proportions of wetland habitat surrounding each wetland they surveyed. The concentric circles show radii of 500, 1,000, and 2,000 m from the focal wetland. After G. Cozzi, C. B. Müller, and J. Krauss, *Landscape Ecology* 23:269–283 (2008).

The landscape surrounding this wetland contains a high proportion of wetland habitat.

The landscape surrounding this wetland contains a much lower proportion of wetland habitat.

(a) *Boloria selene*

(b) *Brenthis ino*

(c) *Boloria titania*

The amount of wetland habitat within a 4,000 m radius best explained the occurrence of *Boloria selene*, a long-range disperser.

The amount of wetland habitat within a 1,000 m radius best explained the occurrence of *Boloria titania*, a poor disperser.

FIGURE 25.8 The scale at which the landscape is measured matters. These analyses show how well the amount of wetland habitat at different distances from a focal wetland explains the occurrence of three butterfly species in northeastern Switzerland. Importance values are goodness-of-fit (r^2) values from correlations between the proportion of wetland habitat at different distances from the focal wetland and the occurrence of butterflies in that wetland; the higher the value, the stronger the correlation. Asterisks indicate the strongest correlations. After G. Cozzi, C. B. Müller, and J. Krauss, *Landscape Ecology* 23:269–283 (2008).

mattered to butterflies with limited dispersal abilities, whereas wetlands within a larger radius mattered to butterflies with greater dispersal abilities. Only by taking a landscape approach and using modern mapping technologies were the researchers able to discern the habitat features that are important to the conservation of each of these endangered butterflies. ▐

Habitat fragmentation can affect species abundance and species richness

When human activities or natural events divide a large, contiguous area of habitat into several smaller habitat patches, we say that the habitat has been *fragmented*. We briefly discussed habitat fragmentation and some of its effects on populations in Chapter 10. Habitat fragmentation can occur in both terrestrial and aquatic habitats. The process of habitat fragmentation produces five effects: (1) the total amount of the habitat decreases, (2) the number of habitat patches increases, (3) the amount of edge habitat increases, (4) the average patch size decreases, and

(5) patch isolation increases. Conversely, as fragmentation increases, the habitat matrix between the fragments (for example, cleared fields between forest fragments) experiences the opposite effects: the total amount increases, the number of distinct patches decreases, the average patch size increases, and the matrix becomes more continuous.

Habitat fragmentation and biodiversity

A question commonly posed by ecologists is, how does fragmentation affect biodiversity? Given the five effects of fragmentation, it turns out that the answer is a bit more complex than it might at first appear. The first effect, a reduction in total habitat area, commonly causes a reduction in species richness. This conclusion should make sense in light of our discussion of species–area relationships in Chapter 20, in which we saw that large oceanic islands contain more species than small islands (see Figure 20.3). Similarly, when Scott Findlay and Jeff Houlahan of the University of Ottawa examined the biodiversity of 30 wetlands across a range of sizes, they found that the smallest wetlands had the fewest species of plants, mammals, birds, amphibians, and reptiles (Figure 25.9). Hence,

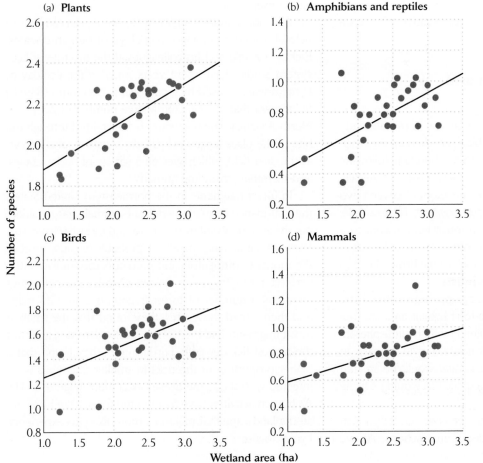

FIGURE 25.9 Species richness is correlated with habitat area. The species richness of (a) plants, (b) amphibians and reptiles, (c) birds, and (d) mammals varies with the size of wetland habitat patches in Ontario, Canada. After C. S. Findlay and J. Houlahan, *Conservation Biology* 11:1000–1009 (1997).

just as we saw in the case of islands, habitat fragments of smaller area generally contain fewer species.

If we were able to examine the effects of fragmentation per se (independent of total habitat size), we could find both positive and negative effects on biodiversity. As we saw in Chapter 12, negative effects can occur when fragments are simply too small to sustain populations and too isolated to receive colonists from other patches. In addition, species living along the edges of habitat fragments may be negatively affected by antagonistic interactions with other species living in the habitat matrix.

In contrast to these negative effects, fragmentation per se can have positive effects when spatial separation of species promotes their coexistence. If prey can disperse to unoccupied patches more readily than their predators, the prey may avoid being driven to extinction throughout the landscape. Similarly, spatial separation may allow two competitor species to coexist in the landscape if one species is a superior competitor but a poor disperser, whereas the other is a poor competitor but can colonize unoccupied patches more rapidly. When all these factors are taken together, it becomes clear that the loss of habitat area associated with fragmentation generally causes a decline in species richness, whereas the increased isolation, smaller patch size, and higher patch numbers that are associated with habitat fragmentation can have both positive and negative effects on species richness.

Habitat fragmentation and species abundances

Although we generally observe a decline in overall biodiversity when a large area of habitat is broken up into fragments, some species become more abundant following habitat fragmentation. The species that benefit are typically those that are specialized to live in ecotones between two habitat types. The brown-headed cowbird, for example, prefers to live in places where a forest lies adjacent to a field. Because such ecotones so often occur at the edges of fragmented habitats, species such as the brown-headed cowbird are termed **edge specialists.**

Consider the effects of splitting up a large area of habitat into smaller fragments. The total habitat area contributed by all the fragments combined might be little altered, but the ratio of edge to interior habitat would increase. In other words, much more edge habitat would be created (Figure 25.10). As a result, edge specialists would be likely to increase in abundance.

Changes in the abundances of edge specialists can have important effects on species interactions. As we

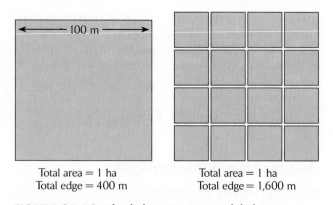

Total area = 1 ha
Total edge = 400 m

Total area = 1 ha
Total edge = 1,600 m

FIGURE 25.10 Edge habitat increases with habitat fragmentation. If 1 hectare of habitat is split into 16 fragments, even with little change in the total habitat area, the ratio of edge to interior habitat increases by a factor of 4.

saw in Chapter 10, increasing forest fragmentation in the eastern and midwestern United States has led to increases in nest parasitism by the brown-headed cowbird, and therefore to decreases in the abundances of other songbirds. Hence, understanding the effects of habitat fragmentation has important implications for species conservation.

An understanding of habitat fragmentation can also be helpful in understanding the ecology of human diseases. Richard Ostfeld and his colleagues at the Institute of Ecosystem Studies have spent years studying the ecology of Lyme disease, caused by a pathogenic bacterium (*Borrelia burgdorferi*) that has infected a large number of people in North America. The bacterium is transmitted through the bite of the black-legged tick (*Ixodes scapularis,* also known as the deer tick), which lives on a variety of birds, reptiles, and mammals, including humans.

In forest fragments in the northeastern United States, the abundance of many vertebrate animals has declined, but the abundance of the white-footed mouse (*Peromyscus leucopus*) has actually increased, probably because most of the mouse's competitors and predators cannot live in the smallest forest fragments. When Ostfeld and colleagues identified a number of forest fragments from GIS maps and then visited those sites, they found that the smallest forest fragments had the highest mouse densities, and thus the highest tick densities. Furthermore, they found that a higher proportion of the ticks in smaller fragments were infected with the Lyme disease bacterium (Figure 25.11). As human activities have fragmented forests, we have created a landscape that makes us more likely to experience Lyme disease.

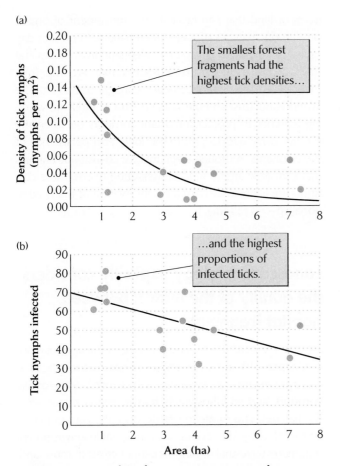

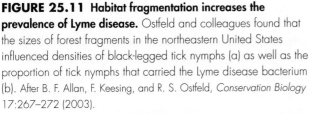

FIGURE 25.11 Habitat fragmentation increases the prevalence of Lyme disease. Ostfeld and colleagues found that the sizes of forest fragments in the northeastern United States influenced densities of black-legged tick nymphs (a) as well as the proportion of tick nymphs that carried the Lyme disease bacterium (b). After B. F. Allan, F. Keesing, and R. S. Ostfeld, *Conservation Biology* 17:267–272 (2003).

Fragment shape and species abundances

Habitat fragments have a variety of sizes and shapes. Fragments of different shapes have different ratios of edge to interior habitat. A circular fragment would contain the minimal amount of edge habitat, whereas a long, slender fragment of the same area would have a much higher proportion of edge (Figure 25.12). How might these differences influence the abundances of species living in fragments?

Ecologists Rick Taylor, Joanne Oldland, and Michael Clarke of LaTrobe University addressed this question in their study of forest fragments in Australia's temperate woodlands. In these woodlands lives a native bird species called the noisy miner. This bird is an aggressive edge specialist that is capable of displacing other small birds. Hence, from a biodiversity conservation point of view, it

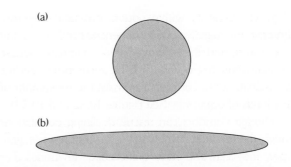

FIGURE 25.12 Fragment shape affects the ratio of edge to interior habitat. (a) A habitat fragment with a circular shape has a minimal amount of edge. (b) A habitat fragment of the same area with a more elliptical shape can have twice the amount of edge.

would be desirable to reduce any habitat that favors the noisy miner. Taylor and his colleagues surveyed 14 woodland fragments in north central Victoria, Australia, looking for patterns that would connect the abundance of noisy miners to forest fragment shapes.

The researchers found that noisy miner densities were highest in "peninsulas" of woodland habitat that projected out into the agricultural matrix. That is, noisy miners strongly prefer woodland habitats that have a high proportion of edge. The researchers suspected that the large amount of cleared land surrounding these woodland peninsulas allows these territorial birds to more easily detect approaching intruders. This habitat preference is of utmost interest to conservation managers because it suggests that by reshaping the woodland fragments to reduce peninsular projections, it should be possible to reduce the abundance of noisy miners and increase the abundances of a variety of other small Australian birds.

◭ DATA ANALYSIS MODULE *Landscape Ecology.* Calculate the future abundance and distribution of northern spotted owls in habitat patches. You will find this module at http://www.whfreeman.com/ricklefs6e.

Habitat corridors and stepping stones can offset the effects of habitat fragmentation

One landscape feature that may lessen the negative effects of habitat fragmentation is habitat **corridors,** which are typically narrow strips of habitat that facilitate the movement of organisms between adjacent habitat fragments. By facilitating movement, corridors increase gene flow and genetic diversity within populations (and therefore counteract the negative effects of genetic bottlenecks and genetic drift), and they permit habitat fragments where local extinction has occurred to be recolonized. As we saw in Chapter 10, corridors can assist the movement of many kinds of organisms (see Figures 10.22 and 10.23).

Although corridors can rescue declining populations through the addition of new colonists bearing new genotypes, they can also have unintended downsides. For example, they can facilitate the movement of predators, competitors, and pathogens among habitat fragments to the detriment of a species of conservation interest. As a result, the costs and benefits of developing corridors among habitat fragments must be examined before resource managers decide to spend their time and limited resources implementing this strategy.

Corridors are probably most important for those organisms that require a continuous connection to move between habitat fragments. However, organisms such as birds and flying insects can pass over stretches of inhospitable habitat matrix and therefore may not need a continuously connected corridor. Other species may be able to move between large patches of favorable habitat if small intervening patches are present where they can stop over to rest or forage. These small intervening patches can be considered **stepping stones** within the matrix.

The importance of stepping stones to dispersal between patches was investigated by Joern Fischer and David Lindenmayer of the Australian National University. In the state of New South Wales, in southern Australia, the landscape was historically composed of temperate woodlands. Today, however, aerial photographs show that these woodlands exist only as fragments. Some of these fragments are very small—even as small as a single tree. Fischer and Lindenmayer hypothesized that these very small woodland fragments might serve as stepping stones for birds traveling between larger fragments. If so, the researchers reasoned, then dispersing birds that fly into a small fragment should arrive from one direction, stop briefly at the fragment to feed and rest, and then continue on their trip in the same direction. By observing the directions of travel of 87 groups of birds that arrived at and departed from small woodland fragments, Fischer and Lindenmayer found that arriving birds either returned to where they had come from or continued on a relatively straight path. These observations suggest that the birds were indeed using the small fragments as stepping stones to move between large woodland fragments.

An appreciation of the role of habitat corridors and stepping stones has spurred major efforts to preserve tracts of land that can facilitate the movement of organisms among patches of fragmented habitat. In present-day India, for example, the Asian elephant lives in several national parks and protected areas that are the fragmented remains of a once larger, contiguous habitat. The World Land Trust and the Wildlife Trust of India are working together to protect important corridors between the protected habitats to ensure the long-term persistence of the elephants. Although elephants are charismatic animals that can draw attention to conservation needs, these corridors are likely to assist in the conservation of many additional species in India.

Landscape ecology explicitly considers the quality of the matrix between habitat fragments

As you will recall from our discussions of metapopulations in Chapters 10 and 12, a landscape can be thought of as a conglomeration of favorable habitat patches, favorable corridors, and an inhospitable habitat matrix surrounding the patches. This model might describe the case of islands surrounded by the ocean, an inhospitable environment that many terrestrial organisms cannot cross. In most landscapes, however, the matrix is composed of habitats that span a range of quality, such that an individual may be able to traverse or even live in some of those less favorable habitats. Indeed, a hallmark of landscape ecology is its emphasis on the matrix as a wonderfully complex mixture of habitats that influence the movement of organisms and materials across the landscape.

Habitat matrix quality and movement between patches

When we think about the matrix between habitat patches being composed of a variety of habitats possessing different characteristics, we arrive at a number of interesting insights. For example, a matrix that contains habitats that are somewhat favorable to a species will promote the species' movement between highly favorable habitat patches. In essence, the entire matrix might serve as a variety of corridors that differ widely in their ability to facilitate movement of organisms among habitat patches. The quality and spatial arrangement of the different habitats in the matrix has thus been termed the **landscape context.**

An example of the importance of the landscape context comes from the investigations of Taylor Ricketts of

Stanford University. In a valley in Colorado, he mapped the locations of meadows, willow thickets, and coniferous forests and then studied the movements of more than 6,000 individual butterflies from 21 species. The butterflies, which feed in the meadow habitat, move between meadow patches by flying through a mosaic of willow thickets and coniferous forests. Ricketts captured butterflies in meadow patches, wrote a unique number on each one with a felt pen, and released it. He then recaptured the butterflies to determine whether they exhibited a preference for dispersing through the willow thickets or the coniferous forests. In four of the six taxonomic groups of butterflies he examined, individuals were 3 to 12 times more likely to move between meadows via willow thickets than via coniferous forests. This study provides strong evidence that the habitats composing the matrix have a substantial effect on the movements of organisms among patches.

The quality of the matrix surrounding a habitat patch can also influence the probability of organisms moving out into the matrix. In fact, the habitat adjoining a patch may be the most important determinant of whether an organism disperses to colonize another patch. Consider, for example, a frog that lives in a patch of mature, moist forest that is favorable to animals such as frogs that are susceptible to dehydration. If the surrounding matrix is a regenerating forest that has moderate amounts of humidity, the frog may easily cross through the matrix. However, if the patch is surrounded by a hot, dry urban area composed of asphalt and concrete, the frog is unlikely even to enter the matrix. Hence, when considering the dispersal of organisms between patches, we need to consider not only the landscape context, but also the **edge context.**

Connectivity and species turnover over time

If the habitat matrix is hospitable enough to allow organisms to move among habitat patches, then the proximity and abundance of habitat patches should influence the proportion of patches that are occupied and the number of species that are present in patches. These landscape characteristics should also influence changes in species composition in a particular patch from year to year. Recall from Chapter 20 that the species composition of a local community is determined by extinctions within the community and colonization from outside the community. Although local conditions determine the probability of extinctions at the local site, the probability of coloniza-

tion depends on characteristics at the regional (landscape) scale. Thus, local conditions and regional conditions should combine to determine the turnover of species in a particular community over time.

Earl Werner and his colleagues, amphibian ecologists at the University of Michigan, tested this prediction. They surveyed 37 wetlands in southeastern Michigan over a 7-year period and recorded the presence of 14 species of larval amphibians. Over the 7-year period, the average wetland was occupied by a cumulative total of 6 of these species, but in any single year, the average wetland was occupied by only 3 species. This finding suggests that the particular assemblage of species occupying a given wetland changes considerably from year to year. How might local and regional factors play a role in these highly dynamic communities?

Werner and colleagues found that the most important local factor determining local extinction was wetland size. Smaller wetlands support smaller amphibian populations, which are inherently more vulnerable to extinction, and smaller wetlands are more likely to dry up before larval amphibians can metamorphose. Following a local extinction, the probability of a species recolonizing a wetland should depend on the number of other wetland habitat patches in the landscape, their distances from the focal wetland, and the number of potential colonists produced by each wetland patch (the researchers assumed that all terrestrial habitats between the wetlands were capable of being traversed). The researchers determined numbers of wetland patches and their distances from each focal patch from aerial photographs, and they determined the number of potential colonists from their survey of the amphibians living in the focal patches. They combined these three factors mathematically into an "index of connectivity" that represents the abundance of potential colonists that could arrive at a given patch. The researchers found that annual species turnover in a focal wetland was positively correlated with the wetland's index of connectivity (Figure 25.13a). They also found that the chance of a species occurring in a focal wetland from year to year was positively correlated with the regional population size for that species (Figure 25.13b).

Thus, while small habitat patches may suffer many local extinctions, those patches can be easily recolonized if they are well connected to nearby populations. This study also shows that it is only by considering local and regional factors simultaneously that we can understand how the species composition of a community changes over time.

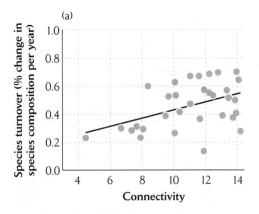

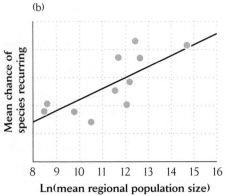

FIGURE 25.13 Habitat patches with greater connectivity have higher species turnover over time. (a) The annual turnover of larval amphibian species in a wetland habitat patch was positively correlated with an "index of connectivity" representing the abundance of potential colonists that could arrive at that patch. (b) The chance of any given species occurring in a wetland patch from year to year was positively correlated with the regional population size for that species, which corresponds to the number of potential colonists. After E. E. Werner et al., *Oikos* 1116:1713–1725 (2007).

Different species perceive the landscape at different scales

It is easy to picture landscape mosaics over very large geographic regions that include a variety of terrestrial and aquatic habitats. While this is certainly a helpful starting point, landscape mosaics can also occur at much smaller spatial scales. For example, the mosaic perceived by a blue whale cruising through thousands of square kilometers of ocean is quite different from the mosaic that a butterfly might experience. Given this difference in scale perception among species, ecologists must consider the scale at which they measure habitat variation in a landscape.

Ecologists have identified two important elements of landscape scale. The first is the degree of resolution at which one views the landscape (termed **grain**). This element is analogous to the resolution of a photograph. If organisms respond to fine-grained habitat variation, then ecologists must measure fine-grained habitat variation, or they will miss the detail that is important to the organism. The second element is the size of the landscape of interest (termed **extent**). If organisms travel over a wide area, then ecologists must examine the landscape mosaic across the entire extent of the population's movements. For example, individual woodland jumping mice use less than 0.004 km^2 of land, so we would need to study fine-grained features of their habitat, including the location of every fallen log, underground burrow, and small grassy area, to determine how the landscape mosaic affects the movements of these mice, but our study could have a relatively small extent. Migrating wildebeests, on the other hand, cover more than 30,000 km^2 each year. In this case, we would have to examine the landscape at a much larger extent, but at a much coarser grain, including the locations of large rivers and of expansive grasslands that have received sufficient seasonal rains to facilitate grass growth. That is not to say that a fine-grained study of the wildebeests' habitat would not be useful, but limited time, money, and personnel means that researchers must arrive at a compromise between grain and extent.

The importance of examining how organisms respond to landscapes at different scales is nicely illustrated in a study of bees by Ingolf Steffan-Dewenter and colleagues at the University of Göttingen in Germany. Using methods similar to that used by Cozzi and colleagues in their study of butterflies in the Swiss Alps, they identified fifteen study sites in Germany that differed in their proportions of cropland, forest, and seminatural habitats such as grasslands and fallow fields (Figure 25.14). Using aerial photographs and GIS, they quantified the proportion of seminatural habitat within different distances from the center of each site. They then placed potted flowering plants at the center of each site and counted the bees that came to pollinate the flowers.

Using these data, the researchers looked at how well the abundances of different types of bees were explained by the proportions of seminatural habitat at various spatial scales. Wild bee abundance was best explained by the proportion of seminatural habitat at small scales (within a radius of 250 m from the potted flowers), whereas bum-

(a)

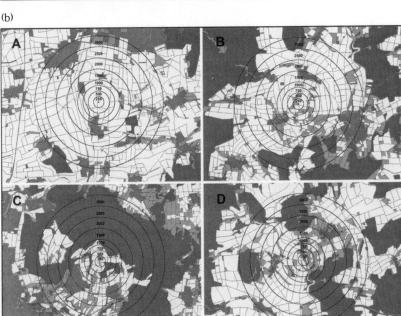

(b)

FIGURE 25.14 Landscape ecologists study how organisms respond to landscapes at different scales. (a) Site map for the study of bee abundance and how it is affected by the proportion of seminatural habitat in the landscape. The circles represent the fifteen study sites. (b) Four study sites (A–D) with varying amounts of seminatural habitat. The concentric circles mark the different spatial scales examined in the study. From I. Steffan-Dewenter et al., *Ecology* 83:1421–1432 (2002).

Legend: Study area 750 m radius; Cropland; Seminatural habitats; Settlement; Woodland/forest; Other land use

blebee abundance was best explained by that at medium scales (750 m), and honeybee abundance by that at large scales (3,000 m) (Figure 25.15). These patterns suggest that the small, solitary wild bees, which fly only short distances and are restricted to natural habitats, are influenced by landscape mosaics at a much smaller spatial scale than are the larger bumblebees and honeybees, which can fly much longer distances to forage and can occupy both natural and agricultural habitats.

These results suggest that if we wish to understand the movements of wild bees, we should examine the landscape mosaic at a smaller extent, and at a finer grain, than we would if we were interested in bumblebees or honeybees. More generally, this study illustrates how examining landscape data at the appropriate grain and extent helps us to understand how landscape mosaics influence organisms and allows us to better manage species that are living in increasingly fragmented landscapes.

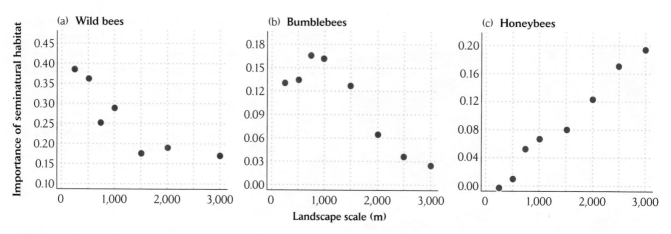

FIGURE 25.15 Abundances of different types of bees are explained by the proportion of seminatural habitat at different scales. For three species of pollinating bees coming to visit potted flowers, correlations between bee abundance and the proportion of seminatural habitat in the landscape have their best statistical fits at different distances. The importance values shown here are goodness-of-fit (r^2) values from correlations between the proportion of seminatural habitats and the abundance of bees at each scale; the higher the value, the stronger the correlation. After I. Steffan-Dewenter et al., *Ecology* 83:1421–1432 (2002).

Organisms depend on different landscape scales for different activities and at different life history stages

As we have seen in previous chapters, organisms face a number of challenges to their survival and reproduction. Animals, for example, must make decisions about mating, feeding, and avoiding predators. These different activities can take the animals across vastly different spatial scales. The German honeybees discussed above cover thousands of meters when foraging for nectar and pollen, but perhaps only a fraction of a meter when feeding their nestmates. Similarly, many birds move little while incubating their eggs, but then travel great distances while collecting food for their new hatchlings. Thus, an animal can experience the landscape at a wide range of different scales depending on its daily activities.

For some organisms, the landscape scale that matters can be very different at different life history stages. For animals such as amphibians and insects that spend the early part of their life cycle in aquatic environments, the relevant landscape mosaic at the larval stage is the range of habitats that is present within a pond or wetland. After metamorphosis, the relevant landscape mosaic encompasses a much wider range of terrestrial and aquatic habi-

tats. The situation of plants is similar. The sedentary lifestyle of most plants means that much of their lives are spent experiencing a small-scale local landscape. When plants reproduce, however, the landscape scale may become much larger. The movement of their pollen and seeds over long distances by wind or animal vectors is dependent on weather patterns, topography, and habitat heterogeneity over large scales. These observations suggest that a landscape perspective may provide tremendous insights, even for species that spend much of their lives within a small area.

This chapter has emphasized the numerous insights that can be gained by applying a landscape perspective to ecology. Modern mapping technologies have improved our ability to quantify the habitat heterogeneity that is such a common feature of the landscape. Embracing this complex heterogeneity has proved to be very helpful in understanding the scales at which different species view their world, how fragmenting the landscape affects biodiversity, and how connectivity via corridors or the habitat matrix assists in recolonization after local extinction. The development of landscape ecology as a field of study not only helps us to understand how natural processes operate in shaping nature, but also offers us predictive power in assessing how our activities are likely to affect the earth's biodiversity in the future.

SUMMARY

1. A landscape is a large area containing a mosaic of heterogeneous habitat types.

2. Landscape mosaics are in part the result of historical processes, including both geologic events, and human activities.

3. Landscape mosaics are also shaped by recent events, including catastrophes such as fires, floods, hurricanes, and tornadoes. The patchy nature of their effects across the landscape is, in part, a function of the landscape itself.

4. Landscape mosaics can be influenced by a number of animals that have a disproportionate effect on their habitat. Humans are the most impressive of these ecosystem engineers.

5. Landscape mosaics over large geographic expanses can be quantified using remote sensing, the Global Positioning System (GPS), and geographic information systems (GIS).

6. When large, contiguous areas of habitat are broken up into smaller fragments, the resulting loss of habitat area typically causes a reduction in biodiversity. Other effects of habitat fragmentation, including increased patch isolation, increased numbers of patches, and decreased patch size, can have either positive or negative effects on species richness.

7. For those species that are edge specialists, increased habitat fragmentation and increased ratios of edge to interior habitat can cause an increase in abundance.

8. Habitat corridors and stepping stones connect patches of fragmented habitat and allow a flow of colonists and genotypes among patches.

9. Landscape ecology explicitly considers the quality of the matrix between habitat patches. It is this landscape context that determines how easily species can move between patches.

10. For habitat patches where local extinctions are frequent, higher connectivity increases the chances of colonization and increases species turnover over time.

11. Whereas we often think of landscapes as habitats over very large regional scales, different species view their worlds at very different scales. Moreover, some species experience very local scales during one portion of their life history, but much larger scales at other times.

REVIEW QUESTIONS

1. Compare and contrast GIS and GPS.

2. Why do many natural catastrophes end up causing patchy effects across the landscape?

3. Why are certain animals considered "ecosystem engineers"?

4. Explain how the fragmentation of a landscape can have both positive and negative effects on biodiversity.

5. Evaluate the importance of creating habitat corridors under two alternative scenarios: (1) the matrix between fragments is inhospitable to a species; (2) the matrix is not the most favorable habitat for a species, but is not uninhabitable.

6. If the habitat loss that results from fragmentation reduces biodiversity, how can some species actually increase in abundance?

7. How can fragment shape affect the abundance of a species?

8. Why do different species perceive the landscape mosaic at different spatial scales?

SUGGESTED READINGS

Allan, B. F., F. Keesing, and R. S. Ostfeld. 2003. Effect of forest fragmentation on Lyme disease risk. *Conservation Biology* 17:267–272.

Cozzi, G., C. B. Müller, and J. Krauss. 2008. How do habitat management and landscape structure at different spatial scales affect fritillary butterfly distribution on fragmented wetlands? *Landscape Ecology* 23:269–283.

Fahrig, L. 2003. Effects of habitat fragmentation on biodiversity. *Annual Review of Ecology and Systematics* 34:487–515.

Findlay, C. S., and J. Houlahan. 1997. Anthropogenic correlates of species richness in southeastern Ontario wetlands. *Conservation Biology* 11:1000–1009.

Fischer, J., and D. B. Lindenmayer. 2002. The conservation value of paddock trees for birds in a variegated landscape in southern

New South Wales. 2. Paddock trees as stepping stones. *Biodiversity and Conservation* 11:833–849.

Murphy, H. T., and J. Lovett-Doust. Context and connectivity in plant metapopulations and landscape mosaics: Does the matrix matter? *Oikos* 105:3–14.

Plue, J., et al. 2008. Persistent changes in forest vegetation and seed bank 1,600 years after human occupation. *Landscape Ecology* 23:673–688.

Ricketts, T. 2001. The matrix matters: Effective isolation in fragmented landscapes. *American Naturalist* 158:87–99.

Steffan-Dewenter, I., et al. 2002. Scale-dependent effects of landscape context on three pollinator guilds. *Ecology* 83:1421–1432.

Sullivan, S. M. P., M. C. Watzin, and W. S. Keeton. 2007. A riverscape perspective on habitat associations among riverine assemblages in the Lake Champlain Basin, USA. *Landscape Ecology* 22:1169–1186.

Taylor, R. S., J. M. Oldland, and M. F. Clarke. 2008. Edge geometry influences patch-level habitat use by an edge specialist in southeastern Australia. *Landscape Ecology* 23:377–389.

Tewksbury, J. J., et al. 2002. Corridors affect plants, animals, and their interactions in fragmented landscapes. *Proceedings of the National Academy of Sciences USA* 99:12923–12926.

Turner, M. G. 2005. Landscape ecology: What is the state of the science? *Annual Review of Ecology, Evolution, and Systematics* 36:319–344.

Werner, E. E., et al. 2007. Turnover in an amphibian metacommunity: The role of local and regional factors. *Oikos* 1116:1713–1725.

CHAPTER 26

Biodiversity, Extinction, and Conservation

The human population has an immense impact on the earth. There are so many of us (the 2008 population of 6.6 billion is increasing at a rate of almost 2% per year), and each individual uses so much energy and so many resources, that our activities influence virtually everything in nature. Most of the land surface of the earth and, increasingly, the oceans have come under the direct control of humankind. Virtually all areas within temperate latitudes that are suitable for agriculture have been brought under the plow or fenced. Worldwide, fully 35% of the land area is used for crops or permanent pastures; countless additional hectares are grazed by livestock. Tropical forests are being felled at the alarming rate of 10 million hectares each year. Semiarid subtropical regions, particularly in sub-Saharan Africa, have been turned into deserts by overgrazing and collection of firewood. Rivers and lakes are badly contaminated in many parts of the world. Gases from chemical industries and the burning of fossil fuels pollute our atmosphere.

We are fouling our nest, and we are still rushing to exploit much of what remains to be taken. If unchecked, this deterioration of the environment will lead to a declining quality of life for all human inhabitants of the earth, as it already has for many. Two-and-a-half billion people, mostly in developing countries, live on less than two dollars a day. The animals and plants with which we share this planet, and on which we depend for all kinds of sustenance, are feeling the impact of the human population even more. They have been pushed aside as we have taken over land and water for our own living space and for the production of our food. We have spoiled their environments with our wastes. Many species have succumbed to habitat destruction, hunting, and other human activities.

This deterioration need not continue. Although many claim that our population exceeds a sustainable level, humans can live in a cleaner and more life-supporting world. This can happen, however, only if we place support for our own population in balance with support for other species and the ecological processes that nurture us. Legislation in many countries has already led to cleaner air and water, more efficient use of energy and material resources, and the rescue of endangered species from further decline. Although the human population will continue to stress natural ecosystems and species survival in the future, there is much we can do to ameliorate the condition of the biosphere and its nonhuman inhabitants.

CHAPTER CONCEPTS

- Biological diversity has many components
- The value of biodiversity arises from social, economic, and ecological considerations
- Extinction is natural but its present rate is not
- Human activities have accelerated the rate of extinction
- Reserve designs for individual species must guarantee a self-sustaining population
- Some critically endangered species have been rescued from the brink of extinction

Although our knowledge is far from perfect, the science of ecology has much to say about rational development and management of the natural world as a sustainable, self-replenishing system. What we have learned about the adaptations of organisms, the dynamics of populations, and the processes that occur in ecosystems suggests simple guidelines for living in reasonable harmony with the natural world.

First, environmental problems will not be brought under control as long as the human population continues to increase. The earth might support more people than it does at present, but their quality of life would be drastically reduced in the short term, and there would be little prospect for sustainability in the long term. Even our present-day human population cannot maintain itself on a sustainable basis. Reforestation cannot keep pace with growing demands for timber, paper, and fuelwood, and so vast amounts of previously uncut forest are being harvested each year. Most of the important fisheries of the Northern Hemisphere have collapsed and now yield only a fraction of their previous production. Large areas of deteriorated farmland are lost to agriculture every year. Fresh water is in critically short supply in many parts of the world. As the human population increases, such demands on the environment will only increase.

Natural populations are controlled by density-dependent factors, which include food shortages, disease, predation, and social strife. These factors reduce fecundity, or increase mortality, or both, as populations grow. If the human population were to come under such external controls, the toll in human suffering—disease, famine, warfare—would be enormous. Thus, maintaining individual quality of life at a high level will require, above all, that humans exhibit a reproductive restraint that defies the entire history of evolution, during which "fitness" has been measured in terms of reproductive success rather than quality of life. Only an appreciation of the negative economic and environmental consequences of overpopulation will cause humanity to value individual human experience over numbers of progeny as the two become increasingly incompatible.

Recently, increased education (especially for women), economic opportunity, and urbanization have combined to reduce birth rates and population growth in most regions of the world. Indeed, many European countries now have negative population growth, and family size in Asia and Latin America has declined from an average of six to fewer than three children during the past 50 years. Only in Africa, with an average family size above five and a population growth rate of 2% per year, have these trends lagged. But although human population growth rates are declining globally, human use of resources is not. Most people aspire to a higher material standard of living, and the aspirations of billions of people will place—indeed, have already placed—a tremendous strain on the earth's resources. One consequence of economic globalization is that the high rates of material consumption and demand for raw materials from wealthy countries adds to environmental stresses in developing nations.

Second, our individual consumption of energy, material resources, and food that is produced at higher trophic levels must be reduced. The earth cannot sustain resource and energy use at the level now enjoyed by affluent citizens of developed countries. Energy consumption in the United States in 2003 was, on average, equivalent to almost 8,000 kilograms of oil per person. This level of consumption was twice the level in England and other European countries and almost 30 times the level in the poorest countries, including Congo, Haiti, and Myanmar. Energy use efficiency can be increased and superfluous consumption reduced without impairing comfort or enjoyment of life. Each individual human can reduce her or his impact by eating lower on the food chain (reducing meat consumption, for example), investing in energy- and resource-efficient technologies, driving smaller cars and occupying smaller living areas, and living closer to equilibrium with the physical world (for example, lowering the thermostat setting in winter and raising it in summer).

Third, although it is inevitable that most of the world will come under human management, ecosystems should be maintained as close to their natural state as possible to keep natural ecosystem processes intact and reduce the costs of water, energy, and materials. As a general rule, the less we alter nature, the easier it will be to sustain the environment in a healthy condition. For example, as we saw in Chapter 24, many areas covered by tropical forests are unsuitable for grazing or agriculture because these activities upset natural nutrient regeneration processes and cause soils to deteriorate. Such areas should be left as forest reserves or recreational areas, or used for sustainable exploitation of forest products. Similarly, deserts can be watered, and they often become tremendously productive for certain types of agriculture. But the costs of maintaining such managed systems can become extremely high as soils accumulate salts from irrigation water and aquifers become depleted. Living with nature is always preferable to, and less costly than, working against it.

Human activities also affect the populations of individual species, either directly through hunting, for example, or indirectly through habitat degradation or the introduction of pathogens. Human actions have led to extinction, or imminent threat of extinction, for many species. In this chapter, we shall consider the challenge of conserving species. In Chapter 27, we shall discuss ways of maintaining natural populations and ecosystem processes so that our generation and future generations will benefit from them. The solutions to all of these problems are informed by basic principles of ecology. We must remember, however, that although solutions can be proposed, implementing them will require concerted social, political, and economic action.

Biological diversity has many components

Nearly 1,500,000 species of plants and animals worldwide have been described and given Latin names. Insects account for about half of these. Many more species, particularly in poorly explored regions of the tropics, await scientific discovery. Indeed, new species are continually being described. For example, a 2006 expedition to Indonesian New Guinea discovered twenty new species of frogs, four new butterflies, five new types of palms, and a new species of honeyeater bird, and rediscovered the golden-mantled tree kangaroo, not previously reported from New Guinea and thought to be extinct. Some experts have estimated that the final global species count could soar to between 10 million and 30 million. Such estimates may be inflated, although the diversity of bacteria and other microorganisms is probably both immense and unknowable. But there is no doubt that we share this planet with several million other kinds of organisms.

Making lists of species names is one way of tabulating diversity, but such lists represent only part of the concept of biodiversity, which includes the many unique attributes of all living things. Although each species differs from every other species in the name that science has assigned to it, it also differs in the way its adaptations define its place in the ecosystem. Different species of plants, for example, have dissimilar tolerances for soil conditions and water stress and disparate defenses against herbivores; they also differ in growth form and in strategies for pollination and seed dispersal. Animals, too, have adaptations that define their place in nature. These variations constitute **ecological diversity.**

Biodiversity results from genetic change, or evolution, which underlies the formation of new species. Because genetic variation is crucial to the evolutionary responses of populations to changes in the environment, **genetic diversity,** both between and within species, is another important component of biodiversity. For many types of organisms, particularly bacteria, the sequencing of genetic material found in samples of soil and water provides the only glimpse of their vast diversity.

All species are related by evolutionary descent from common ancestors, some recent and some in the distant past. **Phylogenetic diversity** takes into account the degree of relationship among organisms, giving greater weight to distantly related forms than to close relatives. Thus, five species of rodents represent less phylogenetic diversity than a mouse, a bat, a deer, a coyote, and a monkey. Ecological diversity and phylogenetic diversity

(a)

(b)

FIGURE 26.1 Many oceanic islands harbor endemic species. (a) The Hawaiian silversword is found only at high elevations on Haleakala Volcano on the island of Maui, Hawaii.

(b) This tortoise is endemic to the Galápagos archipelago, where each island has a distinctive form. Photo (a) by James L. Amos/Peter Arnold; photo (b) by R. E. Ricklefs.

are closely related, but evolutionary convergence and the diversification of species descended from a recent common ancestor (**adaptive radiation**) add complexity to these concepts.

Finally, biodiversity has a geographic component. Different regions have different numbers of species. If diversity were a contest, tropical rain forests and coral reefs would be the clear winners. Equally important, however, is the fact that most regions harbor unique species found nowhere else. Species whose distributions are limited to small areas are called **endemic** species, and regions with large numbers of endemic species are said to possess a high level of **endemism.** Clearly, conservation of global biodiversity is best served by directing efforts toward areas of high endemism as well as high diversity.

Oceanic islands are well known for harboring endemic forms; virtually all the birds, plants, and insects of such isolated islands as the Hawaiian and Galápagos archipelagoes occur nowhere else (Figure 26.1). As a result, when habitat destruction, hunting, or the introduction of alien species results in a loss of local populations in such places, that loss is likely to signify global extinction. Fossil-bearing deposits have shown that more than half of the birds of the Hawaiian Islands have disappeared since human colonization of the islands. Those birds occurred nowhere else; now they are gone forever. So is the dodo, a giant flightless pigeon known only from the island of Mauritius in the Indian Ocean, extinct since the mid-1600s. Steller's sea cow (a giant relative of dugongs and manatees), which was endemic to the Bering Sea, became extinct by

1768, less than 30 years after it was first discovered (and hunted) by Europeans.

ECOLOGISTS IN THE FIELD **Identifying biodiversity hotspots for conservation.** Some relatively small areas of the world support exceptionally large numbers of species. Areas known to be rich in species of large plants, birds, mammals, and reptiles are also likely to be rich in species belonging to less conspicuous groups. Norman Myers, of Oxford University, and his colleagues have identified 25 biodiversity "hotspots" worldwide, which they have proposed for special conservation consideration (Figure 26.2). The boundaries of hotspots are relatively easy to set for such island areas as the West Indies, Madagascar, and New Caledonia. Within continents, hotspot boundaries usually correspond to the edges of important biomes, such as the dry cerrado vegetation of Brazil and the Mediterranean climatic region of southern Europe and northern Africa. To qualify as a hotspot, a region must have a high level of endemism.

The natural vegetation remaining in all the hotspots identified by Myers occupies only 1.4% of the total land area of the earth, yet these hotspots hold as many as 44% of all plant species and 35% of all species of terrestrial vertebrates. They are also regions of rapid habitat destruction where a high proportion of species are threatened with population declines or extinction. Within these hotspot areas, an average of 88% of the natural vegetation has already disappeared.

Myers emphasized that endemism should be a key criterion used to rank the conservation value of an area.

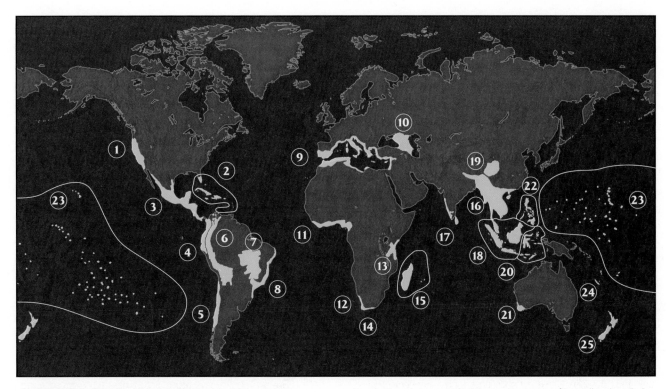

FIGURE 26.2 Twenty-five biodiversity hotspots have been identified worldwide. These areas are receiving special consideration for conservation efforts. Hotspots may include entire regions, such as the West Indies (2) or the Sunda Islands (18), or they may focus on particular environments, such as the South American cerrado (7), a vast region of woodland and savanna, or localized areas of great diversity and endemism, such as the Cape Floristic Province (14) of South Africa. After N. Myers et al., *Nature* 403:853–858 (2000).

In support of this emphasis, an analysis by David Orme, of Imperial College, London, and his colleagues on the distribution of the world's bird species showed that regions of high species richness do not necessarily support the highest numbers of endemic or threatened species (Figure 26.3 on p. 550). For example, the western part of the Amazon basin has the highest species richness of birds (and most other organisms) on earth, but its species tend to be widespread across tropical South America, and relatively few are in danger of extinction. In contrast, the bird species of the Andes and of the Atlantic Forest of Brazil tend to have narrow distributions. In addition to having high proportions of endemics, these areas have been more thoroughly transformed by human activities, whereas large portions of the Amazon basin remain relatively untouched because of their remoteness.

Richard Cincotta and his colleagues at Population Action International have pointed out that Myers's biodiversity hotspots also tend to have above-average human population densities combined with high population growth rates. In 2000, more than 1.1 billion people—nearly 20% of the human population—lived in the 12% of the earth's land area included in the hotspots, and the average growth rate of these populations was 1.8% per year.

Among the most densely populated are southern India and Sri Lanka, the Philippines, and the West Indies. Population growth rates are particularly high in Andean Colombia, Ecuador, and Peru, in Madagascar, and in West Africa. Each of these areas presents special and urgent conservation challenges. The three major tropical wilderness areas—the Amazon basin of South America, the Congo basin of Africa, and New Guinea—have low population densities at present, but high population growth rates, due in large part to immigration. From the standpoint of preserving biodiversity, the hotspots identified by Myers and his colleagues are logical places to focus conservation and management efforts. ∎

The value of biodiversity arises from social, economic, and ecological considerations

The rate of disappearance of certain kinds of species, particularly those most vulnerable to hunting, pollution, and destruction of habitat, is probably now at an all-time high

FIGURE 26.3 Areas of high endemism do not necessarily coincide with areas of high species richness. (a) The global distribution of total bird species richness. (b) The global distribution of bird species threatened with extinction (IUCN categories critically endangered, endangered, and vulnerable). (c) The global distribution of endemic bird species (those occurring in fewer than 30 1° grid squares). The color scale above each map is linear from one species to the maximum shown. After C. D. L. Orme et al., *Nature* 436:1016–1019 (2005).

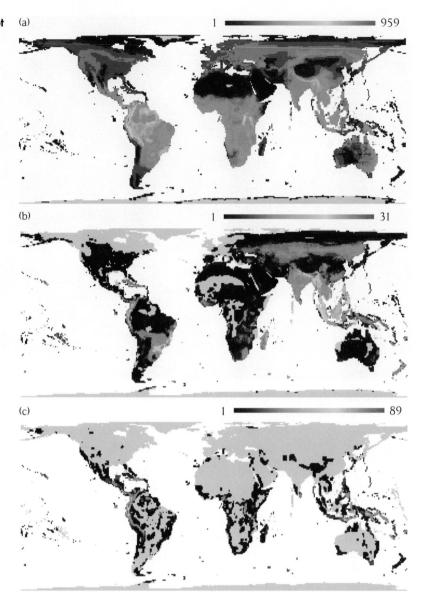

in the history of the earth. Some estimates suggest that more than one species disappears each day, most of them tropical rain forest insects. This accelerated loss of species is directly linked to the growth and technological capacities of the human population.

Why do we care? What concern is it of ours if a species of beetle disappears from a remote corner of South America? Many species are already gone. Do we really miss them? In fact, extinction occurs normally in natural systems. Why should we try to stop it?

Moral responsibility

For many people, extinction raises a moral issue. Some take the position that, because humankind affects all of nature, it is our moral responsibility to protect nature. If

morality derives from a natural law—that is, if morality is intrinsic to life itself—then we may presume that the rights of nonhuman individuals and species are as legitimate as the rights of individuals within human society. Of course, no species is guaranteed a right to perpetual existence, just as no human is guaranteed immortality. But extinction by unrestrained hunting, pollution, habitat destruction, and irresponsible spread of disease is considered by many to be like murder, manslaughter, genocide, and other infringements of individual human rights.

Economic benefits

The value of individual species can be argued from the standpoint of their economic benefits to humankind. Individual species have obvious economic importance as food

FIGURE 26.4 Many species have economic value to humans. Public markets in tropical countries, such as this one in Nairobi, Kenya, offer hundreds of varieties of local plant products, such as fruits, fibers, and medicinals. Many of these products are harvested from natural ecosystems, others from species cultivated locally or throughout the world. Photo by R. E. Ricklefs.

resources, game species, and sources of natural products, drugs, and organic chemicals (Figure 26.4). For example, more than a hundred important medicinal drugs (including codeine, colchicine, digitalin, L-dopa, morphine, quinine, artimesinin, strychnine, and vinblastine) are extracted directly from flowering plants. These drugs account for about one-fourth of all prescriptions filled in the United States.

Some plant and animal species of economic importance have been cultivated or domesticated and then selectively bred to enhance their valuable qualities. These species are not in danger of extinction, but making room for their cultivation on a large scale has often endangered other species that are perceived as having lesser value. An example is the classic conflict between sheep ranchers and wolves, which occasionally kill sheep and other livestock. Because of the economic value of sheep farming, wolves were driven out of most of North America, often with bounties on their heads. Often the result was that herds of deer and other herbivores became so large as to damage the environment, including, ironically, its value for grazing sheep. The point is that assigning economic value to species favors some over others and often does not address the issue of conserving biodiversity in a general sense.

In many systems of accounting, the short-term economic gains of converting natural systems to human uses (such as the conversion of forest to cropland) or of overexploiting natural resources (such as the intensive fishing of Atlantic cod populations) are assumed to outweigh any long-term value of conserving the natural system for sustained economic benefit. The value of conserved species and habitats usually becomes apparent only when the long-term costs of overexploitation or habitat conversion are properly accounted for.

High value may be placed on some individual species because they attract tourists to an area. The practice of visiting an area to see unspoiled habitats and the animals and plants that live in them is referred to as **ecotourism.** Many tropical countries have capitalized on this attraction by establishing parks and support services for tourists. In Latin America, quetzals, macaws, and monkeys draw tourists to areas where those species are protected. Diversity itself is often the attraction in tropical rain forests and coral reefs, with their hundreds of different species of trees, birds, corals, or fish.

In East Africa, lions, elephants, and rhinoceroses have great value because of the tourist dollars, pounds, euros, yuan, and yen they bring into countries that are badly in need of foreign currencies (Figure 26.5). Unfortunately, conflict exists between those few people who poach elephants for their ivory and rhinoceroses for their horn and the many who enjoy watching them. The intensity of poaching was revealed in a study in a national park in Zambia, where the fraction of tuskless female elephants increased from 10% in 1969 to 38% in 1989 as a direct result of selective illegal ivory hunting. Tusklessness in females is a genetic trait, and because poachers kill only individuals with tusks, poaching strongly favors tusklessness in a population. A change in the frequency of a trait from 10% to nearly 40% within one generation suggests strong selection indeed.

Ecotourism has been responsible for the development and maintenance of an increasing number of parks and reserves in many parts of the world. Its impact will expand as more people become aware of the gratification that comes from experiencing nature directly. The capacity of ecotourism to confer value on species is, however, finite. People have limited money to spend, and merely increasing reserve systems will not necessarily generate

FIGURE 26.5 Biodiversity may attract tourists to an area.
The inspiring diversity of wildlife in Africa is as much a part of its attraction as any particular species. Nowhere else can one see Grevy's zebra (*Equus grevyi*, pictured here) or dozens of other large mammals in their natural environments. Many of these species are threatened: the population of Grevy's zebra in Africa declined 70% to fewer than 2,000 individuals between 1977 and 1988, and its recovery is hampered by competition with domestic livestock and overgrazing of pasture land. Photo by R. E. Ricklefs.

more tourism. Furthermore, some areas of immense biological importance, with high diversity and endemism, are not attractive as destinations or are inaccessible to most tourists. Deserts, semiarid regions, many islands, and most marine ecosystems fall into these categories. Moreover, most species simply are not interesting or even perceptible to the general public. Their preservation will depend on their living in association with more highly valued species—so-called **flagship species** such as giant pandas, cheetahs, or leatherback turtles—or habitats.

Indication of environmental quality

Individual species may have considerable value as indicators of broad and far-reaching environmental change. During the 1950s and 1960s, populations of many predatory and fish-eating birds in the United States, particularly the peregrine falcon, bald eagle, osprey, and brown pelican, declined drastically. Several of these species disappeared from large areas, the peregrine falcon from the entire eastern United States. The causes of these population declines were traced to pollution of aquatic habitats by breakdown products (residues) of DDT, a pesticide that was widely used to great immediate benefit to control crop pests and mosquito vectors of malaria after World War II. Worldwide production of the pesticide during the 1970s was about 70,000 tons per year. Unfortunately, the pes-

ticide's residues resisted degradation and entered aquatic food chains, where they accumulated in the fatty tissues of animals and were concentrated with each step in the food chain. The high doses consumed by predatory birds interfered with their physiology and reproduction, making eggshells overly thin and causing the deaths of embryos (Figure 26.6). Breeding success plummeted, and populations followed.

The peregrine population was a sensitive indicator of the general health of the environment. Its demise sounded the alarm to environmentalists; in 1962, Rachel Carson warned of a "silent spring" when no birds would be left to sing. The U.S. Environmental Protection Agency responded by banning DDT and related pesticides in 1972, and chemical companies have since devised alternatives that have less drastic environmental effects. Indeed, DDT had also lost much of its effectiveness in North America and elsewhere, as populations of mosquitoes and other insect pests have evolved resistance to it. Bald eagles and ospreys are becoming familiar sights once again, and, thanks to the helping hands of dedicated biologists, who reared birds obtained from other parts of the geographic range and released them in the eastern United States, peregrine falcons have staged a comeback. This was a major victory, not only for the peregrine and the cause of species conservation, but also for the general quality of our own environment, because the peregrine population could not be saved until DDT, which affects many species, was banned from use. DDT continues to have appropriate local uses—for example, inside houses in tropical countries to control mosquito vectors of malaria—but its use, and that of other pesticides with persistent residues,

FIGURE 26.6 Predatory birds were important bioindicators of the effects of DDT. The shells of these broken eggs in the nest of a brown pelican were thinned by residues of the pesticide DDT. Pelicans feed high on the food chain, on fish in estuaries and other coastal waters. Photo by Betty Anne Schreiber/Animals Animals.

has declined tremendously under tighter regulation and public awareness of its potential dangers.

Maintenance of ecosystem function

Species diversity may have intrinsic value for stabilizing ecosystem function. Several experimental studies suggest that more diverse systems are better able to maintain high productivity in the face of environmental variation. For example, David Tilman and J. A. Downing of the University of Minnesota, using experimental plots of Minnesota prairie containing differing numbers of plant species, demonstrated that severe drought reduced biomass production less on high-diversity plots than on low-diversity plots (Figure 26.7a). Such results can be explained by positing that higher-diversity systems are more likely to include some species that can withstand particular stresses. As the environment changes, different species can take over the roles of predominant producers in an ecosystem. Such switching among species is less likely to occur in less diverse ecosystems. Controlled experiments with artificial communities established in small plots or in laboratory microcosms similarly show that increasing the number of functional guilds among primary producers or increasing the number of links in the food web increases ecosystem productivity and stability in the face of environmental variation (Figure 26.7b).

As we have seen in our discussion of community structure (in Chapter 18), some species, especially top predators, act as "keystone" elements in ecological communities, and their loss can lead to dramatic changes in community structure and ecosystem function. Removal of most of the wolves, bears, and mountain lions in North America, for example, has resulted in an overabundance of deer. In the absence of deer hunting, these herbivores become so numerous that they denude forests of their plant life, putting their own populations in poor condition and having negative effects on other wildlife and natural habitats. One consequence for humans has been an increase in the number of animal-related automobile accidents (over half a million in 2003, resulting in about 100 deaths, 10,000 injuries, and over a billion dollars in vehicle damage) and an increase in the transmission of tick-borne diseases, such as Lyme disease (20,000 cases per year). In extreme cases, removing a keystone predator can lead to the total collapse of an ecosystem.

This general background illustrates why we should value biodiversity and conserve it. But what happens if our conflicting values are not resolved in favor of preserving biodiversity?

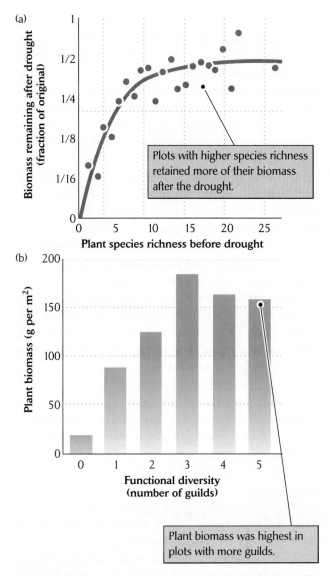

FIGURE 26.7 Diverse ecosystems are more productive and more resistant to perturbation. (a) The effect of plant species richness in prairie plots before a drought on plant biomass after the drought. (b) Plant biomass in artificially seeded experimental plots in relation to the number of plant guilds (early-season annuals, late-season annuals, perennial bunchgrasses, perennial forbs, and nitrogen fixers). After D. Tilman and J. A. Downing, *Nature* 367:363–365 (1994) and D. Tilman et al., *Science* 277:1300–1305 (1997).

Extinction is natural but its present rate is not

Extinction is a major concern of conservationists because it represents the disappearance of evolutionary lineages that can never be recovered. Of course, ecological systems do continue to function after losses of species as ecological feedbacks compensate for changes in production and species interactions. Nonetheless, the present rate of decline and loss of species, whether locally or globally, is a stark indicator that many ecological systems are in fact deteriorating, in ways more fully described in Chapter 27. Humans have already caused the extinction of many species, and the rate of species loss is likely to accelerate. To effectively confront this problem, we need to understand the basic causes of extinction.

It will be useful for us to distinguish three types of extinction. As ecosystems change over thousands and millions of years in response to changes in climate and landscapes, some species disappear and others take their places. This turnover of species, at a relatively low rate, is known as **background extinction.** It appears to be a normal characteristic of natural systems.

Mass extinction refers to the dying off of large numbers of species because of natural catastrophes. Hurricanes, volcanic eruptions, and meteorite impacts happen occasionally, and species that happen to be in their way may disappear. Some of these catastrophes have local consequences; others affect the entire globe.

Anthropogenic extinction—extinction caused by humans—is similar to mass extinction in the number of taxa affected and in its global dimensions and catastrophic nature. Anthropogenic extinction differs from mass extinction, however, in that its causes theoretically are under our control.

Most information on background extinction comes from the fossil record, which reveals appearances and disappearances of species through geologic time. The life spans of species in the fossil record vary according to taxon, but they generally fall within the range of 1 million to 10 million years. The dynamics of speciation and extinction revealed by phylogenetic reconstructions based on DNA sequences of living species generally corroborate these estimates. Thus, on average, the probability that a particular species will go extinct in a single year is in the range of one in a million to one in 10 million. If, as conservative estimates have it, some 1 to 10 million species inhabit the earth, the background extinction rate would amount to about one species extinction per year.

Mass extinctions occupy the other end of the spectrum. Natural catastrophes may cause the disappearance of a substantial proportion of species locally or globally, depending on the severity and geographic extent of the catastrophe. Local catastrophes include prolonged drought, hurricanes of great force, and volcanic eruptions. When Krakatau, a volcanic island in the East Indies, exploded in 1883, not an organism was left alive; any that survived the initial explosion were buried under a thick layer of volca-

nic debris and ash (see Chapter 19). Whether any species disappeared in this catastrophe cannot be known because the biotic diversity of the island had not been well surveyed prior to the explosion, but any species endemic to the island certainly went extinct.

Some mass extinction events apparent in the fossil record are thought to have been caused by the impacts of large comets or asteroids (collectively referred to as *bolides*). Major bolide impacts with global effects have occurred at intervals of 10 million to 100 million years over the history of life. One spectacular example occurred at the end of the Mesozoic era (Cretaceous period). This impact is most famous for causing the extinction of the dinosaurs, but other major groups disappeared as well, including the predatory, nautilus-like mollusks called ammonites (Figure 26.8). The cause of an even greater mass extinction at the end of the Paleozoic era (Permian period), in which perhaps 95% of species and numerous higher taxa disappeared, is less certain. Whatever their exact cause, these extinctions were associated with discrete catastrophic events.

And anthropogenic extinction? Are we to be regarded as a "human bolide" in terms of our impact on biodiversity? Well, not yet. Rates of extinction in many groups (particularly among large animals hunted for food and among island forms) are far above background levels, and

FIGURE 26.8 The earth's second major mass extinction occurred at the end of the Cretaceous period. Dinosaurs and other groups, including ammonites, of which a fossil is pictured above, were extinguished by the global effects of a bolide impact. Photo by Kerry Givens/Bruck Coleman.

many more extinction events have undoubtedly gone unrecorded. Nevertheless, if humankind turns out to be a disaster for global biodiversity, the full force of its impact will come in the future. Most important, such a disaster is preventable. Examining the causes of extinction will enable us to see why this is so.

Human activities have accelerated the rate of extinction

Populations disappear when deaths exceed births over a prolonged period. This much is obvious, but this truth also emphasizes that extinction may result from a variety of mechanisms that influence birth and death processes within a population. It has also been stated that extinction represents failure to adapt to changing conditions, either because the changes occur too rapidly or because a population is unable to respond to them.

The conservation status of species has been evaluated in several ways, most notably in the Red List of the International Union for the Conservation of Nature (IUCN). The IUCN ranking scheme recognizes several levels of extinction risk, ranked from the most imperiled as Critically Endangered, Endangered, and Vulnerable. *Critically endangered* species are those whose numbers have decreased, or are expected to decrease, by 80% within three generations; over 3,000 species worldwide have been placed in this category. *Endangered* species include about 5,000 species at risk of becoming extinct because

their populations are small, their critical habitats are threatened, or they are at imminent risk from predation or disease. At the least critical end of the threatened scale, *vulnerable* species include over 8,000 species likely to become endangered unless the factors threatening their survival and reproduction change.

A survey by Theodore C. Foin, at the University of California at Davis, and his colleagues found a number of causes explaining the population declines of endangered species in the United States. The primary causes were (1) habitat reduction and modification (67% of cases), (2) small population size, (3) overexploitation, and (4) species introductions. Clearly, the conversion of natural habitats for human use and the exploitation of natural resources inadvertently reduce populations of many species to the point at which they can no longer sustain themselves.

Habitat loss and fragmentation

Habitat reduction causes extinction by wiping out suitable places for species to live. Animals of the forest will disappear when all the forest has been cut down. Even where habitat remains, however, conditions within that habitat may deteriorate, causing a population to begin a decline toward extinction.

Reduction of habitat and, especially, fragmentation of habitat into small remnant patches poses a tremendous threat to some kinds of wildlife. For example, the Atlantic coastal forests of Brazil have been reduced to only a small percentage of their former extent (Figure 26.9), critically endangering many endemic birds and mammals, such as

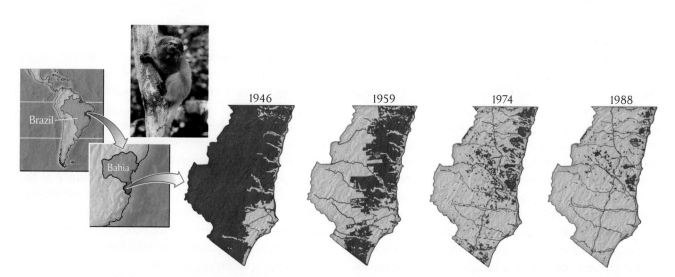

FIGURE 26.9 The Atlantic coastal forests of Brazil have been reduced to a small fraction of their former extent. These maps document the decimation of Atlantic coastal forests in the state of Bahia, Brazil, during the past 60 years. Several endemic species have disappeared from this area, and others, such as the golden lion tamarin, are gravely threatened. Maps by J. R. Mendonça, Projeto Mata Atlântica Nordeste, Convênio CEPLAC/New York Botanical Garden; photo by Tom McHugh/Photo Researchers.

the golden lion tamarin. The remaining fauna is now the focus of intense conservation efforts, but possibly too late for many of the region's endemic species. Even in North America, habitat fragmentation is causing population declines. Tallgrass prairie now exists only in a few isolated refuges, and many populations of prairie plants and animals are locally extinct. Our largest national parks have lost many of their mammal species in the past 50 years, suggesting that these preserves are too small to maintain viable populations.

The effect of reduced habitat area on species richness is a direct extension of the theory of island biogeography. In Chapter 20, we saw that smaller islands support fewer species, in part because smaller populations are at greater risk of extinction due to stochastic processes. In the same way, as the habitat for a species on the mainland is reduced in area, that species' population becomes smaller, and its risk of extinction increases. The dwindling of populations and their eventual extinction takes time, of course, and might not become apparent for many years after a reduction in habitat area.

The disappearance of many native songbird species from small fragments of temperate forest in North America is partly a consequence of small population size and stochastic local extinction. But as we saw in Chapter 25, forest fragmentation has also increased access to forest habitat by predators and nest parasites that are more typical of fields and agricultural lands. Thus, habitat fragmentation also causes a deterioration of habitat quality.

Fragmentation can reduce the dispersal of individuals, turning large populations into metapopulations. As we saw in Chapters 12 and 25, the persistence of a metapopulation depends on the probability of extinction of the individual subpopulations within patches and on the rate of migration of individuals between patches. The smaller the patches, the greater the distance between them, and the more hostile the intervening habitat matrix, the lower the dispersal rate, and the less time the metapopulation can persist.

Over time, changes in climate cause the positions of the major biomes, and of habitats within each biome, to shift, as we saw in Chapter 21. Habitat fragmentation poses a serious threat to the ability of organisms to move with the changing climate. Over the long history of the earth, changes in global climate have been brought about by the drifting of continents and associated changes in oceanic circulation. Where physical barriers to dispersal prevented the distributions of species from following these climatic shifts, local populations became extinct. For example, drastic climate changes during the Ice Age, combined with barriers to dispersal in southern Europe, were responsible for today's impoverished European flora and fauna (Figure 26.10).

The effects of habitat fragmentation may be compounded by the increasing rate of global warming. This anthropogenic change in global average temperatures, which may amount to a rise of 2°C–6°C over the next 50 years, could equal the warming of the earth's climate since the last glaciation, only it is happening 50 times faster (see Chapter 27). It is likely to cause the extinction of many species, particularly plants and sedentary animals with narrow temperature tolerances that cannot shift their range distributions rapidly between habitat fragments.

Small population size

Just by chance, every population experiences variations in birth and death rates. These variations cause what is known as stochastic, or random, variation in population size (see Chapter 12). The magnitude of this variation is inversely related to the number of individuals in a population. Very small populations, such as those in isolated habitat patches, may become extinct just by chance if they suffer a series of very unlucky years. This phenomenon is referred to as stochastic extinction, and although it is relatively unlikely except in the smallest populations, its probability increases with habitat fragmentation. It is a particular threat to species, such as large predators, that typically have low population densities.

Small population size may further increase the probability of extinction by reducing genetic variation in a population (see Chapter 13). A small population contains a smaller proportion of the species' gene pool than a larger population. Furthermore, inbreeding tends to reduce genetic variation. If a population goes through a bottleneck and loses genetic variation, it may not have the capacity to respond to rapid change in the environment. The collared lizard, for example, lives in small populations—generally 20–50 individuals—in isolated glades of xeric habitat on rocky outcrops in the Ozark Mountains of Missouri. This lizard is a resident of southwestern deserts that colonized the Ozarks during a period of hot, dry climate 4,000 to 8,000 years ago. Genetic surveys have shown that the Ozark lizards are genetically uniform within populations, but differ between populations. This is exactly the pattern expected to result from stochastic losses of genetic diversity within small populations and the failure of individuals to disperse between populations.

It is difficult to generalize about problems resulting from population bottlenecks because there are several cases of species that have been reduced to near extinction and lost

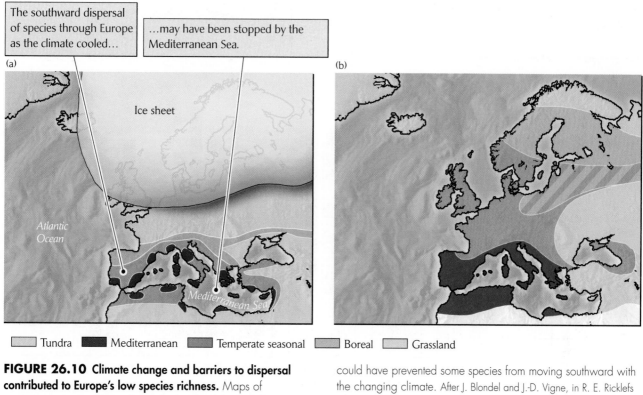

The southward dispersal of species through Europe as the climate cooled...

...may have been stopped by the Mediterranean Sea.

(a)

Ice sheet

Atlantic Ocean

Mediterranean Sea

(b)

☐ Tundra   ■ Mediterranean   ▨ Temperate seasonal   ■ Boreal   ☐ Grassland

FIGURE 26.10 Climate change and barriers to dispersal contributed to Europe's low species richness. Maps of vegetation types in Europe (a) at the height of the last glaciation and (b) today show how the location of the Mediterranean Sea could have prevented some species from moving southward with the changing climate. After J. Blondel and J.-D. Vigne, in R. E. Ricklefs and D. Schluter (eds.), *Species Diversity in Ecological Communities,* University of Chicago Press, Chicago (1993), pp. 135–146.

much of their genetic variation, but have recovered when protected. The northern elephant seal is a case in point (see Figure 11.11). By 1890, hunting had reduced its once numerous population to as few as 100 individuals. Since then, the population has increased explosively to more than 150,000 individuals, distributed throughout much of the species' former range in California and Mexico. This increase represents an annual population growth rate of more than 10%. Several years ago, investigators could not detect any genetic differences between individuals within the species, though they used tests that reveal ample genetic variation in other mammal species. Similarly, one of Africa's large cats, the cheetah (see Figure 13.10), has little detectable genetic variation within its population. The absence of genetic variation might have resulted from a series of population crashes and genetic bottlenecks. Or, the explanation might be found in the metapopulation structure of the cheetah population: cheetahs could have frequently become extinct within isolated habitat patches that were then recolonized by individuals from subpopulations elsewhere. Although cheetahs rarely breed in captivity, the cheetah population appears to be healthy and self-sustaining where it is not hunted by humans. In many other cases, however, populations with reduced genetic variation have experienced serious inbreeding depression, impaired reproduction, and increased mortality.

Overexploitation

Weapons and other harvesting tools, such as kilometer-long drift nets, have made humans such efficient hunters that many species have literally been hunted to extinction. Within recent history, such losses in North America have included the Steller's sea cow, great auk, passenger pigeon, and Labrador duck—all formerly abundant species, all prized for food, all vulnerable, and all hunted until the last individuals were gone. Long-range fishing fleets and improved fishing technology have reduced catches of many fish species and changed the composition of the catch on Georges Bank off New England and maritime Canada, once one of the richest fishing regions in the world (Figure 26.11).

Extinction caused by overhunting and overfishing is not, however, a recent phenomenon. Wherever humans have colonized new regions, some elements of the fauna have suffered. For example, early human populations in the Mediterranean region ate large quantities of tortoises and shellfish, which were easy to catch. As supplies of

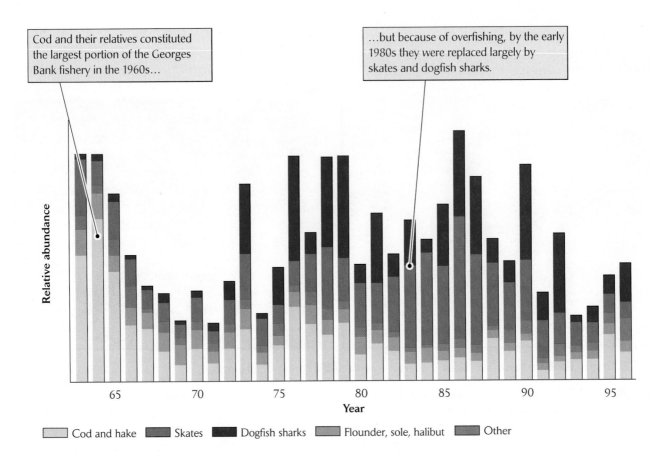

Cod and their relatives constituted the largest portion of the Georges Bank fishery in the 1960s...

...but because of overfishing, by the early 1980s they were replaced largely by skates and dogfish sharks.

☐ Cod and hake   ■ Skates   ■ Dogfish sharks   ☐ Flounder, sole, halibut   ■ Other

FIGURE 26.11 Overexploitation can change the species composition of a community. The composition of the marine community on Georges Bank has been dramatically altered by overfishing. After M. J. Fogarty and S. A. Murawski, *Ecological Applications* 8:S6–S22 (1998).

those foods were depleted, about 30,000 years ago in what is now Italy and about 15,000 years ago in what is now Israel, those human populations were forced to switch to hunting hares, partridges, and other small mammals and birds (Figure 26.12).

Shortly after aboriginal people colonized Australia some 50,000 years ago, several large marsupial mammals, flightless birds, and a tortoise disappeared from the island continent. Madagascar, a large island off the southeastern coast of Africa, received its first human inhabitants only 1,500 years ago, yet their arrival brought the demise of 14 of 24 species of lemurs (mostly large species suitable for food) and between 6 and 12 species of elephant birds, flightless giants found only on Madagascar. A recent analysis suggested that fewer than 1,000 Polynesian colonists of New Zealand hunted 11 species of moas (large flightless birds) to extinction in less than a century at about the same time. Similar extinctions occurred widely on islands in the western Pacific and on the Hawaiian Islands as humans spread throughout the region. In each of these cases, a technologically superior species encountered island popu-

lations unaccustomed to predation pressure. Their lack of defenses and failure even to recognize the danger spelled disaster for these species; lack of restraint on the part of their hunters turned disaster into extinction.

There is no question that human hunters have been responsible for the extinction of many species of large mammals and birds, and have threatened many others. Most of these extinctions occurred years before the development of modern science, so the details have been pieced together from indirect sources, including fossil materials. It is also clear that a more complicated set of circumstances, rather than simple hunting pressure, caused extinctions in some of these cases. For example, after humans arrived in Australia, fires, undoubtedly set by humans, became more prevalent and converted a drought-adapted landscape mosaic of trees, shrubs, and grasslands into fire-adapted desert scrub. Investigators used stable isotopes in carbon-dated eggshell remains to infer the diets of two large flightless birds—the present-day emu and the extinct *Genyornis*. Their results show an abrupt change in the birds' diets at the time humans arrived, particularly a

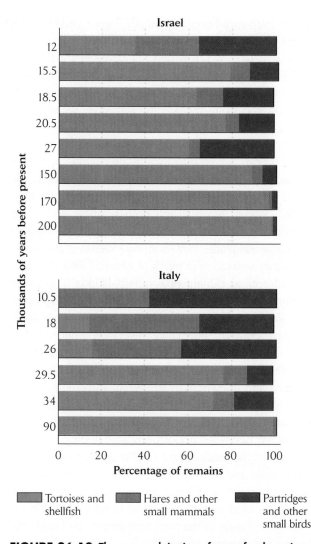

FIGURE 26.12 The overexploitation of some food species has forced people to switch to others. Remains of food items in Paleolithic archeological sites in Israel and Italy document a shift from easily caught prey, such as tortoises and shellfish, to prey requiring more hunting skill as the former prey types were depleted by overexploitation. Data from M. C. Steiner et al., *Science* 283:190–194 (1999).

restriction of the diet to C₃ plants accompanied by a loss of drought-tolerant C₄ plant species (see Chapter 3) from the diet.

The abrupt extinction of many species of large mammals in North America 12,000–13,000 years ago, coinciding with large-scale migrations of humans from Asia, has been a source of controversy for many years. Shortly after the arrival of humans, 56 species in 27 genera of large mammals, including horses, a giant ground sloth, camels, elephants, the saber-toothed tiger, and a lion, disappeared. Explanations for these megafaunal extinctions have ranged from rapid climate change following the retreat of the glaciers, to hunting pressure from the human population,

to epidemic diseases brought with domesticated animals from Asia. Recently, researchers have presented evidence of a bolide exploding in the air over eastern North America about 12,900 years ago. They suggest that the impact could have brought on the Younger Dryas cooling period, resulting in major ecological changes, megafaunal extinctions, and probably a reduction in the size of the human population in the region.

Heike Lotze, at Dalhousie University in Canada, and several colleagues analyzed marine populations in a dozen estuaries and coastal areas in North America, Europe, and Australia. They examined the impacts of several phases of human cultural development on marine populations: prehuman, hunter–gatherer, agricultural, market–colonial establishment, market–colonial development, global market before 1950, and global market at the present. Increasing cultural development, including ship transport, increasingly efficient harvesting technologies, and rising local human population densities, was associated with declines in populations of most species of marine life and other species dependent on marine productivity. It was also associated with declines in habitat quality: nutrient loads increased, accompanied by eutrophication and periods of hypoxia (Figure 26.13). The number of invasive species increased with human cultural development as well. Conservation efforts in these systems have helped to forestall declines in whales, seabirds, and other top predators, but have not significantly altered the deterioration of basic ecosystem structure and function.

Species introductions

Many decreases in habitat quality can be traced to introductions of predators, competitors, or pathogens—that is to say, to biological agents of change. Over the last 200 years, North America has received more than 70 non-native species of fish, 80 species of mollusks, 2,000 species of plants, and 2,000 species of insects. These species may have arrived accidentally—for example, in ship ballast—or may have been deliberately introduced for use as crops, ornamentals, game species, or biological control agents.

One of the surprising results of studies on biological introductions has been that competition from introduced species rarely causes population decline and extinction in native species. For example, the number of non-native species that have successfully colonized and become established on islands is greater than the number of native species that have disappeared from those islands. The critical threats are new predators, competitors, and pathogens, with which the native fauna and flora are often poorly

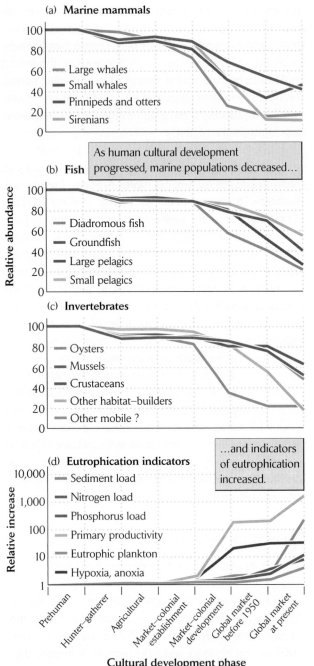

(a) Marine mammals

- Large whales
- Small whales
- Pinnipeds and otters
- Sirenians

As human cultural development progressed, marine populations decreased...

(b) Fish

Realtive abundance

- Diadromous fish
- Groundfish
- Large pelagics
- Small pelagics

(c) Invertebrates

- Oysters
- Mussels
- Crustaceans
- Other habitat–builders
- Other mobile ?

...and indicators of eutrophication increased.

(d) Eutrophication indicators

Relative increase

- Sediment load
- Nitrogen load
- Phosphorus load
- Primary productivity
- Eutrophic plankton
- Hypoxia, anoxia

Cultural development phase:
Prehuman · Hunter–gatherer · Agricultural · Market–colonial establishment · Market–colonial development · Global market before 1950 · Global market at present

FIGURE 26.13 Patterns of decline in marine organisms reflect human cultural development. Changes in populations of (a) marine mammals, (b) fish, and (c) invertebrates, and in (d) indicators of eutrophication, are shown relative to prehuman levels. After H. K. Lotze et al., *Science* 312:1806–1809 (2006).

adapted to cope. The brown tree snake (*Boiga irregularis*), introduced to Guam from Asia, has literally eaten most of that island's endemic land birds to extinction. The Hawaiian Islands have also suffered greatly from species introductions. The native Hawaiian tree snails have fallen prey to introduced predatory snails (Figure 26.14). Major causes of mortality in Hawaiian birds have been malaria and pox virus—which would not have been a problem had the mosquito that transmits these diseases not also been introduced to the islands. Native Hawaiian forests have suffered from invasion by aggressive, weedy plant species, which crowd out native species.

Continental areas have also been vulnerable to introduced species, which may escape the natural controls of predators, parasites, and herbivores that keep their populations in check within their native ranges. Purple loosestrife (*Lythrum salicaria*) and both Japanese and Amur honeysuckle (*Lonicera japonica* and *L. maackii*), introduced as ornamental plants, now dominate wetlands and forest understory vegetation in much of eastern North America. The subtropical climate of southern Florida is ideal for the introduced Australian *Melaleuca* and the Brazilian pepper (*Shinus terebinthifolius*), which cover immense areas and have crowded out native vegetation wherever they have spread. Introduced insects have also naturalized successfully. European honeybees were introduced to North America and elsewhere in the world to pollinate crops. They have been extremely beneficial to agriculture, but have also displaced native bees in many areas. Ironically, naturalized honeybee populations have declined dramatically in recent years owing to the introduced parasitic *Varroa* mite and viruses that it transmits. Other insects, such as fire ants and gypsy moths, have caused drastic changes in local ecosystems, some of which had already been altered early in the twentieth century by Dutch elm disease and chestnut blight.

FIGURE 26.14 Species introductions can have negative effects on native biota. The native tree snails of the Hawaiian Islands are being eliminated by introduced predatory snails like this one. Photo by Bob Gossington/Bruce Coleman.

Aquatic ecosystems seem to be particularly vulnerable to species introductions, perhaps because of the dominance of top-down control by predators and the high rate at which energy and nutrients are processed in aquatic food webs. In the first chapter of this book we saw the devastating effect of the introduced Nile perch on the endemic cichlids of Lake Victoria. As the abundance of cichlids, many of which were zooplankton grazers, decreased, zooplankton abundance increased, and primary production by phytoplankton dropped to a low level, dramatically altering the entire ecosystem of the lake. A large share of the lake's nutrients are now taken up by introduced water hyacinth plants, which form continuous mats over thousands of hectares of the lake surface.

Zebra mussels (*Dreissena polymorpha*) arrived in the North American Great Lakes in 1988 from the Caspian Sea, apparently in ship ballast water. They have now become so abundant in lakes and rivers that they are crowding out native species, severely depleting phytoplankton food sources, and causing billions of dollars in damage to boats, dams, power plants, and water treatment facilities. North America has also supplied its share of invasive species to the rest of the world. The comb jelly (*Mnemiopsis leidyi*) arrived in the Black Sea from North Atlantic coastal waters in 1982. By 1989, this species made up over 95% of the biomass of aquatic organisms in the Black Sea, achieving densities of 400 individuals per cubic meter of water. The species spread to the Caspian Sea in 1999, depleting 75% of the zooplankton there, and has since spread through the Mediterranean to the North Sea and Baltic Sea of northern Europe.

There is no sign that such introductions are slowing down. Sadly, the homogenization of much of the world's biota is occurring at the expense of endemic species.

Emerging diseases

Among the many introduced species are various parasites that cause diseases in their hosts. Such pathogens may be brought to new areas by travelers or their cargo, or may be brought into contact with new potential hosts following habitat fragmentation. In either case, the pathogen may begin to exploit new host species and create an **emerging disease.** The World Health Organization defines an emerging disease as one that has appeared in a population for the first time, or that may have existed previously but is rapidly increasing in incidence or geographic range. Many of the more serious emerging diseases in humans are caused by viruses, including the human immunodeficiency virus (HIV), hantavirus, West Nile virus, and the H5N1 virus (which causes avian flu). Other major human scourges are the protist *Plasmodium falciparum* (which causes malaria) and the bacterium *Yersinia pestis* (which causes bubonic plague).

Animals and plants also suffer from emerging diseases, which can have considerable secondary effects on ecosystems. For example, cattle brought to colonial Africa from Europe and Asia carried a highly infectious viral disease called rinderpest. This disease infected natural populations of several wildlife species, qualifying as an emerging disease for those species. For several decades during the mid-1900s, rinderpest decimated wildebeest populations in the Serengeti ecosystem. No longer controlled by grazers, grasses grew prolifically and, during the dry season, fueled fires that spread across the savannas, killing the scattered trees and altering the structure and functioning of the ecosystem. In the 1960s, widespread use of vaccines controlled rinderpest in cattle, and the disease declined in wildlife populations. The wildebeest population rebounded, largely restoring the Serengeti ecosystem to its earlier state. However, in 1994, a variant of canine distemper virus jumped to the Serengeti lion population, killing a third of the individuals in an epidemic that rapidly spread to other areas. The disease was thought to have originated in domestic dogs living in the many villages surrounding Serengeti National Park and to have spread to lions by way of spotted hyenas, which come into the villages and also feed with lions at kills within the park. The declining populations of these top predators can no longer control herbivore populations to the same degree, so again herbivore populations are on the rise.

Vulnerability to anthropogenic extinction

Why do some species seem more vulnerable to anthropogenic extinction than others? This question has been difficult to answer. Clearly, species that attract the attention of human exploiters are brought under great pressure. In addition, species that have evolved in the absence of hunting (particularly those on remote islands lacking most types of predators) or in the absence of diverse pathogens seem to fare poorly after the arrival of humans. Species with limited geographic ranges, restricted habitat distributions, or small local population sizes are also vulnerable, as one might expect.

But what makes one species rare and locally distributed when a close relative that exhibits superficially similar adaptations is abundant and widely distributed? The difference between success and failure in natural systems may hinge on very small differences in breeding success or longevity—perhaps too small for us to detect in studies of natural populations. Most species persist for a million or more years, so their populations must be fully self-sustaining and capable of recovering from setbacks inflicted by a variable world. The traits that have allowed them to persist over evolutionary time might serve them equally well in our rapidly changing world. The factors that cause a population to embark on a decline to extinction might be very subtle. So far, ecologists have been able to say little on this point. If rare species turned out to be the most specialized ecologically, for example, we would still have to determine what promotes specialization.

By comparing vulnerable species with those that are faring well, ecologists have identified a number of inherent characteristics associated with threatened and declining populations. In an analysis of extinction threats to mammal species, Marcel Cardillo and colleagues at Imperial College, London, determined that small species (less than 3 kg body mass) were vulnerable to anthropogenic extinction primarily because of external factors, including small geographic range size and high human population densities. In contrast, larger species were imperiled primarily because of intrinsic qualities, such as long development periods, low reproductive rates, and low local population densities, that slowed their recovery from population declines. This finding suggests that smaller species would benefit from the general protection of threatened habitats, whereas larger species require closer individual attention to particular factors that influence their survival and reproductive success.

Reserve designs for individual species must guarantee a self-sustaining population

The straightforward way to maintain a population is to guarantee the existence of a sufficient area of suitable habitat that can be kept free of introduced competitors, predators, and pathogens. In practice, the design of nature reserves must take into account the ecological requirements of the species of concern and the amount of space needed to support a **minimum viable population (MVP)**—the smallest population of the species that can sustain itself in the face of environmental variation. In other words, the population must be large enough to remain out of danger of stochastic extinction. The population must also be distributed widely enough that local catastrophes, such as hurricanes and fires, cannot threaten

the entire species. At the same time, some degree of population subdivision may prevent the spread of disease from one part of a population to another.

Guaranteeing suitable habitat becomes more complex when a population has different habitat requirements during different seasons, or when it undertakes large-scale seasonal migrations. In the Serengeti ecosystem of East Africa, patterns of rainfall distribution and plant growth vary seasonally within the region. Huge populations of grazers, such as wildebeests, zebras, and gazelles, undertake long-distance seasonal migrations in search of suitable grazing (see Figure 10.6). It would not be possible to isolate a part of this area as a reserve, because these populations need the entire area of the Serengeti ecosystem at different times of the year. Likewise, thundering herds of buffalo can never be restored to North American prairies, because their seasonal migration routes are now blocked by miles of fencing and agricultural fields. Buffalo survive in a few small reserves in the American West—most notably the Greater Yellowstone Ecosystem—but the natural environment of the buffalo has been irrecoverably lost.

Recognizing the critical importance of large, contiguous areas of habitat for the preservation of wildlife populations and ecosystem functions, a coalition of managers of federal, state, and private lands within an 80,000 km^2 area centered on Yellowstone National Park in Wyoming (Figure 26.15) have developed plans to maintain the region in as close to a natural, self-sustaining condition as possible. The area is called the Greater Yellowstone Ecosystem. The plan includes allowing natural forest fires to burn, as they did in 1988 over half the area of Yellowstone National Park, and restoring populations of top predators, such as the grizzly bear and gray wolf. These top predators will provide natural controls over populations of large grazers, particularly the abundant elk in the ecosystem.

Wolves had been exterminated in the area by 1926, and the explosion of the elk population following its release from predation had considerably altered the vegetation of the area, nearly eliminating aspen and cottonwood groves along rivers. Thirty Canadian wolves were introduced to the Greater Yellowstone Ecosystem in 1995. Their numbers quickly increased, and they immediately started to alter the character of the ecosystem. Populations of elk decreased by half, and the elk shifted their grazing from river courses, where they were vulnerable to predation, to more open, dispersed areas. As a result, aspen and cottonwood trees are now thriving, and beavers have returned to the region. An unexpected consequence of wolf reintroduction was a severe reduction in the population of coyotes, which led to an increase in white-tailed deer in

(a)

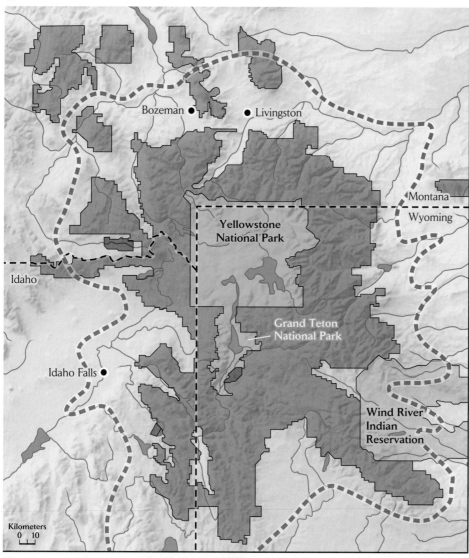

Bozeman ● ● Livingston

Montana
Wyoming

Yellowstone
National Park

Idaho

Grand Teton
National Park

Idaho Falls ●

Wind River
Indian
Reservation

Kilometers
0   10

■ National Wildlife Refuge   □ National Park   ■ National Forest   ▪ ▪ Ecosystem boundary

(b)

FIGURE 26.15 The Greater Yellowstone Ecosystem protects a large, contiguous area of natural habitat. (a) The Greater Yellowstone Ecosystem encompasses national parks, forests, and wildlife refuges, an Indian reservation, and private lands. (b) Although prairies throughout the American West are now fenced and used for crops, this large reserve is one of the few places where buffalo and elk, along with their predators, have access to, and freedom of movement within, large areas of grazing land. Map (a) after Greater Yellowstone Coalition; photo (b) by Fred Bruemmer/DRK Photo.

FIGURE 26.16 Populations that undertake long-distance and seasonal migrations present conservation challenges. *Sandpipers and turnstones feed on eggs laid by horseshoe crabs during May along the shores of Delaware Bay. These eggs are a major food source for migrating shorebirds. This habitat is as necessary for the birds' survival as their breeding and wintering grounds, but horseshoe crab populations are declining owing to environmental pollution and use of horseshoe crab meat as bait in other fisheries.* Photo by John Bova/Photo Researchers.

the region. The abundant carcasses of elk and other prey of the wolves also benefited populations of scavengers such as ravens and golden eagles, which are now common. Overall, the ecosystem is coming into a new, self-sustained natural balance as a result of reintroduction of the wolf.

Long-distance migration poses special problems for the conservation of many types of birds. Some wading birds, such as sandpipers, breed on the Arctic tundra, but maintenance of their populations also depends on conservation of the beaches and estuaries that they use during spring and fall migrations and as wintering grounds (Figure 26.16). Many North American songbirds, whose populations have been declining during recent decades, spend their winters—wisely, it would seem—in forests of Central and South America. Their populations have been placed in double jeopardy by forest fragmentation throughout much of their breeding range in North America and by extensive clearing of forests and spraying of pesticides in Latin America. Birds that migrate between Europe and Africa and between Siberia and Southeast Asia face the same problems.

When threats of extinction come from losses of habitat, conservation strategy is relatively straightforward: the habitat should be preserved. But this may be expensive and politically difficult to achieve. It is also impractical to develop a conservation strategy for every species, and the well-being of the majority will necessarily depend on conservation efforts directed toward a few of the most critically endangered or conspicuous species. As habitat preservation becomes more and more the focus of conservation efforts, it becomes especially important to identify the habitats that are most critical to maintaining

species diversity as a whole and to determine the area of those habitats required to maintain minimum viable populations of most species. Each decision about a species or a habitat will depend on value judgments. What determines which species should be saved? How is their "value" measured?

Critical habitats and geographic areas for conservation

What makes an area critical for conservation? The most valuable areas are those that provide havens for the largest numbers of species not represented elsewhere—such as the biodiversity hotspots described earlier in this chapter. Many types of critical habitats, such as tropical deciduous forests, grasslands, mangrove wetlands, estuaries, and coral reefs, are disappearing at a rapid rate and deserve special attention. Conservation value reflects a combination of local diversity and endemism. As a rule, endemism is highest on oceanic islands, in the tropics, and in mountainous regions. Thus, Madagascar, New Caledonia, and the Hawaiian, Galápagos, and Canary archipelagoes are extremely critical areas for conservation. Extensive surveys of biodiversity in continental regions are beginning to identify critical areas there as well. Planning efforts, however, are continually hampered by lack of detailed information and by conflicting values attached to different components of biodiversity.

The number of reserves we can set aside and the area they can encompass are necessarily limited by economic considerations, so planning for reserves must target habitats and areas of special biological interest. From the standpoint of biodiversity, more is to be gained by setting aside several small reserves spread out over a variety of habitats and areas of high endemism than by preserving an equal area within a single habitat type. Values other than biodiversity may dictate the protection of certain large areas, such as the Greater Yellowstone Ecosystem, and we shall discuss some of them in the next chapter. One thing is certain: the cost of setting aside larger and larger areas of a particular habitat type increases out of proportion to the area itself. This is so simply because land that is least expensive in terms of economic, social, and political values is set aside first. As more land is added to a reserve system, the purchase price invariably increases, as do the potential resources forfeited. It is no accident that most parks and reserves are located in remote, underpopulated areas because establishing a reserve becomes more difficult when it conflicts with economic interests.

One example of such a conflict involved the setting aside in 1968, and expansion in 1978, of a large tract of old-growth redwood forest in northern California as Redwood National Park, which was opposed by the timber industry. In this case, the uniqueness of the redwood habitat and its rapid conversion into managed tree farms greatly increased the value to society of setting aside a large area of this habitat for posterity. Combined with adjacent California state park lands, this reserve covers more than 45,000 ha, including over 15,000 ha of "old-growth" forest. A similar controversy surrounds the old-growth Douglas-fir forests of Washington and Oregon, where the habitat requirements of such unique inhabitants as spotted owls and marbled murrelets have come into conflict with the local timber economy (Figure 26.17).

Many tropical countries, particularly in Central and South America, are in the enviable position of having large tracts of uncut forest and relatively undisturbed tropical habitats of other kinds. These habitats have been protected in the past by their geographic remoteness and by the small size of local human populations. It is still possible to set aside large parks and reserves in such countries as Brazil, Guyana, Ecuador, Peru, and Bolivia, and several governments have moved rapidly during the past decade to preserve tracts of what remains. The problem is complicated, however, by the rapid growth of the human population, by increased exploitation of forest products, and by the conversion of forests to croplands. Such exploitation is justified by a legitimate need to feed people and generate export income for economic development. Thus, the price of conservation is rising rapidly in much of the world, and many developing countries are unable to foot the bill.

Even when lands are set aside "on paper," many countries cannot afford to protect them from squatters, poach-

FIGURE 26.17 Conflicting interests and values present obstacles to conservation. In Washington and Oregon, the habitat requirements of the spotted owl have come into conflict with the interests of the timber industry. Main photo by Tom and Pat Leeson/Photo Researchers; inset photo by Janis Burger/Bruce Coleman.

ers, and politicians who grant mining and logging concessions within protected lands in order to extract short-term profits. For this reason, conservation must be an international effort, and the wealth of the developed countries must be shared globally to protect global biodiversity. One of the most successful recipes for effective protected areas is to involve local people in their design and management so that the benefits of conservation become tangible and economically compelling. A survey of the factors contributing to the success of protected areas of tropical forest in Africa identified a positive public attitude and effective law enforcement, along with large geographic area, low human population density, ecological continuity with other protected areas, and the involvement of nongovernmental conservation organizations (NGOs).

Design of protected areas

In some situations, those who design reserves have some latitude in deciding just how to draw their boundaries. Here, ecologists can provide valuable information by determining the minimum area needed to support a viable population of an endangered species and the critical habitat features that must be present. A helpful tool for planners is **population viability analysis (PVA),** which incorporates demographic information about a particular population into a simulation model to predict the probability that it will avoid extinction within a given period, generally 100–1,000 years. PVA models incorporate information on demographic and environmental stochasticity, including climate change and natural catastrophes, as well as inbreeding depression in small populations. With all of this information in the model, simulations of population change in response to random variation are run many times; the proportion of simulations that avoid population extinction provides an estimate of population viability. A PVA requires considerable data on population changes in the past, as well as many assumptions about population dynamics, so its application is limited to intensively studied species.

Ecological principles derived from metapopulation theory (see Chapter 12) and the theory of island biogeography (see Chapter 20) can also help planners to arrive at the best design. As we saw in Chapter 20, large areas support more species than small areas because large population sizes of individual species reduce the chances of stochastic extinction, promote genetic diversity within populations, and buffer populations against disturbance. When a single type of habitat is preserved, edges should be minimized because the effects of habitat alteration extend for some distance beyond the areas directly altered (see Chapter 10).

According to these considerations, when a reserve is to be carved from an area of uniform habitat, such as a broad expanse of tropical rain forest, (1) larger is better than smaller, (2) one large reserve is better than several smaller reserves that add up to the same total size, (3) corridors connecting isolated reserves are desirable, and (4) circular reserves are better than elongate ones with much more edge. However, when faced with choosing between a single large area of uniform habitat and several smaller areas, each in a different habitat, planners must consider that the smaller areas will often contain a greater total number of species among them because endemic species may be found in one habitat, but not in the others.

Nature reserves must be designed in accordance with the habits of their inhabitants, and requirements for special features (such as nesting sites, water holes, and salt licks) must be taken into account. In mountainous areas, many species undertake altitudinal migrations over the seasonal cycle, so reserves set aside at different elevations must be connected by suitable corridors for travel. Roads and pipelines set in the way of migratory movements or dispersal must be bridged in some manner to allow passage. Similar considerations must go into the design of marine reserves, which have lagged behind those on land. At present, reserves have been set aside and closed to fishing and other exploitation on the Great Barrier Reef of Australia and the Galápagos archipelago of Ecuador, among other areas of particular concern, and there is a growing call for setting aside large areas in international waters of the open ocean to be free of commercial fishing. Clearly, these efforts are making headway in conserving much of the earth's biodiversity. The experience of ecologists, however, including many conservation successes, emphasizes that species preservation will always require considerable research and recovery effort.

Some critically endangered species have been rescued from the brink of extinction

Many species have come so close to extinction that their preservation has required exceptional human intervention. Such efforts, which may cost millions of dollars, are usually directed toward species that appeal to the public. Some may question the wisdom of spending several million dollars (as happened a few years ago) to free three gray whales trapped in Arctic ice. But that incident dramatized the empathy that many people feel for the plight of some other creatures. Even many people who are less than enthusiastic about spending so much to rescue individuals are willing to devote considerable resources to the preservation of species, as shown by the example of the California condor (Figure 26.18).

As the population of California condors in southern California dwindled below 30 individuals in the wild during the 1970s and early 1980s, management personnel made the difficult decision to bring the entire population into captivity. In 1987, the 22 remaining birds were captured and brought to specially constructed breeding facilities located at the Los Angeles and San Diego zoos, where they were protected from the threats that had taken a toll on the wild population: indiscriminate shooting, lead poisoning from slugs left in the deer carcasses on which condors fed, and poisons and traps set out for coyote and rodent control, to which condors were attracted by baits or poisoned carcasses. In captivity, the condors can be induced to lay up to three eggs a year, instead of the usual one, and most of the chicks are reared successfully.

FIGURE 26.18 Some endangered species have been brought back from the brink of extinction. An intensive captive rearing program has begun to re-establish wild populations of the California condor in the western United States. Photo by Tom McHugh/Photo Researchers.

The objective of such a captive rearing program is to produce young that can be reintroduced into their native habitat. Such a program is costly, running into the millions of dollars, and its success ultimately depends on control-ling those mortality factors that threatened the population in the first place, which often requires legislation, land purchases, and public education. In the case of the California condor, the program's success will be properly judged only after 30 or 40 years and the expenditure of tens of millions of dollars. By April of 2008, however, the number of California condors had increased to almost 300 individuals. Of these, 147 are now living in the wild in southern California, Arizona, and Mexico, and the first successful breeding in the wild was reported in 2003.

The California condor can be saved from extinction, as can the Hawaiian crow, the black-footed ferret, and many other species that are now the subject of captive breeding and reintroduction programs. The experience gained through these programs will be useful to similar efforts in the future. The California condor program, like other such programs, has heightened local residents' awareness of conservation issues and has resulted in the preservation of large tracts of habitat in mountainous regions of southern California and elsewhere. Indeed, in May 2008, the Tejon Ranch Corporation agreed to the conservation of 90% of its 270,000-acre (110,000 ha) ranch, one of the largest contiguous areas of natural land in the United States and prime habitat for the California condor. People have also come to understand that as long as care is taken, viable condor populations are compatible with other land uses, such as recreation (as long as human access to nesting sites is restricted), deer hunting (as long as steel rather than lead bullets are used), and ranching (as long as coyote and rodent control programs, if they are to persist at all, are condor-safe). Concessions to condors are neither difficult nor expensive. Making them simply depends on instilling values that acknowledge natural systems as an integral part of the environment of humankind.

SUMMARY

1. Humankind has an immense impact on the earth, managing or otherwise affecting most of its land surface and waters. Human activities have caused deterioration in ecological systems and the extinctions of many species. The repercussions are accelerating as the human population continues to grow and the per capita consumption of energy and resources increases apace.

2. The environmental crisis cannot be fully resolved until human population growth is stopped, consumption of energy and resources declines, and economic development takes ecological values into consideration.

3. Biodiversity encompasses the variety of living organisms on earth. The concept of biodiversity encompasses diversity of ecological roles, genetic diversity within and between populations, phylogenetic diversity, and endemism.

4. The value of individual species is rooted in moral considerations, in the economic benefits we derive from them, and in their role as indicators of environmental deterioration. Species diversity itself may also help to stabilize ecosystem function in the face of environmental variation.

5. Background extinction is the normal turnover of species at a relatively low rate. Mass extinctions, which appear episodically in the fossil record, reflect catastrophic events, including bolide impacts. Anthropogenic extinction is the disappearance of species as a result of human activities.

6. More than 3,000 species are considered critically endangered, meaning that their populations have decreased, or are expected to decrease, by 80% within three generations. Many more species are considered endangered or vulnerable to extinction.

7. Habitat fragmentation may hasten a population's decline toward extinction by reducing its size, by causing deterioration in habitat quality, or by turning large populations into more vulnerable metapopulations.

8. Reductions in size make populations more vulnerable to stochastic extinction, and they reduce genetic variation, thereby impairing the capacity of the population to survive environmental change. The small population sizes of large predators make them especially vulnerable to extinction.

9. Overexploitation by humans has been a major factor in the extinction or decline of many populations, particularly of large mammals and marine organisms.

10. Introductions of non-native species, especially predators, competitors, and pathogens, have led to the decline and extinction of some native populations and to dramatic changes in some ecosystems.

11. Emerging diseases may appear in a population following the introduction of related carrier populations or through habitat changes that bring previously separated populations into proximity. West Nile virus, human immunodeficiency virus, and the avian flu virus are examples of pathogens that have become emergent in the human population in recent years.

12. Although populations of many species are declining, little is known about the factors that make particular populations vulnerable to extinction. Species with small geographic ranges or low population densities are clearly at risk.

13. Increasing attention is being paid to establishing large regional reserves that allow for the migratory movements of large, highly mobile species and the natural regulation of their populations by large predators.

14. Population viability analyses based on simulation models of populations' responses to environmental variation help to predict a species' risk of extinction. Large amounts of demographic data are required to make accurate predictions, however, so this application is limited to well-studied species.

15. Optimally designed protected areas should include a high proportion of endemic species. For a given area of uniform habitat, reserves should consist of a single large area (rather than several small areas) to reduce the chances of stochastic extinction wiping out small populations, and they should have the smallest possible amount of edge. Of major importance, especially in developing countries, are a positive public attitude, effective law enforcement, and participation of nongovernmental conservation organizations.

16. In extreme cases, individual species can be rescued from the brink of extinction by massive recovery efforts that may include captive breeding and reintroduction. Such costly programs, although they are focused on individual species, often highlight more general conservation problems and result in the conservation of large areas of habitat.

REVIEW QUESTIONS

1. Can the global human population continue to grow at the current rate? Explain why or why not.

2. Why should we be concerned not only with preserving species diversity, but also with preserving genetic diversity and phylogenetic diversity?

3. What factors are used to identify biodiversity "hotspots," and why?

4. In what ways can preserving biodiversity provide economic benefits?

5. How does the diversity of an ecosystem affect its ability to withstand environmental variation?

6. Compare and contrast mass extinction and anthropogenic extinction.

7. By what mechanisms can habitat fragmentation lead to the extinction of species?

8. Why are introduced species often a threat to native biodiversity?

9. Why are emerging diseases a concern to species that are not directly affected by them?

SUGGESTED READINGS

Cardillo, M., et al. 2005. Multiple causes of high extinction risk in large mammal species. *Science* 309:1239–1241.

Ceballos, G., and J. H. Brown. 1995. Global patterns of mammalian diversity, endemism, and endangerment. *Conservation Biology* 9:559–568.

Cincotta, R. P., J. Wisnewski, and R. Engelman. 2000. Human population in the biodiversity hotspots. *Nature* 404:990–992.

Daszak, P., A. A. Cunningham, and A. D. Hyatt. 2000. Emerging infectious diseases of wildlife—Threats to biodiversity and human health. *Science* 287:443–449.

Firestone, R. B., et al. 2007. Evidence for an extraterrestrial impact 12,900 years ago that contributed to the megafaunal extinctions and the Younger Dryas cooling. *Proceedings of the National Academy of Sciences USA* 104:16016–16021.

Fogarty, M. J., and S. A. Murawski. 1998. Large-scale disturbance and the structure of marine systems: Fishery impacts on Georges Bank. *Ecological Applications* 8:S6–S22.

Foin, T. C., et al. 1998. Improving recovery planning for threatened and endangered species. *BioScience* 48:177–184.

Hadfield, M. G., S. E. Miller, and A. H. Carwile. 1993. The decimation of endemic Hawaiian tree snails by alien predators. *American Zoologist* 33:610–622.

Haig, S. M., J. R. Belthoff, and D. H. Allen. 1993. Population viability analysis for a small population of red-cockaded woodpeckers and an evaluation of enhancement strategies. *Conservation Biology* 7:289–301.

Harvell, C. D., et al. 2002. Climate warming and disease risks for terrestrial and marine biota. *Science* 296:2158–2162.

Holdaway, R. N., and C. Jacomb. 2000. Rapid extinction of the moas (Aves: Dinornithiformes): Model, test, and implications. *Science* 287:2250–2254.

Kideys, A. E. 2000. Fall and rise of the Black Sea ecosystem. *Science* 297:1482–1484.

Lips, K. R., et al. 2006. Emerging infectious disease and the loss of biodiversity in a Neotropical amphibian community. *Proceedings of the National Academy of Sciences USA* 103:3165–3170.

Marmontel, M., S. R. Humphrey, and T. J. O'Shea. 1997. Population viability of the Florida manatee (*Trichechus manatus latirostrus*), 1976–1991. *Conservation Biology* 11:467–481.

McKinney, M. L. 1997. Extinction vulnerability and selectivity: Combining ecological and paleontological views. *Annual Review of Ecology and Systematics* 28:495–516.

Miller, G. H., et al. 2005. Ecosystem collapse in Pleistocene Australia and a human role in megafaunal extinction. *Science* 309:287–290.

Myers, N., et al. 2000. Biodiversity hotspots for conservation priorities. *Nature* 403:853–858.

Naeem, S. 1998. Species redundancy and ecosystem reliability. *Conservation Biology* 12:39–45.

Naeem, S., D. R. Hahn, and G. Schuurman. 2000. Producer–decomposer co-dependency influences biodiversity effects. *Nature* 403:762–764.

Noon, B. R., and K. S. McKelvey. 1996. Management of the spotted owl: A case history in conservation biology. *Annual Review of Ecology and Systematics* 27:135–162.

Pimentel, D., et al. 2000. Environmental and economic costs of non-indigenous species in the United States. *BioScience* 50:53–65.

Pressey, R. L., et al. 2007. Conservation planning in a changing world. *Trends in Ecology and Evolution* 22:583–592.

Prins, H. H. T., and H. P. Vanderjeugd. 1993. Herbivore population crashes and woodland structure in East Africa. *Journal of Ecology* 81:305–314.

Sax, D. F., S. D. Gaines, and J. H. Brown. 2002. Species invasions exceed extinctions on islands world-wide: A comparative study of plants and birds. *American Naturalist* 160:766–783.

Sharam, G., A. R. E. Sinclair, and R. Turkington. 2006. Establishment of broad-leaved thickets in Serengeti, Tanzania: The influence of fire, browsers, grass competition, and elephants. *Biotropica* 38:599–605.

Struhsaker, T. T., P. J. Struhsaker, and K. S. Siex. 2005. Conserving Africa's rain forests: Problems in protected areas and possible solutions. *Biological Conservation* 123:45–54.

Tilman, D., et al. 1997. The influence of functional diversity and composition on ecosystem processes. Science 277:1300–1302.

Wilcove, D. S., et al. 1998. Quantifying threats to imperiled species in the United States. *BioScience* 48:607–615.

CHAPTER 27

Economic Development and Global Ecology

Humans are major players on the surface of the earth, and the collective activities of nearly 7 billion people have caused profound changes in our environment. To take just one example, the popular press, as well as the scientific literature, has featured much discussion recently of the effects of carbon dioxide emissions from the burning of fossil fuels—coal, oil, and natural gas—on the global climate. Carbon dioxide is a greenhouse gas, and its increase in the atmosphere, by almost 20% in the last 50 years, has contributed to an increase in average global temperatures during the past century. Another source of increased carbon dioxide emissions is forest fires. Devastating fires are increasing in extent and duration due to recent severe droughts in many tropical areas. Some of the worst fires on record have occurred recently in Indonesia, especially on the islands of Sumatra and Borneo, where annual dry season conditions are magnified during years of El Niño events (see Chapter 4). In 1982–1983, more than 3.7 million hectares (37,000 km^2) of rainforest and cropland burned on Borneo. Another 2 million hectares burned during the 1997–1998 El Niño event. The latter fires resulted in over $9.3 billion in economic losses and produced suffocating smoke that blanketed the area for months. These fires released an estimated 0.8 billion–2.6 billion tons of carbon into the atmosphere, an amount corresponding to 13%–40% of all annual global carbon production at that time by the burning of fossil fuels. Little has been done to stem this wasteful loss of resources and the resulting contributions to atmospheric pollution and greenhouse gases. The fires that broke out in 2006 were among the most devastating (Figure 27.1).

Many of the rainforests of this region of Southeast Asia are unusual in that they grow on extensive deposits of nearly pure organic material. These peat deposits can be up to 12 m deep. Fires that begin in these deposits are difficult to extinguish, and widespread cutting of the overlying forest for wood products has dried out the peat, increasing the

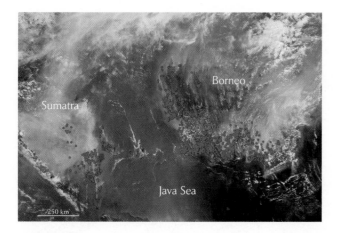

FIGURE 27.1 Smoke from agricultural and forest fires increases atmospheric carbon dioxide concentrations. This satellite image shows fires burning on Sumatra (left) and Borneo (right) in late September and early October 2006. These fires blanketed a wide region with smoke that interrupted air and highway travel and pushed air quality to unhealthy levels. The locations of actively burning fires appear in red. NASA image created by Jesse Allen, Earth Observatory, using data provided courtesy of the MODIS Rapid Response team.

likelihood of fires igniting. Dr. Susan Page, of the Department of Geography, University of Leicester, in the United Kingdom, estimates that these peat deposits are among the largest deposits of organic carbon on earth, rivaling the carbon stored in the boreal forests and tundra of North America and Eurasia. Protecting these forests is important not only to prevent the release of massive amounts of carbon dioxide into the atmosphere, to the detriment of the global climate, but also to conserve the unique biodiversity of the region, which is home, for example, to the largest remaining population of orangutans.

CHAPTER CONCEPTS

- Ecological processes hold the key to environmental policy
- Human activities threaten local ecological processes
- Toxins impose local and global environmental risks
- Atmospheric pollution threatens the environment on a global scale
- Human ecology is the ultimate challenge

We have set aside, or plan to set aside, large areas of the earth's natural environments as reserves to maintain their capacity for supporting species. But what of the other 90% or so that has been, or soon will be, converted to managed systems supporting the human population—devoted to living space, food production, forestry, mineral production, hunting, and recreation? Can these managed ecosystems sustain an expanding human population indefinitely at a high quality of life? Can they serve some of the same functions as natural ecosystems, or will nature reserves stand in stark contrast to completely altered environments dominated by humans and their domesticated species?

A sustainable biosphere is unlikely as long as the human population continues to grow. The earth offers no new regions for humans to colonize. Except for portions of the humid tropics, much of which cannot support dense human populations, most of the habitable areas of the earth have been filled (Figure 27.2). Further population increases will lead to further crowding, tearing not only the fabric of human society but also that of the life-supporting systems of the environment.

Pessimism comes easily in the present environmental climate, but there is also room for optimism. Many programs for cleaning up the environment and protecting endangered species have been undeniable successes. In the United States, the Clean Air Act (1970), the Clean Water Act (1972), and the Endangered Species Act (1973) resulted in considerable protection of the environment. Other countries have adopted similar legislation, and international agreements address such diverse issues as air pollution, carbon dioxide emissions, and trade in endangered species. These successes have not been limited to the developed countries. Relatively straightforward ecological and engineering solutions exist for most environmental problems. Most important, people all over the globe share a deep concern for their environments. The challenge to ecologists is to provide the scientific information needed to develop social consensus, build political commitment, and inform decision making on issues concerning the environment.

FIGURE 27.2 Most of the earth's suitable land area has been converted to croplands, pastures, or rangelands. (a) Potential natural vegetation in the absence of human activity. Areas suitable for tundra, boreal forest, tropical forest, and desert are not suitable for intense exploitation for agriculture, but are sources of wood and minerals, including oil. (b, c) Proportions of land covered by (b) croplands and (c) pastures or rangelands. From J. A. Foley et al., Science 309:570–574 (2005), reprinted with permission from AAAS.

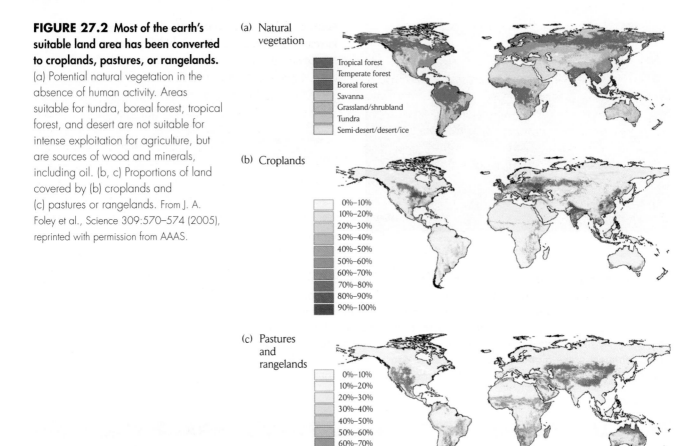

(a) Natural vegetation

Tropical forest
Temperate forest
Boreal forest
Savanna
Grassland/shrubland
Tundra
Semi-desert/desert/ice

(b) Croplands

0%–10%
10%–20%
20%–30%
30%–40%
40%–50%
50%–60%
60%–70%
70%–80%
80%–90%
90%–100%

(c) Pastures and rangelands

0%–10%
10%–20%
20%–30%
30%–40%
40%–50%
50%–60%
60%–70%
70%–80%
80%–90%
90%–100%

Ecological processes hold the key to environmental policy

Throughout this book, we have discussed the processes involved in biological production, the cycling of elements, and the regulation of communities and ecosystems. In the previous chapter, we saw how an understanding of ecological processes can contribute to the conservation of individual populations and species as well as the preservation of biodiversity more generally. Here we will consider the roles of ecological processes in sustaining the functioning of ecosystems and in meeting the needs of the human population. These processes occur in managed as well as natural ecosystems.

Two key aspects of ecosystem function are the harnessing of energy and the continual recycling of materials. In natural ecosystems, the primary source of energy is sunlight. Recycling is accomplished by a variety of regenerative processes, some of them physical or chemical, some of them biological. Any imbalance that leads to the accumulation or depletion of some component of an ecosystem is usually corrected by the ecosystem's self-maintaining dynamic processes. For example, when dead organic matter accumulates within a system, detritivores tend to increase in numbers and consume the excess detritus. When herbivores increase to high numbers and begin to deplete their food resources, density-dependent factors check population growth and tend to restore a sustainable relationship between consumer and resource.

These restorative processes may be physical, but more often they involve biological transformations. From the composition of the atmosphere to the most basic character of many environments, living organisms have greatly modified the conditions of the earth's ecosystems and are responsible for maintaining their suitability for life. When natural ecological processes are disrupted, ecosystems may not be able to maintain themselves. Nowhere is such an outcome more dramatically shown than in the decrease in soil fertility, increase in runoff, erosion, and silting of streams and rivers, and release of carbon dioxide into the atmosphere following the clearing of tropical rain forests (see Chapter 24). Maintaining a sustainable biosphere requires that we conserve the ecological processes responsible for its productivity.

Human activities threaten local ecological processes

All human activities have consequences for the environment. The goal of fishing, for example, is to harvest a food resource for human consumption. However, when we simply maximize short-term returns from a fishery, preferred fish stocks are reduced, or even collapse, and our attention turns to other less desirable but nonetheless exploitable populations. We can see this pattern in the commercial whale fishery, in which one species after another was hunted to near extinction. Humpback, right, bowhead, and gray whale populations were decimated during the nineteenth century. During the twentieth century, a better-equipped whaling industry turned its attention to the more profitable blue whale until the

commercial catch of that species also declined to uneconomical levels during the middle of the century. Whalers moved on to fin whales, whose catch fell off precipitately between 1965 and 1975, and then to less and less profitable species, such as sperm and, finally, sei whales (Figure 27.3). Under current laws, including a total moratorium on commercial whaling since 1985, some populations of whales, such as the gray, bowhead, and humpback, are now increasing at rates of about 2% per year.

Other fisheries, such as the once immensely profitable sardine (*Sardinops*) fishery of western North America, have not recovered from overexploitation. It is possible that sardine stocks were pushed below levels from which they could recover, and that the entire ecosystem has shifted so that the sardine is no longer a prominent link in the food chain. Interestingly, following the collapse of the

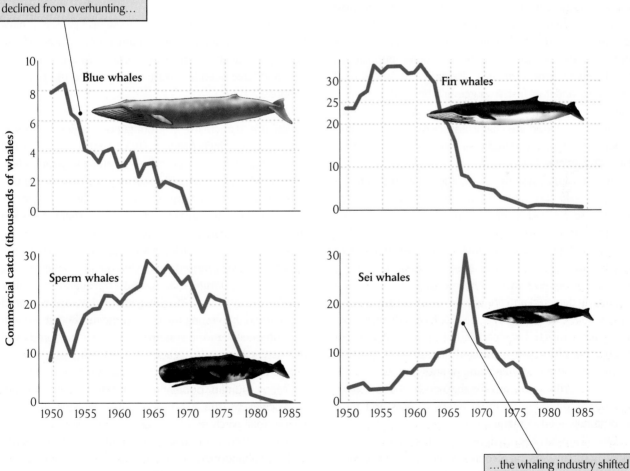

FIGURE 27.3 The whaling industry shifted to new, less profitable species as populations of heavily hunted species dwindled. Commercial catches of four types of whales during the middle of the twentieth century illustrate this shift. After R. Payne, in W. Jackson (ed.), *Man and the Environment*, 2nd ed., W. C. Brown, Dubuque, Iowa (1973), p. 143.

sardine fishery in California in the early 1950s, anchovies (*Engraulis*) increased and are now the basis of an active fishery.

The consequences of human activities are often less direct, however. To provide a relatively straightforward example, the clearing of watershed land for agriculture or timber frequently leads to erosion and deposition of silt downstream from the watershed over long periods. Thus, riverine habitats may be altered and reservoirs behind dams filled in with sediment. Erosion of logged land in Queensland, Australia, is causing damage to streams and to the Great Barrier Reef off the coast. We'll look at some more examples of both direct and indirect effects of human activities as we consider different kinds of threats to local ecological processes.

Overexploitation

Fishing, hunting, grazing, fuelwood gathering, and logging are classic consumer–resource interactions. In most natural ecosystems, such interactions achieve steady states because as a resource becomes scarce, the efficiency of continued exploitation falls off. Consumer populations then begin to decline or seek alternative resources until consumers and the first resource are brought back into balance. Consumer efficiency and the ability of resources to resist exploitation are characteristics of consumers and resources that have evolved over long periods of interaction (see Chapter 17).

In economic systems, consumer–resource interactions may also come into balance because as a resource becomes scarce and its price increases, demand for that resource drops; people either do without or find less expensive alternatives. Nonetheless, because our use of tools has made us such efficient exploiters of natural systems, our resources may not become scarce until they are very nearly depleted and are unable to sustain any level of exploitation. Our technological skills have advanced too rapidly for nature to keep pace. Consequently, many ecosystems that historically supported sparse human populations, such as the vast forests and prairies that were home to native North Americans for thousands of years, have now been converted to more intensive uses and have become degraded in many areas.

Once land has been stripped of nutrients, the human population may lose the resource base it needs to support itself. Where the population can no longer move to other areas or shift to new food sources, the prospect of population control by starvation and its associated diseases, and by social strife, may become reality. Food shortages are already occurring in some parts of the world. In sub-

FIGURE 27.4 In many parts of the world, human land use has left the land unproductive. Overgrazing by goats has removed most of the ground-level vegetation in this woodland in Madagascar. Photo by Walt Anderson/Visuals Unlimited.

Saharan Africa, overgrazing and fuelwood gathering have left little vegetation to support human or any other life. In other parts of the world, vast areas of formerly productive land have been laid waste by local land use practices (Figure 27.4).

Although much of the tropics can sustain intensive agriculture, particularly in mountainous regions with volcanic soils, the high productivity of large portions of the tropics depends on the presence of natural vegetation (see Chapter 24). Because old lowland tropical soils contain few nutrients, the fertility of many tropical ecosystems is maintained by the constant recycling of nutrients between detritus and living plants. Break the cycle by clear-cutting the forest, and the nutrients are lost. Over much of the Amazon basin, forested land cleared for cattle grazing becomes so infertile that it must be abandoned after 3 years of ranching. After vegetation has been cleared and burned, harvesting the cattle removes the last remaining nutrients from the system. To be sure, the forests will regenerate, but many decades or even centuries must pass before the natural fertility of the ecosystem is restored.

Many human populations in the tropics maintain themselves by the practice of "shifting agriculture," in which small patches of forest are cut and the trees burned to release nutrients into the soil, planted for 2 or 3 years, and then abandoned in favor of a new patch (Figure 27.5). A typical patch recovers sufficiently to repeat this process in 50–100 years. Accordingly, as long as only 1%–2% of the forest is cut each year, and thus perhaps 2%–6% is under cultivation at any one time, the land can sustain this practice. This type of agriculture requires little input

FIGURE 27.5 Shifting agriculture can be a sustainable practice. Small subsistence farms, such as this one in Dominica, can provide much of a family's food requirements without overstressing the environment, as long as the human population density is low. Photo by Tom Bean/Corbis.

of labor, materials, or energy and takes advantage of the natural processes of tropical forest development, but it supports only sparse human populations. When the land is cultivated more intensively by tilling, fertilization, watering, and weeding, productivity and long-term sustainability may increase greatly, but so do the inputs of labor and materials required, and so does the cost of the agricultural products.

Many tropical regions lack abundant mineral resources, and their people must rely on agricultural exports to pay for more expensive staple foods and to import the manufactured goods that are considered essential to a high standard of living. Therefore, tropical lands have been cleared to grow cash crops for export to other parts of the world: coffee, sugar, bananas, and beef, to name a few. While this kind of agriculture is sustainable on some soils, for much of the tropics it leads to a decline in the natural productivity of the environment so severe that alternative sustainable land uses may no longer be possible.

There are many ways to reduce these problems. Among the most effective solutions are limiting exploitation of resource populations to the maximum sustainable yield (the number of individuals that can be removed without depressing the resource population's growth rate), implementing alternative sustainable uses of land, increasing agricultural intensity on land that will bear it, and improving the distribution of food from areas of production to areas of need. Most of these solutions carry a price tag. Planning for sustainable use cuts short-term returns, and these losses must be made up elsewhere to maintain the material wealth of a human population. Increasing agricul-

tural intensity requires disproportionately greater inputs of energy, labor, and chemical fertilizers, each contributing its own problems. Relying on crops selected or genetically engineered for high food production can make agriculture more vulnerable to outbreaks of pests and disease.

Providing for larger populations also has human costs. Segments of the human population that are forced by impoverished land to import food must also earn money to buy it; otherwise, they will be reduced to welfare status and impose a burden on other segments of the population. As long as local human population growth does not reflect the abundance of local resources, imbalances between human consumers and their resources will continue to worsen.

Species introductions

Both intentionally and unintentionally, humans have taken other species everywhere they have traveled. Aborigines brought dingoes (semi-domesticated dogs) to Australia; Polynesians brought rats to Hawaii. Of course, the global movements of species by human agency increased immensely after Europeans began colonizing most of the world some 500 years ago. It has been estimated that 50,000 non-native species have been introduced to the United States. These species have included edible and horticultural varieties of plants, and their pests; commercially valuable trees; domesticated animals used for work or meat; familiar backyard animals, especially birds; mammals for sport hunting; human pathogens; and such frequenters of human habitation and transportation as the ubiquitous cockroach and dandelion. The result of this movement of species between continents is a globally distributed flora and fauna of alien species that have in some cases displaced or otherwise altered local biotas. Fully 40% of the species listed under the U.S. Endangered Species Act are threatened because of competition, predation, parasitism, and herbivory by introduced species.

Most of the area of New Zealand now supports a predominantly alien flora and fauna (Figure 27.6). Most of the native forest was cut long ago and replaced by pines from North America and eucalyptus from Australia. Moas (large, flightless birds that grazed vegetation and ate fruit) were killed off by Maori natives before Europeans arrived, and sheep now take their place. Most birds of the countryside are those that were transplanted from England to stave off the homesickness of early colonists. Only at the southern tip of New Zealand do native forests of southern beech (*Nothofagus*) persist in wet and remote mountains. Of New Zealand's 2,500 species of plants, fully 500 are naturalized introductions, and they account for most of

(a)

(b)

FIGURE 27.6 **Few native landscapes remain in New Zealand.** (a) This native forest community is rare in New Zealand today. (b) Most native forests have been replaced by agricultural landscapes featuring plants and animals introduced from Europe. Photos by R. E. Ricklefs.

single keystone consumer, such as the Nile perch, can shift the character of a habitat from one state to a qualitatively different state.

Polynesians arrived on the small (164 km^2) South Pacific island of Rapa Nui (Easter Island) about A.D. 1200. The Polynesian population of Easter Island had increased to perhaps 10,000–15,000 by the 1600s. The island is famous for the gigantic stone heads manufactured by this highly developed culture (Figure 27.7). Before humans arrived, the island supported a lush palm forest. However, by the time the first Europeans reached the island, on Easter Sunday 1722, the forest had disappeared, and the population had declined to only a few thousand

FIGURE 27.7 **Species introductions can cause ecosystems to collapse.** Rapa Nui (Easter Island) was once covered by a lush palm forest, but seed predation by the introduced Polynesian rat, in combination with human overexploitation, severely degraded this ecosystem. Wolfgang Kaehler/Corbis.

the present vegetation. These introduced plants prospered for a variety of reasons. Most of the natural habitat in New Zealand had been greatly disturbed by logging, farming, and ranching, which made invasion by weedy European species, accustomed to disturbance and intensely cultivated landscapes, relatively easy. In addition, because of their comparatively low diversity and simple community structure, island ecosystems tend to be easier to invade than continental ecosystems—there are simply fewer native competitors.

Although introduced species may displace native plants and animals, they do not necessarily disrupt ecosystem function. Introduced species often assume the ecosystem roles of the native species they replace. However, the effects of introduced species are difficult to predict. In aquatic systems in particular, introductions of consumers at higher trophic levels have seriously altered ecosystem function and have caused basic changes in community structure. As we saw in Chapter 26, the introduction of a

FIGURE 27.8 Plowing of prairie land contributed to the creation of the Midwest's Dust Bowl. This farmstead in the midwestern United States was abandoned during the height of the Dust Bowl period in 1937. Courtesy of the U.S. Department of Agriculture, Soil Conservation Service.

individuals. The society and its culture had experienced a catastrophic collapse. UCLA biogeographer Jared Diamond has portrayed Rapa Nui as a dramatic example of the consequences of overexploiting resources in the wake of unchecked population growth. The local population no doubt was responsible for clearing the dominant palm forests, which led to erosion and poor soil fertility, but they also had help. As happened elsewhere in the Pacific, the Polynesians brought a commensal species with them: the Polynesian rat. It turns out that rats love palm seeds. The rat population increased so much on the local feast that few palm seeds escaped predation to germinate and replace the trees cut by the local people. With the

dominant tree unable to regenerate, it is no surprise that the forest, which sustained the productivity of the island, disappeared.

Habitat conversion

Conversion of natural ecosystems to human uses brings its own set of problems. As shown by events on Rapa Nui, altering the basic nature of a habitat often upsets natural processes of regeneration and control of ecosystem function. Cutting a tropical forest on nutrient-poor soil breaks the tight cycling of nutrients that maintains forest productivity. It also greatly alters the physical structure of the soil by exposing it to increased leaching and sunlight. The productivity of the land decreases precipitously, and soil erosion may increase tenfold or more.

Problems associated with habitat conversion are not restricted to tropical forests. On the North American prairies, plowing destroyed the dense root mats of perennial herbs that formerly held the soil together. A prolonged drought in the central United States during the 1920s and 1930s turned former prairies converted to croplands into a "dust bowl" of blowing soil (Figure 27.8). The global inventory of topsoil—the nutrient-rich upper layers of the soil profile—on croplands is about 6,500 gigatons (GT). Erosion from croplands is currently about 100 GT per year, or about 5 times faster than new topsoil is being formed. Thus, at the present rate, the earth's cropland topsoil reserves will, on average, be depleted in 65 years, forcing farmers in many regions to abandon now fertile agricultural regions and become more and more dependent on chemical fertilizers (Figure 27.9).

Other kinds of ecosystems are also being converted at a high rate. Mangrove wetlands provide natural protection

(a)

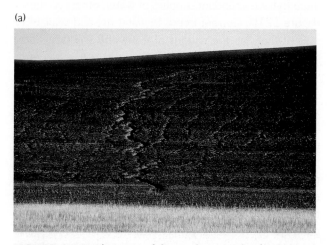

(b)

FIGURE 27.9 About 1% of the earth's topsoil is eroded each year. Soil erosion and gully formation is likely to occur on plowed farmland (a, Whitman County, Washington) and on heavily grazed pasture (b, Shelby County, Tennessee). Photos by Tim McCabe, courtesy of the U.S. Department of Agriculture, Soil Conservation Service.

(a)

(b)

FIGURE 27.10 Irrigation has both benefits and costs.
(a) Irrigation can turn desert into productive farmland, as furrow irrigation of cotton has done in the Imperial Valley of southern California. (b) However, the accumulation of salts in the soil that accompanies irrigation can damage crops, as it has in this irrigated alfalfa field in Colorado. Photos by Tim McCabe, courtesy of the U.S. Department of Agriculture, Soil Conservation Service.

for coastlines in many parts of the tropics. Where they have been cleared for fuelwood, shrimp farming, and land reclamation, coasts have been laid bare to rampaging hurricane-driven floodwaters. Similarly, damming rivers brings the benefits of flood control, increased water supplies, and power generation, but also increases silt transport, blocks fish migrations, alters downstream water conditions, and may even change the local weather.

Irrigation and water use

Water makes the desert bloom. Humankind has employed various irrigation schemes to increase the productivity of land since the beginning of agriculture (Figure 27.10). Only recently, however, has irrigation been applied on immense scales to land that would otherwise be unsuitable for agriculture. The benefits are tremendous, but so are the costs, many of which surface only after years of profitable irrigation. The primary costs are the negative environmental effects of developing the dams, wells, canals, and dike work required to support irrigation: depletion of aquifers; lowered water tables where wells are the source of irrigation water; accumulation of salts in soils as irrigation water evaporates in arid zones; reduction of groundwater quality through the concentration of naturally occurring toxic elements and the introduction of pesticides and fertilizers; and transmission of diseases by aquatic organisms. In most places, the costs of delivering water to crops—including the burden of future environmental problems—are underwritten by the population at large through taxes and other subsidies.

Water looms as one of the most critical limiting resources for the human population. A recent estimate of the fresh water available globally from river runoff and groundwater recharge was on the order of 50,000 cubic kilometers (km^3) per year. However, much of this water, including storm runoff and discharge of rivers in remote areas, is inaccessible. Thus, only about 12,000 km^3 per year are actually available for human use. Human water use currently amounts to about 5,000 km^3 per year: of this total, 3,500 km^3 are for agriculture, 1,000 km^3 for industry (including energy production), and 500 km^3 for domestic use. The demand for water is not uniform from region to region, and does not match the distribution of rainfall. So, while some regions of the world continue to have abundant supplies of water, others go thirsty (Figure 27.11). Current trends in water use and availability suggest that half the nations of the world will face water shortages by 2025, and three-fourths will experience water scarcity by 2050.

Much of this scarcity could be alleviated by several water economies. For example, about 1.5 m^3 of water are required to produce a kilogram of cereal grains, compared with 6–15 m^3 to produce a kilogram of meat or poultry. Thus, eating more grains and less meat reduces pressure on water supplies. Wind and solar power generation require much less water per kilowatt of electricity than fossil fuel and atomic power plants. Water used to generate hydroelectric power remains available for other uses. As the world becomes more industrialized, industrial uses of water will increase its price and reduce its quality by polluting it with industrial wastes. Nonetheless,

FIGURE 27.11 Water use greatly exceeds water supplies in many parts of the western United States. Withdrawals of water as a percentage of available freshwater supplies are shown by county. These withdrawals reflect irrigation, power generation, and domestic water consumption. Human population growth during the coming years will be highest in the most water-stressed areas, particularly the west and the south. After M. Hightower and S. A. Pierce, *Nature* 452:285–286 (2008).

technologies exist for increasing water use efficiencies and reducing wastes in industrial effluents.

Recall from Chapter 24 that runoff from sewage and agricultural fertilizers can cause hypoxic "dead zones" to develop in rivers, lakes, and coastal waters (Figure 27.12). Before water pollution came under strict controls in North America and Europe, large sections of major rivers had become completely anoxic, killing off local fish populations and preventing the migration of other species, such as shad and salmon, between the ocean and their headwater spawn-

FIGURE 27.12 Input of organic wastes can create anoxic conditions in aquatic environments. This fish die-off in a stream near Colorado Springs, Colorado, was caused by oxygen deficiency related to organic pollution. Photo by Shane Anderson/Saturdaze.

ing grounds. In general, cutting off the sources of organic wastes, either by diverting the inputs to larger bodies of water that can absorb them or by improving treatment of sewage, can restore natural conditions. Where these solutions have been implemented, their costs have been more than repaid in the long run by the benefits of enhanced water quality for fisheries, public health, and recreation.

Toxins impose local and global environmental risks

Toxins are poisons that kill animals and plants by interfering with their physiological functions. Many toxins occur naturally, but human activities have increased their accumulation in the environment. Toxic substances can be divided into several classes: acids, heavy metals, organic compounds, and radioactive substances are the most notable. The most efficient way of dealing with anthropogenic toxins is to reduce their entry into the environment through more efficient recovery and recycling. However, many anthropogenic toxins that have already accumulated in the environment will persist for long periods and will require expensive remediation.

Acids

Anthropogenic sources of strong acids are primarily of two kinds. The first is the reduced sulfur compounds exposed to the surface environment at coal mines. Sulfur bacteria oxidize pure and reduced forms of sulfur to sulfates, which may then be converted to sulfuric acid in streams that drain mining areas—hence the term **acidic mine drainage.** In some places, stream water becomes so acidic that it sterilizes the aquatic environment (see Figure 23.18).

The second, and more widespread, problem is **acid rain.** Coal and oil are not pure hydrocarbons; they contain sulfur and nitrogen compounds as well. After all, these fossil fuels are the remains of plants and animals that, when living, contained nitrogen and sulfur in their proteins and other organic molecules. Burning coal and oil, in addition to producing carbon dioxide and water vapor, spews nitrogen oxides and sulfur dioxide into the atmosphere. When these gases dissolve in raindrops, they are converted to acids. In highly industrialized areas, the pH of rain may drop to between 3 and 4, which is 100 to 1,000 times the acidity of natural rain.

The consequences of acid rain have been severe in some regions, such as the northeastern United States, Canada, and Scandinavia (see Chapter 24). Rivers and lakes in these regions tend to be oligotrophic and thus do not contain dissolved bases to buffer acid inputs. Consequently, their pH may drop to as low as 4.0, acidic enough to stunt the growth of, or even kill, fish and other organisms. Acid rain may also lower the pH of soil. At low pH, soil nutrients are leached at a faster rate, and phosphorus compounds in the soil precipitate and become unavailable for uptake by plant roots. The solutions to acid rain pollution are primarily technological and economic: scrubbing offending gases from the effluents of power plants and automobiles, finding alternatives to burning fossil fuels for energy, and reducing total demand for energy. Even when

sulfur and nitrogen emissions are reduced, however, ecosystems may not return to normal for decades or even centuries.

Biologists are now concerned that the increasing carbon dioxide levels in the atmosphere are acidifying the surface waters of the oceans, and that this change will adversely affect many marine organisms that construct calcium carbonate shells or skeletons. Although the oceans are a sink for atmospheric carbon dioxide, the additional dissolved CO_2 will tend to increase their acidity as it forms more carbonic acid (see Chapter 23). Ocean surface waters normally have a slightly basic pH of 8.2. By the late 1990s, the pH had dropped to 8.1, and it has been projected to decrease to 7.8 within a century. These changes may seem small, but increasing concentrations of bicarbonate ions (HCO_3^-) work against the biological formation of calcium carbonate shells by foraminifera, corals, crustaceans, mollusks, and other shelled marine organisms. The full extent of the changes that could result from acidification of marine surface waters is difficult to predict, but most oceanographers and marine ecologists agree that it will cause some shift in the functioning of marine ecosystems.

Heavy metals

Even in low concentrations, mercury, arsenic, lead, copper, nickel, zinc, and other heavy metals are toxic to most forms of life. They are introduced into the environment in a variety of ways, principally as refuse from mining and mineral smelting (Figure 27.13), as waste products of manufacturing processes, as fungicides (such as lead arsenate), and through the burning of leaded fuel (although it is no longer sold in the United States and western Europe,

FIGURE 27.13 Toxic heavy metals are released into the environment by mining and mineral smelting. High concentrations of metals in the waste material from this open-pit copper mine in Bingham, Utah, will prevent recolonization of the area by plants and the redevelopment of natural vegetation. Photo by Gene Ahrens/Bruce Coleman.

leaded gasoline is still used widely in other parts of the world). The effects of heavy metals are varied, but include interference with neurological function in vertebrates.

Many toxic heavy metals, including copper and nickel particulates released into the atmosphere by smelters, eventually accumulate in soils. The concentration of copper averages about 30 parts per million (ppm) in unpolluted temperate soils. Concentrations in excess of 100 ppm adversely affect mosses, lichens, and large fungi. Earthworm abundance drops off dramatically above 1,000 ppm, and most vascular plants cannot tolerate concentrations above 5,000 ppm (0.5%). As fungi die out, decomposition of organic matter and nitrification of organic nitrogen in the soil decrease. In one study in Sweden, in soils with copper levels of 2,000 ppm, fungal populations had fallen to only 20%–30% of their natural levels. With fewer decomposer organisms regenerating mineral nutrients in the soil, the overall productivity of the ecosystem declines.

Concentrations of heavy metals above 1,000 ppm may extend 10–20 km from sites of metal smelting. Taller smokestacks, which distribute wastes over larger areas at lower concentrations, can mitigate these effects, but solving the problem will ultimately require a change in the technology of metal production to reduce toxic by-products.

Organic compounds

Some natural organic compounds, such as nicotine and pyrethrins, are used as agricultural pesticides, but most organic pesticides are far more deadly concoctions produced in the laboratory. Pests have had no previous exposure to these artificial pesticides and no opportunity to evolve resistance. These pesticides include organomercurials (such as methylmercury), chlorinated hydrocarbons (DDT, lindane, chlordane, dieldrin), organophosphorus compounds (parathion, malathion), carbamate insecticides, and triazine herbicides. These compounds do their job in agriculture and pest management, but many accumulate in other parts of the ecosystem, where they adversely affect plant production and wildlife populations.

Modern pesticides and pesticide delivery systems are being designed for efficient use with minimal effects on the environment. Unfortunately, this progress is offset by increasingly widespread and more intensive use of chemicals of all kinds in agriculture. Because insects and other pests may evolve resistance to pesticides, their benefits are often short-lived, and they must be applied in ever larger amounts to achieve continued results (see Chapter

6). Through ecological research, we can assess the vulnerability of natural systems to these pollutants, prescribe safe applications, and—perhaps most important—find suitable alternatives to humankind's chemical warfare with agricultural pests. Some microorganisms—some of them genetically engineered to have special biochemical properties—can be used to metabolize pesticides and other toxic compounds to innocuous by-products. The use of biological agents to clean up the environment and help restore habitats is referred to as **bioremediation.**

Crop plants that are genetically modified to tolerate certain herbicides permit chemical "weeding" of agricultural lands. As a result, productivity is increased with less labor and use of artificial organic compounds. Genetically modified crops are controversial, however, because the price of seed is prohibitive for farmers in poor countries and because of the risk that the genes introduced to these crops might be transferred to non-crop plants, creating additional environmental problems. It is certain, however, that this technology will continue to develop, with exciting possibilities and potential risks.

Oil spills are another type of toxic environmental pollution caused by organic compounds. Crude oil is a complex mixture of hydrocarbons, containing up to 1% nitrogen and 5% sulfur. Oil pollution occurs at the source in areas of oil production, rarely as a result of breaks in the hundreds of thousands of kilometers of oil pipeline in the world, and most frequently in the ocean due to offshore drilling and the wreckage of oil tankers (Figure 27.14). These losses amount to 3 million to 6 million tons of oil annually, or about 0.1%–0.2% of global oil production. Petroleum kills by coating the surfaces of organisms and, because hydrocarbons are organic solvents, by disrupting biological membranes. Over time, oil slicks disperse by evaporation of the lighter fractions, emulsion of other fractions in water, and weathering and microbial breakdown of the rest. But certain types of sensitive ecosystems, such as coral reefs, may take decades to recover fully.

Radioactive substances

Radiation comes in a broad spectrum of energy intensities, ranging from generally harmless long-wavelength radio and infrared radiation, through visible light, to damaging ultraviolet radiation of shorter wavelengths, and the extremely energetic cosmic rays and subatomic particles released by the disintegration of atomic nuclei (radioactive decay). Natural sources create an unavoidable background level of radiation. In some circumstances, natural radioactive substances, such as the radon gas present in

(a)

(b)

FIGURE 27.14 Oil spills occur most frequently in the ocean. (a) Oil slick from a damaged tanker approaching the Caribbean coast of Panama. (b) Fringe of dead mangroves killed by the oil spill. Photos by Carl C. Hansen, courtesy of the Smithsonian Tropical Research Institute.

soils in regions having granitic bedrock, can become concentrated and pose public health hazards. Such dangers are minor, however, compared with the extreme radiation hazards that could result from accidents at nuclear power plants (such as those that occurred at Three Mile Island, Pennsylvania, in 1979 and at Chernobyl, Ukraine, in 1986), from the waste products of nuclear power generation, and from nuclear war.

The possibility of global nuclear war is diminishing as the superpowers dismantle their nuclear arsenals and nations turn their attention to economic, social, and environmental problems. Still, the threat of local nuclear terrorist attacks persists, and radioactive wastes produced by peaceful uses of the atom pose daunting disposal problems. Depending on the waste product, radiation will not decline to harmless levels for thousands or even millions

of years—far beyond the life span of waste containers, not to mention that of the institutions to which their care is entrusted. The waste disposal problem may ultimately limit the use of nuclear power.

Atmospheric pollution threatens the environment on a global scale

Because of the circulation of the atmosphere and oceans, certain types of pollution have global consequences: their effects extend far beyond the sources of the pollution itself. By far the most worrisome of these consequences are the destruction of the ozone layer in the upper atmosphere and the increase in atmospheric concentrations of carbon dioxide and other "greenhouse" gases, which are the major contributors to the current global increase in environmental temperature.

The ozone layer and ultraviolet radiation

Ozone (O_3) is a molecular form of oxygen that is a highly reactive oxidizer, capable of chemically oxidizing organic molecules and destroying their proper functioning. Consequently, ozone is toxic to living organisms even in small concentrations. Near the earth's surface, ozone is produced by the oxidation of molecular oxygen (O_2) in the presence of nitrous oxide (N_2O) and sunlight. Because N_2O is a product of gasoline combustion, ozone can reach high levels in the automobile exhaust fumes that pollute cities, particularly where there is strong sunlight. Los Angeles, for example, was well known for its smog before pollution control measures went into effect. Ozone concentrations in the atmosphere at ground level sometimes reached 0.5 ppm, which is perhaps 20–50 times the normal level and is damaging to human health, crops, and natural vegetation.

Ozone is also produced in the stratosphere, at an altitude of about 25 km, but there it has the beneficial effect of shielding the earth's surface from ultraviolet radiation by absorbing solar radiation of short wavelengths (especially in the range of 200–300 nm). Unfortunately, certain substances, among them chlorine atoms, cause the breakdown of ozone. Levels of chlorine in the upper atmosphere increased over many decades because of the release into the atmosphere of chlorofluorocarbons (CFCs), which were used as propellants in spray cans and as coolants in air conditioning and refrigeration systems. Decreases in stratospheric ozone concentrations of 50% or more—so-called **ozone holes**—have been observed at high latitudes in both hemispheres (Figure 27.15). In

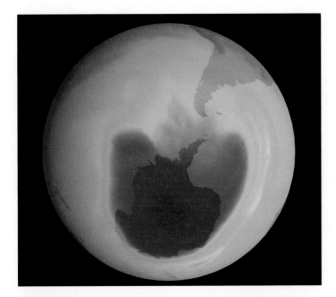

FIGURE 27.15 Ozone holes are areas where the stratospheric ozone concentration has decreased by 50% or more. This false-color satellite image, taken on October 3, 1999, shows the development of a large ozone hole over Antarctica. Courtesy of NASA.

September 2000, a NASA instrument determined that the Antarctic ozone hole, at about 28.3 million square kilometers, was three times larger than the area of the United States—the largest area on record for this hole up to that date; that record was reached again in 2004 and 2006.

There is about a 2% increase in ultraviolet radiation at the earth's surface for every 1% decrease in ozone concentrations in the stratosphere. One expected result of the elevation in UV levels is an increase in the incidence of skin cancer, because ultraviolet radiation between 280 and 320 nm in wavelength (so-called UV-B) causes damage to DNA molecules, which can transform cells to produce skin cancers. Of greater concern is the fact that ultraviolet radiation damages the photosynthetic apparatus of plants and could cause reductions in primary production—the base of the food chain for the entire ecosystem. Such reductions have already been observed in the oceans surrounding Antarctica. The threat is so great that the international community, through the Vienna Convention for the Protection of the Ozone Layer (1985) and the Montreal Protocol (1987), agreed to phase out the use of all CFCs by the end of the twentieth century. CFC molecules can remain in the upper atmosphere for decades, however, so ozone concentrations have not yet begun to recover. Even though it is too early to see the result, we can anticipate that ending the production and use of CFCs and other chlorine-containing gases will reverse the damage that has

already been done and allow atmospheric ozone to return gradually to its natural equilibrium level, perhaps within a century.

Carbon dioxide and the greenhouse effect

Carbon dioxide (CO_2) occurs naturally in the atmosphere. Without it, the earth would be a very cold place, because most of the sunlight absorbed by the earth's surface would be reradiated into the cold depths of space. As it is now, CO_2 forms an insulating blanket over the earth's surface that lets short-wavelength ultraviolet and visible light from the sun pass through, but retards loss of heat as longer-wavelength infrared radiation (see the Global Change box in Chapter 3). The glass in a greenhouse works on the same principle, so this phenomenon is known as the **greenhouse effect.**

At times in the distant past, the concentration of CO_2 in the atmosphere was far greater than it is now (see Figure 23.9), and the average temperature of the earth was correspondingly much warmer. As atmospheric CO_2 declined during the last 50 million–100 million years, the earth experienced a gradual cooling, culminating in the Ice Age of the past million years. The problem we are facing today as atmospheric CO_2 concentrations increase is not that the earth has never been so warm, but that the climate is changing so rapidly that ecological systems will fail to keep up with the changes and suffer losses in biodiversity and productivity.

Before 1850, the concentration of CO_2 in the atmosphere was on the order of 280 ppm (equivalent to 0.028%). During the last 150 years, which have witnessed tremendous increases in the burning of wood, coal, oil, and natural gas for energy production, it has increased to over 380 ppm (Figure 27.16). Half of this increase has occurred during the last 30 years, and the rate of increase appears to be rising. The concentration of carbon dioxide in the atmosphere represents a balance between processes that add CO_2 and those that remove it. Before the Industrial Revolution, the addition of CO_2 to the atmosphere by the respiration of terrestrial organisms (approximately 120 billion metric tons of carbon per year) was balanced by the gross primary production of terrestrial vegetation, and the total amount in the atmosphere was maintained at equilibrium. At present, forest cutting accounts for the addition of about 2 billion tons of carbon to the atmosphere annually; burning of fossil fuels adds about 5 billion tons. Thus, human activities have increased carbon inputs into the atmosphere by about 6%—perhaps more, perhaps less, depending on whose figures we use.

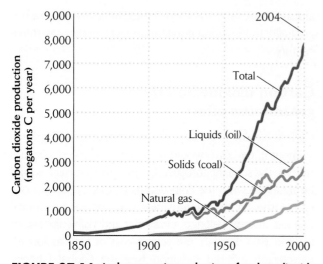
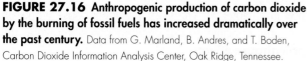

FIGURE 27.16 Anthropogenic production of carbon dioxide by the burning of fossil fuels has increased dramatically over the past century. Data from G. Marland, B. Andres, and T. Boden, Carbon Dioxide Information Analysis Center, Oak Ridge, Tennessee.

The atmosphere exchanges carbon dioxide with the oceans, where excess carbon is precipitated as calcium carbonate sediments (see Chapter 23). It has been estimated that at present, the oceans absorb more carbon than they release to the atmosphere by about 2.4 billion tons annually. In spite of the fact that the oceans are a net carbon sink, their capacity to absorb additional carbon is less than half of the anthropogenic input of carbon into the atmosphere, and part of that capacity may already be used to offset excess production of carbon dioxide by terrestrial systems. No matter how the arithmetic is done, it

shows carbon dioxide concentrations in the atmosphere to be rising rapidly.

How much the surface temperature of the earth will warm as a result of increased atmospheric CO_2 concentrations is uncertain. Moreover, CO_2 is not the only greenhouse gas of concern, and greenhouse gases are not the only factors influencing the heat balance of the atmosphere. Other gases, such as methane, nitrogen oxides, and chlorinated hydrocarbons, which are produced as industrial and agricultural wastes, add to the greenhouse effect (Figure 27.17). Increased ozone in the lower atmosphere is another contributor. Somewhat balancing these factors are increases in the earth's albedo resulting from changing land uses and the reflection of sunlight by **aerosols**—fine solid or liquid particles—in the upper atmosphere. Nonetheless, the current increase in heat input to the atmosphere owing to anthropogenic causes is about 1.5 watts per square meter of the earth's surface.

Warmer temperatures caused by greenhouse gases will have mixed effects on ecosystem productivity. On the positive side, warmer temperatures will lengthen the growing season and speed metabolism, and will thereby tend to enhance production in moist environments. Because plants require carbon dioxide for photosynthesis, productivity in moist environments is likely to be further enhanced by higher CO_2 concentrations in the atmosphere. Balancing this benefit is the likelihood of increasing drought stress in arid environments, which may reduce agricultural production and accelerate the conversion of overused grazing lands and croplands to deserts. Even within wet tropical areas, periodic droughts caused

FIGURE 27.17 Greenhouse gases and other anthropogenic inputs add to the heat budget of the atmosphere. *Radiative forcing* is change in the heat budget of the atmosphere due to absorption or reflection of light. Additional anthropogenic sources of radiative forcing beyond the principal greenhouse gases—carbon dioxide (CO_2) and methane (CH_4)—include increasing ozone in the lower atmosphere (troposphere). Carbon particles deposited on snow absorb light energy (reduce albedo), while land clearing generally increases albedo. Aerosols added to the upper atmosphere (stratosphere) reflect incoming solar radiation and have a net cooling effect. After a figure produced by Leland McInnes based on data from the Intergovernmental Panel on Climate Change (IPCC), Working Group I: The Physical Science Basis of Climate Change, Fourth Assessment Report (2007).

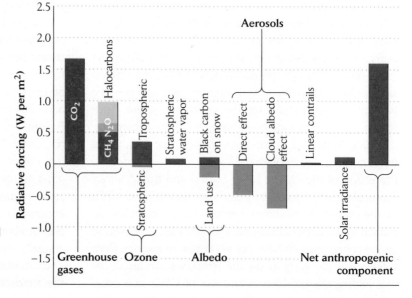

by El Niño events may lead to forest diebacks and contribute additional carbon to the atmosphere. Other potential problems include the inundation of coastal human settlements by rising sea levels fed by melting polar ice caps.

Human ecology is the ultimate challenge

The human population continues to increase worldwide at a rate of almost 2% per year. The highest rates of population increase are in some of the poorest countries. Even if population growth were to stop today, staggering problems would remain. The present human population is consuming resources faster than new resources are being regenerated by the biosphere, all the while pouring forth so much waste that the quality of the environment in most regions of the earth is deteriorating at an alarming rate. If we are to leave a habitable world for future generations, our top priority must be to achieve a sustainable relationship with the rest of the biosphere. This will require putting an end to population growth, developing sustainable energy sources, providing for the regeneration of nutrients and other materials, and restoring degraded habitats.

Ecosystem services

Although most of us live in a world dominated by technology, our well-being depends ultimately on ecosystem services provided by nature. **Ecosystem services** are defined as the benefits provided by ecosystems that support human life. Examples of key ecosystem services are the provisioning of clean, fresh water, food production, pollination of crops, flood control, assimilation of atmospheric carbon dioxide, and outdoor recreation. Some of these services are direct and depend on particular species of animals and plants that provide food, timber, medicines, and other commodities. Other ecosystem services are indirect and depend on the maintenance of intact natural ecosystems, but perhaps not particular species. Among these services are the provisioning of fresh water by forested watersheds and the regeneration of nutrients in soils by detritivores and soil microorganisms. To the extent that we degrade natural environments, we have to make up for the services they provide in other ways, or suffer their loss.

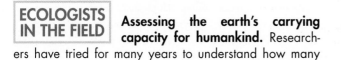

ECOLOGISTS IN THE FIELD **Assessing the earth's carrying capacity for humankind.** Researchers have tried for many years to understand how many people the earth can support. Estimates of the sustainable population vary between about 0.6 billion—less than one-tenth the current population—to several tens of billions. These estimates depend on certain critical assumptions about the limits on human population size and the standard of living for each individual. What is meant by standard of living includes not only material goods, such as housing, televisions, recreational travel, and other measures of quality of life, but also the cost of producing and delivering that most basic human necessity: food. In the United States, for example, every kilojoule of food energy we consume represents 10 kilojoules of fossil fuel energy that have gone into the fertilizers, farm machinery, packaging, and transportation required to put that food on our tables. Most estimates of the earth's carrying capacity for humanity exclude the contributions of fossil fuels, which are nonrenewable resources with high environmental costs. Thus, to be sustainable, human energy needs must be met by renewable sources, such as wind, solar, and hydroelectric power generation and ethanol production from crops.

Mathis Wackernagel, a member of the Task Force on Planning Healthy and Sustainable Communities at the University of British Columbia, developed the concept of the **ecological footprint** based on ethanol as a sustainable energy source for our food and energy needs. The ecological footprint is the amount of land devoted to ethanol production required to sustain a single human individual at his or her material standard of living. Making a generous assumption about how much ethanol can be produced from a hectare of fertile cropland, Wackernagel estimated that the citizens of The Netherlands, for example, would each require 5.3 ha of fertile land to supply their requirements. The area of The Netherlands holds fewer than 2 ha of cropland for each of its 16 million citizens. Clearly, to sustain their current standard of living, which is one of the highest in the world, the Dutch would have to import immense amounts of energy, as they currently do in the form of fossil fuels, or dramatically increase their use of renewable energy resources. The United States has the highest ecological footprint in the world, at 10.3 ha per individual. Thus, even though there are 6.7 ha of productive agricultural land available per person, the population at its present level is not sustainable by domestic renewable resources.

Globally, there are only about 1.7 ha of land of average agricultural productivity available for each of the earth's nearly 7 billion current inhabitants. That is clearly not enough to sustain the present population at a high standard of living, according to Wackernagel's assumptions. What will happen when the human population increases to 10 billion, as it might by the middle of this century? Will the average standard of living decrease as sources of fossil fuel are used up, or will new technologies emerge to keep pace with the growing population? Cornell University ecologist David Pimentel has estimated that the earth can sustain 2 billion people at the standard of living currently enjoyed in the United States. Stanford University ecologist Paul Ehrlich puts the number much lower. According to many ecologists, larger populations, including the present 6 billion, can be sustained only at a lower average standard of living. ▌

In many respects, humankind has stepped beyond the bounds of the usual ecological mechanisms of restraint and regeneration. Our ability to tap nonrenewable sources of energy in the form of coal, oil, and natural gas deposits has temporarily removed ecological limits on our population growth. No longer is most of the human population supported by the land it occupies. Our technological and economic capabilities to reach out for new land and resources have pushed density-dependent population feedbacks into the future—but they remain nonetheless.

Our present course leads in a predictable direction. It is not an inviting one: increasing energy, material, and food shortages; many people living with poverty and disease; a badly polluted environment; escalating social and political strife. These density-dependent mechanisms of population control will inevitably come into play, as they do for every species.

The future need not be like this (Figure 27.18). Where we have escaped natural restraint, we must apply self-restraint. Where we produce waste products that cannot be regenerated by ecological systems, we must find ways to recycle them ourselves. Energy consumption must be scaled back, and production must be based increasingly on renewable energy sources. Achieving these goals will require a consensus among social, economic, and political institutions. The hope for such a consensus lies in making people aware of the global deterioration of the quality of human life and educating them in the basic ecological principles that must form the foundation of a self-sustaining system.

Above all, humankind has the choice of adopting a new attitude toward its relationship with nature. We are a part of nature, not apart from nature. To the extent that our intelligence, culture, and technology have given us the power to dominate nature, we must also use these abilities to impose self-regulation and self-restraint. This is our greatest challenge. We have succeeded spectacularly in becoming the technological species. Our survival now depends on our becoming the ecological species and taking our proper place in the economy of nature.

FIGURE 27.18 Conservation efforts can help us take our proper place in the economy of nature. (a) Well-maintained and productive farmland. (b) A well-planned industrial park. (c) An attractive urban setting with plenty of nature worked in. Photo (a) by Ron Nichols, courtesy of the U.S. Department of Agriculture, Soil Conservation Service; photo (b) by Macduff Everton/CORBIS; photo (c) by Sotographs/The Stock Market.

SUMMARY

1. Maintaining a sustainable biosphere means that we must conserve the ecological processes responsible for its productivity.

2. The principal local threats to ecological processes are overexploitation of resources; species introductions; habitat conversion, particularly deforestation; and overuse and pollution of water supplies.

3. Overexploitation of resources is most intense where the human population cannot easily be sustained. In many places, particularly in the tropics, land has been stripped of its natural fertility.

4. Introduced species may displace native species, introduce new pathogens to humans, crops, and livestock, shift the character of an ecosystem to a different state, or even cause the collapse of an ecosystem.

5. The conversion of natural ecosystems to managed systems may upset natural ecological processes. Conversion to agriculture may cause losses of soil fertility and soil erosion.

6. Water has become scarce in many parts of the world. The greatest use of water by humans is for irrigation, followed by industrial processes and energy production. In dry regions, irrigation can lead to salt deposition in soils. Pollution by industrial effluents and organic wastes are also serious problems.

7. A number of human activities release toxic substances, many of which accumulate in the environment and can adversely affect the functioning of natural ecosystems. Among the most serious problems are acidic mine drainage and acid rain; heavy metals; organic compounds, particularly those used in pesticides; and radioactive substances. Increasing atmospheric carbon dioxide has begun to acidify the surface waters of the oceans and may disrupt calcification by corals and other shell-forming organisms.

8. On a global scale, various airborne pollutants, especially chlorofluorocarbons, have reduced the concentrations of ozone in the upper atmosphere, allowing more damaging ultraviolet radiation to reach the surface of the earth.

9. Increasing concentrations of carbon dioxide in the atmosphere, produced mainly by the burning of fossil fuels, threaten to cause a rapid increase in average global temperatures, with potentially adverse consequences for natural ecosystems and agriculture. In addition, sea levels will rise as polar ice caps melt.

10. The key to survival of the human population is development of sustainable interactions with the biosphere. This will require control of human population growth, an increasing reliance on renewable energy sources, and recycling of material wastes.

11. A major goal must be maintenance of the natural ecosystem services that supply us with food, maintain soil fertility, guarantee adequate supplies of fresh water, pollinate crops, control flooding, assimilate carbon dioxide from the atmosphere, and provide space for outdoor recreation.

12. One way of quantifying the ability of the earth to sustain human populations is to calculate the ecological footprint, the amount of land required to provide for a single individual, assuming that energy is obtained from renewable resources. People in the United States would require an average of 10.3 ha per individual to sustain their current standard of living. In fact, however, the earth has only 1.7 ha of agricultural land to support each individual in today's population.

13. Our environmental crisis is real and will not go away. Yet we can make changes, beginning with controlling population size, reducing energy and material consumption, switching to renewable energy sources, and learning to appreciate and preserve the sustaining value of natural ecosystems.

REVIEW QUESTIONS

1. What does your knowledge of ecology suggest will happen to human populations as they exceed their carrying capacity?

2. Under what conditions would species introductions not affect ecosystem function?

3. What are some of the major costs of using irrigation water?

4. Since many toxins naturally occur in ecosystems, why are we concerned about the current effects of toxins on ecosystems?

5. How can bioremediation be helpful in preserving ecosystem functions?

6. What is thought to be the underlying cause of the thinning ozone layer in the upper atmosphere?

7. How have the production and assimilation of carbon dioxide been altered during the past 200 years?

8. What are some of the expected costs and benefits of global warming?

9. How does the concept of the ecological footprint highlight the impact of humans on the environment?

SUGGESTED READINGS

Arrow, K., et al. 1995. Economic growth, carrying capacity, and the environment. *Science* 268:520–521.

Bongaarts, J. 1994. Can the growing human population feed itself? *Scientific American* 270:36–42.

Clark, D. A. 2004a. Sources or sinks? The responses of tropical forests to current and future climate and atmospheric composition. *Philosophical Transactions of the Royal Society of London* B 359:477–491.

Clark, D. A. 2004b. Tropical forests and global warming: Slowing it down or speeding it up? *Frontiers in Ecology and the Environment* 2:73–80.

Daily, G. C., and P. R. Ehrlich. 1992. Population, sustainability, and Earth's carrying capacity. *BioScience* 42:761–771.

Díaz, S., et al. 2006. Biodiversity loss threatens human well-being. *PLoS Biology* 4:e277.

Dietz, T., E. A. Rosa, and R. York. 2007. Driving the human ecological footprint. *Frontiers in Ecology and the Environment* 5:13–18.

Foley, J. A., et al. 2005. Global consequences of land use. *Science* 309:570–574.

Foley, J. A., et al. 2007. Amazonia revealed: Forest degradation and loss of ecosystem goods and services in the Amazon Basin. *Frontiers in Ecology and the Environment* 5:25–32.

Gattuso, J.-P., D. Allemand, and M. Frankignoulle. 1999. Photosynthesis and calcification at cellular, organismal and community levels in corals reefs: A review on interactions and control by carbonate chemistry. *American Zoologist* 39:160–183.

Halpern, B. S., et al. 2008. A global map of human impact on marine ecosystems. *Science* 319:948–952.

Hitz, S., and J. Smith. 2004. Estimating global impacts from climate change. *Global Environmental Change: Human and Policy Dimensions* 14:201–218.

Holdren, J. P. 2008. Science and technology for sustainable well-being. *Science* 319:424–434.

Houghton, R. A., et al. 2000. Annual fluxes of carbon from deforestation and regrowth in the Brazilian Amazon. *Nature* 403:301–304.

Hughes, T. P. 1994. Catastrophes, phase shifts, and large-scale degradation of a Caribbean coral reef. *Science* 265:1547–1551.

Hunt, T. L. 2006. Rethinking the fall of Easter Island. *American Scientist* 94:412–419.

News Feature on Climate Change. 2007. *Nature* 446:706–707, 716–721.

Orr, J. C., et al. 2005. Anthropogenic ocean acidification over the twenty-first century and its impact on calcifying organisms. *Nature* 437:681–686.

Page, S. E., et al. 2002. The amount of carbon released from peat and forest fires in Indonesia during 1997. *Nature* 420:61–65.

Parmesan, C. 2006. Ecological and evolutionary responses to recent climate change. *Annual Review of Ecology, Evolution, and Systematics* 37:637–669.

Rasmussen, P. E., et al. 1998. Long-term agroecosystem experiments: Assessing agricultural sustainability and global change. *Science* 282:893–896.

Rees, W. E. 1992. Ecological footprints and appropriated carrying capacity: What urban economics leaves out. *Environment and Urbanisation* 4:121–130.

Reynolds, J. F., et al. 2007. Global desertification: Building a science for dryland development. *Science* 316:847–851.

Rousseaux, M. C., et al. 1999. Ozone depletion and UVB radiation: Impact on plant DNA damage in southern South America. *Proceedings of the National Academy of Sciences USA* 96:15310–15315.

Schmer, M. R., et al. 2008. Net energy of cellulosic ethanol from switchgrass. *Proceedings of the National Academy of Sciences USA* 105:464–469.

Special Section on Sustainability and Energy. 2007. *Science* 315:781–813.

Stoddard, J. L., et al. 1999. Regional trends in aquatic recovery from acidification in North America and Europe. *Nature* 401:575–578.

Vitousek, P. M., et al. 1986. Human appropriation of the products of photosynthesis. *BioScience* 36:368–373.

Wackernagel, M., et al. 2002. Tracking the ecological overshoot of the human economy. *Proceedings of the National Academy of Sciences USA* 99:9266–9271.

Walther, G.-R., et al. 2002. Ecological responses to recent climate change. *Nature* 416:389–395.

Weatherhead, E. C., and S. B. Andersen. 2006. The search for signs of recovery of the ozone layer. *Nature* 441:39–45.

Worm, B., et al. 2006. Impacts of biodiversity loss on ocean ecosystem services. *Science* 314:787–790.

Year of Planet Earth. 2008. Feature on global climate change. *Nature* 451:279–300.

GLOSSARY

Acclimatization. A reversible change in the morphology or physiology of an organism in response to environmental change.

Accumulation curve. A method of determining the relationship between species diversity and sample size by plotting the increase in the number of species observed as samples are added. *Compare with* Rarefaction curve.

Acid rain. Precipitation with high acidity (pH < 4) caused by the dissolution of certain gases (sulfur dioxide and nitrous oxide) released into the atmosphere by combustion of fossil fuels.

Acidic mine drainage. Water runoff from mine wastes containing sulfuric acid, which forms when organic sulfur is oxidized upon exposure to the atmosphere.

Acidity. The concentration of hydrogen ions in a solution.

Active transport. Movement of molecules or ions through a membrane against a concentration gradient.

Adaptation. (1) A genetically determined characteristic that enhances the ability of an individual to cope with the conditions of its environment. (2) The evolutionary process by which organisms become better suited to their environments.

Adaptive radiation. Diversification of species descended from a recent common ancestor.

Adiabatic cooling. The decrease in temperature with increasing elevation caused by the expansion of air under decreasing atmospheric pressure.

Aerosols. Fine solid or liquid particles in the atmosphere.

Afrotropical region. The biogeographic region corresponding roughly to the continent of Africa. *Also called* Ethiopian region.

Age structure. The distribution of individuals among age classes within a population.

Albedo. The proportion of the light reaching a surface that is reflected by that surface.

Allele. One of several alternative forms of a gene.

Allelopathy. Interference competition among plants by means of toxic secondary compounds.

Allocation. The division of limited time, energy, or materials among competing functions or requirements.

Allochthonous. Originating outside a system; particularly referring to nutrients and organic matter transported into stream and lake ecosystems. *Compare with* Autochthonous.

Allopatric. Occurring in different places; particularly referring to geographically separated populations. *Compare with* Sympatric.

Alpha diversity. *See* Local diversity.

Alternative stable states. Two or more stable equilibrium points, only one of which is occupied by a population at a given time.

Altruism. Social behavior that benefits the recipient at a cost to the donor.

Ammonification. The chemical transformation of organic nitrogen into ammonia by the metabolic breakdown of proteins and amino acids.

Anaerobic. Without oxygen.

Anoxic. Lacking oxygen; anaerobic.

Anthropogenic extinction. Extinction caused by human activities.

Aphotic zone. The portion of a water body below the depth to which light penetrates. *Compare with* Photic zone.

Aposematism. *See* Warning coloration.

Apparent competition. Competition between two or more species that is mediated by their consumers.

Arbuscular mycorrhizae (AM). Mutualistic associations of fungi with the roots of plants in which fungal hyphae penetrate root cell walls and form vesicles or branched structures in intimate contact with the root cell membranes. *Compare with* Ectomycorrhizae.

Artificial selection. Differential survival or reproduction that results from conscious decisions made by humans concerning desirable phenotypes of domesticated or laboratory animals and crops.

Asexual reproduction. Reproduction without the sexual union of gametes (fertilization). *Compare with* Sexual reproduction.

Assimilated energy. That portion of the energy contained in food that an organism digests and absorbs.

Assimilation efficiency. The ratio of assimilated energy to ingested energy.

Assimilatory. Referring to a biochemical transformation that results in the reduction of an element to an organic form and hence its gain by the organic compartment of the ecosystem. *Compare with* Dissimilatory.

Assortative mating. Preferential mating between individuals having either similar (positive assortative mating) or dissimilar (negative assortative mating) genotypes or phenotypes.

Asymmetric competition. A competitive relationship between two species in which each has an advantage with respect to different limiting factors in the environment.

Australasian region. The biogeographic region corresponding roughly to Australia, New Guinea, and nearby islands.

Autochthonous. Originating within a system; particularly referring to organic matter produced and nutrients cycled within a stream or lake. *Compare with* Allochthonous.

Autotroph. An organism that assimilates energy either from sunlight or from inorganic compounds. *Compare with* Heterotroph.

Autumn bloom. *See* Fall bloom.

Background extinction. Extinction of species or higher taxa at a relatively low rate during periods without rapid environmental change. *Compare with* Mass extinction.

Batesian mimicry. A predator deterrence strategy in which a palatable species (mimic) resembles an unpalatable species (model). *Compare with* Müllerian mimicry.

Benthic zone. The sediments or other substrates at the bottom of a stream, lake, or ocean.

Beta diversity. The difference, or turnover, in species from one habitat to another. *Compare with* Local diversity, Regional diversity.

Bet hedging. Reducing the risk of mortality or reproductive failure in a variable environment by adopting an intermediate strategy or several alternative strategies simultaneously, or by spreading one's risk over time and space (for example, by perennial rather than annual reproduction).

Bicarbonate ion (HCO_3^-). An anion formed by the dissociation of carbonic acid.

Biodiversity. Variation among organisms and ecological systems at all levels, including genetic variation, morphological and functional variation, taxonomic uniqueness, and endemism, as well as variation in ecosystem structure and function.

Biomass accumulation ratio. The ratio of biomass to productivity.

Biome. One of several categories into which communities and ecosystems can be grouped based on climate and dominant plant forms.

Bioremediation. Restoration of natural habitats or ecological processes by use of biological agents.

Biosphere. All the ecosystems of the earth.

Biosphere approach. An approach to ecology that is concerned with phenomena at a global scale.

Bottleneck. *See* Population bottleneck.

Bottom-up control. The influence of producers on the sizes of the trophic levels above them in a food web. *Compare with* Top-down control.

Boundary layer. A layer of still or slow-moving water or air close to the surface of an object.

Browsing. Consumption of a portion of a plant's tissues; generally applied to woody vegetation. *Compare with* Grazing.

C_3 photosynthesis. The most common photosynthetic pathway, in which carbon dioxide is initially assimilated into a three-carbon compound, glyceraldehyde 3-phosphate (G3P).

C_4 photosynthesis. A photosynthetic pathway in which carbon dioxide is initially assimilated into a four-carbon compound, oxaloacetic acid (OAA), in leaf mesophyll cells.

CAM. *See* Crassulacean acid metabolism.

Canopy. The uppermost layer of vegetation in a forest.

Carrying capacity (K). The number of individuals in a population that the resources of a habitat can support.

Caste. Individuals within a social group sharing a specialized form or behavior.

Catastrophe. An unpredictable event that has a strong negative effect on population size.

Cation exchange capacity. The ability of a soil to retain positively charged ions (cations), which provides an index to the fertility of that soil.

Central place foraging. Foraging behavior in which acquired food is brought to a central place, such as a nest with young.

Chaos. A complex, unpredictable pattern of oscillation, as in the sizes of populations with very high intrinsic rates of growth. *Compare with* Damped oscillation, Limit cycle.

Character displacement. Divergence in the traits of two otherwise similar species where their ranges overlap, caused by the selective effects of competition between them.

Chemoautotroph. An organism that oxidizes inorganic compounds to obtain energy for the synthesis of organic compounds. *Compare with* Photoautotroph.

Climate zone. A region in Heinrich Walter's climate classification scheme, defined by the annual cycle of temperature and precipitation.

Climax community. The end point of a successional sequence, or sere; a community that has reached a steady state under a particular set of environmental conditions.

Clone. A group of asexual individuals, all descended from the same parent and bearing the same genotype.

Closed community. A community in which the distributions of several species coincide closely, but are largely separated from those of other sets of species. *Compare with* Open community.

Clumped distribution. A dispersion pattern in which individuals are found in discrete groups. *Compare with* Random distribution, Spaced distribution.

Coalescence time. The period over which all the copies of a particular gene in a population will have descended, just by chance, from a single copy that existed at some time in the past.

Codominant. Referring to alleles that, when heterozygous, produce a phenotype intermediate between their homozygous phenotypes.

Codon. A sequence of three nucleotides in DNA or RNA that specifies which amino acid will be placed at a particular position in a protein.

Coefficient of relationship (r). The probability that one individual shares with another a genetic factor inherited from a common ancestor. *See also* Identity by descent.

Coevolution. The reciprocal evolution in two or more interacting species of adaptations selected by their interaction.

Cohesion–tension theory. The idea that the force required to draw water from the soil and roots of a plant to its leaves is generated by transpiration.

Cohort life table. A life table that follows the fates of a group of individuals born into a population at the same time from their birth to the death of the last individual. *Also called* Dynamic life table. *Compare with* Static life table.

Common garden experiment. A research method in which organisms from different habitats are grown together in the same place to test for genetic differences in the absence of different environmental influences.

Community. An association of interacting populations, usually defined by the nature of their interaction or the place in which they live.

Community approach. An approach to ecology that is concerned with understanding the diversity and relative abundances of different kinds of organisms living together in the same place.

Compartment model. A representation of an ecosystem in which the various components are portrayed as units (compartments) that receive inputs from and provide outputs to other such units.

Compensation point. The depth of water or light intensity at which respiration and photosynthesis balance each other; the lower limit of the photic zone. *Compare with* Saturation point.

Competition. Use or defense of a resource by one individual that reduces the availability of that resource to other individuals, whether of the same species (intraspecific competition) or other species (interspecific competition).

Competitive exclusion principle. The principle that two or more species cannot coexist indefinitely on the same limiting resource.

Conduction. The transfer of heat between substances in contact with one another.

Connectedness web. A food web that emphasizes the feeding relationships among species within a community. *Compare with* Energy flow web, Functional web.

Constancy. A measure of the ability of a system to resist change in the face of outside influences. *Also called* Resistance. *Compare with* Resilience.

Constitutive defense. A defensive structure or compound that is always present. *Compare with* Induced defense.

Consumer-imposed equilibrium. An equilibrium that results when an increase in a resource population results in a corresponding increase in a consumer population, which brings the resource population under control at a point well below its carrying capacity. *Compare with* Resource-imposed equilibrium.

Consumer–resource interaction. Any interaction between species in which one species preys on or otherwise consumes the other.

Continental drift. Movement of continents across the surface of the earth over geologic time.

Continuous-time approach. An approach to population modeling assuming that time flows continuously and that change can occur at every instant. *Compare with* Discrete-time approach.

Continuum. A gradient of environmental characteristics or of change in the composition of communities.

Continuum index. The scale of an environmental gradient based on changes in physical characteristics or community composition along that gradient.

Control. A treatment that reproduces all aspects of an experiment except the variable of interest.

Convection. Transfer of heat by the movement of a fluid, such as air or water.

Convergence. The process by which unrelated organisms evolve a resemblance to each other in response to similar environmental conditions.

Cooperation. Social behavior that benefits both donor and recipient.

Coral reef. A structure built from the skeletons of corals in shallow tropical ocean waters, often the foundation of a rich and productive marine ecological zone.

Coriolis effect. The effect of the earth's rotation on the circulation patterns of the atmosphere and oceans, which causes winds and currents to veer to the right of their direction of travel in the Northern Hemisphere and to the left in the Southern Hemisphere.

Corridors. Narrow strips of habitat that facilitate the movement of organisms between adjacent habitat fragments.

Countercurrent circulation. Movement of fluids in opposite directions on either side of a separating barrier through which heat or dissolved substances can pass.

Crassulacean acid metabolism (CAM). A photosynthetic pathway in which the initial assimilation of carbon into a four-carbon compound occurs at night; found in some succulent plants in arid habitats.

Cross-resistance. Resistance or immunity to one pathogen resulting from infection by another, usually closely related, pathogen.

Crypsis. An appearance that allows an organism to blend into the background to avoid detection by others.

Cyclic climax. A self-perpetuating, repeating sequence of stages produced by ongoing succession, none of which by itself is stable, but which together constitute a persistent pattern.

Damped oscillation. A pattern of oscillations with progressively smaller amplitude, as in the sizes of some populations approaching their equilibrium value. *Compare with* Chaos, Limit cycle.

Defensive mutualism.   A relationship between two species in which one species defends the other against its consumers and usually receives some type of nourishment or living space in return.

Demography.   The study of the structure and growth of populations.

Denitrification.   The reduction of nitrate and nitrite to form molecular nitrogen (N_2), primarily by specialized bacteria.

Density.   (1) Referring to a population, the number of individuals per unit of area or volume. (2) Referring to a substance, the weight per unit of volume.

Density-dependent.   Having an influence on a population that varies with the size of the population. *Compare with* Density-independent.

Density-independent.   Having an influence on a population that does not vary with the size of the population. *Compare with* Density-dependent.

Deoxyribonucleic acid (DNA).   A macromolecule whose sequence of nucleotide subunits encodes genetic information.

Determinate growth.   A growth pattern in which an individual continues to grow only until it reaches maturity. *Compare with* Indeterminate growth.

Deterministic.   Not subject to, or not accounting for, stochastic (random) variation.

Detritivore.   An organism that feeds on detritus.

Detritus.   Freshly dead or partially decomposed remains of organisms and their indigestible excreta.

Developmental response.   A long-lasting, irreversible physical or physiological change in an individual in response to the environmental conditions under which it develops.

Diapause.   Temporary interruption of development or physiological function, usually associated with a period of unfavorable environmental conditions.

Dioecy.   In plants, the occurrence of male and female sexual organs on different individuals. *Compare with* Monoecy.

Diploid.   Having two sets of chromosomes. *Compare with* Haploid.

Directional selection.   Differential survival or reproduction of individuals with a phenotype at one extreme of the population distribution, resulting in an evolutionary shift in the population average toward that phenotype. *Compare with* Stabilizing selection, Disruptive selection.

Discrete-time approach.   An approach to population modeling that uses discrete time intervals, generally corresponding to intervals between reproductive periods. *Compare with* Continuous-time approach.

Dispersal limitation.   The absence of a population from suitable habitat because of barriers to dispersal.

Dispersion.   The spacing of individuals with respect to one another within the geographic range of a population.

Dispersive mutualism.   A relationship between two species in which one species transports pollen or disperses seeds and usually receives some type of nourishment in return.

Disruptive selection.   Differential survival or reproduction of individuals with phenotypes at two or more extremes, resulting in an evolutionary shift of individual phenotypes away from the population average. *Compare with* Directional selection, Stabilizing selection.

Dissimilatory.   Referring to a biochemical transformation that results in the oxidation of the organic form of an element and hence its loss from the organic compartment of the ecosystem. *Compare with* Assimilatory.

Distribution.   The geographic area occupied by a population.

Disturbance.   An event that causes rapid or marked change in a population or community, often thought of as displacing an ecological system from its equilibrium.

DNA.   *See* Deoxyribonucleic acid.

Dominance hierarchy.   The orderly ranking of individuals in a group based on the outcome of aggressive encounters.

Dominant.   (1) Referring to an allele that masks the expression of another (recessive) allele of the same gene. *Compare with* Recessive. (2) Referring to a species that is particularly abundant in or exerts great influence within an ecological system. (3) Behaviorally dominant.

Donor.   The individual in a social interaction that directs behavior toward another individual. *Compare with* Recipient.

Dormancy.   A physiologically inactive state, such as hibernation, diapause, or seed dormancy, usually assumed when conditions do not allow the organism to function normally.

Doubling time.   The time required for a population to grow to twice its size.

Dynamic life table.   *See* Cohort life table.

Ecological diversity.   A measure of biodiversity that takes into account variation in the ecological roles of species.

Ecological efficiency.   The percentage of the energy or biomass produced at one trophic level that is transferred to the next trophic level.

Ecological envelope.   A catalog of ecological conditions at the locations where a species has been recorded, used in ecological niche modeling.

Ecological footprint.   The amount of land required to sustain the energy needs of a single human individual at his or her material standard of living.

Ecological niche modeling.   A method of predicting the distribution of a species from limited information about the range of conditions where individuals of that species are known to occur.

Ecological release.   Expansion of habitat and resource use and increase in abundance by a population in a region of low species richness.

Ecological system.   A regularly interacting or interdependent group of biological entities forming a unified whole that functions in an ecological context.

Ecology.   The study of the interactions of organisms with one another and with their environment.

Ecosystem.   An assemblage of organisms together with their physical and chemical environments.

Ecosystem approach.   An approach to ecology that is concerned with the activities of organisms as well as physical and chemical transformations of energy and materials in the soil, atmosphere, and water.

Ecosystem ecology.   The study of natural systems from the standpoint of the flow of energy and cycling of matter.

Ecosystem engineer.   An animal whose activities have a disproportionately large effect on the landscape.

Ecosystem services.   The benefits provided by ecosystems that support human life.

Ecotone.   A region of rapid turnover of species along a spatial transect or ecological gradient; a zone of transition between communities.

Ecotourism.   Travel for the recreational purpose of observing unusual species or ecological habitats and landscapes.

Ecotype.   A genetically differentiated subpopulation that is adapted to specific habitat conditions.

Ectomycorrhizae (EcM).   Mutualistic associations of fungi with the roots of plants in which the fungus forms a sheath around the outside of the root and penetrates the spaces between the cells of the cortical layer. *Compare with* Arbuscular mycorrhizae.

Ectotherm.   An organism that uses external heat to maintain its body temperature. *Compare with* Endotherm.

Edge context.   The qualities of the habitat matrix adjoining a habitat patch that affect the probability of an organism's leaving that patch.

Edge specialists.   Species that are specialized for living in ecotones.

Effective population size (N_e).   The size of an ideal population that would undergo genetic drift at the same rate as an observed population.

Egested energy.   Energy contained in indigestible components of food that are defecated or regurgitated.

El Niño.   A warm current that appears each winter along the western coast of northern South America; also used to refer to a global climatic pattern seen in years when this current is particularly strong. *Compare with* La Niña.

Emergent.   (1) Referring to trees, rising above the forest canopy. (2) Referring to wetland plants, rooted below the waterline.

Emerging disease.   A disease that appears in a population or species for the first time, or that is rapidly increasing in its incidence or geographic range.

Endemic.   (1) Occurring in a particular region and nowhere else. (2) Referring to disease, present at a low level within a local population.

Endemism.   The state of a species that is restricted to a single region.

Endotherm.   An organism that maintains its body temperature by the metabolic generation of heat. *Compare with* Ectotherm.

Energy equivalence rule.   The principle that populations tend to consume about the same amount of food per unit of area, and thus to have similar effects on population and ecosystem processes, regardless of the size of individuals.

Energy flow web.   A food web in which the connections between species are quantified by the flux of energy between a resource and its consumer. *Compare with* Connectedness web, Functional web.

Environment.   The surroundings of an organism, including the other organisms with which it interacts.

Epilimnion.   The warm, oxygen-rich surface layer of a body of water that lies above the thermocline. *Compare with* Hypolimnion.

Epiphyte.   A plant that grows on another plant and is not rooted in soil, but instead derives its moisture and nutrients from the air and rain.

Equilibrium isocline.   A line on a graph representing the sizes of a predator and a prey population, or of competing populations, designating points at which the growth rate of one of the populations is zero. *Also called* Zero growth isocline.

Equilibrium theory of island biogeography.   The principle that the number of species on an island represents a balance between the process of colonization by new immigrant species and the process of extinction of resident species.

Equilibrium water vapor pressure.   The amount of water vapor in the atmosphere at which the tendency of liquid water to evaporate and the tendency of water vapor to condense back to a liquid state are balanced.

Estuary.   A body of water at the mouth of a river.

Ethiopian region.   *See* Afrotropical region.

Eusociality.   The complex social organization of termites, ants, and many wasps and bees, characterized by sterile adult offspring living and cooperating with one or a few dominant reproductive individuals to produce broods of offspring.

Eutrophic.   Rich in the inorganic nutrients required by organisms. *Compare with* Oligotrophic.

Eutrophication.   Enrichment of an aquatic ecosystem with nutrients; often refers to overenrichment caused by sewage or runoff from fertilized agricultural lands that results in excessive bacterial growth and oxygen depletion.

Evaporation.   The transformation of water from the liquid to the gaseous state with the input of heat energy.

Evolution.   Change in a population's gene pool.

Evolutionarily stable strategy.   A strategy such that, if all members of a population adopt it, no alternative strategy can invade that population.

Excreted energy.   The percentage of assimilated energy that is excreted in the form of nitrogen-containing organic wastes produced when the diet contains an excess of nitrogen.

Experiment.   A controlled manipulation of a system to determine the effect of a change in one or more variables.

Exploitative competition.   Competition between individuals or species by way of their mutual effects on shared resources. *Compare with* Interference competition.

Exponential growth.   Continuous increase (or decrease) in a population at a rate that is proportional to the number of individuals at a given time. *Compare with* Geometric growth.

Extent.   The size of a landscape area of interest. *Compare with* Grain.

Extra-pair copulation (EPC).   Copulation between a paired female and a male that is not her mate.

Extremophile.   An organism that can tolerate extreme environmental conditions.

Facilitation.   A process by which one species increases the probability of another species becoming established, particularly during early succession. *Compare with* Inhibition, Tolerance.

Fall bloom.   The rapid growth of algae in a temperate lake following fall overturn.

Fall overturn.   The vertical mixing of water layers in a temperate lake in fall following the breakdown of thermal stratification as the surface water cools and sinks. *Compare with* Spring overturn.

Fecundity.   The number of offspring produced per reproductive episode.

Fertilization.   In sexual reproduction, the union of male and female gametes to form a zygote.

Field capacity.   The amount of water that soil can hold against the pull of gravity at a matric potential of less than -0.01 MPa.

Fitness.   The genetic contribution by an individual's descendants to future generations of its population.

Fixation.   The loss of all alleles of a gene except one from a population, in which case the remaining allele is said to be *fixed*.

Fixation index (F).   A measure of allele fixation in a population.

Flagship species.   A species highly valued by human society whose preservation results in the preservation of other species living in association with it.

Fluvial.   Of or referring to flowing water.

Food web.   A representation of the various interconnected paths of energy flow through populations in a community, taking into account the fact that each population shares resources and consumers with other populations.

Forbs.   Broad-leaved herbaceous plants.

Founder event.   Colonization of an island or habitat patch by a small number of individuals that possess only a small sample of the genetic variation in the parent population. *Compare with* Population bottleneck.

Frequency.   Referring to alleles, the number of copies of a particular allele of a gene divided by the total number of copies of that gene in the population.

Frequency-dependent selection.   Differential survival or reproduction that depends on the frequency of a phenotype in a population.

Functional response.   A change in the rate of prey consumption by an individual predator as a result of a change in the density of its prey. *See also* Type I, II, III functional response; Numerical response.

Functional web.   A food web in which the importance of each species in maintaining the integrity of a community is reflected in its influence on the growth rates of other species' populations. *Compare with* Connectedness web, Energy flow web.

Fundamental niche.   The range of physical conditions and resources within which individuals of a species can persist. *Compare with* Realized niche.

Gamete.   A haploid cell that fuses with another haploid cell of the opposite sex during fertilization to form a zygote. In animals, the male gamete is called the sperm, and the female gamete is called the egg or ovum.

Game theory.   A method of analyzing the outcomes of behavioral decisions by an individual when those outcomes depend on the behavior of other individuals with which it interacts.

Gamma diversity.   *See* Regional diversity.

Gene.   All the codons that specify the amino acid sequence for a single protein, along with any sequences that regulate its expression.

Gene flow.   The exchange of genes between populations through the movement of individuals, gametes, or spores.

Gene pool.   All the alleles of all the genes of every individual in a population.

Generalization.   A pattern to which a reasonable amount of observation produces few exceptions.

Generation time.   The average period between the birth of an individual and the birth of its offspring.

Genetic diversity.   A measure of the genetic variation in a population, species, or community.

Genetic drift.   Change in allele frequencies in a population due to random variations in fecundity and mortality.

Genetic marker.   Variation of any kind that has a genetic basis and is used to study population processes that influence patterns of genetic variation.

Genotype.   The genetic characteristics that determine the structure and functioning of an organism. *Compare with* Phenotype.

Genotype–environment interaction.   The interaction between the genetic characteristics of an organism and that organism's environment that determines its phenotype.

Genotype–genotype interaction.   Variation in the expression and fitness of genotypes in one species depending on the genotypes of another species with which it interacts.

Geographic information system (GIS).   A computer program that allows researchers to compile diverse sets of geographic information and to quantify characteristics of the landscape.

Geographic range.   The distribution of a population in space.

Geometric growth.   Increase (or decrease) in a population as measured over discrete intervals in which the increment is proportional to the number of individuals at the beginning of the interval. *Compare with* Exponential growth.

Global Positioning System (GPS).   An array of satellites placed in orbit around the earth that send out signals that can be picked up by receivers on earth and used to calculate the latitude, longitude, and altitude of any location,

Gonad.   The primary sexual organ in which male or female gametes are produced.

Gondwana.   The major landmass of the Southern Hemisphere during the early Mesozoic era, made up of present-day South America, Africa, India, Australia, and Antarctica.

Gradient analysis.   The plotting and interpretation of the abundances of species along an environmental gradient.

Grain.   The degree of resolution at which a landscape is viewed. *Compare with* Extent.

Grazing.   (1) Consumption of a portion of a plant's tissues, generally applied to grasses and other herbaceous vegetation. *Compare with* Browsing. (2) Removal of plant or algal cover from a substratum, generally applied to aquatic systems.

Greenhouse effect.   An increase in average global temperatures resulting from an increase in the concentration of carbon dioxide and certain other heat-absorbing gases in the atmosphere.

Gross primary production.   The total energy assimilated by autotrophs through photosynthesis. *Compare with* Net primary production.

Growing season.   The portion of the year during which conditions are suitable for plant growth; at temperate latitudes, defined by the frost-free season.

Growth form.   The physical structure of a plant.

Guild.   A group of species that occupy similar ecological positions within the same community.

Habitat.   The place or physical setting where an organism normally lives, often characterized by a dominant plant growth form or physical characteristic (that is, a stream habitat, a forest habitat).

Habitat matrix.   The habitat types that surround patches of suitable habitat for a particular species.

Habitat patch.   An area of habitat with the resources and conditions necessary for a population to persist.

Hadley cell.   The circulation pattern of rising and falling air within the tropics.

Hadley circulation.   The vertical and latitudinal circulation pattern of air in the atmosphere driven by the warming effect of the sun.

Handicap principle.   The idea that elaborate, sexually selected displays and adornments act as handicaps that demonstrate the generally high fitness of the bearer.

Haplodiploidy.   A sex determination mechanism in which females develop from fertilized eggs and are diploid, and males develop from unfertilized eggs and are haploid.

Haploid.   Having a single set of chromosomes. *Compare with* Diploid.

Haplotype.   The genotype of a haploid organism or a haploid portion of the genome (such as a gamete, mitochondrion, or chloroplast).

Hardy–Weinberg equilibrium (HWE).   The proportions of homozygotes and heterozygotes in a population that meets the assumptions of the Hardy–Weinberg law.

Hardy–Weinberg law.   The mathematical proposition that the frequencies of alleles and genotypes remain unchanged from generation to generation in a population of infinite size in the absence of natural selection, mutation, genetic drift, and assortative mating.

Hawk–dove game.   A game theory analysis of social behavior that compares the fitness outcomes of selfish and cooperative behaviors.

Heat.   The amount of energy transferred between two systems because of a difference in their temperature.

Heat budget.   All the gains and losses of heat by an organism by any mechanism, including metabolism, evaporation, radiation, conduction, and convection.

Herbivore.   An organism that consumes living plants or their parts.

Hermaphrodite.   An individual that has both male and female sexual functions.

Heterotroph.   An organism that uses other organisms or their remains as a source of energy and nutrients. *Compare with* Autotroph.

Heterozygous.   Containing two different alleles of a gene. *Compare with* Homozygous.

Hibernation.   A state of winter dormancy involving lowered body temperature and metabolism.

Holistic concept.   The idea, championed by Frederic E. Clements, that the organisms in a community form a discrete, complex unit analogous to a superorganism. *Compare with* Individualistic concept.

Homeostasis.   Maintenance of constant internal conditions in the face of a varying external environment.

Homeothermy.   Maintenance of a constant body temperature in the face of a fluctuating environmental temperature. *Compare with* Poikilothermy.

Homozygous.   Containing two identical alleles of a gene. *Compare with* Heterozygous.

Horizon.   A layer of soil distinguished by its physical and chemical properties.

Host.   The living organism on or within which a parasite resides.

Hyperosmotic.   Having an osmotic potential (generally, salt concentration) greater than that of the surrounding medium. *Compare with* Hypo-osmotic.

Hypolimnion.   The cold, oxygen-depleted layer of a body of water that lies below the thermocline. *Compare with* Epilimnion.

Hypo-osmotic.   Having an osmotic potential (generally, salt concentration) less than that of the surrounding medium. *Compare with* Hyperosmotic.

Hypothesis.   A conjecture about or explanation proposed for a pattern or relationship in nature embracing a mechanism for its occurrence.

Hypoxia.   The depletion of oxygen in an aquatic environment to the extent that aquatic organisms can no longer survive there.

Ideal free distribution.   The distribution of individuals across habitat patches of different intrinsic quality such that the realized quality of the patches, in terms of fitness for the individuals occupying them, is equalized.

Identity by descent.   The probability that two individuals share copies of a particular gene inherited from a common ancestor; the probability that two alleles are identical copies inherited from a single ancestor. *See also* Coefficient of relationship.

Immobilization.   Removal of inorganic nutrients from the available nutrient pool by incorporation into organic structures.

Inbreeding.   Mating between closely related individuals.

Inbreeding coefficient (*F*).   The departure of the observed frequency of heterozygotes from Hardy–Weinberg equilibrium values.

Inbreeding depression.   The decrease in fitness caused by the high proportion of homozygous genotypes, and the resulting expression of deleterious recessive alleles, among the offspring of matings between close relatives.

Inclusive fitness.   The fitness of an individual plus the fitnesses of its relatives, weighted by the coefficient their of relationship.

Indeterminate growth.   A growth pattern in which an individual continues to grow, usually at a decreasing rate, after maturity. *Compare with* Determinate growth.

Individualistic concept.   The idea, espoused by Henry A. Gleason, that a community is not a discrete unit, but merely a fortuitous association of species whose adaptations and requirements enable them to live together under the physical and biological conditions of a particular place. *Compare with* Holistic concept.

Indomalayan region.   The biogeographic region corresponding roughly to India and Southeast Asia. *Also called* Oriental region.

Induced defense.   A defensive structure or compound that is produced in response to herbivory or predation. *Compare with* Constitutive defense.

Industrial melanism.   The evolution of dark coloration by cryptic organisms in response to industrial pollution, especially by soot, in their environments.

Inflection point.   The point at which a logistic or other sigmoid growth curve changes from its accelerating to its decelerating phase ($N=K/2$).

Infrared (IR) radiation.   Electromagnetic radiation having a wavelength longer than about 700 nm. *Compare with* Ultraviolet radiation.

Inhibition.   The suppression of one species by the presence of another, especially during a successional sequence. *Compare with* Facilitation, Tolerance.

Interference competition.   Direct antagonistic interaction between individuals or species competing for resources, usually by behavioral or chemical means. *Compare with* Exploitative competition.

Intermediate disturbance hypothesis.   The idea that species diversity is greatest in habitats with moderate amounts of physical disturbance, owing to the coexistence of different successional stages.

Interspecific competition.   Competition between individuals of different species. *Compare with* Intraspecific competition.

Intertidal zone.   *See* Littoral zone (2).

Intertropical convergence.   The region at the solar equator where surface currents of air meet and begin to rise under the warming influence of the sun.

Intraspecific competition.   Competition between individuals of the same species. *Compare with* Interspecific competition.

Intrinsic rate of increase.   The growth rate of a population with a stable age distribution.

Inverse density dependence.   *See* Positive density dependence.

Ion.   An electrically charged atom or group of atoms.

Irradiance.   The intensity of the light of all wavelengths impinging on a surface, quantified in watts per square meter.

Iteroparity.   A life history characterized by multiple reproductive episodes. *Compare with* Semelparity.

Jet stream.   A rapidly moving west-to-east air current about 10 km above the earth's surface that forms at the high-altitude junction of two atmospheric circulation cells.

Joint equilibrium point.   The combination of predator and prey population sizes at which the two population sizes do not change; the point at which the equilibrium isoclines of the two populations cross.

Joint population trajectory.   A closed cycle of change in predator and prey population sizes in which increases and decreases in the predator population track those in the prey population.

Keystone consumer.   A species, often a predator, that has a dominant influence on the structure of a community, which may be revealed when that species is removed.

Kin selection.   Differential survival or reproduction that results from variation in genetically based social behavior among closely related individuals.

Lake.   A body of fresh water in any kind of depression.

Landscape.   A large area consisting of a mosaic of diverse habitat types.

Landscape context.   The quality and spatial arrangement of the habitat types in a habitat matrix.

Landscape ecology.   The study of the composition of landscapes and the spatial arrangements of habitats within them, and of how those patterns influence individuals, populations, communities, and ecosystems at different spatial scales,

Landscape model.   A model of spatial population structure in which the effects of differences in habitat quality within the habitat matrix are considered.

La Niña.   A global climatic pattern characterized by strong winds and cool ocean currents flowing westward from the coast of South America into the tropical Pacific Ocean. *Compare with* El Niño.

Laterite.   A hard layer, rich in oxides of iron and aluminum, that may form at the surface of exposed, deeply weathered tropical soils.

Laterization.   The leaching of silicon from soil due to weathering under warm, moist conditions, leaving oxides of iron and aluminum to predominate.

Laurasia.   The major landmass of the Northern Hemisphere during the Mesozoic era, consisting of present-day North America, Europe, and most of Asia.

Legacy effect.   A long-lasting influence of historical processes on a modern-day ecosystem.

Lek.   A gathering of males within a traditional arena to perform courtship displays.

Lentic.   Of or referring to nonflowing fresh waters. *Compare with* Lotic.

Liana.   A climbing woody vine that is rooted in soil.

Liebig's law of the minimum.   The principle that the growth of a population is limited by the resource whose supply is least relative to demand (the limiting resource).

Life history.   The adaptations that constitute the schedule of an organism's life, including traits such as age at maturity, fecundity, and longevity.

Life table.   A summary of the probabilities of survival and the fecundities of the individuals in a population by age.

Lifetime dispersal distance.   The average distance that individuals in a population disperse from their birthplace to where they reproduce.

Life zone.   A more or less distinct belt of vegetation occurring within and characteristic of a particular latitude or range of elevation.

Limit cycle.   A stable pattern of oscillation that alternates between high and low values, as in the sizes of populations with high intrinsic rates of growth. *Compare with* Damped oscillation, Chaos.

Limiting resource.   The resource that is most scarce relative to a population's demand for it and which therefore limits the population growth.

Limnetic zone.   The open water of a lake or pond beyond the littoral zone. *Also called* Pelagic zone.

Littoral zone.   (1) The shallow zone around the edge of a lake or pond within which rooted vegetation is found. (2) The shore of the sea between the highest and lowest tidal water levels. *Also called* Intertidal zone.

Local diversity.   The number of species in a small area of homogeneous habitat. *Also called* Alpha diversity. *Compare with* Regional diversity, Beta diversity.

Local mate competition.   Competition among males for mates that occurs at or near their place of birth, and hence potentially involves competition between close relatives.

Local population.   *See* Subpopulation.

Logistic equation.   The mathematical expression of an exponential rate of population increase that decreases in linear fashion as population size increases.

Longevity.   The life span of an individual.

Lotic.   Of or referring to flowing fresh waters. *Compare with* Lentic.

Lotka–Volterra model.   A continuous-time model that calculates the influence of each of two populations (predator and prey, or competitors) on the abundance of the other.

Macroecology.   The study of patterns in the sizes of the geographic ranges of populations and in the densities and distributions of individuals within those ranges.

Margalef's index.   An index of species richness based on the idea that the number of species in a sample increases in direct relationship to the logarithm of the number of individuals sampled.

Mass extinction.   The abrupt disappearance of a large proportion of species or higher taxa as a result of a natural catastrophe. *Compare with* Background extinction.

Mate guarding.   Close association of males with their female mates to prevent their mating with other males.

Mathematical model.   A representation of a complex system by a set of equations corresponding to the hypothesized relationships of each of the system's components to its other components and to outside influences.

Mating system.   The pattern of matings between males and females in a population, including the number of simultaneous or sequential mates, the permanence of pair bonds, and the degree of inbreeding.

Matric potential.   The water potential generated by the attraction of water to the surfaces of soil particles.

Maturity.   Acquisition of reproductive function, or the age at which this occurs.

Maximum sustainable yield (MSY).   The highest rate at which individuals can be harvested from a population without reducing the size of the population—that is, at which recruitment equals or exceeds harvesting.

Mediterranean climate.   A climatic pattern found at middle latitudes on the western sides of continents, characterized by cool, wet winters and warm, dry summers.

Meiosis.   A series of two divisions by cells destined to produce gametes, involving pairing and segregation of homologous chromosomes and a reduction of chromosome number from diploid to haploid.

Meiosis, cost of.   *See* Twofold cost of meiosis.

Mesic.   Of or referring to habitats with plentiful rainfall and well-drained soils. *Compare with* Xeric.

Metabolic theory of ecology.   A theory stating that temperature has consistent effects on a range of processes important to ecology and evolution at a range of scales.

Metapopulation.   A population that is divided into discrete subpopulations between which individuals move infrequently.

Metapopulation model.   A model of spatial population structure that describes a set of subpopulations occupying habitat patches between which individuals move infrequently, and in which the intervening habitat matrix is considered only as a barrier to the movement of individuals.

Microcosm.   A small, simplified system, often maintained in a laboratory, that contains the essential features of a larger natural system.

Microenvironment.   *See* Microhabitat.

Microhabitat.   A part of a habitat that can be distinguished from other parts by its environmental conditions.

Microsatellite.   A type of non-protein-coding DNA sequence consisting of a tandem repeat of sequences of two, three, or four nucleotides.

Minimum viable population (MVP).   The minimum number of individuals necessary to prevent a population from suffering stochastic extinction.

Minisatellite.   A type of nongenetic repetitive DNA sequence related to the microsatellite.

Mixed strategy.   An evolutionarily stable strategy that results in the persistence of two or more phenotypes within a population, each at an equilibrium proportion set by the frequency-dependent fitnesses of the phenotypes.

Model, mathematical.   *See* Mathematical model.

Monoecy.   In plants, the occurrence of male and female sexual organs in different flowers on the same individual. *Compare with* Dioecy.

Monogamy.   A mating system in which each individual mates with only one individual of the opposite sex, generally involving a strong and lasting pair bond. *Compare with* Polygamy.

Müllerian mimicry.   A predator deterrence strategy in which several unpalatable species adopt a single pattern of warning coloration. *Compare with* Batesian mimicry.

Mutation.   Any change in the genotype of an organism occurring at the gene, chromosome, or genome level; usually applied to changes in the sequence of the nucleotides of DNA.

Mutualism.   An interaction between two species that benefits both.

Mycorrhizae.   Mutualistic associations of fungi and plant roots in the soil that facilitate the uptake of nutrients by the roots.

Natural experiment.   An approach to hypothesis testing that relies on natural variation in the environment to create reasonably controlled experimental treatments.

Natural selection.   Change in the frequency of genetic traits in a population through differential survival and reproduction of individuals bearing those traits.

Nearctic region.   The biogeographic region corresponding to North America.

Negative assortative mating.   Preferential mating between individuals having dissimilar genotypes or phenotypes. *Compare with* Positive assortative mating.

Negative density dependence.   A population response to increasing density that depresses survival and birth rates. *Compare with* Positive density dependence.

Negative feedback.   The action of internal response mechanisms to restore a system to a desired state, or set point, when the system deviates from that state.

Neighborhood size.   The number of individuals in a population included within a circle whose radius is the lifetime dispersal distance of an average individual.

Neotropical region.   The biogeographic region corresponding to South America and tropical Central America and the West Indies.

Neritic zone.   The zone of shallow ocean water from the lowest tidal level to the edge of the continental shelf.

Net ecosystem production.   The balance of energy gain and energy loss in an ecosystem; a measure of net energy accumulation.

Net primary production.   That portion of gross primary production that is accumulated in the tissues of autotrophs. *Compare with* Gross primary production.

Net production efficiency.   The percentage of assimilated energy that is used for growth and reproduction.

Net reproductive rate (R_0).   The number of female offspring produced by an average female during her lifetime.

Neutral mutation.   *See* Silent mutation.

Neutral stability.   A type of dynamic behavior in which a system at equilibrium does not return to that equilibrium after perturbation.

Niche.   The range of conditions a species can tolerate and the ways of life it pursues; the functional role of a species in the community, often conceived as a multidimensional space.

Nitrification.   The oxidation of ammonia by specialized bacteria, yielding nitrite and nitrate.

Nitrogen fixation.   Biological assimilation of atmospheric nitrogen to form organic nitrogen-containing compounds.

Nonrenewable resource.   A resource that is not regenerated and which becomes unavailable when it is used. *Compare with* Renewable resource.

Nucleotide.   Any of several chemical compounds forming the structural units of RNA and DNA and consisting of a purine or pyrimidine base, a ribose (RNA) or deoxyribose (DNA) sugar, and phosphoric acid.

Numerical response.   A change in the size of a predator population as a result of a change in the density of its prey. *See also* Functional response.

Nutrient use efficiency (NUE). The ratio of dry matter production to the assimilation of a particular nutrient, usually expressed as grams per gram.

Observation. A fact, pattern, or relationship seen in nature that invites explanation or speculation.

Oceanic zone. The zone of deep ocean water beyond the neritic zone.

Oligotrophic. Poor in the inorganic nutrients required by organisms. *Compare with* Eutrophic.

Omnivory. Feeding on more than one trophic level.

Open community. A local association of species having independent and only partially overlapping ecological distributions. *Compare with* Closed community.

Optimal foraging. A theoretical concept that seeks to explain foraging behavior in terms of the fitness costs and benefits of each possible alternative behavior.

Optimal outcrossing distance. The distance between mating partners that best balances the risk of inbreeding depression with the risk of reducing the fitness of progeny by passing on genes adapted to different environmental conditions.

Optimum. The narrow range of environmental conditions to which an organism is best suited.

Organism. A living being; the most fundamental unit of ecology.

Organism approach. An approach to ecology that emphasizes the way in which an individual organism's form, physiology, and behavior help it to survive in its environment.

Oriental region. *See* Indomalayan region.

Oscillation. A pattern of regular fluctuation above and below some mean value.

Osmoregulation. The process of maintaining a proper salt balance.

Osmosis. The tendency of water to move from regions of low solute concentration to regions of high solute concentration.

Osmotic potential. The force with which an aqueous solution attracts water by osmosis; usually expressed as a pressure.

Outcrossing. Mating between unrelated individuals; in hermaphrodites, fertilization that takes place between gametes from different individuals. *Compare with* Inbreeding, Selfing.

Ozone hole. A region of severe ozone depletion in the upper atmosphere, usually at high latitudes.

Palearctic region. The biogeographic region corresponding roughly to temperate Asia and Europe.

Pangaea. A supercontinent that existed at the end of the Paleozoic era and which encompassed most of the earth's landmasses, including the future Laurasia and Gondwana.

Parasite. An organism that consumes parts of a living host, usually without killing the host.

Parasite-mediated sexual selection. The idea that females choose male mates on the basis of secondary sexual characteristics that indicate their ability to resist parasite infection.

Parasitoid. An insect whose larvae live within and consume the tissues of a living host, usually another insect.

Parent–offspring conflict. The difference in the optimal level of parental investment in a particular offspring from the viewpoint of the parent and from the viewpoint of that offspring, stemming from the fact that all offspring are genetically equivalent from the point of view of their parents, but siblings have a coefficient of relationship of only 0.50.

Parity. The number of episodes of reproduction in an individual's lifetime.

Parthenogenesis. Reproduction without fertilization by male gametes, usually involving the formation of diploid eggs whose development is initiated spontaneously.

Patch. *See* Habitat patch.

Pathogen. A parasite, especially a microorganism, that causes disease in its host.

Pattern. A set of recurring events or objects, generally repeatable in a predictable manner. *Compare with* Process.

Payoff matrix. In game theory analysis, a chart showing the fitness consequences of the players' behaviors.

Pelagic zone. *See* Limnetic zone.

Per capita. Expressed on a per individual basis.

Perfect flower. A flower having both male and female sexual organs (anthers and carpels).

Periodic cycle. A pattern of fluctuation with regular intervals between high and low values.

Permafrost. A layer of permanently frozen soil found at high latitudes.

pH. A measure of acidity or alkalinity; the negative of the common logarithm of hydrogen ion concentration, measured in moles per liter.

Phenolics. Aromatic hydrocarbons produced by plants, many of which exhibit antimicrobial properties.

Phenotype. The physical expression of an organism's genotype in its structure and function; the outward appearance and behavior of the organism. *Compare with* Genotype.

Phenotypic plasticity. The genetically based capacity of an individual to respond to environmental variation by changing its form, function, or behavior.

Photic zone. The surface layer of a water body in which there is sufficient light penetration for photosynthesis. *Compare with* Aphotic zone.

Photoautotroph. An organism that uses sunlight as its primary source of energy for the synthesis of organic compounds. *Compare with* Chemoautotroph.

Photoperiod. The length of the daylight period in a 24-hour day.

Photorespiration.   Oxidation of carbohydrates to carbon dioxide and water by Rubisco, the enzyme responsible for carbon assimilation, at low CO_2 concentrations, reversing the light reactions of photosynthesis.

Photosynthetically active region (PAR).   The wavelengths of light that are suitable for photosynthesis, ranging between about 400 nm (violet) and 700 nm (red), corresponding to the visible portion of the electromagnetic spectrum.

Photosynthetic efficiency.   The percentage of the energy in sunlight that is converted to net primary production during the growing season.

Phylogenetic diversity.   A measure of biodiversity that takes into account the degree of relationship among organisms, giving greater weight to more distantly related forms.

Phylogenetic effect.   Resemblance between two or more species resulting solely from their common ancestry.

Pleiotropy.   An effect of a single gene on multiple traits.

Podsolization.   The breakdown and loss of clay particles in the acidic soils of cold, moist regions.

Poikilothermy.   Conformity to the external environmental temperature. *Compare with* Homeothermy.

Point mutation.   A substitution of one of the nucleotides in a DNA codon that changes the amino acid that it specifies. *Compare with* Silent mutation.

Polyandry.   A mating system in which a female pairs with more than one male at the same time. *Compare with* Polygyny.

Polygamy.   A mating system in which a male pairs with more than one female (polygyny) or a female pairs with more than one male (polyandry) at the same time. *Compare with* Monogamy.

Polygyny.   A mating system in which a male pairs with more than one female at the same time. *Compare with* Polyandry.

Polygyny threshold.   The degree of variation among territories at which the reproductive success of a female that pairs with a polygynous male holding a high-quality territory is likely to equal that of a female that pairs with a monogamous male holding a lower-quality territory.

Polymorphism.   The occurrence of more than one distinct phenotype or genotype in a population.

Pond.   A body of fresh water smaller than a lake.

Pool.   A stretch of relatively deep, slowly flowing water in a stream. *Compare with* Riffle.

Population.   The individuals of a particular species that inhabit a particular area.

Population approach.   An approach to ecology that is concerned with population dynamics.

Population bottleneck.   An extended period of small population size, during which a population is vulnerable to the loss of genetic diversity through genetic drift. *Compare with* Founder event.

Population genetics.   The study of changes in the frequencies of genes and genotypes within a population.

Population size.   The number of individuals in a population.

Population structure.   Attributes of a population including the density and spacing of individuals within its geographic range and the proportions of individuals in each age and sex class. *See also* Age structure, Spatial structure.

Population viability analysis (PVA).   A method of using simulation models incorporating demographic and environmental information to predict the probability that a population will avoid extinction within a given period.

Positive assortative mating.   Preferential mating between individuals having similar genotypes or phenotypes. *Compare with* Negative assortative mating.

Positive density dependence.   A population response to increasing density that increases survival and birth rates. *Also called* Inverse density dependence. *Compare with* Negative density dependence.

Potential evapotranspiration (PET).   The amount of water that could be transpired by plants and evaporated from the soil, given the local temperature and humidity, if water were not limited.

Prairie.   A temperate grassland in central North America.

Preadaptation.   A trait evolved for one purpose that becomes useful for another purpose in a changed environment.

Predator.   An animal or protist (rarely, a plant) that kills and eats animals or protists.

Prediction.   A logical consequence of a hypothesis or outcome of a model describing some aspect of a system.

Primary consumer.   An herbivore; an organism at the lowest consumer level in a food web.

Primary producer.   A plant or other autotroph that assimilates the energy of sunlight (a photoautotroph) or reduced inorganic compounds (a chemoautotroph) and uses it to synthesize organic compounds.

Primary production.   Assimilation (gross primary production) or accumulation (net primary production) of energy by plants and other autotrophs.

Primary succession.   Succession in a newly formed or exposed habitat devoid of life. *Compare with* Secondary succession.

Priority effect.   The result of an interaction between two species during a successional sequence whose outcome depends on which becomes established first.

Process.   An event that causes change in an ecological system. *Compare with* Pattern.

Programmed death.   Death as part of a semelparous life history.

Promiscuity.   Mating with many individuals of the opposite sex, generally without the formation of strong or lasting pair bonds.

Proximate factor.   An aspect of the environment that the organism uses as a cue for behavior, but which does not directly affect the organism's fitness (such as day length). *Compare with* Ultimate factor.

Pyramid of energy.   The idea that energy flux through the food chain decreases at progressively higher trophic levels.

Q₁₀. The ratio of the rate of a physiological process at one temperature to its rate at a temperature 10°C cooler.

Queen. In a colony of eusocial insects, a fertile, fully developed reproductive female.

Radiation. Energy emitted in the form of electromagnetic waves.

Rain shadow. A dry area on the leeward side of a mountain range.

Random distribution. A dispersion pattern in which individuals are distributed without regard to the positions of other individuals. *Compare with* Clumped distribution, Spaced distribution.

Random walk. The pattern of change in numbers in a population subject to stochastic processes.

Rank-abundance plot. A graph that displays the abundances of species in a community or sample, usually on a logarithmic scale, ranked from the most common to the rarest.

Rarefaction curve. A method of determining the relationship between species diversity and sample size by repeatedly subsampling the pool of individuals in a sample and calculating the average number of species present for a given number of individuals. *Compare with* Accumulation curve.

Reaction norm. The relationship between the phenotype of an individual with a particular genotype and conditions in that individual's environment.

Realized niche. The range of physical conditions and resources within which individuals of a species can persist in the presence of competitors and consumers. *Compare with* Fundamental niche.

Recessive. Referring to an allele whose expression is masked by an alternative (dominant) allele of the same gene. *Compare with* Dominant.

Recipient. The individual in a social interaction toward which behavior is directed. *Compare with* Donor.

Reciprocal transplant experiment. An exchange of individuals of the same species between two different habitats or regions to determine the relative contributions of genotype and environment to the phenotype.

Recruitment. Addition of new individuals to the breeding population by reproduction.

Red Queen hypothesis. The idea that evolutionary change in a population's biological environment, especially in its predators and pathogens, applies continual selection pressure on that population.

Regional diversity. The number of species observed in all habitats within a large geographic area that includes no significant barriers to dispersal. *Also called* Gamma diversity. *Compare with* Local diversity, Beta diversity.

Relative abundance. The proportional representation of a species in a sample or a community.

Remote sensing. The collection of geographic information from a distance, as with images taken from an airplane or satellite.

Renewable resource. A resource that is continually regenerated or renewed. *Compare with* Nonrenewable resource.

Rescue effect. Prevention of the extinction of a declining subpopulation by immigration of individuals from another, more productive subpopulation.

Residence time. The ratio of the size of an ecosystem compartment to the flux through it, expressed in units of time; thus, the average time spent by energy or materials in that compartment.

Resilience. The ability of a system to return to some reference state after a disturbance. *Compare with* Constancy.

Resistance. *See* Constancy.

Resource. Any substance or factor that is consumed by organisms and used for their maintenance and growth, and which supports increased population growth rates as its availability in the environment increases.

Resource-imposed equilibrium. An equilibrium that results when a consumer population becomes limited and a resource population it exploits escapes its control and increases to the carrying capacity set by its own resources. *Compare with* Consumer-imposed equilibrium.

Respiration. Use of oxygen to metabolize organic compounds and release chemical energy.

Respired energy. That portion of assimilated energy that is used to meet metabolic needs, most of which escapes the organism as heat.

Rhizome. An underground, usually horizontal stem of a plant that produces both roots and aboveground shoots and may be modified to store carbohydrate nutrient reserves.

Riffle. A shallow stretch of fast-moving water in a stream. *Compare with* Pool.

Riparian. Of or referring to the bank of a stream.

Riparian zone. An area of terrestrial vegetation bordering a stream that is influenced by seasonal flooding and elevated water tables.

Risk-sensitive foraging. Foraging behavior that is influenced by the presence of predators or risk of predation.

River continuum. The concept of a river system encompassing a continuum of conditions from the headwaters to the mouth, characterized by increasing streambed size, water flow, nutrient and sediment loads, and productivity.

Root pressure. The pressure that is created when the osmotic potential in the roots of a plant draws water from the soil into the plant, and which forces water into the xylem elements.

Rubisco. An enzyme involved in photosynthesis that catalyzes the reaction of RuBP and carbon dioxide to form two molecules of glyceraldehyde 3-phosphate (G3P). *Also called* RuBP carboxylase-oxidase.

RuBP carboxylase-oxidase. *See* Rubisco.

Runaway sexual selection.   Sexual selection by female choice that results in elaborate and costly secondary sexual characteristics in males.

Saturation point.   The light intensity above which the rate of photosynthesis no longer responds to increasing light intensity. *Compare with* Compensation point.

Scale.   The dimension in time or space over which variation is perceived.

Sclerophyllous.   Characterized by tough, hard, often small leaves.

Search image.   A mental image formed by a predator when it encounters prey frequently that helps it to identify and locate suitable prey.

Secondary compound.   A chemical product of plant metabolism that is produced for a purpose other than metabolism, usually as a defense against consumers.

Secondary consumer.   A carnivore; a consumer of primary consumers.

Secondary sexual characteristics.   Traits other than the sexual organs themselves that distinguish males and females.

Secondary succession.   Succession in a habitat that has been disturbed, but in which some aspects of the community remain. *Compare with* Primary succession.

Seed bank.   Dormant seeds in the soil that can germinate when conditions are favorable.

Selfing.   Self-fertilization; a form of sexual reproduction in which a hermaphroditic individual forms both male and female gametes and fertilizes itself. *Compare with* Outcrossing.

Selfishness.   Social behavior that benefits the donor at a cost to the recipient.

Self-thinning curve.   A characteristic relationship between average plant weight and population density found in plant populations that are limited by space or other resources.

Semelparity.   A life history characterized by a single, terminal reproductive episode. *Compare with* Iteroparity.

Semipermeable.   Referring to a membrane that blocks the passage of some, usually large, molecules but not other, usually smaller ones.

Senescence.   Gradual deterioration of physiological function with age, leading to increased probability of death; aging.

Sequential hermaphrodite.   An individual that changes its sex during its lifetime. *Compare with* Simultaneous hermaphrodite.

Sere.   A series of stages of community change in a successional sequence leading toward a stable state.

Sex ratio.   The ratio of the number of individuals of one sex to the number of individuals of the other sex within the progeny of an individual or within a population.

Sexual dimorphism.   A difference in the phenotypes of males and females of the same species.

Sexual reproduction.   Reproduction by means of the union of two gametes (fertilization) to form a zygote. *Compare with* Asexual reproduction.

Sexual selection.   Selection by one sex for specific characteristics in individuals of the opposite sex, usually exercised through mate choice.

Silent mutation.   A substitution of one of the nucleotides in a DNA codon that does not change the amino acid that it specifies. *Also called* Neutral mutation, Synonymous mutation. *Compare with* Point mutation.

Simpson's index (D).   A measure of species diversity weighted by the relative abundance of each species.

Simultaneous hermaphrodite.   An individual that has both male and female sexual functions at the same time. *Compare with* Sequential hermaphrodite.

Sink population.   A subpopulation that would decrease in size owing to high mortality rates, low reproductive rates, or both if it were not maintained by immigration from other subpopulations. *Compare with* Source population.

S-I-R model.   A model of infectious disease transmission in which S represents susceptible individuals, I represents infected individuals, and R represents recovered individuals with acquired immunity.

Social behavior.   Direct interaction of any kind among individuals of the same species.

Soil.   The layer of chemically and biologically altered material that overlies rock or other unaltered material at the surface of the earth; the solid substratum of terrestrial communities.

Soil horizon.   *See* Horizon.

Solar constant.   The intensity of solar radiation reaching the outer limit of the earth's atmosphere, approximately 1,366 W per m^2.

Solar equator.   The parallel of latitude that lies directly under the sun's zenith in a given season.

Solute.   Any substance dissolved in a solvent.

Source population.   A subpopulation that produces an excess of individuals over the number needed to maintain itself and from which there is net emigration. *Compare with* Sink population.

Source–sink model.   A model of spatial population structure in which some subpopulations in high-quality habitat patches (source populations) produce excess offspring that disperse to subpopulations in lower-quality habitat patches that would otherwise decline in size (sink populations).

Southern Oscillation.   A reversal of the typical gradient of atmospheric pressures over the central equatorial Pacific Ocean that triggers an El Niño event.

Spaced distribution.   A dispersion pattern in which each individual maintains a minimum distance between itself and its neighbors. *Compare with* Clumped distribution, Random distribution.

Spatial structure.   The density and spacing of individuals in a population.

Species–area relationship.   The positive relationship between species richness and the area of a habitat patch, island, or region.

Species pool. The entire group of species within a source region from which colonists of an island or habitat patch are drawn.

Species richness. A simple count of the number of species in an area.

Species sorting. A process that determines membership in a local community based on the tolerances of species from the regional species pool for local conditions, their requirements for resources, or their interactions with competitors, predators, and pathogens.

Spitefulness. Social behavior that reduces the fitness of both donor and recipient.

Spring overturn. The vertical mixing of water layers in a temperate lake in spring as surface ice melts and the surface water warms and sinks. *Compare with* Fall overturn.

Stability. In population biology, the achievement of an unvarying equilibrium population size.

Stabilizing selection. Differential survival or reproduction of individuals with phenotypes closest to the average value of a population. *Compare with* Directional selection, Disruptive selection.

Stable age distribution. The proportions of individuals in various age classes in a population that is growing at a constant rate.

Static life table. A life table that follows individuals of various age classes within a population over a given time interval. *Also called* Time-specific life table. *Compare with* Cohort life table.

Steady state. The condition of a system in which opposing forces or fluxes are balanced.

Steppe. A temperate grassland in central Asia.

Stepping stones. Small intervening patches within a habitat matrix where organisms moving between patches of favorable habitat can stop to rest or forage.

Stochastic. Resulting from chance events.

Stoichiometry. The study of the balance between inputs and outputs.

Stomate. An opening in the surface of a leaf through which gas exchange with the atmosphere takes place. *Also called* Stoma.

Stratification. The establishment in a body of water of distinct layers with different temperatures or salinities due to differences in their densities.

Stream. A body of water flowing over the surface of the land.

Subpopulation. A subdivision of a population that exchanges individuals with the remainder of the population infrequently. *Also called* Local population.

Subtropical high-pressure belts. Regions of high atmospheric pressure and dry air centered approximately 30° north and south of the equator, where the air of a Hadley cell sinks.

Succession. A regular sequence of changes in the species composition of a community in a newly formed or disturbed habitat that progresses to a stable state.

Survival (s_x). The probability of an individual living from one age or time period (x) to the next.

Survivorship (l_x). The probability that a newborn individual will be alive at age x.

Switching. A change in diet to favor food items of greater abundance.

Symbiosis. A close physical association between two species, usually coevolved. Symbiotic relationships can be parasitic or mutualistic.

Sympatric. Occurring in the same place; particularly referring to overlapping species distributions. *Compare with* Allopatric.

Synonymous mutation. *See* Silent mutation.

Taiga. A moist coniferous forest found at high latitudes, dominated by spruce and fir trees.

Tannins. Polyphenolic compounds produced by most plants that bind to plant proteins, thereby impairing their digestion by herbivores.

Temperature profile. The relationship of temperature to depth below the surface of water or soil or to elevation above the ground.

Tension–cohesion theory. *See* Cohesion–tension theory.

Terpenoids. Secondary compounds, including essential oils, latex, and resins, produced by plants as defenses against consumers.

Territory. Any area defended by one or more individuals against intrusion by others of the same or different species.

Thermal inertia. The tendency of an object to remain at the same temperature, which is greater in larger organisms owing to their lower surface-to-volume ratio.

Thermocline. The depth in a body of water at which the temperature changes abruptly between an upper layer of warm water (epilimnion) and a lower layer of cold water (hypolimnion).

Thermodynamic principles. A set of principles relating to heat and motion that govern all energy transformations.

Thermohaline circulation. A global pattern of surface and deep-water currents driven by differences in the density of water caused by variations in temperature and salinity.

Thermophilic. "Heat-loving"; able to tolerate high temperatures.

−3/2 (three-halves) power law. A generalization proposing that the relationship between the logarithms of the biomass and the density of a plant population has a slope of $-3/2$.

Time delay. A delay in the response of a population or other system to change in the environment.

Time-specific life table. *See* Static life table.

Tolerance. Referring to succession, a lack of influence by one species on the presence or absence of other species. *Compare with* Facilitation, Inhibition.

Top-down control. The influence of consumers on the sizes of the trophic levels below them in a food web. *Compare with* Bottom-up control.

Torpor. A voluntary, reversible condition of low body temperature and physiological activity.

Trade-off. A consequence of devoting limited time, energy, or materials to one structure, function, or behavior at the expense of another.

Transient climax. A climax community that develops in an ephemeral habitat, such as a temporary pond.

Transpiration. Evaporation of water from leaf cells and other parts of plants.

Transpiration efficiency. *See* Water use efficiency.

Trophic cascade. An indirect interaction in which a consumer–resource interaction influences additional trophic levels of a community.

Trophic level. A position in a food web, determined by the number of energy transfer steps from primary producers to that level.

Trophic mutualism. A symbiotic relationship between two species in which each species supplies the other with a limiting nutrient or energy source that it cannot obtain by itself.

Turnover. (1) The replacement in a population of individuals that die with newborn individuals. (2) *See* Beta diversity.

Twofold cost of meiosis. The 50% fitness cost of sexual reproduction that results from a parent's contributing only one-half of the genetic material of each of its offspring.

Type I functional response. A predator–prey relationship in which an individual predator's rate of food consumption is directly proportional to prey density.

Type II functional response. A predator–prey relationship in which an individual predator's rate of food consumption levels off at high prey densities (predator satiation).

Type III functional response. A predator–prey relationship in which an individual predator's rate of food consumption levels off at high prey densities, and is additionally low at low prey densities owing to lack of a search image or to inefficient prey capture mechanisms.

Ultimate factor. An aspect of the environment that directly affects the fitness of an organism (such as food supply). *Compare with* Proximate factor.

Ultraviolet (UV) radiation. Electromagnetic radiation having a wavelength shorter than about 400 nm. *Compare with* Infrared radiation.

Understory. A layer of vegetation under the canopy of a forest.

Upwelling. Vertical movement of water, usually near coasts where surface currents diverge, that brings nutrients from the depths of the ocean to surface layers.

Vegetative reproduction. An asexual reproductive process by which shoots that sprout from the roots, rhizomes, or leaves of a plant give rise to separate individuals with genotypes identical to that of the parent plant.

Vicariance. The breaking up of a widely distributed ancestral population by continental drift or some other barrier to dispersal.

Virulence. A measure of the capacity of a parasite or pathogen to invade and proliferate in host tissues.

Warning coloration. Conspicuous patterns or colors adopted by unpalatable prey organisms to advertise their noxiousness or dangerousness to potential predators. *Also called* Aposematism.

Water potential. The strength of the attractive forces holding water in the soil or in a cell; usually expressed as a pressure.

Watershed. The drainage area of a stream.

Water use efficiency. The ratio of net primary production to transpiration of water by a plant, usually expressed as grams per kilogram of water. *Also called* Transpiration efficiency.

Weathering. Physical and chemical alteration of rock material near the earth's surface.

Weed. An organism, generally having strong powers of dispersal, that is capable of living in highly disturbed habitats.

Wetland. An area of land consisting of soil that is saturated with water and supports vegetation specifically adapted to such conditions.

Wilting coefficient. The minimum water potential of the soil at which plants can obtain water; generally assumed to be −1.5 MPa. *Also called* Wilting point.

Wilting point. *See* Wilting coefficient.

Xeric. Of or referring to habitats in which plant production is limited by the availability of water. *Compare with* Mesic.

Zenith. The highest position of the sun in a day.

Zero growth isocline. *See* Equilibrium isocline.

Zonation. The distribution of organisms in bands or regions along an environmental gradient (for example, intertidal zonation).

Zygote. A diploid cell formed by the union of male and female gametes during fertilization.

INDEX

Boldfaced page numbers indicate references to boldfaced words or phrases in the text.
Italicized page numbers indicate pages with illustrations.